90 0808280 6

# Proceedings of the Thirteenth International Diatom Symposium

WITH
F
UNIVERSITY
LIBRARY
D1765526

# Proceedings of the Thirteenth International Diatom Symposium

Maratea, Italy, 1st – 7th September 1994

*Edited for the International Society for Diatom Research*

by

**Donato Marino & Marina Montresor**

1994

Biopress Limited
Bristol

© Biopress Ltd., 1995

All rights reserved. No part of this publication may be reproduced, stored in a retrieval system, or transmitted, in any form or by any means, electronic, mechanical, photocopying, recording or otherwise, without the prior permission of the copyright owner.

ISBN 0–948737–35–2

UNIVERSITY OF PLYMOUTH
Item No. 900 382611-0
Date 23 NOV 1998 S
Class No. 589.481 MAR
Contl. No. 0948737352
LIBRARY SERVICES

PUBLISHED BY:

Biopress Ltd.
The Orchard
Clanage Road
Bristol
BS3 2JX
England

*Acknowledgement*

*Many illustrations and diagrams in this volume have been obtained from the publications. Some of the original figures have been slightly modified. In all cases reference is made to the original publication. The full source can be found in the reference list at the end of each chapter. Permission for the reproduction of this material is gratefully acknowledged.*

British Library Cataloguing in Publication Data

A catalogue record for this book is available from the British Library

Printed in Great Britain by Henry Ling Ltd., at the Dorset Press, Dorchester, Dorset

# Contents

**Life histories and their ecological significance**



**Ecology of marine and brackish water diatoms**

**Ecology of freshwater diatoms**



**Morphology and Taxonomy**

**Diatoms as a tool for paleoenvironment reconstruction**



**Workshop**

# Preface

The 13th International Diatom Symposium held at the hotel Villa del Mare, Maratea (Italy) from 1 to 7 September 1994, was convened by the research staff of the Marine Botany and Benthic Laboratories of the *Stazione Zoologica A. Dohrn di Napoli.* This was the first of the International Diatom Symposia to be held in Italy, and thus we had the opportunity of introducing diatomists to our country and to the Stazione Zoologica.

Italy boasts a very old tradition of botanical studies: indeed, the first University botanical gardens were established at the Universities of Padua and Pisa as early as the mid-16th century. In spite of this prestigious tradition and the fundamental work done on diatoms by Castracane degli Antelminelli, De Toni and Forti, studies by Italian investigators on diatoms were, until fairly recently, fragmentary. Now, with the more sophisticated techniques that have become available and with the increasing awareness of the potential of this area of study, diatoms, and their myriad of related aspects, have attracted growing attention. This is clearly reflected in the increasing number of articles on diatoms in the scientific literature and by many Italian researchers attending the 13th International Symposium.

The venue and the scientific programme attracted a total of 162 scientists from 34 countries. During the seven plenary sessions, 38 oral presentations were delivered on the following topics: Ecophysiology of marine and freshwater diatoms; Diatoms as a marker of climate changes and anthropogenic impact; Life histories and their ecological significance; Ecology of marine and brackish water diatoms; Ecology of freshwater diatoms; Morphology and Taxonomy; Diatoms as a tool for paleo-environment reconstruction. In addition, 134 posters were presented and discussed during 15 sessions devoted to the seven topics listed above. An additional poster session was organized to accommodate the numerous abstracts dealing with "Polar diatoms". Three evening workshops and three evening sessions devoted to video presentations completed the scientific programme.

The topics of the sessions were compiled by the Scientific Organizing Committee with a view to including research that focused on the most promising and exciting aspects of diatom studies, and research conducted using classical and new techniques. Visualization and monitoring of motility components in raphid diatoms by image-enhanced video microscopy, application of new transfer functions for paleo-environment reconstructions, application of molecular techniques to define species limits are only a few examples of studies where classical observations and new approaches can be successfully combined.

This volume contains 48 articles that originated from the oral and poster presentations given at the Symposium, as well as a report of a workshop on "CASPIA update on saline lake diatoms". All the papers submitted for publication underwent a rigorous selection procedure akin to the peer review process used by international scientific journals. Each manuscript was appraised by two, in some cases three, experts. Given the large number of papers submitted and the diverse fields of research covered, the selection and editing of papers was not an easy task. We are greatly indebted to the many reviewers whose prompt, in-depth reports on the manuscripts ensured a timely publication of this volume. According to many participants, the Symposium was intellectually stimulating and we hope that this book reflects the enthusiasm that marked the meeting.

We cannot conclude this brief Preface without acknowledging the advice and collaboration of Professor F. E. Round, and the former President, Professor E. F. Stoermer, both of whom generously supported our candidacy to host the 13th Symposium. In addition, we thank Professor Round who acted as Managing Editor of these Proceedings and gave unstintingly of his time, together with G. E. Lockett and all the Biopress staff to produce this book. Our thanks also go to our fellow members of the organizing committee, Dr L. Mazzella and M. C. Buia, who shared with us the difficult task of putting together the Symposium both from the scientific and organizational point of view. Thanks also go to the staff of *Jean Gilder Congressi* and in particular to its Managing Director, Gigi Finizio, who relieved us of much of the worry of organizing a large international meeting. The active interaction between all these persons contributed greatly to the friendly and pleasant atmosphere that prevailed at the Symposium.

Last but not least we express our warmest thanks to Professor G. Salvatore and Dr L. Cariello, President and General Director of the Stazione Zoologica, whose generous financial support made this Symposium possible.

Donato Marino and Marina Montresor

*Stazione Zoologica "A. Dohrn",*
*Villa Comunale, 80121 Napoli, Italy*

*Acknowledgements*

*The Symposium was also supported by grants from Consiglio Nazionale delle Ricerche (Rome) and Regione Basilicata (Potenza, Italy). International Science Foundation (Washington) and the ZEISS company financially supported participants from countries experiencing currency difficulties.*

# Daily changes of reed periphyton composition in a shallow Hungarian lake (Lake Velence)

É. Ács and K. Buczkó

*Botanical Department, Hungarian Natural-History Museum, H–1476 Budapest, Pf. 222, Hungary*

## Abstract

Reed periphyton stucture was investigated in the middle of summer in 1992. Five replicate samples were collected at mid-day over 25 days from living and sterilised reeds. Two weeks before sampling, a 20 cm portion of 150 below water level reed stems were cleaned with a brush. After cleaning they were covered with black nylon and aluminium foil. A day before starting the investigation, some substrata were collected as controls and examined under light microscope after preparation. Just before the experimental period the reed stems were uncovered and 150 pieces each 20 cm long cleaned and sterilised reed stems were placed in a frame. This frame was fixed under the water level to living reeds, at the same sampling point and depth.

The number of living and dead cells was counted and chlorophyll-a content of the periphyton was measured. The ratio of dead:live cells was higher in almost all the samples taken from sterilised reed, and was lowest at the beginning of the experiment. The abundance increased continuously over 11 days on both substrata, then there was a slow-increase period ("steady state"). The same tendency was observed in the chlorophyll-a content of cleaned green reeds. Changes in relative abundance of some diatoms was also examined during the investigation.

## Introduction

One of the key areas in biology is the study of succession. These types of studies are fairly common in the aquatic environment; yet the study of periphytic algal succession has been poorly investigated in Hungarian shallow lakes to date. Succession of shallow-lake phytoplankton in Hungary has been studied mainly by Padisák (e.g. Padisák *et al.* 1988, 1990). The succession of periphyton of the River Danube has been studied by Ács & Kiss (1993). The process of natural succession is often disturbed by human interference, especially environmental pollution. One of the main aims of our study was to investigate the regeneration potential of attached algal communities in

Lake Velence, over a 1 month study period. The relationship between the quality of the substratum and the taxonomic composition of the reed periphyton was also studied.

*Study area*

Lake Velence is the second most important recreation centre in Hungary. It is located in the central part of the country, 45 km to the west of Budapest. The area of the lake is 24.5 $km^2$, the area of the water catchment region is 615 $km^2$. Average depth of the water is 1.2 m and the water is regularly mixed to the bottom. Our sampling station was at Rigya-mellék in the central part of the reed belt of the lake, at a point protected from the wind, where marked wave action occurred only during heavy storms.

## Material and Methods

In the period between 12th July and 5th August 1992 samples were taken daily at the same time (around noon) from the sampling station. Two weeks before the investigations 155 pieces of reed were cleaned by brushing a 30 cm long section of submerged stems. This was then covered with aluminium foil and wrapped in black nylon foil. The reeds remained alive and rooted at their original site but no algal coating could be developed directly on their surface (these will be referred to as "cleaned green" reeds). The extended period in the dark killed any algal cells which survived cleaning.

Just before the first sampling day (day 0) the foils were removed. 5 reed stems were cut for microscopical examination to check their cleanliness. On the same day, 150 pieces of sterilised reed stems were fixed in a frame*. The frame was fixed to rooted reeds at the sampling points, just beside the living cleaned reed, exposing them to identical conditions. The upper parts of the sterilised reed stems were just below the water surface (these specimens will be referred to as "sterilised").

For sampling, the reed stems were cut 5 and 15 cm below the water surface and the resulting 10 cm stem lengths were carefully removed and placed in tubes. 5 replicates of cleaned, uncleaned and sterilised reed stems were sampled at a time. In the laboratory, the samples were washed into water of known volume which was subsequently split into two parts. One half of the sample was used for chlorophyll-a measurement according to the method of Felföldy (1987); the other half was used for taxonomic determinations, using the Utermöhl (1958) method, taking the statistical errors (Lund *et al.* 1958) into consideration. Diatoms were identified after digestion with $H_2O_2$ .

Over the 25 days of the study period (on days 8, 14 and 25), control samples were taken from reed stems which were not previously cleaned. These specimens will be referred to as "uncleaned". "Uncleaned" control pieces were subjected to the same taxonomic study and chlorophyll-a determination.

---

* (These reed stems were cleaned by brushing and sterilised before the experiment).

In the course of the sampling procedure, the following parameters were checked at the sampling station: water temperature, Secchi-transparency, pH and conductivity of the water. 11 samples of water were collected for chemical analyses. The following chemical analyses were performed: suspended matter, KOI (manganatic), $HCO_3$, $CO_3$, $OPO_4$, total-P, $NO_2$, $NO_3$, $NH_3$, total inorganic N, Na, K, Ca, Mg content.

In the course of counting, the proportion of dead cells was also determined.

## Results

75 taxa were identified from the cleaned green stems, 66 from the sterilised reed and 47 from the uncleaned reed, with the following distribution: Cyanophyta 8, 6, 7, Euglenophyta 3, 4, 0, Chrysophyceae 0, 1, 0, Xanthophyceae 0, 1, 1, Bacillariophyceae 49, 42, 27 and Chlorophyta 18, 12, 12.

There was no direct correlation between water chemistry, number of individuals and chlorophyll-a content, thus only the average, minimum and maximum water chemistry parameters are included here (Table I).

Table I. The minimum, maximum and mean values of water chemistry parameters measured during the investigation.

| | min. | max. | mean |
|---|---|---|---|
| pH | 8.8 | 8.92 | 8.86 |
| conductivity ($\mu S\ cm^{-2}$) | 2573.52 | 3532.11 | 3250.02 |
| suspended matter ($mg\ l^{-1}$) | 4.4 | 33 | 19.19 |
| KOI (Mn) | 3.175 | 10.4003 | 8.91 |
| $HCO_3$ ($mg\ dm^{-3}$) | 109.19 | 320.25 | 242.51 |
| $CO_3$ ($mg\ dm^{-3}$) | 132 | 216 | 176.61 |
| $OPO_4$ ($mg\ dm^{-3}$) | 0.0137 | 0.1792 | 0.06 |
| total–P ($mg\ dm^{-3}$) | 0.047 | 0.4026 | 0.14 |
| $NO_2$ ($mg\ dm^{-3}$) | 0.0174 | 0.2646 | 0.06 |
| $NO_3$ ($mg\ dm^{-3}$) | 0.5876 | 1.2219 | 0.78 |
| $NH_3$ ($mg\ dm^{-3}$) | 0.7614 | 1.8889 | 1.41 |
| total anorg. N ($N\ mg\ dm^{-3}$) | 1.5052 | 2.9115 | 2.25 |
| Na ($mg\ l^{-1}$) | 390 | 560 | 461.96 |
| K ($mg\ l^{-1}$) | 50 | 135 | 77.76 |
| Ca ($mg\ l^{-1}$) | 7.5 | 12 | 9.50 |
| Mg ($mg\ l^{-1}$) | 200 | 311 | 247.74 |
| water temperature (C°) | 21 | 28.5 | 25.99 |
| Secchi transparency (cm) | 30 | 49 | 37.59 |

The number of individuals on the sterilised and the cleaned reeds did not reach that of the uncleaned green reeds during the 25 day study period (Fig. 1). The rapid increase in number of individuals slowed down for both substrata after the 11th day. The form of the curve is reminiscent of a saturation curve. The maximum number of individuals on cleaned green reeds was $0.82 \times 10^6$, on the sterilised reed $0.26 \times 10^6$, on the uncleaned green reed $1.86 \times 10^6$ (19th July), $3.89 \times 10^6$ (25th July) and $2.06 \times 10^6$ (5th August) cells $cm^{-2}$, respectively.

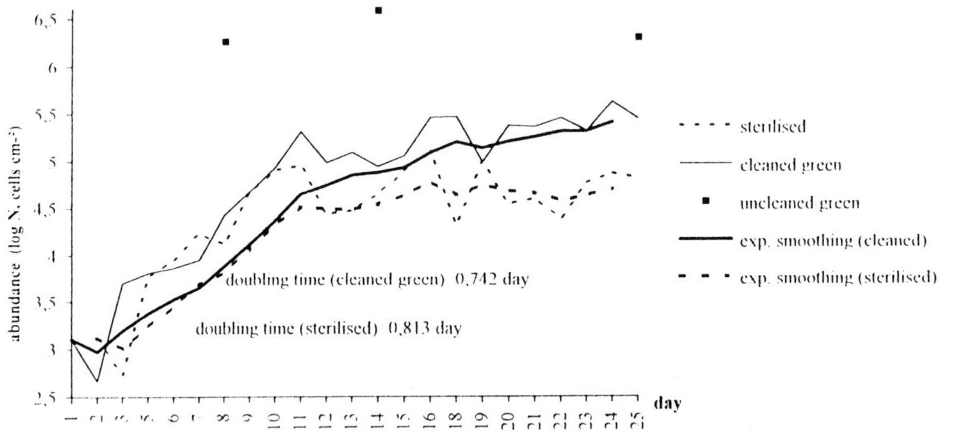


Fig. 1. The changes in algal abundance on cleaned green and sterilised reed stems and their exponential smoothing curve during the investigation. Algal abundance of uncleaned green reed also indicated.

Doubling time was estimated as 0.742 days on the cleaned green reed, and as 0.813 days on the sterilised reed between days 5th and 11th.

2–3% of cells were dead on the cleaned green reed until day 8, until day 6 on the sterilised reed. After this time the percentage ahead increased slowly to an average of 10% (little more on the sterilised reed, see Fig. 2). The proportion of dead cells on the uncleaned green reed was also around 10%. A much higher number of individuals occurred on the cleaned green reed than on the sterilised ones.

Chlorophyll-a content was significantly different on the two types of substratum. On the cleaned green reed it increased more or less gradually until day 19, then reached a "steady state" (Fig. 3). By this time, chlorophyll-a content of the samples matched that of the uncleaned green reeds. Chlorophyll-a content of the sterilised reed was almost constant or decreased slightly over the study period (chlorophyll-a content could only be detected reliably after day 10).

The Fig. 4 shows the cumulative relative abundance of the most abundant species (more than 5% in at least one sample). On cleaned green reed high relative abundance of *Achnanthes minutissima* Kütz. (ACHMIN) and *Gomphonema olivaceum* (Horn.) Bréb. (GOMOLI) was found in almost all samples. *Cymbella lacustris* (Ag.) Cl. (CYMLAC) and *Ctenophora pulchella* (Ralfs) (CTEPUL) had high relative

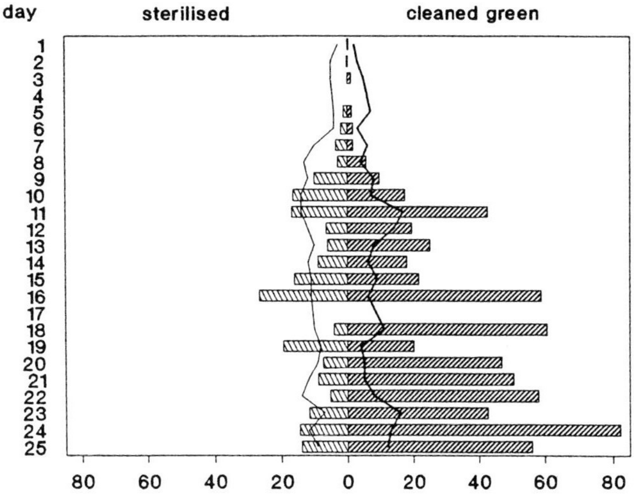


Fig. 2. The changes in periphytic algal abundance (bar, $10^4$ ind. $cm^{-2}$) and dead cells (line, in percents) during the investigation.

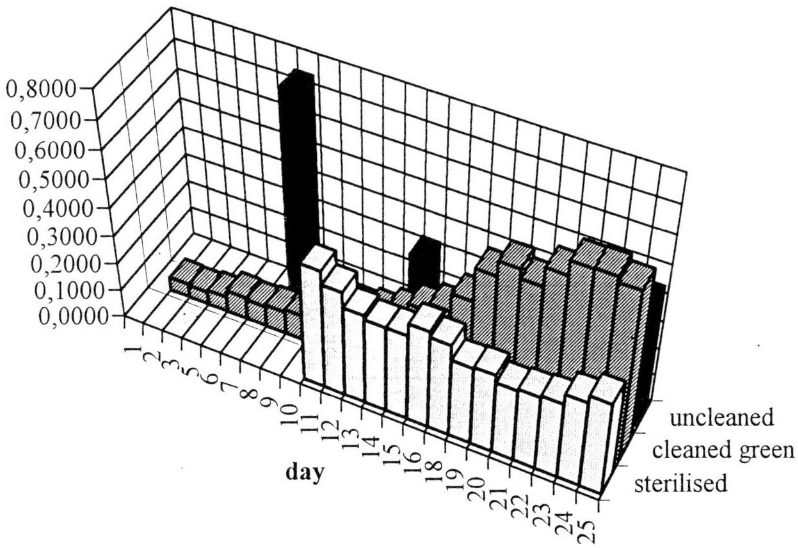


Fig. 3. The changes in chlorophyll-a content of periphyton, as 5 day running average, on cleaned green and sterilised reed stems during the investigation. Chlorophyll-a content of uncleaned green reed also indicated.

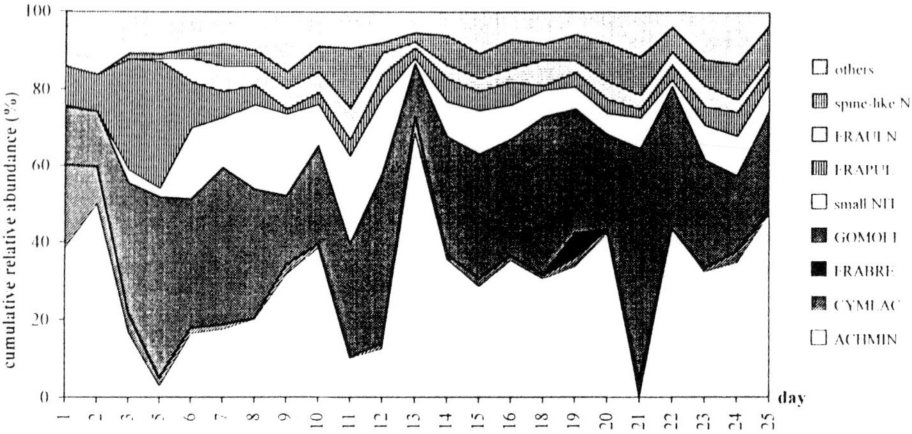


Fig. 4. The change in relative abundance of most abundant species on cleaned green reed stems during the investigation. See abbreviations in text.

abundances at the beginning of the sampling period but decreased later. The relative abundance of a small *Nitzschia* sp. (small NIT), a spine-like *Nitzschia* sp. (spine-like NIT) and *Synedra ulna* (Nitzsch) (SYNULN) was small until day 5, increased later and remained stable. *Fragilaria brevistriata* Grun. (FRABRE) appeared on day 18 and disappeared 2 days later. On sterilised reed, the occurrences displayed by *A. minutissima*, *G. olivaceum*, small *Nitzschia* sp., spine-like *Nitzschia* sp., *pulchella* and *ulna* were similar to those on cleaned green reed (Fig. 5).The relative abundance of *Cocconeis placentula* Ehr. (COCPLA) increased a little after day 18. Occasionally, the relative abundance of *Navicula cryptocephala* Kütz. (NAVCRY) was over 5% but it was completely missing in some samples.

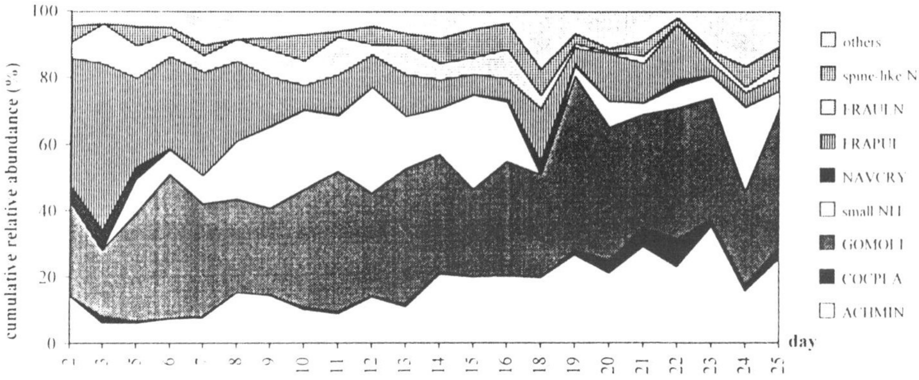


Fig. 5. The change in relative abundance of most abundant species on sterilised reed stems during the investigation. See abbreviations in text.

The species with the highest relative abundances were *A. minutissima* and *G. olivaceum* on both types of substratum. On cleaned green reed the relative abundance of *G. olivaceum* was generally higher than *A. minutissima* until day 12 (Fig. 6; except days 1, 2, 3 and 10), but after day 13 (except on days 15, 18 and 21) the latter was more abundant. In the coating collected from uncleaned reed the relative abundance of *A. minutissima* was always higher. In the sterilised reed coating the relative abundance of *G. olivaceum* was always higher than that of *A. minutissima* (Fig. 7). While the relative abundance of *A. minutissima* surpassed 40% on several occasions on the cleaned green reed, on sterilised reed it never attained this value and generally contributed less than 20%. On the cleaned green reed the relative abundance of *C. lacustris* surpassed 10% in the first two samples, but it decreased below 5% from day 3. On sterilised reed, however, it never exceeded 5%. The relative abundance of *C. placentula* exceeded 5% on the sterilised reed samples by the end of the study period, but was always under 5% on cleaned green reed.

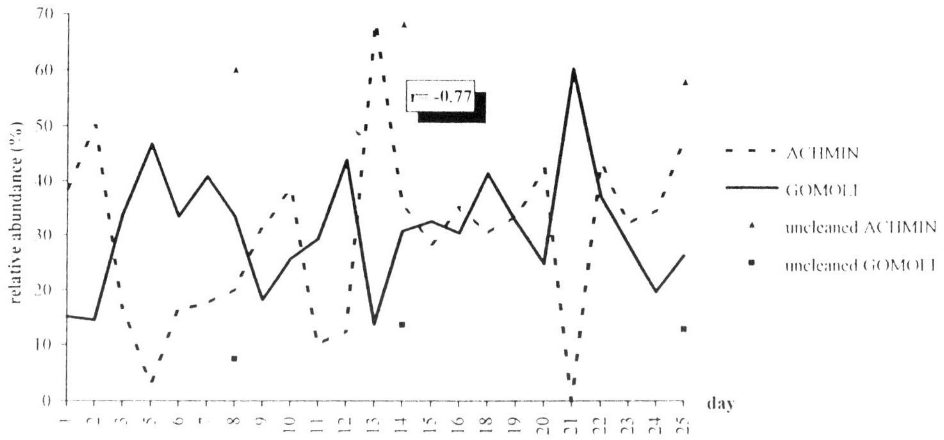


Fig. 6. The changes in relative abundance *of Achnanthes minutissima* (ACHMIN) and *Gomphonema olivaceum* (GOMOLI) on cleaned green reed stems during the investigation; relative abundance on uncleaned green reed stems also indicated.

## Discussion

No direct correlation between the water chemistry parameters and the cell numbers was revealed even for cleaned green reed. This is, however, not surprising because, in spite of using natural substrata, the experimental conditions were artificial. Cleaning the reeds caused severe perturbation. Similarly, in the case of sterilised reed, an empty substratum was colonised while the water chemistry reflected the daily changes of the water body.

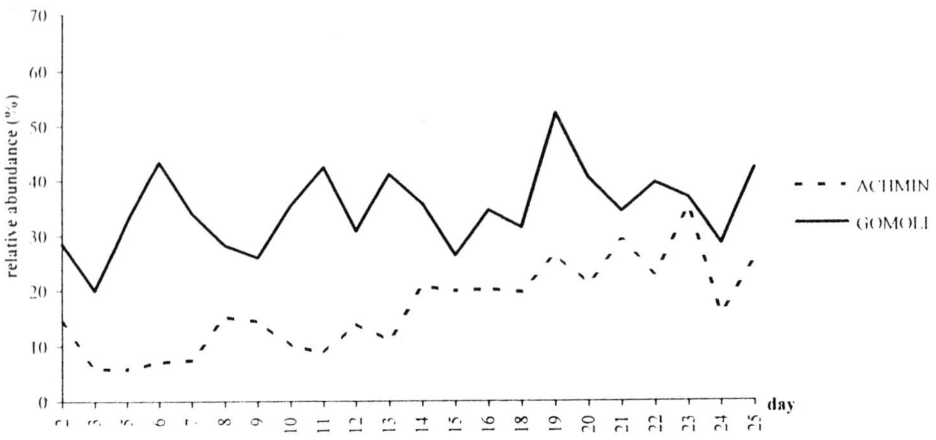


Fig. 7. The changes in relative abundance *of Achnanthes minutissima* (ACHMIN) and *Gomphonema olivaceum* (GOMOLI) on sterilised reed stems during the investigation.

In spite of similar doubling times on both substrata, increase in cell numbers ceased on sterilised reed after the day 11 leading to a "steady state". In the case of the cleaned reeds the number of individuals continued to increase although at a considerably slower pace after day 11. The observed differences can be explained by two hypotheses. One possible explanation invokes nutrient exchange between the living reed and the periphyton formed on it, which may be even more intensive for some species (*Achnanthes minutissima, Cymbella lacustris*). Similar results were reported by Shamess *et al.* (1985) who found that periphyton on a natural substratum (*Typha* sp.) was distinctly different from that on an artificial substratum in two eutrophic lakes. According to Allen (1971), there are metabolic relationships between macrophyte and epiphyton. The macrophytes act as a source of phosphate for their epiphyte, especially in oligotrophic and mesotrophic lakes (Burkholder & Wetzel 1989). The water of Lake Velence was mesotrophic over the study period. Cattaneo & Kalff (1979) found some evidence of nutrient transfer from the host to the epiphyte. Microorganisms on an artificial substratum were more phosphate-limited.

Although environmental conditions were identical for both types of substratum, and there was similar exposure to wave movement for all, we could still observe that, in the case of better developed coating, periphyton on sterilised reed was ready to peel off.

As we did not have any method for studying the compounds excreted by the reed, or the chemical composition of the reed itself, we cannot solve the above problem yet. Both factors probably contribute to the observed phenomena.

The low number of dead cells at the beginning of the study period and their subsequent increase later, indicate that stem cleaning was adequate and that no algae remained on the stems after the cleaning procedure.

While the number of individuals remained about one order of magnitude lower on cleaned green and sterilised reed than on uncleaned green reed, the chlorophyll-a content attained corresponding concentrations to the uncleaned green reed. This is probably explained by the large number of *Achnanthes minutissima* cells in the sample, which, due to their small size, have a lesser effect on the chlorophyll-a content than on cell numbers.

On cleaned green reed, significant negative correlation ($p<10\%$ level) was found between the relative abundances of *A. minutissima* and *G. olivaceum*. It seems that these two species competed in the coating and that *A. minutissima* was more competitive here. On uncleaned green reed there was always significant dominance of *A. minutissima*. On sterilised reed, *A. minutissima* could never "beat" *G. olivaceum*. This may also be explained by nutrient exchange between *A. minutissima* and the host plant.

Several authors (e.g. Goldsborough & Hickman 1991; Otten & Willemse 1988) have pointed out that the physical conditions of the substratum (roughness of the surface, colour, wettability, electrostatic attraction etc.) are important modifiers of accumulation. On hydrophobic substrata the tightly adherent *Cocconeis placentula* was found to be more abundant (Goldsborough & Hickman 1991). In our study the substratum surface properties were the same. We think that *C. placentula* was more competitive under harsh circumstances (e.g. nutrient limitation).

After day 25 the number of individuals on cleaned green and sterilised reed remained lower than on uncleaned green reed, in spite of the calculated doubling time and average mortality rate which would theoretically have allowed this density to be attained within 10 days. This phenomenon can be explained by heavy emigration of the small algae which reproduce rapidly but compete poorly for nutrient (Sommer 1981). This observation is supported by the chlorophyll-a content of the coating which attained that of the control samples (uncleaned green reed) due initially to large (basically filamentous) algae.

## Summary

We found that reed periphyton in Lake Velence could not recover from the perturbation caused by our test over the 25 days of the study period. The number of algal cells did not reach that of the uncleaned reeds. The average doubling time of the periphytic algae (approx. day 10) should be enough to regenerate the coating, but extensive emigration of the rapidly reproducing but poor nutrient competitor small-sized species probably occurred. We can predict from the increase in relative abundance of filamentous algae that the chlorophyll-a content could reach that of the uncleaned reeds.

The increase in the number of individuals on both substrata was quite similar until day 11. After this period, however, growth practically ceased on sterilised reed while it continued at a slower pace on cleaned green reed. The reason for this difference could lie in the metabolic relationship between macrophyte and periphyton, ensuring a phosphate supply for periphyton on the cleaned green reed, while the lower number of individuals of *Achnanthes minutissima* on sterilised reed reflects lower phosphate supply resulted in exfoliation of the coating.

## Acknowledgements

The authors kindly acknowledge the financial support of a research grant OTKA F5251 and would like to express personal thanks to Dr Gyula Lakatos (Univ. L. Kossuth.) and Maria R. Nagy (Environm. Protect. Authority, Székesfehérvár).

## References

Allen, H. L. (1971). Primary production, chemo-organotrophy, and nutritional interactions of epiphytic algae and bacteria on macrophytes in the littoral of a lake. *Ecological Monographs*, **41**, 97–127.

Ács, É. & Kiss, K. T. (1993). Colonization process of diatoms on artificial substrate in the River Danube near Budapest (Hungary). *Hydrobiologia*, **269/270**, 307–315.

Burkholder, J. M. & Wetzel, R. G. (1989). Microbioal colonization on natural and artificial macrophytes in a phosphorus-limited, hardwater lake. *Journal of Phycology*, **25**, 55–65.

Cattaneo, A. & Kalff, J. (1979). Primary production of algae growing on natural and artificial aquatic plants: a study of interaction between epiphytes and their substrate. *Limnology & Oceanography*, **24**, 1031–1073.

Felföldy, L. (1987). A biológiai vízminôsítés (Biological water qualification). Vízügyi Hidrobiológia, **16**, VGI, Budapest.

Goldsborough, L. G. & Hickman, M. (1991). A comparison of periphytic algal biomass and community structure on Scirpus validus and on a morphological similar artificial substratum. *Journal of Phycology*, **27**, 196–206.

Lund, J. W. G., Kipling, C. & Lecren, E. D. (1958). The inverted microscope method of estimating algal numbers and the statistical basis of estimations by counting. *Hydrobiologia*, **11**, 143–170.

Otten, J. H. & Willemse, M. T. M. (1988). First steps to periphyton. *Archiv für Hydrobiologie*, **112**, 177–195.

Padisák, J., G.-Tóth, L. & Rajczy, M. (1988). The role of storms in the summer succession of phytoplankton in a shallow lake (Lake Balaton, Hungary). *Journal of Plankton Research*, **10**, 249–265.

Padisák, J., G.-Tóth, L. & Rajczy, M. (1990). Stir-up effect of wind on a more-or-less stratified shallow lake phytoplankton community, Lake Balaton, Hungary. *Hydrobiologia*, **191**, 249–254.

Shamess, J. J., Robinson, G. G. C. & Goldsborough, L. G. (1985). The structure and comparison of periphytic and planktonic algal communities in two eutrophic prairie lakes. *Archiv für Hydrobiologie*, **103**, 99–116.

Sommer, U. (1981). The role of r- and K-selection in the succession of phytoplankton in Lake Constance. *Acta Oecol. Oecol. Gener.*, **2**, 327–342.

Utermöhl, H. (1958). Zur Vervollkommnung der quantitativen Phytoplankton-Methodik. *Mitteilungen. Internationale Vereinigung für Theoretische und Angewandte Limnologie*, **9**, 1–38.

# A terrestrial epilithic diatom from Roman Catacombs

P. Albertano, L. Kováčik *, P. Marvan* and M. Grilli Caiola

*University of Rome 'Tor Vergata', Department of Biology,*
*via della Ricerca scientifica, I–00133 Rome, Italy*

**Czech Academy of Sciences, Institute of Botany,*
*Dukelská 145, CS–37982, Třeboň, Czech Republic*

## Abstract

The presence of *Diadesmis gallica* W. Smith in terrestrial, light limited environments is recorded and the morphological, ultrastructural and ecophysiological characteristics of the species have been investigated. Following the isolation of several strains from different Roman Catacombs, only a few morphometrical variations were observed among isolates in culture. The results of light-temperature cross-gradient culture of one strain confirmed the adaptation of this species to low irradiances and led to the hypothesis that a competition for light and nutrients can occur *in situ* when prokaryotic and eukaryotic algae are present in the hypogean phototrophic communities.

## Introduction

Up to now the presence of diatoms in terrestrial environments characterized by light limitation has been scarcely recorded (Prát 1925; Krammer & Lange-Bertalot 1986; Claus 1962a, b; Skácelová & Marvan unpublished), and only occasionally have epilithic species been reported for hypogean archaeological sites (Albertano 1993; Albertano & Grilli Caiola 1989; Albertano *et al.* 1991; Altieri *et al.* 1993). The peculiar microclimatic conditions of such sites are due to a limited air circulation, an even temperature throughout the year, a high level of humidity and pronounced light gradients that occur in proximity to entrances or from artificial lamps. Similarities with this type of environment can be found in natural caves where terrestrial algal communities occur and the distribution of phototrophs is also strongly influenced by the low irradiances (Hoffmann 1989). Sciaphilous species are usually the main colonizers of hypogean substrata, and biofilms or crusts of cyanophytes and chlorophytes constitute the major fraction of the lithophytic algal populations which inhabit the exposed surface of rocks, often causing deterioration problems of art works (Albertano 1991).

Among the algal communities found in different Roman hypogea in Rome (Italy), one diatom was recurrently observed in the *in situ* associations. Therefore, morphological and ultrastructural observations and isolation of strains was carried out to identify and characterize the species both in nature and in culture. Moreover, cross-gradient cultures (Albertano *et al.* 1993) of one selected strain were carried out with the purpose of assessing the optimal light and temperature conditions for biomass development in order to determine the growth potentiality of the species.

## Material and Methods

Samples were collected on October 1991 and June 1992 from brick walls, marbles, plasters and frescoes in 4 Roman Catacombs and 1 hypogeum. Sampling sites and microclimatic conditions recorded at the different points are listed in Table 1. Light was measured using a LI–COR LI–185B Quantum/Radio/Photometer equipped with a LI–200 pyranometric sensor, a LI–190SB quantum sensor for photosynthetic available irradiance (PAR), and a LI–210SB photometric sensor. Temperature (°C) and relative humidity (RH %) were recorded by a HI 8564 Hanna Instruments thermo-hygrometer. Aseptically collected samples taken by gently scraping material from the colonized surfaces were either directly prepared for light microscopy and scanning electron microscopy or inoculated in liquid and agarized mineral culture media.

Isolation of strains was achieved in medium Z after Zehnder (Staub 1961) modified by adding 1 ml of $Na_2SiO_3$ solution to a final concentration of 92 mg/l Si. The list of 16 monoalgal strains, which are presently kept in the Culture Collection of Autotrophic Organisms at Treboň (Czech Republic), is reported in Table 1. Cultures of strains were subsequently grown at ±18°C in 16/8 light/dark cycles and analyzed for specific characters and morphometrical differences.

For cross-gradient cultures, 2 ml of uniform algal suspensions of the strain Kováčik & Albertano 1991/1 were inoculated on 48 agar plates and grown in continuous light provided by SHC 400 Watt Tesla lamps as previously reported (Albertano *et al.* 1993). Biomass development at the different irradiances (2–80 $W \cdot m^2$/ 8–440 $\mu mol\ m^{-2} \cdot s^{-1}$) and temperatures (5–31°C) was determined as the increase in dry weight per Petri dish during 10 days from the inoculum. Experiments were made in triplicate.

Light (LM) and electron microscopy (EM) observations were performed either on fresh or treated material. Morphometrical analyses were done using a Leitz Dialux 20 microscope equipped with NPL Fluotar ICT objectives. For electron microscopy, samples fixed in 2.5% glutaraldehyde in 0.2 M phosphate buffer, were post-fixed in 1% $OsO_4$ in the same buffer, dehydrated in ethanol series, gold-coated and observed at 15kV in a DSM 950 Zeiss Scanning Microscope. Permanent preparations of frustules were obtained by hot cleaning in hydrogen peroxide/potassium dichromate (Krammer & Lange-Bertalot 1986) and mounted in Pleurax for LM (Fott 1954) or dehydrated and

Table 1. List of the strains of *Diadesmis gallica* isolated from Roman hypogea.

| Strain (isolators, year/collection n.) | Locality | Sampling site and ecological conditions |
|---|---|---|
| Kováčik & Albertano 1991/1 | Catacombe di Priscilla | 'Criptoportico' corner on the left side of the Cappella Greca, grey-green crusts on plaster near the lamp<br>October 1991: RH 87.4%, T = 17.1°C,<br>light climate 0.3–0.4 μmol · $m^{-2} \cdot s^{-1}$ (0.5–1.0 $W \cdot m^{-2}$/15–20 lux) |
| Kováčik & Albertano 1991/4<br>Kováčik & Albertano 1991/5 | | Entrance corridor, pale-green crusts on the brick wall<br>October 1991 |
| Kováčik & Albertano 1992/8<br>Kováčik & Albertano 1992/9 | Catacombe di S. Agnese | Arch in corridor near the stairs of abside, green crusts on tufo around the lamp<br>June 1992: RH 86.4%, T = 17.5°C,<br>light climate 9–12 μmol · $m^{-2} \cdot s^{-1}$ (7–10 $W \cdot m^{-2}$/200–400 lux) |
| Kováčik & Albertano 1992/2<br>Kováčik & Albertano 1992/3 | | Near the foregoing site at the end of corridor, green films on brick wall around the lamp<br>June 1992: RH 86.4%, T = 17.6°C,<br>light climate 10–12 $\mu mol \cdot m^{-2} \cdot s^{-1}$ (15–35 $W \cdot m^{-2}$/200–500 lux) |
| Kováčik & Albertano 1991/14<br>Kováčik & Albertano 1991/15<br>Kováčik & Albertano 1991/16 | Catacombe di S. Callisto | Soil from a tomb with illumination<br>October 1991 |
| Kováčik & Albertano 1992/6<br>Kováčik & Albertano 1992/7 | | 'Pannello dei simboli' on arch, green and brown film on marble and gypsum<br>June 1992: RH 94.5%, T = 18.1°C, light climate 0.3–0.5 $\mu mol \cdot m^{-2} \cdot s^{-1}$ (0.5–1.2 $W \cdot m^{-2}$/20–30 lux) |
| Kováčik & Albertano 1991/11<br>Kováčik & Albertano 1991/12<br>Kováčik & Albertano 1991/13 | Catacombe di S. Sebastiano | 'Arcosoglio', green soil near the lamp<br>October 1991 |
| Kováčik & Albertano 1991/10 | Basilica Inferiore di S. Clemente | Brown film on brick wall<br>October 1991 |

gold-coated for SEM or negatively stained with 2% uranyl acetate and observed with a CEM 902 Zeiss transmission electron microscope at 80 kV.

## Results

The presence of diatoms was observed in all the Roman Catacombs investigated (see Table 1). Temperature and relative humidity ranged from 16 to 20°C and 86.4–94.5 RH% at the various sampling points therefore showing very uniform values for the various Catacombs. More differences were detectable in the light climates varying from extreme low (0.5 W $m^{-2}$ to low (10 W $m^{-2}$ total irradiance and between 0.3 to 12 µmol $m^{-2}\cdot s^{-1}$ PAR. No apparent preference for one type of substratum was observed in the distribution of the diatom.

From LM and EM observations of natural and isolated material, *Diadesmis gallica* W. Smith 1857 (see Mann in Round *et al.* 1990 for the retention of the original genus *Diadesmis*) (syn: *Navicula gallica* (W. Smith) Lagerstedt 1873) was identified as the only diatom species present *in situ* and in cultures (Figs 1, 2). In natural samples, the cells are frequently arranged into ribbon colonies and interspersed among the mineral fragments of the substratum and other algal components of the microbial communities (Figs 3, 4).

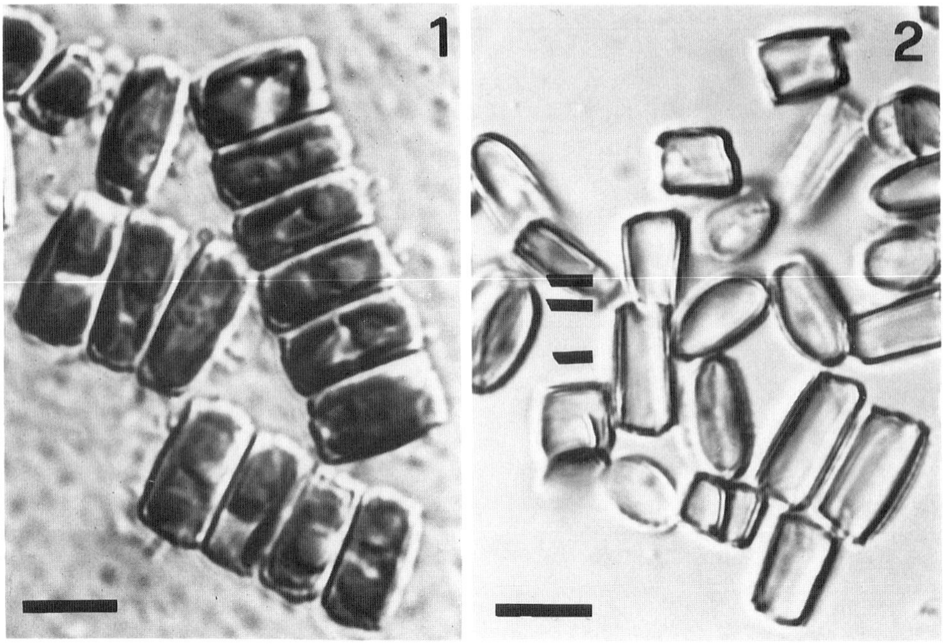


Fig. 1. LM photographs of living colonies of *D. gallica* strain Kováčik & Albertano 1991/1. Fig. 2. LM photographs of cleaned frustules of *D. gallica* strain Kováčik & Albertano 1991/5 in valve and girdle view. Bars = 5 µm.

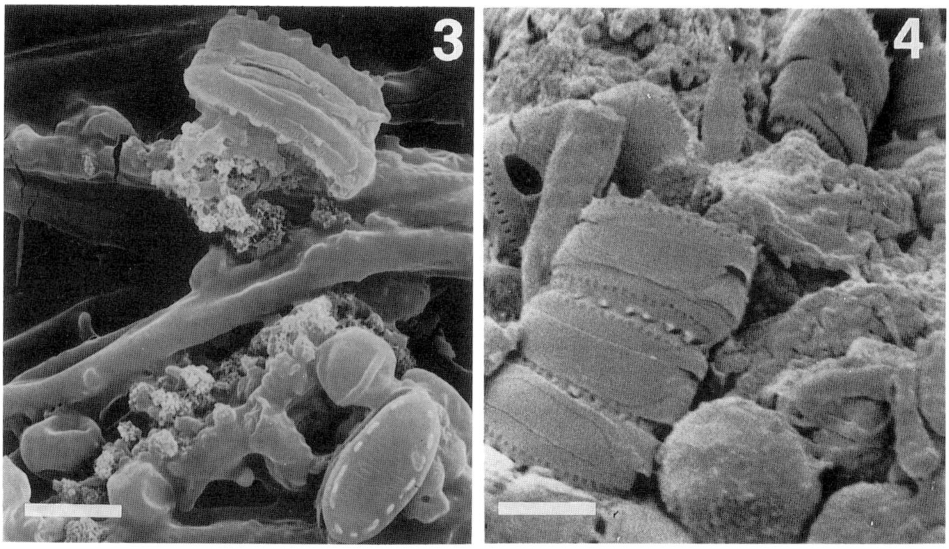


Figs 3–4. SEM micrographs of *D. gallica* in samples collected in June 1992 at the "Catacombe di S. Sebastiano". Single cells and colonies are interspersed among mineral fragments, filamentous cyanophytes (c) and green coccal algae (g). Bars = 2 µm.

*Description of* Diadesmis gallica *from Roman Catacombs*

Cell solitary or in short ribbon colonies (Figs 1, 3, 5) connected one to the other by marginal spines. One parietal chloroplast without pyrenoid. Valves elliptical through linear, with broadly rounded never protracted ends. In LM with hardly visible structure, (2.3) 3.0–9.0 (9.7) µm long and about 2–4 µm wide. Striae more or less shortened, parallel up to slightly radial near the ends, 26–31 in 10 µm, leaving a ± broad lanceolate axial area (Fig. 6). Marginal spines truncated, in girdle view up to 0.5 µm long, about 10 in 10 µm (Figs 6–8). The valve mantle with a row of perforations of the same density as the striae. In LM the raphe is not visible, whereas EM revealed the presence along the apical axis of two simple longitudinal fissures about 2 µm long leaving a central solid area of 1–1.5 µm apparently only in valves lacking marginal spines (Figs 9–11). Simple girdle with one row of perforations more dense than on the valve mantle (Fig. 12).

The LM morphometrical observations of the 16 isolates showed little variation of frustule length in culture. Although there are some differences in cell dimensions, no distinct gap could be detected among the isolates (Fig. 13).

The results of cross-gradient experiments (Fig. 14) indicated growth between 5.5 and 28°C. The best algal development occurred at 14–23°C in the range of light climate 6 to 30 W $m^{-2}$/30–150 µmol $m^{-2} \cdot s^{-1}$, while yield of dry matter decreased at the lowest irradiances even if still influenced by the temperature gradient (Fig. 15).

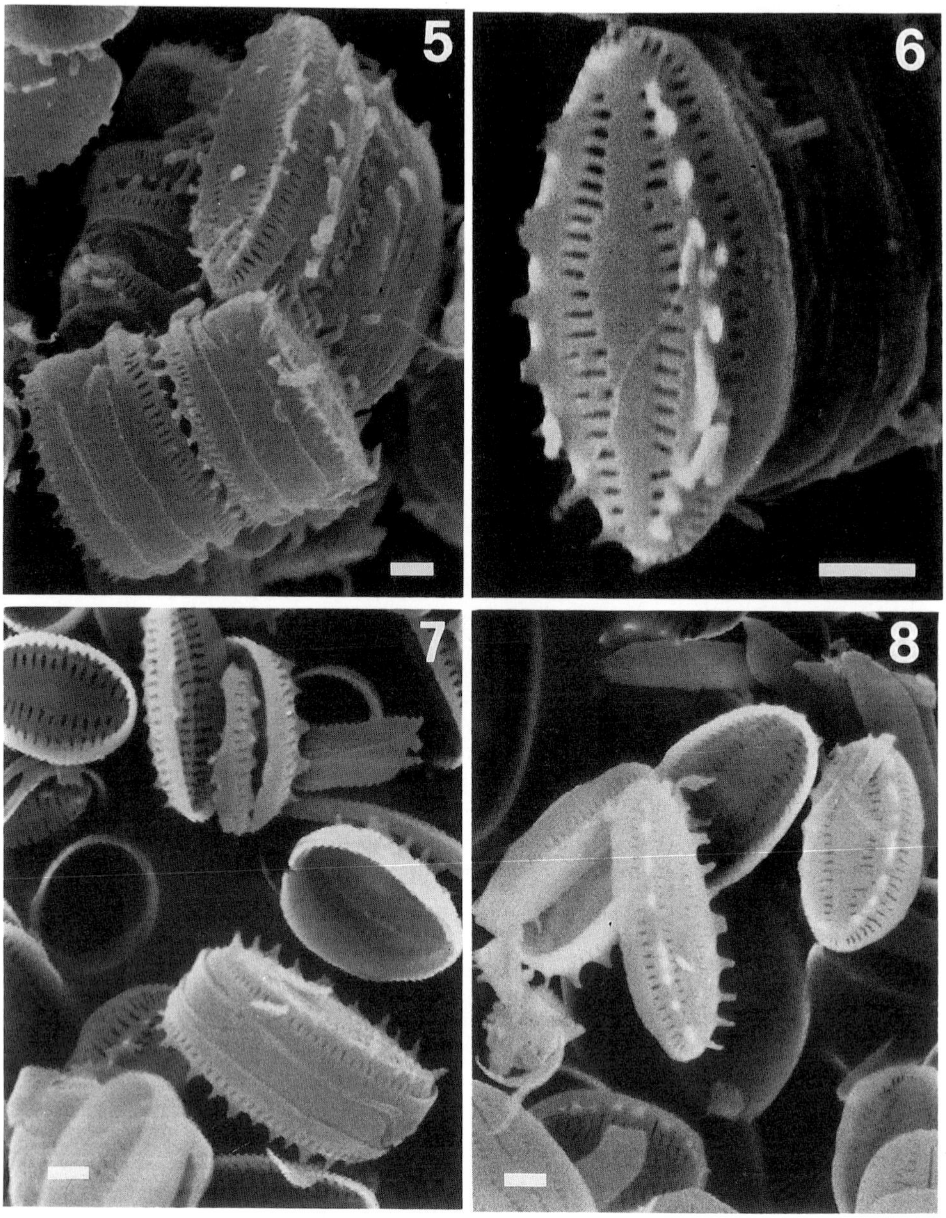


Figs 5–8. SEM micrographs of *D. gallica* strain Kováčik & Albertano 1991/1. Few celled colonies in girdle (Fig. 5) and valve-mantle view (Fig. 6). Cleaned frustules in different external and internal view (Figs 7, 8). Bars = 1 µm.

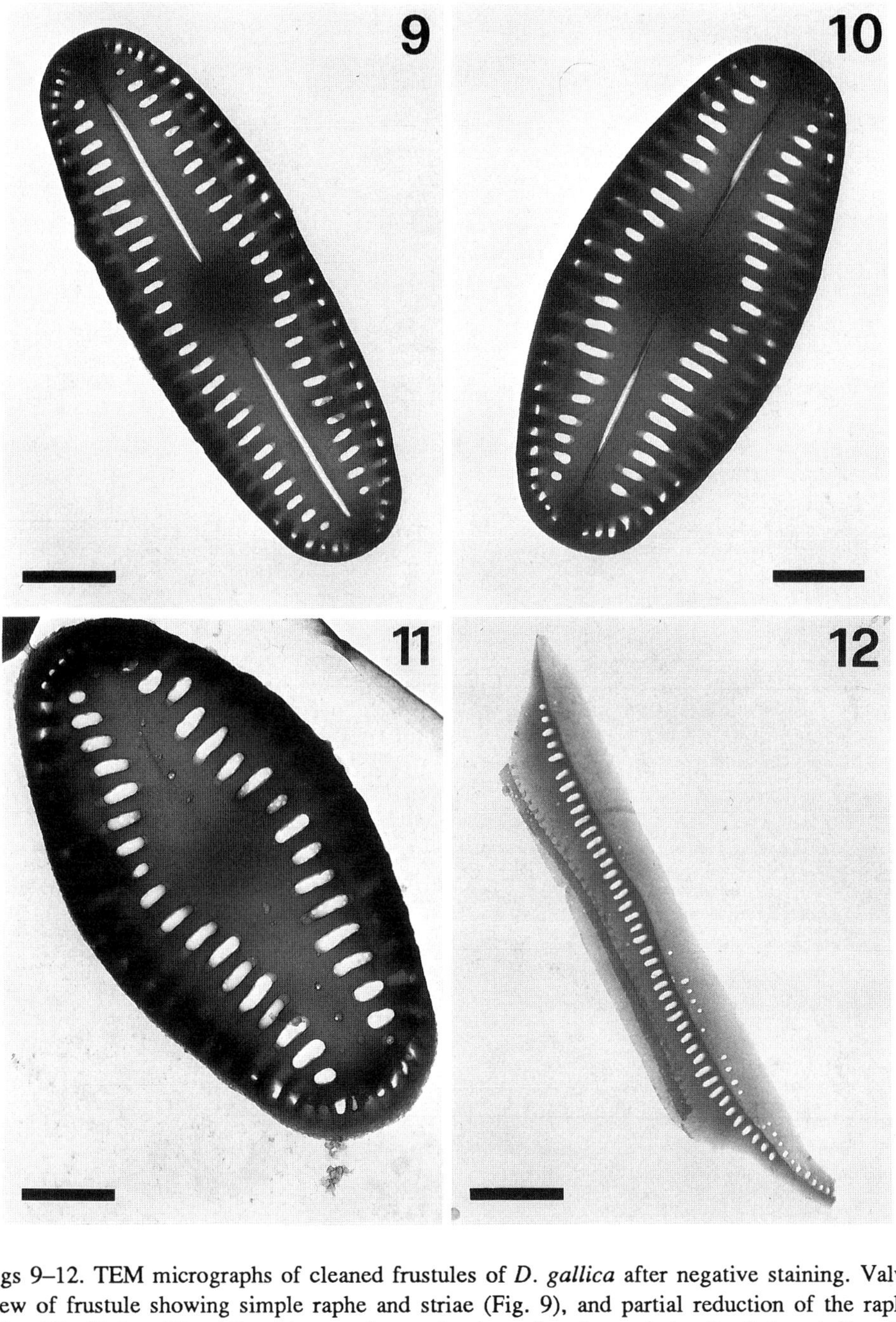


Figs 9–12. TEM micrographs of cleaned frustules of *D. gallica* after negative staining. Valve view of frustule showing simple raphe and striae (Fig. 9), and partial reduction of the raphe (Fig. 10). Valve (Fig. 11) with complete reduction of raphe and detail of the girdle with perforations (Fig. 12). Bars = 1 µm.

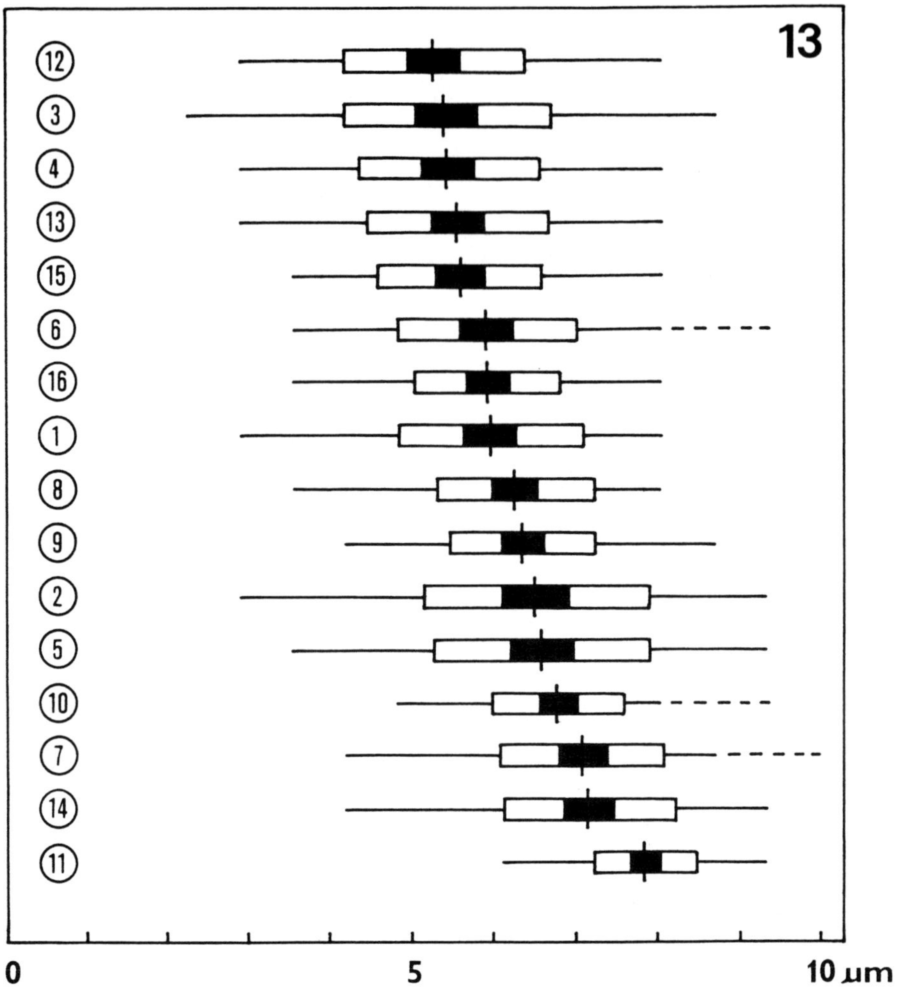


Fig. 13. Cell lengths of the 16 strains of *D. gallica* in culture (see Table 1) arranged according to increasing mean value. Solid lines indicate the whole range of the cell length variation. White rectangulars indicate mean and standard deviation ($\overline{x} \pm s_x$). Black rectangulars indicate 95% confidence interval of mean.

## Discussion

*D. gallica* is the species found in the different hypogean sites investigated. Intermediate forms to other similar taxa as illustrated by Krammer and Lange-Bertalot (1986) for varieties of *Navicula gallica* (W. Smith) Lagerstedt s.l. were never observed. Contrary to the description of *N. gallica* s.s. in Hustedt (1962, p. 207–208), raphe *in situ* and in cultures is almost always fully reduced and the form of axial area resembles that of *N. fragilarioides* Krasske in l.c. fig. 1325. The presence of a simple raphe in

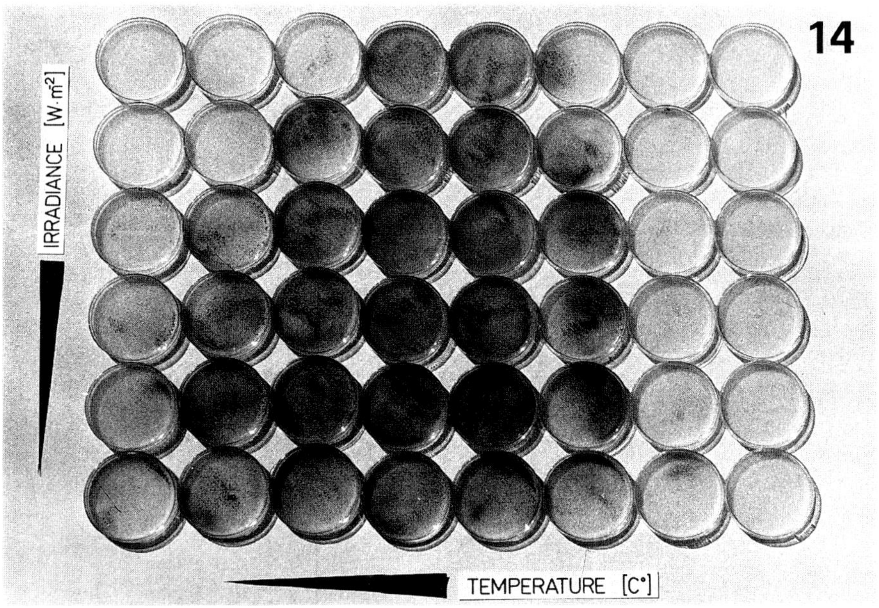


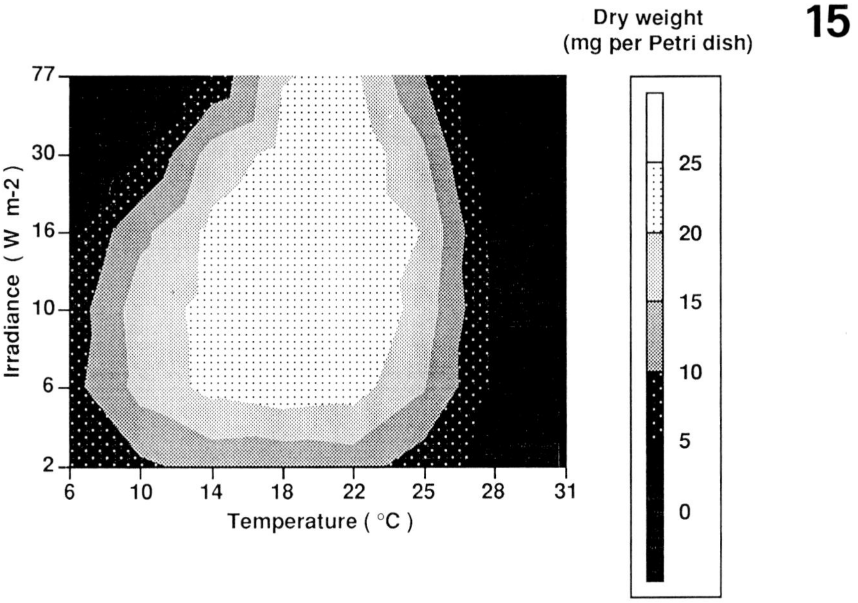


Fig. 14. Cross-gradient cultures of *D. gallica* after 10 days from the inoculation. Growth is visible as dark grey colour in some of the 48 agar plates. Black wedges indicate decrease of temperature and irradiance. Fig. 15. Biomass development (mg of dry weight per Petri dish) of *D. gallica* grown under different combinations of light and temperature.

valves lacking marginal spines may be due to the developmental phase of the frustule or to the adaptation to the colonial life.

As regards light adaptation, on the basis of our results *D. gallica* should be regarded as a sciaphilous species because of its best development in culture at irradiances around 10 W $m^{-2}$/50 µmol $m^{-2} s^{-1}$. These data are partially consistent with those recorded *in situ,* which reported the presence of *D. gallica* also in sites with the most extreme light climates. Nevertheless, both literature (Krammer & Lange-Bertalot 1986) and the present report indicate this species as a typical colonizer of dim light habitats in association with cyanophytes and a few chlorophyte species (Albertano 1993). The same applies to one locality in the Czech Republic where this species grew under extreme low light intensities on the internal walls of a deep well in the Spilberk Castle in Brno (Skácelová & Marvan unpublished).

Therefore, the presence of *D. gallica* in the hypogean phototrophic communities of Roman archaeological sites can be regarded as a distinctive feature of these algal associations which are dominated by filamentous cyanophytes such as *Leptolyngbya* or *Fischerella.* Nevertheless, *D. gallica* is also able to grow at irradiances and temperatures utilized by the cyanophytes (Grilli Caiola *et al.* 1987, Albertano & Grilli Caiola 1988, Albertano *et al.* 1993), and also to develop a greater biomass at the lowest temperatures and highest irradiances tested. Consequently a scarcity of some nutrient or a strong nutrient competition between diatoms and cyanophytes in nature may be hypothesized to explain the successful colonization by the latter. However, increased growth of *Diademis gallica* may also occur as a consequence of possible man-induced variations in light climate or air pollution. Therefore there is a need for evaluation of growth in competition experiments with the other algae present in the association.

## Acknowledgements

We gratefully acknowledge the Pontificia Commissione di Archeologia Cristiana that allowed the samples collection inside the Catacombs of Rome. This work has been supported by a grant of the National Research Council of Italy, C.N.R–CT15.

## References

Albertano, P. (1991). The role of photosynthetic microorganisms on ancient monuments. A survey on methodological approaches. *Journal of European Studies on Physical, Chemical, Biological and Matemathical Techniques applied to Archaeology PACT*, **33**, 151–159.

Albertano, P. (1993). Epilithic algal communities in hypogean environments. *Giornale Botanico Italiano*, **127**, 386–392.

Albertano, P. & Grilli Caiola, M. (1988). Effects of different light conditions on *Lyngbya* sp. in culture. *Archiv für Hydrobiologie, Algological studies*, **50–53**, 47–54.

Albertano, P. & Grilli Caiola, M. (1989). A hypogean algal association. *Braun-Blanquetia*, **3**, 287–292.

Albertano, P., Kovacik, L. & Gardavsky, A. (1993). Cross-gradient cultures of filamentous cyanophytes. *Giornale Botanico Italiano*, **127**, 855–856.

Albertano, P., Luongo, L. & Grilli Caiola, M. (1991). Observations on cell structure of micro-organisms of an epilithic phototrophic community competing for light. *Nova Hedwigia*, **53**, 369–381.

Altieri, A., Pietrini, A.M. & Ricci, S. (1993). Un'associazione di alghe e muschi in un sito archeologico ipogeo. *Giornale Botanico Italiano*, **127**, 611.

Claus, G. (1962a). Data on the ecology of the algae of Peace Cave in Hungary. *Nova Hedwigia*, **4**, 55–79.

Claus, G. (1962b). Beitrag zur Kenntnis der Algenflora der Abaligeter Höhle. *Hydrobiologia (Den Haag)*, **19**, 192–222.

Fott, B. (1954). Pleurax, synthetická priskyrice pro preparaci rozsivek. (Pleurax, a synthetic resin for mounting of diatoms). *Preslia*, **26**, 193–194.

Grilli Caiola, M., Forni, C. & Albertano, P. (1987). Characterization of the algal flora growing on ancient Roman frescoes. *Phycologia*, 26, 387–390.

Hoffmann, L. (1989). Algae of terrestrial habitats. *The Botanical Review*, **55**, 77–105.

Hudstedt, F. (1962). Die Kieselalgen. In: *Dr. L. Rabenhorsts Kryptogamenflora von Deutschlands, Österreichs und der Schweiz* 7/3 (2): 161–348. Akademische Verlagsgesellschaft, Leipzig.

Kramer, K. & Lange-Bertalot, H. (1986). Bacillariophyceae I: Naviculaceae. In: *Süsswasserflora von Mitteleuropa* 2/1 (H. Ettl, J. Gerloff, H. Heynig and D. Mollenhauer, eds), pp. 876. Gustav Fischer Verlag, Jena.

Prát, S. (1925). Das Aèroplankton neu geöffneter Höhlen. *Zentralblatt für Bakteriologie, Parasitenkunde, Infektionkraunkheiten und Hygiene (Jena) II*, **64**, 39–40.

Round, F. E., Crawford, R. M. & Mann, D. G. (1990). *The Diatoms. Biology and morphology of the genera*. 747 pp. Cambridge University Press, Cambridge.

Staub, R. (1961). Ernährungsphysiologisch-autökologische Untersuchungen an der planktonischen Blaualge *Oscillatoria rubescens* DC. *Schweizerische Zeitschrift für Hydrologie*, **23**, 82–198.

# Ecology and autecology of *Tabellaria flocculosa* var. *asterionelloides* Grunow (Bacillariophyceae) with remarks on the validation of the species name

Martin T. Dokulil and Sigrid Kofler

*Institut für Limnologie/Abt. Mondsee,*
*Österreichische Akademie der Wissenschaften,*
*Gaisberg 116, A–5310 Mondsee, Austria*

## Abstract

*Tabellaria fenestrata* Grunow has been an important component of the phytoplankton assemblages of the deep, alpine lake Mondsee in the past. Investigations by LM and SEM on both field material and clonal batch cultures show that the correct species name must be *T. flocculosa* var. *asterionelloides* Grunow.

Long-term records from the sediment and from field material since 1953 indicate that the species was not present during the pre-eutrophication phase and became increasingly abundant as eutrophication progressed. During oligotrophication, biovolume drastically decreased as a consequence of changed N:P and Si:P ratios.

Physiological adaptation to temperatures <20°C, low light intensities, preference for green light, growth under moderatly stable conditions and low loss rates explain the competitive advantage over other species when nutrients are abundant.

## Introduction

The genus *Tabellaria* is a common and widely distributed freshwater alga throughout the world. The few species recorded among the genus, inhabiting the benthos and the plankton of lakes, often form abundant populations and can dominate the diatom assemblages. Because of its widespread apparence, the genus *Tabellaria* has recently been the subject of many morphological and ecological studies (e.g. Knudson 1952, 1953a, b, c, 1954; Lehn 1969; Koppen 1975, 1978; Flower & Batterbee 1985; Kofler 1986; Theriot & Ladewski 1986; Lange-Bertalot 1988). These investigations, however, created new problems regarding the taxonomic position of the species (see Krammer & Lange-Bertalot 1991, page 104 ff.).

Species records from the phytoplankton of many European lakes include *Tabellaria fenestrata* (Lyngbye) Kütz., but this species seems to be widely confused with long-celled forms of *Tabellaria flocculosa* (Roth) Kütz. Since the species is regarded as an important indicator of eutrophication (Nipkow 1920; Huber-Pestalozzi 1942; Grim 1955) clarification of its taxonomic position and ecology is important.

This paper intends to summarise ecological long-term records of *Tabellaria* (Dokulil 1991) from the alpine lake Mondsee in Austria (Dokulil & Skolaut 1986) with observations from clonal cultures (Kofler 1986).

### *Background*

*Tabellaria* first appeared in the phytoplankton counts from Mondsee in the year 1968 (Findenegg 1969) when eutrophication began. This observation is confirmed by diatom fossil records from the sediment (Klee & Schmidt 1987; Schmidt 1991). During the eutrophication period from 1968 to 1982, *Tabellaria* regularly formed large populations every early summer and autumn (Oberrosler 1979; Schwarz 1979; Dokulil 1991; Schmidt 1991) while blooms of *Limnothrix rubescens* dominated during the summer (Dokulil 1987). Restoration measures were introduced in 1973 but significant signs of oligotrophication were not observed until the early 1980´s (Dokulil & Jagsch 1989).

## Material and Methods

### *Field material*

Samples were routinely taken at the mid-lake station from various depths at weekly intervals in the years 1982–1984. In the following years, integrated samples (0–20 m) were collected each month (1985 and 1986) or every two weeks (1987 to 1994). During these years the horizontal and vertical distribution was investigated on several occasions. Phytoplankton counting, cell volume calculations and biomass estimation from cell sizes followed the inverted microscope technique of Lund *et al.* (1958) facilitated by direct input to a computerized plankton counter since 1989 (Hamilton 1990) and sizing on a digitizing tablet.

Standard methods were used for all chemical analyses. Secchi-depth was measured with a white disk of 25 cm diameter, and temperature recorded with a mercury thermometer. Under-water light attenuation was measured with a Li-Cor quantum-radiometer using a $2\pi$-sensor.

### *Culture material*

Clonal cultures of *Tabellaria* were isolated from net plankton samples taken from Mondsee in spring 1983 and kept in a modified Chu-medium (Chu 1942).

All experiments were performed in a light incubator at constant temperature (±0.5°C), illuminated by fluorescent tubes with a light-dark cycle (L:D) of 14:10 hours. Light intensities in the range of 8–260 µE $m^{-2}$ $s^{-1}$ were used at five temperatures between 5° and 25°C.

Effects of light quality on growth were tested at 20°C using coloured plastics screens adjusted to give a total light intensity of 40 µE $m^{-2}$ $s^{-1}$.

All cultures were pre-incubated at defined temperature and light conditions for at least ten days before performing experiments. Average growth rates were calculated from cell counts during the logarithmic phase of 15 parallel inoculations. Chlorophyll-*a* determinations were made in triplicate according to the method of Nusch (1980). Concentration changes of individual algal pigments were estimated from thin layer chromatography (Züllig 1982) during light quality experiments.

*Taxonomic position of the species investigated*

The populations investigated appear as star-shaped to rarely zigzag colonies in the field material, while both forms and all possible intermediate stages occurred at the same time in the clonal cultures. Individual cells have up to 4, in culture sometimes more, intercalary bands. Marginal spines are clearly visible (Fig. 1a). Open intercalary bands are common but do not have regular or rudimentary septa (Fig. 1b, c). The position of the rimoportula is variable but always occurs near to the central inflation (Fig. 1d). Data on dimensions and cell morphology are summarised for the field and cultured material in Table 1. Cells of the cultured strains are on average somewhat shorter than cells from field populations, and hence their length-width ratio and their surface to volume ratio differs too. In some clonal strains extreme size reduction was observed following protoplast contraction to the central area of the cell, as a consequence of strong silica depletion in the culture medium, similar to observations by Kling (1993) on *Asterionella formosa.* These cultures were not used for experiments.

In accordance with Knudson (1953a,b), we conclude from all these characters that the *Tabellaria*-populations in Mondsee should be identified as *Tabellaria flocculosa* var. *asterionelloides*, even though *T. flocculosa* should not have incomplete intercalary bands (Krammer & Lange-Bertalot 1991). This latter feature would be characteristic of *T. fenestrata,* when combined with apparent definite frustule formation and the lack of marginal spines (Koppen 1975). Because specimens from one and the same clonal culture appear with or without incomplete intercalary bands, we reject this as a diagnostic character for delineating between species among the genus *Tabellaria.*

**Results**

Seasonal appearence of *Tabellaria* follows a regular pattern during the period 1982 to 1993. Maximum development of the species occurs in early summer and late fall, except for the years 1984 and 1987 when only a single peak occurred (Fig. 2). Lowest

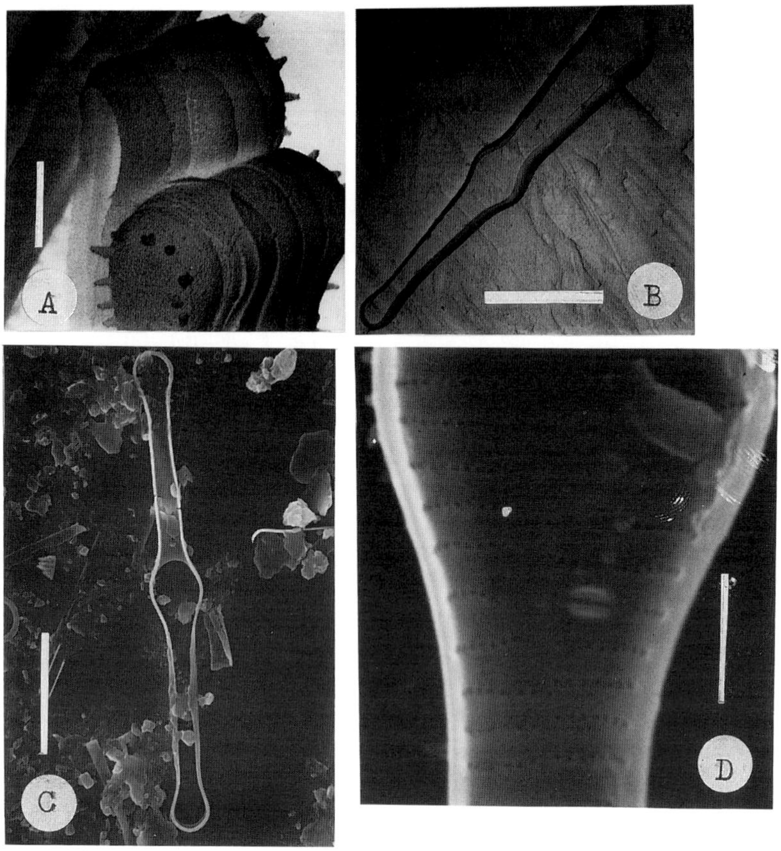


Fig. 1. *Tabellaria flocculosa* var. *asterionelloides*, SEM. A. Cell end with spines and intercalary bands, scale bar = 2 µm. B. Open intercalary band, scale bar = 10 µm. C. Intercalary band with septum, scale bar = 10 µm. D. Central inflation with rimoportula, scale bar = 2 µm.

biovolumes of around 5 × $10^6$ $\mu m^3$ $l^{-1}$ are present throughout the winter while peak biovolumes are in the range 1000–3000 × $10^6$ $\mu m^3$ $l^{-1}$ during the first half of the observation period (1982–1987). A substantial reduction in biovolume of *Tabellaria* was observed from 1987 to 1988. Values for the period 1988 to 1993 are lower by about two orders of magnitude (0.5–20 × $10^6$ $\mu m^3$ $l^{-1}$). In the years 1991 and 1992 only spring peaks developed while *Tabellaria* was sporadically recorded in 1993 and 1994.

The timing of the greatest change in abundance coincides with a substantial increase in the concentration of total inorganic nitrogen and hence a change in the N:P ratio (Fig. 3 and Table 2). Annual average total phosphorus concentrations are significantly lower since 1990 in comparison with previous years (Table 2 and Fig, 3), resulting in a further increase in the N:P ratio. A substantial rise in the Si:P ratio (Table 2) has also occurred due to the steadily increasing dissolved silica concentrations since 1986 (Fig. 4). Water temperature at the surface and Secchi-depth, as a measure of under-water light conditions, do not show any trend (Fig. 4 and Table 2).

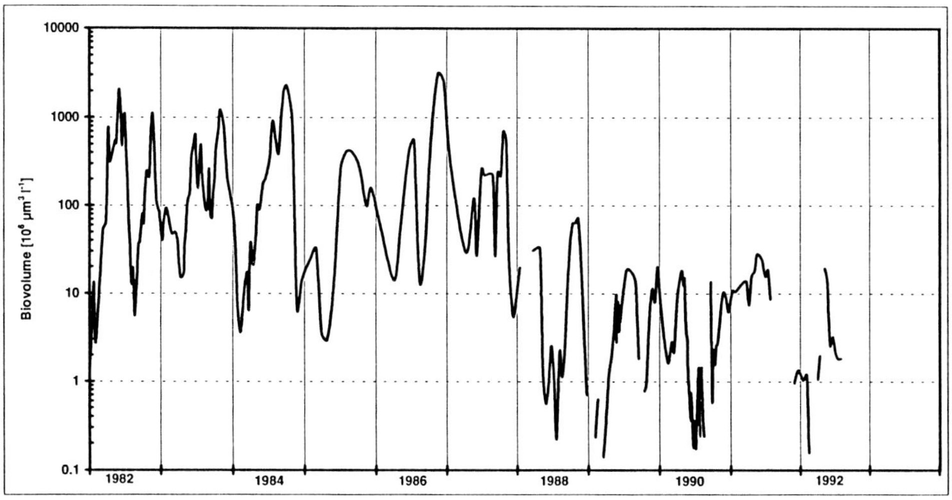


Fig. 2. Time sequence of the average euphotic zone biovolume ($10^6$ $\mu m^3$ $l^{-1}$, log scale) of *Tabellaria flocculosa* var. *asterionelloides* for the years 1982–1993. Abundancies in 1993 are too low and too scarce to show up on the graph.

Table 1. Averages and ranges of the cell dimensions, the cell volume (V), the cell surface (A) and the cell surface to volume ratio (A/V) for the field material from the years 1982–1985, for the clonal strains from 1984 and for a culture from the Windermere algal collection.

| | Field Material (1982–1985) n = 65 | | Strain 1984 n = 37 | | Strain England n = 22 | |
|---|---|---|---|---|---|---|
| Length [µm] | 58.1 | (52 – 57) | 35.6 | (30–39) | 33.9 | (32 – 36) |
| Width [µm] | 3.5 | (3 – 4.5) | 3.8 | ( 3 – 4) | 3.4 | (2 – 4) |
| Depth [µm] | 7.1 | (5 – 8) | 6.5 | ( 4 – 8) | 6.3 | (3 –16) |
| L/W | 16.6 | (12.7 – 17.3) | 9.4 | (9.2 – 10.0) | 9.9 | (9 –16) |
| Surface [$\mu m^2$] | 1057 | (680 – 1200) | 776 | (584 – 936) | 703 | (483 – 1367) |
| Volume [$\mu m^3$] | 1219 | (720 –1452) | 1212 | (591 – 1357) | 861 | (409 – 2182) |
| A/V | 0.87 | (0.83 – 0.94) | 0.78 | (0.70 – 0.98) | 0.85 | (0.62 – 1.01) |

Mean growth rates (µ) from laboratory experiments, temperature and light conditions systematically varied, were highest at 20°C (0.203–0.384) in all but the experiments at the lowest light intensity, where growth rate was highest at 10°C (Fig. 5). Temperatures

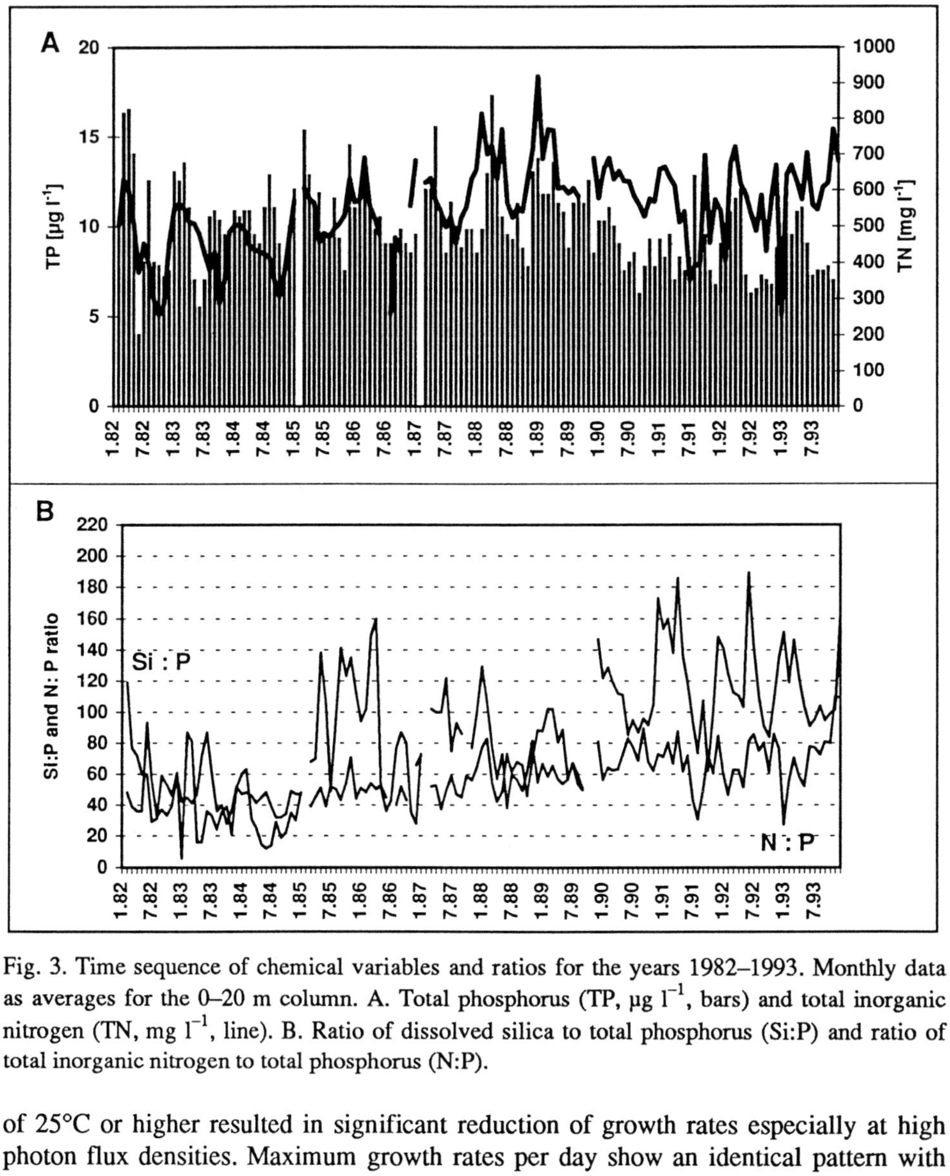


Fig. 3. Time sequence of chemical variables and ratios for the years 1982–1993. Monthly data as averages for the 0–20 m column. A. Total phosphorus (TP, µg $l^{-1}$, bars) and total inorganic nitrogen (TN, mg $l^{-1}$, line). B. Ratio of dissolved silica to total phosphorus (Si:P) and ratio of total inorganic nitrogen to total phosphorus (N:P).

of 25°C or higher resulted in significant reduction of growth rates especially at high photon flux densities. Maximum growth rates per day show an identical pattern with rates about 20% higher. Growth rates become light saturated at 60–80 µmoles of photons per square meter per second (Fig. 5).

Average rates of net population changes (k´) of *Tabellaria* in the epilimnion of Mondsee from the years 1982 to 1984 are compared in Fig. 6 to growth rates derived from culture experiments for the respective mean epilimnetic temperature-light combinations. Missing data in the growth rate curve indicate photon flux densities higher than those tested in our laboratory experiments. In all but one case, net rates of population changes are lower than derived growth rates because these include losses from cell death, sinking, parasitism and grazing. However, the two curves correspond closely between September and May. Differences are greatest during June, July and August. Population changes are mainly controlled by light availability and temperature during most of the year, while nutrients seem to be the major controlling factor during summer.

Table 2. Annual average physico-chemical data for the years 1982–1993, and mean values for the period. Surface temperature, secchi-depth ($Z_{SD}$), dissolved silica concentrations (Si), total phosphorus (TP), total inorganic nitrogen ($TN_{inorg}$) and the ratios TN:TP and Si:P are given.

| Annual Averages | Surface Temp. | $z_{SD}$ | $z_{eu}$ | Si | TP | $TN_{inorg}$ | TN/TP | Si/P |
|---|---|---|---|---|---|---|---|---|
| Year | °C | m | m | mg $l^{-1}$ | µg $l^{-1}$ | µg $l^{-1}$ | | |
| 1982 | 12.71 | 5.32 | 12.68 | 0.51 | 10.21 | 430.73 | 42.19 | 50.11 |
| 1983 | 13.33 | 5.23 | 12.70 | 0.42 | 9.99 | 458.92 | 45.93 | 41.75 |
| 1984 | 9.56 | 5.61 | 14.02 | 0.31 | 10.23 | 428.92 | 41.95 | 30.32 |
| 1985 | 9.86 | 4.81 | 13.50 | 0.48 | 11.32 | 534.82 | 47.25 | 42.25 |
| 1986 | 11.51 | 4.14 | 10.35 | 0.37 | 10.12 | 509.30 | 50.34 | 36.57 |
| 1987 | 13.52 | 4.72 | 11.81 | 0.70 | 10.75 | 566.45 | 52.72 | 65.14 |
| 1988 | 13.73 | 4.51 | 11.28 | 0.73 | 10.95 | 652.42 | 59.56 | 67.05 |
| 1989 | 11.56 | 4.80 | 12.00 | 0.82 | 11.40 | 681.90 | 59.84 | 72.03 |
| 1990 | 15.44 | 4.88 | 12.19 | 0.84 | 8.79 | 606.00 | 68.93 | 95.10 |
| 1991 | 11.98 | 5.64 | 14.10 | 0.91 | 8.54 | 540.08 | 63.23 | 106.89 |
| 1992 | 11.58 | 4.60 | 11.51 | 1.03 | 8.48 | 570.17 | 67.24 | 121.29 |
| 1993 | 10.93 | 5.00 | 12.51 | 1.19 | 8.69 | 604.33 | 69.56 | 137.41 |
| **Mean 82-93** | **12.14** | **4.94** | **12.39** | **0.69** | **9.95** | **548.67** | **55.73** | **72.16** |

Since greatest abundances in the field often extend from the lake surface down to 15 m with maxima usually at 5–7 m, light quality could be of significant importance. Growth rates of both cultured and field material were enhanced when incubated in blue or green light, when compared to white light incubations of the same intensity. Exposure to red light resulted in elevated proportions of dead cells (Kofler 1986). During blue light incubations the chlorophyll-*a* content per cell decreased from 3.39 pg $cell^{-1}$ in white light to 1.30 pg $cell^{-1}$, while it increased to 3.71 pg $cell^{-1}$ in red light and to 4.25 pg $cell^{-1}$ in green light. Thin layer chromatography revealed that green light causes a general pigment increase of up to 4x mainly because of a substantial rise in the chlorophyll-*c* level (approx. 500% of white light concentrations). In contrast, all pigment contents are reduced after exposure to blue light while red light causes stimlulation of total carotenoids, especially fucoxanthin. These experiments and observations confirm that *Tabellaria* is best adapted to grow in green light which is the most penetrating light component in Mondsee.

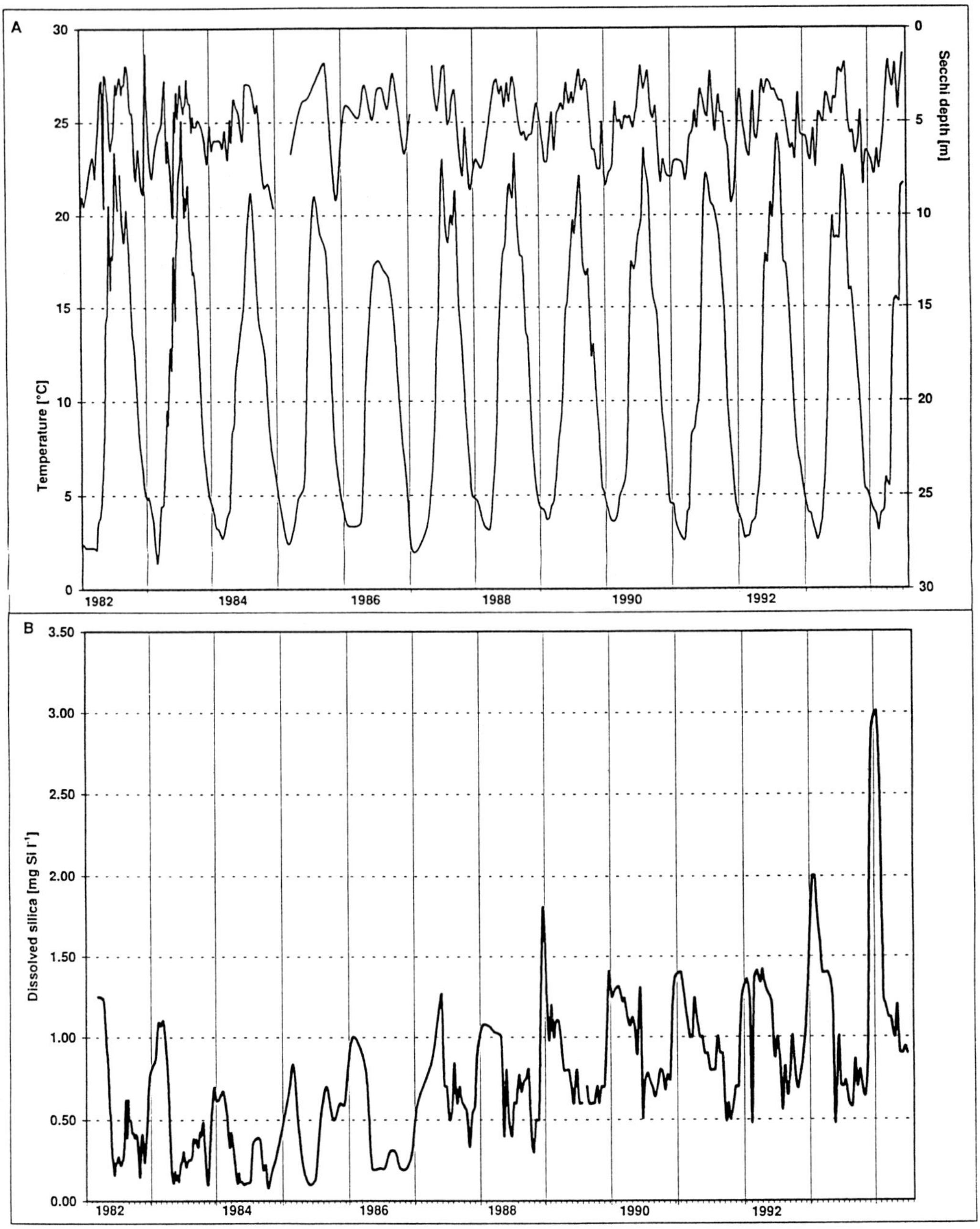


Fig. 4. Time sequence of physico-chemical variables for the years 1982–1994. A. Secchi depth (m) and surface water temperature (°C). B. Dissolved silica (Si, mg $l^{-1}$) averaged over the top 20 m.

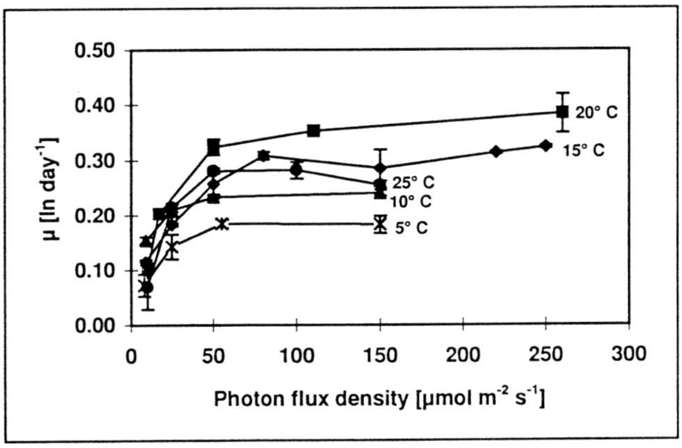


Fig. 5. Growth rates (μ, ln day$^{-1}$) and standard deviation of the mean of laboratory batch cultures at 5, 10, 15, 20 and 25°C versus photon flux density (μmol m$^{-2}$ s$^{-1}$).

Gross photosynthetic rates of cultures of *Tabellaria* expressed as oxygen evolution per cell (Fig. 7) increase in general with temperature. Net rates increase at photon flux densities of >30 μmoles m$^{-2}$ s$^{-1}$ up to 20°C, while rates at 10 μmoles decrease. Both gross and net rates are reduced at 25°C and 150 μmoles of photons due to much greater respiration rates. Comparable results are obtained if rates are expressed per unit chlorophyll-*a*, except for 25°C where rates are depressed at all light intensities. Comparative analysis of carbon uptake and oxygen evolution rates gave photosynthetic quotients of about 0.84. *In situ* estimates of photosynthetic rates of *Tabellaria* were consistent with predictions from laboratory experiments using field data on light intensity and temperature.

## Discussion

Several questions regarding the taxonomic position of the taxon investigated remain open. The combination of characters observed in this study – marginal spines, four to several, frequently incomplete intercalary bands, colonies in the field characteristically "*fenestrata*"-shaped – is to some extent discussed by Lange-Bertalot (1988) when trying to separate *T. fenestrata* from *T. quadriseptata*. Contrary to Koppen (1975) who considers incomplete intercalary bands as typical for definite frustule formers, this was a common feature in our material which is an indefinite frustule former, according to the definition of Knudson (1952). The characters observed, therefore place our taxon somewhere between *T. fenestrata* and *T. flocculosa*.

However, the autecology of the taxon in a moderately alkaline, oligo-mesotrophic lake, the number of intercalary bands, the cell morphology and morphometry, the predominating star-shaped colonies, and the euplanktonic habit allow us to conclude that the taxon is best identified as *Tabellaria flocculosa* var. *asterionelloides* (Knudson

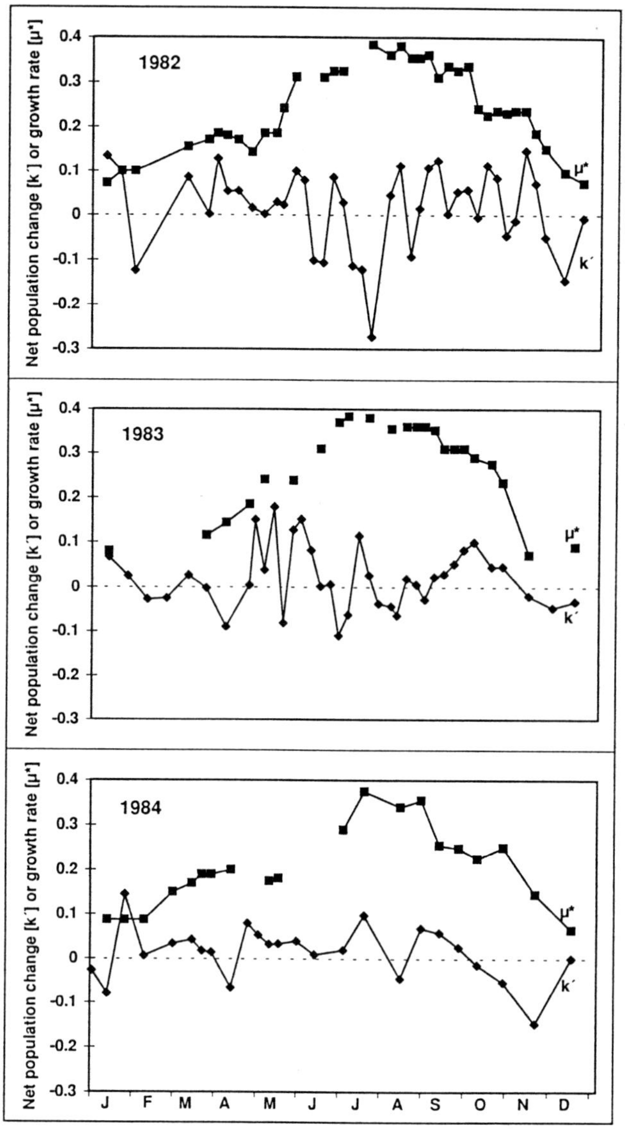


Fig. 6. Time sequence for the years 1982, 1983 and 1984 of average net population change (k´) during the observation interval derived from field data in fig. 2 compared to growth rates (µ*) calculated from data in fig. 5 for the average epilimnetic light and temperature conditions during the interval. Missing µ*-data indicate missing light and/or temperature data, or photon flux densities higher than those tested.

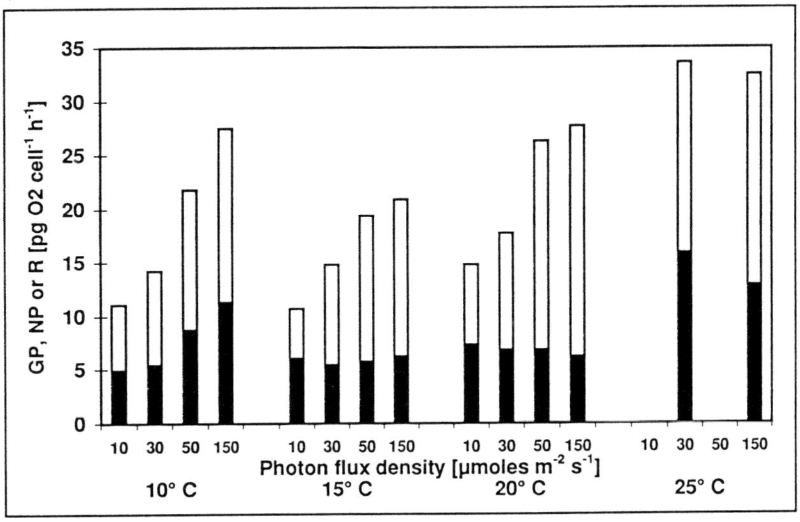


Fig. 7. Rates of gross-photosynthesis (total bar), net-photosynthesis (white part) and respiration (black part) as pg $O_2$ cell$^{-1}$ h$^{-1}$ for laboratory batch cultures of *Tabellaria flocculosa* var. *asterionelloides* for different photon flux densities (µmoles m$^{-2}$ s$^{-1}$) and temperatures (°C).

1953b). We would further conclude that *T. fenestrata* is probably not a good taxon and should be fused with *T. flocculosa.*

Long-term records of *T. "fenestrata"* have been reported by e.g. Knudson (1953a) and Talling (1993) from Windermere, England, and by Lehn (1969) from Lake Constance, Germany. In Windermere the species belongs to the phytoplankton assemblage of the spring peak, while it has characteristically two peaks in Lake Constance. In some years only one maximum of *Tabellaria* occurred (Lehn 1969) similar to our findings. The phenomenon of seasonal dimorphism, with the zigzag form predominating in winter months and the star-shaped form in summer, described by Vollenweider (1950) and by several other authors, and summarised in Knudson (1953a), has not been observed in Mondsee.

Growth rates of *Tabellaria* are largely controlled by water temperature and light intensity during most of the year. Field populations are adapted to rates of effective light climate less than 400 J cm$^{-2}$ day$^{-1}$ and temperatures up to 20°C (Fig. 8) in agreement with laboratory experiments. Average exponential growth rates (0.072–0.384) are in good correspondence with results from elsewhere (Talling 1955; Saraceni 1966; Tilman 1981 for P-limitation). Higher rates of 0.485–0.760 were reported for *Tabellaria* by Tilman (1981) under Si-limitation, and by Løvstad (1984) and Jaworski (cit. In Reynolds 1984) for light intensities higher than those used in the present study. Growth rates of freshwater diatoms are generally in the range of 0.6–2.2 (Eppley 1977; Reynolds 1984). Rates for *Tabellaria* are therefore comparatively low. Other members of the Fragilariaceae, often co-occurring with *Tabellaria,* such as *Fragilaria crotonensis* and *Asterionella formosa* also have considerably higher rates of growth

(Eppley 1977; Reynolds 1984; Hartig & Wallen 1986). Estimates of net population changes from field data are in close agreement with data from Knoechel & Kalff (1978).

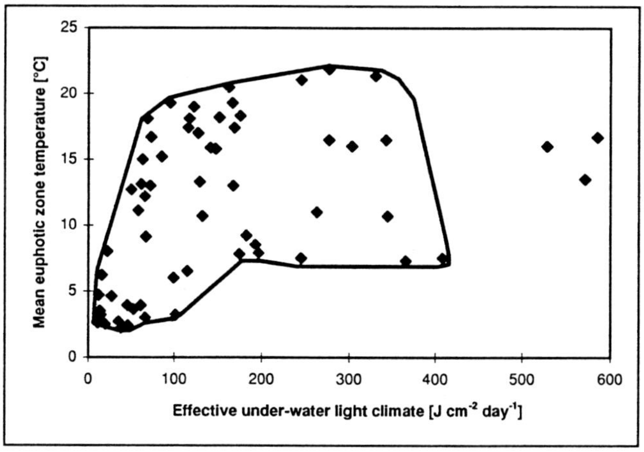


Fig. 8. Positive rates of net population changes from fig. 6 versus mean euphotic zone temperature (°C) and average effective light climate (J $cm^{-2}$ $day^{-1}$) for the observation interval.

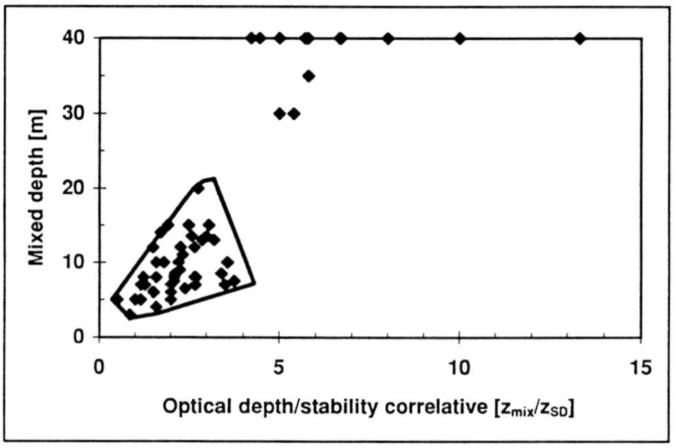


Fig. 9. Positive rates of net population changes from fig. 6 versus mixed depth (m) and the optical depth/stability correlative ($z_{mix}/z_{SD}$).

Competition experiments for nutrients using chemostat cultures (Tilman 1982; Tilman *et al.* 1982) indicate that *Tabellaria* is an inferior competitor for silicate and, to some extent, phosphate when compared to other diatom species. It should therefore theoretically be difficult, if not impossible for *Tabellaria* to persist in lakes (Tilman &

Sterner 1984). Observations using pulsed nutrient supply, summarised in Willén (1991), resulted in *Tabellaria*-dominance when nutrient pulses were infrequent. Periods of very patchy appearance of nutrients under non-equilibrium conditions would then favour *Tabellaria* in nature, as seems to be the case in the field populations of the present study.

Moreover, *Tabellaria* is competitively favoured over other species through increased stability as can be seen from Fig. 9. Net population increase in the field is largely confined to conditions when the mixed zone is less than 20 m and the relation mixed depth to Secchi-depth is less than 4, indicating that mixing is equal or not very much greater than the euphotic zone. Unpublished observations on cell death and sinking rates confirm results on *Tabellaria* by Knoechel & Kalff (1978) indicating considerably lower death and sinking rates than those recorded for other diatoms.

Nutrient availability and column stability can thus explain observed seasonal population changes (Reynolds 1983), while the long-term decline (Fig. 2) is a reflection of altered nutrient availability (Table 2).

## Conclusion

Our results show that *Tabellaria fenestrata* from Mondsee is more correctly identified as *T. flocculosa* var. *asterionelloides*. Appearance of populations in the field is controlled by availability of light, temperature and nutrients. *Tabellaria* is a better competitor under more stable conditions because of adaptation of photosynthesis and growth to lower light intensities, higher temperatures and comparatively lower loss rates. Long-term appearance is governed by changes in nutrient ratios.


## Acknowledgements

The help of many of the staff members of the Institute in collecting and evaluating samples over the years is kindly acknowledged. Especially, we would like to thank the Bundesanstalt für Fischereiwirtschaft, Scharfling for making available unpublished chemical data, Claudia Skolaut for sample analysis, Anne-Marie Schmid and Sybille Wunsam for SEM-micrographs, and to Helen Bannion for correcting the manuscript.


## References


Chu, S. P. (1942). The influence of the mineral composition of the medium on the growth of planktonic algae. Part 1. Methods and culture media. *Ecology*, **30**, 57–63.

Dokulil, M. (1987). Long term occurrence of blue-green algae in Mondsee during Eutrophication and after nutrient reduction with special reference to *Oscillatoria rubescens*. *Schweizerische Zeitschrift für Hydrologie*, **49**, 378.

Dokulil, M. (1991). Populationsdynamik der Phytoplankton-Diatomeen Im Mondsee seit 1957. *Wasser und Abwasser*, **35**, 53–75.

Dokulil, M. & Jagsch, A. (1989). Mondsee. Limnologische Entwicklung. In: *Seenreinhaltung in Österreich. Fortschreibung 1981-1987. Limnologie-Hygiene-Maßnahmen-Erfolge* (Bundes ministerium für Land- und Forstwirtschaft ed.), 91–95. Bundesministerium für Land- und Forstwirtschaft, Wien.

Dokulil, M. & Skolaut, C. (1986). Succession of phytoplankton in a deep stratifying lake: Mondsee, Austria. *Hydrobiologia,* **138**, 9–24.

Eppley, R. W. (1977). The growth and culture of diatoms. In: *The biology of diatoms* (D. Werner, ed.). *Botanical Monographs,* **13**, 24–64. Blackwell Scientific Publications, Oxford.

Findenegg, I. (1969). Die Eutrophierung des Mondsees im Salzkammergut. *Wasser und Abwasserforschung,* **4**, 139–144.

Flower; R. J. & Battarbee, R. W. (1985). The morphology and biostratigraphy of *Tabellaria quadriseptata* (Bacillariophyceae) in acid waters and lake sediments in Galloway, Southwest Scotland. *British Phycological Journal,* **20**, 69–79.

Grim, J. (1955). Die chemischen und planktologischen Veränderungen des Bodensee–Obersees in den letzten 30 Jahren. *Archiv für Hydrobiologie, Suppl.,* **22**, 310–322.

Hamilton, P. B. (1990). The revised edition of a computerized plankton counter for plankton, periphyton and sediment diatom analyses. *Hydrobiologia,* **194**, 23–30.

Hartig, J. H. & Wallen, D. G. (1986). The influence of light and temperature on growth and photosynthesis of *Fragilaria crotonensis* Kitton. *Journal of Freshwater Ecology,* **3**, 371–382.

Huber-Pestalozzi, C. (1942). *Das Phytoplankton des Süßwassers. 2. Teil, 2. Hälfte: Diatomeen.* 549 pp. Schweizerbart'sche Verlagsgesellschaft, Stuttgart

Klee, R. & Schmidt, R. (1987). Eutrophication of Mondsee (Upper Austria) as indicated by the diatom stratigraphy of a sediment core. *Diatom Research,* **2**, 55–76.

Kling, H. J. (1993). *Asterionella formosa* RALFS: The process of rapid size reduction and its possible ecological significance. *Diatom Research,* **8**, 475–479.

Knoechel, R. & Kalff, J. (1978). An in situ study of the productivity and population dynamics of five freshwater planktonic diatom species. *Limnology and Oceanography,* **23**, 195–218.

Knudson, B. M. (1952). The diatom genus *Tabellaria* I. Taxonomy and morphology. *Annals of Botany, N.S.,* **16**, 421–440.

Knudson, B. M. (1953a). The diatom genus *Tabellaria* II. Taxonomy and morphology of the plankton varieties. *Annals of Botany, N.S.,* **17**, 131–155.

Knudson, B. M. (1953b). The diatom genus *Tabellaria* III. Problems of intra-specific taxonomy and evolution in *T. flocculosa. Annals of Botany, N.S.,* **17**, 598–609.

Knudson, B. M. (1953c). The distribution of *Tabellaria* in the English lake district. *Verhandlungen der Internationalen Vereinigung für Limnologie,* **12**, 216–218.

Knudsen, B. M. (1954). The ecology of the diatom genus *Tabellaria* in the English lake district. *Journal of Ecology,* **42**, 345–358.

Kofler, S. (1986). Temperatur und Strahlung als bestimmende Faktoren für Wachstum und Morphologie von *Tabellaria flocculosa* var. *asterionelloides* (Bacillariophyceae) in Kultur. 146 S., Dissertation Universität Wien, Wien.

Koppen, J. D. (1975). A morphological and taxonomic consideration of *Tabellaria* EHR. (Bacillariophyceae) from the north-central United States. *Journal of Phycology,* **11**, 236–244.

Koppen, J. D. (1978). Distribution and aspects of the ecology of the genus *Tabellaria* EHR. (Bacillariophyceae) in the north-central United States. *American Midland Naturalist,* **99**, 383–397.

Krammer, K. & Lange-Bertalot, H. (1991). Bacillariophyceae. 3. Teil: Centrales, Fragilariaceae, Eunotiaceae. In: *Süßwasserflora von Mitteleuropa 2/3* (H. Ettl, J. Gerloff, H. Heynig & D. Mollenhauer, eds) 176 pp., Gustav Fischer Verlag, Stuttgart.

Lange-Bertalot, H. (1988). Die Gattung *Tabellaria* unter Berücksichtigung von *Tabellaria ventricosa* KÜTZING (Bacillariophyceae). *Nova Hedwigia*, **46**, 413–431.

Lehn, H. (1969). Die Veränderungen ds Phytoplanktonbestandes im Bodensee. I. Fluktuationen von *Tabellaria fenestrata* 1890–1967. *Internationale Revue der gesamten Hydrobiologie*, **54**, 367–411.

Løvstad, O. (1984). Competitive ability of laboratory batch phytoplankton populations at limiting nutrient levels. *Oikos*, **42**, 176–184.

Lund, J. W. G., Kipling, C. & Le Cren, E. D. (1958). The Inverted Microscope Method of Estimating Algal Numbers and the Statistical Basis of Estimations by Counting. *Hydrobiologia*, **11**, 144–170.

Nipkow, F. (1920). Vorläufige Mitteilung über Untersuchungen des Schlammabsatzes im Zürichsee. *Schweizerische Zeitschrift für Hydrologie*, **1**, 100–122.

Nusch, E. A. (1980). Comparison of different methods for chlorophyll and phaeopigment determination. *Archiv für Hydrobiologie, Beihefte Ergebnisse der Limnologie*, **14**, 14–36.

Oberrosler, I. E. (1979). Der Einfluß des Phytoplanktons auf die Fütterung von Jungfischen (Karpfen). 137 S., Dissertation, Universität Salzburg, Salzburg.

Reynolds, C. S. (1983). A physiological interpretation of the dynamic responses of populations of a planktonic diatom to physical variability of the environment. *New Phytologist*, **95**, 41–53.

Reynolds, C. S. (1984). *The ecology of freshwater phytoplankton*. 384 pp. Cambridge University Press, Cambridge.

Saraceni, C. (1966). Il fabbisogno in fosforo e ferro nella coltura di tre specie di diatomee planctoniche del Lago Maggiore. *Memorie dell'Instituto Italiano di Irdobiologia*, **20**, 117–131.

Schmidt, R. (1991). Diatomeenanalytische Auswertung laminierter Sedimente für die Beurteilung trophischer Langzeittrends am Beispiel des Mondsees (Oberösterreich). *Wasser und Abwasser*, **35**, 109–123.

Schwarz, K. (1979). Das Phytoplankton des Mondsees 1978. *Arbeiten aus dem Labor Weyregg*, **3**, 83–92.

Talling, J. F. (1955).The relative growth rates of three plankton diatoms in relation to underwater light and temperature. *Annals of Botany, N.S.*, **19**, 329–341.

Talling, J. F. (1993). Comparative seasonal changes, and inter-annual variability and stability, in a 26-year record of total phytoplankton biomass in four English lake basins. *Hydrobiologia*, **268**, 65–98.

Theriot, E. & Ladewski, T. B. (1986). Morphometric analysis of shape of specimens from the neotype of *Tabellaria flocculosa* (Bacillariophyceae). *American Journal of Botany*, **73**, 224–229.

Tilman, D. (1981). Tests of resource competition theory using four species of Lake Michigan algae. *Ecology*, **62**, 802–815.

Tilman, D. (1982). *Resource competition and community structure*. 296pp. Pinceton University Press, Princeton.

Tilman, D. & Sterner, R. W. (1984). Invasions of equilibria: tests of resource competition using two species of algae. *Oecologia (Berlin)*, **61**, 197–200.

Tilman, D., Kilham, S. S. & Kilham, P. (1982). Phytoplankton community ecology: The role of limiting nutrients. *Annual Review of Ecology and Systematics*, **13**, 349–372.
Vollenweider, R. A. (1950). Ökologische Untersuchungen planktischer Algen. *Schweizerische Zeitschrift für Hydrologie*, **12**, 193–263.
Willén, E. (1991). Planktonic diatoms – an ecological review. *Algological Studies*, **62**, 69–106.
Züllig, H. (1982). Untersuchungen über die Stratigraphie von Carotenoiden im geschichteten Sediment von 10 Schweizer Seen zur Erkundung früher Phytoplankton – Entfaltungen. *Schweizerische Zeitschrift für Hydrologie*, **44**, 1–98.

# Interactions of pH and aluminum on cell length reduction in *Asterionella ralfsii* var. *americana* Körner

Robert W. Gensemer[1], Ralph E. H. Smith and Hamish C. Duthie

*Department of Biology, University of Waterloo
Waterloo, Ontario, N2L 3G1, Canada*

## Abstract

Cell length and the relative distribution of different cell length morphotypes of the diatom *Asterionella ralfsii* var. *americana* Körner are thought to vary in response to changes in pH and other related environmental factors. Therefore, we examined the influences of two of these factors (pH and aluminum) on mean population cell length and the variation in cell lengths within individual colonies. *A. ralfsii* was grown in continuous culture to test morphological responses to variation in pH and aluminum concentrations under conditions of silica-limitation. At steady-state, mean population cell lengths were significantly reduced at pH 5 relative to pH 6, whereas cell lengths were unaffected in treatments containing 20 $\mu mol \cdot L^{-1}$ Al at pH 6. The same addition of Al at pH 5 was completely toxic to *A. ralfsii*, so no morphological results could be obtained under these conditions. Cultures grown at pH 5 without Al exhibited greater variation in cell lengths within a single colony, which apparently resulted from rapid cell length reductions occurring after a single cell division. Usually, cell lengths in any given *Asterionella* colony are quite similar, and cell length variance was less than 1.2–2.1 µm for both inoculum cultures and pH 6 treatments. Cell lengths were significantly more variable in any given colony at pH 5, however, and length variance reached 13.3 µm owing to abrupt changes in cell length between adjacent cells in a colony. Thus, pH may influence both mean cell length, and the relative distribution of different cell length morphotypes of *A. ralfsii* populations under conditions of nutrient stress.

[1]Corresponding Author. Present address: Department of Biology, Boston University, 5 Cummington St., Boston, MA, 02215, U.S.A.

## Introduction

Diatoms are important bioindicators of environmental change in freshwaters, and have been used extensively to assess the onset and extent of lake acidification (Davis 1987, Dixit *et al.* 1992). One species, *Asterionella ralfsii* var *americana* Körner, is often abundant in the plankton of acid lakes, and as such is a useful indicator of changes in lake pH (Schindler *et al.* 1985; Duthie 1989; Charles *et al.* 1990; Gensemer *et al.* 1993a). However, correctly interpreting the ecological preferences of *A. ralfsii* for pH (and perhaps other related environmental factors such as trace metal contamination) also may involve an understanding of the ecological significance of morphological variation in the diatom frustule.

A variety of environmental factors may induce changes in frustule morphology in this taxon. For example, in Adirondack lakes *A. ralfsii* var. *americana* is operationally subdivided into two morphological forms: long (>45 µm), and short (<45 µm); their relative distributions are thought to relate to changes in pH and/or dissolved organic carbon concentrations (Charles *et al.* 1990). Lakes in Kejimkujik Park, Nova Scotia, also contain both length varieties of *A. ralfsii,* and their relative distributions vary seasonally or as a function of changing environmental conditions (McIntyre & Duthie 1993). Culture studies have already shown that when Si-limited, the presence of aluminum at acidic pH can reduce the size of "long form" cells to lengths characteristic of "short form" cells (Gensemer 1990). This suggests that different length "forms" of this taxon may be environmentally-induced morphotypes of the same taxon, rather than distinct species or sub-species. Consistent with these observations, Kling (1993) demonstrated that significant cell size reductions can occur in *Asterionella* populations during as little as a single cell division. Kling hypothesized that rapid cell-size reduction could be an adaptation to Si limitation whereby cellular requirements for frustule formation are minimized, but this was not examined experimentally.

The present study investigates environmentally-induced changes in cell morphology in *A. ralfsii* by examining the interactive effects of both Al additions and pH changes on mean cell lengths of populations grown in Si-limited continuous cultures. This was done to compare the effects of both pH and Al stress on cell lengths, as opposed to the effects of Al alone as in earlier studies (Gensemer 1990). Additionally, cell length variance within single colonies was measured to examine whether rapid cell-size reduction occurred under Si-limitation as described by Kling (1993).

## Methods

*Asterionella ralfsii* cf var. *americana* Körner (clone designation PL–10; UTCC clone #261) was grown in continuous culture to examine changes in cell morphology when populations were Si-limited (see Gensemer *et al.* 1993b for details). Continuous cultures were run in continuous light (PFD = 150 $\mu mol \cdot m^{-2} \cdot s^{-1}$) at a single dilution rate of approximately 0.3 $day^{-1}$ to maintain growth rates at less than 50% of maximum observed rates. Nutrients were supplied to all chemostats at nominal concentrations of 12.5 $\mu mol \cdot L^{-1}$ Si(OH) and 5 $\mu mol \cdot L^{-1}$ PO thus providing an influent molar Si:P ratio of 2.5, which was sufficient to Si-limit *A. ralfsii* under these conditions (Gensemer *et al.* 1993b). Cultures contained nominal Al additions of 0 (= no addition) and 20 $\mu mol \cdot L^{-1}$ total Al, and pH levels were set to 5 and 6 so as to compare conditions of high vs. low Al solubility, respectively. After steady-state was established (when at least 3 successive measurements of cell density showed coefficients of variation less than 5%), subsamples were removed for morphological measurements, as well as a variety of chemical and physiological parameters (Gensemer *et al.* 1993b).

To assess the morphological effects of pH and Al variations, cell lengths were measured from chemostat subsamples taken at steady-state (day 25; see Gensemer *et al.* 1993b). Subsamples were preserved in Lugol's solution, then analyzed for cell length in circular (2.3 ml) settling chambers on an inverted microscope at a magnification of 400x. The microscope was fitted with a video camera, and cell lengths were measured using a microcomputer image analysis system ("Image", public domain) which digitizes light microscope images from which screen pixel distances are converted to length in µm. For each sample, all cells from 50 randomly encountered, intact colonies were measured for the total length of each cell (from footpole to headpole), the variance in cell length within a single colony, and the number of cells in each colony (see Gensemer *et al.* 1994 for cell number results). If rapid cell-size reduction occurred in the manner described by Kling (1993), smaller daughter cells would coexist in the same colony as longer parent cells. Colonies containing significantly different cell lengths would, therefore, exhibit greater variance around mean cell lengths in any given colony.

## Results and Discussion

For samples taken at steady-state (day 25), mean population cell lengths were significantly reduced in the pH 5 treatment (no Al added) relative to the inoculum culture (day 1) and both pH 6 treatments (Fig. 1A). Similar to the physiological results from the same experiment (Gensemer *et al.* 1993b), reduced pH had a greater negative impact on *A. ralfsii* when Si-limited than Al additions at higher pH. The number of cells contained in each colony also varied with pH, but was more affected at pH 6 compared to pH 5 (Gensemer *et al.* 1994). In contrast to previous work using *A. ralfsii* cultures (Gensemer 1990), Al additions did not reduce mean population cell length when Si-limited at pH 6.

One explanation for the lack of an Al effect is that this particular isolate (PL–10; UTCC #261) may be inherently more Al tolerant as compared to that previously studied (UTCC #170; Gensemer 1990). Results of batch culture toxicity bioassays using the present isolate (PL–10) are consistent with this hypothesis (Smith, Duthie, Gensemer, McIntyre 1993 – unpublished data), but more complete tests would be required for confirmation. Alternatively, the continuous Si and light supplies used in the present study could have ameliorated the influence of Al on cell morphology. The semicontinuous cultures used previously supplied Si in discrete daily pulses and light on a 14:10 h L:D cycle (Gensemer 1990), thereby potentially affecting several aspects of Si uptake and silicification mechanisms (Paasche 1980; Sullivan & Volcani 1981). Furthermore, the present experiment was only run at a single, low turnover rate (ca. 0.3 $day^{-1}$). This would have produced a similar number of generations by day 25 as those obtained in the low turnover rate treatments (0.22 $day^{-1}$) of Gensemer (1990) in which cell length reductions also were not observed in response to Al additions.

Nevertheless, our present experiments demonstrate that reduced pH can affect cell morphology in *A. ralfsii*, perhaps even to a greater extent than Al additions alone. Changes in *Asterionella* cell length have also been correlated with pH in New England (Rhodes 1991) and Adirondack lakes (Charles *et al.* 1990). However, in natural populations of *Asterionella* in Kejimkujik Park, Nova Scotia, pH was less well-correlated with either cell length or colony structure (McIntyre & Duthie 1993).

Cultures grown at pH 5 also exhibited greater variation in cell lengths within single colonies (Fig. 1B). Usually, cell lengths in any given *Asterionella* colony are similar, and cell length variance was less than 1.2–2.1 µm for inoculum cultures and both pH 6 treatments. Cell lengths were significantly more variable in any given colony at pH 5, however, and length variance reached 13.3 µm in this treatment. A light micrograph of a typical colony containing variable cell lengths (Fig. 2A) shows that the short cells can be roughly 2/3 the length of the rest of the cells in the colony. The longer cells are typical of all the cells contained in colonies for which no size reduction occurred (Fig. 2B). Furthermore, full-length parent cells immediately adjacent to the short cells only possess cell contents in the lower 2/3 of the cell. From this it appears that only the living portion of the longer parent cell participates in forming the shorter sibling cell immediately adjacent. This is consistent with SEM observations by Kling (1993) using both cultured and natural populations of *A. formosa*.

The mechanisms and environmental cues responsible for rapid cell size reduction in this and other diatom taxa are as yet poorly understood, but evidence suggests it may be in part a response to Si limitation. Kling (1993) hypothesized that rapid cell size reduction is an adaptation to growth under Si-limited conditions. Because growth can not occur without sufficient Si to build a new frustule (Paasche 1980), it is logical that growth using smaller cells would minimize Si requirements when external Si supplies are scarce. Results from the present experiment and those of earlier experiments (Gensemer 1990) generally are consistent with this hypothesis, because in both cases, continuous cultures were designed to induce Si limitation.

As Kling (1993) pointed out, however, Al-induced cell size reduction may not have occurred in *A. ralfsii* (Gensemer 1990) when the populations were Si-limited. Because cell sizes only decreased when growth rates and Si cell quotas were high (usually observed during Si-replete conditions; see Kilham 1978; Paasche 1980), Al alone may have induced cell length reduction in these experiments. Enhanced cell quotas alone, however, are not necessarily diagnostic of Si-replete conditions, particularly when populations also are grown under the influence of stress from other

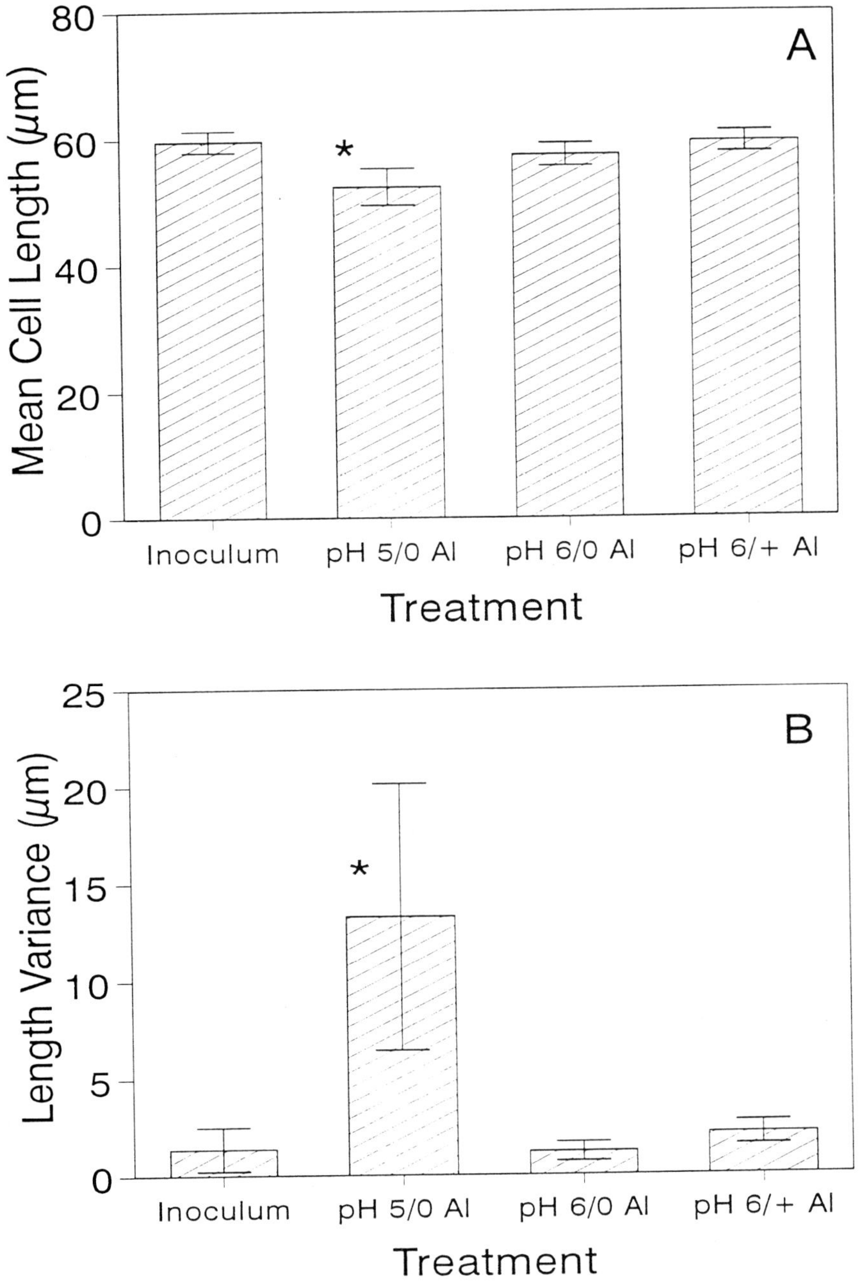


Fig. 1. Effect of chemostat treatments on A) mean population cell lengths, and B) the variance in cell length among all cells in intact *A. ralfsii* colonies (N=50 for each treatment). Asterisks denote treatments that are significantly different from inoculum cultures at $p < 0.05$ using 1-way ANOVA.

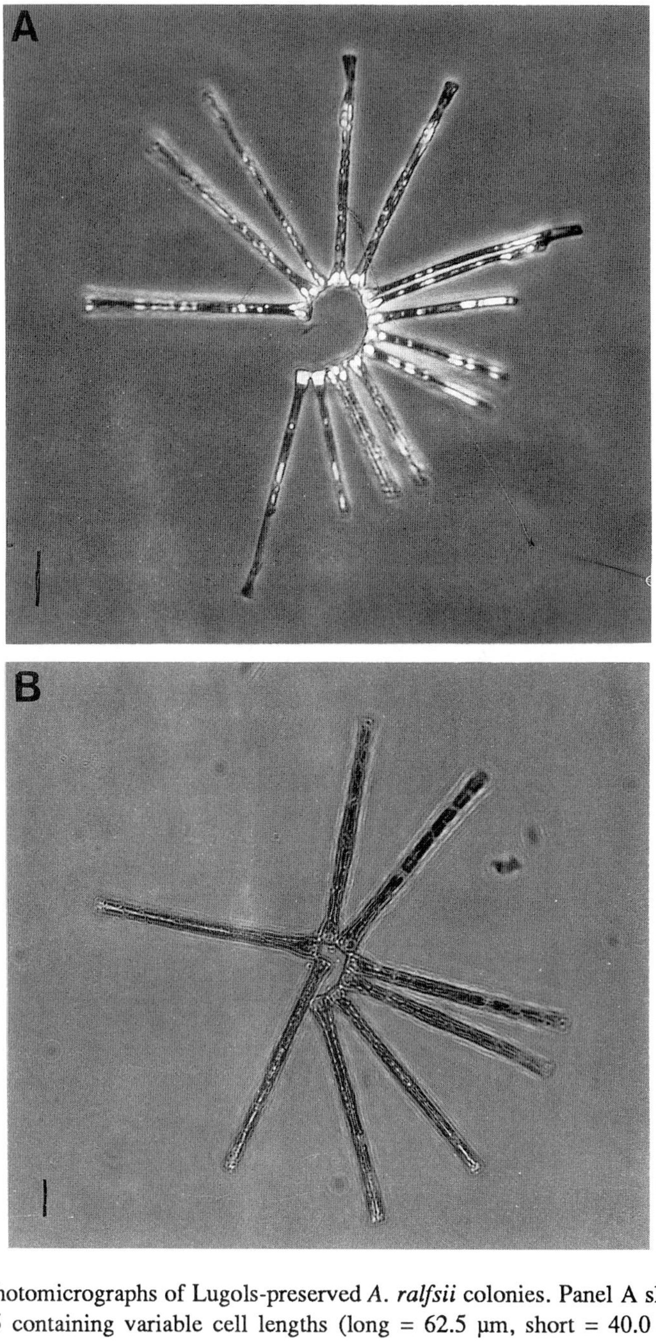


Fig. 2. Light photomicrographs of Lugols-preserved *A. ralfsii* colonies. Panel A shows a colony grown at pH 5 containing variable cell lengths (long = 62.5 µm, short = 40.0 µm). Panel B shows a "healthy" colony containing a normal compliment of long (ca. 60 µm) cells. Scale bars represent 10 µm.

environmental factors (Rhee & Gotham 1981). In the present experiments, Si cell quotas were also highest in the pH 5 treatment where cell length reduction occurred, but Si-limitation was confirmed by the use of lower culture turnover rates and Si:P ratios than those used previously (Gensemer 1990), and the results of nutrient addition bioassays (Gensemer *et al.* 1993b).

In conclusion, *A. ralfsii* populations can respond to environmental stress by undergoing rapid cell size reduction, and this usually occurs when populations are Si-limited. However, given that pH and/or Al were the proximate cues for reducing cell sizes in these experiments, other environmental factors may induce rapid cell size reduction in addition to nutrient limitation *per se*. Taken together with Kling's (1993) study, rapid cell size reduction in *Asterionella* may be a more general response to environmental stress. Whether this is a specific mechanism for minimizing Si requirements during times of nutrient stress, a means by which cells adapt metabolically to other chemical stresses, or a combination of these factors remains to be tested explicitly. Understanding these mechanisms are important in determining whether or not changes in diatom cell length could be diagnostic for nutrient limitation (Kling 1993), metal stress (e.g. Tonolli 1961), pH changes (e.g. Charles *et al.* 1990; Rhodes 1991), or as yet undetermined environmental factors.

**Acknowledgements**

We thank N. S. Hopkins for laboratory assistance, and S. M. McIntyre for assistance with photomicroscopy. This work was funded by grants from the Wildlife Toxicology Fund and the National Sciences and Engineering Research Council of Canada (# STR0045063).

**References**

Charles, D. F., Binford, M. W., Furlong, E. T., Hites, R. A., Mitchell, M. J., Norton, S. A., Oldfield, F., Paterson, M. J., Smol, J. P., Uutala, A. J., White, J. R., Whitehead, D. R. & Wise, R. J. (1990). Paleoecological investigation of recent lake acidification in the Adirondack Mountains, N.Y. *Journal of Paleolimnology*, **3**, 195–241.

Davis, R. B. (1987). Paleolimnological diatom studies of acidification of lakes by acid rain: an application of quaternary science. *Quaternary Science Reviews*, **6**, 147–163.

Dixit, S. S., Smol, J. P., Kingston, J. C. & Charles, D. F. (1992). Diatoms: Powerful indicators of environmental change. *Environmental Science and Technology*, **26**, 22–33.

Duthie, H. C. (1989). Diatom-inferred pH history of Kejimkujik Lake, Nova Scotia: a reinterpretation. *Water Air and Soil Pollution*, **46**, 317–322.

Gensemer, R. W. (1990). Role of aluminum and growth rate on changes in cell size and silica content of silica-limited populations of *Asterionella ralfsii* var. *americana*. *Journal of Phycology*, **26**, 250–258.

Gensemer, R. W., Smith, R. E. H., Duthie, H. C. & Schiff, S. L. (1993a). pH tolerance and metal toxicity in populations of the planktonic diatom *Asterionella*: Influences of synthetic and natural dissolved organic carbon. *Canadian Journal of Fisheries and Aquatic Sciences*, **50**, 121–132.

Gensemer, R. W., Smith, R. E. H. & Duthie, H. C. (1993b). Comparative effects of pH and aluminum on silica-limited growth and nutrient uptake in *Asterionella ralfsii* var. *americana* (Bacillariophyceae). *Journal of Phycology*, **29**, 36–44.

Gensemer, R. W., Smith, R. E. H. & Duthie, H. C. (1994). Interactions of pH and aluminum on cell length and colony structure in a freshwater diatom common to Kejimkujik waters. In: *Proceedings of the Workshop on the Kejimkujik Watershed Studies: Monitoring and Research Five Years after Kejimkujik '88* (C. A. Stacier, M. J. Duggan & J. J. Kerekes, eds), 147–153. Environment Canada, Dartmouth, Nova Scotia.

Kilham, S. S. (1978). Nutrient kinetics of freshwater planktonic algae using batch and semicontinuous methods. *Internationale Vereinigung für Theoretische und Angewandte Limnologie, Mitteilungen*, **21**, 147–157.

Kling, H. (1993). *Asterionella formosa* Ralfs: The process of rapid size reduction and its possible ecological significance. *Diatom Research*, **8**, 475–479.

McIntyre, S. H. & Duthie, H. C. (1993). Morphological variation in populations of *Asterionella ralfsii* W. Smith from Nova Scotia, Canada. *Hydrobiologia*, **269/270**, 67–73.

Paasche, E. (1980). Silicon. In: *Physiological ecology of the phytoplankton* (I. Morris, ed.), 259–284. Blackwell Scientific, London.

Rhee, G. -Y. & Gotham, I. J. (1981). The effect of environmental factors on phytoplankton growth: Temperature and the interactions of temperature with nutrient limitation. *Limnology and Oceanography*, **26**, 635–648.

Rhodes, T. E. (1991). *Effects of late Holocene disturbance and succession on an acidic Maine lake*. 242 pp. Ph.D. Dissertation, University of Maine, Orono, ME.

Schindler, D. W., Mills, K. H., Malley, D. F., Findlay, D. L., Shearer, J. A., Davies, I. J., Turner, M. A., Linsley, G. A. & Cruikshank, D. R. (1985). Long-term ecosystem stress: the effects of years of experimental acidifcation on a small lake. *Science*, **228**, 1395–1401.

Sullivan, C. W. & Volcani, B. E. (1981). Silicon in the cellular metabolism of diatoms. In: *Silicon and Siliceous Structures in Biological Systems* (T. L. Simpson & B. E. Volcani, eds), 15–42. Springer–Verlag, New York.

Tonolli, L. (1961). La polluzione cuprica del Lago d'Orta: comportamento di alcune popolazioni di Diatomee. *Internationale Vereinigung für Theoretische und Angewandte Limnologie, Verhandlungen*, **14**, 900–904.

# Excretion of exudates by a diatom under culture

E. Sdrigotti and L. Talarico

*Department of Biology, University of Trieste, Via Giorgieri 10, I–34127 Trieste, Italy*

## Abstract

Extensive masses of gelatinous aggregates in the Northern Adriatic Sea is mainly due to diatoms. Among these, *Pleurosigma* sp. has been isolated and cultured in *f*/2 medium at 60 µE $cm^{-1}$ $sec^{-2}$ of irradiance (L:D/14:10) at 20°C without aeration, in order to investigate the physiological conditions causing the release of exudate. This occurs during the declining growth phase and continues during the stationary phase. Photosynthetic activity, pigment content and cell number were estimated during the growth cycle and it was demonstrated that increasing population density resulted in nutrient deficiency which inhibits cell division at a time when photosynthesis is at its highest. Cells, being unable to divide, produce a surplus of photosynthate that needs to be discarded since it is produced faster than required for growth. Formation of larger aggregates at high culture density causes $CO_2$ depletion and light attenuation. This will determine the progressive decrease of $O_2$ evolution. Excretion would then occur because division, growth and photosynthesis are unbalanced. Cytochemical tests on the exudates proved their polysaccharide and glycoprotein nature. Charged groups distributed on the molecules suggest that these organic components might form nutrient enriched microhabitats presumably attracting organic and colloidal material by cation exchange.

## Introduction

It is known that diatoms release a great variety of extracellular substances, mainly carbohydrates and proteins, which often play important roles in aquatic food chains and ecosystems (Fogg 1962; Hellebust 1974; Fogg 1983). The biochemical aspects of these products have been investigated for several marine diatoms, e.g. *Chaetoceros*, *Thalassiosira*, *Nitzschia*, *Corethron* (Huntsman & Sloneker 1971; Myklestad *et al.* 1972; Smestad *et al.* 1975; Haug & Myklestad 1976). Chemical analysis of the hollow mucilaginous stalks attaching cells to the substrate was also performed for the freshwater diatom *Gomphonema* (Huntsman & Sloneker 1971) taxon which has been reviewed and split into four "gomphonemoid" clusters, one of which has totally different morphological characters differentiating it from the freshwater species (Medlin & Round 1986). The chemical nature and ecological significance (Hoagland *et al.* 1993) of mucilaginous material have been widely investigated whereas little is known concerning the physiological conditions related to exocellular release. Few papers deal with natural populations of marine phytoplankton under field conditions (Ignatiades 1973; Sharp 1977; Mague *et al.* 1980) and only some of the factors affecting the release, such as nutrient starvation, have been considered in cultural studies (Nalewajko 1966; Myklestad & Haug 1972; Ignatiades & Fogg 1973; Vaulot *et al.* 1987; Myklestad *et al.* 1989).

Formation of large gelatinous aggregates occurred in the Northern Adriatic Sea in 1988 (Herndl & Peduzzi 1988; Honsell & Cabrini 1990–91; Cabrini *et al.* 1992) and it was repeated in 1989 along the Eastern Adriatic coasts though less intense and shorter in duration. Analysis of the organisms associated with these aggregates indicated that benthic species of diatoms were the most abundant (Fanuko *et al.* 1989; Stachowitsch *et al.* 1990). Apart from *Cylindrotheca closterium* (Ehr.) Reimman & Lewin which had been isolated and cultured under controlled conditions (Cortese & Talarico 1993), other diatoms (*Pleurosigma* sp., *Licmophora abbreviata*, *Navicula* sp.) also had been observed to release mucus under the same conditions (Sdrigotti *et al.* 1994; Talarico *et al.* 1994). The present study aimed at investigating the physiological conditions and external factors that may induce the release of exudates by a benthic diatom such as *Pleurosigma* sp. in culture. Cytochemical characterization of exudates was also performed.

## Material and methods

### *Growth and culture conditions*

Samples were periodically harvested from the Gulf of Trieste (Northern Adriatic Sea) at 10 m depth by means of a standard net with a 20 μm mesh size.

A *Pleurosigma* sp. was isolated from phytoplankton samples by a microcapillary pipette (1–5 ml capacity) and subcultured until a monospecific culture was obtained. Cells were grown under nutrient-sufficient conditions in 100 ml conical flasks and in 10 ml cell well plates, without aeration. Culture medium (*f*/2) was prepared by filtering sea-water (35 ‰ salinity) through 0.45 μm pore size Millipore filters and enriching it as described by Guillard & Ryther (1962). Culture media in polycarbonate bottles were autoclaved at 120°C under 2 atm and cooled at room temperature for 24 hrs. Then cells were inoculated, at the logarithmic growth phase, at an initial concentration of ca $2.3 \times 10^4$ cells $ml^{-1}$. Temperature was kept between 19.5 and 20.5°C and culture pH rarely exceeded 8.0. Irradiance of 60 $\mu E\ m^{-2}\ sec^{-1}$, measured with a LI–COR 185A photometer, was provided by 3 Philips TL40w/55 fluorescent tubes, on a light:dark cycle of 14:10 hrs. Diatom suspensions were fixed for 24–48 hrs at 4°C in 4% formaldehyde in sea-water. Growth rate and cell number were estimated daily by cells count, using a Lund chamber (Lund *et al.* 1958): from 200 to 400 cells were counted under an inverted light microscope (Labovert F8 Microscope, Leitz) at × 320. Clumps of cells were disrupted by ultrasonic vibration. Growth rates were calculated using the equation $K = (\log N_2/N_1)(3.322/t_2-t_1)$ (Harrison *et al.* 1990), where $N_1$ is the biomass at time $t_1$ and $N_2$ is the biomass at time $t_2$. The average growth rate was of 0.5 divisions per day. The organisms normally grew as individual sigmoid cells of average size 90 × 20 µm. Student T-test was applied to verify the reliability of cell concentration means.

*Measurement of photosynthesis*

Photosynthetic activity was tested by measuring oxygen production with an Hansatech Clark-type polarographic oxygen electrode (Delieu & Walker 1972). The cuvette for this system held 2 ml of culture medium and a magnetic stirrer and was cooled at 20°C by a circulating water system. Photon flux density of ca 200 µE $m^{-2}$ $sec^{-1}$ was provided by an halogen spot 35W lamp at a distance of 12 cm. The oxygen analyzer was calibrated with sea water, which amounted to 209 µM $O_2$ at 25°C and at normal atmospheric pressure. Diatom suspensions were concentrated by centrifugation. The readings taken before 20 min had elapsed, were found to be the most useful because the later were often invalidated by the formation of oxygen bubbles. Photosynthetic capability was expressed either on a per cell basis ($10^{-10}$ µmol $O_2$ $min^{-1}$ $cell^{-1}$) or normalized to chlorophyll *a* (µmol $O_2$ $min^{-1}$ $mg^{-1}$ Chl *a*). The significance of data has been used by means of Student T-test.

*Pigments*

Cultures for pigment extraction were concentrated by centrifugation at 4,000 × g for 10 min at 20°C. Chlorophylls (Chl) *a*, *c* and carotenoids (Car) were determined spectrophotometrically (Perkin Elmer UV–VIS 554) from 90% acetone extracts after 20 hrs at 4°C in the dark, using the equation of Jeffrey & Humphrey (1975). Total Car concentration was estimated from the equation of Kirk & Allen (1965), using a specific absorption coefficient of 250 $cm^{-1}$ $(g\ l^{-1})^{-1}$ at 480 nm (Davies 1976).

*Light microscopy and cytochemical tests*

Different stages of cellular growth, exudate release and development were observed using an inverted light microscope directly on cell well plates. Chemical composition of the exudate has been tested by the following cytochemical tests: 0.1% Ruthenium Red (RR) (Jensen 1962), 0.01% Toluidine Blue O at pH 0.5 (ATB) and pH 4.4 (BTB) (Chayen *et al.* 1973), 0.1% Alcian Blue 8GX at pH 0.5 (AB) (Pearse 1968), 0.1% Alcian Yellow at pH 2.5 (AY), Sequential Alcian sequence (DA) (Parker & Diboll 1966), 0.1% Fast Green at pH 2.0 (FG) (Ruthmann 1970).

## Results

In *f*/2 medium a single cell (Fig. 1), after a phase of adjustment, gave rise to monospecific cultures. Cells grew and rapidly divided covering the whole bottom of the well plate (Fig. 2). They aligned along rows (Fig. 3) forming filaments (Fig. 4) which became appressed into a loose net-like pattern (Fig. 5) and often packed into some roundish areas (Fig. 6). In all these stages, cells appeared to be highly motile and healthy and started to release exocellular material. Mucus aggregates initially adhered to the bottom and their extension was doubled (up to 400 µm in width) after 24 hrs (Fig. 7). Then they detached progressively from the bottom and floated as small flocks (Fig. 8) into the sorrounding medium, becoming larger and larger. Cells collected from the surrounding medium when re-inoculated died (Fig. 9) whereas those harvested from inside the flocks (Fig. 10) gave rise to new healthy cultures.

Growth of these cultures, confined to a constant volume of medium within a week, followed a sigmoid curve which could be divided into four phases: adjustment (lag) (Fig. 1), exponential (log) (Fig. 5), declining ( Fig. 6) and stationary phase (Figs 8, 11). Adaptation of diatoms to culture conditions required about 24–36 hrs. During this phase the growth rate increased with time ($K = 1.32 \pm 0.2$). Within 48 hrs from inoculation, during logarithmic phase, the cell biomass doubled over each successive and equal time interval and the growth rate was constant ($K = 0.58 \pm 0.3$). On the third day the declining phase commenced with growth rate decreasing with time ($K = 0.26 \pm 0.03$). Stable concentration of cells (stationary phase) was reached four days later and growth rate declined to zero (Table I). Significance of the cell number means was statistically acceptable (more than 95%) between the first and all the remaining days of the culture cycle, and between log (2nd day) and stationary (4th, 5th, 6th days) phases (Table II). During one week the rate of oxygen evolution increased slowly from $8 \times 10^{-10}$ µmol $O_2$ $min^{-1}$ $cell^{-1}$ to a maximum photosynthetic activity ($18 \times 10^{-10}$ µmol $O_2$ $min^{-1}$ $cell^{-1}$) on the third day. Then $O_2$ decreased progressively to a minimum ($4 \times 10^{-10}$ µmol $O_2$ $min^{-1}$ $cell^{-1}$) (Fig. 12). Photosynthetic activity expressed on the basis of Chl *a* content showed the same trend with an initial value of 18 µmol $O_2$ $min^{-1}$ $mg^{-1}$ Chl *a*, a maximum of 33 µmol $O_2$ $min^{-1}$ $mg^{-1}$ Chl *a* and a minimum of 6 µmol $O_2$ $min^{-1}$ $mg^{-1}$ Chl *a* (Fig. 13). Student T-test applied to photosynthetic capacity means showed significant variations (more than 95%) over all the phases if oxygen evolution was normalized to Chl *a*, and only on the third day (declining phase) if oxygen evolution was expressed on per cell basis. All ove r the cell cycle the day which shows the most

---

Figs 1–8. Fig. 1. *Pleurosigma* sp. (W. Smith) originating the culture. Fig. 2. Monospecific culture after 36–48 hours. Cells are covering the whole bottom. Figs 3 & 4. Cells disposed along rows (↑) and forming filaments (⇑). Fig. 5. Cells appressed into a loose net-like pattern (nw) during releasing stage. Figs 6 & 7. Diatoms packed into a roundish area where they start to release exudates (ex). The same mucus aggregate after 24 hours (ag). Fig. 8. Flocks (fl) detached from the bottom are floating in the medium. Fig. 9. Outside the flock a cell destined to die is extruding its protoplast (↑). Fig. 10. Detail of aggregate with embedded healthy cells (hc). Scale bars = 10 µm (Figs 1, 9, 10); 100 µm (Figs 2–8).

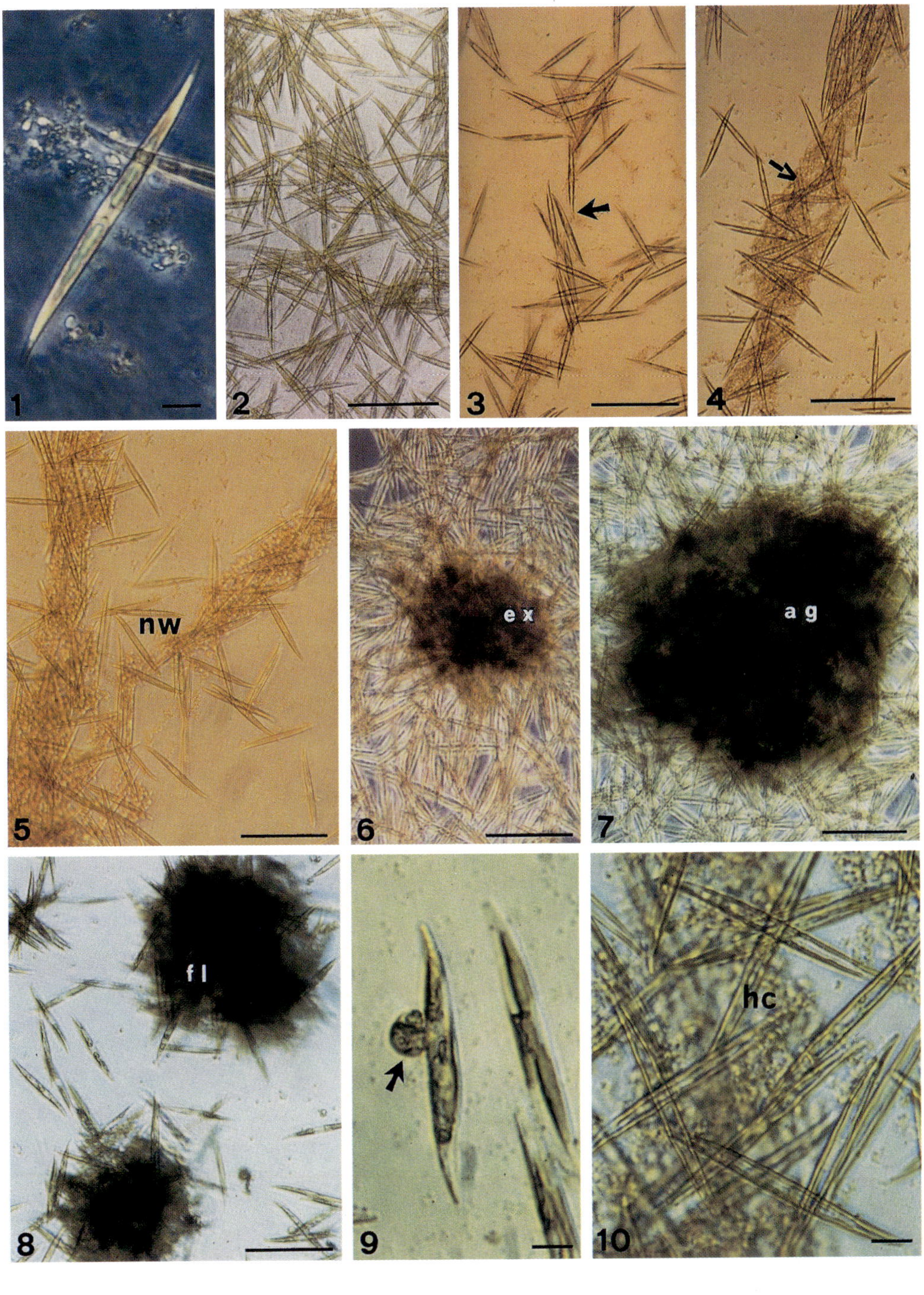
1
2
3
4
5
nw
6
ex
7
ag
8
fl
9
10
hc

Table I. Photosynthetic capability determined either on per cell basis (lane 10) either normalized to chlorophyll a (lane 11). It is evident that variations in chlorophyll a (lane 2) are directly related to cell concentration (lane 7) during a week. Growth rate (K) decreases half by half during the first three days and is about zero from the fourth day (stationary phase). Rate of $O_2$ evolution expressed in µm/min (lane 9) is maximum at the third day when cells start to release exudates. Standard deviations obtained from n measurements are shown in parentheses.

| Days | Chl a ($mg/dm^3$) | Chl c ($mg/dm^3$) | Car ($mg/dm^3$) | Chl c/chl a | Car/chl a | $10^7$ cells/$dm^3$ | K | $O_2$ evolut. (µm/min) | Photos.cap. (µm/min)/ mg chl a | Photos.cap. (µm/min)/ $10^{10}$ cell |
|---|---|---|---|---|---|---|---|---|---|---|
| 1 | 0.10 (0.03) n=6 | 0.038 (0.02) n=6 | 0.057 (0.01) n=6 | 0.38 | 0.57 | 2.3 (0.4) n=6 | 1.32 | 0.018 (0.01) n=6 | 18 (12.9) n=6 | 8 (6.05) n=6 |
| 2 | 0.20 (0.07) n=6 | 0.076 (0.1) n=6 | 0.113 (0.07) n=6 | 0.38 | 0.56 | 5.8 (1.5) n=6 | 0.58 | 0.060 (0.004) n=6 | 30 (7.9) n=6 | 10 (6.7) n=6 |
| 3 | 0.48 (0.04) n=6 | 0.21 0.02) n=6 | 0.242 (0.08) n=6 | 0.43 | 0.50 | 8.7 (3.7) n=6 | 0.26 | 0.157 (0.06) n=6 | 33 (7.6) n=6 | 18 (4.4) n=6 |
| 4 | 0.83 (0.09) n=6 | 0.28 (0.02) n=6 | 0.444 (0.02) n=6 | 0.34 | 0.54 | 11 (1.6) n=6 | 0.0 | 0.111 (0.07) n=6 | 13 (3.6) n=6 | 10 (5.4) n=6 |
| 5 | 0.84 (0.07) n=6 | 0.40 (0.02) n=6 | 0.456 (0.02) n=6 | 0.46 | 0.54 | 11 (1.6) n=6 | 0.09 | 0.082 (0.05) n=6 | 9 (3.6) n=6 | 7 (5.04) n=6 |
| 6 | 0.85 (0.07) n=6 | 0.53 (0.03) n=6 | 0.468 (0.04) n=6 | 0.60 | 0.55 | 11.8 (4.5) n=6 | 0.5 | 0.053 (0.03) n=6 | 6 (3.4) n=6 | 4 (3.1) n=6 |

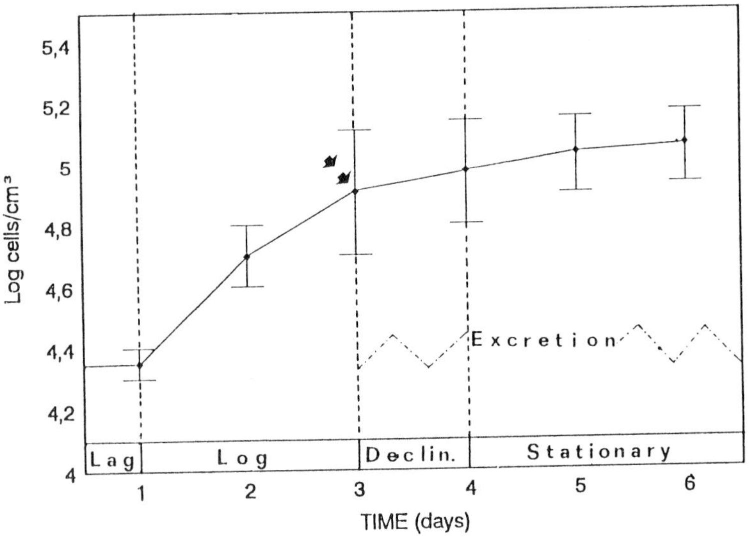


Fig. 11. Growth curve of *Pleurosigma* sp. in batch culture showing four phases. Initial release of photosynthate (arrows) occurs during declining phase. (- -) continuous excretion.

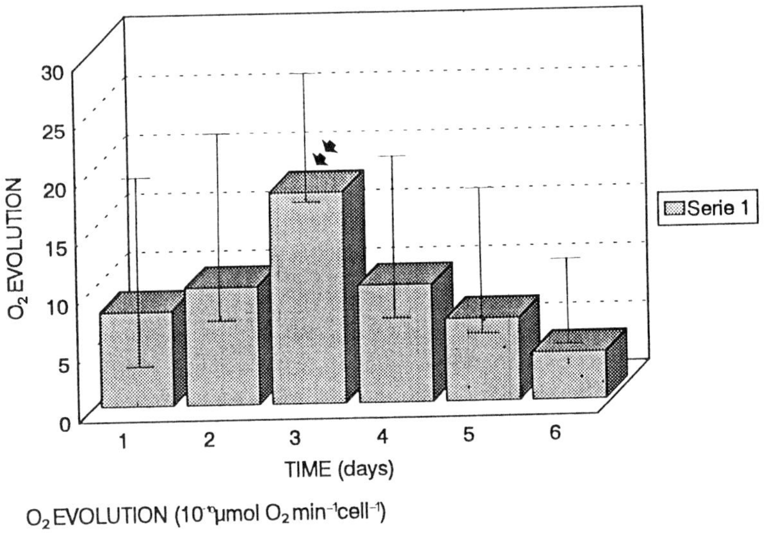


Fig. 12. Photosynthetic capability expressed as $O_2$ evolution on per cell basis within a week. Exudate release (arrows) takes place when $O_2$ evolution is maximum.

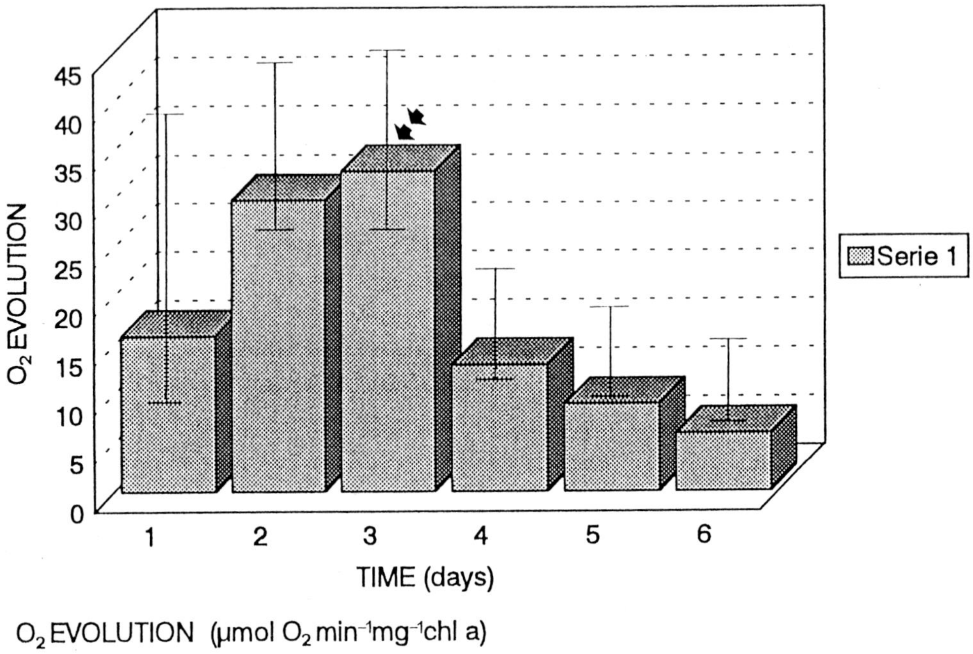


Fig. 13. Photosynthetic capability normalized to chlorophyll *a* content shows the same trend as in Fig. 12.

statistically significant difference appears to be the third one (Table II). This identifies precisely the declining phase when release of exudates starts to take place. Total Chl *a*, *c* and Car, estimated from extracts, increased during log phase (from a minimum of 0.10 to a maximum of 0.82 mg $dm^{-3}$ for Chl *a*; from 0.038 to 0.28 for Chl *c*; from 0.057 to 0.44 mg $dm^{-3}$ for Car) and remained constant during the stationary phase (Table I). On the third day, cells started to excrete material during the declining phase when cell concentration was not at its highest ($8.7 \times 10^4$ cells $ml^{-1}$) while $O_2$ evolution was maximum (Table I). Excretion continued throughout the experiment (Fig. 11). Exudate gave positive reactions to cytochemical tests both for polyanionic polysaccharides and for glycoproteins. RR evenly stained the whole exudate. The inner pectic layer of the cell wall was only weakly stained (Fig. 14). ATB, specific for sulphated groups, gave a metachromatic reaction at the centre of the floc while the cell walls remained uncoloured (Fig. 15). BTB, specific for carboxylated groups, gave a metachromatic reaction at the periphery of the aggregate, staining also the subfrustular layer (Fig. 16). AB and AY confirmed the same distribution of charged groups (Figs 17, 18). FG turned from dark into bright green from the centre to the periphery of the aggregate, staining also part of the cell walls (Fig. 19).

## Discussion

Measurements of photosynthesis indicate that cells are actively photoassimilating during the exponential growth phase and reach a maximum of photosynthesis during the declining phase. This demonstrates that in these phases there is still availability of light, $CO_2$ and nutrients in the culture. Progressive decrease of photosynthesis measured from the fourth day may be due mainly to $CO_2$ depletion owing to lack of aeration and to light attenuation for mucus covering. These results are in agreement with the findings of Nalewajko (1966) on various planktonic algae under culture.

Table II. Student T-test on means (t) and probabiulities of differences of means (p) on the variables: V1, cell number; V2, photosynthetic capacity normalized to Chla; V3, photosynthetic capacity on per cell basis. Significance ≥ 95%.

| | Days | t | p |
|---|---|---|---|
| V1 | 1–2 | 5.522 | 99.9% |
| | 1–3 | 4.212 | 99.8% |
| | 1–4 | 12.92 | 100% |
| | 1–5 | 12.92 | 100% |
| | 1–6 | 5.151 | 99.9% |
| | 2–4 | 5.808 | 99.9% |
| | 2–5 | 5.808 | 99.9% |
| | 2–6 | 3.098 | 98.8% |
| V2 | 1–3 | 3.299 | 97.3% |
| | 2–4 | 0.168 | 99.9% |
| | 2–5 | 0.801 | 100% |
| | 2–6 | 1.947 | 100% |
| | 3–4 | 3.085 | 100% |
| | 3–5 | 4.021 | 100% |
| | 3–6 | 6.405 | 100% |
| | 4–6 | 2.074 | 99.3% |
| V3 | 1–3 | 3.299 | 99.2% |
| | 2–3 | 2.499 | 97.0% |
| | 3–4 | 3.085 | 98.9% |
| | 3–5 | 4.021 | 99.7% |
| | 3–6 | 6.405 | 100% |

Average Chl *a* content is directly correlated to cell growth, since it increases in the log phase and it is nearly constant in the stationary phase in spite of a decreased photosynthetic capability. Nevertheless decrease of photosynthetic rate normalized to

Chl *a* is more pronounced compared with the rate per cell basis. These results seem to suggest that not all the Chl *a*, which is present in the last phase, is photosynthetically active. Comparison between $O_2$ µmol to cell number and $O_2$ µmol to Chl *a* content ratios, highlights possible changes in Chl *a* activity in different phases of the cycle. There is not a direct relationship between photosynthetic capability and Chl *a* content. This becomes more evident in ageing cultures where the highest pigment content does not correspond to the maximum $O_2$ evolution. It might be suggested that functioning Chl *a* is actually less than indicated by the values obtained. This would explain a diminished $O_2$ evolution in spite of the highest Chl *a* total content. This hypothesis is in agreement with the observations of Goedher (1970) who demonstrated that, in ageing cells, Chl *a* in system II, connected to $O_2$ evolution, is less active than Chl *a* in system I. Car have the same trend as Chl *a* and it is presumable that these pigments may lose their photoprotective role in favour of widening the action spectrum when light irradiance is diminished by the shading effect of the exudate and high culture density. As far as excretion is concerned, the whole photosynthate is presumably used to sustain growth when cells are rapidly dividing and nutrients abundant in the culture are actively photoassimilating. The increasing population density, during the declining phase, may result in a nutrient deficiency inhibiting cell division at a time when photosynthetic capability is the highest. Decline of growth together with high photosynthetic activity would determine accumulation and consequent release of photosynthate. That nutrient limitation affects cell division have been largely demonstrated in diatoms both in field and in culture conditions (*Chaetoceros affinis*, Myklestad & Haug 1972; *Thalassiosira weissflogii*, Vaulot *et al.* 1987; Harrison *et al.* 1990). In our experiments, nutrient limitation besides affecting cell division, appears to be related also to photosynthate release, as already suggested by Ignatiades (1973) for *Skeletonema costatum.* Release would occur because cells, being unable to divide owing to depletion of nutrients, would have a surplus of photosynthate. Excretion would then continue at an increasing rate as the $O_2$ evolution decreased during the stationary phase. These considerations are in agreement with the observations of Mague *et al.* (1980) under both field and culture conditions. Excretion would take place when division, growth and photosynthesis are umbalanced, thus producing an overflow mechanism that may become beneficial to the diatom as a secondary effect (Fogg 1983).

---

Fig. 14. Aggregate evenly coloured (⇑) by Ruthenium Red which also weakly stains the inner pectic layer of the cell wall (↑). Exudates at first stage stained by Toluidin Blue at pH 0.5 and at pH 4.4. Metachromatic reaction occurs at the centre (c) and at the periphery (p) of the flock. Note the subfrustular layer (↑) coloured by Toluidin Blue pH 4.4. uc = uncoloured cells. Figs 17 & 18. Alcian Blue and Alcian Sequence confirm the same location of charged groups as in Figs 15 & 16. Note blue at the centre (c) and yellow both in cells (↑) and in exudate periphery (p). uc = uncoloured cells. Fig. 19. Culture stained by Fast Green. Bright green from the centre (c) to the periphery (p) of the flock reveals a diffused distribution of proteins. Part of the cell walls (↑) are also positive. Scale bars = 10 µm (Fig. 15); 50 µm (Figs 16–18); 100 µm (Fig. 14).

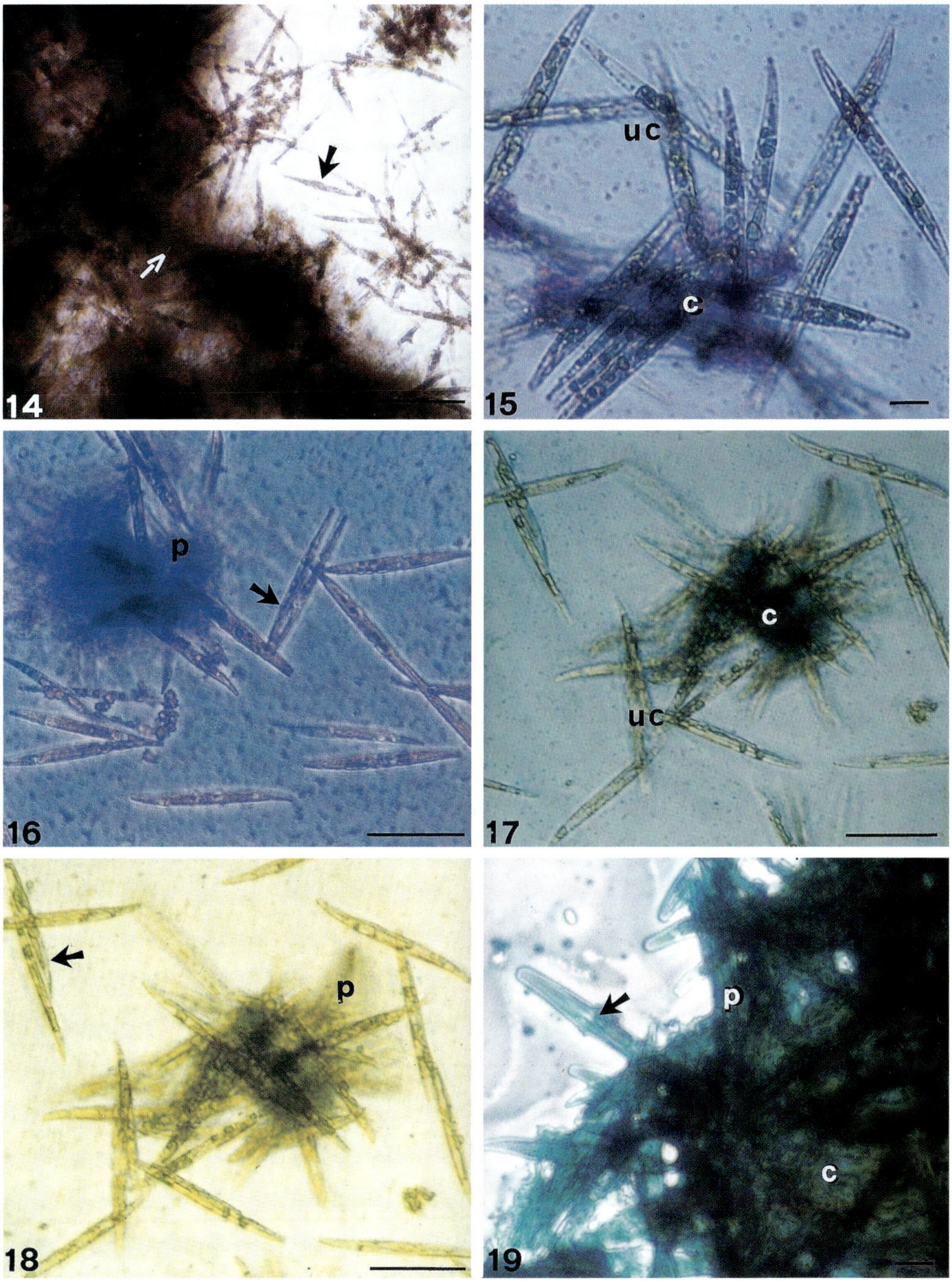
14
15
uc
c
16
p
17
c
uc
18
p
19
p
c

The results of cytochemical tests confirm the presence of carboxylated and sulphated polysaccarides and glycoproteins in exudates, as already found by other authors (Parker & Diboll 1966; Daniel *et al.* 1987; Marchetti *et al.* 1989; Cortese & Talarico 1993). Presence of charged groups with different location within the flock suggests that it might be a center of attraction for organic particles and colloidal material by cation exchange, besides preventing diatoms from dessication (Percival 1979; Daniel *et al.* 1987; Sdrigotti *et al.* 1994). The mucus aggregate would act also as a nutrient enriched microhabitat (Herndl & Peduzzi 1988), since cells inside the flock were more motile and vital than those living outside (Cortese & Talarico 1993) and new healthy cultures originated from inside cells. The chemical nature of the sub-frustular organic layer (also positive to stains specific for polyanionic polysaccharides and proteins), might support the hypothesis that this inner-pectic layer of the cell wall may be related to exudate excretion (Round *et al.* 1990).

In conclusion release of exudate starts to take place at time when photosynthesis is high and cells, still actively metabolizing, are unable to divide owing to mineral deficiency. In the natural environment, where similar physiological responses have been described (Fogg 1977; Sharp 1977), exudate release might be considered a biological advantage. Diatoms may reassimilate these extracellular products that can also be recycled by other microorganisms. In particular aggregates might be considered microhabitats with heterotrophic and autotrophic microorganisms that may function as a cell sufficient-trophic loop (Stachowitsch *et al.* 1990). The formation of flocks may also have a role similar to that played by mucilage structures in colony formation in other algae, particularly with respect to nutrient uptake and light demand. In fact the aggregates would allow the diatom to float, thus favouring wider migrations toward more lightened and nutrient enriched zones. This might be beneficial to the alga to sustain growth when photosynthesis becomes limited (Fogg 1983). As recently suggested by Hoagland *et al.* (1993), also other functions may be attributed to these extracellular polymeric substances such as protection of the cell wall from dissolution (by silica depletion), from chemical and physical (abrasion) attacks and from grazers. Therefore extracellular release may be considered as a normal physiological process, directly related to photosynthetic capability, but it may also play a considerable ecological role for species survival.

**Acknowledgements**

We are very grateful to Dr Anna Maria Schmid, Institute of Plant Physiology, University of Salzburg (Austria) for her kind suggestions and to Dr Frank E. Round, Department of Botany, University of Bristol (UK) for critically reviewing and improving the manuscript.

This research was supported by the Italian CNR (Consiglio Nazionale delle Ricerche) and MURST (Ministero dell'Università e della Ricerca Scientifica e Tecnologica).

## References


Cabrini, M., Fonda Umani, S. & Honsell, G. (1992). Mucilaginous aggregates in the Gulf of Trieste (Northern Adriatic Sea): analysis of the phytoplanktonic communities in the period June–August 1989. *Science of the Total Environment*, **(s.)**, 1–16.

Chayen, J., Bitensky, L., Butcher, R. G. & Poulter, L. W. (1973). *A Guide to Practical Histochemistry*. Lippincot, Philadelphia.

Cortese, A. & Talarico, L. (1993). Observation on gelatinous exocellular products from *Nitzschia closterium* (Ehrenderg) Smith under culture. *Giornale Botanico Italiano*, **127(4)**, 838–840.

Daniel, G. F., Chamberlain, A. H. L. & Jones, E. B. (1987). Cytochemical and electron microscopical observations on the adhesive materials of marine fouling diatoms. *British Phycological Journal*, **22**, 101–118.

Davies, B. H. (1976). Carotenoids. In: *Chemistry and Biochemistry of plant pigments* (T. W. Goodwin, ed.), 38–166. Accademic Press, London.

Delieu, T. & Walker, D. A. (1972). An improved cathode for the measurement of photosynthetic oxygen evolution by isolated chloroplasts. *New Phytologist*, **71**, 201–225.

Fanuko, N., Rode, J. & Draslar, K. (1989). Microflora from the Adriatic mucous aggregations. *Biol Vestn* in Ljubljana, **37**, 27–34.

Fogg, G. E. (1962). Extracellular products. In: *Physiology and Biochemistry of Algae* (R. A. Lewin, ed.), 475–489. Academic Press, N. Y. & London.

Fogg, G. E. (1977). Excretion of organic matter by phytoplankton. *Limnology and Oceanography*, **22 (3)**, 576–577.

Fogg, G. E. (1983). The Ecological Significance of Extracellular Products of Phytoplankton Photosynthesis. *Botanica marina*, **26**, 3–14.

Goedher, J. C. (1970). On the pigment system of brown algae. *Photosynthetica*, **4** (2), 97–106.

Guillard, R. R. L. & Ryther, J. H. (1962). Studies of marine planktonic diatoms. I. *Cyclotella nana* Hustedt and *Detonula confervacea* (Cleve) Gran. *Canadian Journal of Microbiology*, **8**, 229–239.

Harrison, P. J., Thompson, P. A. & Calderwood, G. S. (1990). Effect of nutrient and light limitation on the biochemical composition of phytoplankton. *Journal of Applied Phycology*, **2**, 45–56.

Haug, A. & Myklestad, S. (1976). Polysaccharides of marine diatoms with special reference to *Chaetoceros* species. *Marine Biology*, **4**, 217–222.

Hellebust, J. A. (1974). Extracellular products. In : *Algal Physiology and Biochemistry* (W. D. P. Stewart, ed.), 838–863. Blackwell Scientific Publications.

Herndl, G. J. & Peduzzi, P. (1988). The ecology of amorphous aggregations (marine snow) in the Northern Adriatic Sea: I. General considerations. *Marine Ecology*, **9** (1), 79–90.

Hoagland, K. D., Rosowski, J. R., Gretz, M. R. & Roermer, S. C. (1993). Diatom extracellular polymeric substances: function, fine structure, chemistry, and physiology. *Journal of Phycology*, **29**, 537–566.

Honsell, G. & Cabrini, M. (1990–91). Il fitoplancton durante il "mare sporco" dell'agosto 1988 nel Golfo di Trieste (Adriatico Settentrionale). *Bollettino della Società Adriatica di Scienze in Trieste*, **72**, 7–20.

Huntsman, S. A. & Sloneker, J. H. (1971). An exocellular polysaccharide from the diatom *Gomphonema olivaceum*. *Journal of Phycology*, **7**, 261–264.

Ignatiades, L. (1973). Studies on the factors affecting the release of organic matter by *Skeletonema costatum* (Greville) Cleve in field conditions. *Journal of the Marine Biological Association of the United Kingdom* , **53**, 923–935.

Ignatiades, L. & Fogg, G. E. (1973). Studies on the factors affecting the release of organic matter by *Skeletonema costatum* (Greville) Cleve in culture. *Journal of the Marine Biological Association of the United Kingdom*, **53**, 937–956.

Jeffrey, S. W. & Humphrey, G. W. (1975). New spectrophotometric equations for determinig chlorophylls a, b, $c_1$ and $c_2$ in higher plants, algae and natural phytoplankton. *Biochemie und Physiologie der Pflanzen*, **467**, 191–194.

Jensen, W. A. (1962). *Botanical Histochemistry.* 408 pp. W. H. Freeman and Co., San Francisco.

Kirk, J. T. O. & Allen, R. L. (1965). Dependence of chloroplast pigment synthesis on protein synthesis: effect of actidione. *Biochemistry and Biophysics Research Communities*, **21**, 523–530.

Lund, J. W. G., Kipling, C. & Lecren, E. D. (1958). The inverted microscope method of estimating algal numbers and the statistical basis of estimation by counting. *Hydrobiologia*, **11**, 143–170.

Mague, T. H., Friberg, E., Hughes, D. J. & Morris, I. (1980). Extracellular release of carbon by marine phytoplankton; a physiological approach. *Limnology and Oceanography*, **25 (2)**, 262–279.

Marchetti, R., Iacomini, M., Torri, G. & Tocher, B. (1989). Caratterizzazione preliminare degli essudati di origine fitoplanctonica raccolti in Adriatico nell'estate 1989. *Acqua–Aria*, **8**, 883–887.

Medlin, L. K. & Round, F. E. (1986). Taxonomic studies of marine gomphonemoid diatoms. *Diatom Research*, **1**(2), 205–225.

Myklestad, S. & Haug, A. (1972). Production of carbohydrates by the marine diatom *Chaetoceros affinis var. Willei* (Gran) Hustedt. I. Effect of the concentration of nutrient in the culture medium. *Journal of Experimental Marine Biology and Ecology*, **9**, 125–136.

Myklestad, S., Haug, A. & Larsen, B. (1972). Production of carbohydrates by the marine diatom *Chaetoceros affinis var. Willei* (Gran) Hustedt. II. Preliminary investigation of the extracellular polysaccharide. *Journal of Experimental Marine Biology and Ecology*, **9**, 137–144.

Myklestad, S., Holm-Hansen, O., Varum, M. & Volcani, B. E. (1989). Rates of release of extracellular aminoacids and carbohydrates from the marine diatom *Chaetoceros affinis*. *Journal of Plankton Research*, **11**, 763–773.

Nalewajko, C. (1966). Photosynthesis and excretion in various planktonic algae. *Limnology and Oceanography*, **11 (1)**, 1–10.

Parker, B. C. & Diboll, A. G. (1966). Alcian stains for histochemical localization of acid and sulphated polysaccharides in algae. *Phycologia*, **6 (1)**, 37–46.

Pearse, A. G. E. (1968). *Histochemistry: Theoretical and Applied.* Vol. I. Churchill, London.

Percival, E. (1979). The polysaccharides of green, red and brown seaweeds: their basic structure, biosynthesis and function. *British Phycological Journal*, **14**, 103–117.

Round, F. E., Crawford, R. M. & Mann, D. G. (1990). *The Diatoms. Biology and morphology of the genera.* 747 pp. Cambridge University Press.

Ruthmann, A. (1970). *Methods in Cell Research.* 368pp. Cornell University Press, Ithaca, New York.

Sdrigotti, E., Cortese, A. & Talarico, L. (1994). Preliminary observation on release of extra-cellular products by some diatoms under culture. *Giornale Botanico Italiano*, **128**, 4pp.

Sharp, J. H. (1977). Excretion of organic matter by marine phytoplankton: Do healthy cells do it? *Limnology and Oceanography*, **22**, 381–399.

Smestad, B., Haug, A. & Myklestad, S. (1975). Structural studies of the extracellular polysaccharide produced by the diatom *Chaetoceros curvisetus* Cleve. *Acta Chemica Scandinavica*, **29**, 337–340.

Stachowitsch, M., Fanuko, N. & Ritcher, M. (1990). Mucus aggregates in the Adriatic Sea: an overview of stages and occurrences. *Marine Ecology*, **11 (4)**, 327–350.

Talarico, L., Sdrigotti, E., Cortese, A. & Blasutto, O. (1994). Excretion of photosynthate by some diatoms from the Northern Adriatic Sea (Italy). In: *Fifth International Phycological Congress, Qingdao, Shandong P.R. China*, Abstracts, 39.

Vaulot, D., Olson, R. J., Merkel, S. & Chisholm, S. W. (1987). Cell cycle response to nutrient starvation in two phytoplanktonic species, *Thalassiosira weissflogi* and *Hymenomonas carterae*. *Marine Biology*, **95**, 625–630.

# Lipid fatty acid composition of selected diatom species used in Malaysia aquaculture as live food for penaeid larvae

Lokman Shamsudin

*Faculty of Fisheries and Marine Science, University Pertanian Malaysia, Mengabang Telipot, 21030, Kuala Terengganu, Terengganu, Malaysia.*

## Abstract

The total ash, chlorophyll, phaeopigment, lipid and fatty acid contents of two diatoms (*Chaetoceros calcitrans, Chaetoceros malaysia*) used in tropical Malaysian penaeid mariculture were studied. The axenic laboratory cultures were grown in f-2 medium, while outdoor cultures were grown in a commercial medium designed for optimum nutrition in tropical outdoor aquaculture operations. Considerable amounts (7–14% of the total fatty acid) of the polyunsaturated fatty acid 20:5w3 (*eicosapentaenoic acid*) were present. Lipid content was three to five times higher than chlorophyll–a. There was an increase with culture age in the relative proportion of total C18 and C20 fatty acid components. The diatom contained the w3–polyunsaturated fatty acids (PUFA) necessary for the growth and survival rate of the prawn larvae.

## Introduction

A sufficient quantity and quality of diatoms have to be grown in mass culture under local condition in hatcheries. Subsequently these live food organisms are fed to the early stages of the prawn, *Penaeus monodon* Fabricius, which is an important commercial species in Malaysian shrimp aquaculture. Diatoms have an important role in aquaculture as live food for larval stages of many species of Crustacea and fish, as well as for all stages of bivalves and as food for the zooplankton (*Rotifers, Copepods,* brine shrimps) which are eventually fed to late larval and juvenile fish in hatcheries (Renaud *et al.* 1991). The nutritional value of the diatoms is related to their biochemical composition, especially lipids and fatty acids (Chu & Dupuy 1980; Watanabe *et al.* 1983).

Most marine animals have only a limited ability to synthesise the polyunsaturated fatty acids (PUFA) 20:5w3 (*eicosapentaneoic acid*) and 22:6w3 (*docosahexaenoic acid*) from precursor fatty acids such as linolenic acid (Kanazawa *et al.* 1979). Some aquatic animals may not have an absolute dietary requirement for eicosapentaenoic and docosahexaenoic acids, but growth rates and larval survival usually increase when these fatty acids are included in a diet derived from cultured microalgae (Rodgers and Barlow 1987).

## Material and methods

Strains of *Chaetoceros malaysia* (P. Sayak) and *Chaetoceros calcitrans* were obtained from the Fishery Research Station at Tanjong Demong about 70 km from the University's hatchery research station, which is situated on the east coast of the Peninsular of Malaysia. Axenic strains were used to provide starter cultures, which in turn provided inocula for larger volume cultures for feeding larvae of the prawn *Penaeus monodon.* A line of starter cultures was set up by inoculating one 500 ml flask containing 240 ml of Erdschreiber medium on each of three successive days according to the methods described by Shamsudin (1992). The 200–litre cultures in fibre glass tanks were inoculated at 1000 cells $l^{-1}$ from the 20 l stage and aerated .with 5 l ml–1 of 1% $CO_2$ (Helm *et al.* 1979). Outdoor culture media were prepared using 0.2 µm filtered seawater, enriched with commercial fertilisers (Shamsudin 1992). The diatoms were harvested by centrifugation at various culture ages (2, 8 and 12 d after incubation)

Aliquots were taken during different culture age to determine cell density, pigment (chlorophyll–a, –c and phaeophytin–a), lipid and fatty acid contents. Cell density was determined with a Neubauer haemacytometer using the method of Vollenweider *et al.* (1974); the percentage error in five replicate ranged from 6.4 to 12.2%. The instantaneous growth rate (k) of the exponential phase was estimated according to the methods of Meynell and Meynell (1970). Pigments were extracted from the cell pellet in 90% acetone, followed by sonification (Bransonic S2) (Vollenweider *et al.* 1974; Jeffrey and Humphrey 1975; Shamsudin 1980). The absorbance at 664 nm of the acetone extracts was measured before and after acidification to determine the concentration of chlorophyll–a and phaeophytin–a (Marker & Casey 1982).

For fatty acid analysis, samples of the two diatoms species were collected, weighed, measured, lyophilized and stored at –80°C under nitrogen. Samples were extracted twice with chloroform-methanol and water (1:2:0.8 by volume) according to the Bligh and Dyer (1959) technique as modified by Kates (1972). Lipids were concentrated on a rotary evaporator at 30°C and total lipids mass was determined gravimetrically. The lipid material was methylated with 14% boron trifluoride-methanol (W/V) (Morrison & Smith 1964). Fatty acid methyl ester (FAME) was injected at 60°C onto a single columm Hewlett Packard 5860 series II gas chromatography unit. Initial identification of the FAME was made on a superlcowax fused silica capillary column (30 x 0.53 mm id, 1.0 µm film) with a 40 ml/min flow of nitrogen as the carrier gas as described by Shamsudin (1992) .

## Results and Discussion

Table I shows some of the features of the two diatoms. The growth rate of *C. malaysia* was higher (1.14 div. $d^{-1}$ ) than that of *C. calcitrans* (0.79 div. $d^{-1}$ ). Their lipid contents were 4 to 5 times higher than the chlorophyll–a. The chlorophyll–c contents were less than 10% of their respective chlorophyll–a; however, much lower values were obtained from *C. calcitrans*. The major fatty acid constituents of the diatoms were the saturated fatty acids of the even chain lengths C14–C18, unsaturated monoethylene and polyethylene C16 and C18 acids (Tables II–III). These fatty acid components comprised >90% of the even numbered carbon chain acids in all ages whereas other odd numbered saturated acid types were probably present in small amounts. The fatty acid contents of the two diatoms had high concentrations of 16:1 (palmitoleic acid) and 16:0 (palmitic acid) with variable, but usually, high concentrations of 14:0 (tetradecanoic acid) and 20:5w3 (eicosapentanenoic acid) (Tables II and III). *C. malaysia* contained high amounts of 16:1 acid (30–40%), 14:0 acid (17–22%) and 16:0 acid (10–17%) (Table II). The total saturated acids were higher (37–41%) than those of the polyethylenic unsaturated acids (21–26%). For both diatoms, there was an increase with culture age in the relative proportion of saturated acid components and a decrease of the polyethylenic acids. The ash contents in the diatoms (37–41%) were high; this reflects the high content of the diatom's silicate skeleton. Nalewajko (1966) reported that ash accounted for between 5.3–19.9% of the dry weight of 16 planktonic chlorophytes, but 27–55% for 11 diatom species.

Table I. Some characteristics of the two diatom species used as food for prawn larvae. All values were taken during the log growth phase. (DW, dry weight).

| Algae | Unit | *C. malaysia* | *C. calcitrans* |
|---|---|---|---|
| Mean indiv.cell vol. | $\mu m^3$ | 52.4 | 49.3 |
| Cell growth rate (k) | div. $d^{-1}$ | 1.24 | 0.89 |
| Ash | % DW | 37.1 | 41.4 |
| Lipid | pg $cell^{-1}$ | 0.16 | 0.14 |
| Chl–a | pg $cell^{-1}$ | 0.05 | 0.04 |
| Chl–c | % Chl–a | 8.40 | 9.40 |

Table II. Fatty acid composition, as a percentage of total fatty acids of *Chaetoceros malaysia*. Values are means of duplicate or triplicate analyses. Standard deviations are omitted for clarity, but were normally <5%. (Tr, Trace amounts, less than 0.05%).

| Fatty acid | Culture day | | |
|---|---|---|---|
| | 2 | 8 | 12 |
| *Saturated* | | | |
| 14:0 | 21.7 | 17.9 | 17.4 |
| 16:0 | 10.2 | 14.4 | 17.3 |
| 18:0 | 0.2 | 0.4 | 0.4 |
| Total | 32.1 | 32.7 | 35.1 |
| *Monoethylenic* | | | |
| 16:1 | 34.2 | 34.6 | 37.4 |
| 18:1 | 0.4 | 0.2 | 0.8 |
| 20:1 | – | – | – |
| 22:1 | Tr | Tr | Tr |
| Total | 34.4 | 34.8 | 38.2 |
| *Polyethylenic* | | | |
| 16:3w4 | 0.3 | 2.5 | 2.3 |
| 18:2w6 | 0.2 | 0.4 | 0.2 |
| 18:3w6 | – | 0.4 | Tr |
| 18:3w3 | – | – | – |
| 18:4w3 | 0.2 | 0.4 | 0.4 |
| 20:2w6 | – | – | – |
| 20:4w6 | 11.2 | 4.3 | 3.7 |
| 20:4w3 | – | 0.2 | 0.3 |
| 20:5w3 | 14.1 | 12.4 | 12.7 |
| 22:6w3 | 0.4 | 0.3 | 1.2 |
| Total | 26.4 | 21.0 | 23.7 |
| Total of w3 | 25.9 | 13.3 | 13.4 |
| Totalofw6 | 11.4 | 5.1 | 3.9 |
| w3/w6 | 2.3 | 2.6 | 3.4 |
| Σ unsat 16/16:0 | 3.4 | 2.6 | 2.3 |
| Σ unsat 18/18:0 | 2.0 | 3.0 | 1.5 |
| Σ unsat/Σ sat | 1.9 | 1.7 | 1.8 |

Table III. Fatty acid composition, as percentage of total fatty acids of *Chaetoceros calcitrans*. Values are means of duplicate or triplicate analyses. Standard deviations are omitted for clarity, but were normally <5%. (Tr, trace amounts, less than 0.05%).

| Fatty acid | Culture day | | |
|---|---|---|---|
| | 2 | 8 | 12 |
| *Saturated* | | | |
| 14:0 | 19.8 | 24.4 | 27.9 |
| 16:0 | 14.2 | 17.2 | 16.7 |
| 18:0 | 0.4 | 0.5 | 0.4 |
| Total | 34.4 | 42.1 | 45.0 |
| *Monoethyleic* | | | |
| 16:1 | 29.8 | 27.4 | 24.3 |
| 18:1 | Tr | 1.3 | 1.4 |
| Total | 29.8 | 28.7 | 25.7 |
| *Polyethylenic* | | | |
| 16:3w4 | Tr | 2.7 | 3.4 |
| 18:2w6 | 1.3 | 1.4 | 2.5 |
| 18:3w6 | Tr | 0.2 | Tr |
| 18:3w3 | Tr | 0.1 | 0.5 |
| 18:4w3 | 9.6 | 3.4 | 3.8 |
| 20:2w6 | – | 0.2 | – |
| 20:4w6 | 0.8 | – | – |
| 20:5w3 | 7.4 | 7.7 | 7.9 |
| 20:6w3 | 0.5 | 0.2 | 0.4 |
| Total | 19.6 | 15.9 | 18.5 |
| Total of w3 | 17.5 | 11.4 | 12.6 |
| Total of w6 | 2.1 | 1.8 | 2.5 |
| w3/w6 | 8.3 | 6.3 | 5.0 |
| Σ unsat 16/16:0 | 2.1 | 1.9 | 1.7 |
| Σ unsat 18/18:0 | 24.0 | 10.0 | 14.3 |
| Σ unsat/Σ sat | 1.4 | 1.1 | 1.0 |

The maximum density in the *C. malaysia* culture (17.2 x $10^5$ cells $ml^{-1}$ ) was higher than that of *C. calcitrans* (1.5 x $10^5$ cells $ml^{-1}$). *C. calcitrans* grew poorly under the hatchery culture conditions, as shown by its slow growth rate (0.89 div. $d^{-1}$). Many species of *Chaetoceros* are characterised by their tolerance and adaptability to high temperature (Liao 1980). For instance, *C. gracilis* has an optimal growth at temperature ranging between 25 and 30°C with its maximum toleration temperature at 37°C. Its mass culture in hatcheries is generally performed within the temperature range of 25 to 30°C. Volkman *et al.* (1989) reported that lipids were higher than chlorophyll–a in ten species of microalgae commonly used in mariculture. In the present study, *C. calcitrans* contained lower lipid and chlorophyll–a contents, probably due to its small size. The phaeophytin–a of the diatom ranged from 5–11% during the log growth phase and gradually increased (19–30%) with age of the cultures.

The fatty acid composition of the diatom species used as food is of greatt importance to the growth and survival of aquaculture organisms. Eicosapentaenoic acid (20:5w3) was the major component of both diatoms (7–14% of total acids).These values compared favourably with values for 20:5w3 in other microalgae currently recommended for aquaculture (Volkman *et al.* 1989; Ben-Amotz *et al.* 1987). The docosahexaenoic acid (22:6w3) in both diatoms was low and is comparable with the value reported by Volkman *et al.* (1989). The 18:2w6 linoleic acid was also low in both the *Chaetoceros* species. Changes were observed in the total monoethylenic and polyethylenic acid components with culture age. The total monoethylenic acid tended to increase, whereas total polyethylenic decreased gradually with respect to culture age. Chu and Dupuy (1980) reported that the major fatty acid components comprised 70–93% of the total fatty acids at all ages, whereas odd-numbered saturated and other saturated fatty acids accounted for only 3–18% of the total fatty acids. High ratios of w3 to w6 polyunsaturated fatty acids have been normally used as an index of high nutritional value to aquaculture organisms (Watanabe *et al.* 1983). Various values of the total unsaturated ratio (C–unsat index), the ratio of total unsaturated C16 fatty acids to 16:0 (C–16 index) and the ratio of total unsaturated C18 fatty acids to 18:0 (C–18 index) are some of the other indexes used to denote food value. *C. malaysia* showed high values in total w3 and total w6 indexes but low in C–unsat index. These indices are valuable together with a consideration of which fatty acids are present to indicate the nutritional value of microalgal cells to aquaculture organisms (Volkman *et al.* 1989).

Volkman *et al.* (1989) found that the C16–PUFA were particularly abundant in diatom compared with other algal species. Their major components were identified as 16:2w7, 16:2w4, 16:3w4 and 16:4w1 which were synthesized by further desaturation of 16:1w7 on either side of the 9th carbon bond. The high proportion of these fatty acids accounted for the overall lower abundance of w3 and w6 fatty acids in diatom. In the present studies, the 16:3w4 components of the two diatoms were present in considerable amounts, but the other fatty acid components comprising of 16:2w7, 16:2w4 and 16:4w1 were virtually absent.

**Acknowledgement**

This research was supported by the Malaysian National Research Grant, under Project Code 50366.

## References


Ben Amotz, A., Fisher, R. & Shneller, A. (1987). Chemical composition of dietary species of marine unicellular algae and rotifers with emphasis on fatty acids. *Marine Biology*, **95**, 31–36.

Bligh, E. G. & Dyer, W. J. (1959). A rapid method of total lipid extraction and purification. *Canadian Journal of Biochemistry and Physiology*, **37**, 911–917.

Chu, F. E. & Dupuy, D. J. (1980). The fatty acid composition of three unicellular algal species used as food sources for larvae of the American oyster (*Crassostrea virginica*). *Lipids*, **15**, 356–364.

Helm, M. M., Liang, L. & Jones, E. (1979). The development of a 2001 algal culture vessel at Conwy. *Fisheries Research Technical Report* No. **53**, 1, 2–12.

Jeffrey, S. W. & Humphrey, G. F. (1975). New spectrophotometric equations for determining chlorophylls–a, –b, and–c in higher plants, algae, and natural phytoplankton. *Biochemie und Physiologie der Pflanzen*, **187**, 191–194.

Kanazawa, A., Teshima, S. I. & Kazuo, O. (1979). Relationship between essential fatty acid requirements of aquatic animals and the capacity for bioconversion of linolenic acid to highly unsaturated fatty acids. *Comparative Biochemistry & Physiology*, **63**B, 295–298.

Kates, M. (1972). Techniques of lipidology: Isolation, analysis and identification of lipids. New York, North Holland, pp. 347–353.

Liao, I. C. & Chin, I. P. (1980). Manual on Propagation and Cultivation of Grass Prawn, *Penaeus monodon*. Tungkang Marine Laboratory, Taiwan.

Marker, A. F. H. & Casey, H. (1982). The population and production dynamics of benthic algae in an artificial recirculating hardwater stream. *Philosophical Transactions of the Royal Society of London* **B298**, 265–308.

Meynell, G. G. & Meynell, E. (1970). *Theory and practice in experimental bacteriology*. 2nd edition. Cambridge University Press, Cambridge, pp 370.

Morrison, W. R. & Smith, L. M. (1964). Preparation of fatty acid methyl esters and dimethyl acetate from lipids with boron trifluoride-methanol. *Journal of Lipid Research*, **5**, 600–608.

Nalewajko, C. (1966). Dry weight, ash and volume data for freshwater planktonic algae. *Journal of the Fisheries Research Board of Canada*, **23**, 1285–1288.

Renauld, S. M., Parry, D. L., Thinh, L. V., Kou, C., Padovan, A. & Sammy, N. (1991). Effect of light intersity on the proximate biochemical and fatty acid composition of *Isochrysis* sp.and *Nanochloropsis ocylata* for use in tropical aquaculture. *Journal of Applied Phycology*, **3**, 43–53.

Rodgers, L. J. & Barlow, C. G. (1987). Better nutrition enhance survival of barramundi larvae. *Australian Fish*ery, **46**(7), 30–32.

Shamsudin, L. (1980). The inorganic nutrient contents, photosynthetic values and other related environmental factors of Sungai Terengganu. *Malayan Nature Journal*, **36**, 175–186.

Shamsudin, L. (1992). Lipid and fatty acid composition of microalgae used in Malaysian aquaculture as live food for the early stage of penaied larvae. *Applied Phycology*, **4**, 378.

Volkman, J. K., Jeffrey, S. W., Nichols, P. D., Rogers ,G. I. & Garland, C. D. (1989). Fatty acid and lipid composition of 10 species of micrialgae used in mariculture. *Journal of Experimental Mariculture*, **128**, 219–240.

Vollenweider, R. A., Talling, J. F. & Westlake, D. F. (1974). A manual on methods for measuring primary production in aquatic environment, IBP Handbook No. 12. Blackwell, Oxford, 2nd Edition. pp 255.

Watanabe, T., Kitajima, C. & Jujita, S. (1983). Nutrition value of live organisms used in Japan for mass propagation of fish: a review. *Aquaculture,* **34**, 115–143.

# Are evolutionary tradeoffs evident in responses of benthic diatoms to nutrients ?

R. Jan Stevenson and Yangdong Pan

*Water Resources Laboratory, University of Louisville, Louisville, KY 40292, U.S.A.*

## Abstract

Differences in species growth rates in different nutrient conditions were used to test the hypothesis that there are tradeoffs in species abilities to use different resources. Multiple species diatom communities were developed on clay tiles in incubation chambers supplied with flowing water to simulate stream conditions. A 3 × 2 factorial experiment was conducted to determine the independent and interactive effects of additions of $NH_3$, $NO_3$, and $PO_4$ on species growth rates. Growth rates of most species were significantly affected by $PO_4$ enrichment. Growth rates of *Fragilaria crotonensis* and *Synedra ulna* were independently limited by $NH_3$, $NO_3$, and $PO_4$. In general, the species that had the fastest growth rates in low nutrient regime also grew fastest in the high nutrient regime. Also, the same species had the fastest growth rates in high $NH_3$, $NO_3$, and $PO_4$ concentrations. There was little evidence of adaptive tradeoffs and that competition was a dominant force in natural selection of diatoms in streams.

## Introduction

A variety of different cellular and community features indicate that competition for light and nutrients is an important process in benthic diatom community dynamics. As benthic algal communities develop on substrata, they develop a three-dimensional community matrix of prostrate, motile, apically attached and stalked growth forms (Patrick 1976). The arborescent growth forms of stalked diatoms indicate that nutrient and light availability may be higher in the overstorey of benthic algal communities. Light and nutrient transport through the periphyton matrix is affected negatively by benthic algal density (Hoagland 1983; Stevenson & Glover 1993). Benthic algal growth rates and cellular nutrient concentrations decrease with increasing cell density on substrata (Stevenson 1990; Humphrey & Stevenson 1992).

If competition for nitrogen and phosphorus is important, it may have produced evolutionary strategies for diatoms. Evolutionary strategies are restricted population niches that develop because traits that confer fitness have resource and energetic costs. Such fitness traits may be nutrient uptake abilities, reproduction rates, grazer resistance, or dispersal abilities. During periods of intense competition for a limiting resource, tradeoffs are hypothesized because the species that are evolutionarily successful will have allocated most resources and energy to the fitness trait that enables species survival. Species that allocate energy and resources to several traits would, theoretically, be unsuccessful (competitively excluded) because of their lack of specialization. Therefore, competition for resources may restrict a species' allocation of resources to sets of co-adapted traits that confer fitness during periods of intense competition for a limiting resource (nutrient, space, or light).

Interspecific variation in nutrient uptake abilities may involve tradeoffs in energy and resource allocation. Since nutrient uptake occurs by active transport, it requires cellular investment in the proteinaceous uptake channels in the cell membrane. Energy is also required for active transport of nutrients through transmembrane channels. Therefore, resource and energetic tradeoffs should exist for species abilities to take up and use different nutrients for reproduction. In addition, we could hypothesize that tradeoffs could exist in species abilities to use nutrients in low and high nutrient concentrations. The latter tradeoffs could result from different uptake or cellular resource allocation in low and high nutrient concentrations.

We have studied the intraspecific differences in diatom growth rates in different nutrient regimes to determine whether evidence exists for tradeoffs in fitness traits and thus for evolutionary strategies. We hypothesized that tradeoffs would be evident in different species being able to grow fastest in either high nitrogen and high phosphorus conditions. We also hypothesized that tradeoffs would be evident in different species being able to grow fastest in low and high concentrations of specific nutrients. Experiments to test these hypotheses were conducted in multispecies cultures in artificial stream conditions.

## Methods

Experiments were conducted at the Stream Research Facility of the University of Michigan Biological Station in Pellston, Michigan, USA. Effects of $NH_3$, $NO_3$, and $PO_4$ on diatom growth rates were determined in incubation chambers. These chambers were 25 cm diam. bowls made of clear plastic. Chambers were set about 15 cm under six fluorescent lights (three 40 watt daylight and three 40 watt cool white). These light assemblies supplied between 50 and 100 μmole quanta $m^{-2}$ $s^{-1}$. They could also be raised and lowered for access to the incubation chambers. Incubation chambers were set on magnetic stirrers. A 1 cm magnetic stir bar was used to generate current in the chambers. Evaporation of water from bowls required covering them with clear plastic food wrap that was held onto the bowls with rubber bands. A circular glass plate was under the stir bar at the centre of bowls because stir bars slowly eroded depressions in the acrylic bottoms of the chambers. Air was blown over the chambers in a small room that was cooled by an air conditioner to maintain water temperatures in bowls at 21 ± 2° C, which was the temperature of the stream from which water was taken.

Benthic algal communities were established (pre-colonized) on artificial substrata in the incubation chambers 3 days before measurements of community and population performance. Unglazed, off-white ceramic floor tiles were used (5 cm × 5 cm) as artificial substrata. Clean tiles were placed around the margins of half (12) of the bowls (total = 24). Water from the East Branch of the Maple River was filtered through a 20 µm mesh Nitex screen to remove most algae, larger organisms, and particulate detritus. 2 L of filtered stream water was added to the 12 chambers with tiles. Currents (ca. 5 cm $s^{-1}$) were started with magnetic stirrers and stir bars. Slow current minimized concentric patterns of algal immigration due to decreasing current velocity from centre to edge of bowls. A thick algal suspension for inoculating the chambers and tiles was produced by gathering diatoms from natural substrata in the East Branch of the Maple River. 30 mL of this suspension was added to the 12 chambers, which produced a cloudy algal suspension in the chambers. Three days later, diatom-covered tiles were transferred to clean incubation chambers to start the experiment.

Three or four diatom-covered tiles were placed in each of the 24 incubation chambers. 2 L of filtered stream water was added to each chamber. Water from the East Branch of the Maple River was used because $NO_3$-N and $PO_4$-P concentrations are usually low in this stream during the summer, which was when this experiment was conducted. Concentrated solutions of $NH_4Cl$, $NaNO_3$, and $NaH_2PO_4$ were added to each chamber to manipulate nutrient concentrations. Low and high concentrations of each nutrient were used to produce eight treatments in a 3 × 2 factorial experimental design (3 replicates for each treatment). Target nutrient concentrations in high concentration treatments were 500 µg $NH_3$-N $L^{-1}$, 500 µg $NO_3$-N $L^{-1}$, and 30 µg $PO_4$-P $L^{-1}$. When both $NH_3$ and $NO_3$ were added, volumes of concentrated solutions were halved so that all +N treatments had the same inorganic N concentrations. These concentrations were used because previous experiments showed that they were growth-saturating concentrations (Bothwell 1989; Stevenson submitted).

Every 8 h, 1.0 L of water was drained from the chambers and 1.0 L of filtered stream water was added in a way that is referred to as semi-continuous batch culture (Kilham 1978). When new filtered stream water was added, appropriate amounts of P and N solutions were also added to chambers., This approach was used to maintain relatively constant nutrient conditions in chambers during the 4 day incubation.

Five diatom samples, each comprising three tiles, were collected on day 0 of the experiment (one tile from each of the 15 chambers containing four tiles) to determine initial diatom densities on tiles. After 4 days algae were collected using the three remaining tiles from each chamber. These were preserved as composite samples to minimize variation of algal density due to spatial variation among tiles in a chamber. Densities of cells were determined by enumerating diatoms by species in syrup mounts (Stevenson 1984a) with bright field illumination at 1000× with a research quality Nikon microscope. Only frustules with protoplasm were counted.

Several reproduction characteristics of diatoms were calculated. Growth rates of diatom species ($r_i$) were calculated using the average per-day difference in ln-transformed diatom densities ($N_{i,d}$) during the 4 day incubation period:

$$r_i = (\ln(N_{i,4}/N_{i,0})/4)$$

The stimulation of $i^{th}$ species by specific nutrient enrichment treatments was determined by the change in growth rates of species with a specific nutrient enrichment when the other nutrient (N or P) was saturating:

$$\text{e.g. } r_{i,\text{high } NO_3-\text{high } PO_4} - r_{i,\text{low} NO_3-\text{high } PO_4}$$

Since non-nutrient environmental conditions (light and temperature) may not have been optimal for all species, a standardized growth stimulation was determined by the change in growth rates of species with specific nutrient enrichments divided by the maximum growth rate of the species in high nutrient conditions:

$$\text{e.g. } (r_{i,\text{high } NO_3-\text{high } PO_4} - r_{i,\text{low } NO_3-\text{high } PO_4})/r_{i,\text{high } NO_3-\text{high } PO_4}$$

The hypotheses that tradeoffs between traits and sets of co-adaptive traits occur were tested with Spearman rank correlation analyses (Wilkinson 1989). These hypotheses generally took the form that the ranks of species according to growth rates in high $NO_3$-N and high $PO_4$-P concentrations were negatively correlated. Another hypothesis tested was that ranks of species growth rates in high and low nutrient concentrations were negatively correlated. In these correlations, species were the "cases" or "observations" in the analysis and the variables were ranks of species growth rates or growth stimulation traits in different nutrient treatments.

**Results**

Nutrient concentrations were effectively increased in the incubation chambers (Table 1). Phosphate-P concentrations were increased from control levels of 6.0 to 24.8 µg/L. Nitrate-N was increased from 14.4 to 558 µg/L when no $NH_3$ was added and to 316 µg/L when $NH_3$ was added. Ammonia-N was increased from 32.4 to 572 µg/L when no $NO_3$ was added and to 293 µg/L when $NO_3$ was added.

In general, diatom growth rates were positively affected by P enrichment, but not by N enrichment (Table 2). Only growth rates of *Fragilaria crotonensis* and *Synedra ulna* were significantly ($p<0.05$) stimulated by $NO_3$ and $NH_3$ enrichment alone. Growth rates of both of these taxa were also stimulated by the interaction of both P and N enrichment together. All taxa except *Cymbella microcephala* and *Synedra* spp. were positively affected ($p<0.05$) by P enrichment. Lack of statistical significance of $NO_3$ and $NH_3$ effects for many taxa may have been due to small sample size and not to the magnitude of effects of N enrichment on their growth. Many of these taxa showed positive, but statistically insignificant responses to $NO_3$ and $NH_3$ enrichment (Fig. 1).

Table 1. Nutrient concentrations in incubation chambers.

| Treatment | Nutrient Concentration ($\mu g\ L^{-1}$) |
|---|---|
| Low $PO_4$-P | 6.0 ± 0.10 SE |
| High $PO_4$-P | 24.8 ± 0.74 SE |
| Low $NO_3$-N | 14.4 ± 3.77 SE |
| High $NO_3$–N | 558.1 ± 6.71 SE |
| $NO_3$-N with $NH_3$ | 315.9 ± 1.79 SE |
| Low $NH_3$-N | 32.4 ± 0.50 SE |
| High $NH_3$-N | 572.5 ± 1.75 SE |
| $NH_3$-N with $NO_3$ | 293.4 ± 7.74 SE |

Table 2. Statistically significant (ANOVA, $p<0.05$) responses of species growth rates to changes in nutrient concentrations. A "+" indicates a positive response.

| Species | $PO_4$ | $NO_x$ | $NH_3$ | P+$NO_x$ | P+$NH_3$ |
|---|---|---|---|---|---|
| *Achnanthidium minutissima* | + | | | | |
| *Achnanthidium deflexa* | + | | | | |
| *Cymbella microcephala* | | | – | | |
| *Cymbella minuta* | + | | | | |
| *Cymbella affinis* | + | | | | |
| *Diatoma tenue* | + | | | | |
| *Fragilaria crotonensis* | + | + | + | + | |
| *Gomphonema parvulum* | + | | | | |
| *Navicula radiosa* var. *tenella* | + | | | | |
| *Nitzschia fonticola* | + | | | | |
| *Nitzschia tropica* | + | | | | |
| *Nitzschia palea* | + | | | | |
| *Synedra* spp. | | | | | |
| *Synedra filiformis* | + | | | | |
| *Synedra ulna* | + | + | + | + | |

The fastest growing species in control conditions were *F. crotonensis, Synedra filiformis*, *Cymbella minuta*, and *Nitzschia tropica*, with growth rates ranging from 0.29 to over 0.5 $d^{-1}$ (Fig. 1). Maximum growth rates in high $PO_4$-P concentrations were greater than 1.0 for *Nitzschia tropica* and *Cymbella affinis*. Maximum growth rates in high $NO_3$-N concentrations were 0.5 and 0.7 for *F. crotonensis*, *N. tropica*, *Navicula*

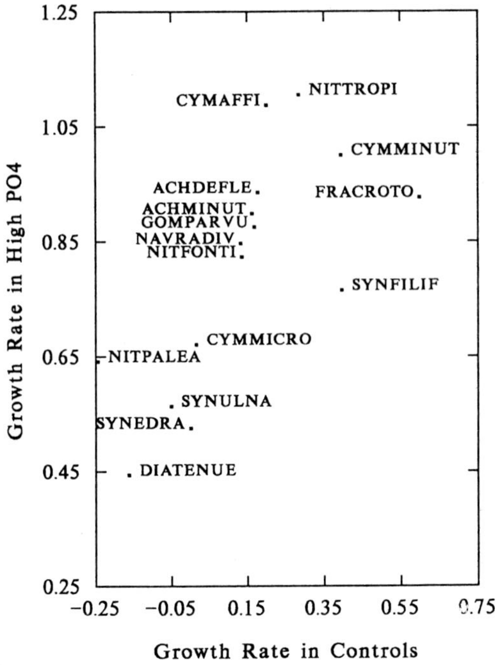


Fig. 1. Growth rates of fourteen diatom species in control and high $PO_4$-P concentrations.

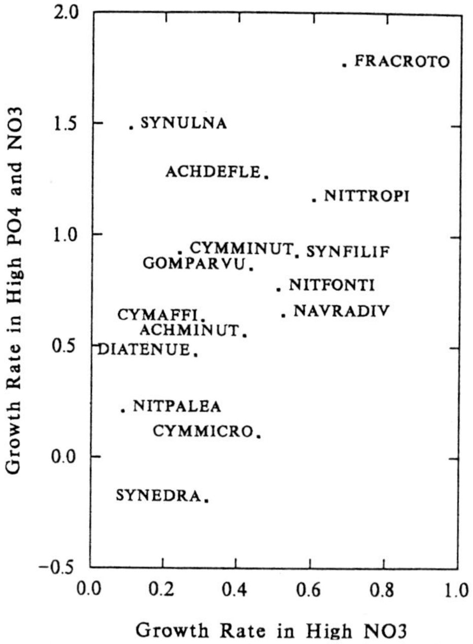


Fig. 2. Growth rates of fourteen diatom species in high $NO_3$-N and high $NO_3$-N+$PO_4$-P concentrations.

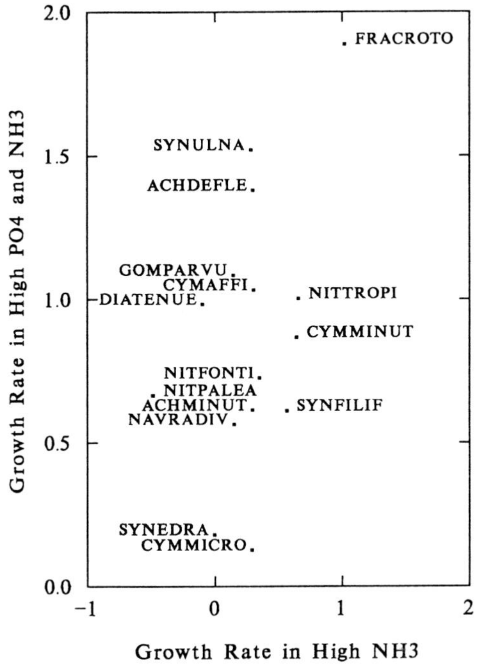


Fig. 3. Growth rates of fourteen diatom species in high $NH_3$-N and high $NH_3$-N+$PO_4$-P concentrations.

*radiosa* var. *tenella*, and *Nitzschia fonticola* (Fig. 2). In high $NH_3$, *Fragilaria crotonensis* had a growth rate of 1.0 and *Synedra filiformis*, *Cymbella minuta*, and *Nitzschia tropica* had growth rates about 0.6 (Fig. 3). *Fragilaria crotonensis* and *Synedra ulna* had the highest growth rates in high P+N, 1.8 and 1.5, respectively (Figs 2, 3).

Patterns in species responses to P and N enrichment indicated that the species that grew fastest in low nutrients also tended to grow fastest in high nutrient concentrations. Ranks of species growth rates were positively correlated between control and either high $NO_3$, $NH_3$, or $PO_4$ treatments (SRC, $p<0.05$, Table 3). For example, *F. crotonensis* had the highest growth rates in control, high $NH_3$, high $NO_3$, and both high P+N treatments (Figs 2, 3). The correlation between ranks of species growth rates in control and high $NH_3$ treatments was the greatest.

Patterns in species responses to P and N enrichment indicated that the species that grew fastest in high P treatments also tended to grow fastest in high N concentrations (Table 3). Ranks of species growth rates in high $NH_3$, $NO_3$, and $PO_4$ treatments were positively correlated ($p<0.05$). Only the relationship between species growth rates in high $NO_3$ and in high $PO_4$ treatments were not significantly ($p>0.05$) positively correlated.

Table 3. Spearman rank correlation coefficients (SRCC) for correlations between species growth rates in different nutrient treatments (p<0;05 for SRCC>0.52, Zar 1974). * Significant correlations.

| Nutrient | Control | $NH_3$ | $NO_3$ |
|---|---|---|---|
| $NH_3$ | 0.882 * | | |
| $NO_3$ | 0.607 * | 0.596 * | |
| $PO_4$ | 0.779 * | 0.721 * | 0.386 |

Patterns in species growth-stimulation by different nutrients (standardized to growth potential in chamber environments), when the other nutrient (either N or P) was saturating, also indicated that the species most stimulated by N tended to be those most stimulated by P. However, few patterns were statistically significant. *Synedra ulna* was the best example of a species stimulated by increases in N or P when the other nutrient was in luxury supply. In fact, the response by *S. ulna* was the greatest of all species. *Synedra ulna* actually had negative growth rates in control conditions, but was one of the fastest growing taxa in high N+P treatments (Figs 1–3). When other nutrients were in luxury supply, correlations between the ranks of species responses were greater than 0, never negative, and seldom statistically significant (Table 4). The only statistically significant ($p<0.05$) relationship was between ranks of species responses to high $NO_3$ when $PO_4$ was abundant and responses to high $PO_4$ when $NO_3$ was abundant.

Table 4. Spearman rank correlation coefficients (SRCC) for correlations between the standardized stimulation of species growth rates by different nutrients when the other nutrient was at growth-saturing concentrations (p<0.05 for SRCC>0.52, Zar 1974). * Significant correlation.

| Nutrient | $NH_3$ (high $PO_4$) | $NO_3$ (high $PO_4$) | $PO_4$ (high $NH_3$) |
|---|---|---|---|
| $NO_3$ (high $PO_4$) | 0.379 | | |
| $PO_4$ (high $NH_3$) | 0.443 | 0.164 | |
| $PO_4$ (high $NO_3$) | 0.450 | 0.596 * | 0.382 |

## Discussion

Substantial interspecific variation occurred in growth rates and responses to nutrients. Per capita growth rates (r) ranged from less than zero (net losses) to 1.8, which is equivalent to a daily 6x ($e^r$) increase in densities. These high growth rates for species such as *Fragilaria crotonensis* and *Synedra ulna* meant that cells divided, on average, more than twice per day in the high N+P treatment (since a 6x increase requires one cell to divide between two and three times to produce 6 cells). All species were not similarly stimulated by the high N+P treatment or $NH_3$, $NO_3$, and $PO_4$ alone treatments. Growth rates varied more among species in high N+P than in the other treatments.

There was little evidence for evolutionary tradeoffs for the ability to grow fast in low and high nutrient regimes, or for the ability to grow fast in different nutrient regimes. A negative correlation was expected in ranks of species growth rates if different species grew fastest in different nutrient conditions. However, in general, the same species were the fastest growing species in different nutrient regimes. In general, the relative ranking of species growth rates stayed the same in different nutrient treatments. Reassessing species' abilities to use nutrients by standardizing growth responses to species growth capacity also indicated that evolutionary tradeoffs did not occur for different nutrient uptake abilities.

*Synedra ulna* was the major exception to the pattern described above. *Synedra ulna* was slightly stimulated by inorganic N (as $NO_3$ or $NH_3$), slightly more by $PO_4$ than inorganic N, but was strongly stimulated by N+P enrichment. This indicated that *S. ulna* required high P and fairly high N for positive growth.

One of the major incentives for this research was to investigate differential species responses to environmental conditions that may autogenically change during benthic algal community development. Supplies of both light and nutrients are known to decrease autogenically (endogenously, or because of the community itself: Hoagland 1983; Humphrey & Stevenson 1992; Stevenson & Glover 1993). The decreases in resources as benthic algal density on substrata increases is hypothesized to cause interspecific exploitative competition for these resources. *Synedra ulna* is one of the few diatom species that has been repeatedly observed to be an early succession species (Stevenson 1984b; Stevenson *et al.* 1991). Both its relative abundance and actual abundance decrease rapidly after the first couple of days of substratum colonization. The observation that *S. ulna* has a high nutrient requirement for growth is a strong indication that exploitative competition is one of the several processes (see Stevenson 1986; Stevenson & Peterson 1991; McCormick & Stevenson 1991) that regulate benthic diatom succession on substrata.

The high growth rates of *Fragilaria crotonensis* are also noteworthy. It grew faster than all other taxa in all treatments, except in the high P treatment. *Fragilaria crotonensis*, a common planktonic species, is usually most successful in mesotrophic lakes. Many pennate, and some centric, diatoms occur as members of both benthic and planktonic habitats. Therefore, its occurrence on substrata was not particularly unusual, but its relatively high performance in this habitat was. Why is this taxon not found more frequently as a benthic diatom in high and low nutrient regimes?

An unexpected result in the research was the significant stimulation of growth by inorganic N and inorganic P independently. Leibig's law of the minimum (Liebig 1840 and Lotka 1925, as cited in Droop 1973) states that growth of a species should be limited by only one nutrient at a time. Two species were stimulated by inorganic N and P enrichment independently. In our research, multispecies cultures may have enabled indirect stimulation of species that were actually limited by one nutrient. Perhaps the non-limiting nutrient (A) indirectly made the more limiting nutrient (B) available by indirect physiological activity of other species. Alternatively, Leibig's law of the minimum may not hold for the species studied. Perhaps increased availability of one nutrient (A) makes use of the other, more-limiting nutrient (B) more effective than when nutrient A is in low supply.

In conclusion, the lack of evidence for tradeoffs between abilities to grow in different resource conditions indicates that competition for nutrient resources has not been an overwhelmingly important evolutionary factor for benthic diatoms in "recent" times. In this context, "recent" refers to within the genetic "memory" of the population. Competition may have been important in the history of benthic diatom evolution and may have caused exclusion of species that were not allocating their resources as efficiently as possible. The example of *Synedra ulna* may be a good example of the ghost of competition past. Thus, competition may still occur on local spatial or temporal scales, but it probably seldom causes competitive exclusion in streams. Perhaps natural, storm-related, disturbances reset community succession and interrupt competitive exclusion. In addition, dispersal of diatom populations is high in streams. Therefore, local genetic bottlenecks could be relieved by dispersal of new genotypes into the local population. For whatever reason, there was little experimental evidence in this research that showed competition for resources to be a driving force in the natural selection of benthic diatoms in streams.

## References


Bothwell, M. L. (1989). Phosphorus-limited growth dynamics of lotic periphytic diatom communities: areal biomass and cellular growth rate responses. *Canadian Journal of Fisheries & Aquatic Sciences*, **46**, 12293–1301.

Droop, M. R. (1973). Some thoughts on nutrient limitation. *Journal of Phycology*, **9**, 264–272.

Hoagland, K. D. (1983). Short-term standing crop and diversity of periphytic diatoms in a eutrophic reservoir. *Journal of Phycology*, **19**. 30–38.

Humphrey, K. P. & Stevenson, R. J. (1992). Responses of benthic algae to pulses in current and nutrients during simulations of subscouring spates. *Journal of the North American Benthological Society*, **11**, 37–48.

Kilham, S. S. (1978). Nutrient kinetics of freshwater planktonic algae using batch and semicontinuous methods. *Mitteilungen. Internationale Vereinigung für Theoretische und Angewandte Limnologie*, **21**, 147–157.

Liebig, J. (1840). Die Chemie in ihreer Anvendung auf Agricultur, und Physiologie. 4th ed., 1847. Taylor and Walton, London.

Lotka, A. J. (1925). *Elements of physical biology*. Williams and Wilkins, Baltimore, reprinted 1956 as *Elements of Mathematical Biology*, Dover, New York. 465 pp.

McCormick, P. V. & Stevenson, R. J. (1991). Mechanisms of benthic algal succession in different flow environments. *Ecology*, **72**, 1835–1848.

Patrick, R. (1976). The formation and maintenance of benthic diatom communities. *Proceedings of the American Philosophical Society*, **120**, 475–484.
Stevenson, R. J. (1984a). Procedures for mounting algae in a syrup medium. *Transactions of the American Microscopical Society*, **103**, 320–321.
Stevenson, R. J. (1984b). How currents on different sides of substrates in streams affect mechanisms of benthic algal accumulation. *Internationale Revue der gesamte Hydrobiologie*, **69**, 241–262.
Stevenson, R. J. (1986). Mathematical model of epilithic diatom accumulation. In: *Proceedings of the Eighth International Diatom Symposium* (M. Ricard, ed.), 209–231. Koeltz, Koenigstein.
Stevenson, R. J. (1990). Benthic algal community dynamics in a stream during and after a spate. *Journal of North America Benthological Society*, **9**, 277–288.
Stevenson, R. J. & Glover, R. (1993). Effects of algal density and current on ion transport through periphyton communities. *Limnology & Oceanography*, **38**, 1276–1281.
Stevenson, R. J. & Pan, Y. (in press). Community dynamics and differential species performance of benthic diatoms along a nitrate gradient. In: A century of Diatom Research in North America: A tribute to the Distinguished Careers of Charles W. Reimer & Ruth Patrick (P. Kociolek & M. Sullivan, eds). Otto Koeltz Publ. (USA), Champaign, Illinois, USA.
Stevenson, R. J. & Peterson, C. G. (1991). Emigration and immigration can be important determinants of benthic diatom assemblages in streams. *Freshwater Biology*, **26**, 295–306.
Stevenson, R. J., Peterson, C. G., Kirschtel, D. B., King, C. C. & Tuchman, N. C. (1991). Density-dependent growth, ecological strategies, and effects of nutrients and shading on benthic diatom succession in streams. *Journal of Phycology*, **27**, 59–69.
Wilkinson, L. (1989). *SYSTAT: The System for Statistics*. Evanston, Illinois, USA. 822 pp.
Zar, J. H. (1974). *Biostatistical Analysis*. Prentice-Hall, Inc. Englewood Cliffs, NJ, USA. 620 pp.

# The effects of silicon deficiency and pH on the composition of lipids of freshwater *Synedra* and *Pinnularia*

Yoko Yamamoto[1] and Hidetaka Tatsuzawa[2]

[1] *Faculty of Agriculture, Meiji University, 1–1–1 Higashimita, Tama-ku, Kawasaki 214, Japan*

[2] *Ebara Research Co., Ltd., 4–2–1 Honfujisawa, Fujisawa 251, Japan*
*Current address: Shimizu Laboratories, Marine Biotechnology Institute, 1,900, Sodeshi, shimizu-shi, Shizuoka 424, Japan*

## Abstract

The effects of silicon deficiency and pH on the composition of lipids in *Synedra ulna*, isolated from a neutral lake, and *Pinnularia braunii*, isolated from a volcanic and acidic lake, was examined. The main lipids found in both diatoms consisted of monogalactosyldiacylglycerol (MGDG), digalactosyldiacylglycerol (DGDG) and sulfoquinovosyldiacylglycerol (SQDG), and also included phosphatidilglycerol (PG). The dominant fatty acid was C16, constituting 50% of all fatty acids. Fatty acids exceeding C20 were localized in glycolipids. As for fatty acid C18, there were either minute amounts or none detected. When cells of *Synedra* were grown under silicon-deficient conditions, there was a tendency for an increase in lipids and unsaturated fatty acids. *Pinnularia braunii* grew well at pH 2 and the lipid content increased when compared to cells grown under pH 6.

## Introduction

The production and storage of lipids by microalgae is regulated by environmental factors such as light irradiance, nutrients and pH. Nitrogen deficiency is known to induce lipid accumulation in several algae, including the Chlorophyceae (Spoer & Milner 1949; Piorreck *et al.* 1984; Suen *et al.* 1987; Chen & Johns 1991) and Rhodophyceae (Cohen *et al.* 1988). The content of TAG in green algae is known to increase with a decrease in the medium's nitrogen content (Piorreck *et al.* 1984).

Silicon deficiency is also known to increase the lipid content of Bacillariophyceae (Shifrin & Chisholm 1981). Acidophilic environmental conditions are known to affect the lipid content and fatty acids in many bacteria (Harwood & Russell 1983). For example, during the growth of *Staphiylococcus aureus* the pH falls due to acid production and there is a rise in the amount of aminoacylphosphatidylglycerol. However, the effect of pH on algae is not fully understood. In this work, we studied the effects of pH on *Pinnularia*, and silicon deficiency for *Synedra*, on lipid and fatty acid compositions.

## Materials and Methods

### *Organisms and culture conditions*

*Synedra ulna* (Nitpch) Ehrenberg was isolated from a neutral lake, and *Pinnularia braunii* var. *amphicephala* (A. Mayer) Hustedt was isolated from Lake Katanuma (pH 2.0), a volcanic and acidic lake, both in Japan.

Cells were cultured in polycarbonate bottles containing CT medium (Table 1), at 20 C under constant illumination provided by fluorescent lamps (40–45 $\mu E \cdot m^{-2} \cdot s^{-1}$).

Table 1. The culture medium for diatoms.

| | |
|---|---|
| $Ca(NO_3) \cdot 4H_2O$ | 15 mg |
| $KNO_3$ | 10 mg |
| $Na_2$–glycerophosphate | 5 mg |
| $MgSO_4 \cdot 7H_2O$ | 4 mg |
| Vitamin$B_{12}$ | 0.01 µg |
| Biotin | 0.01 µg |
| Thiamine HCl | 1 µg |
| PIV metals | 1 ml |
| HEPES or 1,2,3,4–cyclopentatetra acid | 50 mg |
| $NaSiO_3 \cdot 9H_2O$ | 10 mg |
| Distilled water | 100 ml |
| PIV metals | |
| $FeCl_3 \cdot 6H_2O$ | 0.196 g |
| $MnCl_2 \cdot 4H_20$ | 0.036 g |
| $ZnSO_4 \cdot 7H_20$ | 0.022 g |
| $CoCl_2 \cdot 6H_20$ | 0.004 g |
| $Na_2MnO_4 \cdot 2H_20$ | 0.0025 g |
| $Na_2$–EDTA·$2H_20$ | 1 g |
| Distilled water | 1 l |

For studying the effects of silicon-deficiency, cells of *Synedra* were first grown for 70h in silicon-containing media, and then transferred to a silicon-deficient medium and grown for 65h.

For studying the effect of pH on *Pinnularia*, $H_2SO_4$ was added until the required pH was obtained. The pH was checked at frequent intervals during cell growth and adjusted to initial values. Cells were collected after 70h of culturing and growth was monitored by counting with a hemacytometer.

*Lipid extraction and fractionation*

Lipids were extracted from cells with chloroform-methanol-water according to Bligh and Dyer (1959). The crude lipids were evaporated and dissolved in n-hexane. The samples containing about 10mg of lipids were fractionated by silica Set-Pak column (2g silica, Waters, USA) chromatography, using a modified method of Hartmann *et al.* (1986). The following solvents were used; 1) hexane:diethyl ether (99:1 [v/v], 25ml, carotenes), 2) hexane:diethylether (4:1 [v/v], 26ml, TAG), 3) chloroform (40ml, chlorophylls), 4) acetone:chloroform (2:1 [v/v], 30ml) and acetone:methanol (29:1 [v/v], 12ml, MGDG), 5) acetone:methanol (29:1 [v/v], 12ml, DGDG), 6) acetone:methanol (19:1 [v/v], 35ml, SQDG), 7) acetone:methanol (2:1 [v/v], 21ml, PG+PI), 8) methanol (16ml, PE), 9) methanol (41ml, PC), further fractionation for fraction 7, chloroform:methanol (11:10 [v/v], 11ml, PG), and chloroform:methanol (1:4 [v/v], 12ml, PI). Thin-layer chromatography was used to investigate the identity and purity of acyl lipids by placing samples on precoated silica-gel plates (Merck, 5724). The following solvents were used: 1) TAG:heptane-diethyl ether-acetic acid (75:25:4 [v/v], Korte & Casey 1982), and 2) Glycolipids and phospholipids: chloroform-methanol-acetic acid-water (85:15:10:3 [v/v], Nichols 1965). Pure reference substances were co-chromatographed for identification. Spots were detected by the following indicators: cupric acetate (Fewster & Burns 1969) for TAG, α-naphthol for glycolipids (Jacin & Mishkin 1965), molybdenumtrioxide/ sulphuric acid for phospholipids (Dittmer & Lester 1964), Dragendorff reagent for PC (Nichols & Appleby 1969).

*Fatty acid analysis*

Lipids were methanolized with 2% $H_2SO_4$ in methanol for 90min at 90°C. Resultant methylesters were analyzed on a gas-liquid chromatographic system (Shimazu GC–12A) equipped with a fused silica capillary column (0.25mm × 30m; HR–SS–10 ; Shinwa Chemical Ind. Ltd.) and a hydrogen flame-ionization detector. The column temperature was programmed to increase from 175°C to 200°C at a rate of 1°C $min^{-1}$. The injector and detector temperatures were 230°C.

Fatty acid methylesters were identified by comparing their retention times with those of authentic standards (Funakoshi Fatty acid methylester Standards, Japan) and also by gas chromatography-mass spectrometry (GC/MS, TSQ70; Finigan Mat, U.S.A.) attached to the GC. The injection and separation temperatures were 240°C and 250°C.

The ionization voltage, emission current, and multiplier were 70 eV, 200 μA, and 1 kV, respectively. Mass spectra were obtained by both E1 and methan CI modes. Methylesters were quantitatived by comparing with standards.

**Results**

*Growth of diatoms*

The growth rate of *Synedra* at pH 6.5 was $0.033h^{-1}$ (Table 2). The growth rate of *Pinnularia* at pH 3.0 was $0.025h^{-1}$ and $0.014h^{-1}$ at pH 6.0. The optimum pH versus *Pinnularia* growth was 2.0–3.0, corresponding to the results of field observation (pH 2).

Table 2. Physiological characteristics of species examined.

| Species | Cell size (μm) | Specific growth rate ($h^{-1}$)* | Collection site |
|---|---|---|---|
| *Synedra ulna* Ehrenberg G–3 | 110.5 ± 2.39 × 7.7 ± 1.34 | 0.033 (pH 6.5) | Neutral pond (pH 6.8) |
| *Pinnularia braunii* var. *amplicephala* | 19.6 ± 1.98 × 8.3 ± 1.28 | 0.014 (pH 6)<br>0.025 (pH 3) | Acidic lake (pH 2) |

* Cells were cultured at 20°C under light ($45\mu E{\cdot}m^{-2}{\cdot}S^{-1}$)

*Lipids of diatoms*

Table 3 shows the total lipid and fatty acid composition of diatoms. The total lipids in *Synedra* ($113–218pg{\cdot}cell^{-1}$) was greater than that in *Pinnularia* and accumulated in response to silicon deficiency. In contrast, the total lipids in *Pinnularia* grown under a optimum pH 3 were $81pg{\cdot}cell^{-1}$, though this value decreased to 36 $pg{\cdot}cell^{-1}$ at pH 6. No significant difference was observed in the lipid composition of two diatoms. The major fatty acids of the total lipids were 16:0 and 16:1 under all test conditions. The C18 acids were found only in small or trace amounts. Although the 20:4 acid was present at high levels in *Pinnularia*, only small amounts of the same were found in *Synedra* and none in cells grown in silicon-deficient conditions. The 20:5 and 22:6 acids were present in *Synedra*, but absent in *Pinnularia*. There was no significant difference in the C20:5 and C22:6 contents in *Synedra*, regardless of whether the particular condition was silicon-replete or silicon-deficient.

Tables 4, 5 and 6 show lipid classes and fatty acids in each diatom. The major lipids were TAG and MGDG. The TAG content in silicon-deficient cells in *Synedra* was approximately 4–fold of that in silicon-replete cells. In contrast, SQDG and PE reduced in silicon deficiency. Fatty acid composition of each lipid class showed that the C14 and C16 acids were concentrated in TAG, MGDG, DGDG, SQDG, PG and PE, whereas the 22:6 acid was mainly found in PC.

Table 3. Fatty acid composition (mol%) of total lipids, and total cellular content of fatty acid (pg·cell$^{-1}$) in *Synedra ulna* and *Pinnularia braunii* grown at different conditions.

| | *Synedra ulna* | | *Pinnularia braunii* | |
|---|---|---|---|---|
| Fatty acid | Si-replete | Si-deficient | pH 3 | pH 6 |
| 14:0 | 12.9 | 14.7 | 11 | 8.7 |
| 15:0 | 0.6 | 0.6 | 0.9 | 1.6 |
| 16:0 | 18.9 | 15.8 | 19.7 | 22.2 |
| 16:1 | 50.6 | 55.2 | 46.6 | 41.5 |
| 16:2 | 4.0 | 3.6 | 5.4 | 6.1 |
| 16:3 | 4.0 | 2.7 | 5.8 | 7.5 |
| 18:0 | nd | tr | 0.3 | 0.2 |
| 18:1 | 0.8 | 0.6 | 0.9 | 0.5 |
| 18:2 | 0.5 | 0.4 | 1.3 | nd |
| 18:3 | 0.5 | 0.4 | 0.3 | 1.4 |
| 20:4 | 0.7 | tr | 7.9 | 10.3 |
| 20:5 | 5.7 | 5.1 | nd | nd |
| 22.6 | 0.8 | 0.7 | nd | nd |
| Σ FA (pg·cell$^{-1}$) | 112.5 | 217.9 | 81.2 | 36.3 |
| Σ PUFA / Σ SAFA | 2.09 | 2.21 | 2.14 | 2.10 |

Σ FA: Total fatty acid   Σ SAFA: Total saturated fatty acid
Σ PUFA: Total polyunsaturated fatty acid

The lipid fraction in *Pinnularia* was similar to that in *Synedra*, with the main lipids being TAG, MGDG and SQDG. The most significant effects of pH stress on the lipid composition in *Pinnularia* were observed in TAG, MGDG and PG. TAG and MGDG decreased remarkably in cells grown at pH 6 when compared to cells grown at pH 3. PG, containing high proportions of the C18 and C16 saturated fatty acids, showed a tendency to increase in cells grown at pH 6. MGDG in cells grown at pH 6 resulted in a decrease in C16:1 acid and an increase in the 20:4 acid.

Table 4. The composition (mol%) of glycerolipids from *Pinnularia braunii* and *Synedra ulna* grown at different conditions.

| | *Synedra ulna* | | *Pinnularia braunii* | |
|---|---|---|---|---|
| | Si-replete | Si-deficient | pH 3 | pH 6 |
| Triglycerols | 56.5 | 209.6 | 69.1 | 24.4 |
| MGDG | 54.7 | 55.2 | 34.7 | 15.9 |
| DGDG | 9.9 | 3.6 | 1.4 | 1.6 |
| SQDG | 23.7 | 8.1 | 2.2 | 3.3 |
| PG | 12.5 | 12.4 | 2.7 | 7.2 |
| PI | 0.2 | 2.7 | 0.2 | 0.5 |
| PE | 19.9 | 3.3 | 1.4 | 1.3 |
| PC | 3.2 | 6.6 | 1.3 | 0.5 |

**Discussion**

There is some variation in the lipid types in different algal taxa. Although the 18:3 acid was commonly found in green algae (Ben-Amoz *et al.* 1985; Piorreck *et al.* 1984; Spoehr & Milner 1949; Suen *et al.* 1987), the majority of acids in diatoms consisted of the 16:0, 16:1, and C20-polyenoic acids, while the minority constituted the 18 acids. Furthermore, the C20 polyenoic acids in *Synedra* contained the 20:5 and 22:6, and the acid in *Pinnularia* was the 20:4. It is known that silicon-deficiency results in increased lipid accumulation in many diatoms (Laing 1985; Parish & Wangersky 1990; Roessler 1988; Shifrin & Chisholm 1981; Taniguchi *et al.* 1987). The increase in lipids was due to increase in TAG (Parish & Wangersky 1990; Roessler 1988; Sukenik & Carmeli 1990). Our results were also consistent with those reports. Parish and Wangersky (1990) and Roessler (1988) suggested that the production of TAG can be listed as an indicator of physiological stress and further predict that TAG would be a major lipid present in the final stationary phase. Similar observations have been reported by Sukenik and Carmeli(1990). The pH effect on lipids and fatty acids of diatoms has yet to be fully understood. *Pinnularia* is distributed worldwide and could grow in a broad pH range (0.5–6.5). The PG content in cells grown at pH 6 was found to increase and the 16:0 acid became dominant. However, it is unclear whether these results reflected an adaptation to the acidic environment. Further studies will concentrate on the elucidation of this problem.

Table 5. Fatty acid patterns (mol% ) in lipid classes isolated from Si-deficient and Si-replete cells in *Synedra ulna*.

| | TAG | | MGDG | | DGDG | | SQDG | | PG | | PI | | PE | | PC | |
|---|---|---|---|---|---|---|---|---|---|---|---|---|---|---|---|---|
| | +Si | –Si | +Si | –Si | +Si | –Si | +Si | –Si | +Si | –Si | +Si | –Si | +Si | –Si | +Si | –Si |
| 14:0 | 9.0 | 13.8 | 9.3 | 10.1 | 8.3 | 8.1 | 9.3 | 14.7 | 17.6 | 10.9 | 17.3 | 5.8 | 11.5 | 10.2 | 6.7 | 15.9 |
| 15:0 | 0.7 | 0.7 | 1.7 | 0.9 | * | 7.2 | 2.0 | 1.9 | 2.2 | 1.7 | * | 3.0 | 1.1 | 2.0 | 1.6 | 2.6 |
| 16:0 | 18.5 | 14.6 | 28.4 | 20.9 | 24.1 | 22.6 | 21.8 | 21.6 | 36.2 | 26.9 | 82.7 | 28.6 | 19.3 | 23.1 | 13.6 | 16.8 |
| 16:1 | 61.5 | 60.9 | 30.6 | 34.0 | 34.6 | 30.3 | 26.3 | 27.3 | 15.4 | 23.3 | * | 26.0 | 34.4 | 21.9 | 17.3 | 17.7 |
| 16:2 | 1.7 | 2.0 | 4.9 | 7.2 | 5.8 | 9.2 | 4.0 | 13.5 | 3.2 | 9.3 | * | 5.2 | 8.3 | 4.6 | 3.1 | 3.5 |
| 16:3 | 1.5 | 1.4 | 5.6 | 6.7 | 9.1 | 8.4 | 8.0 | 10.9 | 4.8 | 2.7 | * | * | 5.5 | 5.7 | 4.6 | 4.4 |
| 18:0 | 0.4 | 0.3 | 4.2 | 2.4 | 4.4 | 5.9 | 2.3 | 3.2 | 11.8 | 12.2 | tr | 15.3 | 2.0 | 9.9 | 3.9 | 5.3 |
| 18:1 | 0.5 | 0.3 | 4.4 | 0.6 | * | 0.7 | 5.6 | 0.7 | 1.4 | 1.4 | * | 9.1 | 5.2 | 12.5 | 6.2 | 4.0 |
| 18:2 | 0.8 | 0.5 | 1.5 | 0.8 | * | * | 1.0 | tr | * | 0.7 | * | * | 1.5 | * | 8.4 | 3.9 |
| 18:3 | 0.7 | 0.6 | * | 0.5 | * | * | 9.8 | * | * | tr | * | 2.4 | * | * | 1.6 | 4.3 |
| 20:5 | 4.1 | 4.5 | 9.4 | 15.1 | 13.6 | 7.7 | 9.7 | 6.3 | 7.4 | 9.1 | * | 4.8 | 11.2 | 6.3 | 16.7 | 10.5 |
| 22:6 | 0.5 | 0.4 | * | 0.9 | * | * | * | * | * | 1.6 | * | * | * | 3.8 | 16.1 | 11.1 |
| Σ FA | | | | | | | | | | | | | | | | |
| | 31.3 | 69.5 | 30.3 | 18.3 | 5.5 | 1.2 | 13.1 | 2.7 | 6.9 | 4.1 | 0.1 | 0.9 | 11.0 | 1.1 | 1.8 | 2.2 |
| Σ PUFA / Σ SAFA | | | | | | | | | | | | | | | | |
| | 0.33 | 0.32 | 0.49 | 0.91 | 0.77 | 0.58 | 0.92 | 0.75 | 0.23 | 0.45 | – | 0.23 | 0.78 | 0.45 | 1.96 | 0.93 |

+Si: Silicon-replete cells
–Si: Silicon-deficient cells
Σ FA: Total fatty acid

* : not detected
tr : trace
Σ SAFA: Total saturated fatty acid
Σ PUFA: Total polyunsaturated fatty acid

Table 6. Fatty acid patterns (mol%) in lipid classes isolated from cells grown at pH 3 and pH 6 in *Pinnularia braunii.*

| | TAG | | MGDG | | DGDG | | SQDG | | PG | | PI | | PE | | PC | |
|---|---|---|---|---|---|---|---|---|---|---|---|---|---|---|---|---|
| | 3[1] | 6[2] | 3 | 6 | 3 | 6 | 3 | 6 | 3 | 6 | 3 | 6 | 3 | 6 | 3 | 6 |
| 14:0 | 9.7 | 6.2 | 8.8 | 7.8 | 5.8 | 6.9 | 6.9 | 10.9 | 6.6 | 7.5 | 6.0 | 3.3 | 7.2 | 5.0 | 15.3 | 9.4 |
| 15:0 | 0.8 | 1.7 | 0.9 | 2.5 | 1.3 | 1.9 | 1.9 | 3.5 | * | 1.8 | 2.7 | 4.5 | 2.9 | 2.8 | * | 4.2 |
| 16:0 | 19.7 | 21.6 | 19.2 | 14.3 | 16.4 | 19.1 | 18.6 | 24.5 | 27.1 | 34.1 | 34.9 | 46.8 | 27.3 | 31.1 | 28.7 | 52.5 |
| 16:1 | 54.0 | 52.4 | 32.2 | 20.6 | 25.1 | 26.4 | 26.5 | 23.0 | 20.5 | 21.5 | 5.9 | 9.9 | 11.8 | 17.5 | 17.9 | 18.1 |
| 16:2 | 3.1 | 4.1 | 8.6 | 5.1 | 11.7 | 4.7 | 10.0 | 9.4 | 6.7 | 6.4 | 8.4 | 5.7 | 5.1 | 5.1 | 2.9 | * |
| 16:3 | 3.5 | 5.5 | 14.0 | 10.0 | 23.0 | 23.1 | 19.5 | 19.9 | 5.2 | 4.4 | * | * | 4.5 | 7.1 | 3.2 | * |
| 18:0 | 0.3 | 0.9 | 1.0 | 3.7 | 3.2 | 3.4 | 2.6 | * | 12.9 | 11.7 | 19.8 | 26.4 | 10.0 | 12.0 | 9.0 | 7.8 |
| 18:1 | 0.5 | 0.6 | 0.6 | 1.2 | 2.1 | tr | * | * | 2.8 | 0.9 | 5.7 | tr | 21.0 | 2.6 | 12.1 | * |
| 18:2 | 0.5 | 0.6 | 0.4 | 1.6 | 2.0 | * | * | * | * | 1.8 | * | * | * | * | * | * |
| 18:3 | 1.0 | 0.5 | 1.1 | 1.5 | * | * | * | * | * | * | * | * | * | * | * | * |
| 20:4 | 6.8 | 5.9 | 13.2 | 31.8 | 9.6 | 14.5 | 18.2 | 8.9 | 18.2 | 9.9 | 13.5 | 3.4 | 10.3 | 16.8 | 10.9 | 8.0 |
| Σ FA | 59.8 | 44.6 | 32.4 | 29.1 | 1.2 | 3.0 | 1.9 | 6.1 | 2.3 | 13.2 | 0.2 | 0.9 | 1.2 | 2.3 | 1.1 | 0.9 |
| Σ PUFA / SAFA | 0.49 | 0.54 | 1.24 | 1.77 | 1.74 | 1.35 | 1.44 | 0.98 | 0.65 | 0.41 | 0.33 | 0.11 | 0.42 | 0.57 | 0.32 | 0.11 |

1: Cells grown at pH 3
2: Cells grown at pH 6
Σ FA: total fatty acid

* : not detected
tr: trace
Σ SAFA : total saturated fatty acid
Σ PUFA : total polyunsaturated fatty acid

## Acknowledgment

We are grateful to Dr. H. Kobayasi, Tokyo Diatom Institute, for verifying the species identification.

## References

Ben-Amoz, A., Tornabene, T. G. & Thomas, W. H. (1985). Chemical profiles of selected species of microalgae with emphasis on lipids. *Journal of Phycology*, **2**, 72–81.

Bligh, & Dyer, (1959). A rapid method of total lipid extraction and purification. *Canadian Journal of Biochemistry & Physiology*, **37**, 911–917.

Chen, F. & Johns, M. R. (1991). Effect of C/N ratio and aeration on fatty acid composition of heterotrophic *Chlorella sorokiniana*. *Journal of Applied Phycology*, **3**, 203–209.

Cohen, Z., Vonshak, A. & Richmond, A. (1988). Effect of environmental conditions on fatty acid composition of the red alga *Porphyridium cruentum*: correlation to growth rate. *Journal of Phycology*, **24**, 26–332.

Dittmer, J. C. & Lester, R. L. (1964). A simple, specific spray for the detection of phospholipids on thin-layer chromatograms. *Journal of Lipid Research*, **5**, 126–127.

Fewster, M. E. & Burns, B. J. (1969). Quantitative densitometric thin-layer chromatography of lipids using copper acetate reagent. *Journal of Chromatography*, **43**, 170–173.

Hartmann, E., Beutelmann, P., Vandekerkhove, O., Euler, R. & Kohn, G. (1986). Moss cell cultures as sources of arachidonic and eicosapentaenoic acids. *FEBS Letters*, **198**, 51–55.

Harwood, J. L. & Russell, N. J. (1983). *Lipids in plants and microbes*. 64–70 pp. George Allen & Unwin Ltd., London.

Jacin, H. & Mishkin, A. R. (1965). Separation of carbohydrates on borate-impregnated silica gel G plates. *Journal of Chromatography*, **18**, 170–173.

Korte, K. & Casey, M. L. (1982). Phospholipid and neutral lipid separation by one-dimensional thin-layer chromatography. *Journal of Chromatography*, **232**, 47–53.

Laing, I. (1985). Growth response of Chaetoceros calcitrans (Bacillariophyceae) in batch culture to a range of initial silica concentrations. *Marine Biology (Berl.)*, **85**, 37–41.

Nichols, B. W. (1965). Light induced changes in the lipids of *Chlorella vulgaris*. *Biochimica et Biophysica Acta*, **106**, 274–279.

Nichols, B. W. & Appleby, R. S. (1969). The distribution and biosynthesis is of arachidonic-acid in algae. *Phytochemistry*, **8**, 1907–1915.

Parrish, C. C. & Wangersky, P. J. (1990). Growth and lipid class composition of the marine diatom, *Chaetoceros gracilis*, in laboratory and mass culture turbidostats. *Journal of Plankton Research*, **12**, 1011–1021.

Piorreck, M., Baasch, K.- H. & Pohl, P. (1984). Biomass production, total protein, chlorophylls, lipids and fatty acids of freshwater green and blue-green algae under different nitrogen regimes. *Phytochemistry*, **23**, 207–216.

Roessler, P. G. (1988). Effects of silicon deficiency on lipid composition and metabolism in the diatom *Cyclotella cryptica*. *Journal of Phycology*, **24**, 394–400.

Shifrin, N. S. & Chisholm, S. W. (1981). Phytoplankton lipids: Interspecific differences and effects of nitrate, silicate and light-dark cycle. *Journal of Phycology*, **17**, 374–384.

Spoehr, H. A. & Milner, H. W (1949). The chemical composition of *Chlorella*: effect of environmental conditions. *Plant Physiology*, **24**, 129–149.

Suen, Y., Hubbard, J. S. Holzer, G. & Tornabene, T. G. (1987). Total lipid production of the green alga *Nannochloropsis* sp. QII under different nitrogen regimes. *Journal of Phycology*, **23**, 289–296.

Sukenik & Carmeli (1990). Lipid synthesis and fatty acid composition in *Nannochloropsis* sp. (Eustigmatophyceae) grown in a light-dark cycle. *Journal of Phycology*, **26**, 463–469.

Taniguchi, S., Hirata, J. A. & Laws, E. A. (1987). Silicate deficiency and lipid synthesis of marine diatoms. *Journal of Phycology*, **23**, 260–267.

# Sedimentary diatom assemblages in freshwater and saline lakes of the Anatolia Plateau, central part of Turkey: an application for reconstruction of palaeosalinity change during Late Quaternary

Kaoru Kashima

*Department of Earth and Planetary Sciences, Kyushu University, Hakozaki, Fukuoka 812-81, Japan*

## Abstract

Particular emphasis is given to the uses of diatoms as indicators of water salinity. The data indicate great promise for generating predictive relationships useful in reconstructing palaeosalinity fluctuations in closed inland lakes in Turkey.

An important first step in using diatoms as indicators for palaeosalinity is to obtain quantitative data on their ecological characteristics, such as optima and tolerances along environmental gradients. The diatom stratigraphy of sediments in closed-basin lakes in this region then can be used to reconstruct changes in lake salinity and potentially to infer Late-Quaternary climatic change.

## Introduction

In closed inland lakes in arid and semi-arid regions such as Turkey, shifts in effective moisture (precipitation minus evaporation) lead to the concentration or dilution of dissolved salt. Thus changes in palaeosalinity can be related to shifts in climate characteristics such as precipitation and evaporation rates. Few previous studies (Fritz and Battarbee 1989; Fritz *et al.* 1991, 1993; Gasse *et al.* 1990) demonstrate the use of diatoms for reconstruction of salinity change in inland lakes.

In this paper, I investigated living diatom assemblages taken from periphyton scrapes and water samples in freshwater and saline lakes of the central Anatolia Plateau of Turkey (Fig. 1), to obtain quantitative data on their salinity optima and tolerances.

On the basis of a strong relationship between diatom composition and salinity in Turkish lakes (Fig. 2), I defined the diatom-based transfer functions for salinity reconstruction , and then applied it to a Late Quaternary sediment record in Turkey.

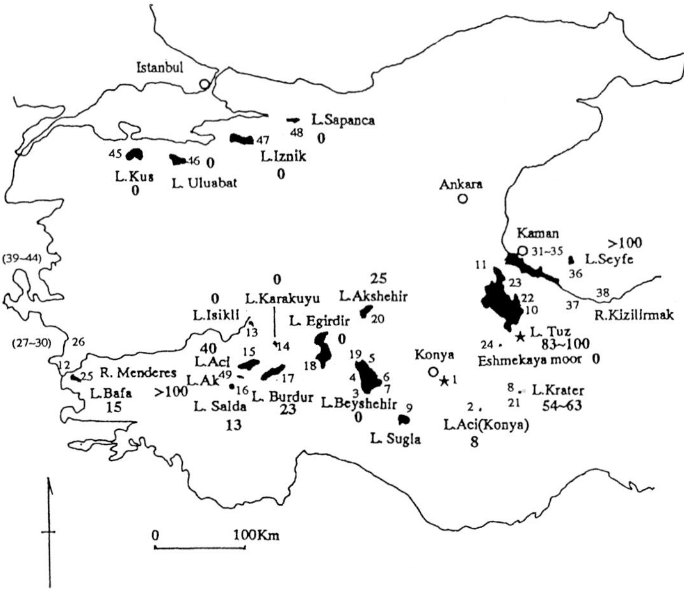


Fig. 1. Sites investigated on the Anatolia Plateau, Turkey in 1991 and 1992. Arabic small numbers indicate the number of sampling sites for living diatoms. Numbers 27-30 and 39-44 were sites for marine diatoms. The 51 samples from 38 sites were used to obtain the data for this paper. Large bold numbers indicate measured salinity values of lakes (‰). Black stars show the locations of drilling surveys.

## *Sample*

The Anatolia Plateau is located in the central and eastern part of Turkey. The height is mostly above 1000 meters. There are many tectonic basins and lake basins that were produced by activities of tectonic movements in the plateau.

It is a suitable region for diatom studies because of the wide range of salinity conditions of the inland lakes in the plateau. 51 samples of living diatom floras were collected from 38 sites at 23 lakes, ponds and rivers in 1991 and 1992 (Fig. 1), for data to be used in a diatom-based transfer function for palaeosalinity reconstruction.

The samples except that from Lake Seife were periphyton scrapes and water samples. At Lake Seife, a surface-sediment sample was taken because the water level of the lake was low. At every sampling site, salinity and pH were measured during sampling. The water samples, except that of Lake Seife were brought back to Japan and were analyzed for chemical data by Prof. Mitamura, Tokai University, Japan (unpublished). There are no detailed chemical data for the river samples.

29 samples were taken from fresh water lakes and rivers, and another 22 samples were taken from saline lakes with range from 8 to over 100 ‰ in salinity.

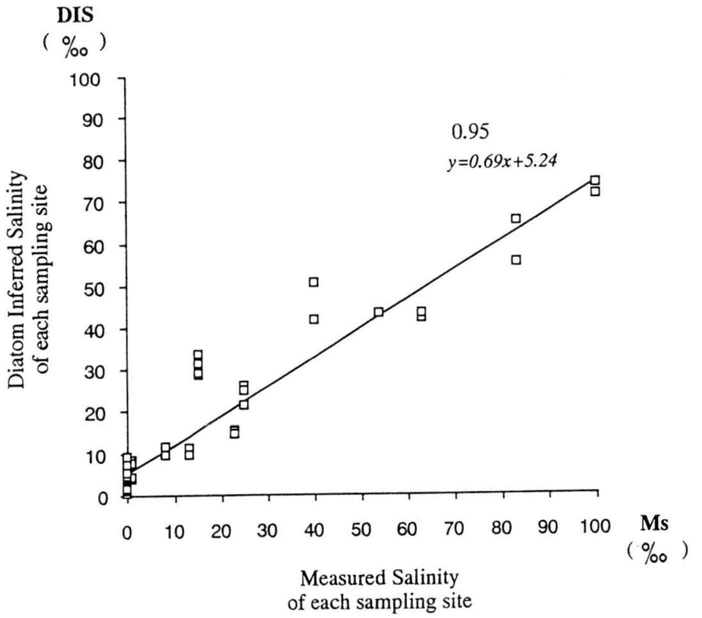


Fig. 2. The relationship between measured salinity (MS) and diatom inferred salinity (DIS). Measured salinity values were taken using a salinometer. Diatom Inferred salinity (DIS) of each sample was obtained from the following function (Fritz *et al.* 1991, 1993).

$$DIS_h = \sum_{i=1}^{n} (AWMS_i \times a_{hi}) / \sum_{i=1}^{n} a_{hi}$$

$DIS_h$ is a DIS value of sample *h*,

$AWMS_i$ is an abundance-weight mean salinity (AWMS) value of taxon *i* (Table 1),

$a_{hi}$ is the abundance (%) of taxon *i* in sample *h*.

*n* is the number of taxa for DIS calculation. In this paper, *n*=61.

### *Diatom-based transfer function for salinity reconstruction*

To provide quantitative reconstructions of lake water salinity, Fritz *et al.* (1991, 1993) refined a diatom-based transfer function for reconstruction of past changes in salinity of lakes in the northern Great Plain region of North America, and applied it to a late-glacial and Holocene sediment record from Devil Lake, North Dakota. Therefore, I applied their methods to the analysis of the samples from Turkey.

Although some taxonomic problems remain, the total number of diatom taxa from the 51 samples came to 126. I listed 61 taxa that occurred with at least 4% abundance in any one sample, and calculated an abundance-weighted mean salinity (AWMS) for each taxon (Table 1). Most of the AWMS of the species are slightly higher than the previous means calculated for other areas (Fritz and Battarbee 1989; Fritz *et al.* 1991, 1993; Gasse *et al.* 1990) . This may be related to the lower number of sampling sites in this study.

Table 1. List of the 61 diatoms for calibration models to infer palaeo-environment.

| Genus | Species | AWMS |
|---|---|---|
| *Nitzschia* | *compressa* | 90.00 |
| *Cyclotella* | *chocwhatcheeano* | 84.12 |
| *Cymbella* | *pusilla* | 81.72 |
| *Gyrosigma* | *strigilis* | 76.12 |
| *Amphora* | *coffeaeformis* | 75.26 |
| *Entomoneis* | *alata* | 73.06 |
| *Navicula* | *cincta ?* | 69.33 |
| *Nitzschia* | *sigma ?* | 63.00 |
| *Cocconeis* | sp. (Krater) | 56.76 |
| *Rhopalodia* | sp. | 47.20 |
| *Pleurosigma* | sp. | 45.93 |
| *Nitzschia* | *constricta* | 43.88 |
| *Stauroneis* | sp. | 40.09 |
| *Synedra* | *tabulata* | 36.84 |
| *Nitzschia* | *obtusa* | 27.23 |
| *Anomoeoneis* | *exilis* | 25.04 |
| *Nitzschia* | *littoris* | 25.00 |
| *Navicula* | *capitata* | 24.25 |
| *Navicula* | *pygmaea* | 23.97 |
| *Navicula* | *protracta* | 23.00 |
| *Synedra* | *pulchella* | 22.43 |
| *Amphora* | *ventricosa* | 15.00 |
| *Mastogloia* | *smithii* | 14.95 |
| *Nitzschia* | *frustulum* | 13.29 |
| *Cymbella* | *lacastris* | 13.00 |
| *Campylodiscus* | *clypeus* | 11.98 |
| *Rhoicosphenia* | *curvata* | 9.88 |
| *Mastogloia* | *elliptica* | 9.64 |
| *Nitzschia* | *palea* | 8.20 |
| *Anomoeoneis* | *sphaerophora* | 8.00 |
| *Cyclotella* | *meneghiniana* | 7.96 |

| Genus | Species | AWMS |
|---|---|---|
| *Achnanthes* | *minutissima* | 7.59 |
| *Navicula* | *cryptocephala* | 6.84 |
| *Diatoma* | *tenuis* | 6.77 |
| *Gomphonema* | *parvulum* | 5.83 |
| *Cymbella* | *minuta* | 5.50 |
| *Cymbella* | *microcephala* | 5.15 |
| *Navicula* | *cryptotenella* | 3.62 |
| *Synedra* | *ulna* | 3.34 |
| *Epithemia* | *adnata* | 2.08 |
| *Cymbella* | *cistula* | 0.98 |
| *Gomphonema* | *gracile* | 0.83 |
| *Fragilaria* | *vaucheriae* | 0.81 |
| *Nitzschia* | *dissipata* | 0.73 |
| *Fragilaria* | *brevistriata* | 0.37 |
| *Amphora* | *libyca* | 0.36 |
| *Fragilaria* | *pinnata* | 0.32 |
| *Cocconeis* | *placentula* | 0.29 |
| *Amphora* | *pediculus* | 0.28 |
| *Rhopalodia* | *gibba* | 0.18 |
| *Epithemia* | *sorex* | 0.12 |
| *Cyclotella* | *comta* | 0.06 |
| *Gomphonema* | *intricatum* | 0.04 |
| *Navicula* | *eliginensis* | 0.03 |
| *Cyclotella* | sp.-1 (Beyshehir) | 0.00 |
| *Opephora* | *martyi* | 0.00 |
| *Navicula* | *rotunda* | 0.00 |
| *Achnanthes* | *lanceolata* | 0.00 |
| *Gomphonema* | *truncatum* | 0.00 |
| *Achnanthes* | *clevei* | 0.00 |
| *Navicula* | *tuscula* | 0.00 |

Abundance-weight mean salinity (AWMS) of each taxa was taken by following function (Fritz *et al.* 1991, 1993).

$$AWMS_i = \sum_{h=1}^{m} (a_{hi} \text{ x } MS_h) / \sum_{h=1}^{m} a_{hi}$$

$AWMS_i$ is an abundance-weight mean salinity (AWMS) value of taxon *i* (Table 1).

$a_{hi}$ is the abundance (%) of taxon *i* in sample *h*.

$MS_h$ is a measured salinity value of sample *h*.

*m* is the number of samples. In this paper, $m$=51.

Then, predictive models developed from the surface sediment study of the Turkish lake samples were used to compute diatom-inferred salinity (DIS) value. The strong relationship between measured and diatom inferred salinity attests to the strength of the diatom-salinity model (Fig. 2). The data indicate a very close agreement between inferred and measured salinity.

***Application of the calibration model to infer palaeosalinity of lake sediments***

In 1991, drilling surveys for lake sediments were made in Lake Tuz, the second largest lake in Turkey, and also in the Konya Basin where a large lake had existed during the Late Quaternary. To reconstruct lake water salinity, transfer functions developed from the living diatom habitats are applied to down-core fossil diatom assemblages preserved in dated lake sediment cores.

*Lake Tuz*

The drilling surveys were done at the two sites (Yasilova and Yenikent) in the south-eastern part of the lake. The DIS data suggest that the lake was much less saline in the past than the present day. The lake had a salinity ranging from 10 ‰ to 20 ‰ in the time interval approximately 10,000–15,000 years BP. Lake level rose and salinity decreased below 10 ‰ from 10,000 years BP through 5,500–3,500 years BP. Thereafter the lake level fell and salinity increased to over 50 ‰ until the present day (Fig. 3). This is also deduced from the height and distribution of lake terraces around the lake.

During the latest Pleistocene and Holocene the lake level has fluctuated as has salinity in response to hydrologic changes controlled in large part by climate, particularly the balance between precipitation and evaporation. Of course, the relationship of salinity to water level and climate is very complex.

*Konya Basin; Palaeo-Konya Lake*

The core samples were taken in the eastern part of the former lake area defined by topographic characteristics. Three alternations from massive calcareous clay to humic clay is assumed to be related to glacial-interglacial climatic change by chemical and isotopic analysis of the core samples (Fig. 4).

DIS values were calculated at the upper humic clay layer, from 790 to 910 cm depth. The data of DIS shows that a highly saline lake was present during the last interglacial period. Analysis of samples from the other layers did not reveal diatom fossils.

**Acknowledgements**

Our research is supported by the International Science Research Program of the Ministry of Education, Science and Culture of Japan. I thank Prof. Yoshinori Yasuda,

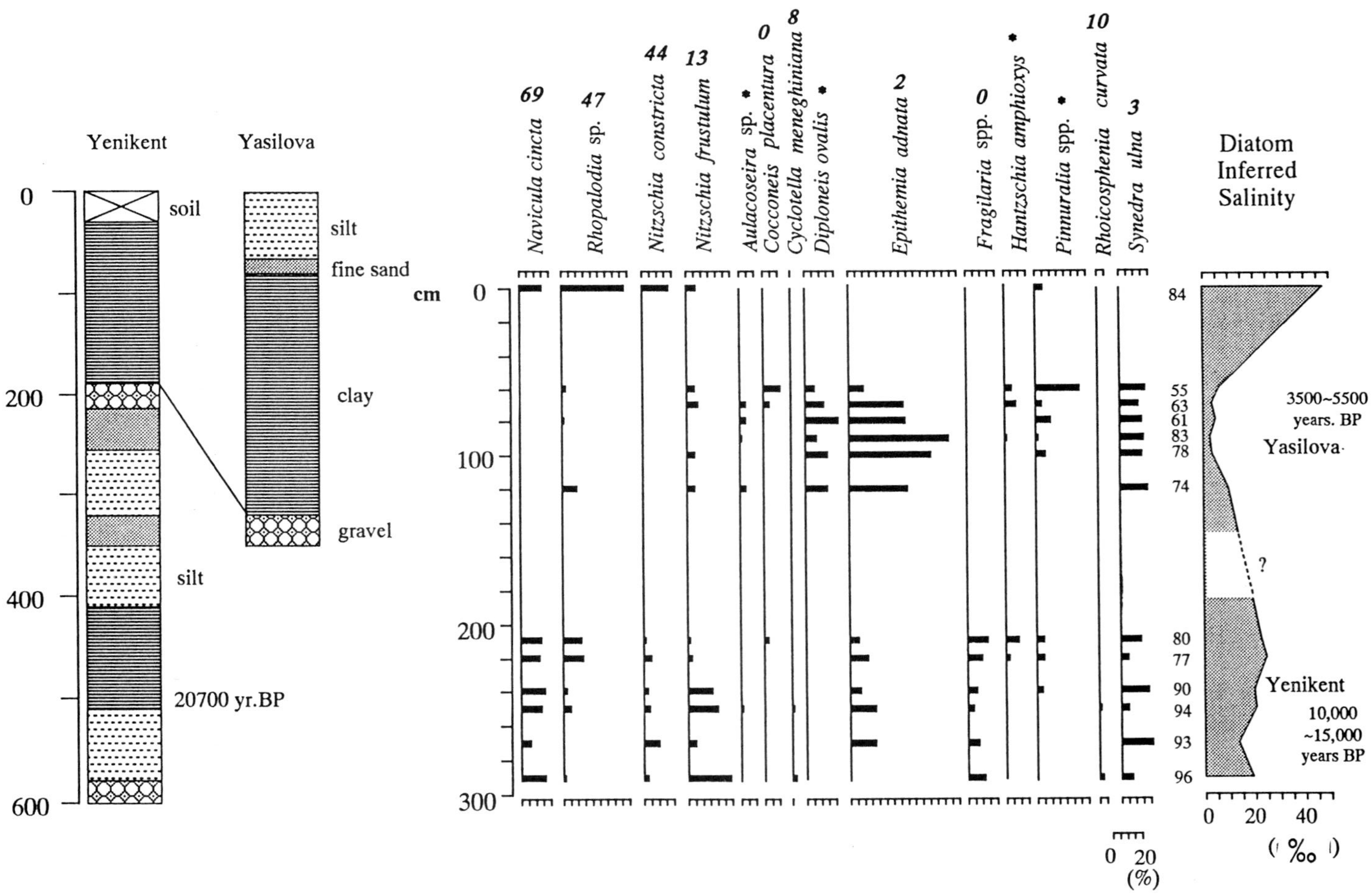


Fig. 3. Diatom assemblages and DIS fluctuation at Lake Tuz, central part of Turkey. The drilling surveys were done at two sites (Yenikent and Yasilova) in the south east part of Lake Tuz. The distance between sites is about 10 km. Diatom assemblages below 200 cm depth were obtained from the data of the Yenikent core samples, and those of 0-120 cm depth from the data obtained from the Yasilova core samples. The left column shows the lithologic change of the coring section. The italic numbers above species names give the abundance-weighted mean salinity (AWMS) values (Table 1). * shows that there is no AWMS value obtained from the living diatom samples. The numbers at the left side of the column of DIS are the cumulative percentages of the 61 taxa (Table 1) in the samples.

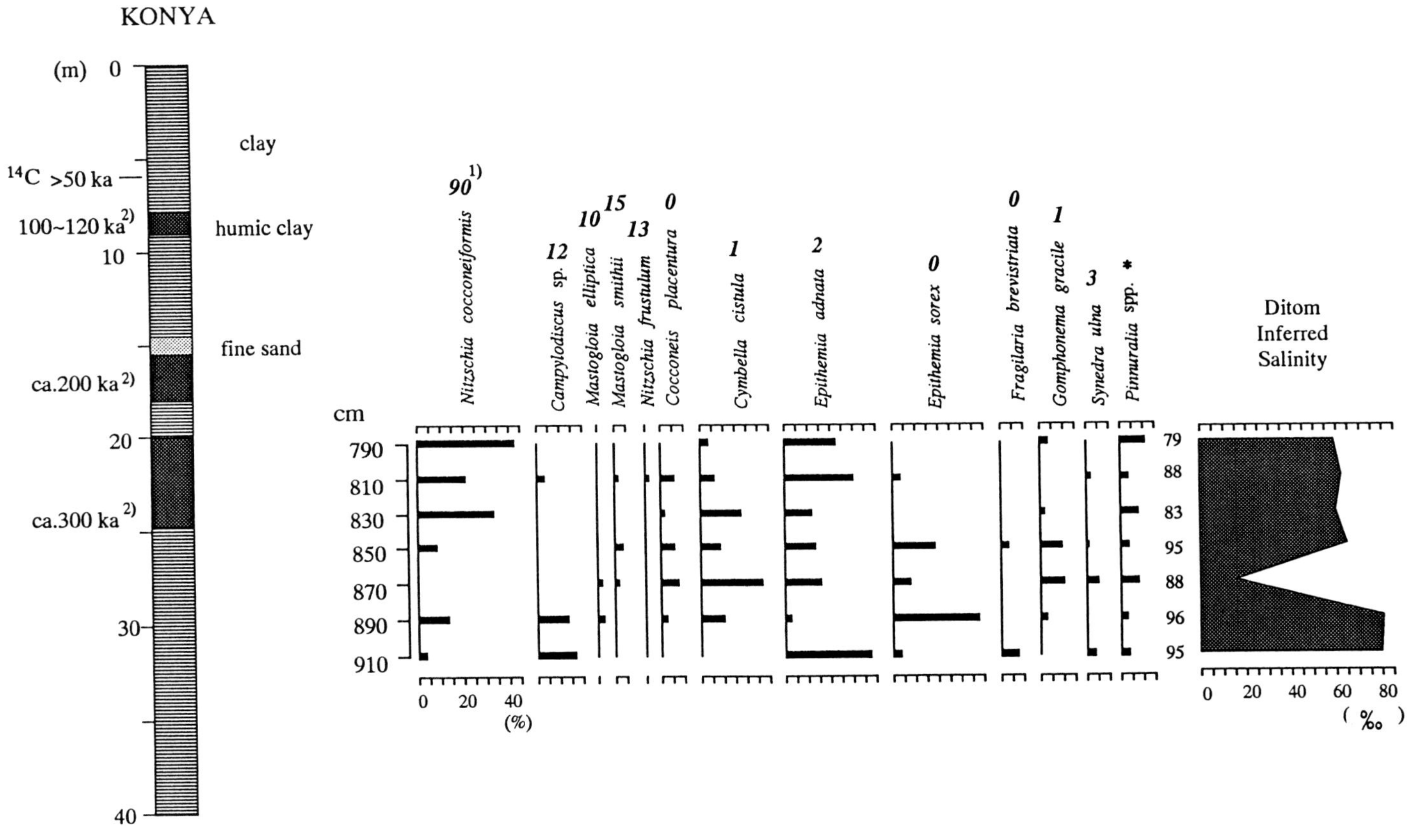


Fig. 4. Diatom assemblages and DIS in the Konya Basin. 1) It is AWMS data of *Nitzschia compressa*. 2) Age data were estimated by the change of chemical component fluctuations. Left column shows the lithology of the coring section. The italic numbers above species names give the abundance-weighted mean salinity (AWMS) values (Table 1). * shows that there is no AWMS value obtained from the living diatom samples. The numbers at left side of the column of DIS gives the cumulative percentages of the 61 taxa (Table 1) in the samples.

Prof. Toshirou Naruse, Dr Hiroyuki Kitagawa, Mr Sachihiro Omura and other members of our research team in 1991 and 1992, for providing facilities, logistical support, discussions, and encouragement during our field studies.

## References

Fritz, S. C. & Battarbee, R. W. (1988). Sedimentary diatom assemblages in freshwater and saline lakes of the Northern Great Plains, North America: preliminary results. In: *Proceedings of the 9th International Diatom Symposium* (F. E. Round, ed.), 265–271. Biopress Ltd. & Koeltz Scientific Books, Koenigstein.

Fritz, S. C., Juggins, S. & Battarbee, R. W. (1993). Diatom assemblages and ionic characterization of lakes of the Northern great Plains, North America: a tool for reconstructing past salinity and climate fluctuations. *Canadian Journal of Fisheries and Aquatic Sciences*, **50**. 1844–1856.

Fritz, S. C., Juggins, S., Battarbee, R. W. & Eingstrom, D. R. (1991). Reconstruction of past changes in salinity and climate using a diatom-based transfer function. *Nature*, **352**.

Gasse, F., Tehét R., Durand, A., Gibert, E. & Fontes, J. R. (1990). The arid-humid transition in the Sahara and the Sahel during the last deglaciation. *Nature,* **346**, 141–146.

# Diatom assemblages from lake sediment samples in Lake Nakaumi and Lake Shinji, coastal lagoons in Japan: an attempt to reconstruct a sensitive high-resolution record of environmental change

Kaoru Kashima

*Department of Earth and Planetary Sciences,*
*Kyushu University, Hakozaki, Fukuoka 812-81, Japan*

## Abstract

Initial analyses of observing the distribution of modern diatom species with regard to salinity levels at 81 localities were done in the brackish lagoons, Lake Shinji and Lake Nakaumi. The results from the modern-distribution analyses were then used in a palaeoecological analysis of diatoms collected from four vertical sections in the two lakes. The data from the diatom stratigraphy of the lagoon sediments yielded a sensitive high-resolution record of environmental change during the last 1,500 years.

## Introduction

Maintaining the quality of natural resources in enclosed coastal seas and lagoons is rapidly becoming an environmental issue of the 1990s. Has the quality changed as a result of anthropogenic activity or as a result of naturally occurring environmental changes, such as climatic change? What was the timing, rate, and extent of environmental change, and how can we infer these aspects?

Diatoms have been extensively used as indicators of environmental change, e.g., eutrophication, acidification, metal contamination, salinification, thermal effluent impact and land use change. The distribution of diatoms preserved in sediment cores can provide a quantitative reconstruction of the aquatic environment.

### *Lake Shinji and Lake Nakaumi*

Lake Shinji and Lake Nakaumi are brackish lagoons located along the coast of the Japan Sea. Two small waterways (Ohashikawa waterway and Sakai waterway) link the two lakes with the sea. A unique aquatic ecosystem under brackish conditions have seen maintained for many years at the two lakes with salinities in the surface waters ranging about 5–10 ‰ at Lake Shinji, and 20–25 ‰ at Lake Nakaumi.

About 30 years ago, the goverment of Japan planned to build a large barrier at the northern part of L. Nakaumi to shut off marine water from the two lakes, the object being to obtain fresh water for agriculture and industry. Because of strong opposition for protection of their natural resources, this project has being abandoned.

As a result, the general public, and consequently government agencies, and environment protection groups have become increasingly concerned to maintain the quality of aquatic resources of the two lakes. Long-term monitoring data are lacking for these aquatic systems.

Environmental conditions can be reconstructed over time spans ranging from a few years to thousands of years based on diatoms. This aspect of the study is necessary because of the lack of detailed data available for palaeoenvironment interpretations.

## Methods

Surface-sediment samples were collected in 1983 and 1986 (Kashima 1990), and undisturbed core samples from the lake bottom in 1989 and 1990 in both lakes (Fig. 1). The core samples were sectioned at intervals of 2.5 cm, and were divided for a number of analyses (e.g. diatom, pollen, shell, foraminifera and chemical analysis).

Diatom valves were separated from sediment matrix by treating it with strong oxidizing acid ($H_2O_2$), then slides are prepared using quantitative methods. Diatoms are mounted using a synthetic resin of high refractive index (Mount media by Wako Pure Chemical Indust.) .

About 200 diatom valves are identified in each sample with high magnification light microscopy and scanning electron microscopy.

The core samples were dated with $^{14}C$ and $^{210}PB$ methods by Brackish Research Center in Shimane University, Japan.

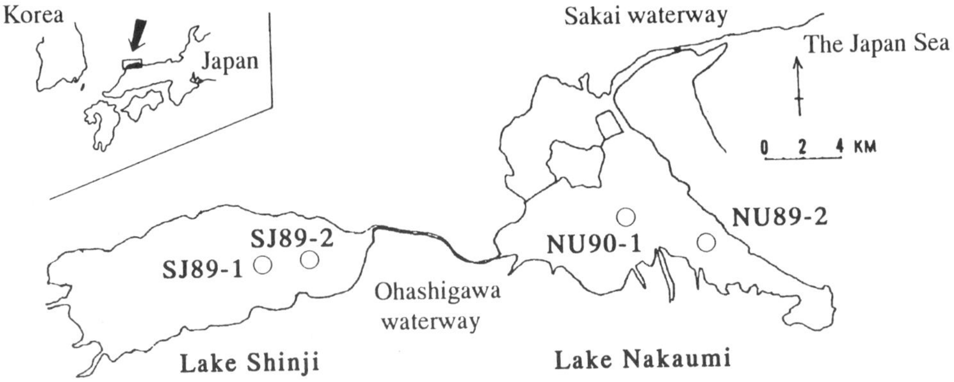


Fig. 1. Location map of studied area. Open circles show the locations of drilling sites.

***Distributions of modern diatom species***

There are apparent differences between diatom assemblages deposited as surface sediments in Lake Shinji and Lake Nakaumi (Fig. 2). At Lake Shinji, *Cyclotella caspia* is dominant and *Diploneis pseudovalis* is co-dominant. The percentages of the two species are above 90% in most of locations in this lake. On the other hand, the flora of diatoms in Lake Nakaumi is highly diversified without any overall dominant species. *Cyclotella caspia, Thalassionema nitzschioides, Fragilaria flavovirens, Thalassiosira* spp. and *Paralia sulcata* appear with abundance ranged less than 50%. Although the different salinity values of the two lakes make the cause of this difference of the modern diatom assemblages, the reason for the low diversity of taxa in L. Shinji has not been ascertained.

***The historic environmental change inferred by diatom fossil assemblages from drilling core samples***

*Lake Shinji* (Fig. 3)

Diatom analysis of a 65 cm-long (SJ89–1, see Fig. 1) showed that a series of oscillations between fresh, brackish and marine episodes has happened during the past 500 years. The lowest part of the core samples (pre 500 years BP) was characterized by marine water taxa, such as *Grammatophora* spp. and *Nitzschia* spp. These marine water species were abruptly replaced 500 years ago ($^{14}C$ method) by *Aulacoseira granulata* and other fresh water species indicating low saline until 50 years ago (by $^{210}Pb$ method). After then, a marked increase in the brackish planktonic taxon *Cyclotella caspia* has happened, and the salinity of lake water has increased again.

*Lake Nakaumi* (Fig. 4)

Two 150 cm long cores were taken at the central and eastern part of Lake Nakaumi. The abundant taxa from the core samples were *Cyclotella caspia, Thalassionema nitzschioides, Fragilaria flavovirens, Thalassiosira* spp., *Paralia sulcata* and *Cocconeis scutellum.* The former five taxa were marine-brackish species very common in the plankton of many brackish lagoons and common in the present Lake Nakaumi (Fig. 2). *Cocconeis scutellum* was benthic species common in present coast of Lake Nakaumi.

The diatom assemblages have hardly changed since 1500 years ago dated by $^{14}C$ method until now.

## References

Kashima, K. (1990). Diatom assemblages in the surface sediments of Lake Shinji and Lake Nakaumi, Shimane Prefecture, Japan. *Diatom (Journal of Japanese Diatom Society),* **5**, 51–58.

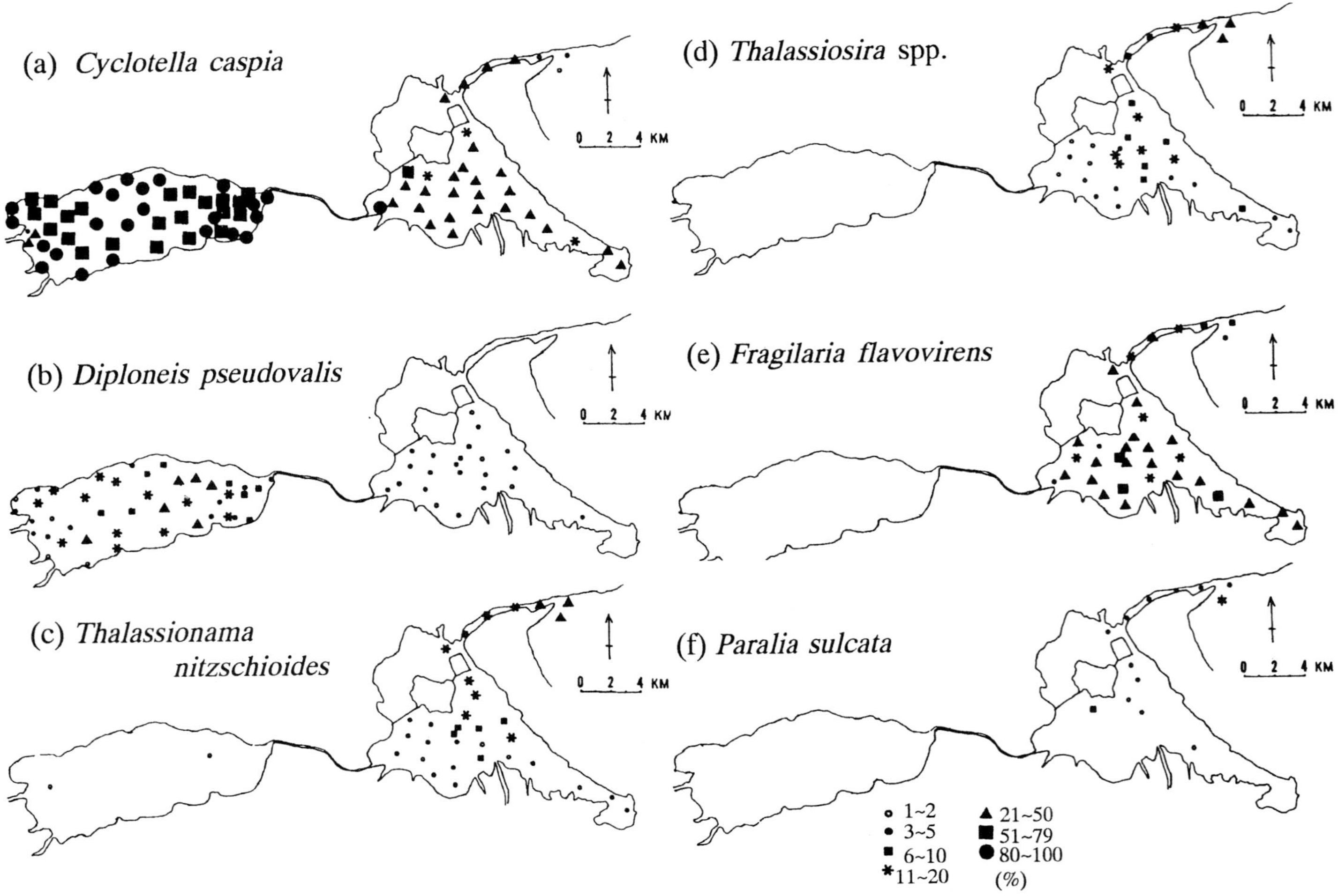


Fig. 2. Distribution maps of diatom taxa from surface-sediment samples in Lake Shinji and Lake Nakaumi.

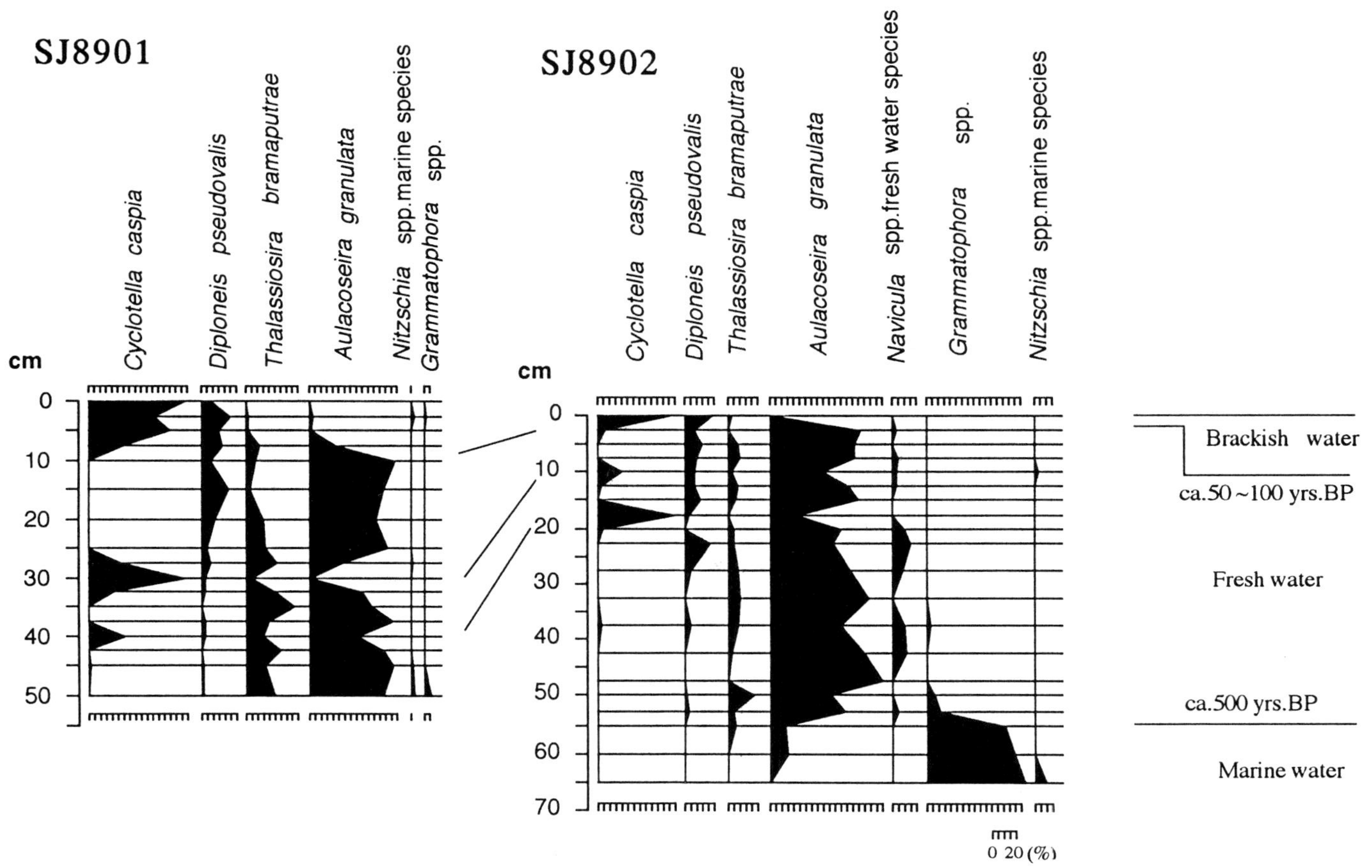


Fig. 3. Composite diatom diagrams of drilling core samples from Lake Shinji.

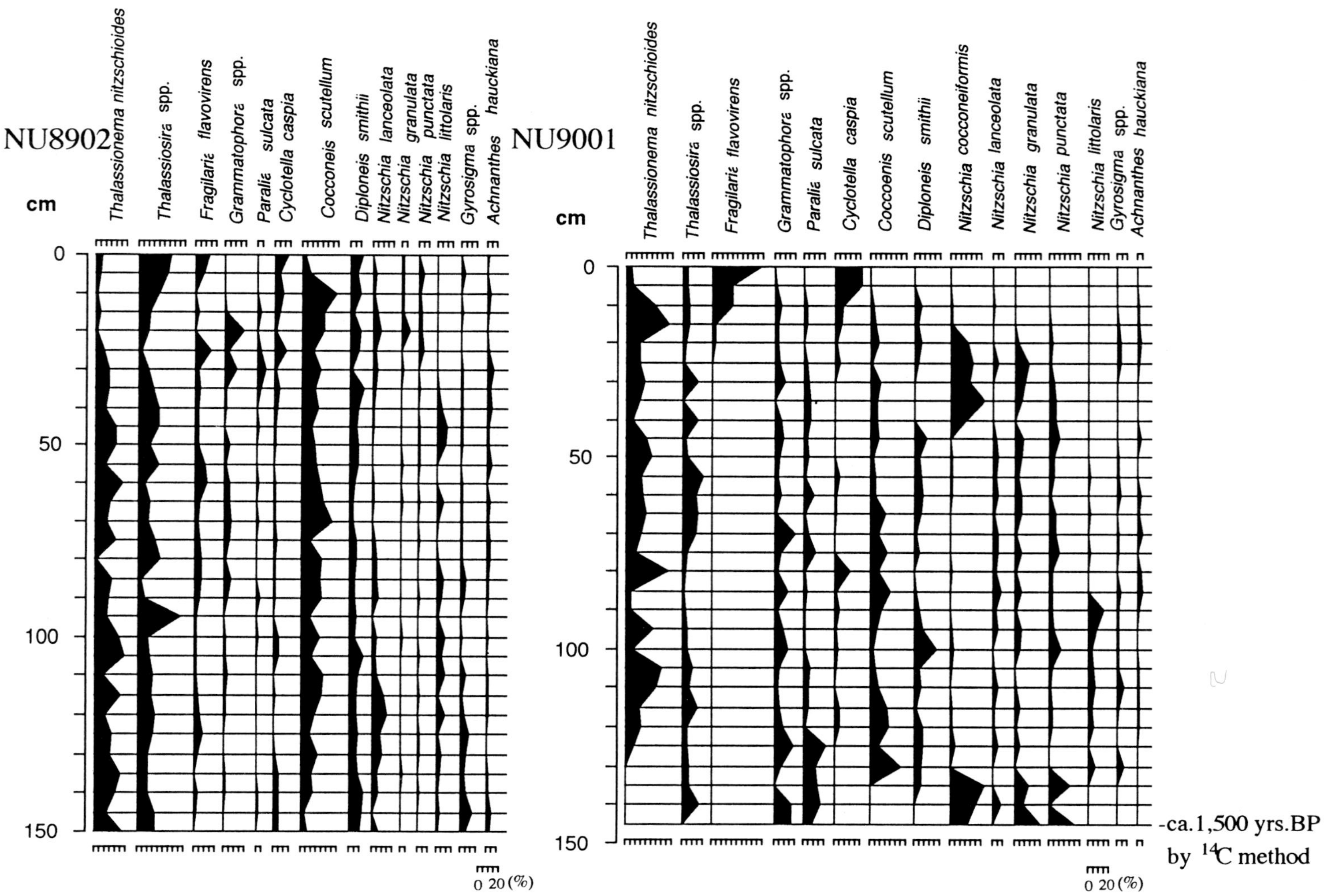


Fig. 4. Composite diatom diagrams of drilling core samples from Lake Nakaumi.

# Use of diatoms for pollution monitoring in the Valira Basin (Andorra)

Vidal Merino, Joan García* and Mariona Hernández-Mariné

*Unitat de Botànica, Facultat de Farmàcia, Universitat de Barcelona, Av. Joan XXIII, s/n. 08028 Barcelona, Spain.*

*Departament d'Enginyeria Hidràulica, Marítima i Ambiental, E.T.S. d'Enginyers de Camins, Canals i Ports, Universitat Politècnica de Catalunya, C/ Gran Capità, s/n. 08034 Barcelona, Spain.*

## Abstract

The water quality of four rivers belonging to the Pyrenean Valira basin (Andorra) has been studied using chemical analyses of the water and the CEE 88 diatom index. The results obtained through both methodologies are in agreement, since the CEE 88 index is strongly related to the chemical parameters that suggest organic pollution. The three sample stations in the lower stretches are the most polluted because of the waste water discharge from the main Andorran towns. The stations in the headwaters have good water quality or are only slightly affected. The changes in water quality cause some modifications in diatom species composition. Thus, *Achnanthes minutissima* and *Hannaea arcus* are the more abundant species in the unpolluted waters, but a shift in the relative frequencies of *Cymbella minuta* and *Nitzschia inconspicua* occurs when the organic pollution increases.

## Introduction

Routine biological monitoring is carried out in many rivers of Europe to complement or supplement chemical monitoring in order to assess water pollution (Whitton 1991). Much effort has been put into studying the animal components (Round 1991b), although diatom studies provide more precise data specially when eutophization and organic pollution are involved (Leclercq 1988; Descy & Coste 1990). Some workers believe that diatoms are difficult to identify, but this is not in fact true since the long tradition of diatom studies has made available good modern floras, so that it is now possible to identify at least the dominant species without much trouble (Round 1991b).

The present study was performed in order to contribute to our knowledge of diatom assemblages in fast-flowing mountain rivers in the Pyrenees and trace the effects of local waste water discharge on the species composition. The diatom communities in the rivers belonging to the Valira basin (Andorra) were studied, and the CEE 88 index (Descy & Coste 1990) was applied to estimate their water quality. Complete chemical analyses of the water were also carried out. The results obtained with both methodologies were compared placing emphasis on the relations between the diatom index and those chemical compounds that indicate organic pollution.

## Material and methods

Two main watersheds form the Valira basin, Valira del Nord and Valira d'Orient, which unite downstream and give rise to the Gran Valira river (Fig. 1). It flows 11.5 km until the border with Catalonia and then runs into the Segre river (a tributary of the Ebro). The Valira del Nord originates at 2600 m.a.s.l. and is 13.9 km long. The Valira d'Orient arises at 2100 m.a.s.l. and flows for 23.3 km. The whole Valira basin drains an area of 559 km$^2$, mainly composed of schists and granitic substrata. Calcareous bedrock is only present in a small section of the Valira del Nord basin. The hydrological regime is typical of the Pyrenean rivers. The maximum flows correspond to spring and early summer because of snow melt in the mountains. Long periods of stable flow characterize the other seasons, with minimum rates in winter.

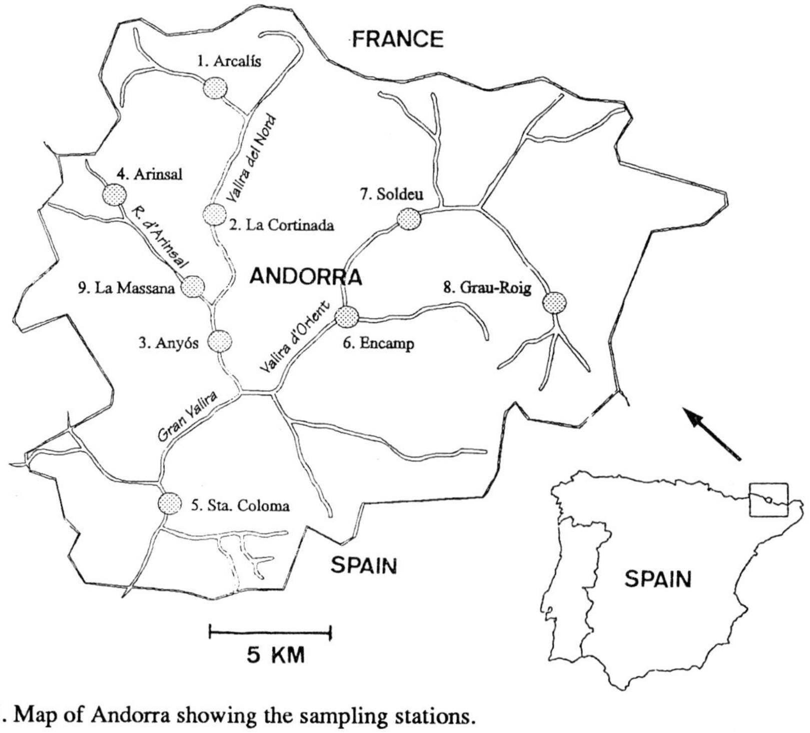


Fig. 1. Map of Andorra showing the sampling stations.

Nine sample stations were set up along the main rivers (Fig. 1). In the headwaters of the Valira d'Orient one station was at Grau Roig (1980 m.a.s.l.) next to the Pas de la Casa-Grau Roig ski resort, where sewage was discharged into the river. Soldeu station (1580 m.a.s.l.) was located in the middle stretch near to the Soldeu-El Tarter ski resort. Soldeu village and a small dam were situated only a couple of kilometres upstream. The dam deflects part of the water for hydroelectrical purposes and reduces and controls the river flow. Encamp station (1205 m.a.s.l.) was located in the more lowland section of the Valira d'Orient river near to the populous town of Encamp, which poured out its waste water to the river. In the Gran Valira river, only one sample station at Santa Coloma (1060 m.a.s.l.) was used. The main Andorran towns (Les Escaldes and Andorra la Vella) were located upstream of Santa Coloma.

Two major rivers run in the Valira del Nord watershed: the Valira del Nord and the Riera d'Arinsal. Along the Riera d'Arinsal river two sample stations were established. Arinsal station (1620 m.a.s.l.) was in the headwaters beside the Arinsal ski resort. La Massana station (1250 m.a.s.l.) was in the outlying area of La Massana town, but before the river crosses the town and enters the Valira del Nord river. Between the stations of Arinsal and La Massana the river runs along one of the Andorran calcareous zones, where there are some villages and gravel extraction installations. Arcalís station (1960 m.a.s.l.) was set up in the headwaters of the Valira del Nord river near to the Ordino-Arcalís ski resort. There are only a few small villages between it and La Cortinada station (1335 m.a.s.l.), which was placed in the middle stretch of the river. Downstream, the Anyós station (1130 m.a.s.l.) was set up beyond the towns of La Massana and the mouth of the Riera d'Arinsal river.

The collection of samples took place seasonally on 11th April 1992, 11th July 1992, 9th October 1992, and 6th February 1993. Diatoms were collected from the epilithon by scraping the upper surface of submerged stones, picked up from the fast flow zones in the river (Round 1991a). They were preserved in 4% formaldehyde, cleaned by boiling in 30% $H_2O_2$ (Renberg 1990), and rinsed in distilled water until neutral. Cleaned material was air-dried onto coverglasses and mounted in microscope slides with Naphrax®. Permanent mounts were observed using Nomarsky optics on a Nikon Optiphot–2 light microscope. 400–500 valves were counted at 1000× magnification for the estimation of relative frequency per sample.

The diatom index used in this study was the CEE 88 (Descy & Coste 1990). This index is aimed to be a standard biological index for management purposes in the countries belonging to the European Union. It has been used in several studies and seems to offer very satisfactory results when compared with the other proposed indices (Descy & Coste 1990). The CEE 88 index was calculated using the program package CO.CA.IN. (Coste 1992). The index scores range from 0 to 10 units, but they were transformed into a 1 to 20 scale. A lower index value means a worse water quality.

At each sample station, water conductivity, pH, dissolved oxygen and temperature were measured by a Crison 523 conductivity-meter, Crison pH-meter and YSI 58 oxygen meter equipped with a thermistor. Alkalinity, $SO_4^{2-}$, $Cl^-$, $Ca^{2+}$, $Mg^{2+}$, $Na^+$, $K^+$, $NH_4^+$, $NO_2^-$, $NO_3$ , SRP (Soluble Reactive Phosphorous), $SiO_2$ and COD (Chemical Oxygen Demand) were analysed in the laboratory using conventional methods (APHA

1989). The results of $NO_3^-$, $NO_2^-$ and $NH_4^+$ were added to give a more general chemical parameter – Total Inorganic Nitrogen (TIN).

A principal component analysis (PCA) was carried out to detect which group of physical-chemical variables were best correlated with the CEE 88 index. The computations were executed before log-transforming the environmental variables (except pH).

**Results**

The values obtained with the CEE 88 index are presented in Table I. The sample stations that had the highest scores were those located in the headwaters (Arcalís, Arinsal and Grau Roig), and in the middle sections of the Riera d'Arinsal river (La Massana) and the Valira del Nord river (La Cortinada). The remaining four stations had index values below 15 units. Three of them were downstream of important towns (Anyós, Encamp and Sta. Coloma). The other (Soldeu) was the station near to the dam. In winter the whole basin showed the best water quality, and in spring it became the worst.

Table I. CEE 88 diatom index values reached at each season by the nine sampling stations. Averages and standard deviations are indicated for both the stations and the seasons.

| Station | Spr. | Sum. | Aut. | Win. | Avg. | Std. dev. |
|---|---|---|---|---|---|---|
| Arcalís | 14.0 | 18.5 | 17.9 | 18.7 | 17.3 | 2.0 |
| La Cortinada | 14.5 | 17.8 | 19.1 | 19.5 | 17.7 | 2.0 |
| Anyós | 12.9 | 12.0 | 12.6 | 14.2 | 12.9 | 0.8 |
| Arinsal | 15.9 | 19.5 | 19.8 | 19.4 | 18.7 | 1.6 |
| Sta. Coloma | 8.8 | 10.5 | 12.3 | 12.6 | 11.0 | 1.5 |
| Encamp | 11.5 | 12.5 | 11.8 | 11.3 | 11.8 | 0.4 |
| Soldeu | 12.8 | 15.3 | 15.4 | 13.6 | 14.2 | 1.1 |
| G. Roig | 14.6 | 19.6 | 14.2 | 19.9 | 17.1 | 2.7 |
| La Massana | 14.1 | 17.8 | 13.8 | 19.2 | 16.2 | 2.3 |
| Avg. | 13.2 | 16.0 | 15.2 | 16.5 | | |
| Std. dev. | 1.9 | 3.3 | 2.9 | 3.3 | | |

The concentrations of TIN and SRP for each sample station are shown in the table II. These results agree with the CEE 88 values, since the downstream stations had the highest TIN and SRP concentrations. However, the Soldeu station had low TIN and SRP values (only SRP concentration in spring was somewhat high, 33 µg/l), which is surprising because of the low CEE 88 values here.

The seasonal CEE 88 scores and the TIN annual average are graphically represented in Fig. 2 (a–c) for the three fluvial axes studied: Valira del Nord-Gran Valira (VN), Valira d'Orient-Gran Valira (VO) and Riera d'Arinsal-Valira del Nord-Gran Valira (RA). The three fluvial axes showed a general pattern of decrease of the CEE 88 index values related to the rise of the TIN content in the four seasons from the headwaters to downstream. In the VN axis (Fig. 2a), the maximum CEE 88 index decrease is located between La Cortinada and Anyós stations, and in this river stretch the TIN showed a maximum. This trend was observed all the year round, though in spring it occurred between Anyós and Sta. Coloma. Upstream, in Arcalís and La Cortinada stations, the CEE 88 index values suggested excellent water qualities (up to 18 units) all the year except spring. The values of the Arcalís station were slightly lower than those of the downstream La Cortinada station. TIN results also reflected this improvement of the water quality. In the RA axis (Fig. 2b), the station placed at the headwaters (Arinsal) had high CEE 88 values (near to 20 units) throughout the year except spring. The index was only slightly lower downstream (La Massana), though an important decrease was recorded in autumn. This decrease did not correspond with the rise in chemical pollutants (see Table II). In the VO axis (Fig. 2c), the most important decrease of the CEE 88 index was usually between the two stations placed upstream (Grau Roig and Soldeu), but a significative chemical deterioration did not accompany it (Table II).

Altogether 82 taxa were found in the epilithon samples, though only 19 of them have relative frequencies greater than 5% (Table III). *Hannaea arcus*, *Achnanthes minutissima*, *Cymbella minuta* and *Nitzschia incospicua* were, in this order, the species that attained the highest abundances. The annual average of their relative frequencies for each sample station is graphically presented in Fig. 3. *A. minutissima* and *H. arcus* were the dominant taxa in the stations with high CEE 88 index values, low TIN and SRP concentrations. Their abundance decreased in the three stations sited downstream. *C. minuta* and *N. inconspicua* (and also *Gomphoneis olivacea*) also had their maximal relative frequencies in these stations. To verify whether this variation in the community composition might be due to a change in the water chemistry, the abundances attained by these four species were compared with the TIN concentrations in each station and season (Fig. 4). Both *A. minutissima* and *H. arcus* attained the highest abundances when TIN concentration was low (200–600 µg N/l). *C. minuta* and *N. inconspicua* seemed to be more tolerant to the organic pollution, since they were most abundant from 500 to 1400 µg N/l TIN. *N. inconspicua* generally had a low abundance (less than 20%), except three times when, surprisingly, TIN concentration was low. However, these results occurred at the three stations placed downstream, in spring, when SPR values were rather high (Table II).

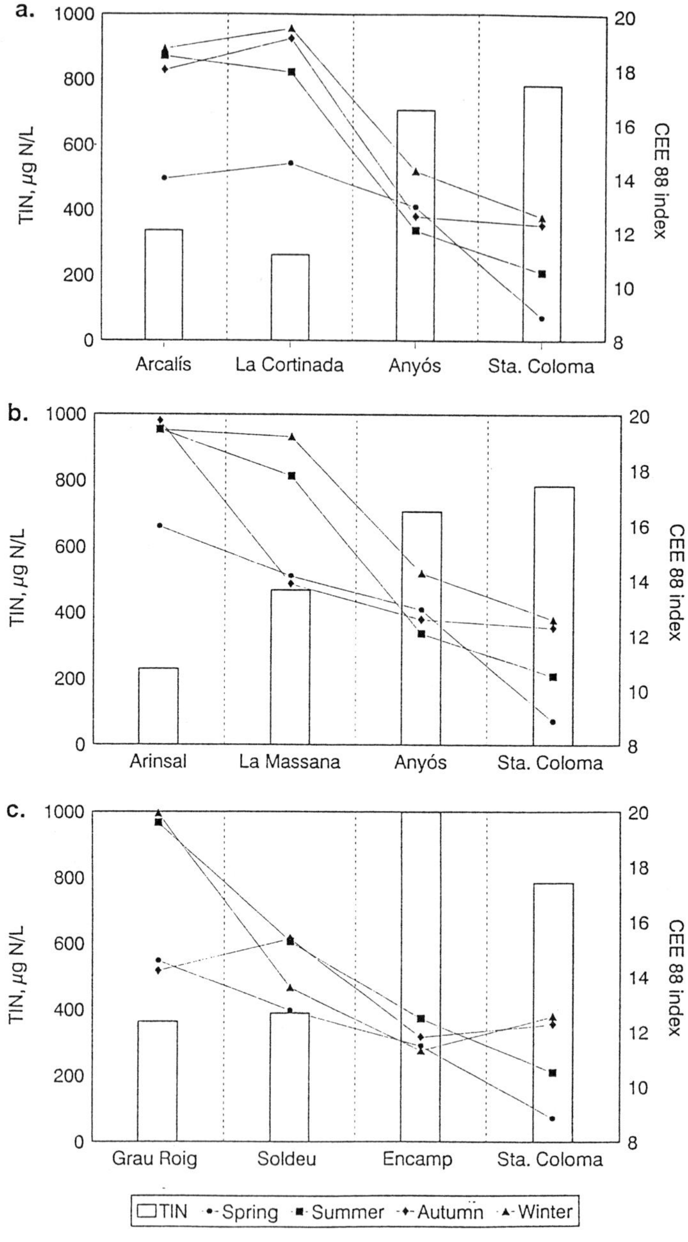


Fig. 2. Seasonal CEE 88 index scores and TIN annual average for each sampling station. **a.** Valira del Nord-Gran Valira fluvial axis, **b.** Riera d'Arinsal-Valira del Nord-Gran Valira fluvial axis, **c.** Valira d'Orient-Gran Valira fluvial axis.

Table II. Concentrations of the Total Inorganic Nitrogen (TIN) and the Soluble Reactive Phosphorous (SRP). Annual averages and standard deviations are indicated for the TIN. (–) means not detected values.

| | Total Inorganic Nitrogen (TIN), µg N/L | | | | | | Soluble reactive Phosphorous (SRP), µg P/L | | | |
|---|---|---|---|---|---|---|---|---|---|---|
| Station | Spr. | Sum. | Aut. | Win. | Avg. | S. dev. | Spr. | Sum. | Aut. | Win. |
| Arcalís | 330 | 260 | 350 | 400 | 340 | 50 | 7 | – | – | 57 |
| LaCortinada | 260 | 210 | 300 | 280 | 260 | 30 | – | – | – | – |
| Anyós | 410 | 490 | 580 | 380 | 700 | 370 | 42 | – | 37 | 175 |
| Arinsal | 250 | 180 | 170 | 330 | 230 | 60 | – | – | – | 5 |
| Sta. Coloma | 560 | 1150 | 690 | 740 | 780 | 220 | 52 | 208 | 87 | 368 |
| Encamp | 510 | 1360 | 650 | 1490 | 1000 | 430 | 72 | 110 | 53 | 288 |
| Soldeu | 390 | 400 | 270 | 490 | 390 | 80 | 33 | – | – | – |
| G. Roig | 390 | 270 | 230 | 570 | 360 | 130 | – | – | – | 43 |
| La Massana | 430 | 420 | 540 | 470 | 470 | 50 | – | – | – | – |

Table III. Epilithic diatom species that attained relative frequencies above 5%.

*Achnanthes biasolettiana* Grunow

*Achnanthes minutissima* Kützing

*Cocconeis placentula* var. *lineata* (Ehrenberg) Van Heurck

*Cymbella minuta* Hilse

*Diatoma ehrenbergii* Kützing

*Diatoma hyemale* var. *mesodon* (Ehrenberg) Grunow

*Fragilaria capucina* Desmazières

*Fragilaria vaucheriae* (Kützing) Petersen

*Gomphoneis olivacea* (Hornemann) Dawson

*Gomphonema pumilum* (Grunow) Reichardt *et* Lange-Bertalot

*Hannaea arcus* (Ehrenberg) Patrick

*Navicula permitis* Hustedt

*Navicula veneta* Kützing

*Nitzschia dissipata* (Kützing) Grunow

*Nitzschia fonticola* Grunow

*Nitzschia inconspicua* Grunow

*Nitzschia palea* (Kützing) W. Smith

*Nitzschia pura* Hustedt

*Reimeria sinuata* (Gregory) Kociolek *et* Stoermer

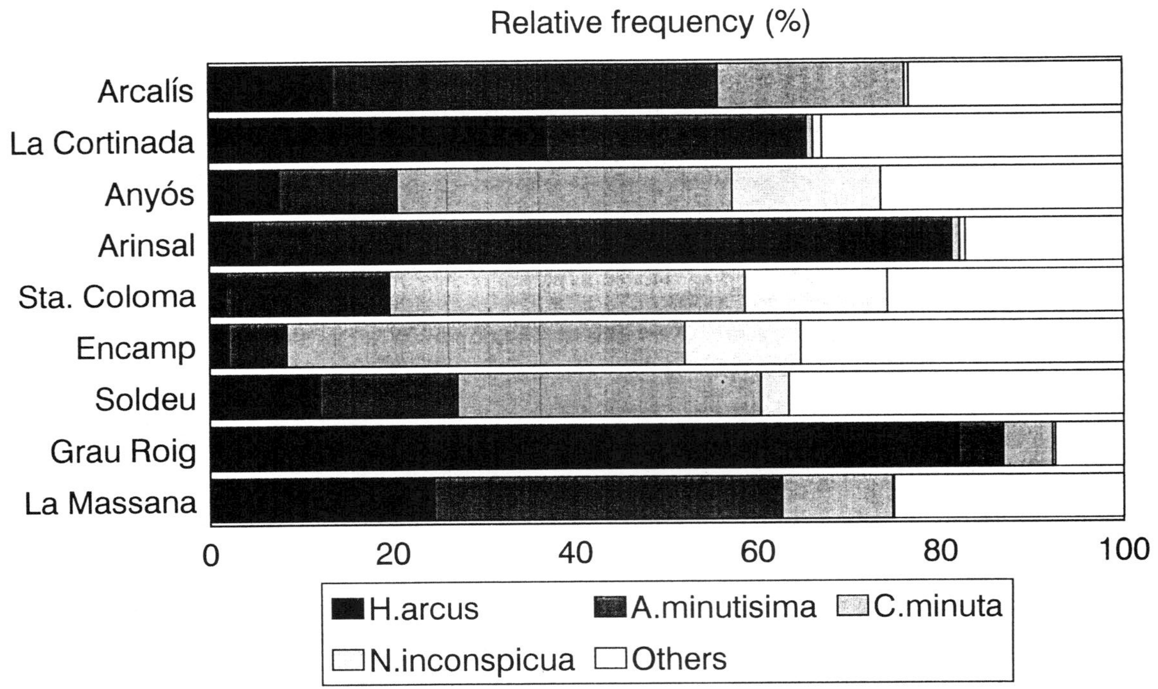


Fig. 3. Annual average of the relative frequencies for the dominant species.

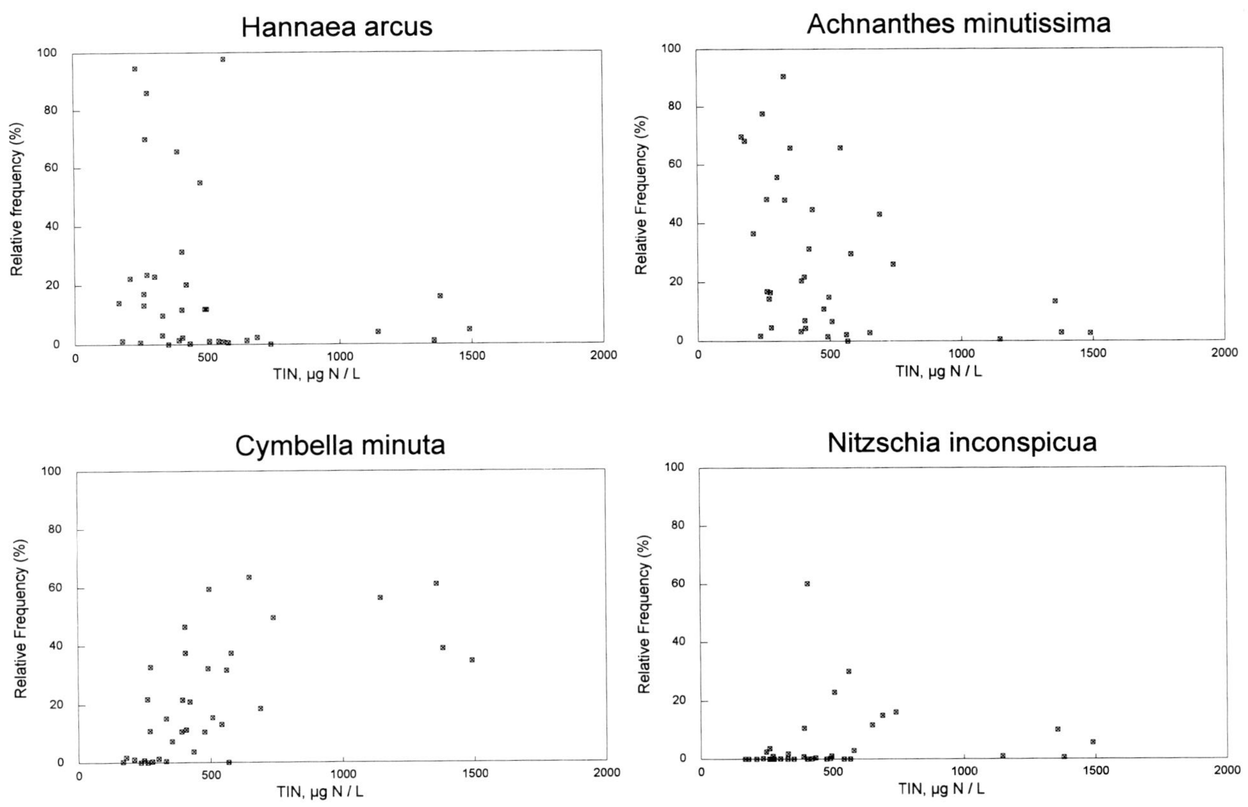


Fig. 4. Plots of the TIN versus the abundances of the dominant species.

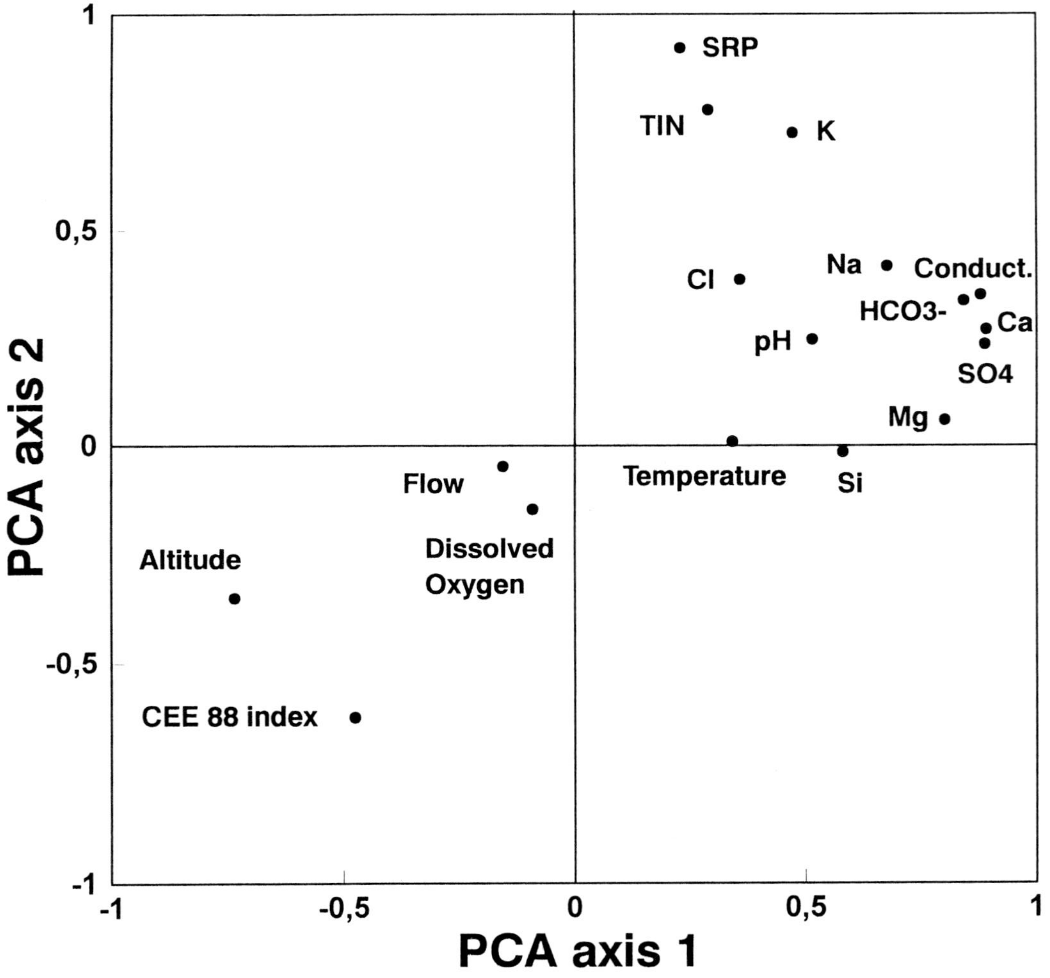


Fig. 5. Principal components correlation plot of 36 samples × 17 chemical and other environmental variables, including the index CEE 88.

The results obtained by PCA to determine the relationship between the CEE 88 index and the chemical parameters are presented in Fig. 5. Three interpretable axes were obtained, which explain 70% of the variance. Only in the first (36% of the variance) and in the second (19%) axis was the CEE 88 index significant. The third component (15%) seemed determined by the highest seasonal changes (temperature, flow, $SiO_2$ and $Cl^-$) and the CEE 88 index showed a poor correlation (0.3). Therefore this axis has not been presented. The first component was strongly correlated to a group of variables that together control the water mineralization (conductivity, alkalinity, $Ca^{2+}$, $Mg^{2+}$, etc.). The sole variable negatively correlated is the altitude. This axis may be interpreted as the spatial organization of the rivers, since water mineralization always increases downstream. The CEE 88 index showed a poor negative correlation (–0.48). The second is determined by the variables associated with waste water discharges: TIN (0.78) and SRP (0.92). The CEE 88 index was negatively correlated (–0.63). This antagonistic position in the space formed by the second component revealed opposite responses to the changes in water quality. When the SRP and TIN concentrations increase as a consequence of waste water discharges, the CEE 88 index decreases. This pattern can be clearly observed in the plot presented in Fig. 6, where the points seemed to fit in a negative exponential curve ($CEE88 = 12.2\ TIN^{-03}$, $r^2=0.97$, $p=0.001$).

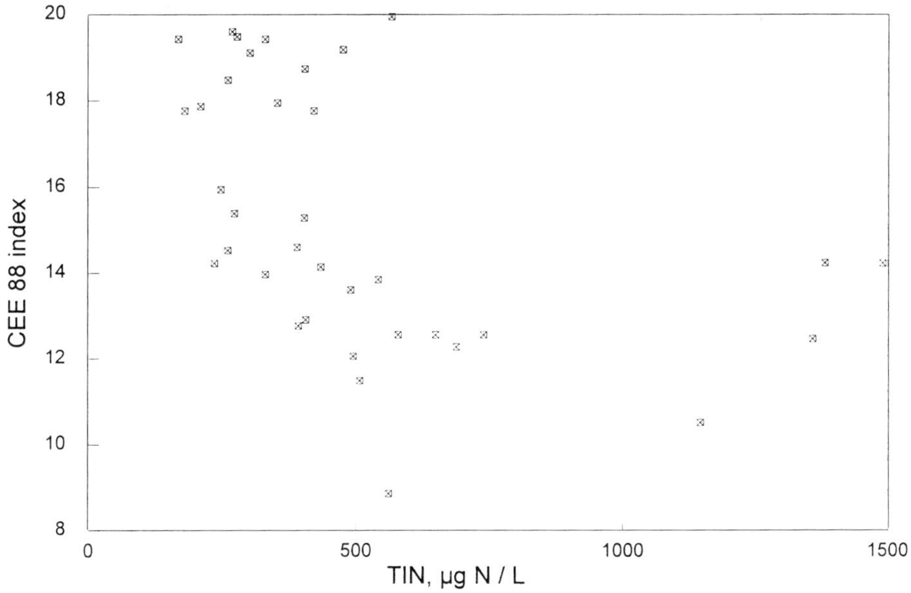


Fig. 6. Plot of the TIN versus the CEE 88 index.

## Discussion

Both the CEE 88 index and the chemical parameters showed that the river waters in the lower sections of the basin were polluted by human activities. The concordance in the results obtained with both methodologies is reasonable because, as seen from the PCA, the CEE 88 index is strongly related to organic pollution. Nevertheless, some discrepancies between them have been observed in the temporal scale. The chemical analyses showed that the river water had the worst quality in winter. Anthropogenic pollutants are generally diluted during the floods of spring and summer, while the diminution of the flow and the tourism caused by winter sports produce the worst quality in the coldest seasons. However, the CEE 88 index indicated the best quality in winter and the worst in spring. This difference might be due to the environmental parameters (e.g. the spring flood) that could also affect the community composition even more strongly than the organic pollution. In these circumstances, the epilithic diatom communities might be more adapted to resist the extreme water flow than the pollution. In winter, the low water temperatures and the ice cover on many river stretches are perhaps also important factors affecting the diatom epilithic assemblages.

A further disagreement between the methodologies was encountered at the Soldeu station. The CEE 88 index values were rather low all the year round (12.8 to 15.4), but TIN and SRP were never high enough to explain them. The flow control caused by the dam located upstream might be responsible for the CEE 88 index decrease, though the presence of unanalysed chemical pollutants might also be involved.

Waste water influence on the diatom flora is difficult to trace in the Valira basin due to minor changes in species composition. However, a trend was observed in the Andorran rivers: *Achnanthes minutissima* and *Hannaea arcus* dominated the epilithic community in unpolluted waters, but a shift in the relative frequencies of *Cymbella minuta* and *Nitzschia inconspicua* occurred when the organic pollution increased. A change in species composition has been previously described for a glacial stream in the Tyrol (Maier & Rott 1988), which had similar environmental conditions as the Andorran rivers.


## Acknowledgements

This work was made possible by a grant from the *Institut d'Estudis Andorrans* in 1992. We would like to express our gratitude to Professor Michel Coste who kindly provided us his computer package CO.CA.IN.


## References


A.P.H.A. (1989). *Standard Methods for the Examination of Water and Wastewater*. American Public Health Association (ed.). New York.

Coste, M. (1992). CO.CA.IN. version 1.3. Logiciel de comptage et calcul d'indices diatomics. Rap. CEMAGREF Bordeaux. 15 pp.

Descy, J. P. & Coste, M. (1990). Utilisation des diatomées benthiques pour l'evaluation de la qualité des eaux courantes. Rapport final. UNECED–NAMUR, CEMAGREF Bordeaux CEE–B. 112 pp.

Leclercq, L. (1988). Utilisation de trois indices, chimique, diatomique et biocénotique, pour l'évaluation de la qualité de l'eau de la Joncquiere, rivière calcaire polluée par le village de

Doische (Belgique, Prov. Namur). *Mémoires de la Societé Royale de Botanique de Belgique,* **10**: 26–34.
Maier, M. & Rott, E. (1988). The effect of local waste-water inflows on the structure of diatom assemblages in fast-flowing streams. In: *Proceedings of the 10th Symposium on Recent and Fossil Diatoms* (H. Simola, ed.), 553–561. Koeltz Scientific Publishers, Königstein, Germany.
Renberg, I. (1990). A procedure for preparing large sets of diatom slides from sediment cores. *Journal of Paleolimnology*, **4**, 87–90.
Round, F. E. (1991a). Use of diatoms for monitoring rivers. In: *Use of algae for monitoring rivers* (B. A. Whitton, E. Rott. & G. Friedrich, eds), 25–32. Innsbruck.
Round, F. E. (1991b). Diatoms in river water-monitoring studies. *Journal of Applied Phycology*, **3**, 129–145.
Whitton, B. A. (1991). Aims of monitoring. In: *Use of algae for monitoring rivers* (B. A. Whitton, E. Rott. & G. Friedrich, eds), 5–8. Innsbruck.

# Effects of intensive forestry and peatland management on forest lake ecosystems in Finland: sedimentary records of diatom floral changes

Heikki Simola, Minna Kukkonen, Jaana Lahtinen and Tarmo Tossavainen

*University of Joensuu, Karelian Institute, Section of Ecology, P.O.Box 111, FIN–80101, Joensuu, Finland*

## Abstract

We have investigated the response of several forest lake ecosystems in eastern Finland to catchment changes caused by intensive forest and peatland management measures, through stratigraphic analyses of sedimentary diatom assemblages. In natural condition, these lakes have been typically polyhumic, oligotrophic and slightly acidic (pH 5–6). Often quite prominent eutrophication, as evidenced by sediment diatoms, coincides with the documented impact of modern forestry practices on the drainage area (sediment dating provided by varve counts in many cases; $^{210}Pb$-dating has also been applied).

In many lakes the eutrophication process is seen as an increase of species and distinct succession of planktonic diatoms up the core (e.g. lakes Haarajärvi, Ilajanjärvi and Polvijärvi). In other cases, however, only the species proportions have changed, but not the species composition (lake Suuri-Rostuvi, in which only the acidophilic *Asterionella ralfsii* has gained over 90% dominance in the sediment representing post-1960 deposition). In the very shallow Lake Petäisjärvi, planktonic diatoms have never been abundant, but clear eutrophication is evidenced by the change in the epiphytic and benthic flora. Such differences reflect differences in flushing rate, nutrient levels and basin morphology.

## Introduction

Modern forestry practices have caused widespread deterioration of surface water quality in Finland, especially since the 1960's. Forest clear-cutting is commonly accompanied by ploughing of the mineral soil, to improve growth conditions for the new trees. Also peatlands have been largely drained (up to 80–90% in southern Finland) for forestry purposes. Soil disturbance may dramatically increase the suspended solids and nutrient loading of forested headwater systems (Holopainen *et al.* 1989; Holopainen & Huttunen 1992; Ahtiainen 1992). Extensive phosphorus-potassium fertilization of managed peatlands has caused widespread and long-lasting eutrophication in the recipient watercourses in Finland (e.g. Ahti 1980; Kenttämies 1981; Rekolainen 1989). Even though the amount of leaching relative to the nutrient reserves in soil and peat need not be very high, the effects may be profound in the naturally very oligotrophic waters (Tossavainen 1991).

This paper summarizes, from the point of view of sedimentary diatom assemblages, several palaeolimnological case studies of forestry impact on small lakes, conducted over the past several years in the province of North Karelia, eastern Finland.

## Material and Methods

The lakes dealt with in this paper are listed in Table I. For palaeolimnological analyses we have obtained short sediment cores usually from the deepest points of the lakes, in most cases with both fresh and freeze coring techniques (Simola *et al.* 1986). Sediment varves (Simola 1992) have provided good dating for many of these lakes, but also $^{210}Pb$ dating has been applied. For other stratigraphic analyses besides diatoms, reference is made to the original case study publications and reports.

All the sites represent very sparsely populated headwater catchments, in which forestry and peatland management are the major human activities. Drainage area management histories of the lakes have been compiled from the archives of the forestry management organizations.

## Results and Discussion

In their natural condition, all the study lakes seem to have been humic, oligotrophic and slightly acidic (pH 5–6), with e.g. *Aulacoseira distans* agg., *Eunotia* spp. and *Frustulia rhomboides* as characteristic diatoms. In the diatom stratigraphies, there are typically slight floral changes dating back to 1800's or even earlier, possibly reflecting effects of slash-and-burn agriculture or early logging activity, but the major changes usually coincide well with the documented impact of modern forestry measures (in some lakes also fuel peat excavation) on the drainage areas (Table I).

Quite typically the eutrophication process is seen as an increase of species and distinct succession of planktonic diatoms. In lake Haarajärvi (Kukkonen 1994), the succession of planktonic taxa appears as follows: *Aulacoseira tenella* → *Cyclotella kuetzingiana* → *Eunotia zasuminensis* & *A. ambigua* → *Asterionella formosa* & *Tabellaria fenestrata* → *Fragilaria crotonensis*. There are also changes in the epiphytic flora, e.g. a decrease of *Navicula soehrensis* and *Achnanthes subatomoides* and the appearance of *Meridion circulare* within the uppermost 5 cm of sediment (Figs 1 & 2). Sediment varves date the onset of prominent change to around 1960.

Rather similar planktonic successions have been reported, e.g in lake Polvijärvi (Simola 1983) and lake Ilajanjärvi (Simola *et al.* 1988). In all these lakes, at least the

Table I. Characteristics of the studied lakes. Water quality parameters calculated as water column mean values from observation data collected by North Karelian Environment Centre. Sampling time and number of sampling occasions (mainly autumn and winter seasons) indicated for each lake.

| Lake (water quality observations) | Elevation m a.s.l. | Surface area, $km^2$ | Max. depth, m | Drainage area, $km^2$ | Drainage area characteristics | Water quality parameters Secchi depth (m) | colour mg Pt $l^{-1}$ | pH | $N_{TOT}$ µg $l^{-1}$ | $P_{TOT}$ µg $l^{-1}$ | $Fe_{TOT}$ µg $l^{-1}$ |
|---|---|---|---|---|---|---|---|---|---|---|---|
| **Haarajärvi** (1990–93; 3) | 111 | 0.52 | 11 | 53 | largely paludified; 40% drained and partly fertilized peatlands; also peat excavation since1986 | 1.1 | 170 | 6.3 | 575 | 17 | 1350 |
| **Ilajanjärvi** (1992–94; 11) | 152 | 7.9 | 14 | 196 | mainly paludified; no lakes; 75% drained, fertilized peatlands; peat excavation on 8 $km^2$ since 1975 | 0.8 | 180 | 6.1 | 625 | 30 | 1700 |
| **Polvijärvi** (1982; 2) | 170 | 1.8 | 35 | 66 | several lakes (including L. Suuri–Rostuvi) on the drainage area; 27% drained, partly fertilized peatland | 1.6 | 120 | 6.0 | 440 | 32 | 1050 |
| **Suuri–Rostuvi** (1985; 1) | 186 | 0.48 | 12 | 12 | small headwater catchment; 40% drained, fertilized peatlands; extensive loggings | 1.0 | 180 | 4.7 | 510 | 58 | 780 |
| **Petäisjärvi** (1987–89; 4) | 166 | 0.46 | 2 | 196 | Large catchment with several lakes; 27% drained and fertilized peatlands, extensive loggings since 1960's | 1.5 | 140 | 6.1 | 550 | 38 | 1350 |

## LAKE HAARAJÄRVI

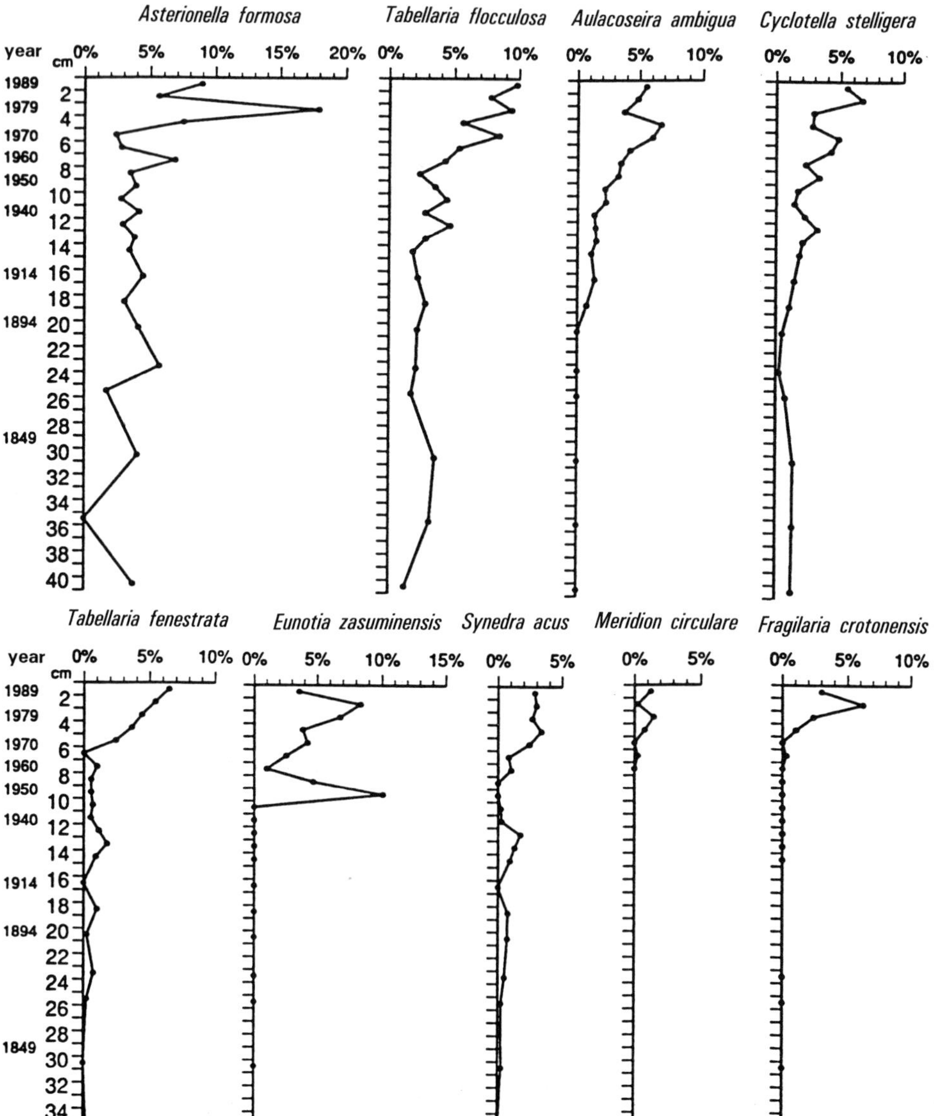


Fig. 1. Stratigraphies of those diatom taxa that clearly increase within the uppermost sediment of Lake Haarajärvi, evidencing eutrophication of the lake due to extensive catchment ditching. Dating provided by sediment varves. Excerpted from Kukkonen (1994).

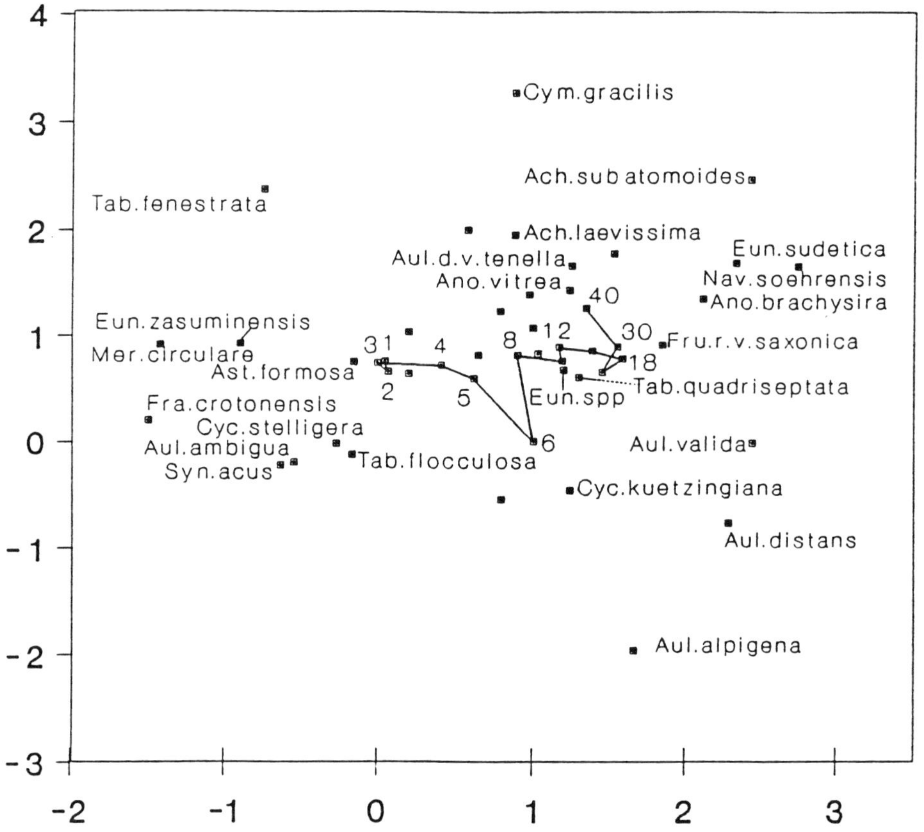


Fig. 2. Ordination by detrended correspondence analysis (DCA; Hill & Gauch, 1980)) of stratigraphic samples (circles connected with a line; sediment depths (cm) indicated) and diatom taxa of stratigraphic significance (squares) in the sediment of Lake Haarajärvi (adopted from Kukkonen 1994). Abbreviations: ACHN LAEV = *Achnanthes laevissima*, ACHN SUBA = *A. subatomoides*, ANOM BRAC = *Anomoeoneis brachysira*, ANOM VITR = *A. vitrea*, ASTE FORM = *Asterionella formosa*, AULA ALPI = *Aulacoseira alpigena*, AULA AMBI = *A. ambigua*, AULA DIST = *A. distans*, AULA TENE = *A. tenella*, AULA VALI = *A. valida*, CYCL KUET = *Cyclotella kuetzingiana*, CYCL STEL = *C. stelligera*, CYMB GRAC = *Cymbella gracilis*, EUNO SPP. = *Eunotia* spp., EUNO SUDE = *E. sudetica*, EUNO ZASU = *E. zasuminensis*, FRAG CROT = *Fragilaria crotonensis*, FRUS RSAX = *Frustulia rhomboides* var. *saxonica*, MERI CIRC = *Meridion circulare*, NAVI SOEH = *Navicula soehrensis*, SYNE ACUS = *Synedra acus*, TABE FENE = *Tabellaria fenestrata*, TABE FLOC = *T. flocculosa*, TABE QUAD = *T. quadriseptata*.

sediment of the period of increasing eutrophication is varved, due to hypolimnetic anoxia, and the corresponding diatom blooms appear as distinct layers in the sediment.

In some lakes, however, only the species proportions have changed, but not the species composition. Such is the case in the very humic Lake Suuri-Rostuvi (Rönkkö & Simola 1986), in which the acidophilic and planktonic *Asterionella ralfsii* has gained over 90% dominance in the sediment representing the post-1960 deposition period (dated by $^{210}Pb$). The absence of more alkaline taxa may be explained by the polyhumic water and acidic bedrock (Rönkkö & Simola 1986).

In some shallow lakes, in which plankton algae do not thrive well, the change may be seen as an increase of alkaliphilous epiphytic and benthic forms, as in Lake Petäisjärvi (Lahtinen 1992). In this very shallow lake planktonic diatoms have never been abundant, but clear eutrophication is evidenced by the epiphytic and benthic flora, e.g. decline of *Anomoeoneis brachysira* and *Eunotia* spp. and increase of *Meridion circulare,* several *Fragilaria* spp., *Achnanthes* spp., *Navicula arvensis* and *N. indifferens.* These changes, dated by $^{210}Pb$, correspond well with the documented management phases of the large drainage area, and with observed and modelled water quality changes (e.g. Tossavainen 1991).

In all lakes the diatom-evidenced eutrophication is accompanied by an increase in sedimentation rate. There are several cases, in which dredging of the inflow channels has caused massive silt deposition (up to several cm per year) near the inflows (e.g. Sandman *et al.* 1990). Lake Haarajärvi (Kukkonen 1994) has received massive silt input at its northern inflow. Stratigraphic diatom analyses were, however, done on samples taken in the southern sub-basin of the lake, in which the sedimentation of inflow solids is not so prominent.

## Conclusions

The limnological consequences of forestry management on the drainage basins are often abrupt and profound in small headwater lakes, even though the actual effects may vary considerably between different lakes. In larger lakes, and in lakes situated at lower reaches of watercourses, the effects may in fact be very difficult to demonstrate (Liehu *et al.* 1986).

Successive small lakes in a catchment system act as ecological filters, retaining much of the nutrient and solid load input from intensive forestry measures. Thus, water quality in the lower reaches of a forested lake-rich drainage system may remain fairly good, and in low-lying larger lakes it is often not possible to discern the effects of forestry practices from other causes of diffuse nutrient loading.

Modern intensive forestry has expanded during the past decades even to the sparsely populated wilderness areas of eastern and northern Finland. This has led to general degradation of water quality in headwater forest lakes and watercourses far away from human settlements. Such lakes are often polyhumic and naturally oligotrophic. Algal blooms, winter anoxia and fish kills have become commonplace phenomena in these former wilderness lakes.

For most of the small headwater lakes water quality observation data is very limited. Therefore palaeolimnology and especially the study of sedimentary diatoms is often the only means of tracing past changes in these environments.

## References


Ahti, E. (1983). Fertilizer-induced leaching of phosphorus and potassium from peatlands drained for forestry. *Communicationes Instituti Forestalis Fennici*, **111**, 1–20

Ahtiainen, M. (1992). The effects of forest clear-cutting and scarification on the water quality of small brooks. *Hydrobiologia*, **243/244**, 465–473.

Hill, M. O. & Gauch, H. G. (1980). Detrended correspondence analysis: an improved ordination technique. *Vegetatio*, **42**, 47–58.

Holopainen, A-L. & Huttunen, P. (1992). Effects of forest clear-cutting and soil disturbance on the biology of small forest brooks. *Hydrobiologia*, **243/244**, 457–464.

Holopainen, A-L., Huttunen, P. & Ahtiainen, M. (1989).Effects of forest practices on water quality and primary productivity in small forest brooks. *Verhandlungen der Internationalen Vereinigung für Limnologie*, **24**, 1760–1766.

Kenttämies, K. (1981). The effects on water quality of forest drainage and fertilization in peatlands. *Publications of the Water Research Institute (Finland)*, **43**, 24–31.

Kukkonen, M. (1994). Haarajärven sedimenttitutkimus: piilevät ja elohopea. *Pohjois-Karjalan Vesi- ja ympäristöpiirin Monisteita*, **16**, 1–20. (Diatoms and mercury in the sediment of Lake Haarajärvi. In Finnish.)

Lahtinen, J. (1992). Nurmeksen Petäisjärven ja Lieksan Suomunjärven stratigrafinen piileväanalyysi ojitusten ja lannoitusten vaikutusten arvioinnissa. Unpublished M.Sc. thesis, Univ. Joensuu, 45 pp. + App. (Stratigraphic studies of diatoms in evaluation of effects of drainage and fertilization on two lakes. In Finnish.)

Liehu, A., Sandman, O. & Simola, H. (1986). Effects of peatbog ditching in lakes: Problems in paleolimnological interpretation. *Hydrobiologia*, **143**, 417–424.

Rekolainen, S. (1989). Phosphorus and nitrogen transport from forest and agricultural areas in Finland. *Aqua Fennica*, **19**, 95–107.

Rönkkö, J. & Simola, H. (1986). Geological control upon the floral manifestation of eutrophication in two headwater lakes. *University of Joensuu, Publications of Karelian Institute*, **79**, 89–96.

Rönkkö, J., Simola, H. and Siira, J. (1988). Effects of forest management on the benthic diatoms of small forest brooks in East Finland. In: *Proceedings of 9th Internation Diatom Symposium, Bristol*, 1986 (F. E. Round, ed.), 102–106. Biopress, Bristol & Koeltz. Scientific Publishers, Koenigstein.

Sandman, O., Liehu, A. & Simola, H. (1990). Drainage ditch erosion history as recorded in the varved sediment of a small lake in East Finland. *Journal of Paleolimnology*, **3**, 161–169.

Simola, H. (1983). Limnological effects of peatland drainage and fertilization as reflected in the varved sediment of a deep lake. *Hydrobiologia*, **106**, 43–57.

Simola, H. (1992). Structural elements in varved lake sediments. In: INQUA workshop on laminated sediments, Lammi, Finland, June 1990 (M. Saarnisto & A. Kahra, eds). *Geological Survey of Finland, Special Paper*, **14**, 5–9.

Simola, H., Huttunen, P. & Meriläinen, J. (1986). Techniques for sediment freezing and treatment of frozen sediment samples. *University of Joensuu, Publications of Karelian Institute*, **79**, 99–107.

Simola, H., Huttunen, P., Uimonen-Simola, P., Selin, P. & Meriläinen, J. (1988). Effects of peatland forestry management and fuel peat mining in Lake Ilajanjärvi, east Finland – a paleolimnological study. Symposium on the Hydrology of Wetlands in Temperate and Cold Regions, Vol. 1. *Suomen Akatemian Julkaisuja*, **4/1988**, 285–290.

Tossavainen, T. (1991). Metsälannoituksen ja metsäojituksen vaikutukset eräiden järvien fosforikuormitukseen sekä puroveden laatuun ja ainehuuhtoutumiin Itä-Suomessa. *Vesi- ja ympäristöhallituksen Monistesarja,* **310,** 1–113. (Effects of forest drainage and fertilization on runoff and phosphorus loading of watercourses in East Finland. In Finnish.)

# Dominance of asexuality in diatom life cycles: evolutionary, ecological and taxonomic implications

Jonathan L. Richardson

*Department of Biology, Franklin and Marshall College, Lancaster, PA 17604–3003, U.S.A.*

## Abstract

Although sexuality has been recognized in most diatom genera (and intensively studied in some), asexual mitotic division is by far the dominant mode of reproduction in Bacillariophyta, as it is in most other one-celled eucaryotes (Protistans). Mutation, not sexual recombination, is almost surely the chief generator of variety within the populations of a diatom species. Indeed, a diatom "population" typically is a collection of clonal lines that for the most part reproduce independently of other clones. New clones may be generated through mutation at any time within the asexual portion of the life cycle, and if a new mutation is dominant and favourable the strain that carries it can multiply rapidly. In a diatom population the relative abundances of both novel mutational clones and older clones originating from auxospores doubtless fluctuate over time as environmental conditions vary. The relatively infrequent episodes of sexual reproduction generate additional clones by genetic recombination, supplementing those generated by mutation. The existence of many clonal varieties, generated both asexually and sexually, provides a diatom population, and the species to which it belongs, with considerable adaptive capability. However, evidence exists that the diversification of higher taxonomic categories within the Bacillariophyta has been relatively slow. Since clones are largely isolated from each other genetically, the population and biological species concepts developed for multicellular organisms may be inappropriate for diatoms. However, the concept of ecotypes that exist *within* species populations clearly is valid and useful when applied to diatoms.

## Introduction

Biologists are amply justified in considering the Bacillariophyta to be an ecologically important, evolutionary enigmatic, highly distinctive, and in many ways highly advanced division within the Protista. Diatoms presently contribute a remarkable 20–25% of earth's total photosynthetic production (Werner 1977). They are the major control of dissolved silica levels in many lakes, and quite possibly in the whole world ocean (Harriss 1967; Gregor 1968). Diatoms appeared relatively late in evolutionary history and have undergone their maximum diversification in the Cenozoic era, like several other dominant components of earth's present biota – e.g. angiosperms, mammals, birds. Exactly when diatoms first evolved remains a matter of debate; it may have been well before the Jurassic fossils that are the oldest diatoms currently known (Philippe *et al.* 1994). The bivalved siliceous skeleton is unique to diatoms, as is the locomotory mode of raphid forms. Also, as is not the case in most other Protistans, the DNA of diatom cells is gathered into "chromosomes as obvious and clear as those in any animal" (Margulis & Sagan 1986), the segregation of which into daughter cells is highly regularized during cell division. Likewise reminiscent of the animal kingdom (but unusual among Protistans) is the thoroughly diploid nature of the diatom life cycle: the only haploid cells in the cycle are the gametes, which must fuse before cell division can resume.

Yet it is the *Protistan* characteristics of diatoms that must be stresssed when one reflects on the nature and adaptability of diatom species and the structure of diatom populations. In the multicellular kingdoms the cells of an individual organism are differentiated into two basic types: (1) somatic cells that divide and build up the tissues of that organism while ordinarily not giving rise to any new organisms, and (2) generative cells, the specialized function of which is the production of new individuals. In unicellular Protistans such as diatoms no comparable distinction exists between nonreproductive somatic cells and reproductive generative cells; instead, every cell, whether it divides mitotically or meiotically, produces a new individual. This capability, together with small size and rapid generation time, allows populations of Protistans to grow far more rapidly under favourable conditions than can the populations of any multicellular organism. Moreover, since asexual, mitotic production of offspring occurs far more frequently than does gamete production and fusion (even in those Protistan groups where sexuality is well known), populations of Protistan species consist largely of a multiplicity of clonal lines that typically remain reproductively isolated from each other for many generations. This papers considers some of the implications for diatom evolution and systematics of these very basic aspects of Protistan biology.

## Sexuality and asexuality in diatom life cycles

Sexuality is widely considered to be integral to diatom life cycles, and the formation of enlarged auxospores after gamete fusion is believed to be the normal mechanism by which cell size is restored after the progressive size reduction that accompanies asexual reproduction in the clonal population initiated by germination of an auxospore. Size reduction occurs in accordance with the well-known Macdonald-Pfitzer Rule (cf. Round *et al.* 1990; Tappan 1980), due to the fact that in each mitotic division the smaller valve (hypotheca) of the parent cell becomes the larger enclosing valve (epitheca) of one of the two daughter cells. When cells in the multiplying clone reach a critically small size, they can be induced by environmental triggers to differentiate sexually and produce gametes (Drebes 1977).

In fact, however, not all diatom species exhibit appreciable size reduction in their clonally dividing lines. This and the fact that sexual reproduction has never been seen in many taxa (Maberly *et al.* 1994; Round *et al.* 1990; Mann 1988) raises at least the possibility that sexuality has been lost from the life cycles of many species. Mann (1988) suggests, however, that a more common reason why sexual stages have gone unobserved in many species is not the loss of sexuality, but simply its intermittent nature and short duration compared to the asexual phase of the diatom cycle.

A diatom life cycle can be defined as the sequence comprising (1) germination of an auxospore to begin a clone; (2) the many subsequent asexual divisions that build the clonal population, ending (3) with auxosporulation of a major proportion of the surviving clonal population (*viz.* those members of the clone that lie within a critical size range required for induction of sexuality). Summarizing the evidence from the few species for which adequate population data exist, Mann (1988) arrived at diatom life cycle lengths as short as 1–2 years (*Cocconeis scutellum* Ehr.; *Stephanodiscus* sp.) and as long as 20 or even 40 years (*Asterionella gracillima* (Hantzsch) Heiberg; *Aulacoseira islandica* O. Müll. v. *helvetica* O. Müll.). Jewson (1992a) believes that the evidence for life cycles longer than 8 years is not convincing, but even the most conservative published estimates of life cycle length imply a great numerical dominance of asexual over sexual reproduction. Except in species that undergo periodic dormancy, mitotic (clonal) multiplication can take place during virtually the whole length of the asexual phase, at rates averaging one division or more per cell per day in optimum times (Tappan 1980). If appreciable sexual reproduction punctuates a process of clonal reproduction as infrequently as Mann's (1988) or Jewson's (1992a, b) calculations indicate, it must follow that natural populations even of diatom species having "short" life cycles are composed overwhelmingly of individuals generated by asexual, clonal reproduction; very few individuals in a population ordinarily are the direct offspring of a mating event (see Table I for sample calculations). It further can be concluded that species populations of diatoms typically consist of a multiplicity of clonal lines leading largely separate existences in a genetic sense, though certainly not in an ecological sense. Thus, asexuality is very much the dominant mode of life in diatom populations, as it seems to be in Protistans generally.

**How do diatom populations adapt?**

To a plant or animal biologist, sexuality in diatoms might appear to be residual or vestigial. Taught to believe that the evolution of a population depends heavily on sexual recombination to generate the individual genetic differences upon which natural selection acts, such a biologist might question how (or whether) diatoms can adaptively evolve. However, it seems very evident that evolution *is* possible even in completely asexual Protistans; this is the conclusion one would draw from the diversity existing within several Protistan divisions in which sexuality is unknown, e.g. the Euglenophyta (Leedale 1971).

Table I. Asexual reproductive potential within diatom life cycles of one to four years' duration.

The calculations below depict how many cells could be produced by clonal division from a single auxospore during the average time interval elapsing before its descendants enter another phase of sexual reproduction. In nature, cell mortality due to grazing, infection or senescence certainly would occur during the phase of clonal reproduction; thus there would actually be fewer clonal descendants than these calculations, based purely on reproductive potential, suggest.

---

1. **Calculation A:** By mitotic division at a rate averaging one division every ten days, a germinating auxospore and its clonal offspring could produce:
   $2^{35}$ cells (= 34,360,000,000 cells) in a one-year life cycle
   or $2^{71}$ cells (= $2.36 \times 10^{21}$ cells) in a two-year life cycle.

2. **Calculation B:** By mitotic division at rates averaging seven divisions per year – the rates calculated by Jewson (1992a, b) for two planktonic species experiencing seasonal light and nutrient limitation – a germinating auxospore and its clonal offspring could produce $2^{21}$ to $2^{28}$ cells in the 3–4 year life cycles of these species (*Stephanodiscus neoastraea* Håkansson & Hickel and *Aulacoseira subarctica* Howarth) in Lough Neagh, Northern Ireland.

3. **Parameter notes:**
   **(a)** Up to 8 divisions/day characterize some diatom species at some times (Tappan 1980); division rates closer to what might be expected as annual averages are used in the sample calculations.
   **(b)** Known diatom life cycles range in length from 1–2 years to 7–8 or possibly even 20–40 years (Jewson 1992a; Mann 1988). Life cycles longer than those used in the calculations theoretically would produce exponentially greater numbers of clonal offspring than the numbers shown above.

---

In diatoms, as in the Euglenophyta, mutation doubtless is the primary vehicle of adaption and diversification. The short generation time of diatoms and the faithful genetic copying that occurs during clonal reproduction tend to ensure that a non-detrimental mutation occurring in one cell during the asexual phase of the life cycle is preserved and transmitted to what often may be millions of clonal descendants. If recessive, the mutation normally will not be expressed, given the diploid nature of diatoms. But if the mutation is dominant, the new mutant line may be morphologically or physiologically different from other strains within the clone. Given the probable frequency of mutation – perhaps once every 50,000 to 1 million divisions at a particular locus (Keeton 1972) – a number of faithfully replicating mutant strains may appear within a clone during a single turn of the life cycle.

Thus we may envision a diatom species population as a collection of true-breeding mutant strains, the reproductive success of which will individually wax and wane as the conditions each strain finds most conducive to growth vary over time. Because *some* members of a strain will not reproduce sexually at times when many other members are doing so, a mutant strain can persist in pure form through the periodic sexual events that may join certain of its members in new genetic combinations involving other strains. Quite possibly this mechanism is largely or wholly responsible for generating the many physiologically and genetically distinct strains that now are being recognized through DNA fingerprinting and culture experiments in species such as *Skeletonema costatum* (Greville) Cleve (Gallagher 1982) and *Fragilara capucina* Dezmazières (Hoagland *et al.* 1993). Likewise, the morphologically distinguishable but highly similar "species flocks" that seem to exist within general such as *Nitzschia* Hassall (particularly in the tropics; *cf.* the studies of Hustedt 1938, 1949) could have originated in this way. Whether these similar taxa should be designated varieties or full species is largely a matter of choice, so long as one realizes how they may have originated and how they probably maintain their identities (cf. Wood & Leatham 1992).

## Further adaptive opportunities through genetic recombination

Despite its infrequency, genetic recombination via sex does not necessarily play an insignificant role in diatom adaptation and evolution. Table II lists several apparent "costs" of sexuality that may have selected negatively against its prominence in the diatom life cycle, but also lists possible benefits that *would* select for retention of sexuality within the cycle. As Lewis (1984) has pointed out, the length of the life cycle of each diatom species presumably represents an evolutionary balance between costs and benefits such as these.

It must be realized that the "life cycle length" of a given diatom species, while it may quite closely characterize all clones of that species, in all cases represents *averages* for those clones. In any clone, the Macdonald–Pfitzer Rule of size reduction ensures that the cycle actually is much longer for a small fraction of the clone (those cell lines that by chance rather consistently receive the epitheca of their parent cells and hence undergo unusually slow size reduction), and considerably shorter for another small fraction within the clone (those that by chance rather consistently receive the hypotheca of their parent cells). Thus the "observed life cycle length" reported by Mann (1988) and others is the statistically most probable time interval that must elapse before a large "middle fraction" of the clone reaches the size range in which sexuality can be induced in many individuals simultaneously. This, the recognized terminus of the clone's life cycle, is the time at which mating and auxospores are most likely to be observed. It is also most likely that the matings that occur at the time will be between clonal siblings. Table II makes clear that benefits derive from such matings, even when the parents are genetically identical.

Table II. Costs and benefits (*cf.* Smith 1978) of sexual reproduction by diatoms.

**Costs:**

1. Sexual parents have lower fitness (defined as an individual's genetic contribution to its offspring) than do asexual parents.
2. Recombination produces some offspring having reduced fitness.
3. "Cost of producing males" (occurs in centric species only: male parents do not bear offspring, only produce gametes)
4. Long interruption of synthesis and growth during meiotic division – a much slower process than mitosis (Lewis 1983).

**Benefits:**

1. New genotypes are produced via genetic recombination (raw material for evolutionary adaptation).
2. Deleterious alleles can be purged. (Even if sibling cells of the same mutant clone are the sexual parents, at least some of their offspring will not inherit the mutation).
3. When siblings of the same clone mate, advantageous but recessive mutations acquired during clonal reproduction will (for the first time) be expressed in at least some offspring. As these "double-recessive" offspring produce new clones of their own, the adaptive value of the recessive mutation may be widely realized in the population.
4. Heterozygosity within the population is increased. (This increase may be minimal except in cases where mating is between members of different clones).
5. Sexual reproduction restores maximal cell size (after repeated asexual division has reduced average cell size within a clone).

Clone members that are sexually out of phase with the majority – i.e. that reach sexually-inducible size much earlier or much later than most of their siblings – are less likely to find a mate among clonal siblings, and more likely (if able to find a mate at all) to pair with a sexually-induced member of some other clone within the population. These cross-clonal matings may be expected to be rare; but when successful, such matings will more radically reorganize the genome than do sibling matings or mutations that occur within a clone during asexual reproduction. In fact, a very few "hybrid matings" have been observed between morphologically distinct varieties within a species (Geitler 1979; Round *et al.* 1990). These instances possibly represent the relatively rare matings that would be predicted between members of different clonal lines. That no inter-species matings have ever been observed suggests that genetic incompatibility barriers might prevent them. Alternatively, interspecies matings may occur, but so infrequently that one would not expect to observe them. Interclonal matings, together with possible interspecies matings, may be important vehicles for the initiation of new lines that are sufficiently radical morphologically to be recognized by systematists as separate species.

## Taxonomic implications

One normally considers a diatom population to be all members of a given species that coexist in a given local habitat. But this collection of individuals does not easily fit the definition of "population" used by most evolutionary biologists. Students of evolution and ecology typically restrict the term "population" to a group of individuals of the same species that can and do interbreed freely. If the reality of a diatom population is that it consists of clones that are largely isolated from each other reproductively, it becomes a matter for argument whether each clone – or even each mutational strain within a clone – is the equivalent of the evolutionist's "subspecies" or "ecotype": a group within a species that is physiologically or morphologically distinct and has relatively little gene exchange with other such groups. In a diatom population, individual clones as well as true-copying mutational strains within these clones do seem to fit the above definition of ecotypes, even though the geographic isolation that usually characterizes multicellular ecotypes often is lacking in the case of diatoms. As earlier suggested, the morphologically distinct "forms" and "varieties" that have been described within recognized diatom species in many cases may be highly successful mutational ecotypes that have arisen during asexual division of a clone, and may be perpetuated asexually for the most part. It is valuable to recognize and name such true-breeding forms since they presumably are often ecologically different, even though – depending on their history – they may be very similar genetically.

## Conclusion

It is widely acknowledged that mutation alone probably can produce sufficient variation in populations of bacteria and other procaryotes to endow them with considerable adaptive and evolutionary potential (Campbell 1993). It does not seem extreme to extend this idea to Protistan eucaryotes having short generation times and life cycles that are largely asexual. The extreme diploidy of diatom life cycles may cause one to wonder whether mutation is an adequate source of variation in this group. Nevertheless, it is premature to accept without question the suggestion of Round *et al.* (1990) that those diatom species (probably relatively few) that have completely abandoned sexuality in their life cycles and become automictic or parthenogenetic "have sacrificed their evolutionary prospects".

In any case, the retention of occasional sexuality within the dominantly asexual life cycle of most diatoms seems to be a very successful life history strategy that preserves successful forms in clonal purity while also providing oppportunities for evolutionary innovation. Lewis (1984) has pointed out that in coupling asexual size reduction during clonal division to the induction of sexuality, diatoms have evolved a unique mechanism by which selection can optimally adjust and regulate the interval between sexual phases. If Philippe *et al.* (1994) are correct in calculating the rate of diatom diversification as somewhat less than that within the chordates, we must regard the life cycle "strategy" of diatoms as evolutionarily quite conservative, but extremely successful.


## Acknowledgements

The author thanks the reviewers of the manuscript for many helpful comments.


## References


Campbell, B. (1993). *Biology*. 3rd ed. xxx + 1190 pp. Benjamin/Cummings Publishing Company, Redwood City, CA.

Drebes, G. (1977). Sexuality. In: *The Biology of Diatoms* (D. Werner, ed.), 250–283. University of California Press, Berkeley & Los Angeles.

Gallagher, J. (1982). Physiological variation and electrophoretic banding patterns of genetically different seasonal populations of *Skeletonema costatum* (Bacillariophyceae). *Journal of Phycology*, **18**, 148–162.

Geitler, L. (1979). On some pecularities in the life history of pennate diatoms hitherto overlooked. *American Journal of Botany*, **66**, 91–97.

Gregor, B. (1968). Silica balance of the ocean. *Nature*, **219**, 275–276.

Harriss, R. (1966). Biologic buffering of oceanic silica. *Nature*, **212**, 360–361.

Hoagland, K., Ernst, S. & DeNicola, D. (1993). Physiological and genetic variation in *Fragilaria capucina* clones along a latitudinal gradient. In: *Abstracts of 12th North American Diatom Symposium* (L. Goldsborough, ed.), 17–18. University of Manitoba, Winnipeg.

Hustedt, F. (1938). Systematische und ökologische Untersuchungen über die Diatomeen-Flora von Java, Bali und Sumatra. 1. Systematischer Teil. *Archiv für Hydrobiologie, Supplement-Band*, **15**, 131–177, 187–295, 393–506.

Hustedt, F. (1949). *Süsswasser-Diatomeen.* Exploration du Parc National Albert, Mission H. Damas (1935–1936), Fascicule **8**. 199 pp. Institut des Parcs Nationaux du Congo Belge, Bruxelles.

Jewson, D. (1992a). Life cycle of *Stephanodiscus* sp. (Bacillariophyta). *Journal of Phycology*, **28**, 856–866.

Jewson, D. (1992b). Size reduction, reproductive strategy and the life cycle of a centric diatom. *Philosophical Transactions of the Royal Society of London* **B, 336**, 191–213.

Keeton, W. (1972). *Biological Science*. 2nd ed. xiv + 888 pp. W. W. Norton & Company, New York.

Leedale, G. (1971). *The Euglenoids*. 16 pp. Oxford Biology Reader, **5**. Oxford University Press, London.

Lewis, W. Jr. (1983). Interruption of synthesis as a cost of sex in small organisms. *The American Naturalist*, **121**, 825–834.

Lewis, W. Jr. (1984). The diatom sex clock and its evolutionary significance. *The American Naturalist*, **123**, 73–80.

Maberly, S., Hurley, M., Butterwick, C., Corry, J., Heaney, S., Irish, A., Jaworski, G., Lund, J., Reynolds, C. & Roscoe, J. (1994). The rise and fall *of Asterionella formosa* in the South Basin of Windermere: analysis of a 45–year series of data. *Freshwater Biology*, **31**, 19–34.

Mann, D. G. (1988). Why didn't Lund see sex in *Asterionella*? A discussion of the diatom life cycle in nature. In: *Algae and the Aquatic Environment* (F. E. Round, ed.), 384–412. Biopress Limited, Bristol.

Margulis, L. & Sagan, D. (1966). *Origins of Sex.* xi + 258 pp. Yale University Press, New Haven, CT.

Philippe, H., Sorhannus, U., Baroin, A., Perasso, R., Gasse, F. & Adoutte, A. (1994). Comparison of molecular and paleontological data in diatoms suggests a major gap in the fossil record. *Journal of Evolutionary Biology*, **7**, 247–265.

Round, F. E., Crawford, R. M. & Mann, D. G. (1990). *The Diatoms. Biology and morphology of the genera.* 747 pp. Cambridge University Press, Cambridge.

Smith, J. (1978). *The Evolution of Sex.* x + 222 pp. Cambridge University Press, Cambridge.

Tappan, H. (1980). *The Paleobiology of Plant Protists.* xxi + 1028 pp. W. H. Freeman & Company, San Francisco.

Werner, D. (1977). Introduction with a note on taxonomy. In: *The Biology of Diatoms* (D. Werner, ed.), 11–17. University of California Press, Berkeley.

Wood, A. & Leatham, T. (1992). The species concept in phytoplankton ecology. *Journal of Phycology*, **28**, 723–729.

# Sexual reproduction in *Coscinodiscus granii* Gough in culture: a preliminary report *

Anna-Maria M. Schmid

*Univ.-Salzburg, Inst. f. Pflanzenphysiologie,*
*A–5020 Salzburg, Hellbrunnerstr. 34, Austria*

*** Dedicated to the late Prof. H. A. v. Stosch on the Occasion of his 85th Birthday**

## Abstract

Sexual reproduction was elicited in the heteropolar *Coscinodiscus granii* Gough by increase of the salinity of the medium and subsequently enhanced by daily alternation of the salinity and by increasing the light intensity. The clone proved to be monoecious (=homothallic), generating both female and male gametangia. With respect to the valves, *C. granii* exhibits a sex dimorphism. Sex-determination appeared to be phenotypic and is suspected to be temporally correlated with sex-induction. Formation of the male gametangium is coupled to a hitherto undescribed cell tetrad. At the two cell stage the two inside valves are structurally connected close to the narrow pole (6 o'clock). From this tetrad, two cells develop into spermatogonangia, each of which produces about 128 uniflagellated sperms. There is also a pattern anomaly here at 6 o'clock, providing the first opportunity among the diatoms to recognise a valve of a male gametangium in a population. The other two cells undergo at least one further vegetative division. This type of spermatogonangium formation delays sperm release so that it is spread over time thus increasing the chance of fertilising a mature oogonium. Isolated tetrads lead to a purely male subclone, indicating that male determination appears fixed prior to tetrad formation. Oogonia can be distinguished from vegetative cells by their longer pervalvar axis and the centrally located nucleus. They develop directly from a vegetative cell and never from tetrads. Sperms enter at the broad pole (12 o'clock) through a fertilisation pore, attaching with their flagellar tip. Auxospores possess a scale case with two different scale types: normal ones and scales with a slit, may be rudiments of the rimoportulae. The diameter of initial cells varies between 220–400 µm, independent from the size of the parent cells. Ten (10) months after the first induction of sexual stages the clone is still fully inducible.

## Introduction

The specific mode of cell division in diatoms usually leads to a progressive reduction in the average size of the cell walls, and thus the cells, in a population. The expression of this phenomenon is known as the McDonald-Pfitzer rule ( refs. in Round *et al.* 1990). Restoration of the species-specific maximum size is commonly brought about through sexual reproduction and the expansion of the zygote into the auxospore, which ultimately leads to the vegetative development of a new clone (refs in Drebes 1977b; Pickett-Heaps *et al.* 1990; Round *et al.* 1990). Since the discovery that diatoms not only multiply by binary fission, but also reproduce (at intervals) sexually (refs in Geitler 1932, 1952; v. Stosch 1951, 1954; Drebes 1977b), researchers who were able to grow diatoms in the laboratory have tried to induce formation of gametes and auxospores. A variety of environmental cues seemed able to elicit or enhance formation of sexual stages, but only when the cells were of a species-specific permissive size class (refs in Dring 1974; Drebes 1977b; and see below). Lewis (1984) suggested that in nature, environmental scenarios ocurring once a year, would be perfect "clocking" factors for organisms with intermittent sex, when sensitive stages of the organism respond to such factors. In diatoms it appears that only cells within a certain "size window" are inducible. This relation between cell size and sensitivity to external factors may be based on the cell's altered physiology and biochemistry due to physical constraints caused by the continuous decrease in cell size (Werner 1971a, b, c).

The factors active in sexualisation, often in combination, are sudden changes in light (daylength and intensity: Drebes 1964, 1966; v. Stosch & Drebes 1964; v. Stosch *et al.* 1973; Armbrust & Chisholm 1992), temperature (v. Stosch & Drebes 1964; Drebes 1966; Holmes 1966, 1967; Rozumek 1968; Werner 1971c), nutrients, specific ions in and salinity of the medium (Geitler 1932; Steele 1965; Holmes 1966; Drebes 1964, 1966; Schultz & Trainor 1968, 1970; Werner 1971c; Hoops & Floyd 1979; Roemer & Rosowski 1980; French & Hargraves 1985; Idei & Chihara 1992; Jewson 1992b; Perez-Martinez *et al.* 1992; refs. in Drebes 1977b; refs in Schmid 1979, 1984, 1990). For some species, an increase in the intensity of the factor has been found to be effective, and for others a decrease.

*Coscinodiscus granii*, the object of this study, is a marine centric diatom of worldwide distribution, can be easily kept under laboratory conditions and is therefore well studied. The characteristic heteropolar form of the cell wall and its ultrastructure has been investigated by Brooks (1975), and cell division was cinematographically documented by Drebes (1975). Drebes (1974) also pictured several stages of sexual reproduction, i.e. of spermiogenesis and initial cells. Sexual stages occur around Helgoland from July to September. Drebes' investigations (1974) revealed that natural populations of *C. granii* consist of 2 subpopulations which are capable of cross-fertilisation: the "major" form (100–400 µm) and the "minor" form (40–200 µm). The clone used in this study belongs to the "major" form. The present paper is a preliminary report of a study into sexual reproduction of this species, carried out by a combination of light- and immuno-fluorescence microscopy, and scanning and transmission EM of intact cells and cleaned cell walls. A detailed series of communications is in preparation (Schmid 1994b, and in prep).

## Material and Methods

A culture of the heteropolar *Coscinodiscus granii* (isolated from the North Sea and provided by Dr R. M. Crawford in March 1993) was subcloned and grown in a Rumed incubator for 15 months in 100 ml Erlenmeyer flasks and artificial seawater medium of 900 mOsmol salinity (for details of the medium see Franz & Schmid 1994), at 18 ± 1°C, and 12:12 h LD regime, exposed to 10 $\mu E.m^{-2}sec^{-1}$ cool white light.

A re-isolate of *C. granii* became sexualized by increasing the salinity to 950 mOsmol (same ion-balance as the 900 mOsmol medium). At first only spermatogonangia occurred, and only in a low percentage. The harvest was increased by means of a daily transfer to fresh medium with salinities of 900 and 950 mOsmol, which alternated at each transfer, and by increasing the light intensity to 50 $\mu E.m^{-2}$ $sec^{-1}$, and resulted in the development of about 20% spermatogonangia. Photoperiod, light quality and temperature were kept constant. This treatment also triggered the formation of oogonia. The daily harvest of a 60 ml culture was ca 600 spermatogonangia and 20–25 oogonia for the duration of 35 days. Comparable results were achieved by application of this method to *C. wailesii, Odontella regia,* and *O. sinensis* (in prep.). Sudden increase in light intensity alone, or coupled to an increase in temperature (25°C) proved to be unsuccessful for all of the species investigated.

Both oogonia and mature spermatogonangia have been isolated and incubated together in a 3 cm petri-dish to enhance the possibility of fertilisation. No medium change was performed at this stage, i.e., some of the auxospores developed at 900, and others at 950 mOsmol salinity. Zygotes were transferred to a small 10 ml vial which was set into a rotor with 8 rpm, to avoid settling and thus damage, and illumination was reduced to normal. After the first division of the inital cell, all cells were transferred to 900 mOsmol-medium. The experiment was stopped after 35 days, because of exhaustion not of the culture, but the observer, since an internal clock (perhaps induced by the light cycle) caused sperm maturation of the majority of the cells between 0100 and 0500 hours. Both pre- and post auxospore cells are still in culture.

To follow the development of tetrads, i.e., the prestages of spermatogonangia, they have been isolated several times (either as single tetrads/triplets in 3 cm Petri-dishes, or as a population of 30 tetrads in a 6 cm Petri-dish) and grown under the same conditions as the other cultures.

Light microscope observation was by means of a Reichert Zetopan with Nomarski optics and a Leitz Labovert. Photographs were taken from living cells, or SDS-cleaned and toluidine-blue stained cell walls. EM preparation will be described elsewhere.

**Note added at revision:** To test whether auxospores could still be induced in the population conditions, triggering sexual stages were applied to the pre-auxospore cells at least once every second month, with positive results even 10 months after the first induction (end of May 1994 and March 1995 respectively).

## Results

*The vegetative cells*

In girdle view the cells are asymmetric with a broad and a narrow pole respectively. Nuclei are located in the cortical cytoplasm at the centers of the epivalves and cytoplasmic strands traverse the vacuole towards the hypovalves. In preprophase in each cell, the nucleus migrates to the broad pole for mitosis (Figs 1, 3) and migrates during cleavage to the centre of the new hypovalve, and then back into its interphase position (Drebes 1975). The mitotic position of the nucleus of *C. granii* is identical with that of *C. wailesii* (Schmid & Volcani 1983). In the latter species, this position was marked as 12 o'clock, when a clock face is placed over the valve face, and the two macrorimoportulae (=mLP) within the marginal ring of LPs are set to 4 and 8 o'clock (Schmid 1994a). Also in *C. granii* (Fig. 5), *C. asteromphalus, C. centralis* (AMS, unpubl.) and probably others, mitosis is at 12 o'clock, with respect to the mLPs. During interphase the two sibling cells remain enclosed by the parent cingulum, put separate in preprophase, when the last hypothecal band is formed.

*Cell tetrads and formation of spermatogonangia (=gametangia)*: 130–170 µm

In *C. granii, C. wailesii* and *O. regia*, a modified vegetative cell quickly divides twice to produce a chain of four cells (Figs 2, 5–12). Commonly two of these remain vegetative during the next division, and two become spermatogonangia. During the first division establishing this tetrad of *C. granii,* a tubular silicified link between the sibling hypovalves is formed, as a rule, at the 6 o'clock position (Figs 6, 8). Alternatively it is found nearer to the valve centre, and is the result of an incomplete cleavage. In SEM (not illustrated) a clear disturbance in the valve pattern can be seen at this site. During the formation of the tetrad stages several cells have been observed – unfortunately not before they were fixed – in which the nucleus was at 6 o'clock instead of 12 prior to division. Whether these were the parent cells of tetrads is not yet clear. The connected cells attract attention because in the normal situation cells become separated long before such an increase of the pervalvar axis (Figs 2, 5–7). The second division leading to the tetrad appears normal in LM, with the nuclei in the centre of the epivalves (Figs 9–10). When the outermost cells of the tetrad become detached, a pair of cells remain with their nuclei at the internal sibling (now epi-) valves. This gives the initial impression of an incorrect nuclear position (i.e. in the hypovalves).

The spermatogonangial formation is variable: a) the outermost cells may be differentiated into gametangia and the two cells on either side of the intercellular bridge remain dividing, forming new tetrads, or b) the outer cells and one of the connected inner cells are vegetative, and only one of the connected cells becomes a spermatogonangium, or c) two cells on one side of the intercellular bridge remain vegetative and the two on the other side differentiate into gametangia. A single observation revealed four spermatogonangia derived simultaneously from a tetrad.

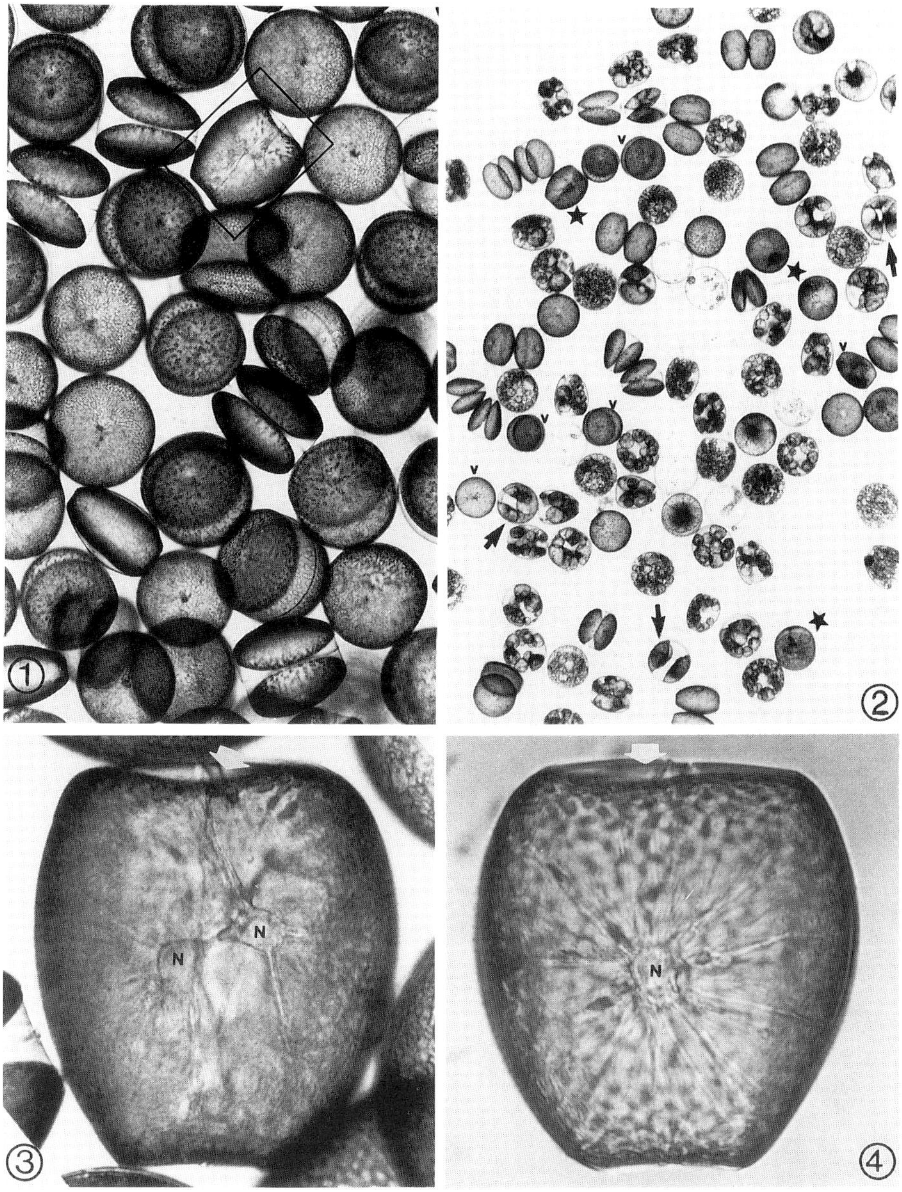


Figs 1–4. *C. granii*. Fig. 1. Vegetative cells in interphase; one (box) dividing. ×120. Fig. 2. A few vegetative cells (v) among sexual stages. Oogonia are marked by asterisks, all others are males from tetrad to sperm-release; arrows indicate cells performing second division of the spermatogonangium (6 to 12). ×50. Fig. 3. Detail of Fig. 1 dividing cell from 12 o'clock, telophase nuclei (due to an optical illusion the cell looks narrower at the lower part of the photograph). ×400. Fig. 4. Oogonium with long pervalvar axis, nucleus in the cell centre. ×400.

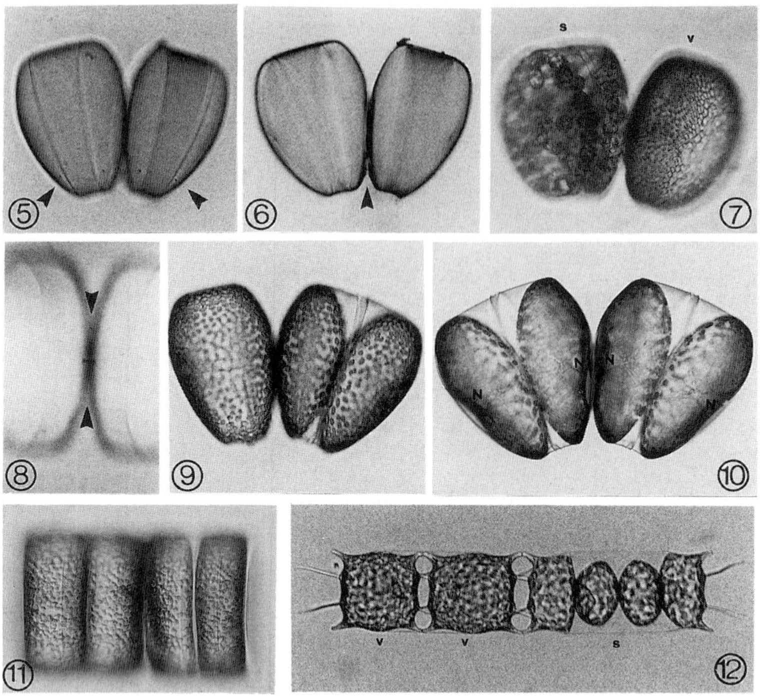


Figs 5–10. *C. granii*, tetrad formation. Fig. 5. SDS-cleaned cell wall of connected cells; 4 mLPs aligned between the arrows. ×200. Fig. 6. Same cell, focus on the silicified connection (arrow). Fig. 7. Connected cells, 1 vegetative (right), 1 spermatogonangium (left). ×200. Fig. 8. Connected valves from 6 o'clock: silified tube (arrows). ×250. Fig. 9. Left cell in prophase, right divided. ×200. Fig. 10. Tetrad; nuclei in the centre of epivalves. ×200. Fig. 11. *C. wailesii*, tetrad. ×170. Fig. 12. *O. regia*; chain of two vegetative cells (v) and one spermatogonangium (s) with two half-spermatogonangia each with 2 spermatogonia. ×250.

Isolation of tetrads (and triplets, e.g. Fig. 9) revealed that vegetative cells once derived from the cell chain can be converted into spermatogonangia after two or three vegetative divisions, so that finally all cells can transform into gametangia. The final number of gametangia deriving from a tetrad is dependent on a) the timing of gametangium formation and their initial number, i.e. how many "vegetative" cells remain; and b) the number of vegetative cycles of the remaining cells prior to their conversion into gametangia.

The 30 tetrads, separated into a petri-dish and grown under the same conditions as the remaining part of the clone, ceased sperm production after 14 days and the rate of vegetative divisions declined, while the cells in the other vessels were still actively producing both gametangia and auxospores. Many of the separated cells looked unhealthy and subsequently died. From the surviving tetrads, usually only one, usually peripheral, cell survived. During the 10 months of experimental induction, no oogonium was ever seen to differentiate from a tetrad. These results suggest that male determination takes place in the tetrad-parent cell. According to Drebes (1974) spermatogonangial size in nature varies from below 260 up to 300 μm.

*Spermatogonia, spermatocytes, sperms*

A series of successive "depauperizing mitoses and divisions" (a term coined by v. Stosch & Drebes 1964) divide the protoplast of the gametangium into 32 small diploid spermatogonia which remain enclosed by the parental cell wall. As in other centric diatoms, meiosis then occurs in the final stages of gametogenesis (refs. in Drebes 1977b).

The first division of the male gametangium resembles a vegetative division only insofar as the nucleus is at 12 o'clock for mitosis and the cleavage furrow cuts through the cell in the valvar plane to produce two spermatogonia. Their pervalvar axes are, however, shorter than in vegetative daughter cells, and to judge from the apparent decrease in pigment content, growth and division of chloroplasts ceases at this stage. The spermatogonia do not form rudimentary valves as described for *Stephanopyxis* (v. Stosch & Drebes 1964; Drebes 1964, 1966, 1969), *Chaetoceros* (v. Stosch *et al.* 1973) or *Odontella regia* and *O. sinensis* (Fig. 12, and in prep.; Drebes 1974, p. 86/87). It is not yet known whether they only lack silica deposition or whether they also fail to form an SDV.

The valves of the male gametangia also exhibit a disturbance at the 6 o'clock region with one LP either absent or reduced, or placed nearer to the centre (Fig. 13). This is perhaps, a prerequisite for the second division of the spermatogonia, which is not in the valvar plane as is the first one, but pervalvar from 6 to 12 o'clock (arrows in Figs 2, 14, 15). The protoplast contracts locally, but remains connected with fine cytoplasmic strands to the LPs of the valves (Fig. 14). Antibodies against α–tubulin have been used to demonstrate the presence of microtubules in these strands towards what are considered to be anchor points (in prep; and see discussion in Schmid 1987, 1994a). The third division is from the cell centre and oriented in the direction of 3- and 9 o'clock. The divisions then become irregular with respect to cell wall symmetry (Figs 2, 16–18), but the spermatogonia are still anchored to the LPs. During the "swelling phase", coupled to meiotic prophase I (Drebes 1974, 1977b), the spermatogonia take up water and the increased turgor causes a very balanced pushing force acting against the valves, causing the gametangial thecae to slide apart but not to open completely (cf. v. Stosch 1954; Drebes 1969, 1974; Manton *et al.* 1969; Pickett-Heaps 1995). The cytoplasm, with the chloroplasts arranged in a ring around its periphery, sits as a cap on one side of a large vacuole (Figs 16, 17; cf. also Pickett-Heaps 1995). The vacuoles are then extruded, together with most of the chloroplasts (Fig. 18). Meiosis I results in 64 biflagellated spermatocytes devoid of vacuoles and with incomplete flagella (not illustrated). Chloroplast extrusion is continued (Fig. 18). Following meiosis II, 128 uniflagellated sperms result (see also Drebes 1974). The latter stages are not always synchronous. The anterior flagellum helps to find the exitpore for the first sperms to reach maturity. The thecae usually split apart through sperm activity at meiosis II.

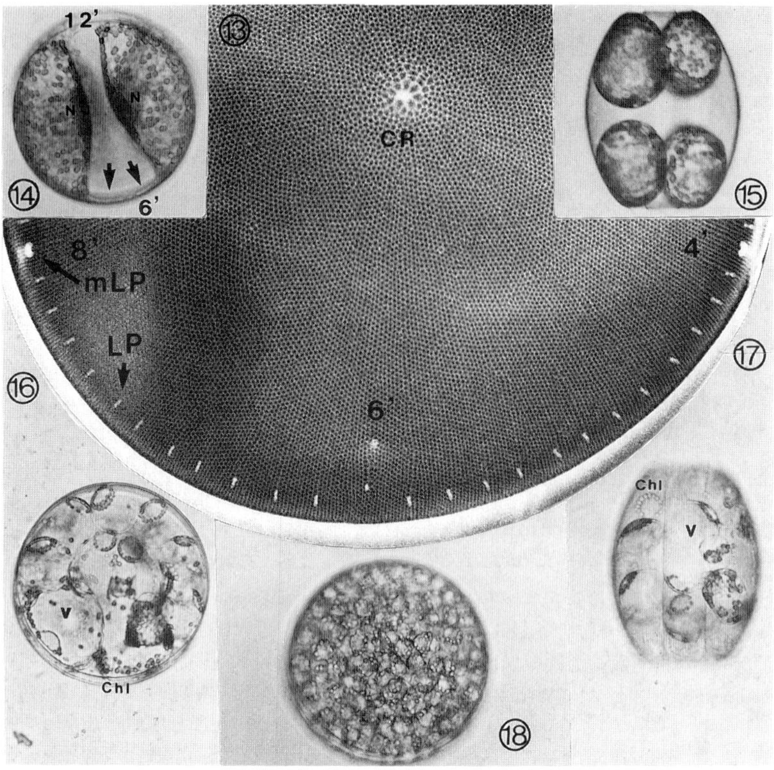


Figs 13–18. *C. granii*, spermatogenesis. Fig. 13. SEM, internal view of spermatogonangium valve demonstrating LP-anomaly at 6 o'clock; mLPs at 4 and 8 o'clock; CR = central rosette; LP = normal rimoportulae. ×950. Figs 14–15. Completed second division of spermatogonangium pervalvarly from 6 to 12 o'clock, resulting in 4 naked spermatogonia. ×270. Fig. 14. Valve view; note cytoplasmic connection (arrows) of spermatogonia to LPs in 6 and 12 o'clock. Fig. 15. Girdle view from 12 o'clock. Figs 16–17. Swelling phase of spermatogonia at onset of meiosis I. ×270. Fig. 16. Valve view; Fig. 17. Girdle view. Note cytoplasmic caps on top of vacuoles (v) and circular arrangement of chloroplasts (chl). Fig. 18. Valve view; spermatocytes after meiosis I. Vacuoles have disappeared, chloroplasts are extruded. ×270.

Sperm liberation occurs some 20 to 25 hours after the first spermatogonial division. As in *Attheya decora* (Drebes 1976a, b, 1977a), in *Odontella regia* (in prep.), and perhaps also in *Melosira moniliformis* (Idei & Chihara 1992), the sperm attaches to the oogonium or egg at first with the flagellar tip and then with the cell body (in prep.).

*Oogonia*: 130–170 µm

Under the conditions provided, the production of female cells as compared to males in our clone, is very low: ca 25 vs. ca 600 cells per day. Oogonia appear to

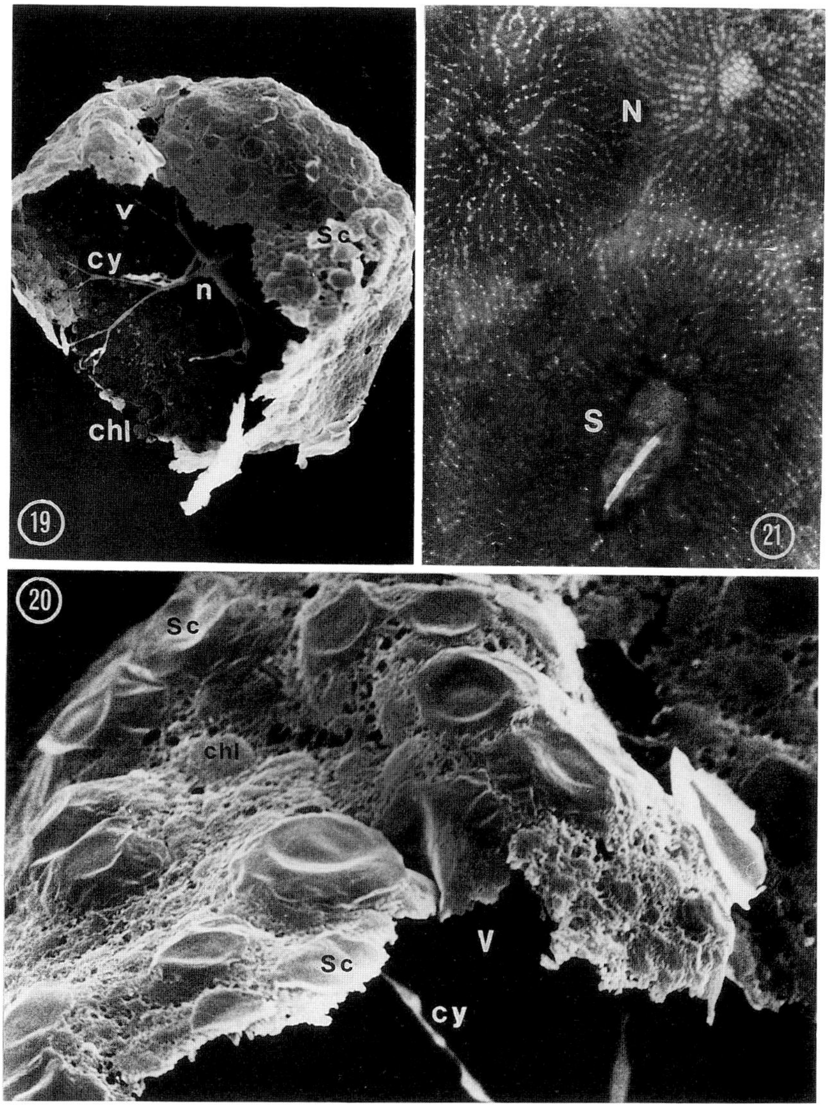


Figs 19–21. *C. granii*, auxospore formation. Figs 19, 20. SEM; Fig. 21 TEM. Fig. 19. Young auxospore mechanically liberated from oogon wall after CPD-preparation and therba cut ope. Huge vacuole (V); central nucleus (N) with radiating cytoplasmic strands (Cy); extremely thin cytoplasmic cortex covered with newly exocytosed scales in their organic coat (Sc); Chloroplasts (chl). ×600. Fig. 20. Detail from 19. Close up of first synthesiszed scales (Sc) in different sizes. Plasmamembrane torn during preparation; chloroplasts (chl); cytoplasmic strands (Cy). ×4,000. Fig. 21. SDS cleaned case of mature auxospore with scales *in situ*; note difference between normal and slit-scales. ×18,000.

develop directly from vegetative cells and have never been seen to develop from tetrads. They produce a single egg and function as auxospore mother cells (see also Drebes 1974). The pervalvar axis is longer than in a vegetative cell undergoing division (cf. Figs 3 and 4) and the nucleus is in the geometric cell centre held in position by numerous cytoplasmic strands. It moves here directly from the normal interphase position at the epivalve centre, and not through a detour, e.g. via 12 o'clock. Oogonial valves do not possess the exceptional LP anomalies at 6 o'clock found in tetrad and spermatogonangial valves, and therefore *C. granii* gives us the earliest opportunity to recognise sex-dimorphism in the valves. The sperm enters through a fertilisation pore in the girdle (not shown) at the broad pole (12 o'clock), to which the egg nucleus also migrates. Then, perhaps as a reaction to a signal from the sperm, all cytoplasmic strands disappear. This stage is impossible to fix for EM, without major ruptures of the cytoplasm. The resulting zygote contracts, disconnects from the anchors (LPs) in the oogonial wall, and eventually escapes from it by osmotic swelling (in prep.).

*Auxospores / Initial cells:* 220–400 µm

Isometric swelling is mediated by a scaly auxospore envelope (Figs 19–21) within which the initial cell for a new clone is laid down (48 h). The auxospore envelope, a single layer of overlapping scales in an organic matrix, is composed of two types of scales (Schmid 1994b): the normal ones, known from other species, e.g. *Melosira varians* (Reimann 1960), *M. varians & M. nummuloides* (Crawford 1974), *Ellerbeckia arenaria, Biddulphia sp.* (Round & Crawford 1981), *Thalassiosira* (Schmid 1984), and mixed among them, around 60–65 scales per auxospore which are larger and possess a clear slit within their annulus area (Fig. 21). At the early developmental stages, when the scale-layer is not yet complete (Figs 19, 20), the cell is easily distorted by even careful manipulation during preparation for electron microscopy. The nucleus in the auxospore is situated in the cell centre (Fig. 19) only during the early swelling phase, and thereafter located in the cortical cytoplasm. Several strands traverse the vacuole from the nucleus to other places in the cytoplasmic cortex (figs 19, 20), possibly to the position of the slit scales. Formation of the first initial valve is preceded by an acytokinetic mitosis and accompanied by a local contraction (termed a "localized plasmolysis" in previous studies, e.g. Hoops & Floyd 1979). The result of this process is pictured in Drebes (1974). Heteropolarity of the whole clone with respect to the valve, i.e. position of the mLPs at 4 and 8 o'clock (Schmid 1994a) is manifested at this stage, and heteropolarity with respect to the cell (i.e. broad pole at 12 and narrow pole at 6 o'clock) is created with the first division by means of asymmetric valves and girdle bands. Both initial valves of *C. granii* appear symmetric in gross-morphology. During the final events in initial cell formation the auxospore investment ruptures, probably at predetermined sites, and is shed.

During 10 months of sex-induction, initial cells of this clone varied considerably in size: 220–400 µm in diameter. This variation is independent of the size of the parents, and greater than indicated by Drebes (1974) for the natural situation (280–400 µm).

## Discussion

*Induction of gamete formation: laboratory results vs. field studies; environmental cues.*

Sexual reproduction in plants is commonly a response to changing or adverse environmental conditions (e.g. "Notblüte" or "vernalisation" of spermatophytes). Sexuality may be both the motor of evolution and of adaptation to changing environments; progressive radiation into new habitats. This may be especially relevant for diatoms, because of a rapid dispersal of newly created genotypes via mitotic cell multiplication (Alberts *et al.* 1990; Armbrust & Chisholm 1992). Provided the cells are of the correct size, sexual reproduction in diatoms can be induced in the laboratory only when one or more environmental factors are suddenly changed (see refs. above and in the Introduction). This also depends on the species-specific range of sensitivity to a given trigger: see also Armbrust & Chisholm (1992).

Besides resting spore formation, sexual reproduction appears as one of the versatile strategies available in the diatoms' repertoire, for survival as a population and for interspecific competition in changing habitats. Both strategies are obligately coupled in the unique diatom *Leptocylindrus danicus*, where the resting spore is directly formed within the auxospore (French & Hargraves 1985). A comparable behaviour was suggested for *Cyclotella ocellata* (Perez-Martinez *et al.* 1992). In *Chaetoceros diadema* nitrogen deprivation has been shown to induce not only resting spore formation, but simultaneously also sexuality, when cells are at the narrow end of species specific diameter scale (French & Hargraves 1985). Both strategies mean disruption of the mitotic cycle and cell multiplication; but while resting spore formation appears a fairly safe mechanism (Hargraves 1976) with low costs, sexual reproduction is a time and energy consuming process with the implicit danger of failure (e.g. Lewis 1984). When cells are triggered to undergo meiosis, they are not viable for a longer lifespan. So if sexual fusion fails, then the cells are lost (see also Jewson 1992a, b). *C. granii*, can easily cope with the loss of cells producing oogonia, since under our optimal conditions oogonia are less than 1% of the total population. The wasted cells are those producing sperms to ensure that sufficient gametes are produced to optimise encounter. In *Corethron* for instance, the majority of the population seems to be sacrificed for the sake of sperm production, as concluded from the high amount of empty thecae in the ocean (Crawford 1995).

In older literature (rev. by Drebes 1977b) osmotic changes have often been held responsible for sexualisation, but to these workers change in salinity was commonly synonymous for dilution of the medium. Transfer of old stationary cultures into fresh medium enhanced sexual reproduction in *Coscinodiscus asteromphalus* (Werner 1971c), *Stephanopyxis turris* (v. Stosch & Drebes 1964), *Cyclotella meneghiniana* (Hoops & Floyd 1979) and others (see refs in Introduction). Schultz & Trainor (1968, 1970) found that sex-induction of *Cyclotella* caused by increase in salinity is apparently correlated with an increase in the $Na^+$ concentration of the medium. This possibility may also account for *C. granii*, since the media of both osmolalities were balanced, i.e $Na^+$ concentration was also alternating each transfer.

Sperm liberation between 0100 h and 0500 h for the majority, but not all, cells of *C. granii* indicates, that the time course of spermatogenesis is correlated with an internal clock. Although detailed experiments as carried out in *Chaetoceros curvisetum* (Furnas 1985) have not been done, the observations during this study do not indicate such a strict dependency on the diel LD-rhythm as found in *Ch. curvisetum.* The influence of light intensity, quality (i.e. phytochrome system; blue-light effects), and photoperiod is reviewed by Dring (1974), who concludes that auxospore formation is normally controlled by the total light energy incident upon the cultures, rather than by photoperiodicity, as seems to be the case in *C. granii* too (see also v. Stosch *et al.* 1973). Furthermore, this would resemble sexualisation in *Thalassiosira weissflogii,* where spermatogenesis was induced in an apparently subdioecious clone by exposure to two days darkperiod followed by continuous light (Vaulot & Chisholm 1987). *Aulacoseira subarctica* on the other hand, responded in nature to the low light conditions during winter (Jewson 1992a).

The role of light in sex determination was studied by v. Stosch (1954), who found that continuous high light intensities resulted preferentially in females, dim light in a LD cycle resulted in sperms, and a LD cycle with strong light induced both sexes in a clone of *Lithodesmium* in the inducible size range. *C. granii* behaved similarly during this study, indicating that sex determination is phenotypic, but sudden increase in light intensity alone or even in combination with an increase in temperature, as successful for *St. turris* and *St. palmeriana* (v. Stosch & Drebes 1964; Drebes 1966) proved unsuccessful. The primer for sexualisation in *C. granii* (as well as *C. wailesii, Odontella regia* and *O. sinensis*) was clearly the increase in salinity (or possibly the $N^+$ content).

The conditions provided during this study lead to a relatively high percentage of ca 20% of spermatogonangia in the clone of *C. granii.* But meanwhile, the majority of cells multiply vegetatively, providing a pool of reserve cells, in case sexual reproduction is not successful. Since some cells may remain theoretically for years in the inducable size-class, due to the specific mode of division, the population would not die out, and sexual recombination could occur several times. Cushing (1955) observed auxospore formation in *O. (Biddulpia) sinensis* twice in quick succession. Comparable results were obtained in laboratory cultures of *T. weisflogii* (Armbrust & Chisholm 1992). Sexual capacity of our *C. granii* clone is retained even after 10 months.

The essential element in all of the laboratory observations in diatoms is that a sudden quantitative change of a factor leads within two or three days to sexualisation of a diatom clone or population apparently as a physiological response to a "stress situation". Factors inducing sex in *C. granii* are salt- and light-stress. Light-, temperature, and salt-stresses are known to interfere with the protein metabolism ("stress-proteins"), and all are coupled to water-stress (e.g. Kirst 1990; Borkird 1991; Burton 1991; Hagemann *et al.* 1991; Herskovitz *et al.* 1991). *C. granii* is now inducible for 10 months. Each time the external salinity and light intensity are raised, the cellular answer is, as a rule, gamete formation. This behaviour is reminiscent of some intertidal brown algae (reviewed by Dring 1974), where the periodicity of their sexual activity seems to be correlated with the pattern of exposure in the natural habitat, indicating that also here the osmotic stress acts as a trigger for reproductive activity. *C. granii's* response to fluctuations in the intracellular water potential closely resembles that of *C. wailesii, O. regia* and *O. sinensis* (in prep.), but not of *Thalassiosira eccentrica,* since in the latter species sexualisation was not inducible by increasing the external salinity. Instead, *T. eccenctrica* responded as if exposed to microtubule inhibitors, with reorganizing its morphogenetic machinery and forming scales instead of valves (Schmid 1984). Scale formation in both *C. granii* and *T. eccentrica,* however, can be correlated with a reorganisaion of the microtubule cytoskeleton, which in turn, might be the result of a change in proteinsynthesis due to water potential deficiency. Studies investigating adaptive changes in the metabolism of diatoms with biochemical methods have just started. The pennate diatom *Phaeodactylum tricornutum* revealed differences in protein synthesis and post-translational modifications upon exposure to different culture conditions coupled to a change of the morphotype (Gutenbrunner *et al.* 1994 and refs. therein). A preliminary study with *Biddulphiopsis titiana* revealed a short term production of a 65kD protein after salt-stress exposure (Kranewitter, Thalhamer & Schmid, unpubl.).These results may indicate that diatoms respond to changing environments in a similar way as other organisms. Just how the external signal is transmitted and transformed, perhaps via synthesis and release of hormones, so that ultimately gamete and auxospore formation occurs, is unknown.

Knowledge about correlation of auxospore formation and triggers in the natural habitat is limited. In freshwater plankton diatoms, auxosporulation is frequently observed during spring (*Stephanodiscus alpinus:* Hickel & Håkansson 1993; *Cyclotella pseudocomensis* Scheffler 1994; *C.* cf. *comensis-ocellata-krammeri,* E. Hegewald pers. comm; *St. alpinus, C.* cf. *radiosa (comta),* Grabensee, Salzburg, AMS unpubl.) or autumn holomixis of a lake (*Stephanodiscus* cf. *neoastraea*: Round 1982; Jewson 1992b; *St. alpinus* Hickel & Håkansson 1993). Jewson (1992b) convincingly demonstrated correlation of auxospore formation in *Stephanodiscus* with the increase of the nitrate-N above a certain treshold due to watermixis. In marine environments almost all reports of sexual auxosporulation deal with neritic diatoms, i.e., occupants of a habitat with periodic fluctuations in physico-chemical parameters (e.g. Drebes 1974; French & Hargraves 1985, and refs therein), while reports on auxospores in the open ocean are rather scarce. Crawford's (1995) observation of a *Corethron* bloom under-going a mass sexual phase at the polar frontal zone, can perhaps, also be correlated with a mixis of waterbodies of different physico-chemical parameters. More studies are required to evaluate whether extrapolation of laboratory results to field situations is permitted. Jewson's (1992b) correlation of sex-induction in *Stephanopyxis* with the increase in the N-level as apparent trigger suggests that this is possible. Also, periodic alteration in salinity as applied during this study could be visualized to have been the elicitor for sex induction in Hofker's (1928) *Coscinodiscus* of the Zuiderzee (NL), where auxospores occurred only in the area influenced by the freshwater inflow of Eem and Ysel, i.e. in an environment of fluctuations in salinity and chemistry.

*Cell biology of gametogenesis and auxospores*

*Tetrads / Subdioecy*

One of the most striking observations during this study is the hitherto undescribed formation of tetrads as pre-stages of spermatogonangial differentiation. The siliceous tubes connecting the innermost cells of a tetrad as a result of an incomplete division, are reminiscent of "intercellular bridges" present during spermatogenesis in vertebrates (Alberts *et al.* 1990). In this case it is thought that transfer of substances can readily occur between the connected cells. It is also thought that these bridges are responsible for the precise synchrony of the developmental events occurring in joined cells during spermatogenesis (Alberts *et al.* 1990). At present it is not clear whether this is also true for *C. granii.*

In delaying gametogenesis in one to three cells of a tetrad until the next generation, *C. granii* may be exhibiting a survival strategy. The "vegetative" offsprings of the tetrad divide further while the spermatogonangia are evacuated upon maturation of the sperms. Tetrad formation has been found also in *C. wailesii* and *O. regia* (Figs 11, 12), indicating that such a "safety mechanism" may have a wider distribution among centric diatoms.

Monoecy has also been reported for *C. granii* by Drebes (1974) from Helgoland but the same author (1968) had previously found several clones which behaved unisexually, producing either male or female gametes. In light of the present study the "subdioecious" behaviour could be explained by the probability of male sex determination prior to tetrad formation. Since part of the offsprings of the tetrad remain vegetatively dividing, isolation of such a cell in combination with sex-induction would give rise to male gametes only. In addition, Drebes' (1968) observation indicates, that the female sex must also be determined prior to oogonium formation in the diploid state, when cells are still able to produce vegetatively. While male induction may be correlated with the unusual position of the nucleus in 6 o'clock (in prep), female induction cannot be detected prior to oogonium-formation, which is again correlated with an unusual position of the nucleus in the geometric cell centre instead of the epivalve centre. An unfertilised haploid egg has never been seen dividing again.

The cessation of sperm production in a population of isolated tetrads, and the reduction of viability in their "vegetative" offsprings, gives the impression that some kind of chemical communication exists within the cells of a clone or population. This coincides with the observation of Rozumek (1968), that in the dioecious pennate *Rhabdonema adriaticum* gametes were formed only when male and female clones were cultured in the same vessel.

*Spermatogenesis*

The LP-anomaly at 6 o'clock of the spermatogonangial walls (Fig. 13) is probably a consequence of the abnormal cell division and valve formation during tetrad formation. This LP-anomaly helps to orientate the second divison in the spermatogonangium, which is pervalvar from 6 to 12 o'clock (Fig. 14). This correlation between the dislocated LP and the direction of cleavage also demonstrates that LPs are not only unoccluded passages through the wall, but are anchor and reference points for developing tensile forces (see fig. 30 in Franz & Schmid 1994), and for intracellular orientation in general (Schmid 1994a). Detailed ultrastructural investigations are in progress.

Spermatogenesis in centric diatoms has been known since the last century ("microspores": refs. in Hofker 1928; Schmidt 1931), but correct interpretation (as oogamy) of the observations were provided first by Geitler (1952) and v. Stosch (1951, 1954), and their co-workers, and documented in spectacular films first by Drebes (1969a, b, 1976a, b) and subsequently by Pickett-Heaps (1995).

The sequence of events of spermatogenesis in *C. granii* is comparable to that described for other marine species (Drebes 1964, 1966, 1977a; v. Stosch 1977; v. Stosch & Drebes 1964; v. Stosch *et al.* 1973; refs in Pickett-Heaps *et al.* 1990; Round *et al.* 1990); sperm formation itself resembles that in *Lithodesmium* (e.g. Manton & v. Stosch 1966; Manton *et al.* 1969). These authors first described the lack of the 2 central microtubules in the sperm flagellum, which was subsequently confirmed by Heath and Darley (1972); Idei (1995, this volume); and is true also for *C. granii*: (AMS in prep). TEM and immuno-fluorescence preparations have been done to study in more detail the cytological changes occurring during spermiogenesis and auxospore formation of *C. granii* (in prep.). From light microscope it appears that chloroplast division ceases, and their morphology becomes reduced prior to extrusion. It seems that chloroplasts are maternally inherited in *C. granii*. Decrease of chlorophyll fluorescence over time was monitored by flow cytometry also during spermatogenesis in *Thalassiosira weissflogii* (Vaulot & Chisholm 1987). The physiological changes that accompany sexual reproduction, were first studied by Werner (1971a, b, c).

Although interpretation of observed stages of spermatogenesis was rarely correct in early publications (e.g. Hofker 1928; Schmidt 1931), the described sequence of events was roughly comparable to the detailed LM reports of v. Stosch, Drebes, and co-workers (refs above), indicating a mechanism that is basically similar operating in those few species investigated so far. The existence, let alone the mechanism, of auxospore formation is unknown for the majority of diatom species, and mechanisms may vary under the influence of environmental constraints. A very peculiar mode of sperm formation requiring a more detailed discussion, was described for *Stephanodiscus neoastraea* (Jewson 1992). According to the author, "the intact protoplast of "a male-determined cell" expanded, formed a sphere that gradually opened up the frustule which fell away after ca 30 min. Spermatozoa became increasingly active inside the sphere [termed now a "gametangium"], which finally burst after 45–75 min since the first expansion." (Jewson 1992b). If Jewson had not reported subsequent sperm release from such a cell, one would be inclined to interpret this description, as well as the light micrographs (Jewson fig. 5a–c) and the diagram (Jewson fig. 9) as an early development of an auxospore, where the protoplast eventually degenerates. The time schedule given is also unique, proposing that the protoplast of the "male-determined cell" would have formed a spherical gametangium, and would have completed all spermatogonial divisions and meiosis I and II within 75–95 min, whereas spermatogenesis in *Stephanopyxis, Chaetoceros, Lithodesmium* (see refs. above), *Coscinodiscus* (this paper), *O. regia* and *O. sinensis* (in prep) takes 15 to 25 hours according to the species. Such a contrast with spermatogenesis of other diatoms requires additional studies to establish whether this type of formation is specific for *Stephanodiscus*.

*Auxospores / Initial cells*

The most interesting discovery with respect to auxospore formation was the detection of a limited number of "slit scales" (= LP scales, Fig. 21) in the auxospore envelope. They appear slightly stronger silicified and are usually larger than the normal type, bearing a clear slit within the the area circumscribed by a pronounced annulus (Fig. 21). Circumstantial evidence suggests that these are rudimentary rimoportulae (Schmid 1994b)

Reinforcement of the organic auxospore envelope of *C. granii* is by a single layer of overlapping scales (Figs 19–21; Schmid 1994b), which are only slightly silicified, perhaps even less than those in *Cyclotella meneghiniana* (Hoops & Floyd 1979). The siliceous components of auxospore walls of a variety of species have been studied through the light-, and later the electron microscope by v. Stosch and co-workers (reviewed in v. Stosch 1982), but the first classical EM study on auxospore and initial cell formation in a centric diatom is that of Crawford (1974, 1975). He was able to demonstrate formation of scales in their own SDV, attached to the plasmamembrane (confirmed by Hoops & Floyd in 1979). Thus, scale formation in diatoms was found to resemble that in testate amoebae rather than that in Chrysophyceae and Synurophyceae, where the scales are moulded and silicified along cisternae of the cytoplasmic or chloroplast associated endoplasmic reticulum (refs. in Crawford 1974; Pickett-Heaps *et al.* 1990). The only other detailed TEM study concerning auxospore and initial cell formation has been published on *C. meneghiniana* (Hoops & Floyd 1979).

Reinforcement of the auxospore envelope by siliceous scales or bands is apparently not universal in centric diatoms, since the envelope in two species of *Stephanodiscus* (Round 1982; Edlund & Stoermer 1991), and in *Aulacoseira herzogii* (Jewson *et al.* 1993) appears purely organic, lacking any siliceous component. By contrast, *Cyclotella meneghiniana,* living in a similar environment, does form scales (Hoops & Floyd 1979). These reports encourage a broader study into auxospore formation of a variety of species and genera, including TEM investigations, in order to understand the ecological, phylogenetic and cell biological meaning of such variations.

Variation in the size of auxospores is not specific to *C. granii.* In *Stephanodiscus niagarae* the size of initial valves produced in culture covers more than 40% of the taxon's size range (Edlund & Stoermer 1991). That this is not a cultural anomaly was demonstrated by the authors correlating the data from their culture with H. L. Smith's plankton collection from the same sampling site, but the last century. The results of *C. granii*, whose initial cells vary between 280–400 µm in nature (Drebes 1974) and between 220 and 400 µm within a single clone, apparently without any size relation to the parent's size, corroborates their observations. Also in *Thalassiosira weissflogii* (Armbrust & Chisholm 1992) the maximum mean cell size in originally genetically identical populations is not constant. All these results are not congruent with the generally accepted concept of diatom life cycles, that initial cells display always maximum size (see also discussion in Schmid 1990; Edlund & Stoermer 1991; Davidovich 1994). The development of the auxospores in lower and higher salinity in response to altered culture conditions suggests that salinity may be a factor controlling auxospore expansion. However, genetic diversity due to genetic recombination cannot be excluded and this matter merits further intensive research. In *C. granii* also the inducible size range varies enormous. Pre-auxospore cells during this study varied from 130–170 µm, while Drebes (1974) found spermatogonangia from below 260 up to 300 µm, indicating a clear overlap of the "inducible size window" of 130–300 µm with "maximum size" of 220–400 µm. Investigations of a variety of genera would be highly rewarding.

## Acknowledgements

I wish to thank Dr R. Crawford for the gift of a culture of *C. granii,* and Prof. J. Pickett-Heaps for the gift of a video-copy of his unpuplished Video-Disc on "Diatoms"; I also gratefully appreciate their continuous discussions and share of results, and their help in improving the English of this paper. Financial support by the Austrian FFWF Projekt P8727–Bio is also acknowledged.

## References

Alberts, B., Bray, D., Lewis, J., Raff, M., Roberts, K. & Watson, J. D. (1990). Molekularbiologie der Zelle. Kap. 15. Keimzellen und Befruchtung, 998–1045. VCH–Verlagsgesellschaft mbH, Weinheim, Germany, pp. 1490.

Armbrust, E. V. & Chisholm, S. W. (1992). Patterns of cell size change in a marine centric diatom: Variability evolving from clonal isolates. *Journal of Phycology*, **28**, 146–156.

Borkird, C., Claes, B., Caplan, A., Simoens, C. & Montagu, v. M. (1991). Differential expression of water-stress associated genes in tissue of rice plants. *Journal of Plant Physiology*, **138**, 591–595.

Brooks, M. (1975). Studies on the genus *Coscinodiscus*. III. Light, transmission and scanning electron microscopy of *C. granii* Gough. *Botanica Marina,* **18**, 29–39.

Burton, R. S. (1991): Regulation of proline synthesis in osmotic response: Effects of protein synthesis inhibitors. *Journal of Experimental Zoology*, **259**, 272–277.

Crawford, R. M. (1974). The auxospore wall of the marine diatom *Melosira nummuloides* (Dillw.) C. Ag. and related species. *British Phycological Journal,* **9**, 9–20.

Crawford, R. M. (1975). The frustule of the initial cells of some species of the diatom genus *Melosira* C. Agardh. In*: Proceedings of the Third Symposium on Recent and Fossil Marine Diatoms*, Kiel, 1974 (R. Simonsen, ed.), 37–50. J. Cramer, Vaduz.

Crawford, R. M. (1995). The role of sex in the sedimentation of a marine diatom bloom. *Limnology & Oceanography*, **40** (1) 200–204.

Cushing, D. H. (1955) Production and a Pelagic Fishery. *Fishery Investigations, Series II*, **18** (7), pp. 104.

Davidovich, N. A. (1994). Factors Controlling the Size of Initial Cells in Diatoms. *Russian Journal of Plant Physiology*, **41** (2), 220–224.

Drebes, G. (1964). Über den Lebenszyklus der marinen Planktondiatomee *Stephanopyxis turris* (Centrales) und seine Steuerung im Experiment. *Helgoländer wissenschaftliche Meeresuntersuchungen*, **10**, 152–153.

Drebes, G. (1966). On the life history of the marine plankton diatom *Stephanopyxis palmeriana*. *Helgoländer wissenschaftliche Meeresuntersuchungen*, **13**, Nr. 1–2, 101–114.

Drebes, G. (1968). Subdiözie bei der zentrischen Diatomee *Coscinodiscus granii*. *Naturwissenschaften*, **55**, 236.

Drebes, G. (1969a–f). Asexual und sexual reproduction in *Stephanopyxis* – 6 b/w 16mm cine films, Cat. Nos E1341–1344, Institut für Wissenschaftlichen Film, Göttingen.

Drebes, G. (1974). *Marines Phytoplankton*. Eine Auswahl der Helgoländer Planktonalgen (Diatomeen, Peridineen). Georg Thieme Verlag, Stuttgart. 186pp.

Drebes, G. (1975). *Coscinodiscus granii* (Centrales). Vegetative Vermehrung. b/w 16mm cine film, Cat.No. E1840, Institut für Wissenschaftlichen Film, Götttingen.

Drebes, G. (1976a,b). *Attheya decora* (Centrales). Vegetative Vermehrung und Geschlechtliche Fortpflanzung. b/w 16mm cine films, Cat.Nos. E2275, E2383, Institut für Wissenschaftlichen Film, Göttingen.

Drebes, G. (1977a). Cell structure, cell division, and sexual reproduction of *Attheya decora* West (Bacillariophyceae, Biddulphiinae). In: *Proceedings of the 4th International Diatom Symposium* (R. Simonsen, ed.), 167–178, O. Koeltz, Koenigstein.

Drebes; G. (1977b). Sexuality. In: *The Biology of Diatoms* (D. Werner, ed), *Botanical Monographs*, **13**, 250–283.

Dring, M. (1974). Reproduction. In: *Algal Physiology and Biochemistry. Botanical Monographs*, Vol. **10** (W. D. P. Stewart, ed.), 814–837. University California Press, Berkeley & Los Angeles.

Edlund, M. B. & Stoermer, E. F. (1991). Sexual reproduction in *Stephanodiscus niagarae* (Bacillariophyta). *Journal of Phycology*, **27**, 780–793.

Franz, S. M. & Schmid, A. M. (1994). Cell-cycle and phenotypes of *Biddulphiopsis titiana*. *Diatom Research*, **9** (2), 265–288.

French, F. W. & Hargraves, P. E. (1985) Spore formation in the life cycles of the diatoms *Chaetoceros diadema* and *Leptocylindrus danicus*. *Journal of Phycology*, **21**, 477–483.

Furnas, M. J. (1985). Diel Synchronisation of sperm formation in the diatom *Chaetoceros curvisetum* Cleve. *Journal of Phycology*, **21**, 667–671.

Geitler, L. (1932). Der Formwechsel der pennaten Diatomeen. *Archiv für Protistenkunde*, Bd. **78**, 1–226.

Geitler, L. (1952). Oogamie, Mitose, Meiose und metagame Teilung bei der zentrischen Diatomee *Cyclotella*. *Österreichische Botanische Zeitung*, **99**, 506–520.

Geitler, L. (1963). Alle Schalenbildungen der Diatomeen treten als Folge von Zell-oder Kernteilungen auf. *Berichte der Deutschen Botanischen Gesellschaft*, **75**, 393–396.

Gutenbrunner, S., Thalhamer, J. & Schmid, A. M. (1994). Proteinaceous and immunochemical distinctions between the oval and fusiform morphotype of *Phae*odactylum tricornutum (Bacillariophyceae). *Journal of Phycology*, **30**, 129–136.

Hagemannn, M., Techel, D. & Rensing, L. (1991). Comparison of salt- and heat induced alterations of protein synthesis in the cyanobacterium *Synechocystis sp.* PCC6803. *Archives of Microbiology*, **155**, 587–592.

Hargraves, P. E. (1976). Studies on Marine Plankton Diatoms. II. Restingspore Morphology. *Journal of Phycology*, **12**, 118–128.

Heath, B. I. & Darley, M. W. (1972). Observations on the ultrastructure of the male gametes of *Biddulphia laevis* EHR. *Journal of Phycology*, **8**, 51–59.

Hershkovitz, N., Oren, A., Post, A. & Cohen, Y. (1991). Induction of water-stress proteins in cyanobacteria exposed to matric- or osmotic- water stress. *FEMS Microbiology Letters*, **83**, 169–172.

Hickel, B. & Håkansson, H. (1993). *Stephanodiscus alpinus* in Plußsee, Germany. Ecology, morphology and taxonomy in combination with initial cells. *Diatom Research*, **8**, 89–98.

Hofker, J. (1928). Die Teilung, Mikrosporen- und Auxosporenbildung von *Coscinodiscus biconicus*. *Ann. Protistologie*, Paris, **1**, 167–194.

Holmes, R. W. (1966). Short-term temperature and light conditions associated with auxospore formation in the marine centric diatom, *Coscinodiscus concinnus* W. Smith. *Nature*, Lond., **209**, 217–218.

Holmes, R. W. (1967). Auxospore formation in two marine clones of the diatom genus *Coscinodiscus*. *American Journal of Botany*, **54**, 163–168.

Hoops, H. J. & Floyd, G. L. (1979). Ultrastructure of the centric diatom *Cyclotella meneghiniana*: vegetative cell and auxospore development. *Phycologia*, **18**, 424–435.

Idei, M. & Chihara, M. (1992). Successive Observations on the Fertilisation of a Centric Diatom *Melosira moniliformis* var. *octagona*. *Botanical Magazine Tokyo*, **105**, 649–658.

Jewson, D. H. (1992a). Size reduction, reproductive strategy and the life cycle of a centric diatom. *Philosophical Transactions of the Royal Society, London*, B 335, 191–213.

Jewson, D. H. (1992b). Life cycle of *Stephanodiscus sp* (Bacillariophyta). *Journal of Phycology*, **28**, 856–866.

Jewson, D. H., Khondker, M., Rahman, M. H. & Lowry, S. (1993). Auxosporulation in the freshwater diatom *Aulacoseira herzogii* in Lake Banani, Bangladesh. *Diatom Research*, **8**, 403–418.

Lewis, W. M. (1984). The diatom sex clock and its evolutionary significance. *The American Naturalist*, **123**, 73–80.

Manton, I. & v. Stosch, H. A. (1966) Observations on the fine structure of the male gamete of the marine centric diatom *Lithodesmium undulatum*. *Journal of the Royal Microscopical Society*, **85**, 119–134.

Manton, I., Kowallik, K. & v. Stosch, H. A. (1969) Observations on the fine structure and development of the spindle at mitosis and meiosis in a marine centric diatom (*Lithodesmium undulatum*).II. The early meiotic stages in male gametogenesis. *Journal of Cell Science*, **5**, 271–298.

Perez-Martinez, C., Cruz-Pizarro, L. & Sanchez-Castillo, P. (1992). Auxosporulation in *Cyclotella ocellata* (Bacillariophyceae) under natural and experimental conditions. *Journal of Phycology*, **28**, 608–615.

Pickett-Heaps, J. D. (1995) Sexual reproduction in the centric diatom *Ditylum brightwellii*. In: *Diatoms: Cell Division, Motility and Sexual Reproduction* (J. D. Pickett-Heaps & J. Pickett-Heaps, eds), Video-Disc: Cytographics-Production (in press).

Pickett-Heaps, J. D., Schmid, A. M. & Edgar, L. (1990). The cell biology of diatom valve formation. In: *Progress in Phycological Research* (Round & Chapman, eds), Vol. **7**, 1–168. Biopress Ltd., Bristol.

Reimann, B. E. F. (1960). Bildung, Bau und Zusammenhang der Bacillariophyceenschalen (elektronenmikroskopische Untersuchungen). *Nova Hedwigia*, **2**, 349–373.

Roemer, S. C. & Rosowski, J. R. (1980). Valve and band morphology of some freshwater diatoms. III. Pre- and postauxospore frustules and the initial cell of *Melosira roeseana. Journal of Phycology* **16**, 399–411.

Round, F. E. (1982). Auxospore structure, initial valves and the development of populations of *Stephanodiscus* in Farmoor Reservoir. *Annals of Botany*, **49**, 447–459.

Round, F. E. & Crawford, R. M. (1981). The lines of evolution of the Bacillariophyta. I. Origin. *Proceedings of the Royal Society of London* B **211**, 237–260.

Round, F. E., Crawford, R. M. & Mann, D. G. (1990). *The Diatoms. Biology and morphology of the genera.* 747 pp. Cambridge University Press, Cambridge.

Rozumek, K. E. (1968). Der Einfluß der Umweltfaktoren Licht und Temperatur auf die Ausbildung der Sexualstadien bei der pennaten Diatomee *Rhabdonema adriaticum* Kütz. *Beitr. Biol. Pfl.*, **44**, 365–388.

Scheffler, W. (1994). *Cyclotella pseudocomensis* nov. sp. (Bacillariophyceae) aus norddeutschen Seen. *Diatom Research*, **9**, 355–369.

Schmid, A. M. (1979). Influence of environmental factors on the development of the valve in diatoms. *Protoplasma*, **99**, 99–115.

Schmid, A. M. (1984). Wall morphogenesis in *Thalassiosira eccentrica*: Comparison of auxospore formation and the effects of MT-inhibitors. In: *Proceedings of the 7th International Diatom Symposium* (D. G. Mann, ed.), 47–70. O. Koeltz, Königstein.

Schmid, A. M. (1987). Morphogenetic forces in diatom cell wall formation. In: *Cytomechanics* (Bereiter-Hahn, Anderson & Reif, eds), 183–199. Springer, Berlin.

Schmid, A. M. (1990). Intraclonal variation in the valve structure of *Coscinodiscus wailesii* Gran et Angst. *Beiheft zur Nova Hedwigia*, **100**, 101–119.

Schmid, A. M. (1994a). Aspects of morphogenesis and function of diatom cell walls with implications for taxonomy. *Protoplasma*, **181**, 43–60.

Schmid, A. M. (1994b). Slit-scales in the auxospore scale case of *Coscinodiscus granii*: the rudiments of rimoportulae? *Diatom Research*, **9** (2), 371–375.

Schmid, A. M. & Volcani, B. E. (1983). Wall morphogenesis in *Coscinodiscus wailesii* Gran et Angst. I. Valve morphology and development of its architecture. *Journal of Phycology*, **19**, 387–402.

Schmidt, P. (1931). Die Reduktionsteilung bei der Mikrosporenbildung von *Coscinodiscus apiculatus* Ehrenberg und andere Ergebnisse zur Biologie dieser Diatomee. *Internationale Revue der gesamten Hydrobiologie*, **25** 68–101.

Schultz, M. E. & Trainor, R. R. (1968). Production of male gametes and auxospores in the centric diatoms *Cylotella meneghiniana* and *C. cryptica. Journal of Phycology*, **4**, 85–88.

Schultz, M. E. & Trainor, R. R. (1970). Production of male gamestes and auxospores in a polymorphic clone of the centric diatom *Cyclotella. Canadian Journal of Botany*, **48**, 947–951.

Steele, R. L. (1965). Induction of sexuality in two centric diatoms. *Bioscience*, **15**, 298.

Stosch, H. A. von (1951). Entwicklungsgeschichtliche Untersuchungen an zentrischen Diatomeen. I. die Auxosporenbildung von *Melosira varians. Archiv für Mikrobiologie*, **16**, 101–135.

Stosch, H. A. von (1954). Die Oogamie von *Biddulphia mobiliensis* und die bisher bekannten Auxosporenbildungen bei den Centrales. *Rapport, VIIIth Congrés International de Botanique*, 58–68.
Stosch, H. A. von (1977). Observations on *Bellerochea* and *Streptotheca* including descriptions of three new planctonic diatom species. *Beiheft zur Nova Hedwigia*, **54**, 113–166.
Stosch, H. A. von (1982). On auxospore envelopes in diatoms *Bacillaria*, **5**, 127–156.
Stosch, H. A. von & Drebes, G. (1964). Entwicklungsgeschichtliche Untersuchungen an zentrischen Diatomeen. IV. Die Planktondiatomee *Stephanopyxis turris* – Ihre Behandlung und Entwicklungsgeschichte. *Helgoländer Wissenschaftliche Meeresuntersuchungen*, **11**, 209–257.
Stosch, H. A. von, Theil, G. & Kowallik, K. V. (1973). Entwicklungsgeschichtliche Untersuchungen an zentrischen Diatomeen. V. Bau und Lebenszyklus von *Chaetoceros didymum* mit Beobachtungen über einige andere Arten der Gattung. *Helgoländer Wissenschaftliche Meeresuntersuchungen*, **25**, 384–445.
Vaulot, D. & Chisholm, S. (1987). Flow cytometric analysis of spermatogenesis in the diatom *Thalassiosira weissflogii* (Bacillariophyceae). *Journal of Phycology*, **23**, 132–137.
Werner, D. (1971a). Der Entwicklungszyklus mit Sexualphase bei der marinen Diatomee *Coscinodiscus asteromphalus*. I. Kultur und Synchronisation von Entwicklungsstadien. *Archiv für Mikrobiologie*, **80**, 43–49.
Werner, D. (1971b). II. Oberflächenabhängige Differenzierung während der vegetativen Zellverkleinerung. *Archiv für Mikrobiologie*, **80**, 115–133.
Werner, D. (1971c). III. Differenzierung und Spermatogenese. *Archiv für Mikrobiologie*, **80**, 134–146.

# The relationship between available silicate and marine diatoms in the Gulf of Trieste

M. Cabrini, M. Celio, S. Fonda Umani and I. Pecchiar

*Laboratory of Marine Biology, Str. Costiera 336, S. Croce, 34010 Trieste, Italia*

## Abstract

Phytoplankton samples were collected and environmental parameters measured in the Gulf of Trieste from March to November 1992. Temperature and salinity data were organised in a matrix on which cluster analysis was computed with the aim of identifying the different water bodies present in each month. New matrices were organised for each identified water body using averages of temperature, salinity and nutrients. Correlation between salinity and silicates is evident only in the thin surface lens. No significant correlation appears between dissolved silicates and diatom abundance. This is because of the faster utilization of this nutrient in the more coastal station, in which the diatom bloom appears well developed in comparison with the offshore station where the dissolved silicate has a higher concentration and the diatom bloom is beginning.

*Keywords:* diatom, silicate, water column structure, Gulf of Trieste.

## Introduction

The growth and behaviour of plankton diatom populations are subject to two predominant special influences. The first is the availability of dissolved silicates which the cells need to make their frustule. The second is the tendency for diatom cells to sink and is strongly related to the stability of the water column (Round *et al.* 1990). In the estuarine and coastal waters the input of nitrogen and phosphorus is increased by human activities whereas silicate input is decreased. The availability of dissolved silicates decreases due to eutrophication and the presence of river dikes causing an increase in silt deposition which prevents silicate resuspension. Furthermore silicon is not recycled as rapidly as other nutrients (Egge & Aksnes 1992). In the Gulf of Trieste

(Northern Adriatic Sea) nutrient concentrations are strongly correlated with river inputs and consequently with salinity (Burba *et al.* 1994).

The purpose of this work was the identification of the water column structure in the Gulf of Trieste over a period of months, with the aim of highlighting the partitioning of each water body by its thermohaline characteristics and the distribution of nutrient concentrations. Diatom abundance in the various water bodies and the role of dissolved silicates in their distribution were then analyzed.

## Study Area, Material and Methods

The Gulf of Trieste is the northernmost part of the Adriatic Sea and is characterized by shallow waters (maximum depth about 25m). The water mass circulation in the Gulf of Trieste can be represented by a three layer model. The bottom layer, below about 10m depth with characteristic salinity of 37.5 ± 0.3 psu, and the intermediate one (from 5m to 10m) flow almost permanently counter clockwise in the NE part of the Gulf. The surface layer, with a thickness of about 5m, commonly flows clockwise towards Trieste (Stravisi 1983). Temperature is at a maximum of about 26°C in summer: July, August and September (Stravisi 1987). During the period of stratification from April to October, the pycnocline separates two or three different water masses in which the growth of diatoms can vary significantly.

From March to November 1992 environmental data and biological samples were collected at four stations located along a coastal offshore transect in the Gulf of Trieste (Fig. 1). The hydrological data were collected using an Idronaut CTD vertical profile. Profiles of pressure, temperature, salinity recorded with a 20 dbar pressure step were averaged on a per metre basis. The estimated parameters were obtained by the application of algorithms for the computation of fundamental seawater properties (UNESCO 1983). Chemical and phytoplankton samples were collected at 0m, 5m, 10m and at the bottom (from 18 to 22m). Chemical analyses were carried out simultaneously on ammonia-nitrite and nitrate-nitrogen, reactive phosphate-phosphorus and orthosilicate-silicon, by means of an Alleance Autoanalyser, according to Strickland & Parson (1972). Phytoplankton samples were fixed in 4% buffered formaldehyde and analysed by the Utermöhl (1958) method.

Monthly temperature and salinity data were organised in a matrix and cluster analysis using a similarity ratio where

$$SIMij = \frac{\sum_{r=1}^{R} Xr,i * Xr,j}{\sum_{r=1}^{R} (Xr,i)^2 * \sum_{r=1}^{R} (Xr,j)^2 - \sum_{r=1}^{R} Xr,i * Xr,j}$$

with a complete linkage method, was computed. We organised new matrices using the average temperature, salinity and nutrient values for each identified water body. P.C.A. based on the correlation coefficient were computed on these matrices by Matedit

(Burba *et al.* 1992). A ratio of the total number of diatoms to the maximum total number of diatoms (X/Xmax) was calculated for each water body identified by cluster analysis.

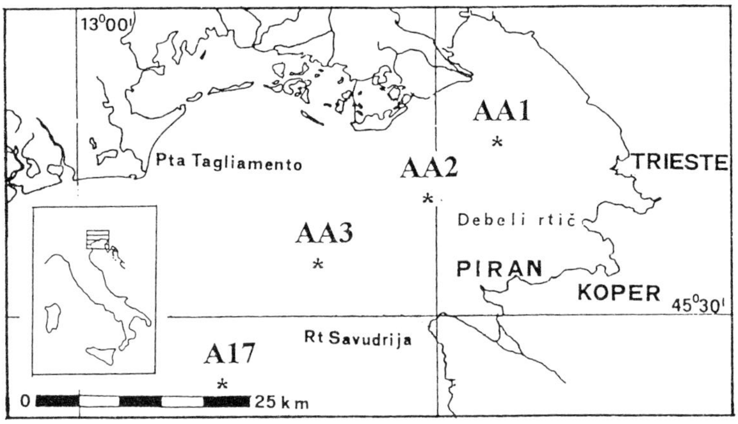


Fig. 1. Sampling stations located along a NE–SW transept in the Gulf of Trieste.

## Results

Cluster analyses identified (at 70–80% of similarity) three to seven water bodies throughout the year:

4–5 March: The water column was homogenous. Only a small surface lens and a small residual nucleus appeared.

6 and 13 April: A thin low salinity surface lens floated above distinct underlying water bodies (Fig. 2a).

7–8 May: Haline stratification reached its maximum depth (5m) in AA3 – A17 and was less evident in stations closer to the coast. In the station 6 miles offshore (AA2) a small surface lens was still present. Two different bottom water bodies were present and characterized by high salinity and low temperature. The southern body was the result of south eastern water advections linked to the general anticlockwise circulation of the North Adriatic (Franco *et al.* 1982), which tend to substitute the deep water residual body which originated during the previous winter and is still present in the most western part of the basin.

28–29 May: A clear stratification in which cluster analyses identified five different water bodies was evident; the surface layer appeared less diluted than in the previous months.

15–16 June: A small diluted surface lens reappeared during stratification (Fig. 2b).

13–14 July: Stratification reached the maximum depth. A surface lens was again present at the St. AA2.

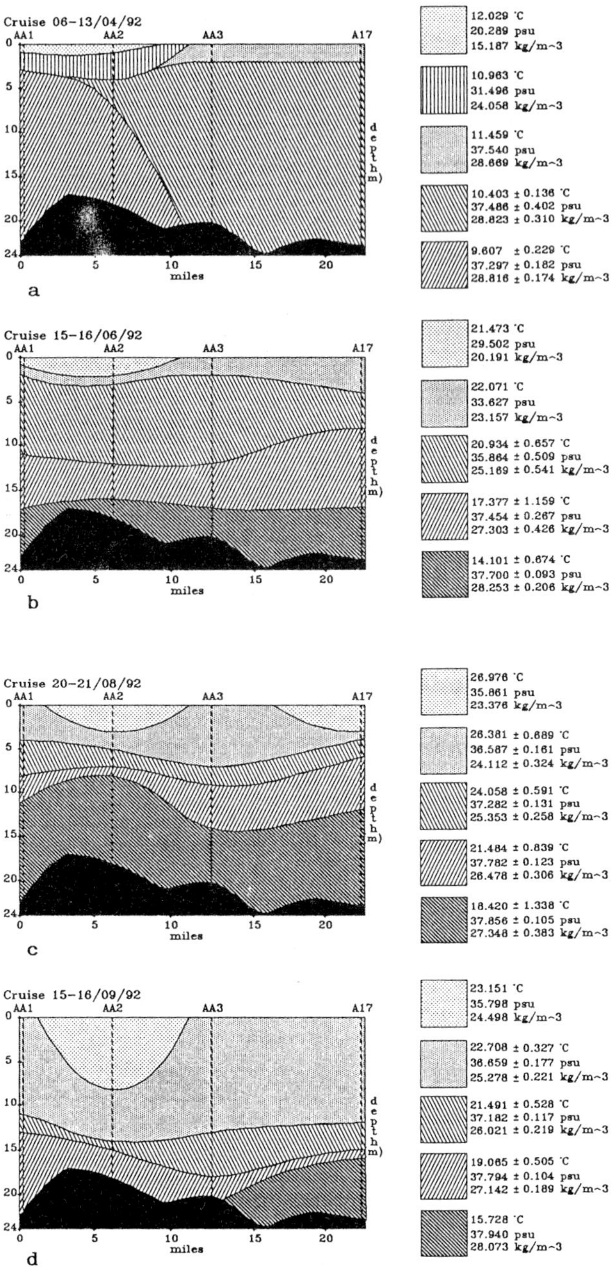
Cruise 06-13/04/92
AA1
AA2
AA3
A17
depth (m)
miles
a
12.029 °C
20.289 psu
15.187 kg/m-3
10.963 °C
31.496 psu
24.058 kg/m-3
11.459 °C
37.540 psu
28.869 kg/m-3
10.403 ± 0.136 °C
37.486 ± 0.402 psu
28.823 ± 0.310 kg/m-3
9.607 ± 0.229 °C
37.297 ± 0.182 psu
28.816 ± 0.174 kg/m-3
Cruise 15-16/06/92
b
21.473 °C
29.502 psu
20.191 kg/m-3
22.071 °C
33.627 psu
23.157 kg/m-3
20.934 ± 0.657 °C
35.864 ± 0.509 psu
25.169 ± 0.541 kg/m-3
17.377 ± 1.159 °C
37.454 ± 0.267 psu
27.303 ± 0.426 kg/m-3
14.101 ± 0.674 °C
37.700 ± 0.093 psu
28.253 ± 0.206 kg/m-3
Cruise 20-21/08/92
c
26.976 °C
35.861 psu
23.376 kg/m-3
26.381 ± 0.689 °C
36.587 ± 0.161 psu
24.112 ± 0.324 kg/m-3
24.058 ± 0.591 °C
37.282 ± 0.131 psu
25.353 ± 0.258 kg/m-3
21.484 ± 0.839 °C
37.782 ± 0.123 psu
26.478 ± 0.306 kg/m-3
18.420 ± 1.338 °C
37.856 ± 0.105 psu
27.348 ± 0.383 kg/m-3
Cruise 15-16/09/92
d
23.151 °C
35.798 psu
24.498 kg/m-3
22.708 ± 0.327 °C
36.659 ± 0.177 psu
25.278 ± 0.221 kg/m-3
21.491 ± 0.528 °C
37.182 ± 0.117 psu
26.021 ± 0.219 kg/m-3
19.065 ± 0.505 °C
37.794 ± 0.104 psu
27.142 ± 0.189 kg/m-3
15.728 °C
37.940 psu
28.073 kg/m-3

20–21 August: Two different surface lenses were present at 6 (St.AA2) and 24 (St. A17) offshore miles, the water reached maximum temperatures (26.976°C at the surface and 18.420°C at the bottom). A conspicuous bottom advection invaded the Gulf from the southeast (Fig. 2c).

15–16 September: Fresh water input, which became more pronounced at St.AA2 determining a surface lens exceeding 5 meters in depth. The bottom intrusion was confined to the southern region (A17) (Fig. 2d).

10–11 November: A frontal system which probably arose in October, appeared well defined. At St.AA2 a small surface lens was still present.

For each identified water body all the related data are reported following the results of cluster analyses in Figs 3, 4, 5 and 6. Diatom abundance was reported as X/Xmax and its sequence. We considered only the temperature, salinity and nutrient data for highlighting any possible temporal partitioning of the water masses in order to follow their temporal evolution. In each of these identified groups we subsequently analysed the diatom content, trying to define which of the considered parameters might more strictly influence the diatom temporal and spatial patterns.

*Surface lens:*

The first axis (61.62%) can be interpreted as a decreasing gradient for salinity and an increasing one for $Si-SiO_4^{4-}$, $N-NO_3^-$, $N-NO_2^-$. The second one (20.83%) as an increasing axis for temperature and $N-NH_4^+$. The best inverse correlation appeared to be among salinity and $Si-SiO_4^{4-}$, $N-NO_3^-$, $N-NO_2^-$. Station group 1 is composed of AA1 and AA2 in April and June. These samples are characterized by high values for all nutrients analysed, particularly for $Si-SiO_4^{4-}$ and by the highest concentration of diatoms (1,668,919 cell/l). In these communities, *Skeletonema costatum* prevails. Group 2 is primarily represented by stations AA1 and AA2 (only one A17 in August). Intermediate values of salinity and nutrients were found in these samples. The diatoms are still present in high numbers (652,289 cell/l, Fig. 3). In the summer samples *Chaetoceros* spp. prevails, while from September *Leptocylindrus danicus* takes its place. No significant correlation appears between salinity or silicates and diatom abundance because the high values of diatoms correspond to the more coastal station (AA1) where silicates were already utilized, while high values of silicates were often found in the more offshore station (AA2) where the diatom bloom was just starting. This situation was particularly evident in March, April and June.

---

Fig. 2 a–d. Some examples of water column structure (2a– April, 2b– June, 2c– August and 2d– September) in which are reported water masses identified by cluster analysis. For each of them temperature (°C) and salinity (PSU) average, density excess as gt and standard deviation (number of data above 10) are reported.

*Surface layer:*

The first axis (45.79%) is an increasing gradient for $N\text{-}NO_3^-$ and $P\text{-}PO_4^{3-}$; the second one (17.58%) an increasing gradient for salinity and a decreasing one for temperature. Station group 1 is composed of AA1 and AA3 mainly in the summer period. The diatom abundance is relatively high with an average of 181,462 cell/l. In these samples *Thalassiosira* sp. and *Chaetoceros* sp. are prevalent. Station group 2 is mainly composed of A17 from spring to summer and by AA1, AA2 and AA3 samples in May and September. The abundance of diatoms is slightly lower (176,065 cell/l, fig. 4). Besides *Thalassiosira* sp., *Pseudonitzschia* sp. appears prevalent. In this case silicates do not seem to influence the partitioning of water bodies and consequently the distribution of diatom abundance.

*Intermediate layer:*

The first axis (49.64%) is an increasing axis for $N\text{-}NO_2^-$, temperature, $N\text{-}NO_3^-$ and $Si\text{-}SiO_4^{4-}$. The second one (21.10%) an increasing gradient for salinity and a decreasing one for $P\text{-}PO_4^{3-}$. The best correlation, in this case, is between temperature, $Si\text{-}SiO_4^{4-}$, $N\text{-}NO_2^-$, and $N\text{-}NO_3^-$. Group 1 stations are composed of samples from June and July. Diatom abundance showed intermediate values (278,076 cell/l) typical of the end of the diatom bloom, when *Leptocylindrus danicus* and *Cyclotella glomerata* clearly prevail. Station group 2 consists of May samples and is characterized by low diatom abundance and the dominance of *Pseudonitzschia* sp. Station group 3 is composed of three August samples identified by the highest values of $Si\text{-}SiO_4^{4-}$ among this group. Diatom abundance is of the same magnitude as the first group and may be due to a slight increase in silicates (Fig. 5). The prevailing species is *Pseudonitzschia* sp.

*Bottom layer:*

The first axis (48.92%) shows an increasing gradient for $Si\text{-}SiO_4^{4-}$, $N\text{-}NO_3^-$ and temperature; the second one (19.10%), an increasing gradient for $P\text{-}PO_4^{3-}$ and $N\text{-}NH_4^+$. The best correlation appears to be between temperature and $Si\text{-}SiO_4^{4-}$. Station group 1 is composed of samples from May through July and one in September. Silicate values show an increase although the diatom abundance is still low. They are mainly constituted by *Pseudonitzschia* sp. and *Leptocylindrus danicus*. Station group 2 is mainly composed of August samples and show higher values of silicates. Diatom abundance in this group is similar to group 1 (Fig. 6) *Leptocylindrus danicus* still prevails but also *Hemiaulus hauckii* and *Navicula* sp. are abundant. It appears that silicate regeneration is evident mainly in August, when, because of the presence of a well defined pycnocline, no mixing with the upper layers takes place.

Table 1. The characteristic of the two groups of samples referred to the "surface lens" identified by P.C.A. are reported.

| GR. | ST. | °C | PSU | PO4 | NO3 | NO2 | NH3 | SiO2 | Cell.*dm^-3 | X/Xmax | N. |
|---|---|---|---|---|---|---|---|---|---|---|---|
| | AA104L | 11.290 | 20.230 | 0.17 | 19.22 | 0.28 | 1.62 | 23.60 | 5487506 | 1.00 | 1 |
| I | AA204L | 12.767 | 20.347 | 0.25 | 47.15 | 0.35 | 1.07 | 39.90 | 303743 | 0.06 | 7 |
| | AA106L | 21.479 | 30.373 | 0.24 | 25.79 | 0.29 | 2.02 | 9.98 | 748192 | 0.14 | 5 |
| | AA206L | 21.469 | 29.067 | 0.18 | 33.42 | 0.31 | 1.14 | 11.20 | 136236 | 0.02 | 11 |
| mean I | | 16.751 | 25.004 | 0.21 | 31.40 | 0.31 | 1.46 | 21.17 | 1668919 | | |
| | AA103L | 8.198 | 33.907 | 0.06 | 6.33 | 0.07 | 0.00 | 1.61 | 2451272 | 0.45 | 2 |
| | AA203L | 8.068 | 35.551 | 0.03 | 10.24 | 0.26 | 0.00 | 1.58 | 245673 | 0.04 | 9 |
| | AA15BL | 18.339 | 35.782 | 0.15 | 1.98 | 0.03 | 0.66 | 0.63 | 1322179 | 0.24 | 3 |
| | AA211L | 16.966 | 37.074 | 0.14 | 0.79 | 0.01 | 0.12 | 1.09 | 8475776 | 0.15 | 4 |
| II | AA205L | 14.349 | 35.233 | 0.17 | 1.17 | 0.03 | 0.31 | 0.00 | 109434 | 0.02 | 13 |
| | AA207L | 23.226 | 30.048 | 0.19 | 9.37 | 0.02 | 0.65 | 6.26 | 125068 | 0.02 | 12 |
| | AA208L | 26.964 | 35.781 | 0.09 | 3.93 | 0.18 | 1.07 | 2.70 | 341707 | 0.06 | 6 |
| | A1708L | 26.987 | 35.941 | 0.10 | 0.95 | 0.07 | 1.34 | 0.45 | 274705 | 0.05 | 8 |
| | AA209L | 23.151 | 35.798 | 0.00 | 0.95 | 0.03 | 0.06 | 1.64 | 152984 | 0.03 | 10 |
| mean II | | 18.472 | 35.013 | 0.10 | 3.97 | 0.08 | 0.53 | 1.77 | 652289 | | |
| | | | | | | | | | | | |
| mean | | 17.943 | 31.933 | 0.14 | 12.41 | 0.15 | 0.82 | 7.74 | 965098 | | |
| st. dev. | | 6.329 | 5.549 | 0.07 | 14.22 | 0.13 | 0.61 | 11.27 | 1453215 | | |

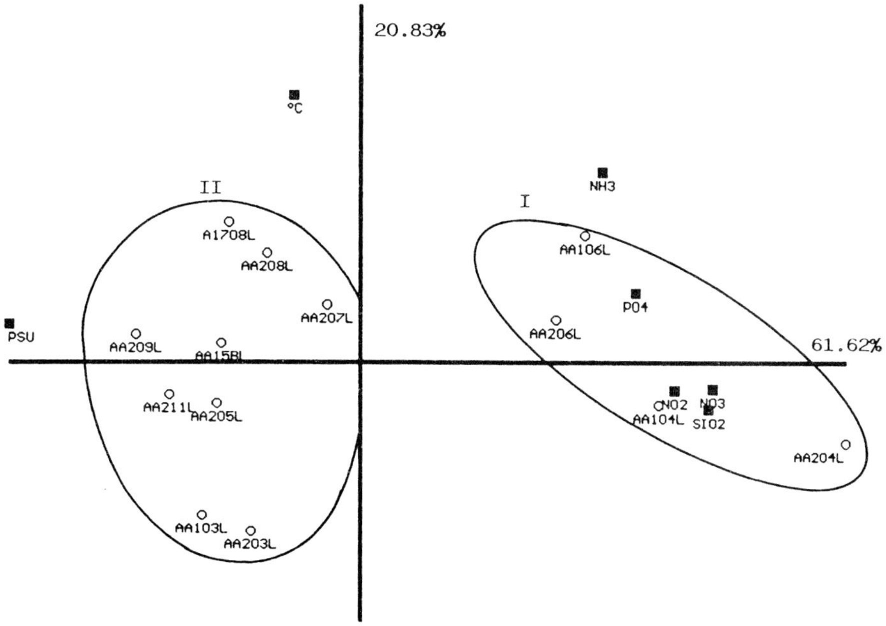


Fig. 3. Ordination of the samples shown by P.C.A. on physical and chemical parameters for the "surface lens". In the scattergram the distribution of the parameters and the percentage of the total variance are also shown.

Table 2. The characteristic of the two groups of samples referred to the "surface layer" identified by P.C.A. are reported.

| GR. | ST. | °C | PSU | PO4 | NO3 | NO2 | NH3 | SiO2 | Cell.*dm^-3 | X/Xmax | N. |
|---|---|---|---|---|---|---|---|---|---|---|---|
| | AA306S | 22.382 | 33.131 | 0.19 | 5.43 | 0.13 | 1.33 | 0.94 | 303741 | 0.53 | 2 |
| | AA308S | 26.541 | 36.615 | 0.10 | 1.09 | 0.08 | 1.29 | 0.97 | 231154 | 0.40 | 4 |
| | AA108S | 26.282 | 36.536 | 0.21 | 1.55 | 0.10 | 1.21 | 1.84 | 75935 | 0.13 | 14 |
| I | AA307S | 23.506 | 34.492 | 0.30 | 3.58 | 0.05 | 0.50 | 2.34 | 238972 | 0.42 | 3 |
| | AA107S | 22.620 | 34.520 | 0.24 | 3.49 | 0.02 | 1.37 | 1.98 | 96034 | 0.17 | 13 |
| | AA304S | 11.378 | 37.489 | 0.10 | 2.95 | 0.06 | 1.46 | 2.46 | 142935 | 0.25 | 11 |
| mean I | | 22.118 | 35.464 | 0.19 | 3.06 | 0.07 | 1.19 | 1.76 | 181462 | | |
| | A1709S | 22.651 | 36.613 | 0.00 | 0.79 | 0.01 | 1.57 | 0.06 | 221104 | 0.39 | 5 |
| | AA309S | 22.604 | 36.719 | 0.00 | 0.37 | 0.01 | 1.41 | 0.46 | 178668 | 0.31 | 8 |
| | AA209S | 23.223 | 36.561 | 0.00 | 0.43 | 0.00 | 0.39 | 0.83 | 192070 | 0.34 | 7 |
| | AA109S | 22.727 | 36.637 | 0.00 | 0.56 | 0.01 | 0.67 | 1.21 | 206958 | 0.36 | 6 |
| | AA25BS | 19.043 | 35.798 | 0.13 | 1.07 | 0.04 | 0.67 | 0.39 | 160801 | 0.28 | 9 |
| II | A175BS | 18.897 | 36.194 | 0.02 | 0.31 | 0.00 | 0.42 | 0.27 | 142937 | 0.25 | 10 |
| | AA35BS | 19.395 | 36.096 | 0.08 | 0.19 | 0.01 | 0.30 | 0.00 | 22334 | 0.04 | 17 |
| | A1705S | 15.562 | 37.203 | 0.17 | 0.39 | 0.02 | 0.78 | 0.00 | 69424 | 0.12 | 15 |
| | AA305S | 15.216 | 37.403 | 0.10 | 0.60 | 0.00 | 0.39 | 0.00 | 62534 | 0.11 | 16 |
| | A1706S | 22.092 | 33.778 | 0.21 | 0.49 | 0.04 | 0.07 | 0.01 | 571752 | 1.00 | 1 |
| | A1704S | 11.539 | 37.591 | 0.06 | 0.58 | 0.09 | 0.23 | 1.05 | 102735 | 0.18 | 12 |
| mean II | | 19.589 | 36.338 | 0.08 | 0.73 | 0.03 | 0.73 | 0.50 | 176065 | | |
| mean | | 20.333 | 36.081 | 0.11 | 1.40 | 0.04 | 0.87 | 0.87 | 177652 | | |
| st. dev. | | 4.421 | 1.285 | 0.09 | 1.47 | 0.04 | 0.46 | 0.82 | 122521 | | |

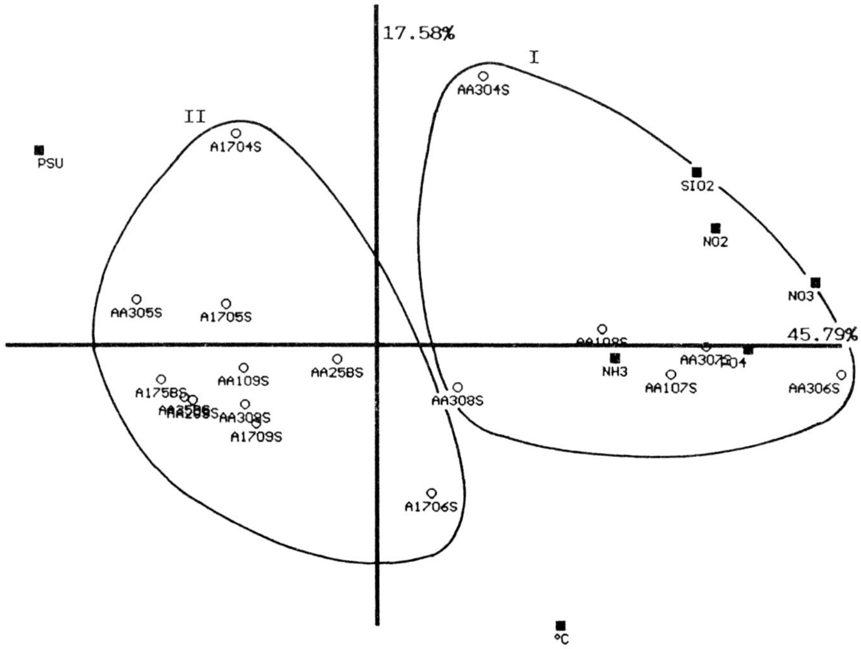


Fig. 4. Ordination of the samples shown by P.C.A. on physical and chemical parameters for the "surface layer". In the scattergram the distribution of the parameters and the percentage of the total variance are also shown.

Table 3. The characteristic of the two groups of samples referred to the "intermediate layer" identified by P.C.A. are reported.

| GR. | ST. | °C | PSU | PO4 | NO3 | NO2 | NH3 | SiO2 | Cell.*dm^-3 | X/Xmax | N. |
|---|---|---|---|---|---|---|---|---|---|---|---|
| I | AA107M | 22.772 | 35.720 | 0.20 | 2.19 | 0.08 | 1.10 | 0.89 | 558352 | 1.00 | 1 |
| | AA306M | 20.958 | 35.944 | 0.14 | 0.49 | 0.07 | 0.92 | 0.01 | 385262 | 0.69 | 2 |
| | AA206M | 21.103 | 35.887 | 0.08 | 0.83 | 0.03 | 0.77 | 0.12 | 166388 | 0.30 | 10 |
| | AA106M | 20.888 | 35.940 | 0.10 | 0.96 | 0.03 | 0.98 | 0.16 | 76081 | 0.14 | 13 |
| | A1706M | 20.383 | 35.627 | 0.16 | 0.81 | 0.06 | 1.53 | 0.01 | 341709 | 0.61 | 3 |
| | AA207M | 22.759 | 35.897 | 0.12 | 0.66 | 0.09 | 0.39 | 1.24 | 297042 | 0.53 | 4 |
| | AA307M | 22.534 | 36.661 | 0.15 | 0.35 | 0.04 | 0.42 | 1.04 | 232271 | 0.42 | 6 |
| | A1707M | 23.166 | 36.948 | 0.16 | 1.20 | 0.10 | 0.28 | 1.19 | 167504 | 0.30 | 9 |
| mean I | | 21.820 | 36.078 | 0.14 | 0.94 | 0.06 | 0.80 | 0.58 | 278076 | | |
| II | A175BM | 18.534 | 36.566 | 0.02 | 0.17 | 0.00 | 0.13 | 0.27 | 187606 | 0.34 | 8 |
| | AA35BM | 18.731 | 36.785 | 0.08 | 0.22 | 0.01 | 0.31 | 0.13 | 11167 | 0.02 | 15 |
| | AA25BM | 17.091 | 37.112 | 0.09 | 0.08 | 0.00 | 0.42 | 0.25 | 125069 | 0.22 | 12 |
| | AA15BM | 16.932 | 37.426 | 0.10 | 0.19 | 0.01 | 0.36 | 0.02 | 55839 | 0.10 | 14 |
| mean II | | 17.822 | 36.972 | 0.07 | 0.17 | 0.01 | 0.31 | 0.17 | 94920 | | |
| III | AA108M | 23.903 | 37.321 | 0.08 | 1.05 | 0.09 | 1.03 | 2.18 | 272474 | 0.49 | 5 |
| | AA208M | 23.834 | 37.230 | 0.07 | 1.92 | 0.12 | 1.01 | 2.66 | 194300 | 0.35 | 7 |
| | A1708M | 24.619 | 37.313 | 0.00 | 1.46 | 0.05 | 1.47 | 0.71 | 160800 | 0.29 | 11 |
| mean III | | 24.119 | 37.288 | 0.05 | 1.48 | 0.09 | 1.17 | 1.85 | 209191 | | |
| mean | | 21.214 | 36.558 | 0.10 | 0.84 | 0.05 | 0.74 | 0.73 | 215458 | | |
| st. dev. | | 2.389 | 0.637 | 0.05 | 0.62 | 0.04 | 0.43 | 0.79 | 135712 | | |

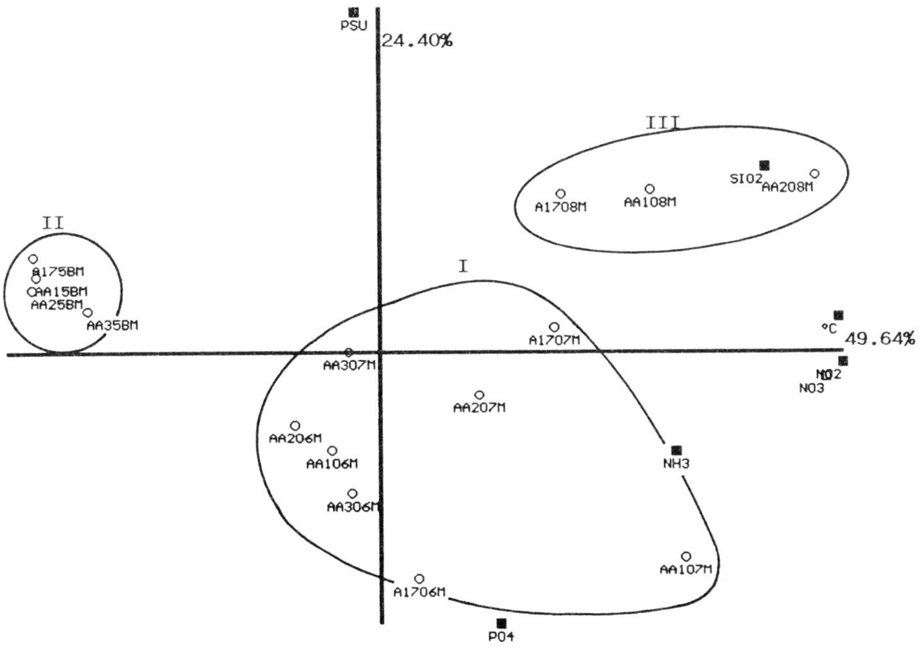


Fig. 5. Ordination of the samples shown by P.C.A. on physical and chemical parameters for the "intermediate layer". In the scattergram the distribution of the parameters and the percentuage of the total variance are also shown.

Table 4. The characteristic of the two groups of samples referred to the "bottom layer" identified by P.C.A. are reported.

| GR. | ST. | °C | PSU | PO4 | NO3 | NO2 | NH3 | SiO2 | Cell.*dm^-3 | X/Xmax | N. |
|---|---|---|---|---|---|---|---|---|---|---|---|
| I | AA207F | 17.922 | 37.575 | 0.00 | 0.67 | 0.02 | 2.25 | 3.19 | 471246 | 1.00 | 1 |
| | AA307F | 17.270 | 37.646 | 0.00 | 0.38 | 0.01 | 0.54 | 3.74 | 109433 | 0.23 | 6 |
| | AA25BF | 13.217 | 37.647 | 0.11 | 0.29 | 0.02 | 1.31 | 1.29 | 42433 | 0.09 | 11 |
| | AA306F | 14.619 | 37.683 | 0.07 | 0.56 | 0.02 | 1.21 | 1.93 | 13398 | 0.03 | 16 |
| | AA35BF | 13.151 | 37.785 | 0.07 | 0.12 | 0.03 | 1.51 | 0.46 | 11167 | 0.02 | 17 |
| | A175BF | 12.037 | 37.833 | 0.02 | 0.21 | 0.08 | 0.33 | 0.99 | 29033 | 0.06 | 14 |
| | A1709F | 15.729 | 37.940 | 0.11 | 0.64 | 0.01 | 1.73 | 4.70 | 109433 | 0.23 | 7 |
| | A1707F | 17.707 | 37.991 | 0.03 | 1.02 | 0.04 | 1.03 | 2.22 | 104966 | 0.22 | 8 |
| | A1706F | 13.527 | 37.746 | 0.36 | 1.60 | 0.02 | 2.21 | 0.10 | 122836 | 0.26 | 4 |
| | AA15BF | 12.773 | 37.731 | 0.42 | 0.13 | 0.00 | 2.30 | 1.78 | 49134 | 0.10 | 10 |
| mean I | | 14.795 | 37.758 | 0.12 | 0.56 | 0.03 | 1.44 | 2.03 | 106308 | | |
| II | AA107F | 17.098 | 37.671 | 0.08 | 1.27 | 0.07 | 1.72 | 5.11 | 303740 | 0.64 | 2 |
| | AA108F | 18.750 | 37.726 | 0.09 | 1.31 | 0.08 | 2.19 | 7.84 | 22332 | 0.05 | 15 |
| | AA208F | 19.289 | 37.822 | 0.05 | 1.67 | 0.15 | 3.51 | 7.43 | 30148 | 0.06 | 13 |
| | AA106F | 14.104 | 37.668 | 0.17 | 1.40 | 0.06 | 3.03 | 4.10 | 125066 | 0.27 | 3 |
| | AA206F | 14.659 | 37.691 | 0.23 | 2.35 | 0.07 | 2.97 | 4.49 | 37965 | 0.08 | 12 |
| | A1708F | 16.928 | 37.970 | 0.00 | 2.43 | 0.07 | 2.21 | 7.03 | 120602 | 0.26 | 5 |
| | AA308F | 19.013 | 37.925 | 0.03 | 2.63 | 0.07 | 6.13 | 4.88 | 75933 | 0.16 | 9 |
| mean II | | 17.126 | 37.782 | 0.09 | 1.87 | 0.08 | 3.11 | 5.84 | 102255 | | |
| mean | | 15.775 | 37.768 | 0.11 | 1.10 | 0.05 | 2.13 | 3.60 | 104639 | | |
| st. dev. | | 2.327 | 0.123 | 0.12 | 0.80 | 0.04 | 1.30 | 2.35 | 114381 | | |

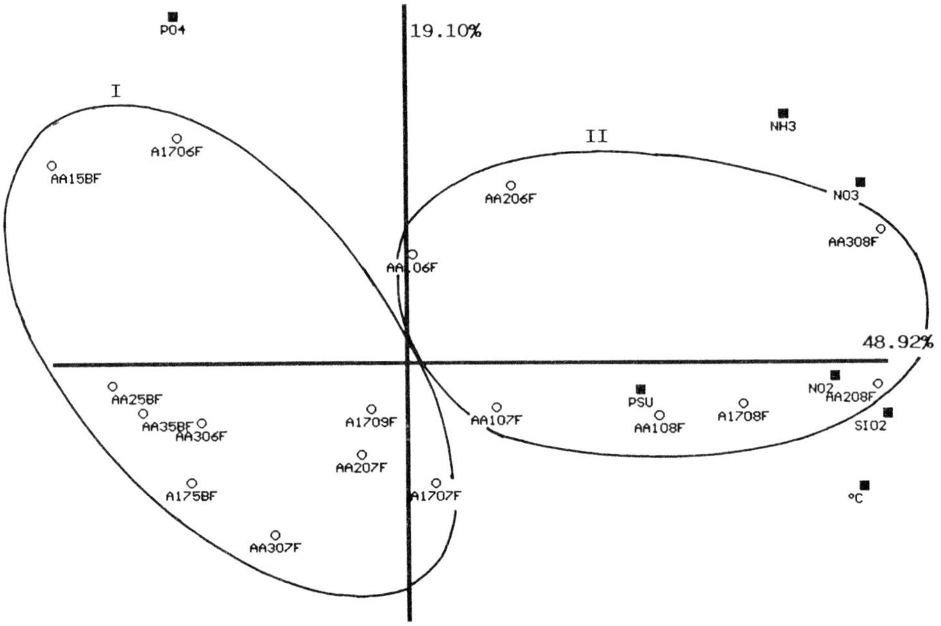


Fig. 6. Ordination of the samples shown by P.C.A. on physical and chemical parameters for the "bottom layer". In the scattergram the distribution of the parameters and the percentuage of the total variance are also shown.

## Conclusion

Clear evidence of a relation between salinity and silicates exists only in the surface lens during the period of stratification, confirming the strong dependence of their concentration on river inputs. In this water body the highest values of diatom abundance are also detected in spring and early summer even if no clear correlation appears between their density and salinity or dissolved silicates. The highest values of $Si - SiO_4^{4-}$ were detected in the more offshore stations where the diatom bloom was just starting or perhaps did not appear owing to some other nutrient limitation, while in the more coastal stations the dissolved silicates might already be partially taken up by diatoms. The surface layer shows intermediate values of diatom abundance and silicates, mainly due to sinking and diffusion processes respectively. The intermediate layer generally corresponds to the layers of density discontinuity. It generally shows intermediate values of diatom abundance and at the end of May the values decrease sharply in correspondence with the lowest mean values of dissolved silicates. At the bottom layer the increase of silicate is due to regeneration processes enhanced by the high temperatures because the regeneration of particulate biogenic silicate is strongly temperature-dependent (Conley & Malone 1992). This availability is not completely utilized by diatoms because of growth inhibition due to photolimitation at the end of summer.

It appears that silicate inputs in the Gulf of Trieste are confined only to the surface lens which strongly depend on river inputs over a short period determining a very limited spring diatom bloom. In the subsurface and intermediate layers only a small amount of silicates seems to be available thus determining small populations of diatoms. Regenerated silicate cannot enhance diatom growth due to the existence of photolimitation at these depths at the end of summer.

Analysing an annual cycle of dissolved silicates in Chesapeake Bay, Conley & Malone (1992) found that the spatial and temporal distribution of dissolved silicates was dominated by river flow during winter–spring in the upper and mesohaline Bay (which may correspond to our surface lens), and by benthic regeneration during summer in the lower Bay as happens in our bottom water body. They also assessed that the dissolved silicates might control the magnitude of diatom production during the spring bloom and the flux of phytoplankton biomass to the benthos. In our case, besides $N - NO_3^-$, which basically shows the same pattern as dissolved silicate, these two nutrients could at least share control of the diatom abundance.

Finally, as regards the difference in taxonomic composition among the various water bodies, it appears interesting to highlight that the species dominant only in the surface lens (*Skeletonema costatum*) requires 0,56–2 pg Si/cell (Paasche 1973) while for example one of the dominant species at the intermediate layer (*Cyclotella glomerata*) needs 100 pg Si/cell (Einsele & Grim 1938). These differences in the rate of dissolved silicate uptake may be the explanation for the relevant differences observed in the highest number of cells reached by each of the prevalent species.

## Acknowledgements

The authors are very grateful to Dr N. Burba for statistical help, Dr S. Predonzani for chemical analyses and Mr E. Cociancich for technical assistance. They want to thank also the two anonymous reviewers for their helpful comments.


## References


Burba, N., Cabrini, M., Del Negro, P., Fonda Umani, S. & Milani, L. (1994). Variazioni stagionali del rapporto N/P nel Golfo di Trieste. *Atti del 10° Congresso A.I.O.L.*, 333–344.

Burba, N., Feoli, E., Malaroda, M. & Zuccarello, V. (1992). Un Sistema Informativo per la Vegetazione. *Manuale di utilizzo dei programmi.* Gorizia, Collana quaderni C.E.T.A., **2**, 1–78.

Conley, D. J. & Malone, T. C. (1992). Annual Cycle of dissolved silicatess in Chesapeake Bay: implications for the production and fate of phytoplankton biomass. *Marine Ecology Progress Series*, **81** (2), 121–128.

Egge, J. K. & Aksnes, D. L. (1992). Silicate as regulating nutrient in phytoplankton competition. *Marine Ecology Progress Series*, **83**, 281–289.

Einsele, W. & Grim, J. (1938). Über den Kieselsäuregehalt planktischer Diatomeen und dessen Bedeutung für einige Fragen ihrer Ökologie. *Z. Bot.*, **32**, 545–90.

Franco, P., Jeftic, L., Malanotte Rizzoli, P., Michelato, A. & Orlic, M. (1982). Descriptive model of the Northern Adriatic. *Oceanologica Acta*, **5** (3), 379–389.

Paasche, E. (1973). The influence of cell size on growth rate, silica content and some other proprieties of four marine diatom species. *Norwegian Journal of Botany*, **20**, 197–204.

Round, F. E., Crawford R. M., Mann, D. G. (1990). *The Diatoms. Biology and morphology of the genera.* Cambridge University Press, Cambridge. 747 pp.

Stravisi, F. (1983). The vertical structure annual cycle of the mass field parameters in the Gulf of Trieste. *Boll. Oceanol. Teor. Appl.*, **1** (3), 239–250.

Stravisi, F. (1987). Interannual climatic variation in the Northern Adriatic Sea. In: *Jellyfish blooms in the Mediterranean.* UNEP II Workshop "Jellyfish in the Mediterranean". *Map Technical Reports Series* No. **47**, Athens, 1991, 175–187.

Strickland, J. D. H. & Parsons, T. R. (1972). A pratical handbook of seawater analysis. *Bulletin of the Fisheries Research Board of Canada*, 1–167.

UNESCO (1983). Algorithms for computation of fundamental properties of seawater. *UNESCO Technical Papers in Marine Science*, **44**, 1–53.

Utermöhl H. (1958). Zur Vervolkommung der quantitativen Phytoplankton methodik. *Mitteilungen. Internationale Vereinigung für Theoretische und Angewandte Limnologie*, **9**, 1–38.

# Diatoms from hypersaline coastal environments in La Trinitat salt works (Spain)

E. Clavero*, V. Merino, J. Grimalt* and M. Hernández-Mariné

**Departament Química Ambiental, CID–CSIC, Jordi Girona 18, 08034 Barcelona, Spain.*

*Unitat de Botànica, Facultat Farmàcia, Universitat de Barcelona. Av. Joan XXIII, s/n. 08028 Barcelona, Spain.*

## Abstract

A taxonomic study of the benthic and epiphytic diatoms of a coastal salt works (La Trinitat, Ebre Delta, Spain) has been made. Samples were collected in three ponds, which covered a salinity range between 44 ‰ and 108 ‰. The diatoms identified were species frequently encountered in littoral marine environments. Diversity was low at all stations, but decreased with increased salinity. As salinity increased, the dominant species were replaced by others. Only some species of the genera *Amphora*, *Nitzschia* and *Surirella* were present throughout all the ponds.

## Introduction

Of the environmental factors that influence the distribution of diatoms, salinity is probably the strongest (Battarbee 1986). As in many others taxonomic groups, diatoms range through the salinity spectrum from 0 to 100 ‰ (Remane 1971), and they completely disappear when concentrations reach 130–140 ‰ (Noël 1984); some workers have investigated waters at even higher salinities: 168 ‰ (Rincé & Robert 1983), 180 ‰ (Ehrlich 1978), 205 ‰ (Ehrlich & Dor 1985). Most diatoms are more or less tolerant to salinity, but it is still possible to recognize the limits within which species achieve their maximum development (Tomás 1988). Thus, they are of great use mainly to palaeoecologists as indicators of salt concentration (Battarbee 1986; Schiefele & Schreiner 1991). Nevertheless, it is never a simple task. Salinity not only affects the community composition but also the diatom morphogenesis. Changes in salinity can induce in some species a variety of phenotypes genetically determined or environmentally induced, and be the cause of many systematic confusions (Tomás 1988; Carperlan 1978).

Coastal salt-works (salinas) are man-made environments that are exploited to obtain halite (NaCl) for human consumption and industrial purposes. The process is based on the evaporation of brines using the sun and the wind as the energy sources. Thus, they are restricted to areas with climates characterized by periods during which evaporation exceeds precipitation such as the Mediterranean region. Halite is obtained by passing seawater to an array of evaporation ponds. The water flows through this system, evaporates and increases in salinity through successive ponds. Salt-works are a typical example of extreme environments in the sense that only a few taxonomic groups are present (de Witt & Grimalt 1992). Fishes, macrophytes and macroalgae are found only in low salinity ponds (up to 60 ‰), and the diversity of invertebrates species decreases as the salinity increases (Britton & Johnson 1987). The reduced activities of macrophytes and invertebrate animals allow the proliferation of benthic cyanobacteria and bacteria resulting in the formation of microbial mats (de Witt & Grimalt 1992).

The most effective exploitation of the salinas depends on the constancy of the salinities of the different ponds. Therefore, the salinity in the ponds is checked repeatedly and the water flow adjusted to reduce salinity fluctuations. The salt content is not the sole variable that determines the community composition since salinity fluctuation is strongly involved (Carpelan 1978). We can avoid the latter environmental factor when dealing with salt-works, because of the fairly stable salinity conditions that characterize them. It makes the salinas an excellent environment to study the ranges of salinity tolerance for diatoms (Noël 1984).

For the geologists the salinas represent a model system to study salt precipitation and related biogeochemical processes in shallow evaporitic environments. The knowledge of the diatom communities that thrive in the current evaporitic environments is extremely valuable to understand the genesis of the ancient evaporite deposits, since many species are also found in the diatomites (Noël 1984).

The major objective of the present study was to examine the diatom flora presently existing in the array of ponds with different salinities. Preliminary information about the autoecology of some species is also presented, paying special attention to the ranges of salinity tolerance. The most evident trends observed in this man-made system are also described, and compared with those present in the literature for similar environments. Finally, some taxonomical comments are presented.

## Material and methods

La Trinitat salt-work is located near the mouth of the Ebro river, on the Ebro delta (40°35'N, 0°40'E), Catalonia, Spain. Sea water is pumped from Els Alfacs Bay and runs through three consecutive large and shallow lagoons called depositors (Fig. 1). They are arranged in array and connected by several sluices. Coarse particles settle in these ponds and, because of evaporation, the brines become more and more concentrated (approximately from 40 ‰ to 150 ‰). Calcium carbonate also mainly precipitates here. The depositor 2 attains the highest salinity and its brine is transferred to a following series of ponds: the heaters. There are ten heaters having water of salinity from 150 ‰ to 280 ‰. In such salinities, calcium sulfate precipitates, therefore the heaters enclose the evaporitic gypsum domain (Ortí *et al.* 1984). Finally, the brine enters to the crystallizers where salinities become higher than 300 ‰ and halite is harvested. The salt-works are opened from March–April to October–November. When they are closed and no seawater is pumped in, neither the deposits nor the heaters become entirely dry.

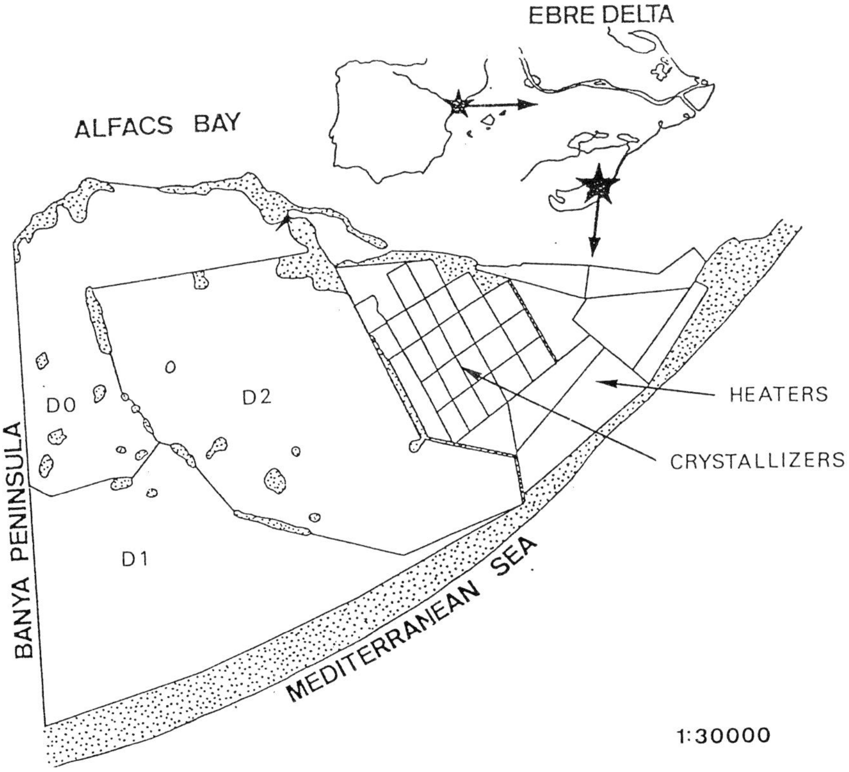


Fig. 1. Map of La Trinitat salt-works showing the sampling stations.

Only the three depositors (D0, D1, D2) were periodically sampled, since they had the lowest salinity water and diatoms thrived abundantly in them. The samples were obtained monthly from March (before the opening of the salt works) to July 1994. Several cores (about 0.5–1 cm) were obtained from the sediment in each sample station using a top-cut syringe. They were placed in Petri dishes and some coverglasses were placed on the sediments. After about 24 hours, the living diatoms that adhered to the coverglasses were transferred to test tubes, and preserved with several drops of formaldehyde. The fixed material was rinsed with distilled water, cleaned with 5% hydrochloric acid and then boiled in 30% $H_2O_2$ until cleared of the organic matter.

After repeated rinsings, the cleaned material was air-dried onto coverglasses and mounted on microscope slides using Naphrax®. The diatoms were identified using Nomarsky optics and a Nikon Optiphot–2 light microscope. For SEM observation, the material was prepared as above and air-dried on membrane filters. They were placed onto aluminium stubs and coated with gold. The SEM pictures were taken with a Hitachi S–2300. At each sampling, salinity was obtained by a densitometer and also by the standard method of titration with silver nitrate using seawater as a chlorinity standard.

**Results**

Throughout the period of study, salinity in each depositor remained fairly stable, slightly increasing from winter to summer (Fig 2). A remarkable change appeared in June after some days of a heat wave. The salinity of the three ponds became closer maybe due to a seawater intake justified by the increased evaporation.

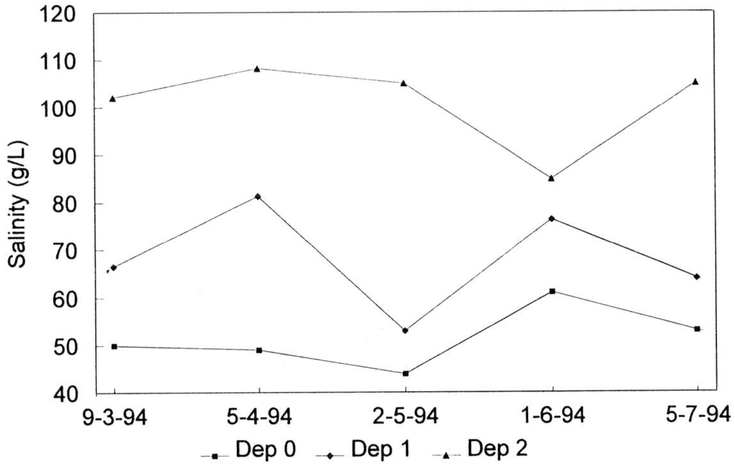


Fig. 2. Salinities recorded in the three stations during the sample period.

Altogether twenty-eight taxa were identified, which represented the most common diatoms growing in the three lagoons (Table II). The largest variety of taxa occurred at depositor 0 where the lowest salinity values were recorded. Macroscopic growth of diatoms was observed at all of the stations, but no attempt was made to obtain quantitative data.

In depositor 0 salinity usually ranged from 44 ‰ to 53 ‰, though there was a peak of maximum salinity (61 ‰) on May (Table I). The benthic diatom community thrived on a sandy sediment and the low salinity did not allow microbial mat establishment. In April, *Nitzschia* cf. *vidovichii* (Fig. 18) was the dominant benthic diatom but it was replaced by *Amphora* cf. *hyalina* (Fig. 19), *Nitzschia frustulum* (Fig. 9) and *N. sigma* in May. In July there was a great development of the aquatic

Angiosperm *Ruppia maritima* L., and the green algae *Chaetomorpha* sp. and *Cladophora* sp. which allowed the growth of epiphytic species such as *Achnanthes brevipes* (Figs 16, 17), *Striatella unipunctata* (Figs 12, 13), *Synedra crystallina* but mainly *Cocconeis placentula* var. *euplypta* (Fig. 14).

Table I. Data of the salinities recorded in the three stations during the sample period.

| | Salinity (‰) | | |
|---|---|---|---|
| Date | Dep.0 | Dep. 1 | Dep. 2 |
| 9/3/94 | 50 | 66 | 102 |
| 5/4/94 | 49 | 81 | 108 |
| 2/5/94 | 44 | 53 | 105 |
| 1/6/94 | 61 | 76 | 84 |
| 5/7/94 | 53 | 64 | 105 |

In depositor 1, salinity varied from 53 ‰ to 80 ‰ (Table I). *Pleurosigma elongatum* (Fig. 8) was the dominant species throughout the period of study. It appeared abundantly associated with filaments of cyanophyta constituting a thin layer that recovered the sediment surface. It also appeared on sporadically wet sediments and forming floating lumps with *Lyngbya aestuarii* Liebm. Other abundant species were, from March to May, *Amphora* aff. *ostenfeldi*, *Scoliopleura tumida* (Fig. 11) and *Gyrosigma spencerii*; and in July, *Surirella striatula* (Fig. 15). When in June *Cladophora* sp. strongly developed, it appeared surrounded by *Cocconeis placentula* var. *euglypta*, *C. bardawilensis* and *Pleurosigma elongatum*. Observations of the deep sediment by SEM showed that the diatom frustules did not maintain their structure but dissolved and disappeared a few millimetres (1–2 mm) below the surface.

In depositor 2, salinity averaged 100 ‰, except for the unusual value of 85 ‰ attained in June (Table I). *Nitzschia lembiformis* (Fig. 7) was the most abundant all through the period sampled. *Nitzschia closterium*, *N. sigma*, *Amphora coffeaeformis* (Figs 5, 6) and *Surirella striatula* were also frequent.

**Discussion**

All the diatoms found in La Trinitat salt works are frequent species of the littoral marine environment. Many similarities were found between La Trinitat and other Mediterranean salt-works (Noël 1982, 1984).

The lagoons showed a low diatom diversity which decreased as salinity increased. In the least saline lagoon, 28 species were recorded but only 11 species were found in the highest salinities. The dominance of *Nitzschia* cf. *vidovichii* (at 44 ‰–53 ‰) gave way

Table II. Diatom taxa and corresponding salinity ranges in La Trinitat salt works. * common, ** abundant, *** dominant.

| Taxa | Depositor 0 44 – 61 ‰ | Depositor 1 61.1 – 81 ‰ | Depositor 2 84.9–108.2 ‰ |
|---|---|---|---|
| *Achnanthes brevipes* Ag. | * | | |
| *Amphora acutiuscula* Kütz. | * | | |
| *Amphora* cf. *ostenfeldi* Hust. | * | *(*) | * |
| *Amphora arcus* Greg. | * | * | |
| *Amphora coffeaeformis* Ag. | * | * | ** |
| *Amphora* cf. *tenerrima* Hust. | * | * | * |
| *Amphora* aff. *hyalina* Kütz. | ** | * | * |
| *Cocconeis bardawilensis* Ehrlich | * | * | |
| *Cocconeis placentula* var. *euglypta* (Ehr.) Cl. | *(*) | *(*) | |
| *Gyrosigma spencerii* (Quekett) Grif. *et* Henfr. | * | *(*) | |
| *Mastogloia braunii* Grun. | * | | |
| *Navicula cincta* (Ehr.) Ralfs | * | * | |
| *Navicula complanata* (Grun.) Grun. | * | * | |
| *Navicula ramosissima* (Ag.) Cl. | * | * | |
| *Nitzschia acicularis* W. Smith | * | * | * |
| *Nitzschia closterium* (Ehr.) W. Smith | * | * | ** |
| *Nitzschia frustulum* (Kütz.) Grun. | ** | * | * |
| *Nitzschia lanceolata* var. *minor* (W. Smith) Van Heurck | * | * | * |
| *Nitzschia lembiformis* Meister | * | * | **(*) |
| *Nitzschia sigma* (Kütz.) W. Smith | ** | * | ** |
| *Nitzschia* cf. *vidovichii* (Grun.) Peragallo | ** | * | |
| *Pleurosigma elongatum* W. Smith | * | *** | |
| *Rhopalodia musculus* (Kütz.) O. Muller | * | | |
| *Scoliopleura tumida* (Breb.) Rabenhorst | * | *(*) | |
| *Striatella unipunctata* (Lyngb.) Ag. | * | | |
| *Surirella fastuosa* (Ehr.) Kütz. | * | | |
| *Surirella striatula* Turpin | * | *(*) | *(*) |
| *Synedra crystallina* (Ag.) Kütz | * | | |

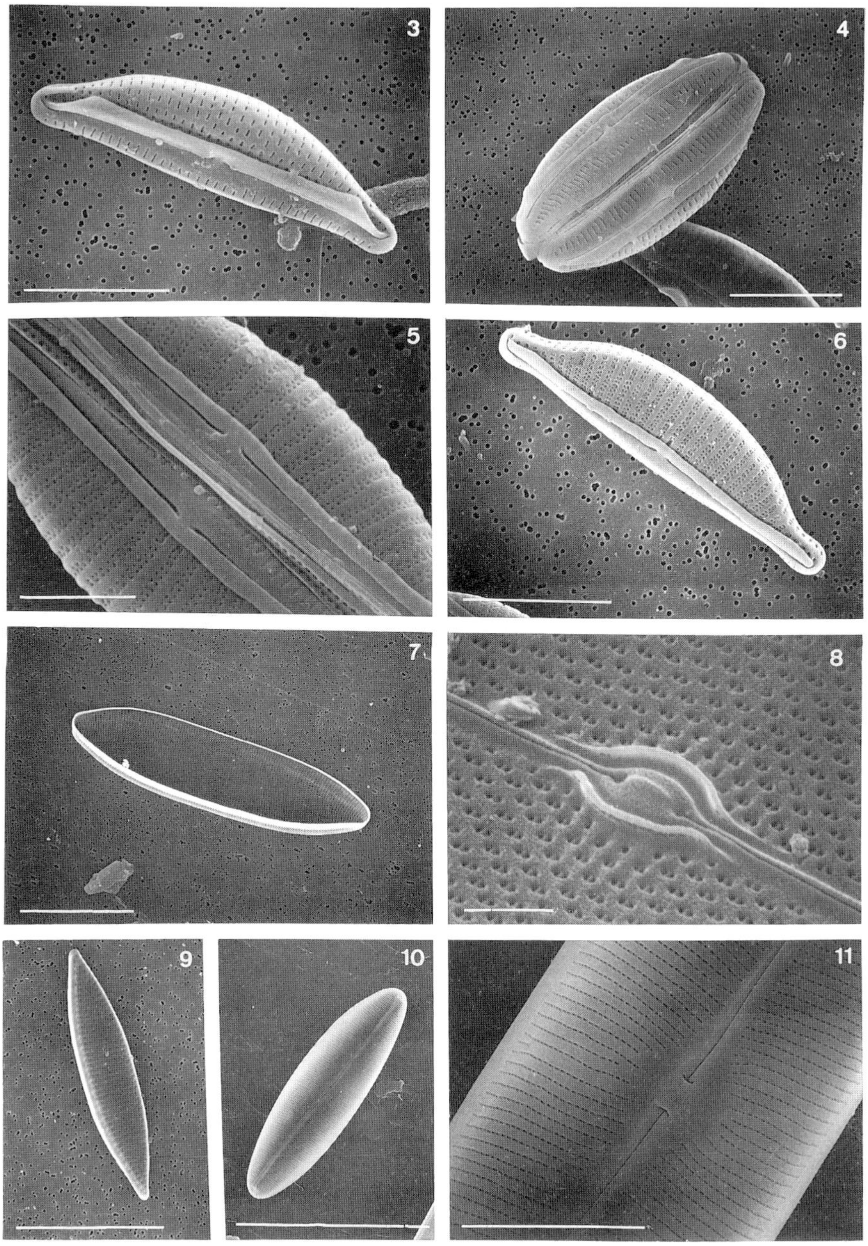


Figs 3–11. SEM micrographs of some diatoms from La Trinitat salt works. Figs 3, 4. *Amphora* cf. *ostenfeldi*. Figs 5, 6. *Amphora coffeaeformis*. Fig. 7. *Nitzschia lembiformis*. Fig. 8. *Pleurosigma elongatum*. Fig. 9. *Nitzschia frustulum*. Figs 10, 11. *Scoliopleura tumida*. Scale bars = 2 μm (Figs 5, 8), 5 μm (Figs 3, 4, 6), 10 μm (Figs 7, 9, 11), 50 μm (Fig. 8).

to *Pleurosigma elongatum* at salinities of 53 ‰ to 80 ‰, and then by *Nitzschia lembiformis* at the highest salt concentrations (85 ‰–108 ‰). A few species of the genera *Amphora*, *Nitzschia* and *Surirella*, were present throughout the entire area, which agrees with the data of Noël (1984) for the Bras del Port salt-works. They thrived at salinities from 44 to 108, and are therefore considered here as holoeuryhaline (*Amphora coffeaeformis*, *A.* cf. *tenerrima*, *Nitzschia sigma*, *Surirella striatula*, etc.).

The limits of salinity tolerance of the identified diatoms generally agreed with those indicated by the literature (Ehrlich & Dor 1985; Noël 1982, 1984; Rincé & Robert 1983; Tomás 1988). But the upper limit of salt tolerance of some diatoms in La Trinitat salt-works is difficult to establish, since the maximum salinity recorded was lower than the limit expected. Nevertheless, the maximum salinity tolerated by some species did not fit with previous data. *Surirella fastuosa* is considered a marine or brackish species, but we found it up to salinities of 61 ‰. This coincided with another study of Mediterranean coastal areas (Tomás 1988) that also reported it alive in highly saline environments. The upper tolerance limit of *Cocconeis placentula* var. *euglypta* (Fig. 14) has been established at 130 ‰ (Ehrlich & Dor 1985). In the present study it only occurred in salinities from 44 ‰ to 80 ‰. But as its presence seems to be related to macroalgae (*Cladophora*, *Chaetomorpha*) (Tomás 1988), its absence in higher salinities might be associated to the absence of a suitable substrate.

Some species seemed to prefer the upper salinities of the range were they thrived. *Surirella striatula* (Fig. 15) is a cosmopolitan species, common in both coastal and at athalasiohaline environments (Tomás 1988). Hustedt (1957) found it in several rivers of the Weser basin (Germany) and records it as (α)-mesohalobe. Felix and Rushforth (1979) cite it in the plankton of the Great Salt Lake (USA) at high salinities (113 ‰–129 ‰). Drum & Weber (1966) report it at salinities from 0 ‰ to 22 ‰ in a Massachusetts salt marsh. In Bras del Port salt-works (Noël 1984) it appeared, with different morphologies, between 64 ‰ and 128 ‰. We recorded it from 44 ‰ to 108 ‰, mainly in the higher salinities of the range.

Another species that proliferated best in high salt concentrations was *Nitzschia lembiformis*. Ehrlich & Dor (1985) also point to this behaviour in the Gavish Sabkha. Noël (1982) also records it at salinities of 52 ‰ and 164 ‰, but she identifies it as *Nitzschia pusilla* (Kütz.) Grun. In our opinion every report might simply be the same species, but a taxonomical study is required.

The proliferation of *Striatella unipunctata*, *Synedra crystallina* and *Achnanthes brevipes* that occurred over *Ruppia* and on the sediment where this angiosperm grew, was also observed by Noël (1984) in the Giraud salt works.

The upper limit of salinity tolerance showed by *Pleurosigma elongatum* (80 ‰), was the same as that established by Tomás (1988). The constitution of a thin layer on the sediment by species of the same genera is also reported by Van Heurk (1880–1885)

The degradation of the diatom frustules below the sediment surface appears to be a usual phenomenon in some extreme saline environments (Noël 1982; Ehrlich 1978). The frustules, instead of fossilizing, dissolve and play an important role in the silicon cycle of the brines and interstitial waters (Noël 1982). This dissolution of the frustules

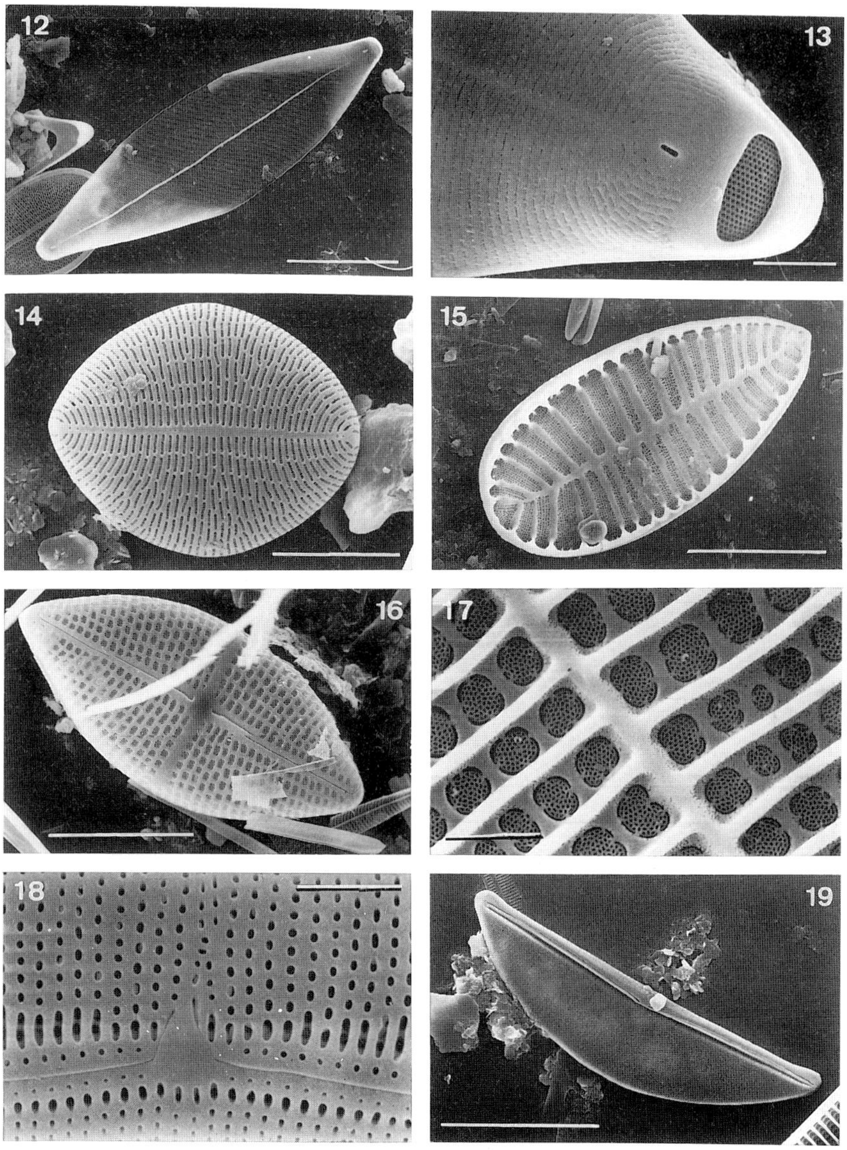


Figs 12–19. SEM micrographs of some diatoms from La Trinitat salt works (cont.). Figs 12, 13. *Striatella unipunctata*. Fig. 14. *Cocconeis placentula* var. *euglypta*. Fig. 15. *Surirella striatula*. Figs 16, 17. *Achnantes brevipes*. Fig. 18. *Nitzschia* cf. *vidovichii*. Fig. 19. *Amphora* aff. *hyalina*. Scale bars = 2 µm (Figs 13, 17, 18), 5 µm (Fig. 14), 10 µm (Fig. 19), 20 µm (Figs 12, 15, 16).

in hypersaline environments questions the mechanisms involved in their conservation in the ancient evaporitic environments.

## Acknowledgements

We thank the Spanish Ministry of Education (DGICYT PB93–0190–C02–01) for financial support. The Electron Microscopy Service of the University of Barcelona made available the electron microscopes for this study and provided technical help.,

## References

Battarbee, R. W. (1986). Diatom analysis. In: *Handbook of Holocene Palaeoecology and Palaeohydrology* (B. E. Berglund, ed.), 527–570.

Britton, R. H. & Johnson, A. R. (1987). An ecological account of a Mediterranean Salina: The Salin de Giraud, Camargue (S. France). *Biological conservation*, **42**, 185–230.

Carpelan, L. H. (1978). Evolutionary euryhalinity of diatoms in changing environments. *Nova Hedwigia*, **29** (3 + 4), 489–526.

de Witt, R. & Grimalt, J. O. (1992). Microbial ecosystems in Spanish coastal salinas; an ecological and geochemical study of biomarkers. *Limnetica*, **8**, 205–212.

Drum, R. W. & Webber, E. (1966). Diatoms from a Massachusetts salt marsh. *Botanica Marina*, **9** (1/2), 70–77.

Ehrlich, A. (1978). The diatoms of the hypersaline Solar lake (NE Sinai). *Israel Journal of Botany*, **27**, 1–13.

Ehrlich, A. & Dor, I. (1985). Photosynthetic microorganisms of the Gavish Sabkha. *Ecological Studies*, **53** (*Hypersaline Ecosystems*), 295–321.

Felix, E. A. & Rushforth, S. R. (1979). The algal flora of the Great Salt Lake, Utah, U.S.A. *Nova Hedwigia*, **31**, 163–195.

Hustedt, F (1957). Die diatomeenflora des Fluss-systems der Weser in Gebiet der Hansestadt Bremen. *Abhandlungen herausgegeben vom Naturwissenschaftlichen Verein zu Bremen*, **34** (3), 181–441.

Noël, D. (1982). Les diatomées des saumures des marais salants de Salin-de-Giraud (Sud de la France). *Géologie Mediterranéenne*, **9**, 413–446.

Noël, D. (1984). Les diatomées des saumures et des sédiments de surface du Salin de Bras del Port (Santa Pola province d'Alicante, Espagne). *Rev. Inv. Geol.*, **38/39**, 79–107.

Orti, F., Pueyo, J. J., Geisler-Cussey, D. & Dulau, N. (1984). Evaporitic sedimentation in the coastal salinas of Santa Pola (Alicante, Spain). *Rev. Inv. Geol.*, **38/39**, 169–220.

Remane, A. (1971). Ecology of brackish water. In: *Die Binnengewasser* (A. Remane & C. Schlieper, eds), vol. **22**, part I, 1–210. Stuttgart.

Rincé, Y. & Robert, J. M. (1983). Évolution des peuplements de diatomées planctoniques et benthiques d'un marais salant lors des variations printanieres de salinité. *Cryptogamie*, Algologie, **4**, 73–87.

Schiefele, S. & Schreiner, C. (1991). Use of diatoms for monitoring nutrient enrichment, acidification and impact of salt in rivers in Germany and Austria. In: *Use of algae for monitoring rivers* (B. A. Whitton, E. Rott & G. Friedrich, eds), 103–110. Innsbruck.

Tomás, X. (1988). Diatomeas de las aguas epicontinentales saladas del litoral mediterráneo de la Península Ibérica. (Tesis Doctoral, Universidad de Barcelona), 714 pp.

Van Heurk. (1880–1885). *Synopsis des Diatomées de Belgique*. Texte, 235p.

# The distribution of psammic algae on a marine beach at Praia Azul, Brazil

Marinês Garcia-Baptista

*Universidade Federal do Rio Grande do Sul, Departamento de Botânica, Av. Paulo Gama s/n °, Porto Alegre, Brazil, CEP 90040–060*

## Abstract

Psammic microalgae including numerous diatoms species were collected monthly along the backshore and swash zones on a dissipative sandy beach at Praia Azul. Data on species composition, horizontal distribution, abundance and comments on similarity among the Praia Azul community and other countries are presented. Three species groups, influenced by conductivity values which are related to land topography, were revealed by Reciprocal Averaging.

## Introduction

Praia Azul (municipality of Arroio do Sal) is a small village located on the coast in the northern part of the state of Rio Grande do Sul (29°29'S, 49°49'W) (Fig. 1 A and B).

The coastal region of Rio Grande do Sul is notable for a large (> 600 km) sandy beach stretching from the rocks of Torres to Arroio Chuí. More information is presented in Garcia-Baptista (1993), Toldo Jr. *et al.* (1993) and Calliari & Klein (1993).

The sediments of Praia Azul are well sorted, subround to round, very fine quartz (2.52 to 2.62 diameter) sand.

On the Rio Grande do Sul coast, the metereological tides, which are caused mainly by the wind, are more important than astronomic tides. The shore is subject to microtides which have an amplitude up to 2 m and because of this, the upper stations near the dunes are exposed to the air day and night.

The purpose of this work is to characterize the species composition of the psammic community in relation to seasonal variation, abundance, distribution and related environmental conditions.

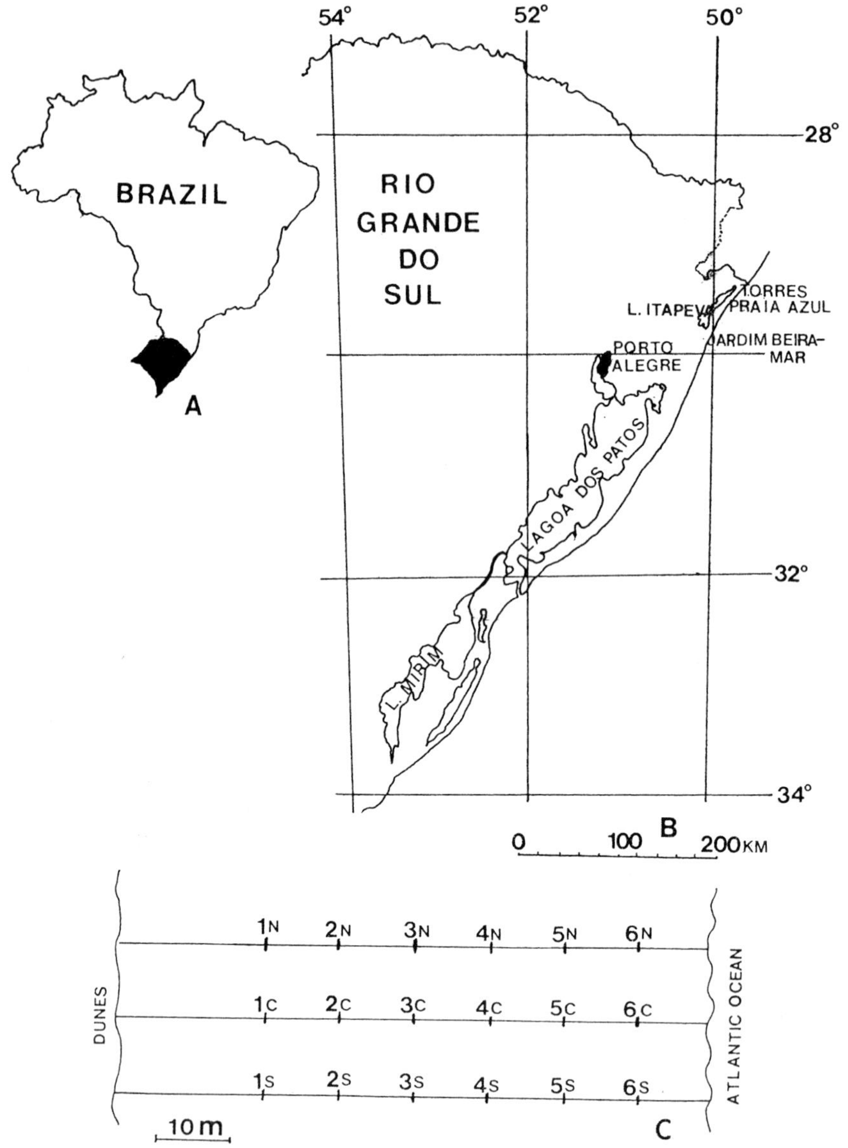


Fig. 1. Map showing the location of Praia Azul and the distribution of the sampling sites.

## Material and Methods

The samples were collected monthly during a 13 month period, from April 1 1990 to April 7 1991, between 8:30 a.m. to 12:30 p.m. The methodology employed follows.

A line on the beach was chosen where there were no freshwater, open drains or domestic sewage. Stakes (10) were planted every 10 m in a straight line, parallel with the foredunes line. One of the stakes was chosen at random and used as a fixed point of reference throughout the study. From that point, all collecting activities were developed. A transect, perpendicular to the foredunes line, was recorded monthly, starting 20 m from the foredunes, extending from the backshore to the swash. The latter was variable depending on sea level. On this transect, stakes were placed every 10 m marking the stations. Two other transects, parallel to the first and 10 m apart were also marked. The collecting stations, marked by stakes, were designated according to their positions on the beach with a number defining the station and a letter defining the transect (N = north, C = central, S= south); this allowed the collecting of duplicates. (Fig. 1 C)

At each sampling station a 0.50 m quadrat was subdivided into 16 small squares. Three of the squares were randomly chosen to sample algae and the sand of the other squares was used for chemical analyses.

The samples for algal analyses were taken with a 3.7 cm diameter PVC tube, in which a fine cut was made 1 cm in the middle. To collect the samples the PVC tube was pressed down vertically into the sediment. Sand from the first centimeter was collected by inserting a small knife into the ring. The sand from the three subsquares were pooled in a flask with 60 ml of 0.3% Lugol solution.

Samples were homogenized by shaking. Subsamples (1 to 10 ml) were withdrawn with a pipette and transferred to a Utermöhl chamber. The counting was done in random fields. Due to the great variability both in diversity and abundance of algae in the samples, the number of the fields in each chamber was determined in relation to minimal area and species-area curve. Cain & Castro (1959) criteria were used to determine the point on the curve at which it has 10% increase of the total number of species to a maximum of 10% of the total sample area. The counting was done with at least 100 organisms in each chamber. Only algae possessing chloroplasts were counted.

The number of algae counted was converted to density (number per $cm^3$) by the following formula (Taasen & Høisæter 1981), with some adaptations.

$$y = \frac{x.(60{,}400 + 32{,}256)}{z.32{,}256}.10^3 \text{ where:}$$

$y$ = is the number of algae per $cm^3$ in the top 1 cm layer,
$x$ = is the number of algae counted in each subsample,
$z$ = is the volume ($mm^3$) of sediment suspension counted each time,
60,400 = is the dilution volume ($mm^3$), and
32,256 = is the volume ($mm^3$) of sediment pooled.

Species were designated as abundant when present in numbers greater than the medium number of individuals per sample. Many species were indistinguishable at 400× and therefore they were counted in aggregates and listed in Table 1.

The weather readings taken from the Weather Station of Torres (29° 29'S, 49° 43' W), refer to 5 days before the sampling.

Ammonium, nitrite/nitrate, conductivity, sodium and available phosphorus concentration was determined by the Soil Laboratory of Faculdade de Agronomia (UFRGS) according to Tedesco *et al.* (1985).

The data were divided into two matrices: the species abundance data matrix and the chemical data matrix. To examine affinities among samples in relationship to abundance of their constituent taxa, both samples and species were ordered along axes by reciprocal averaging (RA). To facilitate interpretation, the vectors of samples ordination and scores derived from the species abundance data were correlated with each variable in the chemical data matrix. Species that have at least one occurrence with the relative abundance greater than or equal to 4% are included in RA. (Field *et al.* 1982)

The data matrix processed has a dimension of 52 species versus 60 samples. Three replica average (C,S and N) were made in order to reduce the matrix size.

Multivariate analysis was performed in NTSYS-pc program version 1.7 (Rohlf, 1992).

Table 1. Aggregate species list.

| |
|---|
| *Anabaena fertilissima* / *Anabaena iyengarii* |
| *Chlamydomonas* sp.1 / *Chlamydomonas* sp.2 |
| *Chloromeson parva* / *Chromulina* sp. |
| *Hantzschia distinctepunctata* / *Hantzschia longiareolata* |
| *Hantzschia psammicola* / *Hantzschia distinctepunctata* |
| *Navicula cryptocephala* / *Navicula consentanea* / *Navicula gregaria* / *Navicula* sp. |

## Results and Discussion

### *Species composition and species richness*

The Bacillariophyta accounted for the highest number of species (59.8%), followed by Cyanophyta and Dinophyta. In addition, the Bacillariophyta often appear in dense patches containing many species in great abundance, e.g. *Nitzschia palea, Navicula* cf. *tripunctata, Navicula cancellata, Navicula cryptocephala, Nitzschia* aff. *lanceolata* var. *pygmaea, Navicula salinaroides, Diploneis litoralis, Nitzschia arenosa, Campylosira cymbelliformis* and *Asterionellopsis glacialis* (Table 2). Cyanophyta and Euglenophyta were not important in number but can sometimes thrive, changing the colour of the sand. *Amphidinium britannicum* was the most important species of the Dinophyta, *Nodularia harveyana* and *Aphanothece castagnei* of the Cyanophyta and *Euglena viridis* of the Euglenophyta. The most numerous taxa were mobile.

Table 2. Species list distributed according to frequency (number of samples in which the species were observed). * Density greater than $10^5$ org./$cm^3$, ** Density greater than $10^6$ org./$cm^3$.

| Species | Frequency | Species | Frequency |
|---|---|---|---|
| ***Nitzschia palea* (Kütz.) W. Sm. | 167 | *Nitzschia epithemioides* Grun. | 31 |
| * *Navicula cancellata* Donk. | 155 | Gymnodiniales sp.2 | 29 |
| * *Navicula* cf. *tripunctata* (O.F. Müll.) Bory. | 154 | *Nitzschia gandersheimiensis* Krasske | 27 |
| **Navicula cryptocephala* Kütz. | 141 | *Chloromeson parva* Carter | 26 |
| ***Nitzschia* aff. *lanceolata* W. Sm. var. *pygmaea* Cl. | 138 | *Amphiprora alata* (Ehr.) Kütz. | 25 |
| *Hantzschia distinctepunctata* (Hust.) Hust. | 127 | *Pinnullaria cruciformis* (Donk.) Cl. | 21 |
| **Navicula salinaroides* Cholnoky | 123 | **Cylindrotheca fusiformis* Reim. & Lewin var. *fusiformis* | 19 |
| **Diploneis litoralis* (Donk.) Cl. | 119 | **Mastogloia exigua* Lewin | 17 |
| ***Nitzschia arenosa* Garcia-Baptista | 115 | *Navicula flagellifera* Hust. | 15 |
| *Campylosira cymbelliformis* (A. Schmidt) Grun. | 105 | *Aphanothece castagnei* (Bréb.) Rab. | 13 |
| ***Asterionellopsis glacialis* (Cast.) Round | 106 | *Phormidium* aff. *arcuatum* Skuja | 15 |
| *Chroomonas salina* (Wislouch) Butcher | 100 | *Skeletonema costatum* (Greville) Cl. | 15 |
| *Amphora laevis* Greg. var. *laevissima* (Greg.) Cl. | 74 | *Cylindrotheca signata* Reim. & Lewin | 10 |
| **Amphidinium britannicum* (C. Herdman) Lebour | 71 | **Merismopedia punctata* Meyen | 10 |
| *Pinnularia intermedia* (Lag.) Cl. | 66 | *Microcystis* sp. | 10 |
| **Nitzschia archibaldii* Lange-Bertalot | 62 | *Hantzschia amphioxys* (Ehr.) Grun. var. *vivax* (Hant.) Grun. | 9 |
| *Petroneis humerosa* (Brébisson) Stickle & Mann | 60 | *Nostoc paludosum* Kütz. | 10 |
| **Euglena viridis* Ehr. | 60 | *Cylindrospermum* sp.2 | 8 |
| *Nitzschia marginata* Hust. | 57 | *Nostocales* sp.1 | 8 |
| *Tropidoneis lepidoptera* Greg. var. *minor* Cl. | 56 | *Anabaena variabilis* Kütz. | 7 |
| *Hantzschia psammicola* Garcia-Baptista | 53 | *Anabaena oscillarioides* Bory | 7 |
| *Cylindrotheca closterium* (Ehr.) Reim. & Lewin | 44 | **Oscillatoria nigroviridis* Thwaites | 7 |
| *Navicula soodensis* Krasske | 38 | *Cylidrospermum* sp.1 | 6 |
| *Nitzschia spathulata* W. Sm. | 38 | *Anabaena fertilissima* Rao | 4 |
| *Dimerogramma hyalinum* Hust. | 38 | Euglenales | 4 |
| **Chlamydomonas* sp.1 | 35 | *Chrysococcus* cf. *ornatus* Pascher | 3 |

Both planktonic freshwater algae (*Aulacoseira ambigua* (Grun.) Simonsen) and marine species (*Skeletonema costatum, Chaetoceros pelagicus, Leptocylindrus danicus* Cl., ?*Stephanopyxis* sp., *Chaetoceros* aff. *affinis* Lauder, Centrales sp. 1 and sp. 2) were observed but not in great number, probably because they did not adapt to this environment.

The number of taxa observed at each sample ranged from 0 to 33. The driest stations had in general the lowest diversity values. No algae were detected in those stations where, in March soft, dry sand had been deposited by wind action. The highest numbers of algae were observed in April 1990 in stations with high humidity and low conductivity and where the sand became green. (Fig. 2 and Fig. 3)

The greatest number of species was found in April 1990 (68), in November (57) and in March (54). This coincides with wide beaches that have berms and depressions which permit microenvironments to emerge (Fig. 3 and Fig. 4).

Fig. 2. Horizontal distribution of mean organisms/ $cm^3$ (columns) and number of taxa (dots) in each station in each month.

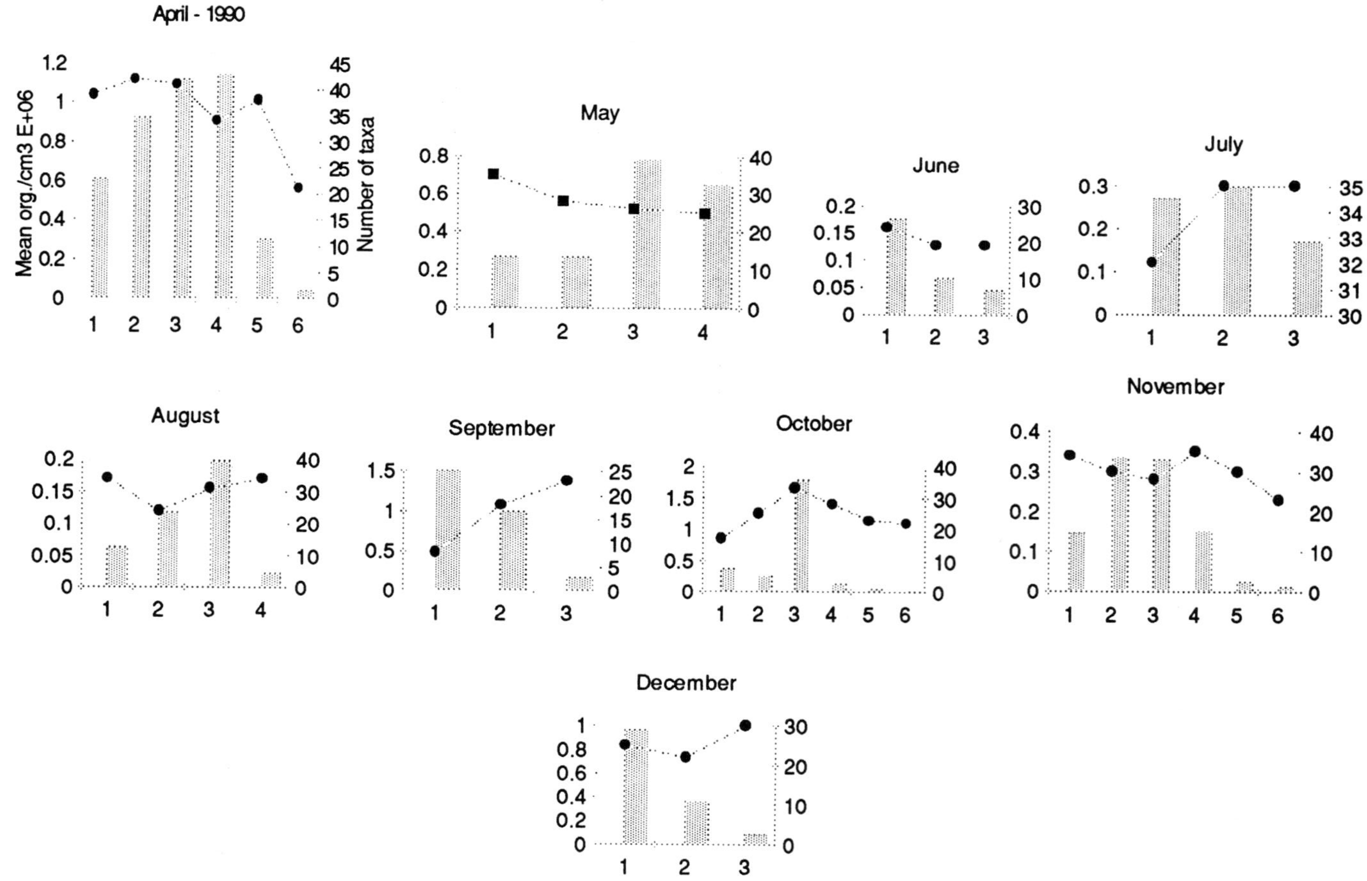

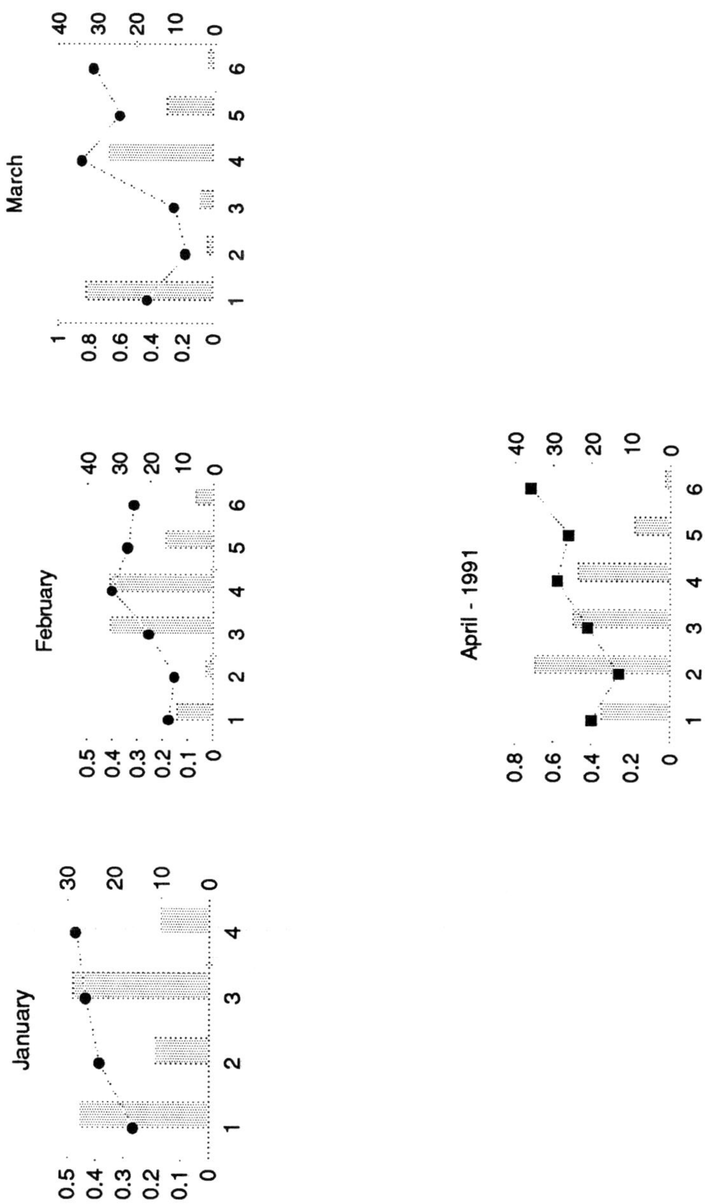
January
February
March
April - 1991

*Similarity*

From the flora found in Jardim Beira-Mar (beach 30 km south of Praia Azul) by Garcia-Baptista & Baptista (1992), only *Diploneis interrrupta* Kütz. and *Navicula pupula* Kütz. were absent in Praia Azul.

The diatoms *Asterionellopsis glacialis, Campylosira cymbelliformis, Chaetoceros* aff. *affinis, Cylindrotheca closterium, Dimerogramma hyalinum, Eunotogramma dubium* Hust./*Eunotogramma laeve* Grun., *Leptocylindrus danicus, Navicula cancellata* and *Skeletonema costatum* were all found in plankton during this study. The presence of *Dimerogramma hyalinum, Eunotogramma dubium / Eunotogramma laeve, Cylindrotheca closterium* and *Navicula cancellata* in Praia Azul plankton, lead me to consider the possibility of a continous interchange between plankton and sand.

The present study indicates that diatom assembleges in Praia Azul differ in at least two features from the ones found in other countries. Epipsammic diatoms were not found in Praia Azul unlike the situation described by Amspoker (1977) for an exposed beach. This is presumably related to sediment grain size since Amspoker & McIntire (1978) revealed an increase in the epipsammic taxa as the mean size of the sediment increased. Furthermore, *Achnanthes, Cocconeis, Gyrosigma* and *Pleurosigma,* found in great diversity or abundance by Aleem (1950), Round (1960), Riznyk (1973), Rao & Lewin (1976), McIntire (1978), Hay *et al.*(1993) and others, in benthic habitats, were never observed in Praia Azul. *Navicula* and *Nitzschia* are as common as in any other marine littoral habitats studied e.g. by Hustedt (1939), Brockmann (1950), Hendey (1964), Cook & Whipple (1982) and Oppenheim (1988).

*Horizontal distribution*

The distribution of some taxa along the transect and number of taxa in each station in each month is presented in Fig. 5. They are the result of the mean of different counts for each taxa or aggregates in each station during the study.

The highest number of organisms/$cm^3$ was found generally near the dunes (stations 1, 2 and 3). It probably resulted from substrate disturbance by wave action in stations 4, 5 and 6. Station 4 often supported the highest number of taxa. These results disagree with that of Riznyk & Phinney (1972) for an Oregon estuary where they found the greatest biomass in the lower intertidal zone.

The majority of the species had a wide distribution at the stations. Because of the sand grain homogeneity in relation to mean size and particle roundness, an homogeneous algal distribution could be expected, but the distribution througout the stations was not as uniform as the sediment.

The species with high frequency and abundance (see Table 2) thrived in stations 3 and 4, except *N. salinaroides* which was more abundant in station 1.

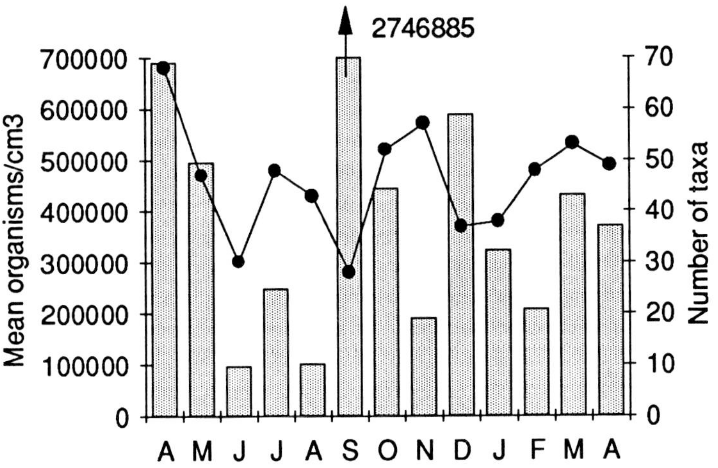


Fig. 3. Changes in the organisms/cm³ number (columns) and number of taxa (dots) for the period April 1990 to April 1991.

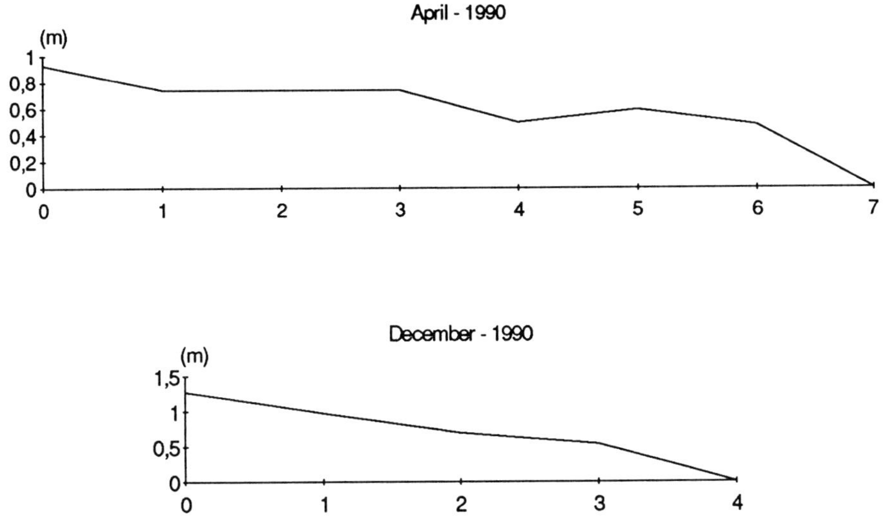


Fig. 4. Land topography in Praia Azul.

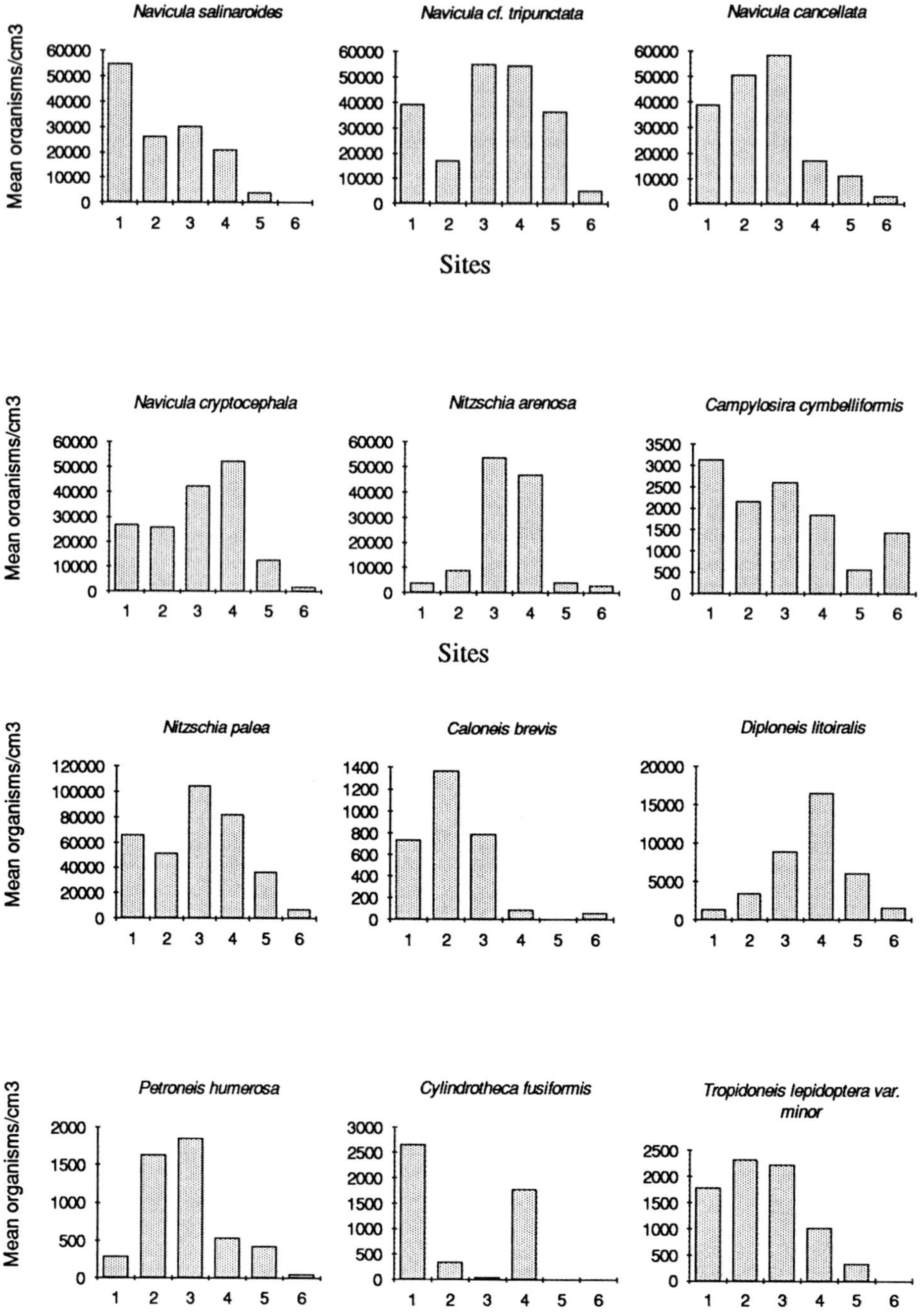
Navicula salinaroides
Navicula cf. tripunctata
Navicula cancellata
Mean organisms/cm3
Sites
Navicula cryptocephala
Nitzschia arenosa
Campylosira cymbelliformis
Mean organisms/cm3
Sites
Nitzschia palea
Caloneis brevis
Diploneis litoiralis
Mean organisms/cm3
Petroneis humerosa
Cylindrotheca fusiformis
Tropidoneis lepidoptera var. minor
Mean organisms/cm3

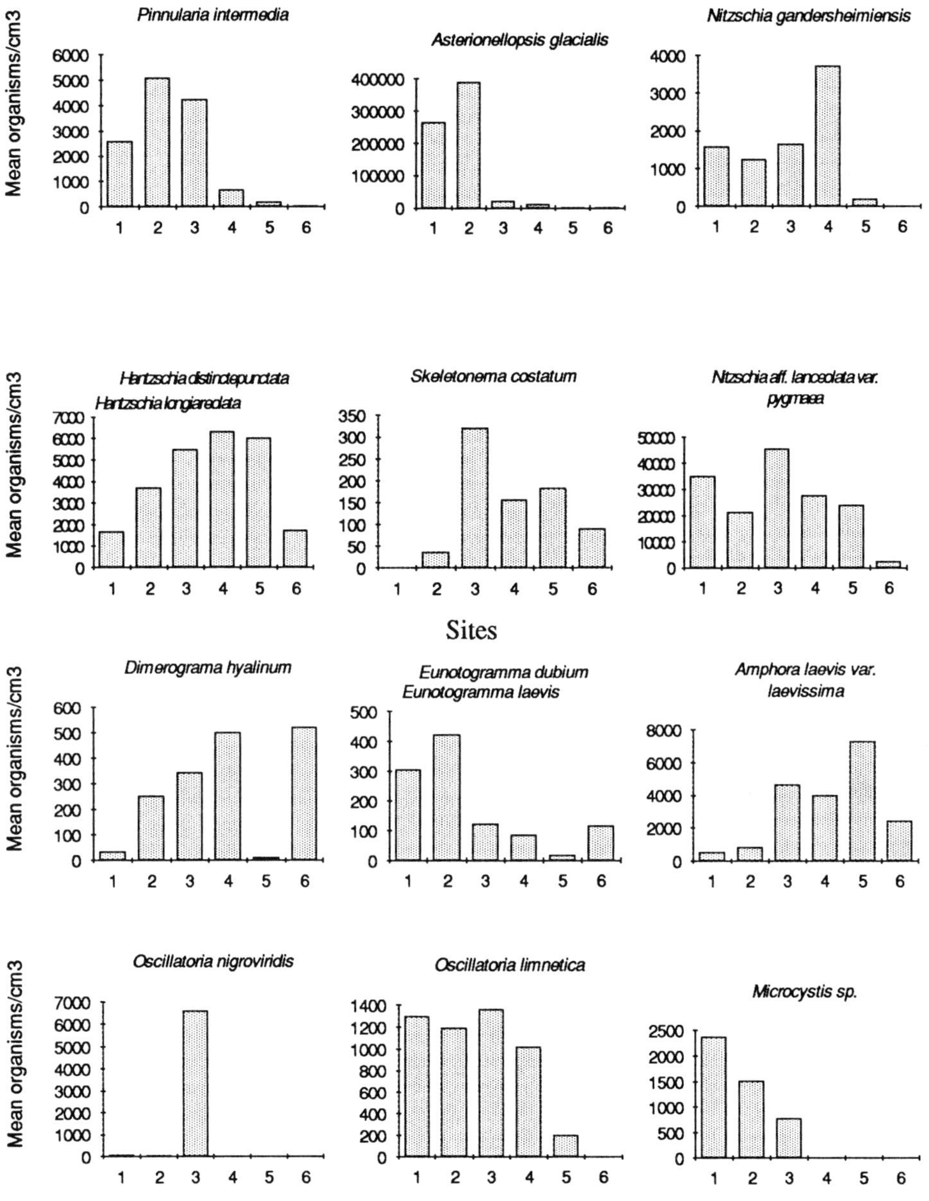
Mean organisms/cm3
Pinnularia intermedia
Asterionellopsis glacialis
Nitzschia gandersheimiensis
Hantzschia distinctepunctata
Hantzschia longiareolata
Skeletonema costatum
Nitzschia aff. lanceolata var. pygmaea
Sites
Dimerograma hyalinum
Eunotogramma dubium
Eunotogramma laevis
Amphora laevis var. laevissima
Oscillatoria nigroviridis
Oscillatoria limnetica
Microcystis sp.

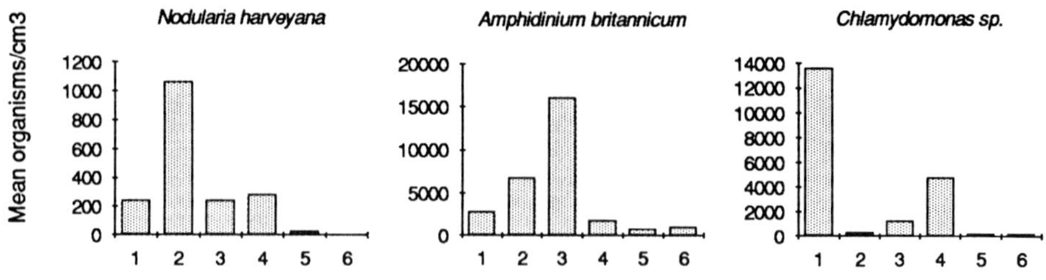


Fig. 5. Distribution of several species along the transect.

The Bacillariophyta had a wide distribution in the area, occurring frequently in one or more collecting stations. The Cyanophyta, Dinophyta and Euglenophyta thrive at a single station (except *Oscillatoria limnetica*).

Some taxa were restricted to two or three stations (1, 2 and 3) especially Cyanophyta and Chlorophyta like: *Oscillatoria nigroviridis, Scenedesmus acuminatus* (Lag.) Chod., *Chroococcus limneticus* Lemm., *Chroococcus turgidus* (Kütz.) Nag., *Anabaena variabilis, Dysmorphococcus varibilis* Takeda, *Mesotaenium macrococcum* var. *micrococcum* (Kütz.) W. & W., and *Scenedesmus acutus* Meyen, and the ones related with the planktonic habitat like *Cylindrotheca closterium, Leptocylindrus danicus, Chaetoceros pelagicus* (stations 5 and 6). Cook & Whipple (1982) found blue-green algal mats in a station where the tide action was absent just like it occurred in the collecting stations near the dunes, since the sea did not reach this point very often.

Besides being a planktonic alga, *Asterionellopsis glacialis* was found in all collecting stations probably related to the height of the tide. It was rarely observed in stations 1 and 2, and when found the abundance was low. The greatest number observed (Fig. 5) is due to a storm in September when it was brought in by the waves.

Several species, like *Navicula salinaroides, Navicula* cf. *tripunctata, Navicula cryptocephala* and *Nitzschia palea* seem to adapt to different environmental conditions of the beach, growing in humid or dry sand (0.1% humidity) under low and high conductivity values (20–5180 µS/cm). Low humidity did not prevent the growth of algae but the number of organisms was reduced.

The presence of algae in samples with low humidity (0.1 or 0.4%) is a surprising fact, though Aleem (1950) observed that several benthic diatoms could withstand desiccation in the laboratory for periods varying from 3 to 6 months and Williams (1964) refers to some benthic marine diatoms that can survive under large and rapid variation of salinity. This may occur with Praia Azul diatoms since they grow in different stations and under different conditions. Humidity was not a primary factor in reducing the number of diatoms on a salt marsh studied by Round (1960).

Other features should be considered in the algae distribution. First, land topography which permits sea or rain water retention as Brockmann (1950) observed. This factor was evident in April when a berm was formed and isolated the backshore from the sea, permitting freshwater accumulation and minimizing the wind action. Thus many freshwater species of Chlorophyta and Cyanophyta thrived (Fig. 4).

Another feature that may influence this distribution could be the presence of psammic invertebrate herbivores found in Praia Azul. Many herbivores were observed in the stations where there was a great diversity and number of algae. This confirms data of Amspoker (1977) and Griffiths & Griffiths (1983).

*Seasonal variations and abundance*

All algal groups are generally found throughout the year but some species are more abundant at certain times, Table 3 and Fig. 6.

Table 3. Distribution of diatoms with season.

| | |
|---|---|
| Winter abundant: | Summer abundant: |
| *Amphidinium britannicum* | *Amphiprora alata* |
| *Pinnularia intermedia* | *Chlamydomonas* sp.1 |
| | *Cylindrotheca closterium* |
| Spring abundant: | *Chroomonas salina* |
| *Amphora laevis* var. *laevissima* | *Hantzschia distinctepunctata* |
| *Asterionellopsis glacialis* | *Euglena viridis* |
| | *Navicula* cf. *tripunctata* |
| | *Nitzschia epithemioides* |
| | *Nitzschia palea* |
| | *Oscillatoria limnetica* |

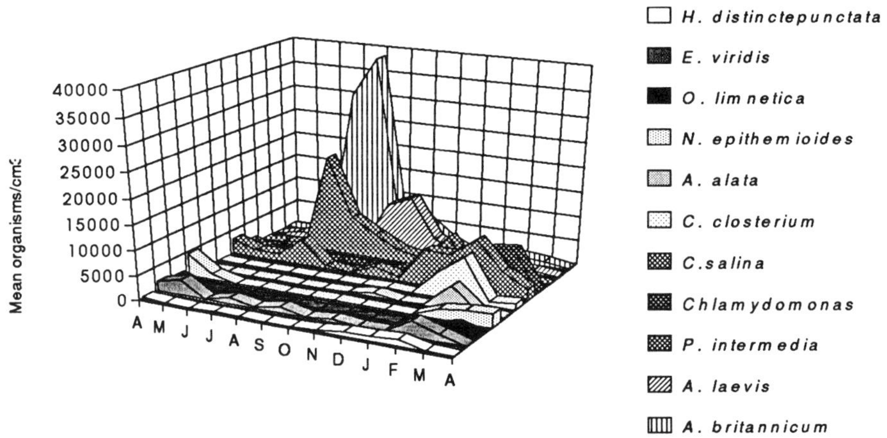


Fig. 6. Seasonal growth of some species for the period the April 1990 to April 1991.

Aleem (1950), Hendey (1964) and Riznyk (1973), observed a similar situation. On the other hand Admiraal *et al.* (1982) noted a great growth of Cyanophyta and Euglenophyta during the summer and Sullivan (1975) found Cyanophyta in the summer and Chlorophyta in the winter.

The Bacillariophyta comprise the highest number of organisms/cm$^3$ throughout the study (Fig. 7). The overall dominance of diatoms is shown in Table 2 and Fig. 7 which reveals that diatoms form over 90% of the flora. Diatoms contribution in other benthic habitats have already been emphasized e.g. by Round (1960) and Burkholder *et al.* (1965).

Blooms with sand discoloration were observed many times along the transects. Colour, stations, dates and main species are listed below.

– **green** – 1, 2, 3 and 4 – April 1990 – *Navicula cryptocephala, Oscillatoria nigroviridis, Aphanothece castagnei, Oscillatoria limnetica* Lemm., *Nostoc paludosum, Nitzschia archibaldii, Microcystis* sp. and *Mastogloia exigua.*

– **brownish yellow** – 3 – May 1990 – *Navicula cancellata, Navicula* cf. *tripunctata, Nitzschia arenosa, Nitzschia palea* and *Nitzschia* aff. *lanceolata* var. *pygmaea.*

– **brownish yellow** – 1N and 2N – June 1990 – *Navicula cancellata, Nitzschia* aff. *lanceolata* var. *pygmaea, Nitzschia palea, Asterionellopsis glacialis* and *Campylosira cymbelliformis.*

– **brownish yellow** – 2N and 3 – August 1990 – *Amphidinium britannicum, Navicula cancellata, Navicula cryptocephala, Nitzschia arenosa* and *Pinnularia intermedia.*

– **brownish yellow** – 3 and 4 – October 1990 – *Nitzschia* aff. *lanceolata* var. *pygmaea, Nitzschia arenosa, Navicula* cf. *tripunctata, Amphora laevis* var. *laevissima, Nitzschia palea* and *Navicula* cf. *tripunctata.*

– **brownish yellow** – 1C – December 1990 – *Navicula salinaroides, Nitzschia palea* and *Navicula cancellata.*

– **green** – 4C and 4S – February 1991 – *Nitzschia palea, Euglena viridis, Navicula* cf. *tripunctata, Chlamydomonas* sp1.and *Chroomonas salina.*

– **green** – 1S – March 1991 – *Chlamydomonas* sp1., *Navicula* cf. *tripunctata, Nitzschia palea* and *Navicula salinaroides.*

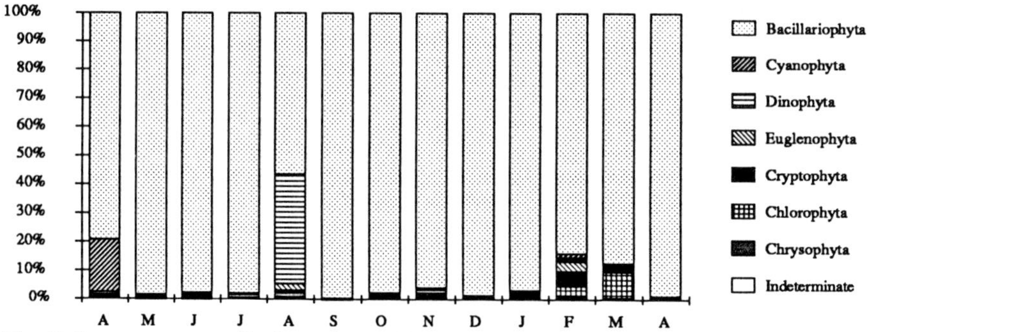


Fig. 7. Percentage contribution of the different algal groups for the period April 1990 to April 1991.

The mean number of organism/cm$^3$ found is similar to those recorded by Taasen & Høisæter (1981) and Riznyk & Phinney (1972) though a higher number of organisms can be found.

Fig. 3 shows that every peak is followed by a decrease in the number of organisms. This feature was detected by Round (1960) studying the changes in diatom flora of upper to the lower marsh on the river Dee (U.K.).

The number of organisms/cm$^3$ (Fig. 3) found on the beach in one month seems not to have any relation with the width and morphology of the beach, because in a narrow beach like that in December 1990 (Fig. 4) it is possible to find as many organisms as in a wide beach as in April (1990) or in greater number when compared with the beach of October 1990 and April 1991 which are similar to April 1990.

Seasonal variation is very difficult to understand since some species seem to have an opportunistic behaviour. For example, station 1 (July 1990), where there was low conductivity and low sodium values (Fig. 8), *Nitzschia archibaldii*, *Microcystis* sp., *Nitzschia gandersheimiensis* Krasske and *Navicula capitata* Ehr. var. *hungarica* (Grun.) Ross grew, but not in densities as high as the ones observed in April 1990, when the temperature and sand humidity were higher .

*Environmental conditions*

All environmental variables presented a marked seasonal and spatial variation. Fig. 8 shows the variation of conductivity, sodium, phosphorus, ammonium, humidity and nitrite/nitrate values.

The lowest average, minimum and maximum temperature (13.5°C; 9.8°C; 17,7°C) were recorded in August and the highest average, minimum temperature (25.1°C and 22.8°C) in February and the highest average maximum temperature (27.2°C) in March. The highest relative humidity recorded was of 99% in July and August when a strong mist lay on the beach. The lowest relative humidity was never less than 50%, permitting a humid environment on the beach throughout the year, May and February had no rain but in September the rainfall was the highest (72.1mm).

Fig. 8 shows that conductivity values increased throughout the year. The lowest sodium values and conductivity were observed in April 1990 ranging from 4 to 34 ppm and 17.5 to 63 µS/cm respectively. All other months had higher concentrations. January and March, (1620–1836 ppm) had higher sodium value and conductivity. The conductivity ranged from 4250 to 5180 µS/cm in stations 2C and 2N in January 1991.

Chomenko & Schäfer (1984) recorded the conductivity in many coastal lagoons and shallow lakes of Rio Grande do Sul and concluded that when values lower than 1000 µS/cm existed, water can be considered as freshwater. This criterion suggests that the sand microenvironment provides freshwater in its interstices.

When comparing the phosphorus (Fig. 8) and the number of organisms/cm$^3$ (Fig. 3) the same trends can be detected. This relationship could be attributed to the stability of phosphorus in the sediment since it is not leached by precipitation (Moniz 1975) unlike nitrite and nitrate. The highest concentration of phosphorus was observed in October and November near to sea and nitrite/nitrate (December and April 1991) and

Conductivity

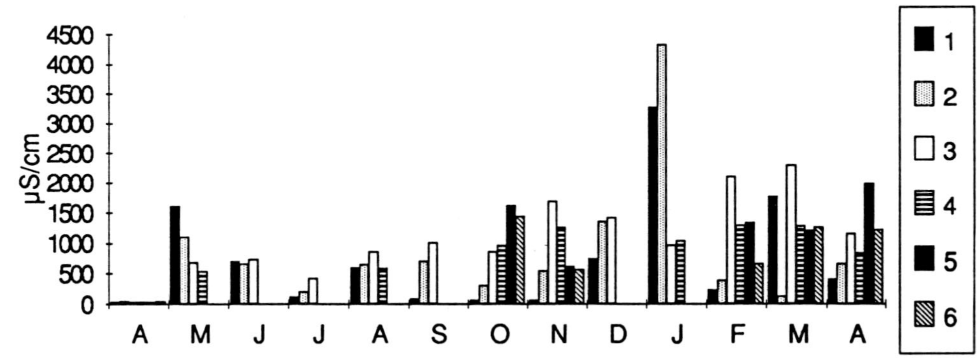


Sodium

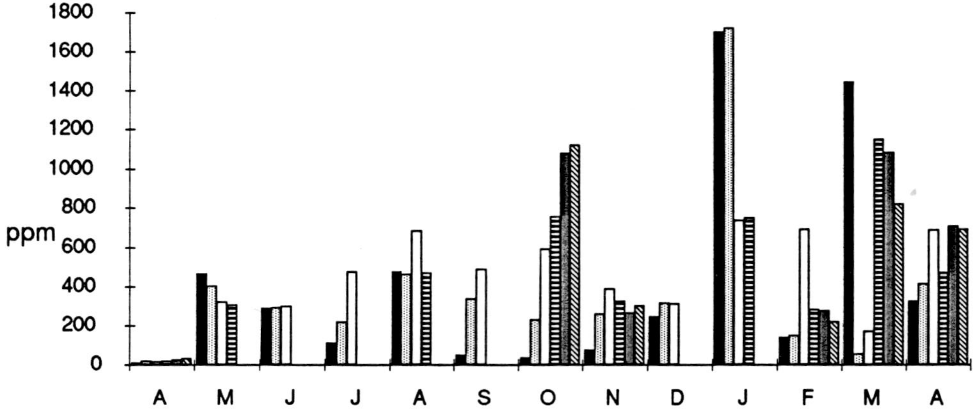

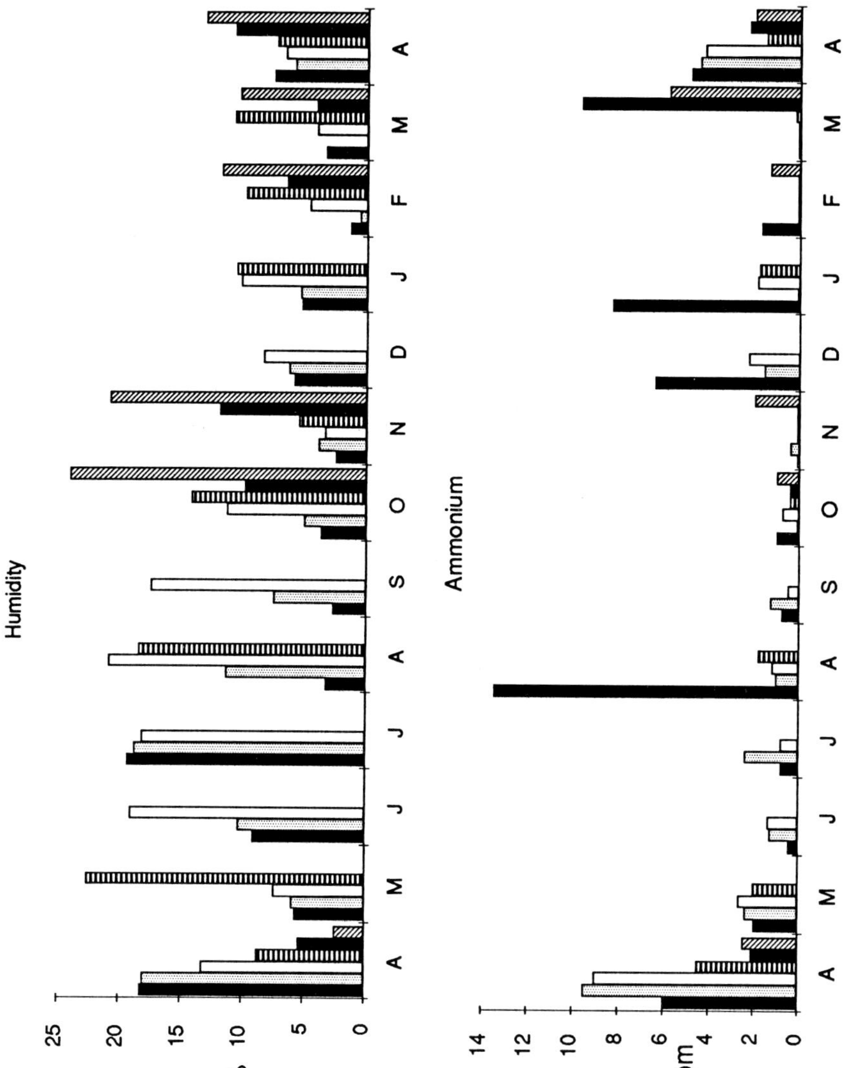

Humidity
%
0
5
10
15
20
25
A M J J A S O N D J F M A
Ammonium
ppm
0
2
4
6
8
10
12
14
A M J J A S O N D J F M A

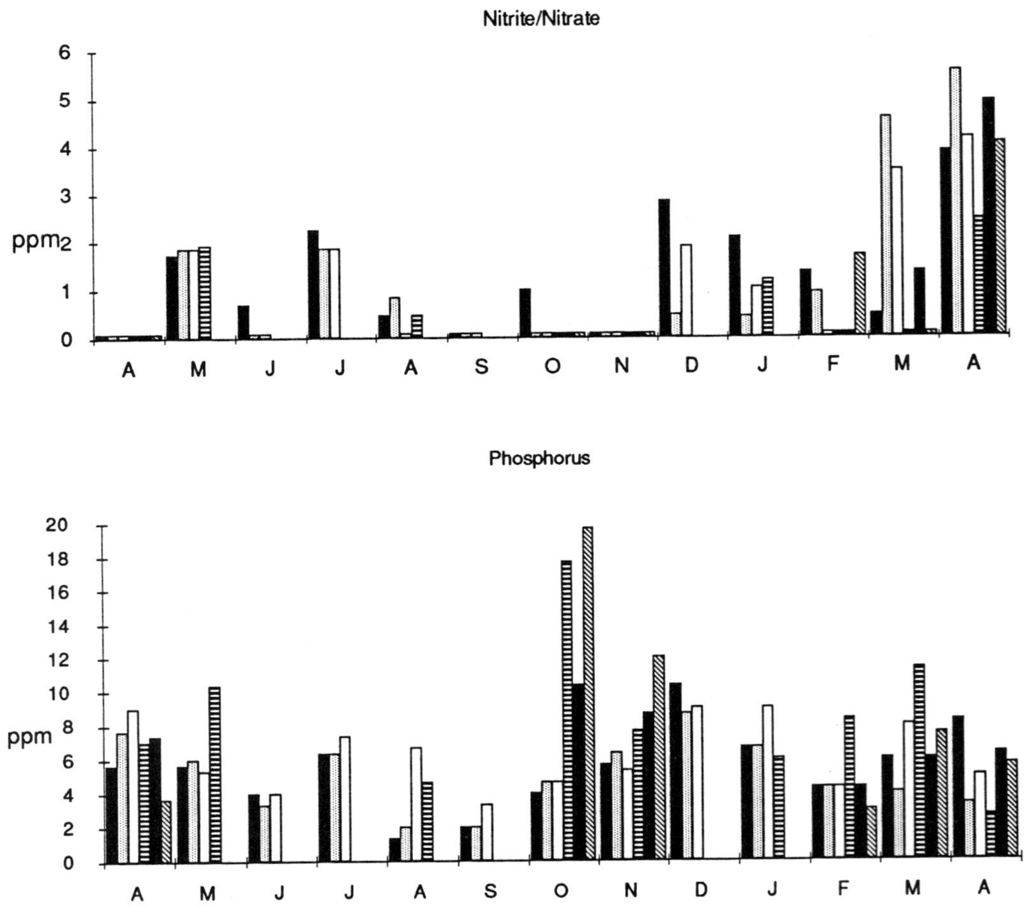


Fig. 8. Seasonal variation of conductivity, sodium, humidity, phosphorus, nitrite/nitrate and ammonium in each sampling station.

ammonium (August and January) near the dunes. Ammonium was always present. The great number of dead invertebrates that are deposited on the beach every day or stay there because of the tide enriches the sediment with ammonium and nitrite/nitrate.

The stations near the sea were the most humid, except in April and July 1990 where the highest humidity was detected near the dunes.

*Community ordination*

The reciprocal averaging procedure generated species ordinations (Fig. 9) and sampling ordinations (Fig. 10) using species abundance matrix. There is a correspondence between the sample and species ordination which clearly shows three groups of samples and species.

Four samples of April 1990, station 1 (November), station 1 (February) station 2 (March) and station 5 (April 1990), form Group 1, which is dominated by *Navicula cryptocephala, Aphanothece castagnei, Oscillatioria nigroviridis, Nostoc paludosum, Nitzschia archibaldii, Microcystis* sp., *Nitzschia gandersheimiensis; Mastogloia exigua* is abundant; *Dysmorphoccocus variabilis, Mesotaenium macrococcum* var. *micrococum, Scenedesmus acutus, Spirulina subtilissima* Kütz., *Fallacia pygmaea* (Kütz.) Stickle & Mann and *Nodularia harveyana* (Thw.) Thuret were present but restricted to April 1990.

The largest group (Group 2) has samples with different chemical characteristics and many species are involved.

In Group 3, samples that were under a strong marine influence with high density of *Asterionellopsis glacialis* occur.

To examine relationships between species groups and the chemical data, the species abundance data were partitioned in 3 subsets. For each subset, reciprocal averaging ordination were generated and ordinations scores for axes 1 and 2 of each subset were correlated with the corresponding chemical variables. (Table 4).

Group 1, the first axis is correlated with sodium values (r=-0.96) and conductivity (r=-0.96). The second axis is correlated with humidity (r=-0.89). This reveals that these species can occur in stations which are almost dry but with interstitial freshwater.

Group 2 is related to conductivity (r=-0.66) showing that these speices grow better when conductivity is low. Ammonium (r=0.68) has an influence on the species growth.

Group 3 shows a correlation with humidity (r=0.53 on the first axis and r=-0.60 on the second) demonstrating that *Asterionellopsis glacialis* can be found in humid sand, when recently deposited, or in dry sand when deposited up to 15 days before.

Salinity (Amspoker & McIntire 1978) and sediment proprieties (Whiting & McIntire 1985) are the most important factors to explain floristic discontinuity in sediment-associated marine diatoms. Combinations of environmental variables (Oppenheim 1991) may also operate. In Praia Azul the most important factors for floristic discontinuity were the salinity related to land topography, storm periods and ammonium.

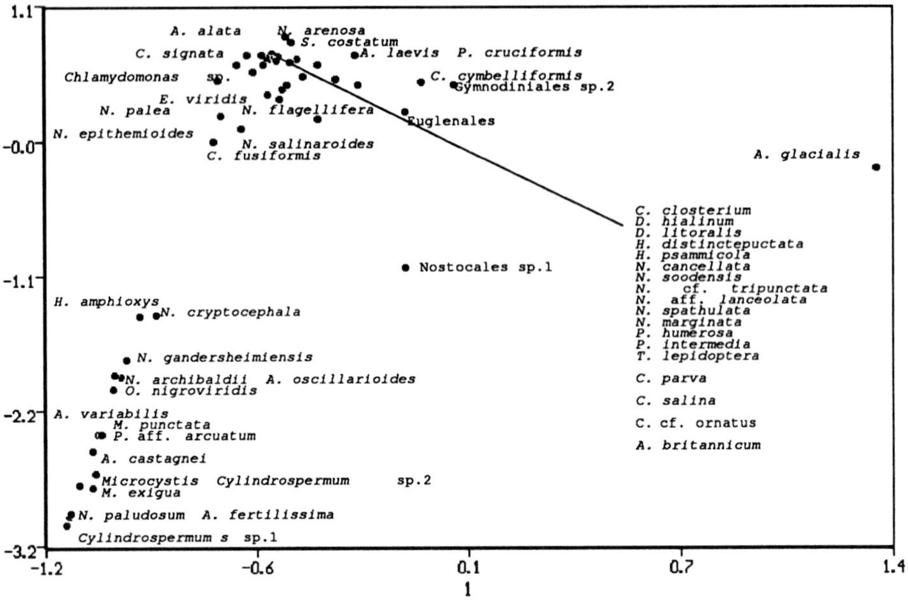


Fig. 9. Reciprocal averaging ordination of 52 taxa.

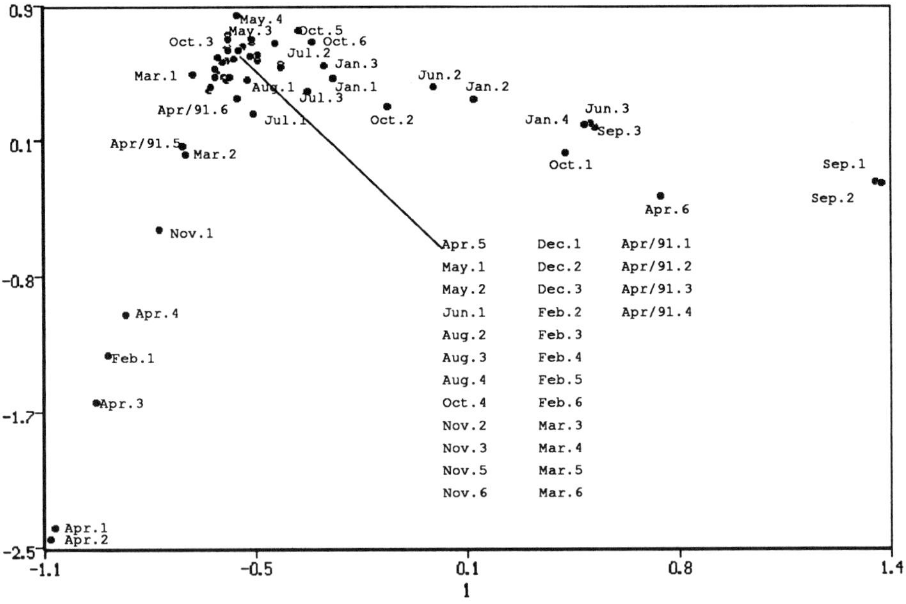


Fig. 10. Reciprocal averaging ordination of 60 samples (month.station).

Table 4. Pearson coefficients of correlation for relationships between environmental variables and two RA axes gerated from the three subsets.

| | Group 1 | | Group 2 | | Group 3 | |
|---|---|---|---|---|---|---|
| | RA1 | RA2 | RA1 | RA2 | RA1 | RA2 |
| Conductivity | -0,96 | -0,44 | -0,66 | -0,35 | -0,04 | -0,03 |
| Sodium | -0,96 | -0,38 | -0,45 | -0,37 | 0,05 | -0,15 |
| Humidity | 0,16 | -0,89 | 0,16 | -0,35 | 0,53 | -0,60 |
| Phosphorus | 0,10 | -0,53 | -0,51 | 0,01 | 0,47 | -0,38 |
| Ammonium | 0,43 | -0,60 | 0,53 | 0,68 | 0,05 | 0,31 |
| Nitrite/Nitrate | -0,74 | 0,11 | -0,03 | -0,14 | -0,14 | 0,51 |

## Conclusion

The Bacillariophyta is the most important group in number of species and abundance. Cyanophyta and Dinophyta were important during the summer and winter, respectively, when they changed the colour of the sand. The majority of the species had a wide distribution in the area. The highest number of organisms/cm$^3$ was generally found near the dunes.

Praia Azul differs from other benthic habitats especially in the absence of epipsammic species.

Factors related to land topography such as slight depressions that permit water accumulation from the rain or open drains lowered the conductivity values to < 155 µS/cm and then a distinctive flora grew with many freshwater species. The most important factors for floristic discontinuity were the salinity and ammonium.

This work demonstrates the importance of studing the whole benthic algal community because the diatoms can be better understood when analysed together with other algal groups. In some groups it is possible find species with an homogeneous distribution, e.g. *Merismopedia punctata* in the Group 1 at Praia Azul.


## Acknowledgements

I want to thank Professor F. E. Round and anonymous reviewers for their suggestions to improve this paper. This work was supported by FAPERGS (Fundação de Amparo à Pesquisa do Rio Grande do Sul): Grants no. 91.00506.0 and 92.60816.7


## References


Admiraal, W., Peletier, H. & Zomer, H. (1982). Observations and experiments on the populations dynamics of epipelic diatoms form an estuarine mudflat. *Estuarine, Coastal and Shelf Science*, **14**, 471–487.

Aleem, A. A. (1950). The diatoms community inhabiting the mud-flats at Whitstable. *The New Phytologist*, **49**, 174–188.

Amspoker, M. C. (1977). The distribution of intertidal epipsamic diatoms on Scripps Beach, La Jolla, California, U.S.A. *Botanica Marina*, **20**, 227–232.

Amspoker, M. C. & McIntire, C. D. (1978). Distribution of intertidal diatoms associated with sediments in Yaquina estuary, Oregon. *Journal of Phycology*, **14**, 387–395.

Brockmann, C. (1950). Die Watt-Diatomeen der schlesswig-holsteinischen Westküste. *Abhandlungen der senckenbergischen naturforschenden Gesellschaft*, **478**, 1–26.

Bulkholder, P. R., Repak, A. & Sibert, J. (1965). Studies on some Long Island Sound littoral communities of microorganisms and their primary productivity. *Bulletin of the Torrey Botanical Club*, **92**, 378–402.

Cain, S. A. & Castro, G. M. O. (1959). *Manual of vegetation analysis*. 325 pp. Harpey, New York.

Calliari, L. J & Klein, A. H. (1993). Características morfodinâmicas e sedimentológicas das praias oceânicas entre Rio Grande e Chuí, RS. *Pesquisas*, **20**, 48–56.

Chomenko, L. & Schäfer, A. (1984). Interpretação biogeográfica da distribuição do gênero *Littoridina* (Hydrobiidae) nas lagoas costeiras do Rio Grande do Sul, Brasil. *Amazoniana*, **9**, 127–146.

Cook, L. L. & Whipple, S. A. (1982). The distribution of edaphic diatoms along environmental gradients of a Louisiana salt marsh. *Journal of Phycology*, **18**, 64–71.

Field, J. G., Clarke, K. R. & Warwick, R. M. (1982). A practical strategy for analysing multispecies distribution pattterns. *Marine Ecology Progress Series*, **8**, 37–52.

Garcia-Baptista, M. (1993). Psammic algae from Praia Azul. *Bibliotheca Phycologica*, **94**, 167 pp.

Garcia-Baptista, M. & Baptista, L. R. M. (1992). Algas psâmicas de Jardim Beira-Mar, Capão da Canoa, Rio Grande do Sul. *Revista Brasileira de Biologia*, **52**, 325–342.

Griffiths, C. L. & Griffiths, R. J. (1983) . Biology and distribution of the littoral rove beetle *Psamathobledius punctatissimus* (Le Conte) (Coleoptera: Staphylinidae). *Hybrobiologioa*, **101**, 203–214.

Hay, S. I., Maitland, T. C. & Paterson, D. M. (1993). The speed of diatom migration through natural and artificial substrata. *Diatom Research*, **8**, 371–384.

Hendey, N. I. (1964). *An introductory account of the smaller algae of British coastal waters*, Part 5*: Bacillariophyceae (Diatoms)*. 317pp. Her Majesty's Stationery Office, London.

Hustedt, F. (1939). Die Diatomeenflora des Küstengebietes der Nordsee vom Dollart bis zur Elbemündung I. Die Diatomenflora in den Sedimenten der unteren Ems sowie auf den Watten in der Leybucht, des Memmert und bei der Insel Juist. *Abhandlungen der Naturwissenschaftlichen Verein zu Bremen*, **31**, 572–677.

McIntire, C. D. (1978). The distribution of estuarine diatoms along environmental gradients: a canonical correlation. *Estuarine and Coastal Marine Science*, **6**, 447–457.

Moniz, A. C. (1975). *Elementos de pedologia*. 320pp. Livros Técnicos e Científicos Editora S.A., Rio de Janeiro.

Oppenheim, D. R. (1988). The distribution of epipelic diatoms along an intertidal shore in relation to principal physical gradients. *Botanica Marina*, **31**, 65–72.

Oppenheim, D. R. (1991). Seasonal changes in epipelic diatoms along an intertidal shore, Berrow flats, Somerset. *Jounal of the Marine Biological Associations of the United Kingdom*, **71**, 579–596.

Rao, V. N. R. & Lewin, J. (1976). Benthic marine diatom flora of False Bay, San Juan Island, Washington. *Syesis*, **9**, 173–213.

Riznyk, R. Z. (1973). Interstitial diatoms from two tidal flats in Yaquina estuary, Oregon, U.S.A. *Botanica Marina*, **16**, 113–138.

Riznyk, R. Z. & Phinney, H. K. (1972). The distribution of intertidal phytopsammon in an Oregon estuary. *Marine Biology*, **13**, 318–324.

Rohlf, F. J. (1992). NTSYS-pc, Numerical Taxonomy and Multivariate Analysis system. Exeter Software, Setasuket.

Round, F. E. (1960). The diatom flora of a salt marsh on the river Dee. *The New Phytologist*, **59**, 332–348.

Sullivan, M. J. (1975). Diatoms communities from a Delaware salt marsh. *Journal of Phycology*, **11**, 468–475.

Taasen, J. P. & Høisæter, T. (1981). The shallow-water soft-bottom benthos in Lindåspollene, western Norway. 4. Benthic marine diatoms, seasonal density fluctuations. *Sarsia*, **66**, 293–316.

Tedesco, M. J., Volkweiss, S. J. & Bohnen, H. (1985). Análise de solo, plantas e outros materiais. *Boletim Técnico de Solos*, **5**. Faculdade de Agronomia, Universidade Federal do Rio Grande do Sul, Porto Alegre. 98 pp.

Toldo Jr, E. E., Dillenburg, S. R., Almeida, L. E. S. B., Tabajara, L. L., Martins, R. R. & Cunha, L. O. B. P. (1993). Parâmetros morfodinâmicos da Praia de Imbé, RS. *Pesquisas*, **20**, 27–32.

Whiting, M. C. & McIntire, C.D. (1985). An investigation of distributional patterns in the diatom flora of Netarts Bay, Oregon, by correspondence analysis. *Journal of Phycology*, **21**, 655–661.

Williams, R. B. (1964). Division rates of saltmarsh diatoms in relation to salinity and cell size. *Ecology*, **45**, 877–880.

# Fine-scale distribution of dominant *Chaetoceros* species in the Gruž and Mali Ston Bays (Southern Adriatic)

Nenad Jasprica and Marina Carić

*Institute of Oceanography and Fisheries, Laboratory of Plankton Ecology, P. O. Box 39, HR–50001 Dubrovnik, Croatia*

## Abstract

Seasonal and vertical distribution of *Chaetoceros* species were investigated in the Gruž and Mali Ston Bays with biweekly or monthly sampling during two one-year cycles (1984–1985 and 1988–1989). Twenty-eight species of *Chaetoceros* were identified. *Chaetoceros compressus* Lauder, *Ch. danicus* Cleve, *Ch. decipiens* Cleve and *Ch. diversus* Cleve constituted more than 80% of the microphytoplankton cells present during spring diatom blooms. Seasonal alternations between mixing and stratification of the water column in Mali Ston Bay affected *Chaetoceros* dynamics on a seasonal scale. Vertical distribution of dominant *Chaetoceros* spp. depended on the pycnocline position in the water column. Highest cell numbers of the dominant *Chaetoceros* in the water column was recorded above the pycnocline depth in the Mali Ston Bay, and in Gruž Bay at the pycnocline depth.

## Introduction

Planktonic diatoms in the southern Adriatic were investigated as a part of phytoplankton composition studies (Viličić 1984, 1985; Jasprica 1989, 1994). Diatoms with their high species diversity, form the most important fraction of the phytoplankton in the southern Adriatic coastal waters, both qualitatively and quantitatively. Ecology of the most common diatom genus *Chaetoceros*, has not been investigated in detail. This paper describes vertical and seasonal distribution of dominant *Chaetoceros* species in the Mali Ston and Gruž Bays.

The Mali Ston Bay (Usko Station, 12 m maximum depth, Fig. 1) is sparsely populated and is a relatively unpolluted area suitable for oyster and mussel farming. Its major nutrient sources are dense vegetation, fresh water from the river Neretva in the outer part of the Bay and submarine springs in the inner part of the Bay. The station

Gruž (25 m maximum depth) is located in the Dubrovnik Harbour in the Gruž Bay. It is influenced by the open sea water in the bottom layer and by freshwater from the river Ombla in the surface layer. The river Ombla estuary brings the largest quantity of fresh water to the area, with maximum values from November to December. The nutrients in the Gruž Bay are mostly derived from stream runoff and sewage from the town of Dubrovnik. However, since 1985, the bulk of sewage effluents are being discharged towards the open sea. According to the annual phytovolume distribution, both bays have been included in the category of "moderately eutrophicated ecosystem" (Viličić 1989).

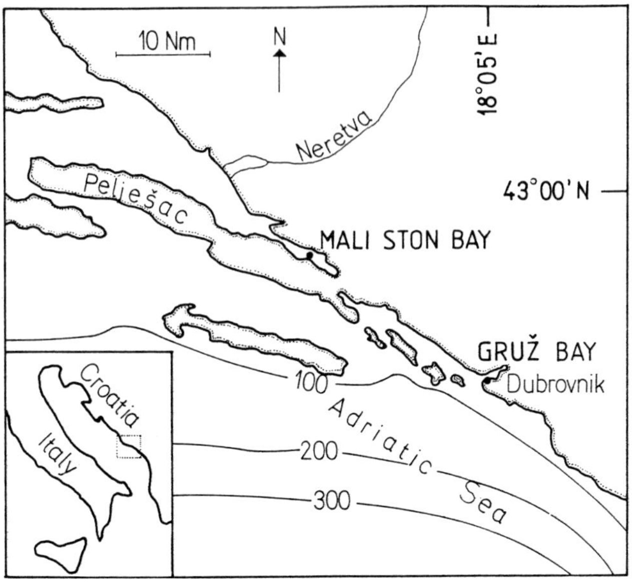


Fig. 1. Location of stations.

## Materials and Methods

During 1984 and 1985 biweekly phytoplankton samples were taken by hand pump (30 l $min^{-1}$ – intake tube diameter 32 mm) to 10 m depth, at depth intervals of 1 meter. From 1988 to 1989 monthly samples were taken by 5 l Niskin bottles at two meter depth intervals in the Mali Ston Bay, and at 0.5, 5, 10 and 20 meters depth in the Gruž Bay. In total, 576 samples were collected and analysed. All the samples were preserved in 2% neutralized formaldehyde solution. Phytoplankton cell number was determined using the inverted microscope method (Utermöhl 1958). Counting of microplankton (>20 μm) cells was performed at a magnification of 100×. Nanoplankton (<20 μm) cells were counted in 20–30 randomly selected fields of view along the counting chamber at a magnification of 320×. Precision of the counting method was ±10 per cent. The counts were also transformed into cell volume using the method described by Smayda (1978). Cell volume of various species was determined according to the cell model (geometrical bodies) constructed by means of photomicrographs and drawings.

To minimize a possible error in total cell volume estimations because of the cell size variability, the measuring of cell dimensions was performed simultaneously with cell counting. Hair-like spines and diatom setae were not included in cell volume values. Total phytoplankton organic carbon was calculated from total cell volume using equations of Strathmann (1967) as modifed by Eppley *et al.* (1970). Physico-chemical parameters (temperature, salinity, sigma-t, reactive silicate, nitrate, total inorganic nitrogen and reactive phosphorus) were determined by standard oceanographic methods (Strickland & Parsons 1972). Nutrients were measured during 1988–1989. Statistical analysis was done using STATGRAPHIC and SYSTAT software packages. All variables were logarithmically transformed (log ($x$+1)) before statistical analysis to improve correlations among variables (Cassie 1962).

The similarity index of Jaccard (Legendre & Legendre 1978) was chosen for quantifying species associations. This index is based on presence/absence of a species rather than on the actual numbers and it is easily computed. It is expressed as:

$$S= c/a + b + c$$

where $a$ and $b$ indicate total occurrence of species in locations a and b, while $c$ is the number of joint occurrences. Double absences were not considered.

## Results

In the period 1984–1985 the temperature range at the surface was from 9.1 to 25.3°C in the Gruž Bay and 7.5–25.2°C in the Mali Ston Bay. The spring temperature began to rise in March and continued rising into summer before decreasing again in September. Low surface salinity values ($21.97 \times 10^{-3}$–$29.10 \times 10^{-3}$), caused by strong precipitation, occurred in the winter-spring period in Gruž Bay and in the early summer in Mali Ston Bay. In the period of 1988–1989 hydrographic conditions were similar to those of the period 1984–1985 and are given in Fig. 2. In the Gruž Bay, a strong density stratification within the upper 0–2 m of the water column persisted throughout the greater part of the year. In the Mali Ston Bay, on the contrary, two periods were recognized according to the hydrodynamic characteristics of the water column: mixing (October–April) and stratification (May–September) periods. A steep pycnocline was present at 6–8-m depth, generated by both salinity and temperature.

The ranges of nutrient concentrations in the Gruž and Mali Ston Bays were for reactive silicate 0.44–13.08 and 0.21–7.15, nitrate 0.01–21.90 and 0.01–9.73, total inorganic nitrogen 0.09–22.40 and 0.14–10.70, and reactive phosphorus 0.01–2.26 and 0.01–0.33, respectively. All values were higher in the Gruž Bay. In the Mali Ston Bay, all nutrients, except reactive silicate, had the lowest range, mean and standard deviation during the stratification period. As regards the parameters above, at and below pycnocline, maximum values were found above pycnocline depth.

Two or three maxima of phytoplankton biomass and cell number were noted throughout the year at both bays (Fig. 3). Maximum of total phytoplankton biomass, microphytoplankton and nanophytoplankton cell numbers in the period 1984–1985 was

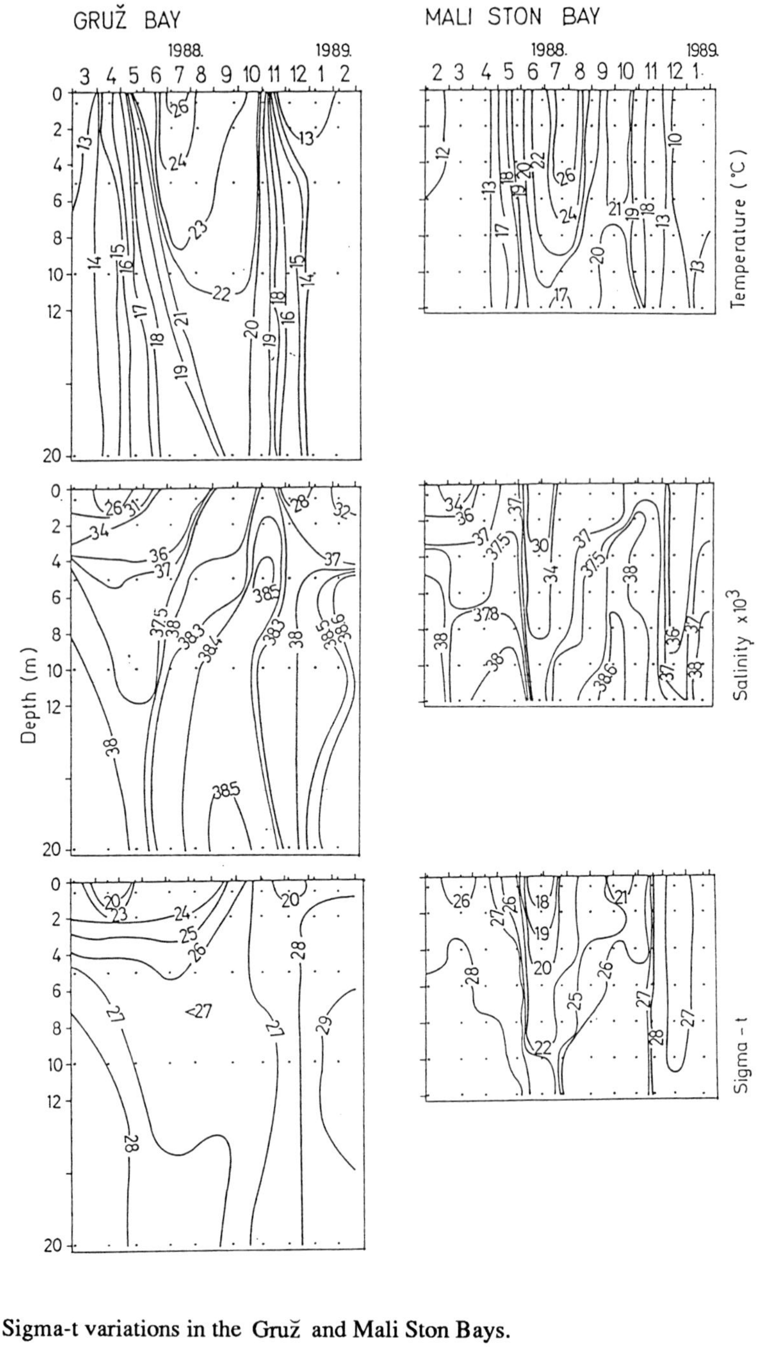


Fig. 2. Sigma-t variations in the Gruž and Mali Ston Bays.

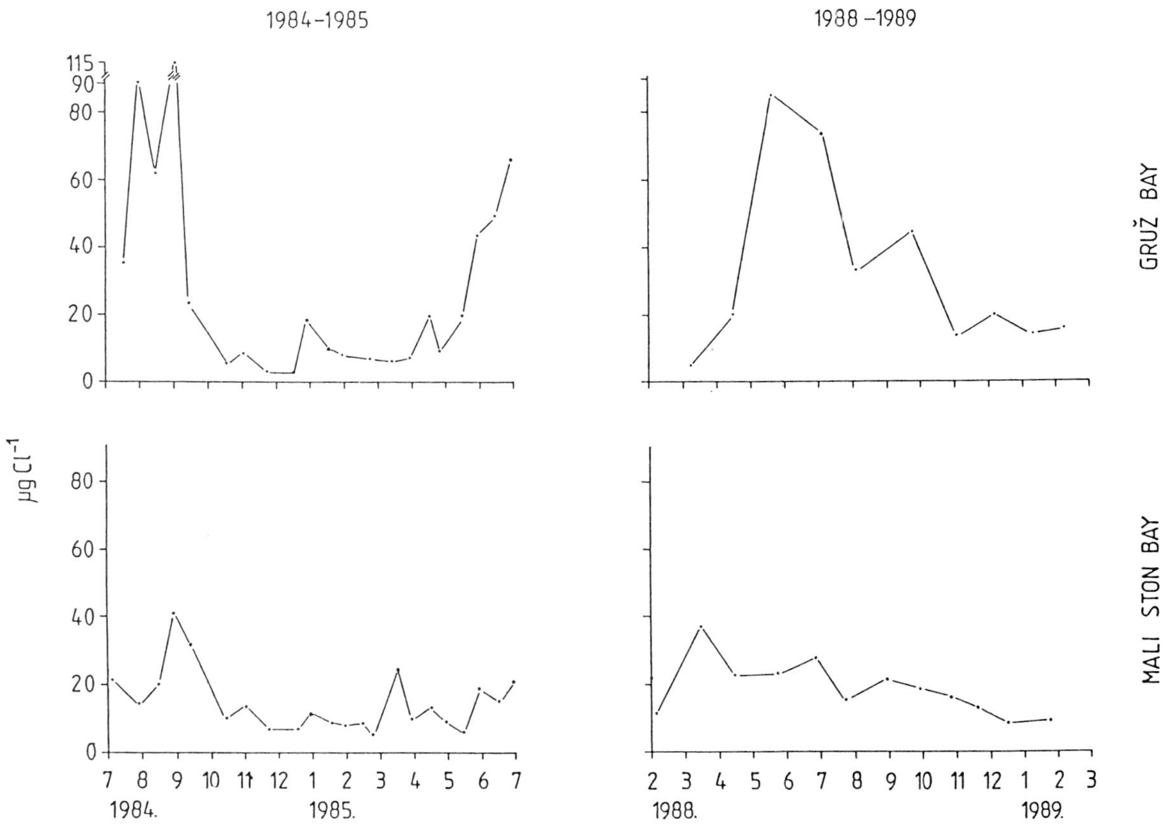


Fig. 3. Seasonal variations of total phytoplankton biomass (μg C $l^{-1}$), microphytoplankton and nanophytoplankton cell numbers ($10^6$ cells $l^{-1}$). Values are expressed as water column mean.

noted in summer (August, 115 μg C$l^{-1}$, 1.9 x $10^5$ cells $l^{-1}$, 3.3 x $10^6$ cells $l^{-1}$ in the Gruž Bay; 42 μg C$l^{-1}$, 1.8 x $10^5$ cells $l^{-1}$, 3.3 x $10^6$ cells $l^{-1}$ in the Mali Ston Bay) and in the period 1988–1989 in spring (March–May, 38 μg C$l^{-1}$, 4.6 x $10^5$ cells $l^{-1}$, 5.6 x $10^6$ cells $l^{-1}$ in the Gruž Bay; 84 μg C$l^{-1}$, 6.0 x $10^5$ cells $l^{-1}$, 5.2 x $10^6$ cells $l^{-1}$ in the Mali Ston Bay). The second peak was during September–December period. The spring–summer peak was composed of diatoms (70–85%) and dinoflagellates (14–28%). The autumn–winter peak consisted of diatoms (72–78%), dinoflagellates (15–26%) and coccolithophorids and silicoflagellates (1–7%). Microphytoplankton in Mali Ston Bay was dominated by the diatoms throughout the year, but in Gruž Bay dinoflagellates caused late summer outbursts of growth and biomass (Fig. 4).

A total of 112 Bacillariophyceae were identified during the two one-year cycles. Community similarity coefficients between Gruž and Mali Ston Bay diatom flora were in the range of 0.45–0.73. The greatest similarity coefficients were noted during autumn–winter months. Chain-forming diatoms were the dominant morphological type among diatoms throughout the year. The most abundant (population density greater than $10^4$cells$^{-1}$) diatom species were found between March and August. In the Gruž Bay they were *Skeletonema costatum* (Greville) Cleve, *Pseudonitzschia delicatissima* (Cleve) Hasle, *Rhizosolenia stolterfothii* Péragallo and *Thalassionema frauenfeldii* (Grunow) Hallegraeff and in Mali Ston Bay *Chaetoceros compressus* Lauder, *Leptocylindrus danicus* Cleve, *Pseudonitzschia delicatissima* (Cleve) Hasle and *Rhizosolenia stolterfothii* Péragallo. The genus *Chaetoceros* comprises the highest number of species; 28 during 1984–1985 and 24 during 1988–1989. *Chaetoceros* contributed the most significant amount to the total diatom cell number in both annual cell cycles in Mali Ston Bay as compared to Gruž Bay (Fig. 4). The species *Chaetoceros compressus* Lauder, *Ch. danicus* Cleve, *Ch. decipiens* Cleve and *Ch. diversus* Cleve constituted more than 80% of the microphytoplankton cell number during spring–summer diatoms blooms. In both bays, *Chaetoceros compressus* Lauder was the most abundant of the *Chaetoceros* species (Fig. 5a,b). During spring–summer bloom its contribution to the total diatom cell number was over 95%. The species *Ch. danicus* and *Ch. decipiens* were abundant during winter. *Ch. diversus* in the 1988–1989 period has not been found in Gruž Bay. In both periods, the dominant *Chaetoceros* species, except *Ch. danicus*, were more abundant in the Mali Ston Bay than in the Gruž Bay. A summary of the dominant *Chaetoceros* species is given in Table I.

The vertical distribution of the *Chaetoceros* depended on the pycnocline position in the water column. Highest cell numbers of *Chaetoceros* in the water column are recorded at the pycnocline depth in Gruž Bay, but above the pycnocline depth in the Mali Ston Bay. Correlation between *Chaetoceros* cell numbers and physical-chemical parameters were computed. All dominant *Chaetoceros* in both bays were significantly correlated with temperature. Salinity, sigma-t and nutrients, did not show a significant relationship to *Chaetoceros* with the exception of *Ch. diversus* during 1988–1989 in the Mali Ston Bay.

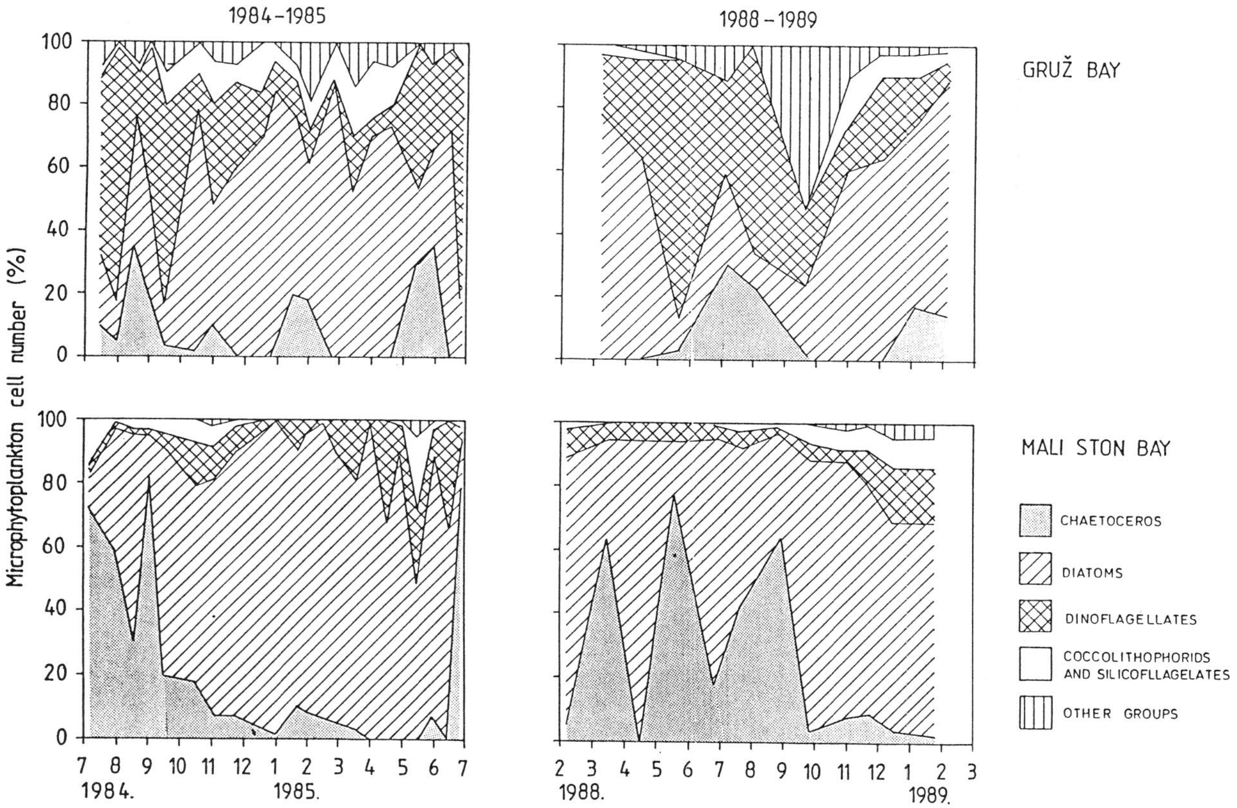


Fig. 4. Percentage contribution of *Chaetoceros* species and different groups to the microphytoplankton cell number. Values are expressed as water column mean.

Table I. List of *Chaetoceros* from the Gruž and Mali Ston Bays.

| Species | 1984–1985 Gruž | 1984–1985 Mali Ston | 1988–1989 Gruž | 1988–1989 Mali Ston |
|---|---|---|---|---|
| *Chaetocers affinis* Lauder | * | * | * | * |
| *Ch. anastomosans* Grunow | * | * | * | * |
| *Ch. atlanticus* v. *neapolitanus* (Schröder) Hustedt | * | * | * | * |
| *Ch. brevis* Schütt | * | * | * | * |
| *Ch. coarctatus* Lauder | . | * | . | . |
| *Ch. compressus* Lauder | * | * | * | * |
| *Ch. constrictus* Gran | * | * | * | * |
| *Ch. convolutus* Castracane | * | * | * | * |
| *Ch. costatus* Pavillard | * | * | * | * |
| *Ch. curvisetus* Cleve | * | * | * | * |
| *Ch. dadayi* Pavillard | * | * | * | * |
| *Ch. danicus* Cleve | * | * | * | * |
| *Ch. debilis* Cleve | * | * | * | * |
| *Ch. decipiens* Cleve | * | * | * | * |
| *Ch. delicatulus* Ostenfeld | . | * | . | . |
| *Ch. densus* (Cleve) Cleve | * | . | . | . |
| *Ch. didymus* Ehrenberg | * | * | * | * |
| *Ch. diversus* Cleve | * | * | * | * |
| *Ch. lauderi* Ralfs in Pritchard | * | * | * | * |
| *Ch. lorenzianus* Grunow | * | * | * | * |
| *Ch. messanensis* Castracane | * | * | * | * |
| *Ch. peruvianus* Brightwell | . | * | . | . |
| *Ch. rostratus* Lauder | * | * | * | * |
| *Ch. simplex* Ostenfeld | * | * | * | * |
| *Ch. tetrastichon* Cleve | * | * | * | * |
| *Ch. tortissimus* Gran | * | * | * | * |
| *Ch. vixvisibilis* Schiller | * | * | * | * |
| *Ch. wighamii* Brightwell | * | * | * | * |

## Discussion

Our results emphasize the variable nature of phytoplankton populations in a coastal habitat. During 1984–1985 maximum phytoplankton biomass development occurred in August at the time of the pronounced vertical temperature and salinity

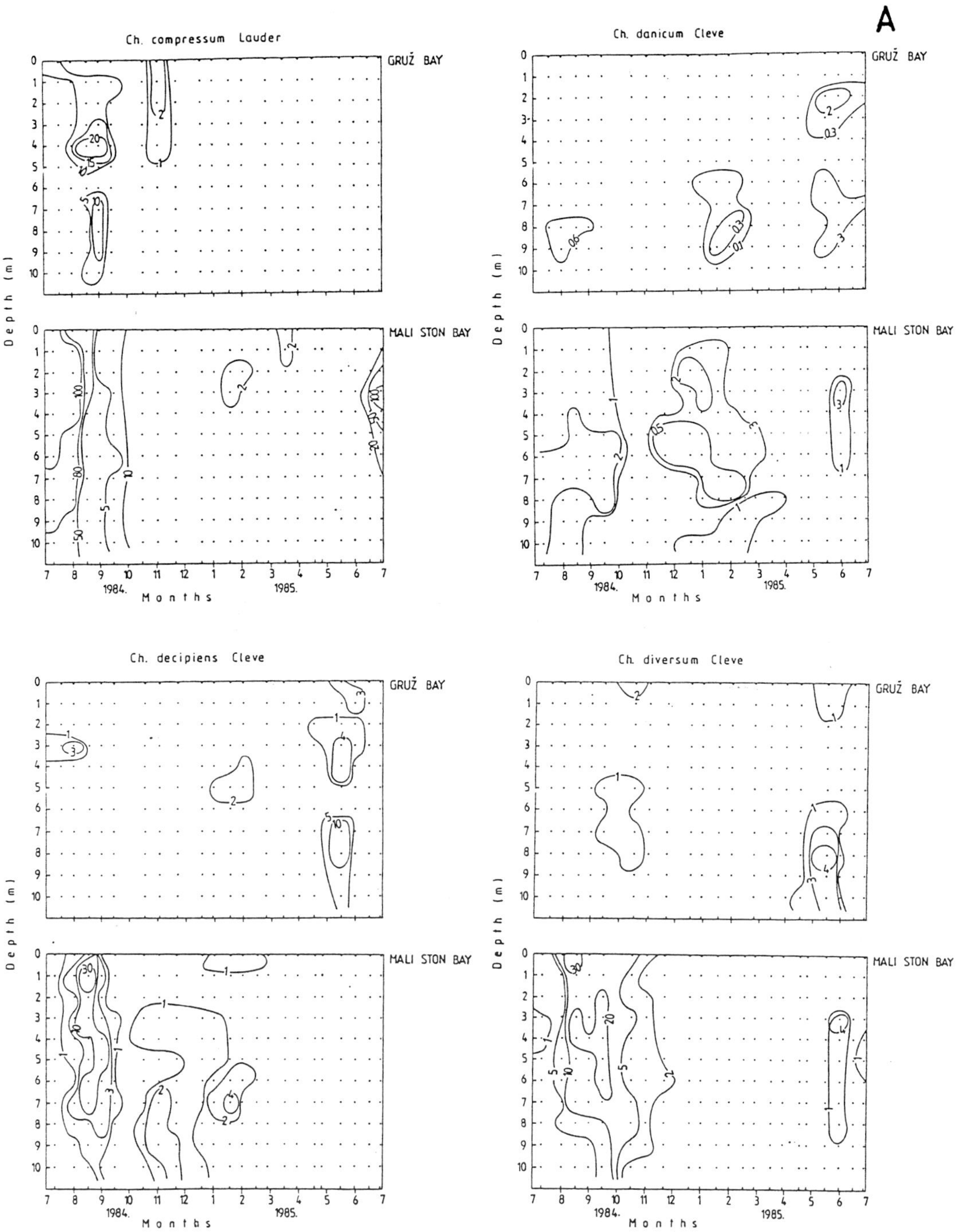


Fig. 5A. Distribution of dominant *Chaetoceros* species ($10^3$ cells $l^{-1}$) in the Gruž and Mali Ston Bays in 1984–1985

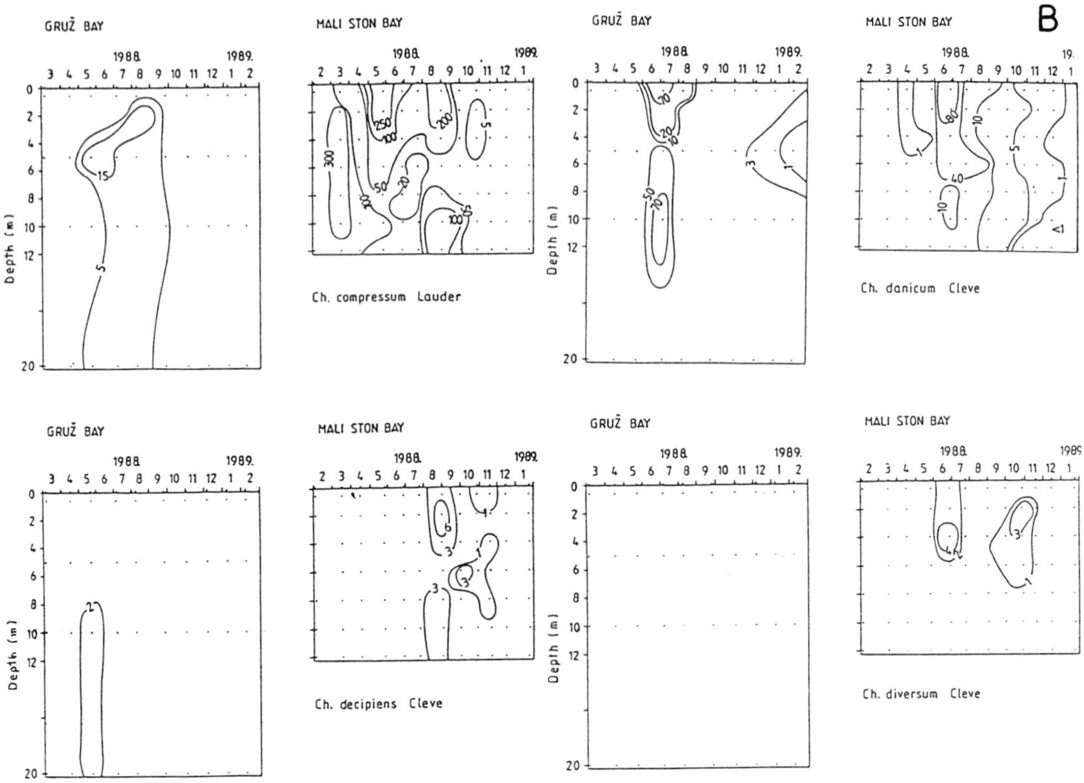


Fig. 5B. Distribution of dominant *Chaetoceros* species ($10^3$ cells $l^{-1}$) in the Gruž and Mali Ston Bays in 1988–1989.

gradient. Previously established patterns of seasonal variation of phytoplankton quantity in the middle (Pucher-Petković 1966) and the southern Adriatic (Viličić 1985), as well as data obtained during 1988–1989 were not in agreement with the data from 1984–1985 cycle. This is presumably due to specific precipitation events in that particular year. In the case of diatoms a large number of periodically dominating species demonstrated large inter-annual variations, both in average annual number and in timing of the periods of maxima. This is a commonly observed phenomen (Smetacek 1985, Bakker *et al.* 1990).

The most abundant diatoms were small, chain-forming species, capable of rapid growth response to favourable conditions. The dominance in number of species of *Chaetoceros* and *Rhizosolenia* in temperate areas has been previously noted (Nikolaides & Moustaka-Gouni 1990; Chiu *et al.* 1994). Among the dominant *Chaetoceros* spp., *Ch. compressus* and *Ch. danicus* showed an increase in number during the 1988–1989 summer period as compared to 1984–1985. It is not known whether these differences are within the normal range of seasonal variability expected for these populations or if these growth patterns are a response to some atypical environmental factor. Although several environmental variables are correlated, transparency should not be ignored as a determining factor in distribution patterns. In summer (stratification period) Secchi disc visibility in both bays decreased (<5 m) (Jasprica 1989), and in the reduced light regime, i.e. below the pycnocline depth, lower cell numbers were observed.

We did not find a correlation between *Chaetoceros* and nutrients. Nutrients could not be directly associated with the observed differences in distribution patterns of the examined diatom species. Even, an intensive development of phytoplankton preceding the stratification period caused a decrease in concentration of most nutrients. An increase in reactive silicate concentration was caused by haline stratification, namely a freshwater influx. Unfortunately, for comparison we do not have nutrient data for the period 1984–1985. The highest values for nitrate, total inorganic nitrogen, reactive phosphorus and reactive silicate have been found in the layer above the pycnocline in Mali Ston Bay, as well as in pycnocline layer in Gruž Bay. These are layers where the largest cell numbers of dominant *Chaetoceros* were found.

Multiple and changing combinations of examined environmental variables, in addition to phytoplankton biomass control by zooplankton and cell sinking, are factors which are likely to regulate diatoms and total phytoplankton abundance and diversity in both bays.

## References


Bakker, C., Herman, P. M. J. & Vink, M. (1990). Changes in seasonal succession of phytoplankton induced by the storm-surge barrier in the Oosterschelde (S.W. Netherlands). *Journal of Plankton Research*, **12**, 947–972.

Cassie, R. M. (1962). Frequency distribution models in the ecology of plankton and other organisms. *Journal of Animal Ecology*, **31**, 65–92.

Chiu, H. M. C., Hodgikiss, I. J. & Chan, B. S. S. (1994). Ecological studies of phytoplankton in Tai Tam Bay, Hong Kong. *Hydrobiologia*, 273, 81–94.

Eppley, R. W., Reid, F. M. H. & Strickland, J. D. H. (1970). The ecology of the plankton off La Jolla, California, in the period April through September 1967. III. Estimates of phytoplankton crop size, growth rate and primary production. *Bulletin, Scripps Institute of Oceanography*, **17**, 534–551.

Jasprica, N. (1989). Distribution of phytoplankton population density and volume-biomass in the Mali Ston and Gruž Bays (the southern Adriatic). *Ekologija*, **24**, 83–96.

Jasprica, N. (1994). The phytoplankton biomass in the south Adriatic coastal and open sea waters. *Ph. D. thesis. University of Zagreb*, 129 pp.

Legendre, L. & Legendre P. (1978). Associations. In: *Phytoplankton mannual* (A. Sournia, ed.), 261–272. UNESCO, Paris.

Nikolaides, G. & Moustaka-Gouni, M. (1990). The structure and dynamics of phytoplankton assemblages from the inner part of the Thermaikos Gulf, Greece. I. Phytoplankton composition and biomass from May 1988 to April 1989. *Helgoländer Meeresunters.*, **44**, 478–501.

Pucher-Petkovíc, T. (1966). Vegetation des Diatomees pelagiques de l'Adriatique moyenne. *Acta Adriat.*, **13**, 1–98.

Smayda, T. J. (1978). From phytoplankters to biomass. In: *Phytoplankton mannual* (A. Sournia, ed.), 273–279. UNESCO, Paris.
Smetacek, V. (1985). The annual cycle of Kiel Bight plankton: a long-term analysis. *Estuaries*, **8**, 145–157.
Strathmann, R. R. (1967). Estimating organic carbon content of phytoplankton from cell volume or plasma volume. *Limnology & Oceanography*, **12**, 411–418.
Strickland, J. D. H. & Parsons, T. R. (1972). A practical handbook of seawater analysis. *Bulletin. Fisheries Research Bd Canada*, **167**, 1–311.
Utermöhl, H. (1958). Zur Vervollkommung der quantitativen Phytoplankton Methodik. *Mitteilungen. Internationale Vereinigung für Theoretische und Angewandte Limnologie*, **9**, 1–38.
Viličić, D. (1984). Phytoplankton communities of the south Adriatic in the greater vicinity of Dubrovnik, *Acta Bot. Croat.*, 43, 175–189.
Viličić, D. (1985). Phytoplankton study of southern Adriatic waters near Dubrovnik for the period from June 1979 to July 1980. *CENTRO*, **1/2**, 35–56.
Viličić, D. (1989). Phytoplankton population density and volume as indicators of eutrophication in the eastern part of the Adriatic Sea. *Hydrobiologia*, **174**, 117–132.

# Microhabitat and morphology of a *Navicula* species from a hypersaline lake in Western Australia

Jacob John

*School of Environmental Biology, Curtin University of Technology, GPO Box U1987, Perth WA 6001, Australia*

## Abstract

More than 10% of the 1,900 ha area of Rottnest Island, located in the Indian Ocean, close to Perth, the capital of Western Australia, is occupied by salt lakes. The deepest of these lakes (Government House Lake), ranges in salinity from 130 to 150 ppt. Phytoplankton are sparse in this lake, most of the primary producers being the benthic microbial communities. Diatoms and Cyanobacteria were the dominant components of the benthic microbial mats. *Amphora coffeaeformis* and a *Navicula* species were most dominant throughout the year in the microbial mats. The floor of the lake was covered by a microbial mat and had living undulous stromatolites. Most of the year, a species of *Artemia*, a crustacean was abundant on the floor of the lake. The setae of the detached limbs and caudal rami of the dead animal were colonised by the *Navicula* sp. The internal space of these setae was used by the *Navicula* species as dwelling tubes.

As the animals die and undergo decay, the diatoms appeared to migrate to the internal opening of the setae of the detached limbs and caudal rami. The varying internal space available in the different types of setae appears to impose size restrictions on the migration of diatom cells – the smallest occupying the narrow filter setae. As the detached limbs with their setae are light with large surface area, they easily float in water. The diatom cells thus were functioning as "plankton" using the transparent exoskeleton as a "aft". In a nutrient poor hypersaline environment, the unusual behaviour of the diatom colonising the dead *Artemia* limb parts can be interpreted as of adaptive significance. The morphology and ultrastructure of the *Navicula* species was investigated by light and scanning electron microscopy.

## Introduction

During a study on the microbial communities of a hypersaline lake on Rottnest Island in Western Australia, a *Navicula* species was observed to be abundant. The species occurred predominantly in a microhabitat not previously reported. Rottnest Island is located in the Indian Ocean, 27 km west of Perth, the capital of Western Australia (32°S, 115°E). More than 10 percent of the total area (1,900 ha) of the island is occupied by salt lakes. The island was once part of the mainland of Western Australia. The fall and rise of sea level during the glaciation and interglaciation periods coupled with the earth movement in south-western part of Australia have been responsible for the emergence of the Island and the formation of the salt lakes (Playford 1983).

Rottnest Island became separated from the mainland about 6,500 years ago, and due to subsequent fall in sea level, the island was uplifted to the present level. The salt lakes on the island, which were formed as a result of the collapse of a cave system, remained connected to the sea for some 3,000 years. Eventually the salt lakes became isolated from the sea by beach bridges and sand dunes and became hypersaline due to evaporation (Playford 1983).

The largest and deepest of the salt lakes is Government House Lake (8 m deep and 1,600 $m^2$ in area) and has a salinity range of 130 to 150 ppt. Government House Lake is noted for its meromixis and has benthic microbial mats of 5 to 10 cm thickness covering parts of its floor. In some areas, the microbial mats form living undulous stromatolites, growing to a height of 5–20 cm.

During winter, when most of the rainfall occurs, the water level in these lakes rises up to the sea level, but falls to 0.4 m to 0.8 m below sea level during summer. Edward (1983) recorded 43 species of invertebrates in the salt lakes on Rottnest Island. One of the most predominant species is the crustacean *Artemia.* belonging to the order Anostraca (closely related to *Artemia salina*). The *Artemia* sp. (red-brine-shrimp) occurs in large numbers predominantly in winter in the Government House Lake on the benthic microbial mats, though it is present throughout the year (Fig. 1). *Artemia* is not found in the natural salt lakes of the mainland, although ten species of native Anostraca are recorded from the mainland (Geddes 1981; Geddes *et al.* 1981). It is believed that the *Artemia* sp. was introduced in the form of "eggs" through salt cleaning machinery imported from Europe to the Government House Lake when it was used as a salt works in the last century until the 1950s. The microbial communities composed of bacteria, cyanobacteria and diatoms act as producers and decomposers, sustaining the invertebrates. The objective of the present paper is to record the microhabitat and morphology of a species of *Navicula* found in Government House Lake probably related to *Navicula duerrenbergiana* Hustedt.

## Materials and methods

Samples of microbial mats were collected from Government House Lake during winter, spring and summer in 1991. Planktonic samples were collected using a 10 µm plankton net during the same period. Temperature, salinity and pH were measured at the time of collection. Artificial glass substrates (10 cm × 5 cm) were used for studying colonisation by microbial communities. The samples were examined fresh and algal species identified. Diatoms were cleaned and processed into permanent slides by traditional methods (John 1983). Both fresh and cleaned material was examined by light microscopy and photographed using a Vanox photomicroscope. The ultra structure of the *Navicula* species was investigated by scanning electron microscopy (JEOL JSM 35C).

## Observations

The *Navicula* species occurred commonly in the microbial communities along with *Amphora coffeaeformis* (Ag.) Kütz., *Navicula elegans* (W. Smith), *Entomoneis tenuistratia* John, *Synedra acus* Kütz., *Nitzschia communis* Rabh. and *Nitzschia rostellata* Hust. Both coccoid and filamentous Cyanobacteria constituted the bulk of the microbial communities. *Aphanothece halophytica* and *Spirulina subsalsa* were the dominant cyanobacteria in the microbial communities. The glass substrates were mostly dominated by *Amphora coffeaeformis* and sparsely by other diatoms and Cyanobacteria.

The mean lake temperature recorded was 18°C for the winter and 32°C for the summer, and pH ranged from 8 to 8.1 for both periods. The salinity recorded was a mean of 130 ppt for the winter and 150 ppt for the summer.

The setae of the limbs and caudal rami of dead animals in a state of decay were colonised by the *Navicula* species which was actively mobile. Planktonic samples collected in winter and spring were full of detached limb appendages of *Artemia* with the setae occupied by the *Navicula* species. Often the water was cloudy with dense floating detached limbs of *Artemia*. These limbs, and other parts of the exoskeleton of the animal, formed a loose top layer on the microbial mats, which at the slightest disturbance, dislodged and floated into the water column.

The cylindrical body of *Artemia* has a trunk with 11 pairs of limbs and a posterior with six segments terminating in a caudal furca of two rami (Fig. 1). Each limb at its base has several narrow elongated curved filter setae of more or less uniform diameters of 3–4 µm (Fig. 3). The detached filter setae were internally colonised by smaller cells of the *Navicula* species.

The setae attached to the distal endite and the exopodite of the upper segments of the limbs were about 18 to 20 µm wide at the base, narrowing up to 4 µm or less. The *Navicula* species were observed to move into the tubes and often overcrowding was noticed at the opening of the setae (Figs 2 to 7) at the point of attachment to the limb.

*Valve morphology* (Figs 8 to 18 and 19 to 22)

The valve outline typically is linear-lanceolate, rarely lanceolate or linear with barely protracted rounded apices. Length of the valve ranges from 14 µm to 45 µm and width 4 to 6 µm. There are two chloroplast plates located diagonally opposite, closer to the girdle margin, connected by a cytoplasmic bridge. There is a libroplast located close to each plastid plate (Figs 8, 9 & 10). The axial area is linear and narrow, expanding slightly towards the centre (Fig. 21). The two central raphe endings are

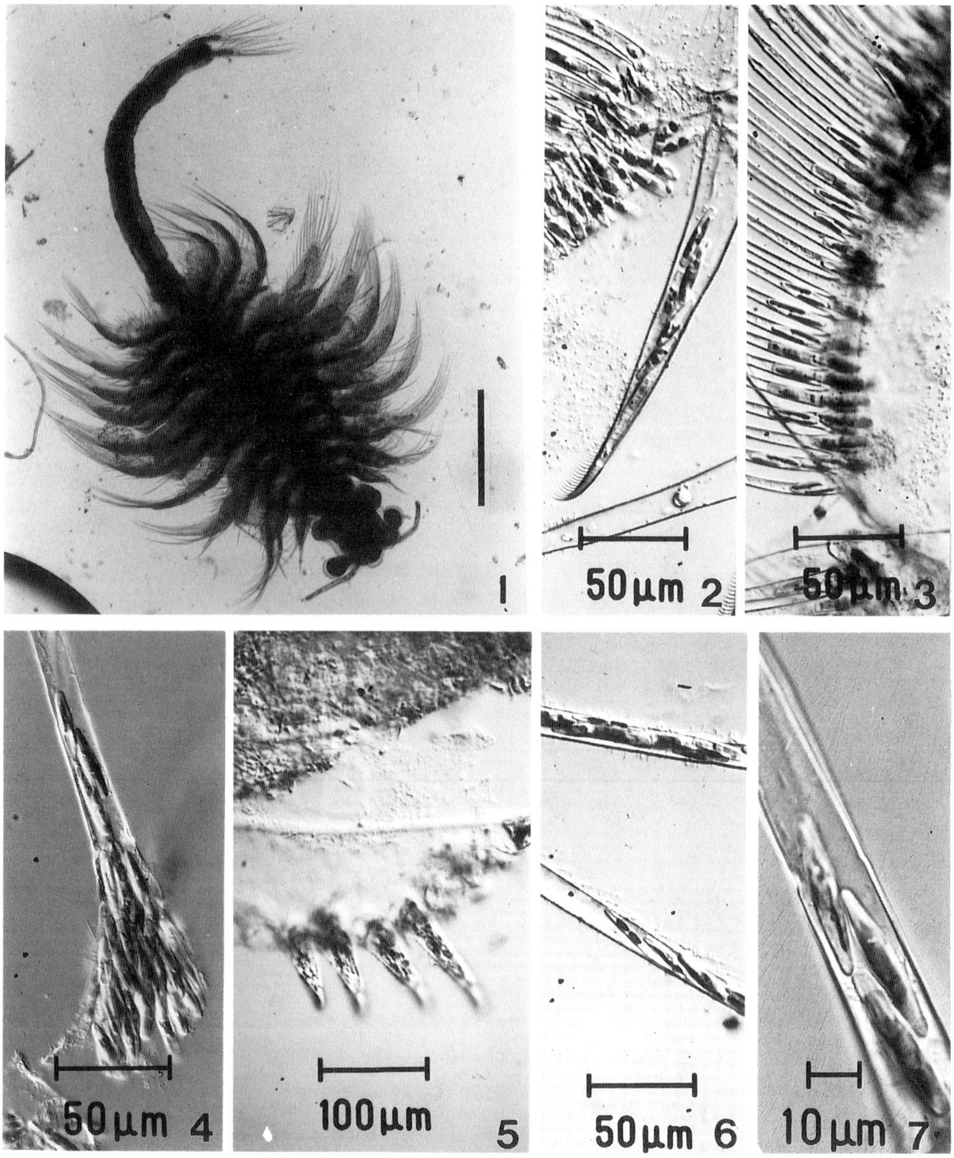


Fig. 1. *Artemia* sp. scale = 1 mm. Fig. 2. Seta from the endite (mid part of limb) with living diatoms inside. Fig. 3. Narrow long curved filter setae from the base of the limb being colonised by *Navicula* cells; note the crowding at the point of entry. Fig. 4. Migrating diatom cells inside a seta. Fig. 5. Short seta of the mid segment (endite). Fig. 6. Filter setae with their narrow lumen colonised by the diatom cells. Fig. 7. Migrating cells in a broad seta.

close together, with slightly expanded pores at the external surfaces; the pores are slightly bent towards the primary side. The two polar terminal fissures are hooked towards the secondary side (Figs 11 & 19). A narrow raphe-sternum is evident internally (Fig. 22). The striae are uniseriate, parallel, 11 to 13 in 10 µm (Figs 11 to 17), the poroids are apically elongate and linear (Figs 19 to 21), 30 to 35 in 10 µm arranged in longitudinal rows. Internal terminal fissures end in a conspicuous helictoglossa (Fig. 20). There are four girdle bands without any markings (Fig. 18).

**Discussion**

The microhabitat of the *Navicula* species reported in this paper adds to a wide variety of microhabitats of diatoms already known. Like many other naviculoid species, this *Navicula* species is a highly mobile, epipelic diatom coexisting with *Amphora coffeaeformis* in large numbers. However, its microhabitat appears to be exclusive to the species as no other diatoms in the lake community were observed to colonise the setae, although there were many other mobile raphid pennate diatoms in the microbial communities. But none displayed any trend in migration to the internal tubes of the setae of *Artemia*.

There are several well documented cases of epizoic associations in diatoms. Shells of many molluscs and crustaceans, skins of dolphins and whales, and feathers of birds all form substrates for diatom growth. Often these associations are specific (Round *et al.* 1990). Host specific diatom associations with pelagic copepods are reported by several authors (Simonsen 1970; John 1991). These associations are often by diatoms which grow attached to the animals by stalks or pads. However, the association of the present species appears to be the result of a specific migratory behaviour of the diatoms. As the setae differ in their internal dimensions, the migrating cells need to select tubes according to their size, the smallest cells (4 µm wide) moving to the filter setae and the larger ones to other setae. The overcrowding at the entrance of the opening of the setae (Fig. 3) indicates directional movement. The organic material within the limbs after decomposition would form a valuable source of nutrients for the diatoms, especially in an environment such as the Government House Lake which has very little available phosphate and nitrate (Edward 1983; John 1984 unpublished).

The exoskeletons of *Artemia* of which the setae and limbs form parts, are transparent to light penetration. The detached exoskeleton limbs with their long narrow setae are light with large surface area and are carried easily into the water; the diatoms which are basically epipelic seem to exist as pelagic "pseudoplankton" using the setae as "rafts". The large amount of such "rafts" that can be collected by a 10 µm plankton net indicates that the diatom is able to thrive and move protected by their artificial dwellings constantly exposed to well oxygenated water.

In shape, size and nature of raphe, it appears similar to *Navicula duerrenbergiana* described by Hustedt in 1934 and *Navicula stundlii* described by Hustedt in 1959. Krammer & Lange-Bertalot (1986 p. 119, Figs 39, 6–11) synonymised the latter with

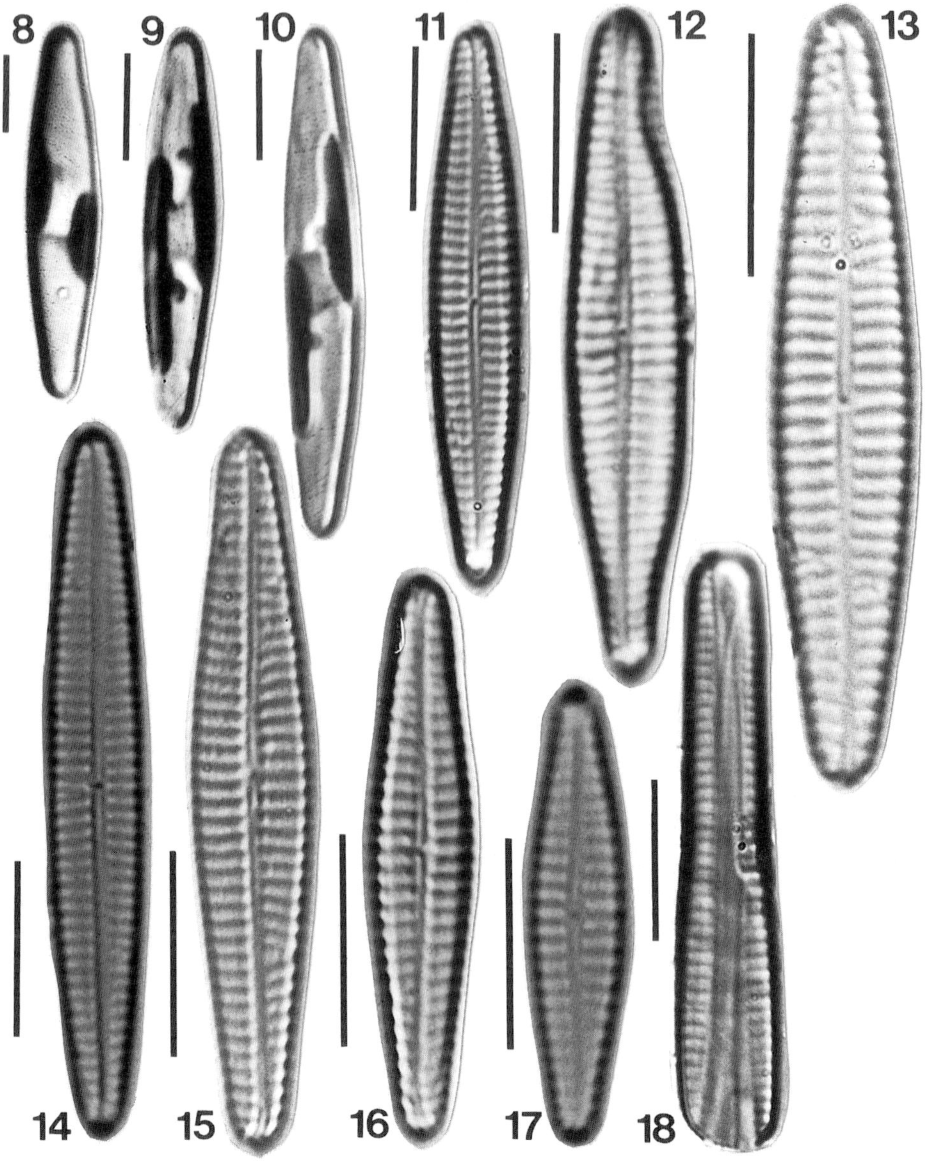


Figs 8–18. *Navicula* sp. LM. Scale bar = 10 μm. Figs 8–10. Fresh cells showing chloroplastid plates connected by the cytoplasmic bridge, note the libroplasts close to each plate. Figs 11–17. The valve ranges from linear-lanceolate to lanceolate and linear with barely protracted rounded apices. Fig. 18. Girdle view showing girdle bands.

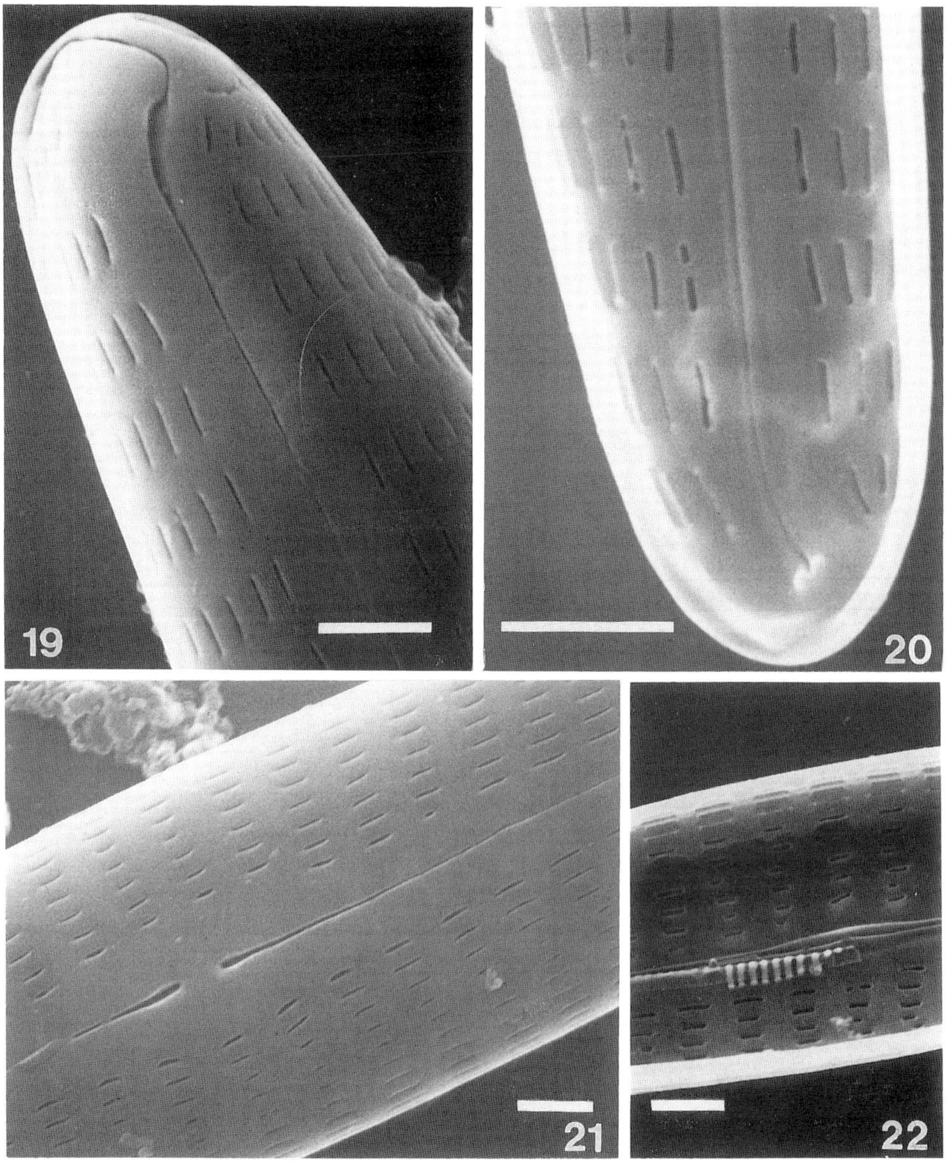


Figs 19–22. *Navicula* sp. SEM. Scale bar = 1 μm. Fig. 19. External surface of a valve apex showing linear poroids and curved external terminal raphe fissures. Fig. 20. Internal surface of a valve apex showing the internal terminal raphe fissure with a helictoglossa. Fig. 21. External surface of the central area of the valve showing the closely located central pores, the narrow central and axial areas and apically linear poroids. Fig. 22. Internal surface of the central area. Note the central raphe endings and the narrow raphe-sternum.

*N. duerrenbergiana*. Both species were found in saline waters, in Europe and Israel. Interestingly enough, Hustedt reported *N. duerrenbergiana* from the salt works of Durrenberg on the Neusiedler See on the Spanish mediterranean coast and in the desert lakes at Negev-Wüste in Israel. While the above two species are linear-lanceolate with obtuse apices, the *Navicula* species from the Government House Lake at Rottnest Island shows a vast range of variation in shape from linear to linear-lanceolate with obtuse apices, many with slightly produced apices. The diatom is smaller than *N. duerrenbergiana* (length 20–60 µm, width 4–6 µm). The striae and poroid density is less in the present species compared to *N. duerrenbergiana* which has 14–15 striae in 10 µm and 45 poroids in 10 µm. No information is available on the chloroplasts of *N. duerrenbergiana*. Typically *Navicula* has two chloroplast plates located directly opposite to each other, each plate being closer to the girdle (Cox 1979; Round *et al.* 1990), although there are a few species of *Navicula* with diagonally displaced plastid plates as observed in *N. gregaria* Donkin by Cox (1987).

As the *Artemia* sp. is believed to have been imported into Australia in the last century through salt cleaning machinery, speculation on the origin of the diatom species becomes interesting. Is this an indigenous species or an introduced species? I have encountered this species with considerable variation in valve size and shape, in a few hypersaline lakes in the mainland of Western Australia in its epipelic habitat.

Further work is required to establish the distribution of the diatom in other salt lakes in Western Australia. Could the diatom have also been imported with the *Artemia* sp.? Are there such associations in Europe? It is worthwhile looking for such an association in Europe. If the diatom species is proved to be native to Australia, it has obviously adapted in a relatively short period to a unique life-style with the immigrant *Artemia*. Do the species in the mainland salt lakes form similar associations with native members of Anostraca? This is yet to be investigated.

### Acknowledgements

I with to thank Mr Clinton Syers for assistance with the field work associated with the present project. I am grateful to Dr Pauli Snoeijs for providing me with a slide of *N. duerrenbergiana*.

### References

Cox, E. J. (1979). Taxonomic studies on the diatom genus *Navicula* Bory: the typification of the genus. *Bacillaria*, **2**, 137–153.

Cox, E. J. (1987). Studies on the diatom genus *Navicula* Bory VI, The identity, structure and ecology of some freshwater species. *Diatom Research*, **2**(2), 159–174.

Edward, D. H. D. (1983). Inland waters of Rottnest Island. *Journal of Royal Society of Western Australia*, **66**, 41–47.

Geddes, M. C. (1981). The brine shrimps *Artemia* and *Parartemia*: comparative physiology and distribution in Australia. *Hydrobiologia*, **81**, 169–179.

Geddes, M. C., De Deckker, P., Williams, W. D., Morton, D. W. & Topping, M. (1981). On the chemistry and biota of some saline lakes in Western Australia. *Hydrobiologia*, **82**, 201–222.

John, J. (1991). The epizoic diatom *Falcula hyalina* Takano on a pelagic copepod from the Swan River estuary, Western Australia. *Journal of Royal Society of Western Australia*, **73**(3), 73–75.

John, J. (1983). The diatom flora of the Swan River Estuary. *Bibliotheca Phycologica*, **64**, 360 pp.

Krammer, K. & Lange-Bertalot, H. (1986). Bacillariophyceae 1 Teil: *Naviculaceae Süsswasser-flora von Mitteleuropa* (Begrundet von A. Pasher) (H. Ettl, J. Gerloff, H. Heynig & D. Möllenhauer, eds), **2**(1). Fischer Verlag, Stuttgart and New York.

Playford, P. E. (1983). Geological research on Rottnest Island. *Journal of Royal Society of Western Australia*, **66**, 10–15.

Round, F. E., Crawford, R. M. & Mann, D. G. (1990). *The Diatoms. Biology and morphology of the genera*. Cambridge University Press. 747 pp.

Simonsen, R. (1970). *Protoraphidaceae*, eine neue Familie der Diatomeen. *Nova Hedwigia, Beiheft* **31**, 377–394.

# Diatom flora in oval faecal pellets from Terra Nova Bay (Antarctica)

Donato Marino, Marina Montresor,
Lucia Mazzella and Vincenzo Saggiomo

*Stazione Zoologica "A. Dohrn",*
*Villa Comunale, 80121 Napoli, Italy*

## Abstract

The concentrations of oval faecal pellets at a coastal station of Terra Nova Bay (Antarctica), frequently exposed to drifting ice, were measured and their algal content was studied. Diatom species, both whole cells with intact cytoplasm and empty frustules, dominated the samples; dinoflagellate resting cysts and chrysophyte statospores were also present. Small sized species, mainly *Fragilariopsis curta*, dominated the diatom assemblages in all samples. The presence of a double-layered peritrophic membrane, surrounding the pellets and the dominance inside the pellets of sea-ice/planktonic species, suggested that a planktonic crustacean, either an animal able to scrape food off the underside of ice or a filter feeder, could be the producer organism. Taking into account the high concentrations of the oval faecal pellets, their constant presence in the water column during the sampling period and the high number of diatom cells in the pellets (from $5.6 \cdot 10^4 \cdot m^{-2}$ to $29.2 \cdot 10^4 \cdot m^{-2}$), the contribution of these pellets to the total silica sinking should range between 6 and 27%, thus confirming their importance in the sinking of organogenic material in this area.

## Introduction

In recent years there has been renewed interest in studies on biogeochemical cycles and vertical fluxes in the oceans. These studies originated mainly from strategic projects designed to evaluate global climatic changes. An increasing number of investigations have focused on the Southern Ocean, where macronutrients do not appear to be limiting for phytoplankton growth, and the "biological pump" could be enhanced by the additional $CO_2$ in the atmosphere (Longhurst 1991; Tréguer & Quéguiner 1991; Priddle *et al.* 1992).

In the complex mosaic of subsystems of the Southern Ocean, the ice edge systems play a crucial role in the ecological processes that promote phytoplankton bloom formation during spring and summer. The structure of the sea ice microflora that is dominated by diatoms, and the physical and biological factors that regulate the seeding effect of ice-diatoms during ice melt is receiving increasing attention (Palmisano & Sullivan 1983; Riebesell *et al.* 1991). A key area for these studies is the Ross Sea, which is characterized by peculiar physical, chemical and biological properties (Jaques 1991). In this area, primary production is relatively high and also the *f* ratio (new production/total production) is rather high not only in spring but also in summer, mainly due to the ice-edge systems which are still found during this season (Nelson & Smith 1986). Most of the new production comes from recurrent phytoplankton blooms dominated by small diatom species, mainly *Fragilariopsis curta* (Carbonell 1985; Nuccio, pers. comm.), which create the main organic flux and are quantitatively important for the annual deposition of silica in sediments (Kellog & Truesdale 1979; Truesdale & Kellog 1979). A large part of the vertical flux is controlled by grazing activity of zooplanktonic organisms (Priddle *et al.* 1992) but the role of their faecal pellets in the transfer of carbon and biogenic silica to the sediments is still disputed (Dunbar 1984; von Bodungen *et al.* 1986). The factors that regulate the sinking velocity and the dissolution/sedimentation ratio of faecal pellets are their shape, size, the presence of a fairly resistant peritrophic membrane, and the content of the pellets (Cadée *et al.* 1992).

Due to the relevance of krill in the Antarctic food webs, detailed information are available on their faecal strings mainly obtained by drifting or moored sediment traps (see references quoted in Karl *et al.* 1991). In recent years attention has turned also to non-krill faecal pellets: round, small pellets produced by heterotrophic protists (Nöthig & von Bodungen 1989; Buck *et. al* 1990; González 1992), copepod cylindrical pellets (Granéli *et al.* 1993) and oval pellets produced by unknown planktonic organisms (Fischer *et al.* 1988; Nöthig & von Bodungen 1989; Bathmann *et al.* 1991; Cadée *et al.* 1992; González 1992). These authors group under the term "oval faecal pellets", faecal material of a rather wide size range which share the common features of the oval outline, the presence of a resistant peritrophic membrane and their content which is mainly diatoms.

We have studied oval faecal pellets and their algal content obtained from selected net samples collected during the 1989–1990 Italian expedition in Terra Nova Bay (Ross Sea). In an attempt to cast some light on the producer organism, we examined the planktonic ostracod *Alacia belgicae* and the thecosome *Limacina helicina* because of their relative abundance in the area (Benassi *et al.* 1992; Guglielmo *et al.* 1992). Using the carbon and silica fluxes estimated by Nelson & Smith (1986) for the same area, we also infer a quantitative evaluation of the role of these pellets in the sinking of organic and organogenic material in this "eutrophic" area of the Southern Ocean.

## Material and Methods

Sampling was conducted on board a small vessel during the course of the Italian Antarctic expedition on five dates from 15–1–1990 to 1–2–1990. The station (74° 41',33 S, 164° 07',15 E) was located 250 m from the coast (60 m depth) in Terra Nova

Bay (Ross Sea). In the 1989–1990 austral summer the opening of the Ross Sea to the ocean waters, due to ice melting, was completed at the end of December. During the sampling period the coastal waters of the Bay were frequently exposed to drifting sea ice.

Faecal pellets were retrieved from net (diameter 1.13 m, length 4 m, mesh size 200 μm) samples collected by vertical haul (0–50 m) and fixed on board with 4% formaldehyde. At the same station, phytoplankton samples were examined and chlorophyll *a* concentrations and primary production were determined as described elsewhere (Innamorati *et al.* 1991; Saggiomo *et al.* 1992).

Faecal pellet concentration was estimated by counting 5 replicates of subsamples obtained with a Stempel pipette. Thirty faecal pellets of each sample were isolated in a settling chamber and measured on a Zeiss IM35 inverted microscope. Their volume was calculated by matching them to a rotational ellipsoid with ellipse-shaped cross-section. The algal cells and empty diatom frustules contained in a number of pellets from each sample were counted. Single faecal pellets were measured, their volume was calculated and then they were transferred with a micropipette to a slide, gently squashed under a coverslip and observed on a Zeiss Axiophot microscope. Algal cells were counted and identified, whenever possible, at the species level; single valves were considered as half of a cell and broken valves as a quarter of a cell. The biovolume of the most abundant diatom species was calculated by measuring the size of many cells (Edler 1979), and converted into carbon using the formula suggested by Strathmann (1967). Biogenic silica content was calculated on the basis of the Si/C ratio (0.57 by moles) measured by Nelson & Smith (1986) in the same study area.

Electron microscopic observations on entire pellets and on their microalgal content were carried out. About 100 pellets were washed in distilled water, dehydrated in serial dilution of ethanol in a centrifuge tube, transferred to propylene oxide and finally embedded in Epon. After polymerization at 70°C for 24–35 h, the blocks were cut with a Reichert Ultracut ultramicrotome. Ultrathin sections were stained by uranyl acetate and observed using a Philips EM 400 microscope. About ten faecal pellets for each sample were dissected under a dissection microscope, placed on a Nuclepore filter, rinsed with distilled water, dehydrated with serial dilutions of ethanol, critically point dried, sputter-coated with gold and examined in a Philips 505 scanning electron microscope. For transmission electron microscope studies of diatom frustules, about 100 pellets from each sample were prepared. Pellets were isolated in small centrifuge tubes, rinsed with distilled water, and cleaned overnight with diluted (10%) permanganate. The samples were centrifuged, acidified with few drops of HCl, gently heated, centrifuged again and rinsed several times with distilled water. A small drop of clean sample was placed on a grid covered with Formvar film and observed with a Philips EM 400 microscope.

Selected specimens of *Alacia belgicae* from samples collected with a "Bioness" net in the same area (Guglielmo *et al.* 1992) and of *Limacina helicina* from our net samples were dissected and their gut content was observed on a Zeiss Axiophot microscope. Ostracod specimens were prepared for scanning electron microscope observations using the same procedure as for faecal pellets.

## Results

All samples contained an extremely high number (from 5.6 to 29.2 $10^4 \cdot m^{-2}$) of oval faecal pellets (Table 1). Pellets were brown and oval (Fig. 1), and surrounded by a very resistant peritrophic membrane, which appeared double-layered in the trans-mission electron microscope (Fig. 2). Their maximum diameter was between 200 and 500 µm (mean value: 309 µm; n = 150) and their volume between 2.6 and $21.5 \cdot 10^6$ $\mu m^3$ (mean value $6.7 \cdot 10^6$ $\mu m^3$). More than 65% of the observed pellets were comprised between $4 \cdot 10^6$ and $8 \cdot 10^6$ $\mu m^3$ and less than 5% had a volume larger than $16 \cdot 10^6$ $\mu m^3$ (Fig. 4).

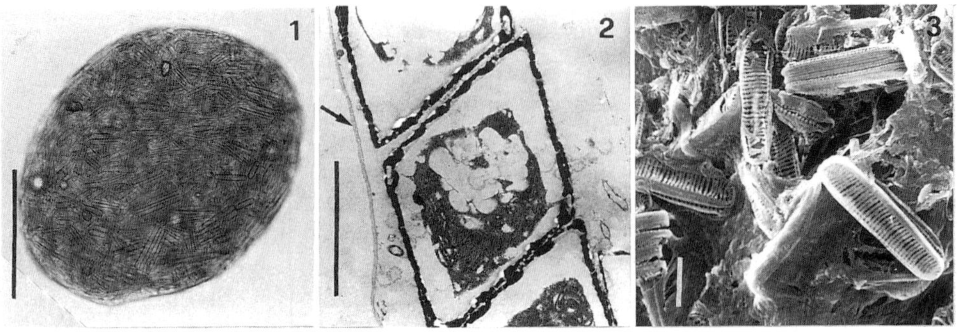


Figs 1–3. Fig. 1. Light micrograph of an oval faecal pellet. Scale bar = 100 µm. Fig. 2. Section of a short chain of whole cells of *Fragilariopsis curta* contained in a faecal pellet. The arrow indicates the peritrophic membrane. TEM micrograph. Scale bar = 5 µm. Fig. 3. SEM micrograph of the gut content of *Alacia belgicae*. Scale bar = 10 µm.

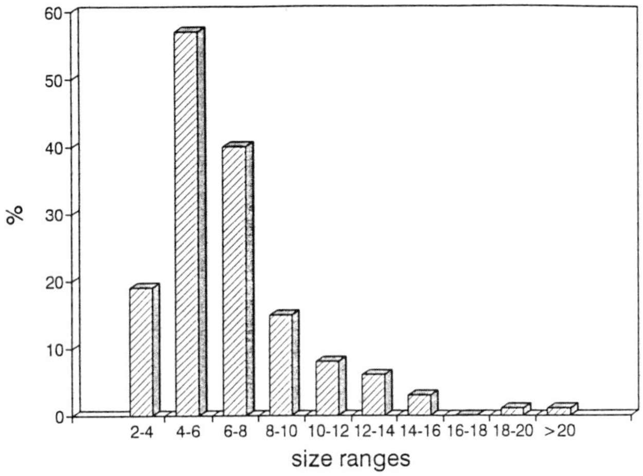


Fig. 4. Size class distribution of the volumes ($\cdot 10^6$ $\mu m^3$) of faecal pellets from all samples (n=150).

Table 1. Volumes, abundances, cell concentrations, total carbon and silica contents of the faecal pellets from the different samples.

| Sampling dates | 15. 01. 90 | 18. 01. 90 | 20. 01. 90 | 25. 01. 90 | 01. 02. 90 |
|---|---|---|---|---|---|
| Faecal pellet mean volume $\cdot 10^6$ $\mu m^3$ (± s.d.). | 8.02 (± 3.71) | 6.08 (± 2.79) | 6.67 (± 3.45) | 6.80 (± 2.76) | 5.91 (± 2.14) |
| n° faecal pellets $\cdot 10^4 \cdot m^{-2}$ | 5.6 | 11.1 | 29.2 | 7.5 | 9.6 |
| n° cells $\cdot 10^6 \cdot m^{-2}$ | 62.51 | 93.68 | 270.19 | 66.57 | 78.44 |
| n° *Fragilatiopsis curta* $\cdot 10^6 \cdot m^{-2}$ | 57.31 | 85.89 | 247.71 | 61.03 | 71.91 |
| n° cells with cytopl. $\cdot$ $10^6 \cdot m^{-2}$ | 19.27 | 28.88 | 83.28 | 20.52 | 24.18 |
| mg C $\cdot m^{-2}$ | 3.48 | 5.21 | 15.03 | 3.70 | 4.36 |
| mg Si $\cdot m^{-2}$ | 4.63 | 6.93 | 20.00 | 4.93 | 5.81 |

The content of all the pellets was constant throughout the sampling period. Diatoms, mainly pennate species, and a small number of dinoflagellate resting cysts and chrysophyte siliceous statospores were recognized in the pellets (Table 2); remnants of planktonic animals were never observed. The volumes occupied by algal cells constituted from 6% to 19% (mean value 10%) of the faecal pellet volumes. The major part of the diatom frustules were unshredded, and whole, undigested cells with an intact cytoplasmic content were rather abundant (from 20% to 41% of the total cell number). The colonial species *Fragilariopsis curta* (Figs 5, 6) largely dominated the diatom assemblages (from 74% to 97% of the total cell number) and appeared frequently in short chains if undigested (Fig. 2). This species was followed in dominance by *F. cylindrus* (Figs 7, 8), *F. sublinearis* (Fig. 9) and *Pseudo-nitzschia turgiduloides* (Fig. 10). Other poorly represented diatom species are illustrated in Figures 11–19. Such large species as *Pseudo-nitzschia turgiduloides*, *Thalassiothrix antarctica* and *Trichotoxon reinboldii* were always present as fragments.

Besides diatom species, a number of resting stages were also observed. The most common were the spiny resting cyst of a dinoflagellate and some chrysophyte statospores. The dinoflagellate cyst had a clearly recognizable paracingulum and a number of aciculate spines covering the epi- and hypo-cyst (Fig. 20). It exactly fits the description given by Buck *et al.* (1992) for spiny cysts collected in the Weddell and Scotia Seas and in our sampling area, probably produced by an athecate small dinoflagellate species. *Litheusphaerella spectabilis* (Fig. 21) and *Archeomonas areolata* (Fig. 22) were the most common chrysophyte statospores identified in our samples, and are reported here using their paleontologic name because of the uncertainty about the free-living producer organisms.

Table 2. Species recorded in the oval faecal pellets.

| |
|---|
| Diatoms |
| *Achnantes brevipes* Agardh |
| *A. clevei* Grunow |
| *A. vincenti* Manguin |
| *Achnantes* sp. |
| *Actinocyclus actinochilus* (Ehrenberg) Simonsen |
| *Asteromphalus hookeri* Ehrenberg |
| *Asteromphalus parvulus* Karsten |
| *Chaetoceros dichaeta* Ehrenberg |
| *Cocconeis fasciolata* (Ehrenberg) Brown |
| *C. schuettii* Van Heurck |
| *Fragilariopsis curta* (Van Heurck) Hustedt |
| *F. cylindrus* (Grunow) Krieger |
| *F. rhombica* (O' Meara) Hustedt |
| *F. sublinearis* (Van Heurck) Heiden |
| *Melosira adeliae* Manguin |
| *Navicula* sp. |
| *Odontella litigiosa* (Van Heurck) Hoban |
| *Pseudogomphonema* sp. |
| *Pseudo-nitzschia turgiduloides* (Hasle) Hasle |
| *Stellarima microtrias* Hasle & Sims, |
| *Thalassiosira gracilis* v. *gracilis* (Karsten) Hustedt |
| *T. lentiginosa* (Janish) Fryxell |
| *Thalassiothrix antarctica* (Grunow) Cleve |
| *Trichotoxon reinboldii* (Van Heurck) Reid & Round |
| Archaeomonads |
| *Archeomonas areolata* Deflandre |
| *Litheusphaerella spectabilis* Deflandre |

The examination of the gut content of the planktonic ostracod *Alacia belgicae* revealed that the largest part of its diet was constituted by *Fragilariopsis curta* (Fig. 3); other less common species found in faecal pellets were also present in the gut of *Alacia belgicae*. The alimentary canal of most of the specimens of *Limacina helicina* was almost completely empty; *Fragilariopsis curta* cells were found only in a few specimens.

## Discussion

This is the first record of oval faecal pellets in the Ross Sea. If we exclude the pellets described by Bathmann *et al.* (1991) which are smaller (80–180 µm in diameter) and almost round, the pellets we observed (200–500 µm in diameter) fit the descriptions provided by other authors (Fischer *et al.* 1988; Nöthig & von Bodungen 1989; Cadée *et al.* 1992; González 1992) for other areas of the Southern Ocean. All the oval pellets recorded contained diatoms except those found by González (1992) in the Scotia and Weddell Seas, which differed from the others because of the presence of crustacean remnants and almost completely repackaged and shredded diatom frustules. The authors hypothesize that these pellets were produced by an omnivorous and/or coprophagous organism that also feeds on krill faecal strings. In the present instance this hypothesis cannot be sustained because of the evidence that the producer organism was a typical herbivore.

The content of the pellets that we found in the Ross Sea was mainly made up by relatively small *Fragilariopsis* species *(F. curta, F. cylindrus, F. tugiduloides)*, which dominated the planktonic communities (Nuccio, pers. comm.), but which are also very common in sea ice; a few benthic and sea-ice diatom species (e.g. *Cocconeis fasciolata, C. schuettii, Achnanthes brevipes*) were also observed. The pellets invariably contained dinoflagellate resting cysts and chrysophyte statospore, which are typical of sea ice biota (Buck *et al.* 1992; Mitchell & Silver 1982). Taking into account that our sampling station was close to the coast, in an area where drifting sea ice was present throughout summer, phytoplankton communities were probably seeded continuously by ice species and resting stages. We therefore suggest that the organisms producing the pellets were either an animal provided with appendages designed for scraping food off the underside of ice or a plankton feeder.

It has been suggested that scraper ostracods could be the organisms producing oval faecal pellets in the Southern Ocean (Angel *in litt.* and in Cadée *et al.* 1992). The high numbers of *Fragilariopsis curta* we found in the gut of the ostracods *Alacia belgicae* seem to confirm this hypothesis. The total numbers of planktonic ostracod recorded in the area (up to 665 ind.·100 $m^{-3}$, Benassi *et al.* 1992) do not justify the number of pellets found; however the ostracods were collected in sea waters relatively far from our sampling station and drifting sea ice. Moreover, the scarcity of ostracods in our samples could reflect both the patchy distribution of these animals and the inadequacy of the sampling device.

The abundance of the thecosome *Limacina helicina* in the coastal surface waters of the Ross Sea (Guglielmo *et al.* 1992) and the large volume of its mucous feeding web (80 ml, Gilmer & Harbison 1986) first suggested it could be the producer organism, but the subsequent gut content analysis did not support this hypothesis. Moreover, to our knowledge, among Gastropods, a true peritrophic membrane has been described only for *Patella* species (see Peters 1992 for a review).

Finally, it cannot be excluded that the organism that produced the oval pellets could be a typical filter feeder crustacean grazing on phytoplankton cells. In this

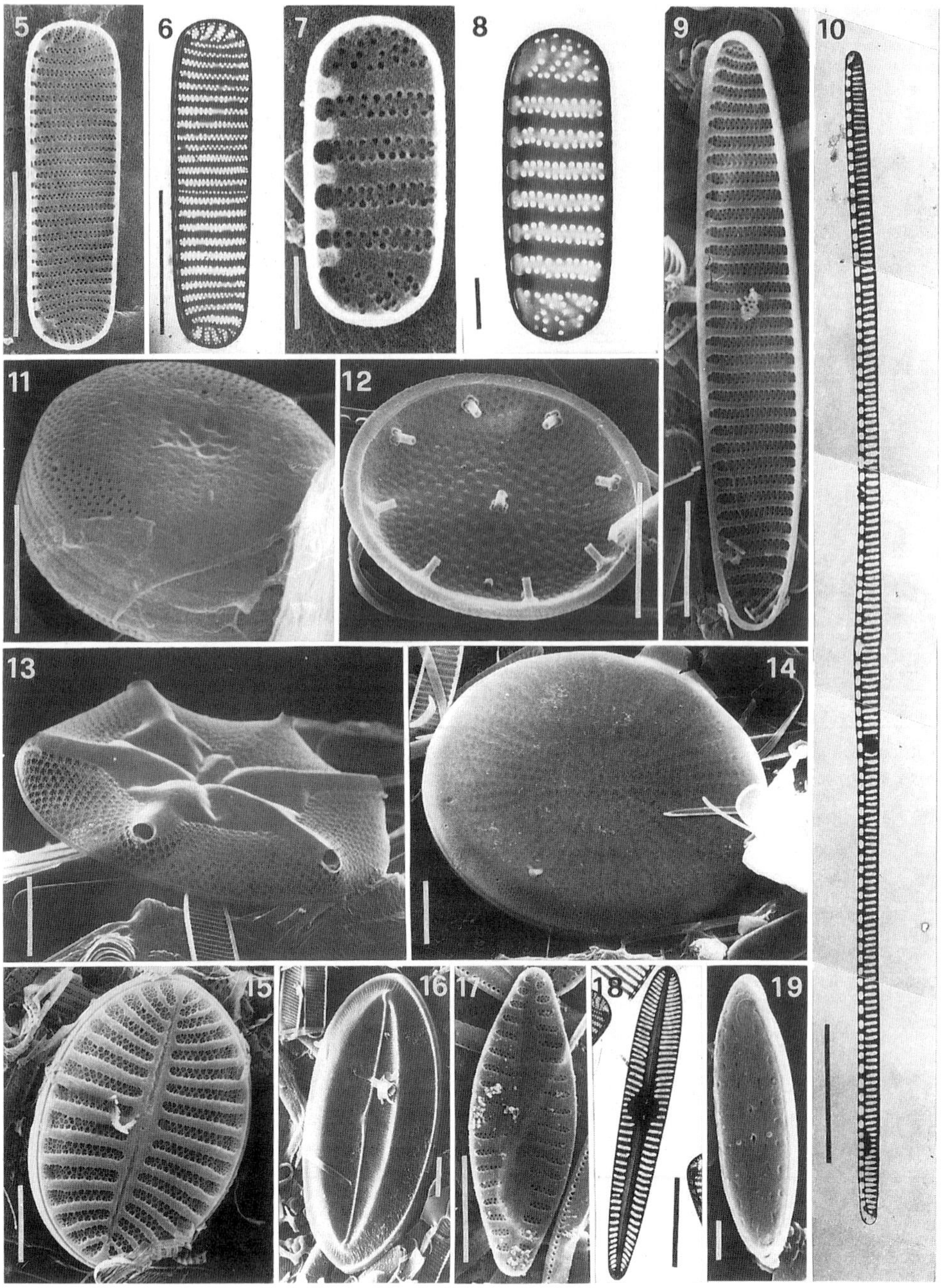
5
6
7
8
9
10
11
12
13
14
15
16
17
18
19

instance, the finding of benthic or typically sea-ice species in the faecal pellets could be explained both by the ice melting and the water column turbulence.

Although data on the distribution of oval pellets are rather difficult to compare due to differences in sampling devices, the concentrations we measured are among the highest recorded so far. From sediment trap experiments in the Weddell Sea, Fischer *et al.* (1988) calculated an annual flux of $94 \cdot 10^3$ oval pellets$\cdot m^{-2}$. In the same area the presence of oval pellets down to a depth of 1000 m and a non homogeneous vertical distribution were observed in discrete water samples by Nöthig & von Bodungen (1989) who also estimated the contribution (64% in volume) of the oval pellets to the total faecal flux in sediment trap material. The uneven distribution and the importance of oval pellets were confirmed by González (1992) who recorded concentrations between 2 and 456 pellets$\cdot 30\ l^{-1}$ in samples collected at different depths from the surface to 300$\cdot m^{-2}$ and a maximum flux of 631mg dry weight$\cdot m^{-2}$ in the Scotia and Weddell Seas.

The relatively high concentrations of oval faecal pellets we found, their constant presence in the water column during the sampling period, and the scarcity of information available on faecal material in the Southern Ocean, other than krill and copepod faeces, prompted us to provide a crude estimation of the contribution of oval pellets to silica fluxes. The standing stock of organic carbon in faecal pellets, obtained from biovolumes of diatom cells, ranged between 3.48 and 15.03 mgC$\cdot m^{-2}$. Assuming a sinking velocity of 150 m$\cdot day^{-1}$ (Cadée *et al.* 1992), oval pellets produced in the surface layer will sink through the water column (50 m) in 8 hours (10.44–45.09 mg C$\cdot m^{-2} \cdot day^{-1}$). In the case of a continuous and homogenous production of pellets during the day in all layers of the water column, these values must be doubled to obtain the daily flux of carbon (20.88–90.18 mg C$\cdot m^{-2} \cdot day^{-1}$). Using a Si/C ratio of 0.57 (by moles) (Nelson & Smith 1986), the total flux of biogenic silica attributable to oval pellets is between 27.8 and 120 mg Si$\cdot m^{-2} \cdot day^{-1}$.

The revised silica budget proposed by Nelson *et al.* (1991) for ice-edge blooms in the Ross Sea gives a net production rate of 12.4 mmol Si$\cdot m^{-2} \cdot day^{-1}$ (336mg Si$\cdot m^{-2} \cdot day^{-1}$) with a dissolution rate of 63.5% within the upper 50 m. If our estimation of silica flux via oval faecal pellets is realistic, the contribution of these pellets to the net production rate of silica should be between 6% and 27%. These values should be

Figs 5–19. Diatom species found inside faecal pellets. Scale bars = 1 µm (Figs 7, 8, 19), 5 µm (Figs 11, 12, 18), 10 µm (Figs 5, 6, 9, 10, 13, 14, 15, 16, 18). Fig. 5. SEM micrograph of *Fragilariopsis curta.* Fig. 6. TEM micrograph of *Fragilariopsis curta.* Fig. 7. SEM micrograph of *Fragilariopsis cylindrus.* Fig. 8. TEM micrograph of *Fragilariopsis cylindrus.* Fig. 9. SEM micrograph of *Fragilariopsis sublinearis.* Fig. 10. TEM micrograph of *Pseudo-nitzschia turgiduloides.* Fig. 11. TEM micrograph of *Thalassiosira gracilis* v. *gracilis.* External view. Fig. 12 SEM micrograph of *Thalassiosira gracilis* v. *gracilis.* Internal view. Fig. 13 SEM micrograph of *Asteromphalus parvulus* Fig. 14. SEM micrograph of *Actinocyclus actinochilus.* Fig 15. SEM micrograph of *Cocconeis fasciolata.* Fig. 16. SEM micrograph of *Cocconeis schuettii.* Fig. 17. SEM micrograph of *Achnantes vincentii.* Fig. 18. TEM micrograph of *Pseudogomphonema* sp. Fig. 19. SEM micrograph of *Navicula* sp.

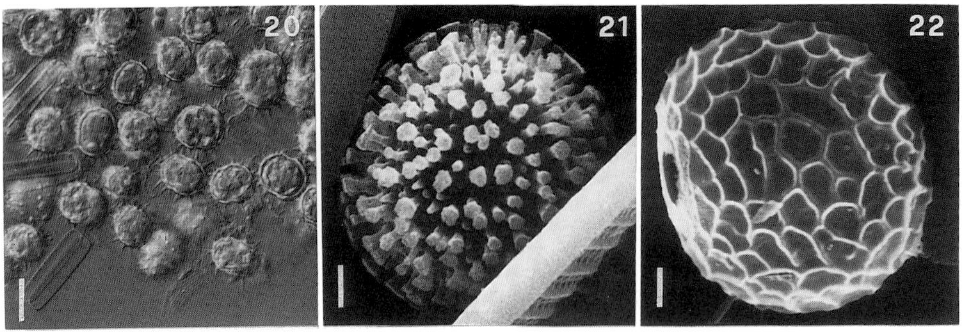


Figs 20–22. Algal species found inside faecal pellets. Fig. 20. Light micrograph of dinoflagellate resting cysts. Scale bar = 10 µm. Fig. 21. SEM micrograph of *Litheusphaerella spectabilis*. Scale bar = 1 µm. Fig. 22. SEM micrograph of *Archeomonas areolata*. Scale bar = 1 µm.

viewed with caution because the dissolution rate of silica is highly variable and depends on the sinking velocity of diatom aggregates (Nelson *et al.* 1991). For instance, Cadée *et al.* (1992) verified that krill faecal strings are easily destroyed and dissolved during their sinking thus contributing to the regenerated production; by contrast oval faecal pellets have a higher probability of reaching intact the sediments.

Although we cannot evaluate the contribution of oval pellets to the carbon flux because we studied only diatom biovolumes, the relevant percentage of whole cells found in the pellets indicate that they may contribute significantly to the transfer of organic carbon from the pelagic to the benthic food-web in the Ross Sea.

## Conclusions

Studies performed so far on carbon and silica fluxes in the Ross Sea have focused on the ecology of krill organisms and on the fate of senescent diatom blooms (Quéguiner *et al.* 1991). A large part of the siliceous sediments in this area for the past 18,000 years is constituted by *Fragiliariopsis curta* (Kellog & Truesdale 1979; Truesdale & Kellog 1979), and this taxon is still the dominant diatom species in the plankton (Carbonell 1985; Nuccio, pers. comm.). The results of our observations suggest that oval non-krill faecal pellets could play an important role in the sinking of organic and organogenic material in the Ross Sea at least in the areas exposed to ice covering by the drifting sea ice.


## Acknowledgements

This work was conducted within the frame of the Italian National Programme for Antarctica. The authors are particularly grateful to Prof. I. Ferrari and Dr M. Angel for their help and fruitful discussions on ostracod ecology and to Dr Buck for his help in determining the chrysophyte statocysts.

## References


Bathmann, U., Fischer, G., Müller, P. J. & Gerbes, D. (1991). Short-term variations in particulate matter sedimentation off Kapp Norvegia, Weddell Sea, Antarctica: relation to water mass advection, ice cover, plankton biomass and feeding activity. *Polar Biology*, **11**, 185–195.

Benassi, G., Ferrari, I., Gentile, G., Menozzi, P. & McKenzie, K. G. (1992). Planktonic Ostracoda in the Southern Ocean and in the Ross Sea: 1989–1990 campaign. *National Scientific Commission for Antarctica, Oceanographic Campaign 1989–1990, Data Report* **II**, 247–300.

Buck, K. R., Bolt, P. A., Bentham, W. N. & Garrison, D. L. (1992). A dinoflagellate cyst from Antarctic sea ice. *Journal of Phycology*, **28**, 15–18.

Buck, K. R., Bolt, P. A. & Garrison, D. L. (1990). Phagotrophy and fecal pellet production by an athecate dinoflagellate in the Antarctic sea ice. *Marine Ecology Progress Series*, **60**, 75–84.

Cadée, G. C., González, H. & Schnack-Schiel, S. B. (1992). Krill diet affects faecal string settling. *Polar Biology*, **12**, 75–80.

Carbonell, M. C. (1985). *Phytoplankton of an ice-edge bloom in the Ross Sea, with special reference to the elemental composition of Antarctic diatoms*. 133 pp. M.Sc. Thesis, Oregon State University, Corvallis, USA.

Dunbar, R. B. (1984). Sediment trap experiments on the Antarctic continental margin. *Antarctic Journal of the United States*, **19**, 70–71.

Edler, L. (1979). *Recommendations for marine biological studies in the Baltic Sea. Phytoplankton and chlorophyll*. 38 pp. Baltic Marine Biologists Publication 5, National Swedish Environment Protection Board, Stockholm, Sweden.

Fischer, G., Fütterer, D., Gersonde, R., Honjo, S., Ostermann, D. & Wefer, G. (1988). Seasonal variability of particle flux in the Weddell Sea and its relation to ice cover. *Nature*, **335**, 426–428.

Gilmer, R. W. & Harbison, G. R. (1986). Morphology and field behavior of pteropod molluscs: feeding methods in the families Cavoliniidae, Limacinidae and Peraclididae (Gastropoda: Thecosomata). *Marine Biology*, **91**, 47–57.

González, H. E. (1992). The distribution and abundance of krill faecal material and oval pellets in the Scotia and Weddell Seas (Antarctica) and their role in particle flux. *Polar Biology*, **12**, 81–91.

Granéli, E., Granéli, W., Rabbani, M. M., Daugbjerg, N., Fransz, G., Cuzin-Roudy, J. & Alder, V. A. (1993). The influence of copepod and krill grazing on the species composition of phytoplankton communities from the Scotia-Weddell Sea. An experimental approach. *Polar Biology*, **13**, 201–213.

Guglielmo, L., Costanzo, G., Zagami, G., Manganaro, A. & Arena, G. (1992). Zooplankton ecology in the Southern Ocean. *National Scientific Commission for Antarctica, Oceanographic Campaign 1989–1990, Data Report* **II**, 301–467.

Karl, D. M., Tilbrook, B. D. & Tien, G. (1991) Seasonal coupling of organic matter production and particle flux in the western Bransfield Strait, Antarctica. *Deep-Sea Research*, **38**, 1097–1126.

Kellog, T. B. & Truesdale, R. S. (1979). Late Quaternary paleoecology and paleoclimatology of the Ross Sea: the diatom record. *Marine Micropaleontology*, **4**, 137–158.

Innamorati, M., Lazzara, L., Mori, G., Nuccio, C. & Saggiomo, V. (1991). Phytoplankton ecology. *National Scientific Commission for Antarctica, Oceanographic Campaign 1989–1990, Data Report* **I**, 141–252.

Jaques, G. (1991). Is the concept of new production-regenerated production valid for the Southern Ocean? *Marine Chemistry*, **35**, 273–286.

Longhurst, A. R. (1991). Role of the marine biosphere in the global carbon cycle. *Limnology and Oceanography*, **36**, 1507–1526.

Mitchell, J. G. & Silver, M. W. (1982). Modern archaeomonads indicate sea-ice environments. *Nature*, **296**, 437–439.

Nelson, D. M., Ahern, J. A. & Herlihy, L. J. (1991). Cycling of biogenic silica within the upper water column of the Ross Sea. *Marine Chemistry*, **35**, 461–476.

Nelson, D. M. & Smith, W. O. Jr. (1986). Phytoplankton bloom dynamics of the western Ross Sea ice edge – II. Mesoscale cycling of nitrogen and silicon. *Deep-Sea Research*, **33**, 1389–1412.

Nöthig, E-M. & Von Bodungen, B. (1989). Occurrence and vertical flux of faecal pellets of probably protozoan origin in the southeastern Weddell Sea (Antarctica). *Marine Ecology Progress Series*, **56**, 281–289.

Palmisano, A. C. & Sullivan, C. W. (1983). Sea Ice Microbial Communities (SIMCO). 1. Distribution, abundance, and primary production of ice microalgae in McMurdo Sound, Antarctica in 1980. *Polar Biology* **2**, 171–177.

Peters, W. (1992). *Peritrophic membranes*. 238 pp. Zoophysiology, vol. 30. Springer-Verlag, Berlin, Germany.

Priddle, J., Smetacek, V. & Bathmann, U. (1992). Antarctic marine primary production, biogeochemical carbon cycles and climatic change. *Philosophical Transactions of the Royal Society of London B*, **338**, 289–297.

Quéguiner, B., Tréguer, P. & Nelson, D. M. (1991). The production of biogenic silica in the Weddell and Scotia Seas. *Marine Chemistry*, **35**, 449–459.

Riebesell, U., Schloss, I. & Smetacek, V. (1991). Aggregation of algae released from sea ice: implications for seeding and sedimentation. *Polar Biology*, **11**, 239–248.

Saggiomo, V., Massi, L., Modigh, M. & Innamorati, M. (1992). Size-fractionated primary production in Terra Nova Bay (Ross Sea) during the Austral summer (1989–1990). In: *Oceanografia en Antartica* (V. A. Gallardo, O. Ferretti & H. I. Moyano eds), 289–294. Centro EULA, Conception, Chile.

Strathmann, R. R. (1967). Estimating the organic carbon content of phytoplankton from cell volume or plasma volume. *Limnology and Oceanography*, **12**, 411–418.

Tréguer, P. & Quéguiner, B. (1991). Biochemistry and circulation of water masses in the Southern Ocean. Preface. *Marine Chemistry*, **35**, xi–xii.

Truesdale, R. S. & Kellog, T. B. (1979). Ross Sea diatoms: modern assemblage distributions and their relationship to ecologic, oceanographic and sedimentary conditions. *Marine Micropaleontology*, **4**,13–31.

Von Bodungen, B., Smetacek, V. S., Tilzer, M. M. & Zeitschel, B. (1986). Primary production and sedimentation during spring in the Antarctic Peninsula Region. *Deep-Sea Research*, **33**, 177–194

# Biodiversity of epiphytic diatom community on leaves of *Posidonia oceanica*

L. Mazzella, M. C. Buia and L. Spinoccia

*Laboratorio di Ecologia del Benthos, Stazione Zoologica "A. Dohrn" di Napoli, 80077 Ischia (Na), Italy*

## Abstract

The composition and the structure of the diatom community epiphytic on leaves of the endemic Mediterranean seagrass *Posidonia oceanica* were studied using Scanning Electron Microscopy. The shoots were collected in a meadow around the Island of Ischia (Gulf of Naples – Italy) at 5 and 22 m during the winter and summer seasons. Species were identified, counted and also classified into growth forms.

A total of 56 species representing 4 growth forms were identified, all belonging to the group of *Pennatae*. The *Naviculaceae* was most species-rich, including both free and erect forms. However, prostrate species of *Cocconeis* (9 species), strongly adherent to the the substratum, were the most frequent and abundant at all depths and seasons. What differentiated the community was the erect stratum (e.g. *Gomphonemopsis*). The factor structuring diatom communities seems to be the leaf age and competition both within diatom species and between diatom and encrusting macro-algae: in general, the highest abundance was recorded on leaf apices (2965 cell/$mm^2$), or when species belonging to different growth forms co-occurred or when encrusting calcareous algae (e.g. *Fosliella–Pneophyllum* spp.) were absent.

## Introduction

The crucial role of seagrasses as "structural species" in coastal ecosystems has been emphasized for many temperate and tropical seas, in particular for the Mediterranean Sea. In the *Posidonia oceanica* ecosystem, endemic to the Mediterranean basin, energy matter are not transferred directly from seagrasses to higher trophic levels (Mazzella *et al.* 1992). Rather this role is played by the macro- and micro-floral epiphytes on leaves and rhizomes. The diatom community, in particular, represents the main food for herbivores such as Molluscs and amphipod Crustaceans (Mazzella & Russo 1989; Scipione & Mazzella 1992).

In the framework of an integrated project aimed at identifying the functional organization and the pathways of energy flow of the *P. oceanica* ecosystem in the Mediterranean Sea, the species composition of the leaf diatom community was investigated.

The seagrass leaf blades represent a unique substratum, with a gradient in age from the meristematic base to the more mature apex. Therefore, the leaves represent a gradient in time available for colonization by epiphytes, as well as a gradient in light intensity and hydrodynamic forces (Mazzella *et al.* 1982).

Moreover, colonization patterns of epibenthic diatoms are greatly influenced by the growth forms and functional characteristics of the diatoms themselves. Many papers have described the importance of diatom motility in relation to a species life cycle and ecological requirements (Edgar & Pickett-Heaps 1984). The importance of the relationship between form and function has also been stressed also for phytoplankton communities (Sournia 1982). Hudon & Legendre (1987) and Kawamura & Hirano (1992) showed that growth form, adherence to the substrate, and motility are features of paramount importance in relation to nutrient and light availability, colonization patterns, and grazing pressure. Due to the importance of the diatom microflora for the grazers inhabiting the *Posidonia* beds, the epiphytic diatom community was analyzed by growth form. In addition, a method to quantify diatom abundance by using Scanning Electron Microscopy was developed.

## Material and Methods

Samples were collected by SCUBA diving in a *Posidonia oceanica* bed, located along the coast of the Ischia Island (Gulf of Naples – Italy ) and described previously by Buia *et al.* (1982). Sampling was in January and July (1989) at 5 m and 22 m, where plant primary production was also followed over one year (Buia *et al.* 1992). The longest leaf in the shoot was selected for studies of the diatom microflora and was fixed in 4% gluteraldehyde. In the two months of investigation, the ribbon-like leaves reached the maximum length of 67 and 59 cm in summer, and 41 and 18 cm in winter, respectively for 5 and 22 m samples (Buia *et al.* 1992).

Qualitative and quantitative analyses were conducted to identify the diatom microflora and to estimate their abundance (Mazzella & Spinoccia 1992). Qualitative analyses were made by scraping the entire leaf and cleaning the material according to Hasle (1978). For the quantitative analyses, three portions of the blade tissue were taken at three different levels (basal, intermediate, and apical); each portion was about 1 cm long and 0.9–1.0 cm width. They were prepared for SEM analyses as follows: 1. portions were gently washed in distilled water; 2. dehydrated in ethanol series, and 3. critical point dryed. The samples were then sputter-coated and mounted in an upright position allowing the observations of both sides of the leaf. SEM analysis was on two transects lines, from which 20 points (each of 0.0625 $mm^2$ area) were selected and photographed for a total surface area of 1.25 $mm^2$, on each side (Fig. 1). In each photograph, diatom cells were counted and categorized by growth form. Higher magnifications were used for identification to the species level. Simonsen (1979) classification was used for identification at genus level. Growth categories were identified on the basis of several characters which were important mainly in relation to the potential grazing pressure, such as posture and adherence to the substrate (prostate vs. erect forms, motile vs. attached forms) (Hudon and Legendre 1987).

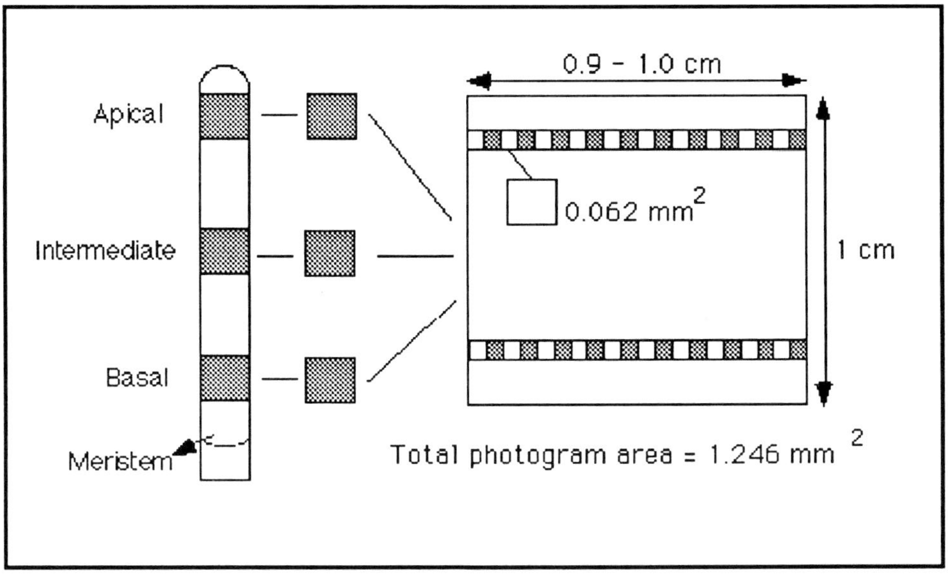


Fig. 1. Illustration of the leaf portions observed by SEM.

## Results

### *Algal colonization along the leaf axes*

Distinct colonization of both micro- and macroalgae was recorded from leaf portions of different age. In young (basal) tissues, bacteria are the dominant colonizers (Fig. 2a). On the intermediate portions of the leaf, diatoms were the dominant flora (Fig. 2c and d); towards the leaf apices, diatoms tended to be replaced by the macroflora, represented in winter by the encrusting algae (*Fosliella–Pneophyllum* spp., *Myrionema orbiculare*) (Fig. 2e and f), and by both encrusting and erect forms in summer. Evidence of herbivory was common (Fig. 2 b).

The difference between the diatom flora at the 5 and 22 m stations was striking, particularly in the winter season. While the microflora was dominated by different species of *Cocconeis* at 5m, these coexisted with *Gomphonemopsis* spp. at 22 m.

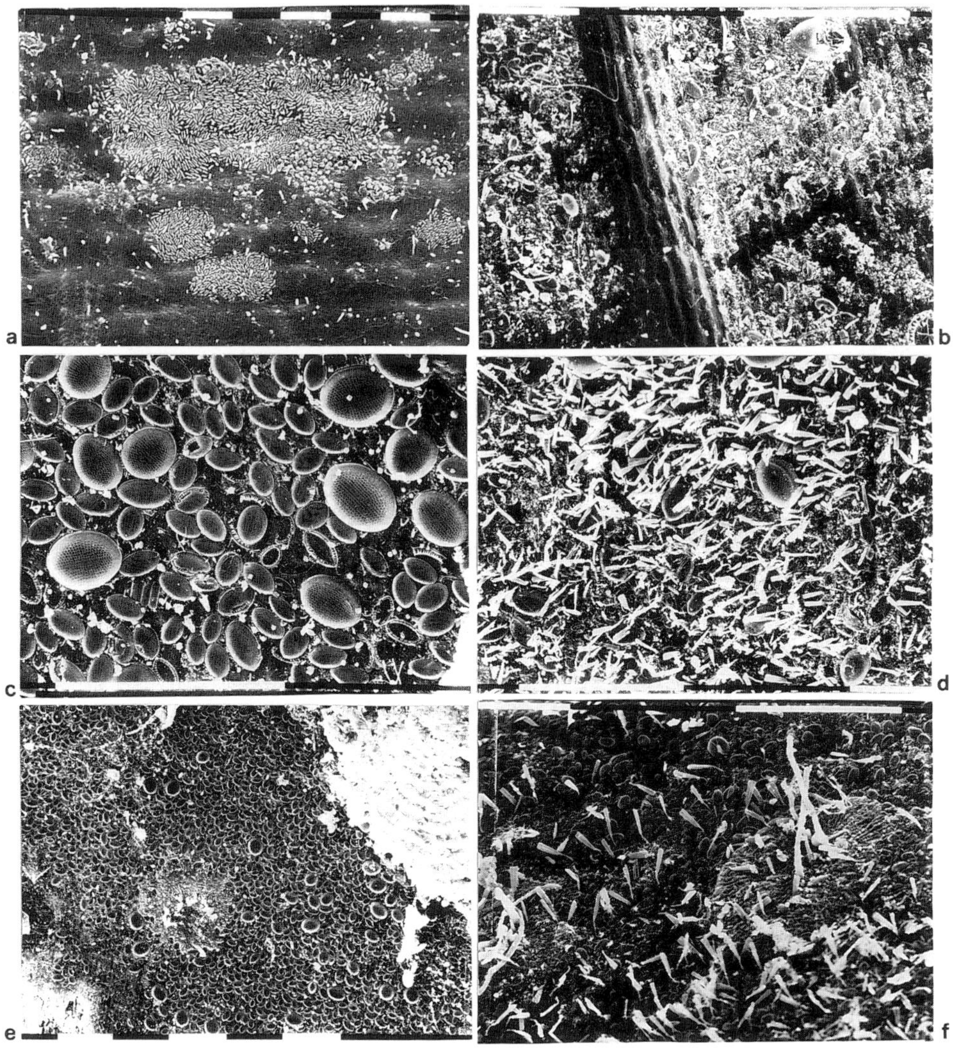


Fig. 2. (a) Bacteria on young leaf tissue; (b–d) diatoms on intermediate leaf tissue; (e) diatoms and *Fosliella–Pneophyllum* spp.; (d) diatoms and *Myrionema orbiculare*. All bar scales = 0.1 µm.

*Species composition and classification by growth forms (morphofunctional groups)*

In the two seasons, at the two depths, a total of 56 diatom species were found along the leaf axes, all belonging to the group *Pennatae*. The highest number of species was recorded in the winter samples, at both stations (31 species at 5 m, 25 at 22 m depth), while in summer, plants at the deep station accounted for a higher species richness (17 species at 5 m vs 21 at 22 m). The family *Naviculaceae* accounted for the highest number of species (22 species) including both free living forms such as *Amphora* spp., *Navicula spp.*and *Mastogloia* spp., and attached forms such as *Gomphonemopsis* spp. The family *Achnanthaceae* was present with a higher number of species; 9 species of *Cocconeis* (e.g. C. *placentula* var. *euglypta* (Ehrenberg) Cleve, *C. pediculus* Ehrenberg, *C. pseudomarginata* Gregory, *C. scutellum* Ehrenberg, *C. scutellum* var. *parva* (Grunow) Cleve, *C. stauroneiformis* (W. Smith) Okuno) and two species of *Achnanthes*.

The family *Diatomaceae* was present with *Grammatophora angulosa* Ehrenberg and *G. marina* (Lyngbye) Kützing, *Fragilaria* and *Synedra* spp. Lastly, *Rhopalodia gibba* (Ehrenberg) Müller and *Nitszchia panduriformis* Gregory were also found but were generally rare.

Four types of growth forms were identified:

**Group A**, characterized by sessile species, in an upright position, attached to the leaf blade by means of a mucopolysaccharide stalk, solitary or colonial (e.g. *Gomphonemopsis* sp., *Grammatophora* sp.);

- **Group B**, characterized by sessile prostrate species strongly adherent to the substrate by the raphe valvar side, scarcely mobile, (e.g. *Cocconeis sp.*) (Figs 3 and 4);
- **Group C**, characterized by species adherent to the subsrate but having a good mobility (e.g. *Amphora* sp., *Rhopalodia* sp.) (Figs 5 and 6);
- **Group D**, characterized by motile species not adherent to the substrate but entrapped among the macroalgae (e.g. *Mastogloia sp.*, *Navicula sp.*, *Nitszchia sp.*) (Figs 7and 8).

Group A was represented by 5 species, Group B by 12 species of which 9 were *Cocconeis*; 22 species were grouped in the category D and only a few species in the Group C.

*Quantitative analyses: abundance of growth forms along the leaf axis*

Cell density was calculated only for the groups A and B, since the remaining groups, were generally not abundant. In winter, at the shallow station (Fig. 4a), diatoms classified into Group A were scarcely represented, while the group B was very abundant especially towards the apical portions; a side-effect (inner vs. outer) was remarkable in the distal portions of the leaf tissue, where the macrophytic covering was often also different (Fig. 4a). The high density of Group B was due maily to *Cocconeis* species; a lower abundance of *Cocconeis* were recorded only when another species of the same group, e.g. *Fragilaria* sp., accounted for the highest numbers of cells. For the deep station (Fig. 4b), the Group A showed a remarkable difference between the internal and external side of the leaf, at the apical portions on the latter reaching values of 1680 cells/mm$^2$. There was a gradual increase in the abundance of Group B on both sides from the meristem to the apices, with consistent high values (e.g. 1285 cells/mm$^2$).

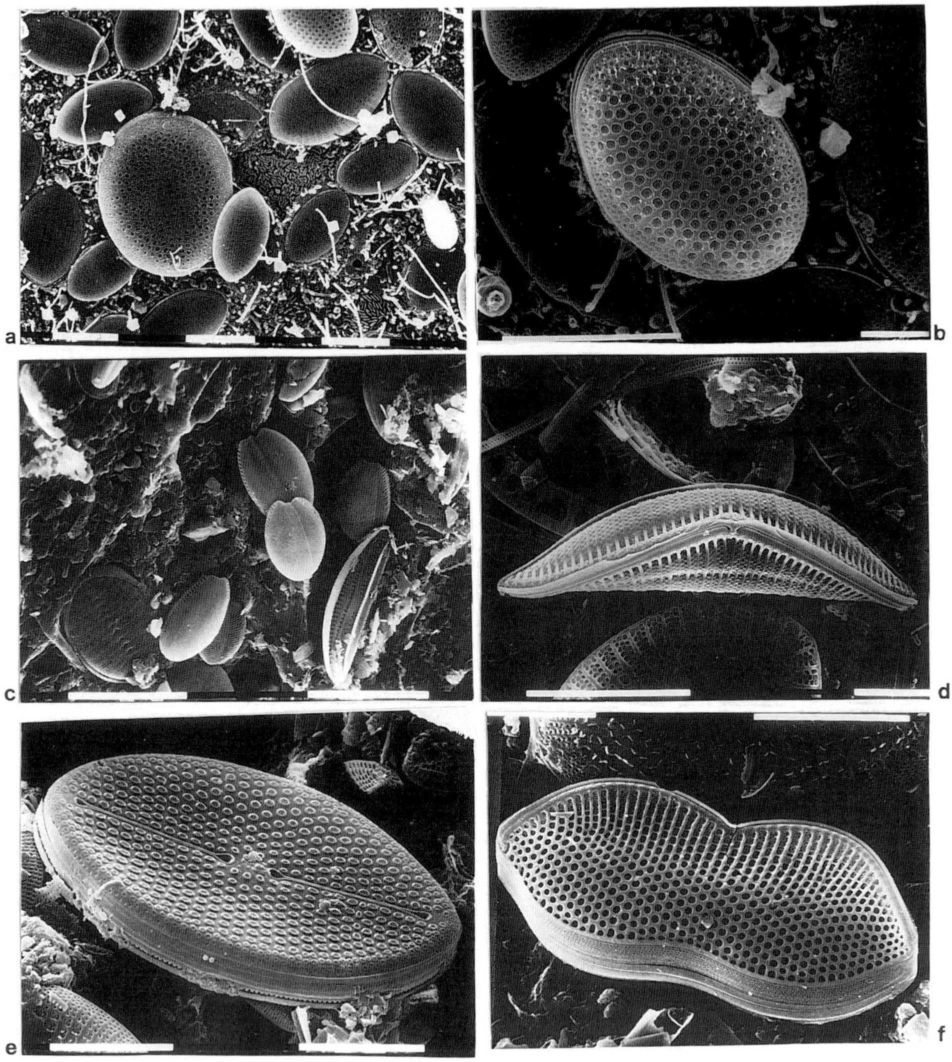


Fig. 3. (a) *Cocconeis stauroneiformis*; (b) *Cocconeis scutellum*; (b) *Amphora* sp.; (d) *Rhopalodia gibba*; (e) *Mastogloia fimbriata*; (d) *Nitzschia panduriformis*. All bar scales = 10 µm.

The summer samples from 5m depth, again showed low densities for the Group A, while the Group B was very abundant, (Fig. 4c). The summer samples from the deep station (Fig. 4d), on the contrary, differ from the winter samples by the virtual absence of Group A diatoms. A side effect has observed for Group B colonization: the external and distal portions recorded the highest values (Fig. 4d).

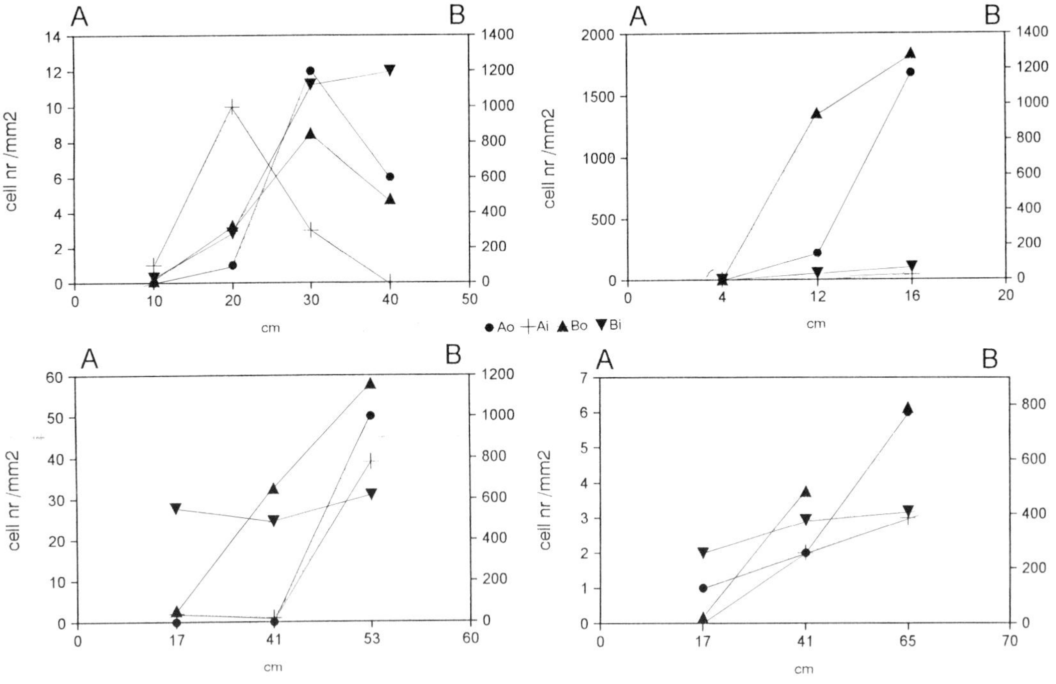


Fig. 4. Cell abundance along the leaf axis of Group A and B in winter (up) and in summer (down) at 5 m (left) and 22 m (right). (o = outer leaf side; I = inner leaf side).

*Diatom microdistribution and relationships with encrusting macroalgae*

Along the leaf axis, the diatom community exhibited a distribution characterized by high patchiness, depending on the age of leaf tissue, on the co-occurrence of diatoms from different growth groups, and by the presence of macroalgae. On old leaf tissue from the shallow and deep stands, in fact, the diatom density seems to be greatly influenced by the presence of the calcareous encrusting *Fosliella–Pneophyllum* species, which seem to inhibit diatom settlement. The soft encrusting *Myrionema* sp. did not appear to influence diatom colonization (Fig. 5a and b).

**Discussion**

The seagrass leaves constitute a suitable settlement surface for a characteristic microalgal community. The species found on the Mediterranean seagrass *Posidonia oceanica* belong to well defined systematic groups (Mazzella 1983; Mazzella & Spinoccia 1992). Species diversity is found within relatively few genera, all belonging to the group *Pennatae*. The same species or congeneric diatom species found on *P. oceanica*, have also been found epiphytic on other seagrasses, distributed both in tropical and temperate waters (*Cocconeis*, *Mastogloia*, *Gomphonemopsis* spp., e.g. by Sieburth & Thomas 1973; Ferreira & Seeliger 1975; Sullivan 1977; Stephen & Gibson 1979; Jacobs & Noten 1980; Sullivan 1982; Medlin & Round 1986). This indicates that the substratum offered by the leaf could be "*per se*" an important factor structuring the epiphytic diatom community, regardless of geographic location. The substratum could be favourable to particular growth forms, such as the prostrate species that are strongly adherent. This could explain the presence of *Cocconeis* spp., in particular *C. stauroneiformis* and *C. scutellum;* as a continuous film found also on *Zostera marina* blades (Siebur t & Thomas 1973; Jacobs & Noten 1980). The quantitative analysis allowed an estimation of the colonization along the leaf axis, which can be considered as a gradient in age and exposure. The classification of growth forms along with the quantitative analyses indicated that the community in both seasons and stations is characterized by prostrate species, dominated by *Cocconeis spp.* What mainly differentiates the two stations are the erect forms and the motile species, the latter being less abundant or rare.

Competition for space is evident between species belonging to the same growth form; *Cocconeis* sp. and *Fragilaria* sp., in particular, when co-occurring, seem to exclude each other. This is not found for *Cocconeis* spp. (prostrate) and *Gomphonemopsis* spp. (erect): in the deep station, in the same leaf tissue, they can co-occur at high abundances (up to 2965 cells/mm$^2$). Competition for space has been found also between the adherent *Cocconeis* spp. and the macroalagal species such as the crustose calcareous *Fosliella–Pneophyllum* spp.

Another factor which likely influences the distribution of diatoms along the leaf axes is the grazing pressure by herbivores or herbivore-deposit feeders which in *Posidonia oceanica* meadows are the most abundant type among the invetebrates (Mazzella *et al.* 1992). Previous investigations on the feeding behaviour have shown that diatoms are the dominant components in the diet of Molluscs and amphipod Crustaceans (Mazzella & Russo 1989; Scipione & Mazzella 1992). However, only some Molluscs can feed on the strongly adherent *Cocconeis spp.*, while the other grazers are able to remove the more mobile species, such as *Amphora sp.* and *Navicula sp.*. This indicates that *Cocconeis spp.*, unless the most abundant, are not preferred; this selectivity could influence the abundance. On the other hand, species diversity seems to be very important both in the substrate colonization patterns and as food source for grazers.

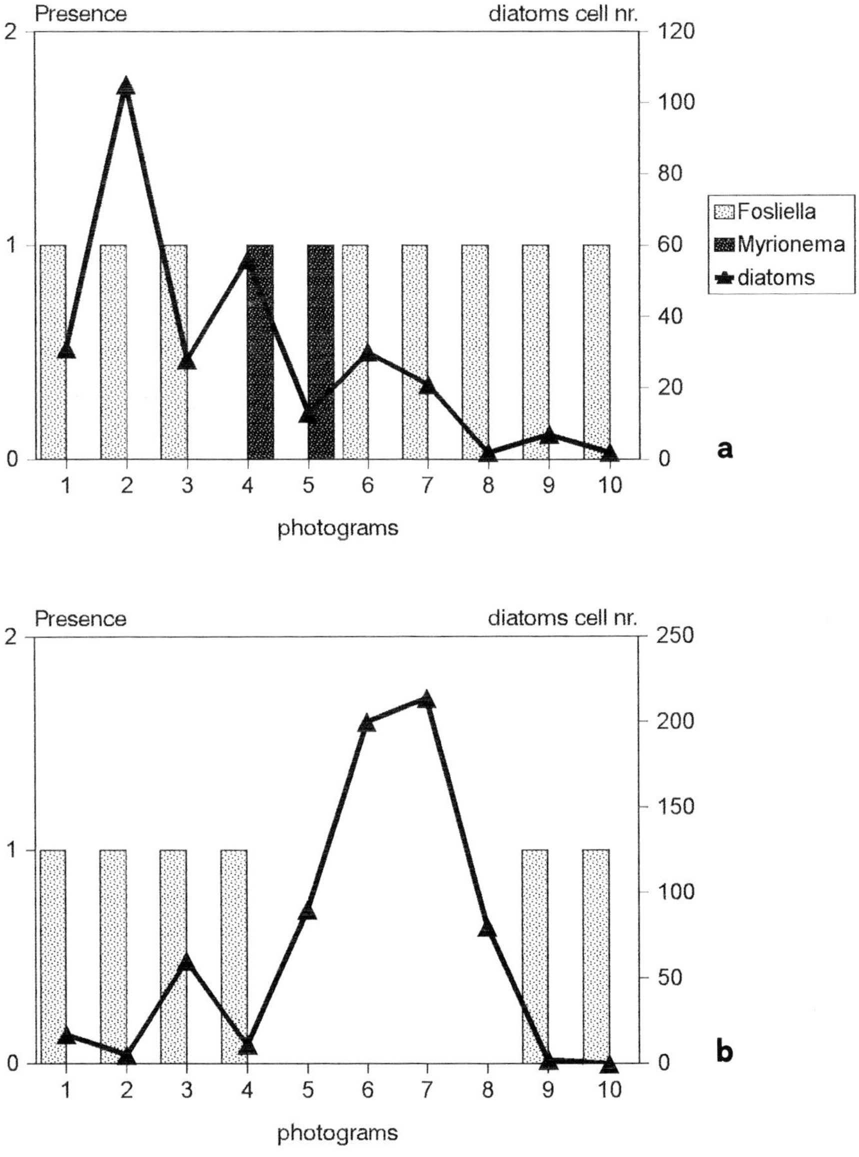


Fig. 5. Cell abundance along a transect line for 5 m (a) and 22 m (b) in winter.

In conclusion, the diatom microflora of *Posidonia oceanica* leaves is well structured; the age gradient along the leaf axes seems to be the training factor controlling the diatom community composition. The age gradient is concurrent in competition for space, grazing pressure, exposure to light and hydrodynamic forces.

The prostrate diatom growth forms constitute a continuous layer which is comparable to that of macroalgal encrusting algae; the prostrate growth form could be considered as a survival strategy against grazing.

## Acknowledgement

Work partly supported by the European Community Commission (ENV. Prog. N.EV4V–0139–b; STEP–0063–C).

## References

Buia, M. C., Zupo, V. & Mazzella, L. (1992). Primary production and growth dynamics in *Posidonia oceanica* P.S.Z.N.I: *Marine Ecology*, **13** (1), 2–16.

Edgar L. A. & Pickett-Heaps, J. D. (1984). Diatom locomotion. In: *Progress in Phycological Research* (F. E. Round & D. J. Chapman eds), **3**, 47–88. Biopress Ltd, Bristol.

Ferreira S. & Seeliger, U. (1985). The colonization process of algal epiphytes on *Ruppia maritima* L. *Botanica Marina*, **28**, 245–249.

Jacobs, R. P. W. M. & Noten, T. M. P. A. (1980). The annual pattern of the diatoms in the epiphyton of eelgrass (*Zostera marina* L.) at Roscoff, France. *Aquatic Botany*, **8**, 355–370.

Hasle, G. R. (1978). Some specific preparations. Diatoms. In: *Phytoplankton Manual (*A. Sournia, ed.), UNESCO *Monographs on Oceanographic Methodology*, 136–142.

Hudon, C. & Legendre P. (1987). The ecological implications of growth forms in epibenthic diatoms. *Journal of Phycology*, **23**, 434–441.

Kawamura, T. & Hirano, R. (1992). Seasonal changes in benthic diatom communities colonizing glass slides in Aburastubo Bay, Japan. *Diatom Research*, **7** (2), 227–239.

Mazzella, L. (1983). Studies on the epiphytic diatoms of *Posidonia oceanica* (L.) Delile leaves. *Rapport Commission International Mer Mediterranée*, **28** (3), 123–124.

Mazzella, L. & Russo G. F. (1989). Grazing effect of two *Gibbula* species (Mollusca, Archaegastropoda) on the epiphytic community of *Posidonia oceanica* leaves. *Aquatic Botany*, **53**, 357–373.

Mazzella, L. & Spinoccia, L. (1992). Epiphytic diatoms of leaf blades of the Mediterranean seagrass *Posidonia oceanica* (L.) Delile. *Giornale Botanico Italiano*, **126** (6), 752–754.

Mazella, L., Wittmann, K. & Fresi, E. (1982). La comunità epifita e suo ruolo nella dinamica dell'ecosistema *Posidonia*. In: *Atti Convegno Unità Operative Afferenti ai Sottoprogetti Risorse Biologiche e Inquinamento Marino*, 215–224. CNR, Rome, Italy.

Mazella, L., Buia, M. C., Gambi, M. C., Lorenti, M., Russo, G., Scipione, M. B. & Zupo, V. (1992). Plant–animal trophic relationships in *the Posidonia oceanica* ecosystem of the Mediterranean Sea: a review. In: *Plant–animal interactions in the Marine Benthos* (D. M. John *et al.* eds), 165–187. Clarendon Press, Oxford.

Medlin, L. K. & Round, F. E. (1986). Taxonomic studies of marine gomphonemoid diatoms. *Diatom Research*, **1**, 205–225.

Scipione, M. B. & Mazzella, L. (1992). Epiphytic diatoms in the diet of Crustacean Amphipods of *Posidonia oceanica* leaf stratum. *Oebalia*, suppl .**17**, 409–412.

Sieburth, J. McN. & Thomas, C. D. (1973). Fouling on eelgrass (*Zostera marina* L.). *Journal of Phycology*, **9**, 46–50.

Simonsen, R. (1979). The diatom system: Ideas on Phylogeny. *Bacillaria* (R. Simonsen, ed.), 9–72. Bremerhaven.
Sournia, A. (1982). Form and function in marine phytoplankton. *Biological Review*, **57**, 347–394.
Stephen, F. C. & Gibson, R. A. (1979). Ultrasructural studies on some Mastogloia (Bacillariophyceae) species belonging to the group Ellipticae. *Botanica Marina*, **22**, 499–509.
Sullivan, M. J. (1977). Structural characteristics of a diatom community epiphytic on *Ruppia maritima*. *Hydrobiologia*, **53**, 81–86.
Sullivan, M. J. (1982). Community structure of the epiphytic diatoms from the Gulf Coast of Florida, U.S.A. In: *Proceedings of the Seventh International Diatom Symposium* (D. G. Mann, ed.), 373–384. O. Koeltz Science Publisher, Philadelphia, U.S.A.

# The epipelic and epipsammic diatom communities at a sandy coastal site on the South China Sea

Lokman Shamsudin and Kartini Mohamad

*Faculty of Fisheries and Marine Science, University Pertanian Malaysia, Mengabang Telipot, 21030 Kuala Terengganu, Terengganu, Malaysia.*

## Abstract

The epipelic flora on the sandy surface of the coast of the South China sea on the east coast of the penisular of Malaysia is composed predominantly of diatoms (*Navicula* spp., *Nitschia* spp., *Surirella* spp. and *Diploneis* spp.). The epipsammic diatom flora, on the other hand, is composed of an association of relatively (< 20 µm) small "diatoms" (mainly *Amphora* spp., *Achnanthes* spp. and *Fragilaria* spp.) attached quite firmly to the surface of sand grains in the littoral zone. The standing crop per unit area ($m^2$) of the epipelic diatom population was estimated using cell counts and chlorophyll–a content. A good correlation was found between the chlorophyll–a content and the cell counts.

## Introduction

The epipelic algal flora of the tropical coast of South China Sea is composed predominantly of diatoms which move freely over the surface of the sandy sediments. These diatoms are usually small, with an average size of 50 µm and are mainly motile (such as *Navicula* spp., *Nitzschia* spp.); a small amount of filamentous blue-green algae (mainly *Lyngbya* spp. and *Anabaena* spp.) also occurs. The epipelic and the attached epipsammic diatoms can be separated from the detrital material simply by repeated swirling of the sediments with water (Round 1965). The unattached material is swept into suspension and can be decanted off while episammic diatoms remain, attached to the sand grains. During the study period (January 1991 to January 1992) the standing crop of the epipelic algae was measured using cell counts and chlorophyll–a. Cell counts of the epipelic algae give a certain amount of taxonomic information as well as an indication of crop size.

## Methods

On each (monthly) sampling date, 5 to 6 samples of the surface sediment were taken from a site along the coast of the South China sea near the vicinity of the University Pertanian marine station. A sample of the top 5mm of sediment was obtained using a cylinder which was pushed into the sediment (enclosing an area of about 60 $cm^2$ ). The sample thus obtained was allowed to settle in the dark for 5 hours after which the supernatant liquid was removed. The sediments was then thoroughly mixed by shaking and from this, a petri dish was filled to a depth of about 0.5 cm. Half of the dish was covered with a double layer of white lens tissue while a square piece of lens tissue (1 $cm^2$) was placed on the other half. The dish was left overnight and between 09.00 hrs and 10.00 hrs the next morning, the epipelic algae, which by now had moved up into the tissue, was collected. The small (1 $cm^2$) tissue containing the epipelic algae was placed on a slide and teased apart and a cover glass, 19 mm in diameter, was placed over it. The epipelic algae lying along a standard transverse plane was counted under a microscope using the high-power objective.

To determine the chlorophyll–a content, the larger pieces of lens tissue were first placed on a piece of filter paper and allowed to air-dry in the dark for 2 hours at room temperature. A standard volume (25 ml) of 90% acetone was used to extract the pigments of the epipelic algae enmeshed in the tissue. Degradation of the pigments during the extraction process was prevented by the addition of 0.3 g dry magnesium carbonate. Grinding of the algae was not necessary for complete extraction which was accomplished in less than 24 hours. The chlorophyll–a and pheophytin–a contents were than measured using the absorption spectrophotometric method adopted by Moss (1967a, b), the sand remaining in the petri dish was dried at 120°C to a constant weight. After the petri dish sample preparation on the day of sampling, a portion of the remaining sediment was air dried in the dark on filter paper, and then weighed. After exrtacting the pigments using 90% acetone in the presence of magnesium carbonate, the pigment content of the sediment was determined using the method described earlier for the epipelic algae. A further portion of the sediment was washed by repeated swirling with water to remove all non-epipsammic material. It was then air-dried and a certain amount (1–2 g) was taken to extract the pigments using 90% acetone.

## Results

Fig. 1 shows that the chlorophyll–a content in epipelic and epipsammic algae had peak values in February 1991 (immediately after the monsoon season) and November 1992 (before the start of the monsoon season). In view of the relatively long period between these sampling dates, it is not possible to attempt a detailed analysis of this discrepancy. Neither is it possible to etablish a correlation between the chlorophyll–a values and environmental factors. As seen in Fig. 2, the results from the epipelic cell counts show a tendency to follow those of chlorophyll–a content. Judging from the pigment values, the standing crop of epipsammic algae was, on the whole, considerably higher than that of epipelic algae. The counts obtained for the epipsammic diatom population are 2 to 3 times those of the epipelic algae (unpublished data). This study also showed that the pigment content of raw sediment was higher than that of algae on washed sand (Table 1). It has been suggested that the washing process might possibly remove cells from the sand grains, especially the dead cells. Eaton and Moss (1966) showed that there was no significant removal of live cells from sand grains. However, they found that large numbers of dead cells occur freely in the sediment. In addition, they discovered that live cells of the chain-forming genera such as *Fragilaria* can be removed by mechanical agitation of the grains. The high levels of pigment found in untreated sediment can be attributed to the presence of dead algal cells as well as fragments of higher plant material in the sediment. The results also showed that the phaeophytin–a content of untreated sediment was higher than that of washed sand.

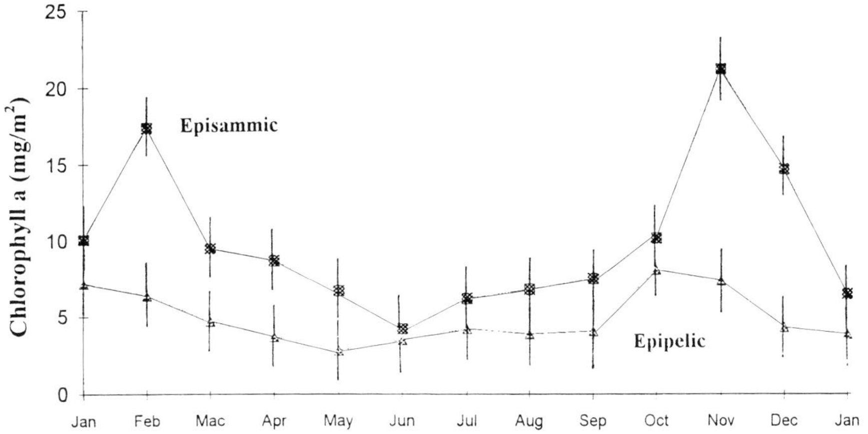


Fig. 1. Variation in chlorophyll– a (mg/m$^2$) of epipelic and epipsammic diatom populations at a sandy coastal site on the South China Sea from January 1991 to January 1992 (Vertical bars at 95% confident limit).

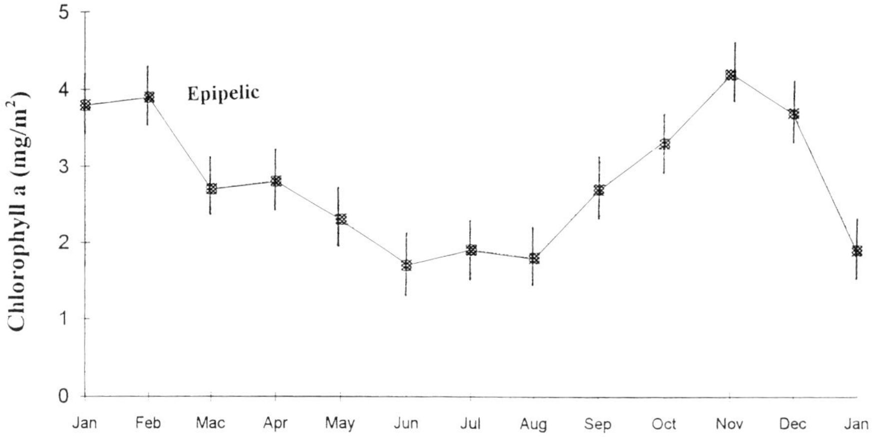


Fig. 2. Variation in epipelic diatom counts (cell number/m$^2$) at a sandy coastal site on the South China Sea from January 1991 to January 1992 (Vertical bars at 95% confident limit).

Table I. Pigment content (chlorophyll–a and phaeophytin–a) of diatom and detritus in washed and untreated sediment (x = mean ; SD standard deviation). All values are in mg pigment/$m^2$ of sediment. At least three samples were averaged for each estimation. The mean percentage degration (phaeophytin–a/ phaeophytin in a + chlorophyll–a) was 39.4 % for washed sand and 84.3 % for untreated sediment.

| Date | Washed | Untreated sediment | | |
|---|---|---|---|---|
| | x ±(SD) | x ± (SD) | x ± (SD) | x ± (SD) |
| Jan 91 | 6.11 (.54) | 10.4 (.94) | 12.2 (.94) | 20.4 (.94) |
| May 91 | 4.21 (.42) | 8.22 (.71) | 9.14 (.56) | 15.2 (.93) |
| July 91 | 2.32 (.24) | 6.42 (.52) | 6.34 (.32) | 10.4 (.95) |
| Sept 91 | 5.17 (.41) | 11.4 (.94) | 10.4 (.54) | 18.4( 2.2 ) |
| Jan 92 | 7.12 (.62 ) | 16.4 (.91) | 15.3 (.87) | 31.4(2.91) |

## Discussion

The sediment of the coast of South China sea has a vast population of burrowing invertebrates and thus a constant loss of epipelic cells can be expected. All along the shore, water movement causes the surface layer of sand to be thrown into ridges and furrows. Disturbance of the algal inhabitants occurs as a direct result of this. It has been suggested that a portion of the algal cells is invariably lost due to the grinding action of the moving sand grains. Wind-induced water movement is frequent on the sea shore and as such the free-living cells on the sediment surface are constantly subjected to agitation and loss. The attached community, especially the epipsammic diatoms (such as *Amphora* spp. and *Opephora* spp.) is probably less subject to removal compared to the free-living and as such will give a higher standing crop value than that of the epipelic algae.

## Acknowledgements

The authors wish to express their gratitude to Universiti Pertanian Malaysia for providing the necessary funds for the project. They also to thank the laboratory technicians for their kind assistance.

## References

Eaton, J. W. & Moss, B. (1966). The estimation of numbers and pigment content in epipelic algal populations. *Limnology & Oceanography*, **11**, 584–595.

Moss, B. (1967a). A spectrophotometric method for the estimation of percentage degradation of chlorophylls to pheo-pigments in extract of algae. *Limnology & Oceanography*, **12**, 335–340.

Moss, B. (1967b). A note on the estimation of chlorophyll a in freshwater algal commmunities. *Limnology & Oceanography*, **12**, 340–342 .

Round, F. E. (1965). The epipsammon; a relatively unknown freshwater algal association. *British Phycological Bulletin*, **2**, 456–463.

# The epiphytic and microphytobenthic diatoms in Estonian coastal waters (the Baltic Sea)

Sirje Vilbaste

*Institute of Zoology and Botany, Estonian Academy of Sciences, Str. Vanemuise 21, EE 2400 Tartu, Estonia*

## Abstract

The species composition of epiphytic and microphytobenthic diatom assemblages collected in two different areas of Estonian shallow coastal waters in midsummers 1990, 1991 and 1993 are discussed.

All in all 169 taxa of diatoms were recorded. The number of taxa varied from 49 to 87 in different samples. The diversity index H' varied between 2.66 and 4.55 and the evenness index I' between 0.52 and 0.82 for individual assemblages. The Sørensen index of similarity between microphytobenthic and epiphytic diatom assemblages varied from 57 to 84%.

The greater species composition of microphytobenthic assemblages better expresses the whole diatom flora in coastal waters compared with the corresponding data from epiphytic assemblages.

## Introduction

Studies of diatom species living in the Baltic Sea go back at least a century; see references in Snoeijs (1989) and Sakson & Miller (1993). There are however only a few floristic studies on diatoms from the Estonian coast, although some of them date from the middle of the previous century (Eichwald 1849, 1852; Weisse 1861). The author of the present paper has published a study (Vilbaste 1987) and now Estonian diatomists are taking part in the intercalibration of diatom species in the Baltic Sea (Snoeijs 1993).

The aim here is to provide some data about epiphytic and microphytobenthic diatom assemblages in Estonian coastal waters. The analysis is restricted to nineteen samples and for that reason the results are preliminary.

## Material and Methods

Käsmu Bay is a part of the Gulf of Finland and it is located on the north coast of Estonia. Väike Väin strait lies between Saaremaa and Muhu islands in the West-Estonian archipelago (Fig. 1).

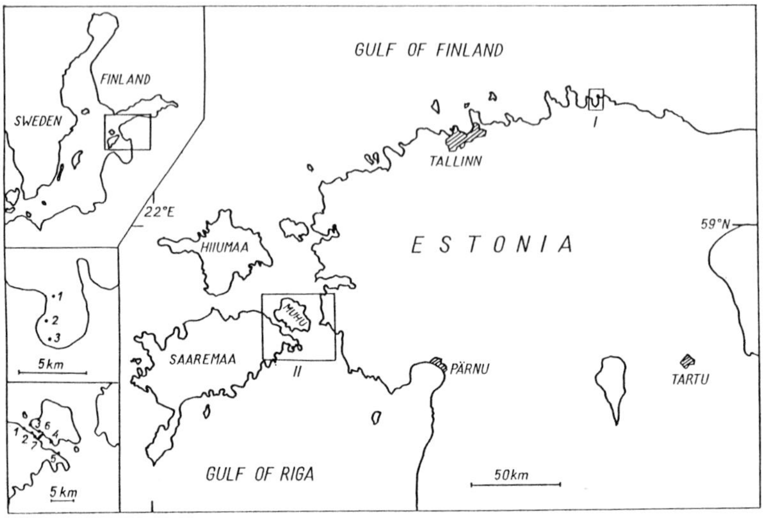


Fig. 1. Study area: I Käsmu Bay, II Väike Väin strait and the sampling sites.

The nature of the two study areas is quite similar. They are non-tidal shallow waters (Käsmu Bay <20 m; Väike Väin <5 m) with sandy sediments predominant. Salinity of the water is 5–6 ‰. In many places the bottom is unevenly covered with stones, which are coated by filamentous algae.

The material from Käsmu Bay was collected in July 1990 and 1991, whereas the material from Väike Väin strait is a part of an ecological study and was gathered together with Elina Leskinen during Estonian–Finnish collaborative work in August 1993 (Vilbaste & Leskinen 1994).

The microphytobenthic samples were taken by piston corer (2.4 cm diameter) in a water depth of 0.5 m. Pieces of macrovegetation and filamentous algae were collected at the same depth.

The diatom slides were prepared after sulphuric acid treatment and mounted in Hyrax or Clophen Harpiks. A bright field Leitz Laborlux III microscope was used. At least 500 valves were counted per sample. The remaining valves were also identified and recorded.

The floras of Pankow (1976) and Krammer & Lange-Bertalot (1986–1991) were used for diatom species identification. The nomenclature was mainly updated with the help of Round *et al.* (1990).

## Results

The number of diatom species was high at all sampling sites and varied from 49 to 87. The diversity index H' varied between 2.66 and 4.55 and the evenness index I' between 0.52 and 0.82 for individual assemblages (Table I).

169 taxa of diatoms were distinguished, 14 of them were taxonomically problematical and 26 were sporadic species; 12 taxa belonged to the Eupodiscales and 157 to the Bacillariales (Table II).

Table I. Main characteristics of microphytobenthic (B) and epiphytic (E) samples investigated, total number of taxa observed (in brackets the number of taxa of counted valves), diversity index (H') and evenness index (I').

| Area | Year | Sample | | Site | Counted valves | No. of taxa | H' | I' |
|---|---|---|---|---|---|---|---|---|
| Käsmu | | | | | | | | |
| | 1990 | 1 | B | 1 | 584 | 68 (44) | 3.89 | 0.71 |
| | | 2 | | 2 | 622 | 87 (56) | 4.53 | 0.78 |
| | | 3 | | 3 | 579 | 74 (62) | 4.47 | 0.75 |
| | | 4 | E | 1 | 552 | 49 (29) | 3.20 | 0.66 |
| | | 5 | | 2 | 557 | 65 (43) | 4.43 | 0.82 |
| | | 6 | | 3 | 564 | 70 (46) | 4.19 | 0.76 |
| | 1991 | 7 | B | 1 | 566 | 61 (46) | 4.14 | 0.75 |
| | | 8 | | 2 | 589 | 73 (54) | 4.54 | 0.79 |
| | | 9 | | 3 | 547 | 69 (46) | 4.04 | 0.73 |
| | | 10 | E | 1 | 577 | 60 (34) | 3.59 | 0.71 |
| | | 11 | | 2 | 563 | 67 (47) | 4.23 | 0.76 |
| | | 12 | | 3 | 573 | 68 (46) | 4.28 | 0.78 |
| Väike Väin | | | | | | | | |
| | 1993 | 13 | E | 1 | 566 | 66 (34) | 3.42 | 0.67 |
| | | 14 | | 2 | 568 | 73 (51) | 3.55 | 0.63 |
| | | 15 | | 3 | 578 | 62 (35) | 3.97 | 0.77 |
| | | 16 | | 4 | 564 | 62 (30) | 2.92 | 0.59 |
| | | 17 | | 5 | 573 | 56 (35) | 2.75 | 0.54 |
| | | 18 | | 6 | 540 | 73 (34) | 2.66 | 0.52 |
| | | 19 | | 7 | 622 | 75 (50) | 4.55 | 0.81 |

Table II. The list of diatom taxa. An estimation of the relative abundance of each taxon at the sampling site (the number of sample is the same in Table I) is indicated with p = predominant (50–20%), f = frequent (20–5%), c = common (5–1%), r = rare (1–0.2%), + = noted, – = not noted.

| Taxon | Number of sample | | | | | | | | | | | | | | | | | | |
|---|---|---|---|---|---|---|---|---|---|---|---|---|---|---|---|---|---|---|---|
| | | B | | | E | | | B | | | E | | | | | E | | | |
| | 1 | 2 | 3 | 4 | 5 | 6 | 7 | 8 | 9 | 10 | 11 | 12 | 13 | 14 | 15 | 16 | 17 | 18 | 19 |
| Eupodiscales (Centrales) | | | | | | | | | | | | | | | | | | | |
| *Actinocyclus octanarius* Ehrenberg | – | – | – | – | – | – | – | + | – | + | – | – | – | – | – | – | – | – | – |
| *Aulacoseira granulata* (Ehrenberg) Simonsen | – | – | – | – | – | – | – | – | – | – | – | + | – | – | – | – | – | – | – |
| *Chaetoceros* sp. | + | – | – | – | – | – | – | – | – | – | – | – | – | – | – | – | – | – | – |
| *Cyclotella comta* (Ehrenberg) Kützing | – | – | – | – | – | – | – | – | – | – | – | – | – | – | – | – | – | – | + |
| *C. meneghiniana* Kützing | + | + | – | – | + | – | – | + | – | – | – | + | + | r | – | – | + | + | r |
| *Melosira lineata* (Dillwyn) C. A. Agardh | – | – | – | – | – | – | – | r | – | – | – | r | + | + | + | – | – | + | + |
| *M. moniliformis* (O. F. Müller) C. A. Agardh | – | – | – | – | – | – | – | – | – | – | – | – | – | f | + | – | + | + | r |
| *M. nummuloides* (Dillwyn) C. A. Agardh | – | – | – | – | – | – | – | – | – | – | – | – | + | + | – | – | + | – | r |
| *Stephanodiscus hantzshii* Grunow | + | + | – | + | + | + | + | r | c | r | c | c | – | – | – | – | – | – | – |
| *S. neoastraea* Håkansson & Hickel | + | + | – | – | – | – | – | – | r | – | – | – | – | – | – | – | – | – | – |
| *Thalassiosira baltica* (Grunow) Ostenfeld | + | – | – | – | + | + | + | + | – | + | + | + | – | – | – | – | + | – | – |
| *T. bramaputrae* (Ehrenberg) Håkansson & Locker | – | – | – | – | – | – | – | – | – | – | – | – | – | – | + | – | – | – | – |
| Bacillariales (Pennales) | | | | | | | | | | | | | | | | | | | |
| *Achnanthes amoena* Hustedt | – | + | c | r | – | r | + | + | c | – | c | r | – | + | – | – | – | – | + |
| *A. brevipes* C. A. Agardh | – | – | – | – | – | – | – | – | + | – | – | – | – | + | + | – | – | – | + |
| *A. delicatula* (Kützing) Grunow | f | f | p | r | f | f | c | c | p | r | c | f | r | r | r | – | r | + | + |
| *A. exigua* Grunow | – | – | c | r | c | c | r | – | c | r | – | – | – | – | – | – | – | – | – |
| *A. lanceolata* (Brébisson) Grunow | – | + | – | – | – | – | – | – | c | – | – | – | – | – | – | – | – | – | – |
| *A. lemmermannii* Hustedt | r | r | c | – | c | r | r | c | c | + | r | c | – | – | – | – | – | + | + |
| *A. minuscula* Hustedt | + | – | c | r | – | c | – | – | c | – | – | – | – | – | – | – | – | – | – |

| | | | | | | | | | | | | | | | | | | | |
|---|---|---|---|---|---|---|---|---|---|---|---|---|---|---|---|---|---|---|---|
| *A. minutissima* Kützing | + | – | + | + | r | r | + | + | c | + | – | – | – | – | – | – | – | + | – |
| *A. taeniata* Grunow | + | – | r | – | – | + | c | + | r | r | r | r | – | – | – | – | – | + | – |
| *Amphora coffeaeformis* (C. A. Agardh) Kützing | c | c | c | + | c | c | c | c | c | + | r | r | r | f | c | r | + | r | c |
| *A. copulata* (Kützing) Schoeman & Archibald | – | + | + | – | + | – | – | + | + | – | + | + | – | – | + | – | + | – | – |
| *A. delicatissima* Krasske | – | – | – | + | + | – | – | – | – | – | – | – | – | r | – | – | – | – | – |
| *A. holsatica* Hustedt | – | + | + | – | + | – | + | – | + | – | – | – | – | – | + | + | + | + | + |
| *A.* cf. *hybrida* Grunow | + | – | r | – | + | + | – | – | + | + | + | + | – | – | – | – | – | – | – |
| *A.* cf. *lineolata* Ehrenberg | – | – | – | – | – | – | – | – | – | – | – | – | + | + | – | + | + | – | r |
| *A. ovalis* Kützing | + | + | – | – | – | + | – | + | + | – | + | – | + | + | + | + | – | – | – |
| *A. pediculus* (Kützing) Grunow | r | r | c | r | r | c | + | r | c | r | r | r | + | r | + | – | – | + | r |
| *Aneomastus tusculus* (Ehrenberg) Mann & Stickle | – | – | – | – | – | – | – | – | + | – | – | – | – | – | – | – | – | – | – |
| *Anomoeoneis sphaerophora* (Ehrenberg) Pfitzer | – | + | – | – | – | – | – | – | – | – | – | – | – | – | – | – | – | – | – |
| *Bacillaria paxillifer* (O. F. Müller) Hendey | – | – | – | – | + | + | – | – | – | – | – | – | – | + | + | – | – | + | r |
| *Berkeleya rutilans* (Trentepohl) Grunow | c | r | + | r | c | r | c | r | – | c | c | c | r | r | c | + | r | + | c |
| *Brebissonia lanceolata* (C. A. Agardh) Mahoney & Reimer | – | + | – | – | – | – | – | + | – | – | – | – | – | + | – | – | – | – | – |
| *Caloneis amphisbaena* (Bory) Cleve | – | – | + | – | – | – | – | + | – | – | – | + | – | – | – | – | – | – | – |
| *C. amphisbaena* fo. *subsalina* (Donkin) Van der Werff & Huls | – | – | – | – | + | – | – | – | + | – | – | – | – | – | – | – | – | – | – |
| *C. bacillum* (Grunow) Cleve | – | – | + | – | – | – | – | + | – | – | – | – | – | – | – | – | – | – | – |
| *Cavinula cocconeiformis* (Gregory ex Greville) Mann & Stickle | – | + | – | – | + | + | – | – | – | – | – | – | – | – | – | – | – | – | – |
| *Cocconeis neodiminuta* Krammer | – | – | c | r | r | c | – | – | c | – | + | c | + | + | r | + | + | + | + |
| *C. pediculus* Ehrenberg | r | r | r | c | r | r | c | r | + | c | f | c | r | r | f | p | + | + | c |
| *C. placentula* Ehrenberg | + | – | + | + | + | + | + | + | – | c | r | + | + | + | + | r | – | + | r |
| *C. placentula* var. *euglypta* (Ehrenberg) Cleve | + | + | + | + | – | – | + | – | + | + | – | – | + | – | – | – | + | + | + |
| *C. scutellum* Ehrenberg | – | – | – | r | – | – | – | – | + | – | – | – | – | – | r | + | – | + | + |
| *Ctenophora pulchella* (Ralfs ex Kützing) Williams & Round | + | c | r | – | r | + | c | c | – | + | c | r | f | r | f | c | r | c | + |
| *Cylindrotheca closterium* (Ehrenberg) Reimann & Lewin | + | – | – | – | – | – | – | – | – | – | – | – | – | – | – | – | – | – | – |
| *Cymbella amphicephala* Naegeli | – | + | + | – | – | – | – | – | – | – | – | – | – | – | – | – | – | – | – |
| *C. helvetica* Kützing | – | – | – | – | – | – | + | – | – | – | – | – | – | – | – | – | – | – | – |
| *C. microcephala* Grunow | – | + | + | – | – | – | – | – | – | – | – | – | – | – | – | – | – | – | – |
| *C. pusilla* Grunow | – | + | + | – | + | – | – | + | – | – | + | + | + | + | – | c | – | + | c |
| *Diatoma moniliformis* Kützing | f | c | c | p | f | c | f | f | c | f | f | f | f | + | f | c | c | c | + |
| *D. tenuis* C. A. Agardh | r | + | r | c | c | + | + | + | – | – | c | r | r | r | c | c | + | + | r |
| *D. vulgaris* Bory | + | + | – | + | + | + | – | – | – | + | – | – | – | – | – | – | – | – | + |
| *Diploneis* cf. *bombus* Ehrenberg | – | – | – | – | – | – | – | – | – | – | – | + | – | – | – | – | – | – | – |

| Taxon | Number of sample | | | | | | | | | | | | | | | | | | |
|---|---|---|---|---|---|---|---|---|---|---|---|---|---|---|---|---|---|---|---|
| | B | | | E | | | B | | | E | | | E | | | | | | |
| | 1 | 2 | 3 | 4 | 5 | 6 | 7 | 8 | 9 | 10 | 11 | 12 | 13 | 14 | 15 | 16 | 17 | 18 | 19 |
| *D. didyma* (Ehrenberg) Cleve | – | – | – | – | – | – | – | – | – | – | – | – | + | – | – | – | – | – | – |
| *D.* cf. *oblongella* (Naegeli) Cleve-Euler | – | – | – | – | – | – | – | – | – | – | – | – | – | – | – | – | – | + | + |
| *D. smithii* (Brébisson) Cleve | – | + | + | – | – | – | – | + | + | – | + | – | + | + | – | + | + | + | – |
| *D.* cf. *stroemi* Hustedt | – | + | – | – | – | – | – | – | – | – | – | – | – | – | – | – | – | – | – |
| *Encyonema caespitosum* Kützing | – | – | + | – | – | + | + | + | + | + | + | + | – | – | – | – | – | – | – |
| *Entomoneis paludosa* (W. Smith) Reimer | – | + | + | – | – | – | + | – | + | – | – | – | + | + | + | + | – | + | r |
| *E. paludosa* var. *subsalina* (Cleve) Krammer | – | + | – | – | – | – | – | – | + | – | – | – | – | – | – | | – | – | – |
| *Epithemia adnata* (Kützing) Brébisson | – | + | – | – | – | – | – | – | – | – | + | + | + | r | + | + | r | r | r |
| *E. sorex* Kützing | + | c | r | + | r | r | + | c | + | – | c | r | r | + | + | c | f | f | f |
| *E. turgida* (Ehrenberg) Kützing | + | r | + | – | – | – | – | + | – | + | + | + | + | f | – | + | + | + | c |
| *Fallacia cryptolyra* Brockmann | + | – | + | – | + | r | – | + | + | – | + | + | + | + | – | – | + | + | – |
| *F. pygmaea* Kützing | – | r | – | – | + | – | – | + | + | + | – | – | + | + | + | – | – | – | + |
| *F.* cf. *tenera* Hustedt | – | – | – | – | – | – | – | – | – | – | – | – | – | r | – | + | – | – | – |
| *Fragilaria capucina* Desmazières | – | – | – | – | – | – | – | – | – | – | – | – | – | – | – | r | – | + | + |
| *F. construens* var. *binodis* (Ehrenberg) Hustedt | – | – | – | – | – | – | – | + | – | – | – | – | – | – | – | – | – | – | – |
| *F. construens* var. *subsalina* Hustedt | – | – | – | – | – | – | – | – | – | – | – | – | – | – | – | + | – | – | + |
| *F. construens* var. *venter* (Ehrenberg) Grunow | + | f | f | c | f | f | c | f | f | c | c | f | + | c | + | c | r | + | r |
| *F. cylindrus* Grunow | + | – | + | – | – | – | + | – | – | + | + | – | – | – | – | – | – | – | – |
| *F. elliptica* Schumann | + | c | c | r | c | r | – | r | – | – | + | + | – | + | – | + | + | + | + |
| *F. hyalina* (Kützing) Grunow | – | – | – | – | – | – | – | – | – | – | – | – | r | – | f | – | + | + | – |
| *F. mutabilis* (W. Smith) Grunow | + | c | + | + | + | + | + | – | – | + | – | + | – | – | – | – | – | – | – |
| *F. pinnata* Ehrenberg | – | – | – | – | + | – | – | – | – | – | – | – | – | – | – | – | – | – | – |
| *Frustulia vulgaris* (Thwaites) De Toni | – | – | – | – | – | – | – | – | – | + | – | – | – | – | – | – | – | – | – |
| *Gomphonema exiguum* Kützing | – | – | – | + | + | r | r | – | – | r | – | r | + | – | – | – | + | – | – |
| *G. olivaceum* (Hornemann) Brébisson | c | + | r | + | r | + | c | r | – | + | r | r | + | – | – | – | – | + | + |
| *Grammatophora oceanica* (Ehrenberg) Grunow | – | – | – | – | – | – | – | – | – | – | – | – | + | + | + | – | – | + | + |
| *Gyrosigma acuminatum* (Kützing) Rabenhorst | – | – | – | – | – | – | – | – | – | – | – | – | – | – | – | – | – | – | + |
| *G.* cf. *scalproides* (Rabenhorst) Cleve | – | – | – | – | – | – | – | – | – | – | – | – | + | – | + | – | – | – | + |
| *Hyalosira delicatula* Kützing | – | – | – | – | – | + | – | – | – | – | – | – | r | – | r | – | r | c | – |

| | | | | | | | | | | | | | | | | | | | |
|---|---|---|---|---|---|---|---|---|---|---|---|---|---|---|---|---|---|---|---|
| *Licmophora communis* (Heiberg) Grunow | + | + | + | – | – | + | + | + | – | r | f | c | f | – | r | – | – | – | – |
| *L. debilis* (Kützing) Grunow | – | – | – | – | – | – | – | – | – | – | – | – | p | + | f | p | + | + | + |
| *L. gracilis* (Ehrenberg) Grunow | – | – | – | – | – | – | – | – | – | – | – | – | – | – | – | + | – | – | – |
| *L. rhombica* Ravenko & Tynni | – | – | – | – | – | – | – | – | – | – | – | – | + | – | + | – | – | – | – |
| *Martyana atomus* (Hustedt) Snoeijs | – | c | r | + | c | r | + | c | – | + | + | r | – | – | – | – | + | – | + |
| *M. schulzii* (Brockmann) Snoeijs | c | r | + | r | r | + | + | r | – | – | + | + | + | – | – | – | + | + | – |
| *Mastogloia baltica* Grunow | – | – | – | – | – | – | – | – | – | – | – | – | – | – | – | + | + | + | + |
| *M. braunii* Grunow | – | – | – | – | – | – | – | – | – | – | – | – | – | + | + | r | + | + | c |
| *M. elliptica* (C. A. Agardh) Cleve | – | c | + | – | – | + | – | r | – | – | r | + | + | + | – | r | + | + | c |
| *M. exigua* Lewis | – | r | – | – | – | – | – | – | – | – | – | – | + | + | + | + | + | r | f |
| *M. pumila* (Grunow) Cleve | – | r | – | + | + | + | – | – | – | – | – | – | r | c | r | c | + | r | c |
| *M. smithii* Thwaites | – | c | + | + | r | + | – | c | – | – | r | r | + | r | + | c | c | c | f |
| *Meridion vernale* C. A. Agardh | – | + | + | – | – | – | – | + | – | – | + | + | – | – | – | – | – | – | – |
| *Navicula* cf. *arenaria* Donkin | – | + | – | – | – | – | – | – | – | – | – | – | – | – | – | – | + | – | – |
| *N. capitata* Ehrenberg | + | r | – | – | – | – | – | c | – | + | + | – | – | – | – | – | – | – | – |
| *N. capitata* var. *hungarica* (Grunow) Ross | + | + | – | – | + | – | + | + | + | – | – | – | – | – | – | – | + | + | – |
| *N.* cf. *cincta* (Ehrenberg) Ralfs | – | – | – | + | – | – | + | – | – | – | – | – | – | – | – | – | – | – | – |
| *N. clementis* Grunow | r | r | r | – | r | + | – | + | + | + | + | + | + | r | + | – | + | + | – |
| *N. crucicula* (W. Smith) Donkin | + | + | – | – | – | + | – | – | – | – | + | – | – | + | + | – | – | + | – |
| *N. crucifera* Grunow | f | c | f | – | – | + | + | c | r | + | r | r | – | – | – | – | – | – | – |
| *N. cryptocephala* Kützing | c | r | c | + | r | r | r | c | r | + | r | r | + | + | + | + | + | + | r |
| *N. digitoradiata* (Gregory) Ralfs | – | – | + | – | – | – | – | – | – | – | – | + | – | – | – | + | – | + | + |
| *N.* cf. *elginensis* (Gregory) Ralfs | – | – | – | – | – | – | – | + | – | – | – | – | – | – | – | + | – | – | – |
| *N.* cf. *erifuga* Lange-Bertalot | – | – | – | – | – | – | – | – | – | – | – | – | – | – | – | + | – | – | – |
| *N. gregaria* Donkin | c | c | c | r | c | c | c | c | c | c | r | c | r | r | + | + | + | + | r |
| *N. halophila* (Grunow) Cleve | + | + | + | + | + | + | + | – | + | + | – | + | – | – | – | – | – | – | – |
| *N. lanceolata* (C. A. Agardh) Ehrenberg | r | + | – | + | – | – | r | – | – | + | + | – | – | – | – | – | – | – | – |
| *N. menisculus* Schumann | r | r | r | – | + | r | + | + | r | + | r | + | – | – | – | – | – | + | – |
| *N. meniscus* Schumann | – | + | r | – | – | – | – | c | + | – | + | – | – | – | – | – | – | – | – |
| *N. peregrina* (Ehrenberg) Kützing | + | + | – | – | – | + | – | + | + | – | – | – | – | – | – | – | – | – | – |
| *N. perminuta* Grunow | f | f | f | p | f | f | f | f | f | p | f | f | + | r | c | – | r | + | c |
| *N. phyllepta* Kützing | c | r | c | r | c | c | c | r | c | r | r | r | f | c | f | + | r | + | c |
| *N. protracta* (Grunow) Cleve | – | c | r | – | + | + | + | + | + | + | – | – | + | + | – | – | – | – | – |

| Taxon | Number of sample | | | | | | | | | | | | | | | | | | |
|---|---|---|---|---|---|---|---|---|---|---|---|---|---|---|---|---|---|---|---|
| | | B | | | E | | | B | | | E | | | | | E | | | |
| | 1 | 2 | 3 | 4 | 5 | 6 | 7 | 8 | 9 | 10 | 11 | 12 | 13 | 14 | 15 | 16 | 17 | 18 | 19 |
| *N. radiosa* Kützing | + | – | – | – | – | – | + | + | – | – | – | – | – | r | + | + | – | + | – |
| *N. reinhardtii* Grunow | – | + | – | – | – | – | – | – | – | – | – | – | – | – | – | – | – | – | – |
| *N. rhynchocephala* Kützing | r | c | c | – | r | + | – | f | c | – | r | r | + | + | + | – | + | + | + |
| *N. salinarium* Grunow | c | r | + | + | r | + | + | + | + | + | + | + | – | + | – | + | – | + | r |
| *N. scutelloides* W. Smith | – | – | – | – | – | – | – | + | – | – | – | – | – | – | – | – | – | – | – |
| *N. spicula* (Hickie) Cleve | – | – | – | – | – | – | – | – | + | – | + | – | + | c | – | + | – | + | r |
| *N.* cf. *tenelloides* Hustedt | – | – | – | – | – | – | – | – | – | – | – | – | – | + | – | – | – | – | + |
| *N. tripunctata* (O. F. Müller) Bory | – | – | – | r | – | – | r | – | – | – | + | – | – | + | – | – | – | – | – |
| *Neidium iridis* (Ehrenberg) Cleve | – | – | – | – | – | – | – | – | + | – | – | – | – | – | – | – | – | – | – |
| *Nitzschia acicularis* (Kützing) W. Smith | – | – | r | – | – | – | r | – | – | – | – | r | – | – | + | – | + | – | – |
| *N. dissipata* (Grunow) Hasle | + | – | r | + | r | + | r | – | r | – | + | r | – | + | – | + | – | + | + |
| *N. dubia* W. Smith | + | – | – | – | – | – | + | – | – | – | + | – | – | – | – | – | – | – | – |
| *N. frustulum* (Kützing) Grunow | – | – | – | – | – | – | – | + | – | + | – | – | – | – | – | – | – | – | – |
| *N. inconspicua* Grunow | r | – | r | f | r | c | f | + | + | f | r | c | – | – | – | – | + | – | + |
| *N. linearis* (C. A. Agardh) W. Smith | – | – | – | – | – | – | – | – | – | – | – | – | + | + | – | + | – | – | + |
| *N. microcephala* Grunow | c | f | c | c | c | f | c | c | c | f | f | f | c | c | c | r | c | r | f |
| *N. palea* (Kützing) W. Smith | r | c | c | – | c | c | c | c | c | r | – | + | + | c | r | + | – | – | r |
| *N. paleacea* Grunow | f | r | c | r | c | c | c | c | c | c | r | c | f | p | f | – | r | + | f |
| *N. perminuta* (Grunow) M. Peragallo | r | r | r | f | + | c | – | – | + | c | r | + | r | r | r | + | r | + | c |
| *N. reversa* W. Smith | + | + | + | – | – | r | + | – | + | r | – | – | + | c | + | + | – | – | + |
| *N. sigma* (Kützing) W. Smith | + | + | – | – | – | + | – | + | + | – | – | – | – | + | + | – | – | + | c |
| *N. sigmoidea* (Nitzsch) W. Smith | – | + | – | – | + | – | + | – | + | + | – | – | – | – | – | + | – | – | – |
| *N. umbonata* (Ehrenberg) Lange-Bertalot | – | – | – | – | – | – | – | – | – | – | – | – | – | – | – | + | – | – | – |
| *N. valdestriata* Aleem & Hustedt | – | – | – | c | r | r | c | – | – | c | + | – | + | f | c | – | r | + | c |
| *Opephora olsenii* Möller | + | f | c | r | c | r | r | f | c | r | c | c | + | r | + | r | r | + | r |
| *Petroneis humerosa* (Brébisson) Stickle & Mann | + | + | + | – | – | – | – | + | + | – | + | + | – | – | – | – | – | – | – |
| *P. marina* (Ralfs) Mann | – | + | – | – | – | – | – | – | – | – | + | + | – | – | + | – | – | – | – |
| *Pinnularia globiceps* Gregory | – | + | + | – | – | – | – | – | – | – | – | – | – | – | – | – | – | – | – |
| *Pleurosigma angulatum* (Quekett) W. Smith | – | – | + | – | – | + | – | – | + | – | – | – | – | – | – | – | – | – | – |

| | | | | | | | | | | | | | | | | | | | |
|---|---|---|---|---|---|---|---|---|---|---|---|---|---|---|---|---|---|---|---|
| *P. elongatum* W. Smith | – | + | – | – | – | – | – | – | – | – | – | – | + | + | + | + | – | + | – |
| *P. normanii* Ralfs | – | – | – | – | – | – | – | – | – | – | – | – | + | + | – | – | – | – | – |
| *P. salinarium* Grunow | – | – | – | – | – | – | – | – | + | – | – | + | – | – | – | – | – | – | – |
| *Pseudostaurosira brevistriata* (Grunow) Williams & Round | – | + | – | – | – | – | – | – | – | – | – | – | – | – | – | – | – | – | – |
| *Rhoicosphenia abbreviata* (C. A. Agardh) Lange-Bertalot | f | r | r | f | r | c | c | r | + | c | r | r | r | c | f | c | f | p | c |
| *Rhopalodia gibba* (Ehrenberg) O. Müller | – | r | – | + | + | + | – | r | – | – | + | + | + | + | + | + | c | c | r |
| *R. musculus* (Kützing) O. Müller | – | – | – | – | – | – | – | – | – | – | – | – | + | – | + | + | – | + | r |
| *Sellaphora bacillum* Ehrenberg | – | – | – | – | + | – | – | – | – | + | – | + | – | – | – | – | – | – | – |
| *S. pupula* (Kützing) Mereschkowsky | – | + | – | + | – | – | – | – | – | – | – | – | – | – | – | – | – | – | – |
| *Stauroneis anceps* Ehrenberg | – | – | r | – | – | – | – | + | – | – | – | – | – | – | – | – | – | – | – |
| *S. phoenicentron* (Nitzsch) Ehrenberg | + | – | – | – | – | – | – | – | – | – | – | – | – | – | – | – | – | – | – |
| *S.* cf. *tackei* (Hustedt) Krammer & Lange-Bertalot | – | – | – | – | – | – | – | – | – | – | – | – | – | – | – | + | – | – | – |
| *Surirella* cf. *brebissonii* Krammer & Lange-Bertalot | – | + | – | – | – | – | + | – | + | + | r | + | – | – | – | + | – | + | – |
| *S. minuta* Brébisson | – | + | + | – | – | + | – | + | + | + | – | – | – | – | – | – | – | – | – |
| *S. ovalis* Brébisson | + | c | + | – | + | + | – | c | + | + | – | – | + | + | + | + | + | + | – |
| *S. subsalsa* W. Smith | – | – | – | – | – | – | – | – | – | – | – | – | + | – | – | – | – | – | – |
| *Tabularia fasciculata* (C. A. Agardh) Williams & Round | f | c | r | r | c | c | f | f | r | f | f | f | f | f | f | c | p | p | c |
| *T. tabulata* (C. A. Agardh) Snoeijs | + | r | + | – | + | + | + | c | – | + | r | r | f | + | c | r | c | f | + |
| *T. waernii* Snoeijs | + | + | – | + | + | + | r | r | r | c | c | c | – | + | + | r | r | + | + |
| *Tropidoneis dannfeltii* Cleve-Euler | + | – | + | – | – | + | – | – | + | + | – | – | – | – | – | – | – | – | – |
| *T.* cf. *vanheurckii* (Grunow) Cleve | – | – | – | – | – | – | – | – | + | – | – | – | – | + | – | – | – | – | – |
| *Tryblionella gracilis* W. Smith | – | – | – | – | – | – | – | – | – | – | – | – | – | – | – | + | – | – | – |
| *T. hungarica* Grunow | + | – | – | – | – | – | + | – | – | – | – | – | + | + | + | + | – | + | + |
| *T. levidensis* W. Smith | – | + | – | – | – | – | – | – | – | – | – | – | – | – | – | – | – | – | + |
| *T. punctata* W. Smith | – | – | – | – | – | – | – | – | – | – | + | – | – | – | – | – | – | – | – |
| number of taxa (total 169) | 68 | 87 | 74 | 49 | 65 | 70 | 61 | 73 | 69 | 60 | 67 | 68 | 66 | 73 | 62 | 62 | 56 | 73 | 75 |

The Sørensen index of similarity between microphytobenthic and epiphytic assemblages varied from 57 to 84% (Fig. 2 I, II, IV, V, VI).

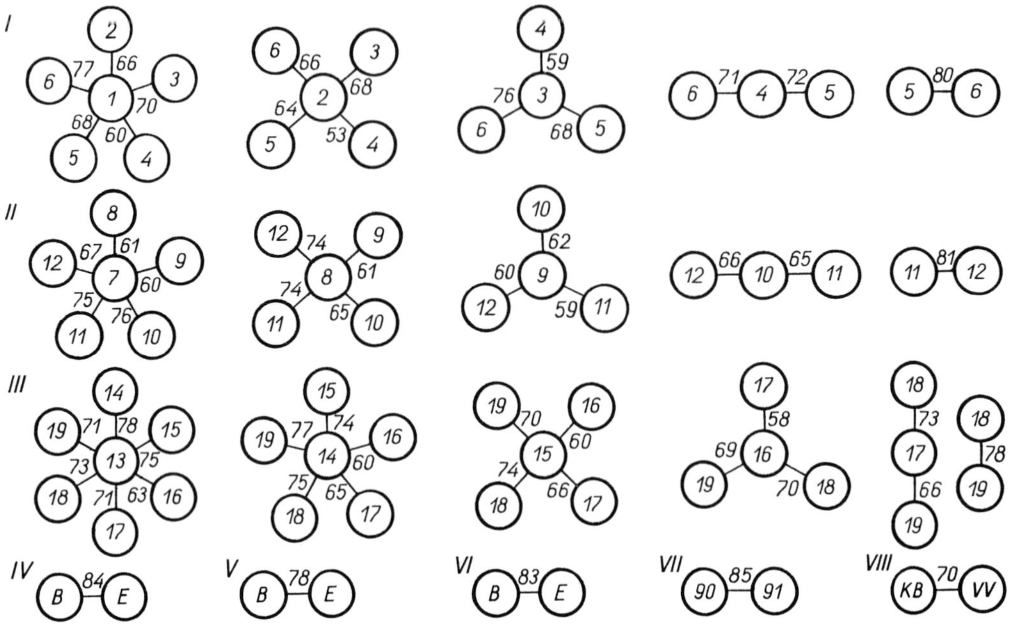

Fig. 2. Sørensen index of similarity (%) *between various samples* (number of sample is the same in Table I): I Käsmu Bay in 1990; II The same in 1991; III Väike Väin strait in 1993; *between microphytobenthic (B) and epiphytic (E) diatoms*: IV Total in Käsmu Bay; V Käsmu Bay in 1990; VI The same in 1991; *between diatom species*: VII Different years (1990 and 1991) in Käsmu Bay; VIII Käsmu Bay (KB) and Väike Väin (VV) strait.

## Discussion

The diatom flora in Estonian coastal waters is very similar to that along the Swedish east coast (Snoeijs 1989) and the Finnish south coast (Leskinen & Hällfors 1990) of the Baltic Sea.

The resemblance of epiphytic diatom assemblages to microphytobenthic ones was great. Although some species were determined only in microphytobenthic assemblages (*Caloneis bacillum*, *Cymbella amphicephala*, *C. microcephala*, *Stauroneis anceps*), there was no one species found only in epiphytic assemblages (if the sporadic taxa are excluded). At the same sampling site the diversity and the number of diatom taxa of microphytobenthos are higher compared with the corresponding indices of epiphytic assemblage. The epiphytes are deposited in greater quality in the epipelon than on the contrary. Consequently the species composition of microphytobenthic assemblages expresses more adequately the whole diatom flora in Estonian coastal waters.

**Acknowledgements**

This study was carried out in the Institute of Zoology and Botany, Estonian Academy of Sciences. The author is obliged to Dr Elina Leskinen (Uppsala University) for co-operation in the field work and encouragement to write the paper. For critically reading the manuscript I thank my colleague Dr Tiiu Trei.

## References

Eichwald, E. (1849). Zweiter Nachtrag zur Infusorienkunde Russlands. *Bulletin de la Société Impériale des Naturalistes de Moscou*, **22**, 2, 400–548.

Eichwald, E. (1852). Dritter Nachtrag zur Infusorienkunde Russlands. *Bulletin de la Société Impériale des Naturalistes de Moscou*, **25**, 1, 388–536.

Krammer, K. & Lange-Bertalot, H. (1986–1991). Bacillariophyceae. Teil 1–4. *Süßwasserflora von Mitteleuropa*, **2/1**, 876 pp., **2/2**, 596 pp., **2/3**, 576 pp., **2/4**, 437 pp. VEB Gustav Fischer Verlag, Jena, Stuttgart, New York.

Leskinen, E. & Hällfors, G. (1990). Community structure of epiphytic diatom in relation to eutrophication on the Hanko Peninsula, South coast of Finland. In: *Proceedings of the 10th International Diatom Symposium* (H. Simola, ed.), 323–333. Koeltz Scientific Books, Koenigstein.

Pankow, H. (1976). Algenflora der Ostsee. II Plankton. 493 pp. VEB Gustav Fischer Verlag, Jena.

Round, F. E., Crawford, R. M. & Mann, D. G. (1990). *The Diatoms. Biology and morphology of the genera.* 747 pp. Cambridge University Press, Cambridge.

Sakson, M. & Miller, U. (1993). Diatom assemblages in superficial sediments from the Gulf of Riga, eastern Baltic Sea. *Hydrobiologia*, **269/270**, 243–249.

Snoeijs, P. (1989). A check-list of the benthic diatoms at Forsmark (northern Baltic Sea). 1. Epilithic and epiphytic taxa. *Annales Botanici Fennici*, **26**, 427–439.

Snoeijs, P. (ed.), (1993). *Intercalibration and distribution of diatom species in the Baltic Sea.* I. 129 pp. Opulus Press, Uppsala.

Vilbaste, S. (1987). The investigation of microphytobenthos of Rame Bay by the glass slide methoes. In: *Proceedings of the Estonian Academy of Sciences. Biology*, **36**, 44–52 (in Russian).

Vilbaste, S. & Leskinen, E. (1994). Species composition and biomass of epiphytic microalgae at Väike Väin, Western Estonia. In: *Ecological studies in the aquatic environment of Väike Väin strait in West Estonia.* 32–41. Yliopistopaino, Helsinki.

Weisse, J. (1861). Die Diatomaceen des Badenschlammes von Arensburg und Hapsal, wie auch des sogenennten Mineralschlammes der Soolen-Badeanstalt in Staraja-Russa. *Mélanges Biologiques Bulletin de l'Académie Impériale des Sciences des St.-Pétersbourg*, **3**, 4, 357–362.

# Type and quality of river waters in central Finland described using diatom indices

Pertti Eloranta

*Department of Limnology and Environmental Protection/ Section Limnology, P.O.Box 27, E-house, Viikki, FIN-00014 University of Helsinki, Finland*

## Abstract

Biological river monitoring in Finland needs new methods, tested and adapted for the environmental conditions in the country. In central Finland 56 fast flowing rivers were sampled in July 1986 for diatom analyses. Most of the rivers were only weakly loaded by agriculture and fish farming, but they differed widely in the content of humic substances, pH, conductivity and colour. Results of diatom analyses were processed using "Omnidia" software (Lecointe *et al.* 1993) which calculates besides other variables also eight different pollution indices. Results showed that all rivers were rather unpolluted but index values did not correlate with C.O.D. nor nutrient concentration in the rivers. Some indices had significant negative correlations to water pH and conductivity. Some indices reacted in opposite ways to nutrient concentration and humic substances. Index values for different diatom communities varied both between communities and indices for the same community.

## Introduction

Biological river monitoring in Finland has been much overlooked due to the lack of methods tested and adapted for the environmental conditions in the country. During recent years a program involving the use of algae for monitoring river waters has been carried out by the author's working group. The program studies diatoms and filamentous algae and their seasonality and variability among epilithic, epiphytic and epipelic communities. Later works have been carried out in rivers with sewage and agricultural loads but this paper presents the results from more unpolluted rivers. The aim of this study was to test the usefulness and sensitivity of different diatom indices in Finnish rivers with differing humus concentrations, conductivity and pH.

## Methods

In central Finland, 56 fast flowing rivers were sampled for diatom analyses. All samples were taken in July 1986 by washing epilithic diatoms from stones. During sampling, current velocity, water pH, conductivity and colour were measured to describe the river type. Other water quality data were taken from the national water quality data base.

After collecting, the samples were cleaned by normal hot acid combustion and mounted using Hyrax as a medium. Microscopical analyses used interference contrast optics and ca. 200–400 frustules were counted and identified from each sample. The diatom analyses were done using the newest version of the "Omnidia" software (Lecointe *et al.* 1993) to calculate eight different diatom indices. These were IPS or Specific Pollution Sensitivity Index (SPI) (Coste in CEMAGREF 1982), Sládeček's index (1986), GENRE or Generic Diatom Index (GDI) (Coste & Ayphassorho 1991), CEE (CEC) index (Descy & Coste 1991), Descy's index (1979), Leclercq & Maquet's index (1987), Schiefele & Schreiner's index (1991) and Watanabe's index (Watanabe *et al.* 1990). All values were transformed in the "Omnidia" program to a scale from 1 to 20 for comparison. The indices used variable numbers of indicator species from 106 (Descy's index) to ca. 2000 (IPS and GENRE). From these eight indices, IPS, GENRE, Descy's index, Sládeček's index and Leclercq & Maquet's index are based on sensitivity classes (i) and indicator value groups (v) of each taxa and their abundance (a) as in Zelinka & Marvan (1961):

$$\mathrm{ID} = \frac{\sum_{j=1}^{n} a_j i_j v_j}{\sum_{j=1}^{n} a_j v_j}$$

CEE is based on the biotic index grids containing 208 taxa grouped into 8 groups with taxa ranked according to decreasing tolerance of pollution and into 4 subgroups of taxa with higher indicator value (see Descy & Coste 1988, 1991). Watanabe's index (DAIpo) (Watanabe *et al.* 1990) is calculated using relative abundances of saprophilous (Sj), and saproxenous taxa (Xi):

$$\mathrm{DAIpo} = 50 + \frac{1}{2}\left(\sum_{i=1}^{p} Xi - \sum_{j=1}^{q} Sj\right).$$

For comparative calculations all index values were transformed in the "Omnidia" program to the scale 0–20.

## Results

### *Areal comparisons*

The rivers were grouped according to their river basin and limnological water quality into five main groups (Fig. 1). Three groups were characterised by brown water and rather low conductivity, whereas two other groups were mostly oligohumic and with a little higher conductivity and pH (Table 1). All rivers were rather unpolluted or they received some organic and nutrient load from fish farms. In the brown water group the indices IPS, Sládeček and GENRE gave increasing values with decreasing water colour, whereas CEE, Descy's and Watanabe's indices gave rather similar results (Fig. 2). In general Watanabe's index varied very little and did not show any clear trend with water quality. Leclercq & Maquet's index values decreased with decreasing colour and C.O.D. values and increasing pH and conductivity values. Although the river groups IV and V consisted of most clear and oligotrophic rivers, only the IPS index gave to them the highest index values, when e.g. Leclercq & Maquet's index gave the lowest values.

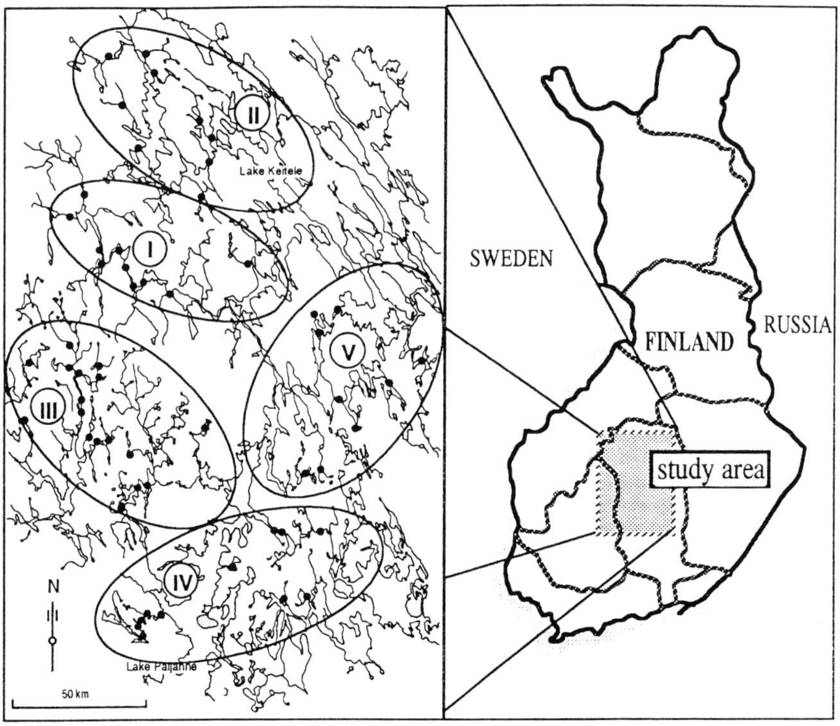


Fig. 1. Study area and the location of sampling stations. The subareas are circled and indicated with roman numbers I–V.

Table 1. Average values of some river characteristics and water quality variables in the five groups of the studied rivers.

| Group | Curr.vel. cm $s^{-1}$ | Colour mg $l^{-1}$ Pt | pH | Tot.P µg $l^{-1}$ | Width m | Conduct. mS $m^{-1}$ | C.O.D. mg $l^{-1}$ $O_2$ |
|---|---|---|---|---|---|---|---|
| I | 89 | 146 | 6.3 | 34 | 17 | 3.6 | 21 |
| II | 93 | 127 | 6.2 | 21 | 11 | 3.1 | 21 |
| III | 66 | 110 | 6.4 | 24 | 9 | 3.4 | 16 |
| IV | 118 | 44 | 6.7 | 19 | 4 | 4.4 | 8 |
| V | 91 | 68 | 6.8 | 28 | 12 | 4.5 | 12 |

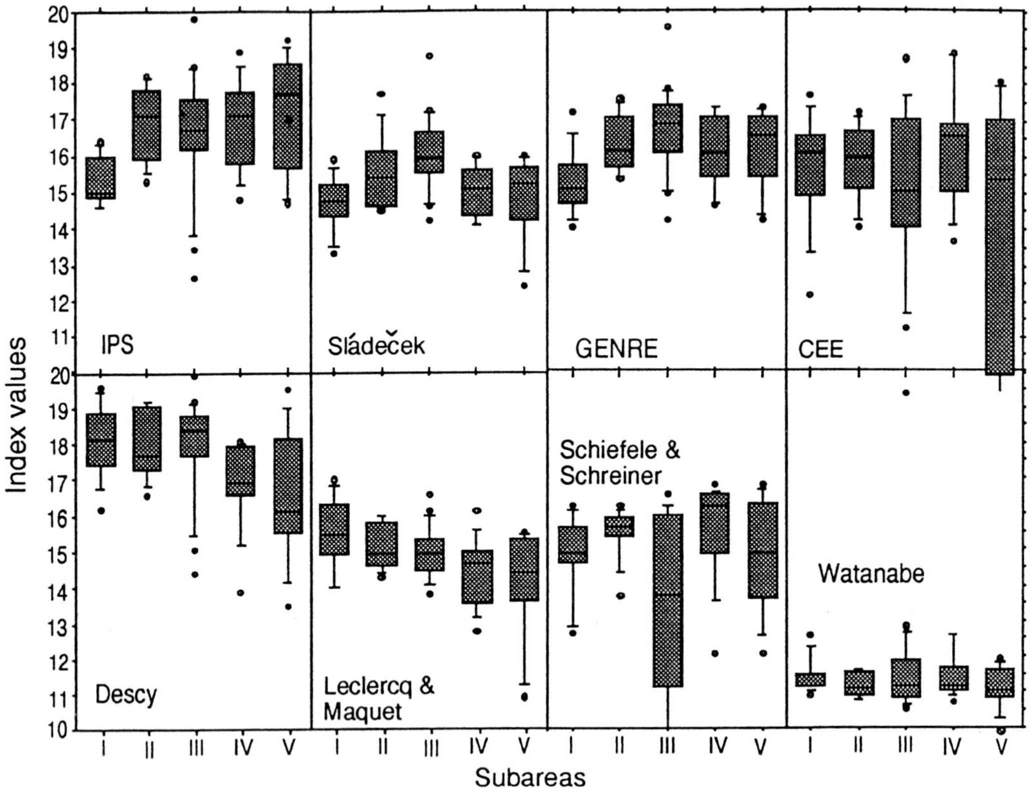


Fig. 2. Box plots of different diatom indices for the five subareas (see Fig. 1).

*Diatom indices vs. water quality variables*

The calculated diatom index values were correlated with water quality variables indicating organic load in the water (C.O.D. and colour) and with the total phosphorus concentration as an indication of trophic level. IPS, CEE and Watanabe's index did not correlate significantly with any of these variables in calculations (Table 2). Sládeček's index and GENRE index had significant negative correlations to water pH and conductivity, Schiefele & Schreiner's index correlated weakly and positively with current velocity. Only Descy's index correlated significantly with C.O.D. and water colour, but positively, showing better water quality along increasing organic content (Fig. 3). The same index correlated negatively with water pH and conductivity. Descy's index also showed an increasing trend with increasing phosphorus concentration (Fig. 4). In general, the variation of all index values obtained from sites of different C.O.D. of the water was wide and without any clear relation (Fig. 5).

Table 2. Correlations between different diatom indices and some river characteristics and water quality variables (** $p < 0.01$; * $p < 0.05$; Le&Ma = Leclercq & Maquet's index, Sc&Sc = Schiefele & Schreiner's index).

| | IPS | Sládeček | Descy | Le&Ma | GENRE | CEE | Sc&Sc | Watan. |
|---|---|---|---|---|---|---|---|---|
| Current veloc. | n.s. | n.s. | n.s. | n.s. | n.s. | n.s. | .294* | n.s. |
| Colour | n.s. | n.s. | .364** | n.s. | n.s. | n.s. | n.s. | n.s. |
| C.O.D. | n.s. | n.s. | .360** | n.s. | n.s. | n.s. | n.s. | n.s. |
| pH | n.s. | -.353** | -.381** | -.415 | -.277* | n.s. | n.s. | n.s. |
| Conductivity | n.s. | -.495** | -.404** | -.424 | -.262* | n.s. | n.s. | n.s. |
| Tot. P | n.s. | n.s. | n.s. | -.290* | n.s. | n.s. | n.s. | n.s. |

*Index values in different diatom communities*

Diatom results were processed using clustering analyses (Pearson correlation and average linkage method) to distinguish different community types. This method produced nine groups. *Fragilaria virescens*, *Eunotia minor* and *Synedra tenera* communities occurred in most brown waters (Table 3), whereas *Achnanthes minutissima* and *Cocconeis placentula* communities were found in more clear, oligotrophic and neutral waters. Most indices gave the highest index values for the *Fragilaria virescens* community, whereas *Gomphonema* spp., *Synedra tenera* and *Cocconeis placentula* community achieved the lowest index values (Fig. 6, Table 3). IPS, CEE and Schiefele & Schreiner's index values were rather high for the *Achnanthes minutissima* community. The average values of all indices to all communities, however, did not differ greatly (Table 3).

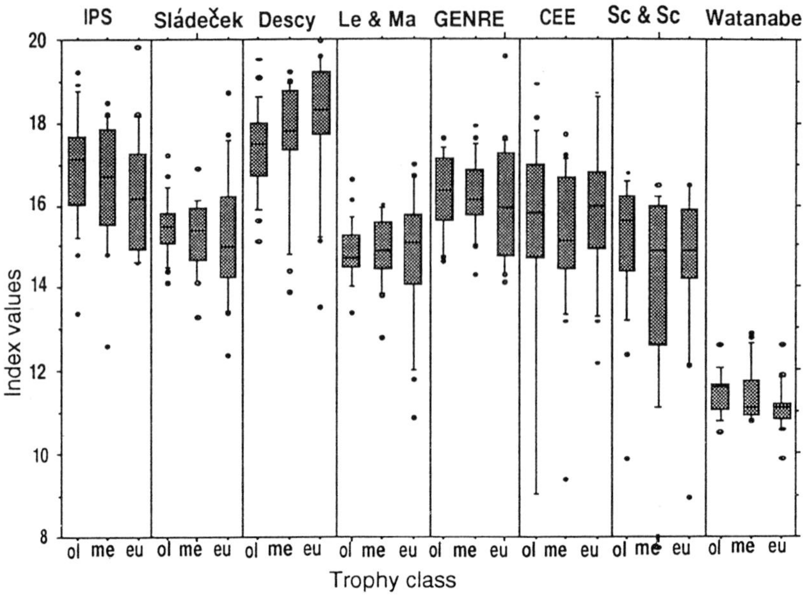


Fig. 3. Box plots of different diatom indices in rivers with different trophy (ol = oligotrophy; tot. P ≤ 15 µg $l^{-1}$; me = mesotrophy; tot. P 16 – 30 µg $l^{-1}$; eu = eutrophy; tot. P > 30 µg $l^{-1}$).

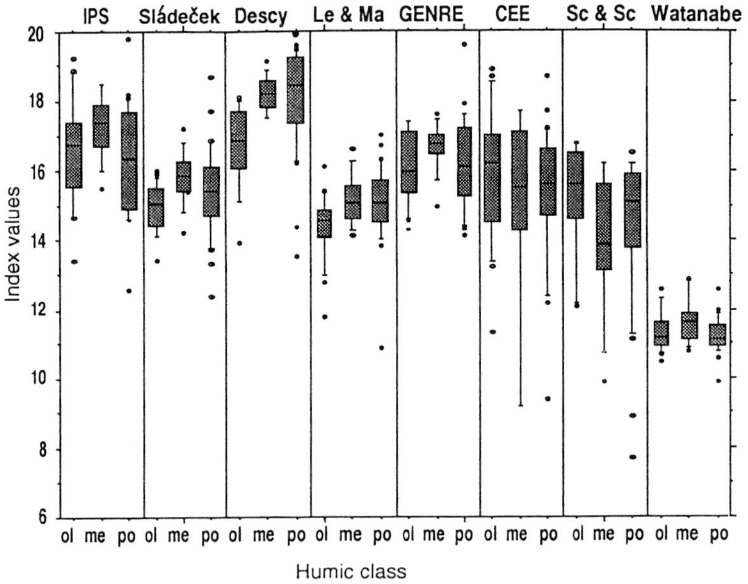


Fig. 4. Box plots of different diatom indices in rivers with differing degrees of dystrophy (ol = oligohumic; colour – 50 mg $l^{-1}$ Pt; C.O.D. ≤ 9 mg $l^{-1}$; me = mesohumic; colour 5–100 mg $l^{-1}$ Pt; C.O.D.≤ 9.1 – 15 mg $l^{-1}$; po = polyhumic; colour > 100 mg $l^{-1}$ Pt; C.O.D. > 15 mg $l^{-1}$).

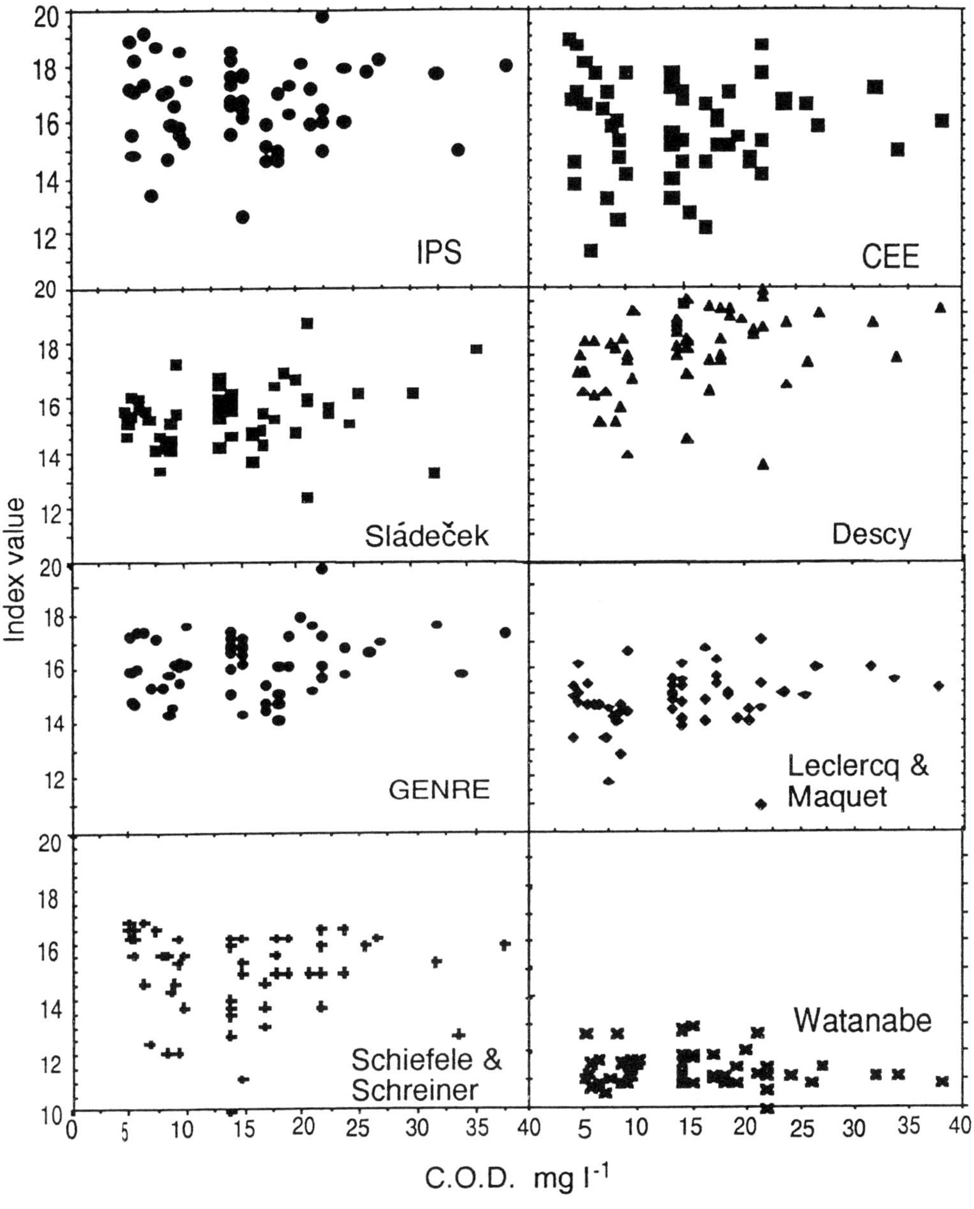


Fig. 5. Scatter plots of different diatom indices and C.O.D. of the river waters.

Table 3. Average values of the diatom indices, number of taxa, diversities and some water quality variables in rivers with different diatom communities indicated by dominant taxa (Fra vi = *Fragilaria virescens*, Eun min = *Eunotia minor*, Ach min = *Achnanthes minutissima*, Tab flo = *Tabellaria flocculosa*, Fra ca lan = *Fragilaria capucina* var. *lanceolata*, mix = mixed community without clear dominant, Coc pla = *Cocconeis placentula*, Go spp. = *Gomphonema* spp., Syn ten = *Synedra tenera*).

| Community | n | Taxa | Divers. | Av.diatom index | Colour mg $l^{-1}$ Pt | pH | Tot.P µg $l^{-1}$ | Cond. mS $m^{-1}$ | C.O.D. mg $l^{-1}$ $O_2$ |
|---|---|---|---|---|---|---|---|---|---|
| Fra vir | 3 | 48 | 3.22 | 16.51 | 190 | 6.1 | 43 | 2.8 | 26.3 |
| Eun min | 7 | 52 | 4.04 | 15.84 | 143 | 6.2 | 25 | 3.2 | 20.1 |
| Ach min | 17 | 66 | 3.78 | 15.59 | 68 | 6.5 | 17 | 3.9 | 11.6 |
| Tab flo | 9 | 56 | 4.32 | 15.20 | 116 | 6.4 | 27 | 3.5 | 17.7 |
| Fr ca lan | 3 | 46 | 3.91 | 14.94 | 117 | 6.5 | 31 | 3.8 | 16.3 |
| mix | 6 | 60 | 4.40 | 14.61 | 59 | 6.8 | 39 | 4.6 | 10.7 |
| Coc pla | 1 | 63 | 4.25 | 14.46 | 40 | 7.1 | 16 | 5.7 | 8.8 |
| Go spp. | 6 | 45 | 3.75 | 14.35 | 89 | 6.5 | 17 | 3.4 | 12.9 |
| Syn ten | 4 | 49 | 3.62 | 14.14 | 125 | 6.6 | 32 | 4.1 | 19.6 |

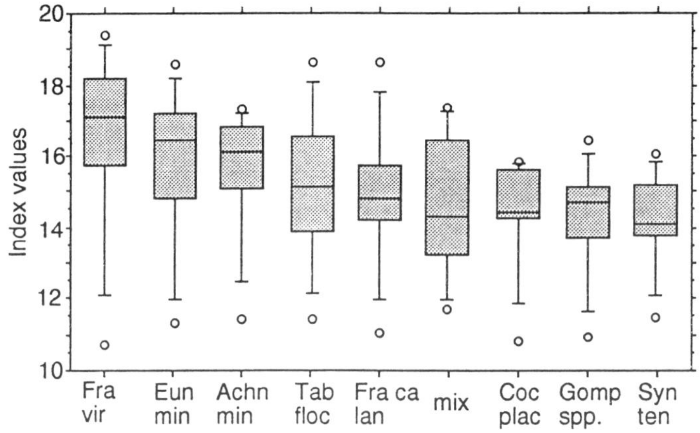


Fig. 6. Box plots of different diatom index values for different diatom communities (Fra vir = *Fragilaria virescens*, Eun min = *Eunotia minor*, Achn min = *Achnanthes minutissima*, Tab floc = *Tabellaria flocculosa*, Fra ca lan = *Fragilaria capucina* var. *lanceolata*, mix = mixed community, Coc pla = *Cocconeis placentula*, Gomp spp. = *Gomphonema* spp. and Syn ten = *Synedra tenera*).

## Conclusions

The diatom indices used did not reflect the differences in nutrient and humus content in the unpolluted rivers. Diatom taxa dominating in polyhumic brown waters with low pH and conductivity are in general classified to be xeno- to oligosaprobic, thus the effects of organic humic compounds were not parallel with the effects from organic wastes, e.g. sewage. Thus the index values decreased with increasing pH and conductivity, although these waters were the best in water quality in the region. Also the differences in nutrient concentrations were not large enough to be seen in the index values. When looking at the general view of the results, IPS index seemed to give the best results when compared to the general water quality in the different areas, whereas the values of the Watanabe's index were more responsive in a specific type of river. Further tests with more polluted and loaded waters are needed to see which index is the most suitable for monitoring of Finnish rivers. Those are already under way and will be published later.

## Acknowledgements

The author will thank Mr. Kurt Andersson for diatom preparations and Lic. Sc. Anssi Eloranta for collecting water chemistry data from National Water Quality data bank.

## References

CEMAGREF (1982). Etude des méthodes biologiques quantitatives d'appréciation de la qualité des eaux. *Rapport Division Qualité des Eaux Lyon – Agence financière de Bassin Rhone – Méditerranée – Corse, Pierre-Bénite,* 218 pp.

Coste, M. & Ayphassorho, H. (1991). Etude de la qualité des eaux du Bassin Artois-Picardie à l'aide des communautés de diatomées benthiques (Application des indices diatomiques). *Rapport Cemagref, Bordeaux – Agence de l'Eau Artois-Picardie,* 227 pp., Douai.

Descy, J. P. & Coste, M. (1988). Utilisation des diatomées benthiques pour l'évaluation de la qualité des eaux courantes. *3éme rapport d'avancement. Contrat CEE B–71–23.* 49 pp. FNUDP, Namur/CEMAGREF, Bordeaux.

Descy, J. P. & Coste, M. (1991). A test of methods for assessing water quality based on diatoms. *Verhandlungen der internationalen Vereinigung für theorethische und angewandte Limnologie,* **24**, 2112–2116.

Descy, J. P. (1979). A new approach to water quality estimation using diatoms. *Nova Hedwigia,* **64**, 305–323.

Leclercq, L. & Maquet, B. (1987). Deux nouveaux indices chimique at diatomique de qualité d'eau courante. Application au Samson et à ses affluents (bassin de la Meuse belge). Comparaison avec d'autres indices chimiques, biocénotiques et diatomiques. *Institut Royal des Sciences Naturelles de Belgique, document de travail,* **28,** 113 pp.

Lecointe, C., Coste, M. & Prygiel, J. (1993). "Omnidia": software for taxonomy, calculation of diatom indices and inventories management. *Hydrobiologia,* **269/270**, 509–513.

Schiefele, S. & Schreiner, C. (1991). Use of diatoms for monitoring nutrient enrichment, acidification and impact of salt rivers in Germany and Austria. In: *Use of algae for monitoring rivers* (B. A. Whitton, E. Rott & G. Friedrich, eds), 103–110. Institut für Botanik, Universität Innsbruck, Innsbruck.

Sládeček , V. (1986). Diatoms as indicators of organic pollution. *Acta hydrochimica et hydrobiologica,* **14**, 555–566.

Watanabe, T., Asai, K. & Houki, A. (1990). Numerical simulation of organic pollution in flowing waters. In: *Encyclopedia of Environmental Control Technology. 4. Hazardous Waste Containment and Treatment* (P. N. Cheremisinoff, ed.), 251–284. Gulf Publishing Company, Houston.

Zelinka, M. & Marvan, P. (1961). Zur Präzisierung der biologischen Klassifikation der Reinheit fliessender Gewässer. *Archiv für Hydrobiologie*, **57**, 389–407.

# Benthic diatom communities in two acidic mountain lakes (Sudeten Mts, Poland)

Janina Kwandrans

*Karol Starmach Institute of Freshwater Biology, Polish Academy of Science, Sławkowska 17, 31–016 Crakow, Poland*

## Abstract

Benthic diatom communities in two mountain lakes with low water pH (4.6–5.8) were analysed. Lakes Mały and Wielki Staw lie on the granite in the Karkonosze range of the Sudeten Mts at 1183 m and 1225 m a.s.l. in the upper montane and mountain-pine zones. The lakes are fed by permanent and temporary streams, rainfall and numerous trickles from the slopes and moraines as well as flow from peat bogs. Stone and sediment samples were collected in June and September 1986 by SCUBA diving along depth transects from the shore to the lake deep.

The most abundant taxa were acidophilous taxa such as *Achnanthes marginulata* Grunow, *Aulacoseira alpigena* (Grun.) Krammer, *A. distans* (Ehr.) Simonsen, *A. lirata* (Ehr.) Ross, *A. pelgrabra* (Oestr.) Haworth, *A. pfaffiana* (Reinsch) Krammer, *Eunotia incisa* Gregory, *E. pseudopectinalis* (Dillwyn) Rabenh., *E. rhomboidea* Hust., *E. tenella* (Grun.) Hust., *Tabellaria flocculosa* (Roth) Kützing and acidobiontic *Eunotia exigua* (Bréb.) Rabenh. Although several species were common to all sites, the diatom communities varied in composition, number of species, structure of dominance and share of ecological groups, especially between the two lakes, but also between shallow and deeper sites in both lakes.

Measured pH values were close to those calculated from benthic diatom communities by some equations. However four different equations gave significantly different pH values

The study confirms that benthic diatom communities are sensitive indicators of the water pH, as is well documented by several authors.

## Introduction

Diatoms are sensitive indicators of many environmental factors and for this reason are frequently used to asses the state of, as well as ecological and chemical changes in, the aquatic environment (Van Dam 1993; Kawecka & Eloranta 1994).

The increase of anthropogenic acidity of lakes has given rise to numerous neo-and paleolimnological studies, concentrating on the diatom communities especially with a view to determining the acidity of the lakes on the grounds of their ecological preferences in relation to the pH of the water. This relationship between diatoms and pH, discovered by Hustedt (1937–39), became the basis for the development of numerical methods of assessing the pH of the water and its changes in aquatic ecosystems (e.g. Nygaard 1956; Meriläinen 1967; Renberg & Hellberg 1982; Charles 1985; Eloranta 1990; Huttunen & Turkia 1990; Håkansson 1993). This approach proved particularly useful in paleolimnological investigations aimed at reconstruction of the history of the lakes on the basis of sedimentary diatoms. They were successfully applied in the investigations based on periphytic diatoms and surface sediment diatoms in recent ecological studies (Jones & Flower 1986; Eloranta 1988, 1990; Charles 1985; Huttunen & Turkia 1990).

In the present work the structure and differences in the benthic diatom communities along a depth transect in two acidic mountain lakes were investigated. A further aim was to assess the relationship between the measured and calculated pH values based on diatom pH-groups, using different equations.

*Study area*

The subject of the investigation were Lakes Mały and Wielki Staw, situated in the eastern part of the Karkonosze Mts, which are the highest mountain range of the Sudeten Mts, extending along the south-western frontier of Poland (Fig. 1). The Karkonosze Mts are built of granite rocks, and their eastern part represents a fragment of metamorphic rocks of Upper Proterozoic age. The soils are mostly brown soils, strongly acidic with a mixture of peat soil. Higher parts are covered by soil-free areas, so-called stone fields. In spite of the relatively moderate altitude of the whole mountain massif, its climate is typical for high mountain regions with high precipitation and long-lasting snow cover, low temperature in summer, and frequent changes of the weather. The area studied lies in the zone of an upper montane and mountain-pine zone consisting mainly of complexes of spruce and thickets of the mountain pine. Within the whole region of the Karkonosze Mts there occur peat bog complexes, occurring frequently especially in the area of the upper montane zone and above, especially on the slopes near Lake Wielki Staw.

The investigated area lies within the territory of the Karkonoski National Park. A detailed description of this area is given by Jahn (1985).

Mały and Wielki Staw are typical, post-glacial lakes from the period of the Baltic glaciation which were formed in the basins of local mountain glaciers, now enclosed by frontal moraines.

Both lakes are fed by permanent and temporary streams, rainfall and numerous trickles from the slopes and also by direct precipitation and from water percolating from the lakes shores. Wielki Staw is fed mainly by temporary streams from melting snow, while on the western slopes there are numerous peat bogs; seepage springs on the slopes were also found.

The water chemistry of two lakes (Table 1) is characterized by being poorly buffered, with low conductivity values and low pH (ranging from 4.6–5.8), and a high

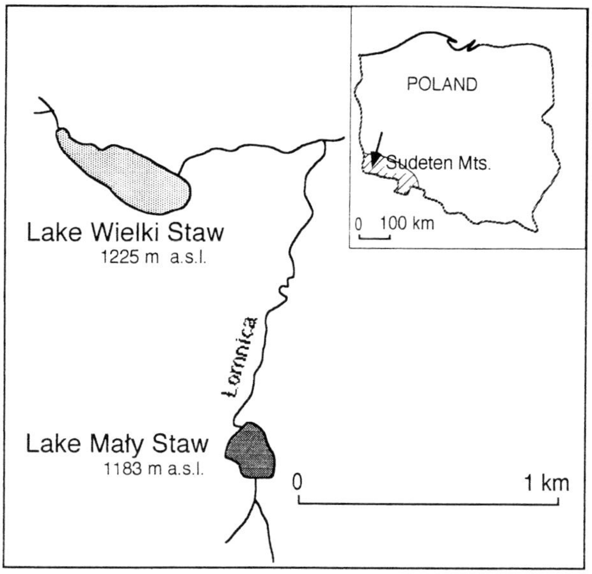


Fig. 1. Location of the studied lakes.

level of aluminium. The description of two lakes is supplemented by Table 1, which contains their more important morphometric features and the physico-chemical data concerning the water.

## Materials and methods

The material was collected in June and September 1986 by SCUBA diving. Samples from Wielki Staw were collected once only (in September ) along the transect at 0.5m, 2m, 5m, 7m, 10m, 12m, 17m. From Mały Staw the samples were collected twice (in June and September ) from 0.5m, 1m, 2m, 4m, 5m, 7m. The stones from the bottom were placed under the water into plastic bags and on shore the algae were washed carefully from their surface. Sediment samples were collected using a plexiglass sediment corer and 0.5 cm of surface layer was taken The material was preserved in a 4% solution of formalin. After standard preparation, described in detail by Kawecka (1980), *ca* 400 frustules were counted from each sample in randomly chosen microscopic fields. If the species was present in the slides but not found during counting it was given the agreed value 0.1. Species attaining at least a 10% share in the community were regarded as dominants. Species with a 5–10% share were described as abundant, and the remaining ones as rare. Diatoms were identified mainly according to Krammer & Lange-Bertalot (1986–1991). Diatom taxa were divided into ecological and pH-groups according to ecological data from Hustedt (1937–39), Cholnoky (1968), Lowe (1974), Van Dam *et al.* (1981), Håkansson (1993). pH calculations were made from equation in Renberg & Hellberg (1982), Eloranta (1990), Huttunen & Turkia

(1990) and Håkansson (1993). Group analysis was done by Pearson correlation and average linkage clustering.

The results of chemical analysis of the water of Mały and Wielki Staw from the Report of the Czechoslovakian Head Office of the Karkonosze National Park were utilized in the study (Table 1). In addition, some environmental parameters were measured in the field during the investigated period; temperature, transparency, oxygen, and pH of the water (by colorimetric method).

Table 1. Characteristics of the studied lakes. Morphometic data according to Komar (1985), chemical data according to Report of the Head Office of the Karkonoski National Park.

| | | L Mały Staw | L. Wielki Staw |
|---|---|---|---|
| Surface area | ha | 2.88 | 8.32 |
| Volume | $10^3$ m | 9.9 | 790.6 |
| Max. depth | m | 7.3 | 24.4 |
| Mean depth | m | 3.4 | 9.5 |
| Secchi disc v. | m | >7 | 8.7 |
| Temperature | °C | | |
| surface | 8.8.86 | 14.0 | 14.7 |
| bottom | 8.8.86 | 13.0 | 7.5 |
| surface | Sept. 86 | 7.5 | 9.1 |
| $0_2$ surface | mg $l^{-1}$ | 10.6 | 8.9 |
| Conductivity | µS $cm^{-1}$ | 18 | 27 |
| Hardness | °dH | 0.28 | 0.37 |
| pH | 8.7.86 | 5.4 | 4.6 |
| | Sept. 86 | 5.8 | 5.2 |
| Alkalinity | µeq $l^{-1}$ | 0 | 0 |
| Ca | mg $l^{-1}$ | 1.4 | 1.4 |
| Mg | mg $l^{-1}$ | 0.37 | 0.28 |
| $SO_4^{2-}$ | mg $l^{-1}$ | 4.50 | 6.80 |
| $NO_3$- | mg $l^{-1}$ | 0.65 | 2.16 |
| Al | µg $l^{-1}$ | 100 | 240 |
| Fe | µg $l^{-1}$ | 50 | |

## Results

### *Characteristics of diatom communities*

In Lakes Mały and Wielki Staw altogether 96 diatom taxa were identified, most of which occurred rarely or sporadically. The number of species in a community varied from 45 to 68 in Mały Staw, and from 36 do 47 in Wielki Staw. The number of species was very similar at all depths in Wielki Staw (Fig. 2).

Among the communities distinguished there were many species common to both lakes, though the diatom communities varied in composition and the dominant species differed (Fig. 3).

In the Mały Staw the following species were dominant: *Aulacoseira alpigena* (Grun.) Krammer (11.4–30.6%), *A. lirata* (Ehr.) Ross (14.7–25.5%), *Tabellaria flocculosa* (Roth) Kützing (32.3–40.7), *Aulacoseira distans* (Ehr.) Simonsen (12%), and *Eunotia tenella* (Grun.) Hust. (10.2%).

In the Wielki Staw the dominant species were *Achnanthes marginulata* Grunow (13.7–45.8%), *Eunotia incisa* Gregory (10.4–37.3 %), *Aulacoseira pfaffiana* (Reinsch) Rabenh. (18.7–27.0 %), *A. alpigena* ( 10.8–13. 6 %), *A. perglabra* (Oestr.) Haworth (10.2 %), *Eunotia exigua* (Breb.) Rabenh.10.8–12.4 %), *E. pseudopectinalis* (Dillwyn) Rabenh. (10.7–12.7) and *E. rhomboidea* Hust. (20.7 %).

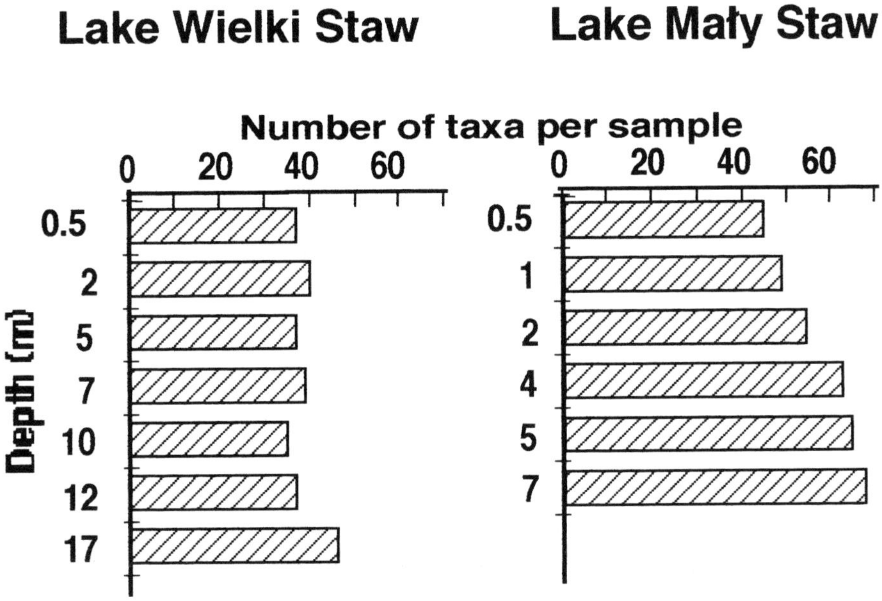


Fig. 2. Number of diatom taxa per sample at different depths.

Acidophilous diatoms were the dominant group along the entire depth transect in both lakes (Fig. 4). The percentage share of diatom pH-groups showed a dependence on the pH of the water . With a higher pH of the water in Mały Staw (Table 1), the acidophilous diatoms constituted about 75–82% of the total communities, in Wielki Staw, with a lower pH, the share of this group was 84–96.5%, and the number of acidobiontic forms was also higher.

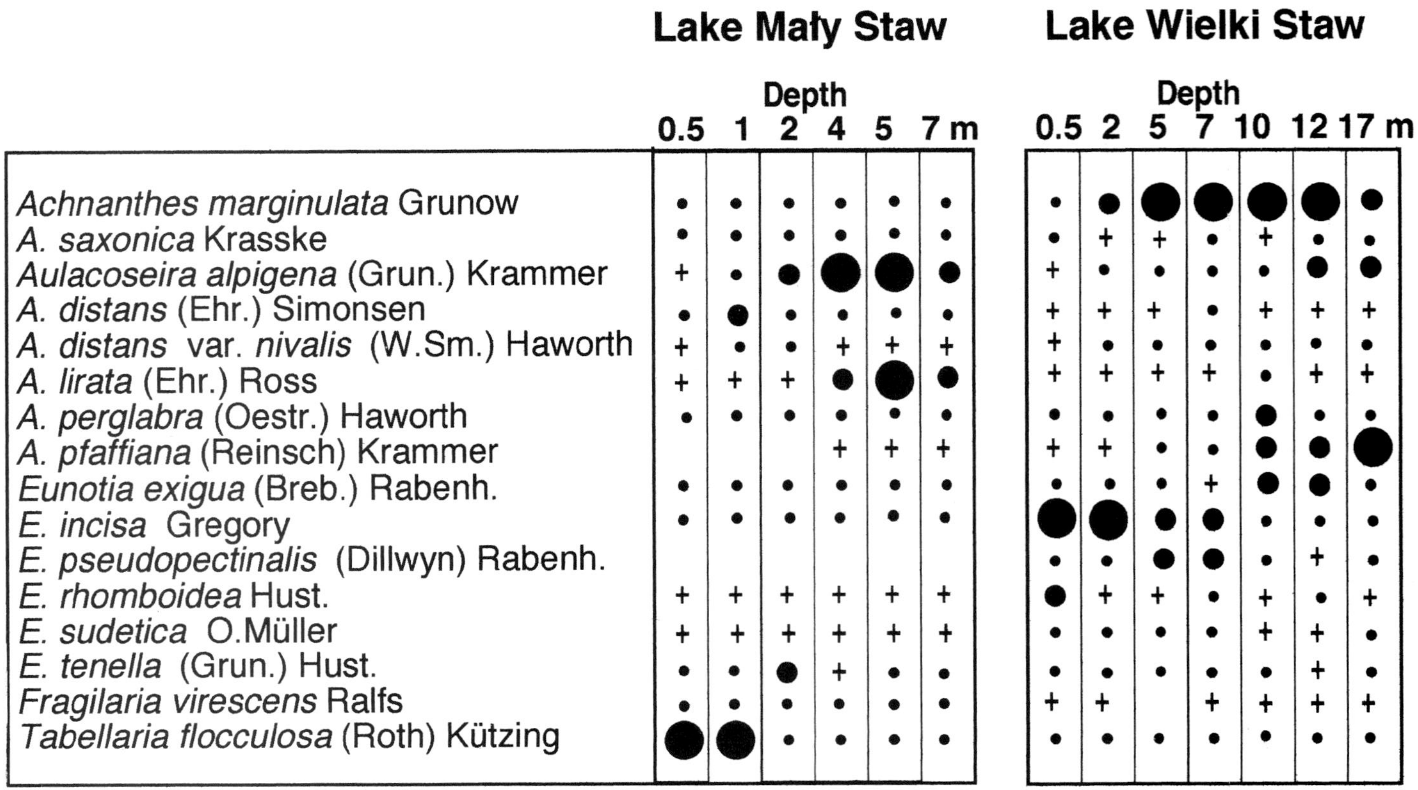


Fig. 3. More abundant diatom species (>5%) found at the different depths in the studied lakes. + < 5%, • 5–10%, ● 10–25%, ● 25–50%.

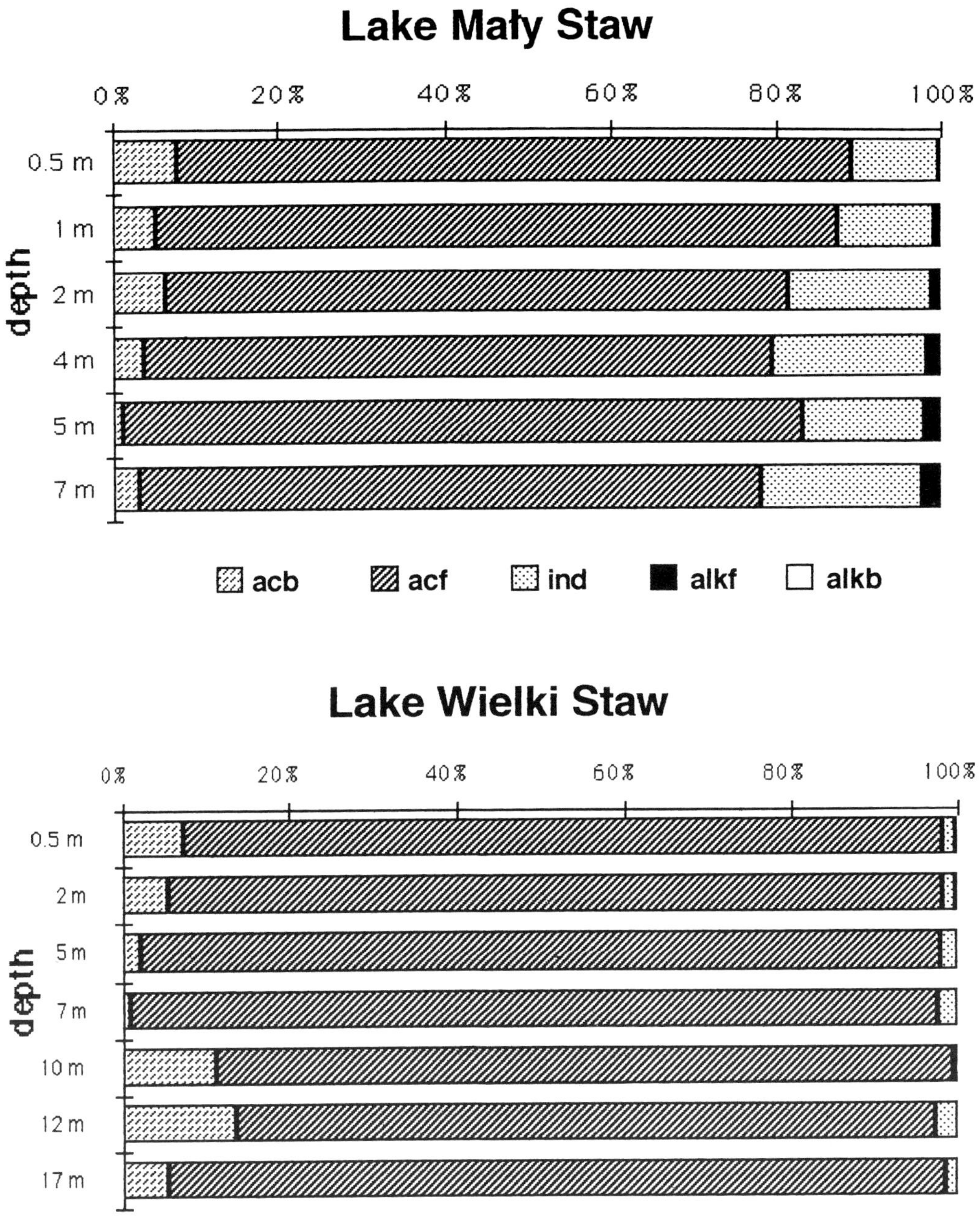


Fig. 4. Relative abundance of diatom pH groups at the different depths in the studied lakes.

The bottom communities of diatoms in both lakes consisted mainly of epilithic and planktonic taxa, while epipelic taxa were present to a smaller degree (Fig. 5). The proportion of these ecological groups changed with depth. The ratio of epilithic to planktonic taxa was higher in Wielki Staw than in Mały Staw. Moreover the population of the epipelic species in the lake Mały Staw was higher along the entire depth transect.

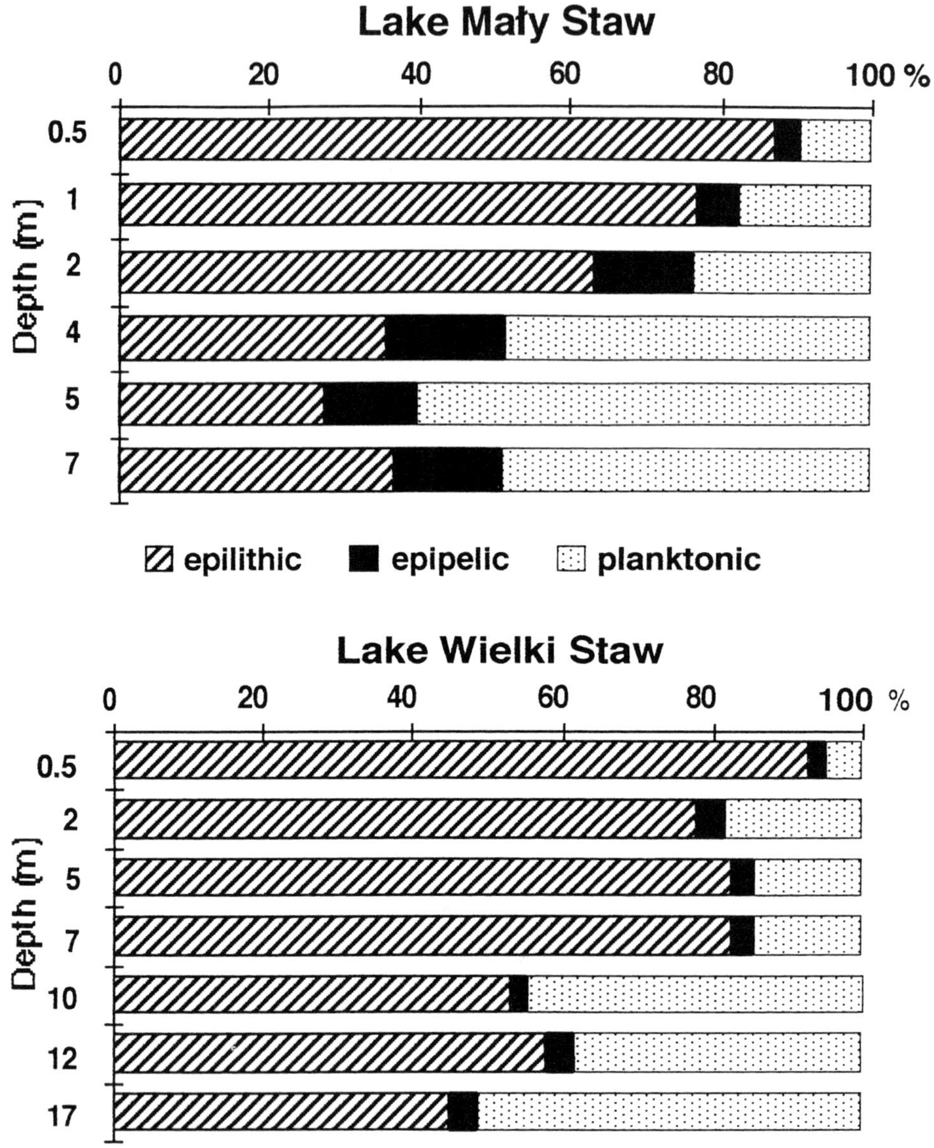


Fig. 5. Relative abundance of ecological groups at the different depths in the studied lakes.

*Cluster analysis of diatom communities*

The results of the cluster analysis showed that the communities of the two lakes differ greatly from each other, whereas in each of the lakes, along the depth transect, the communities show some similarities (Fig. 6). Based on cluster analysis the diatom communities can be divided into 3 main groups in both lakes (Fig. 6).

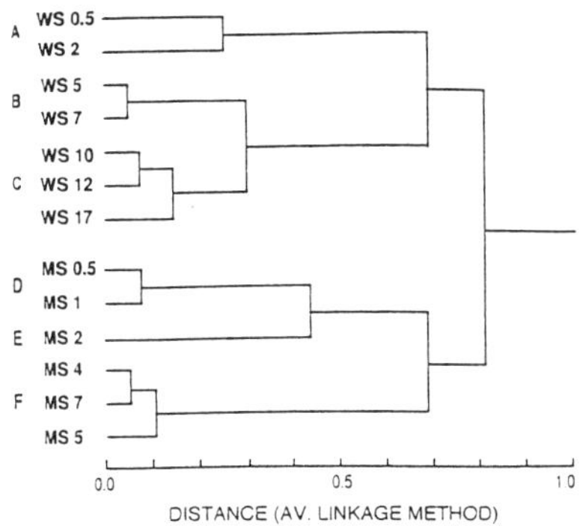


Fig. 6. Comparison of diatom communities in the studied lakes according to Pearson correlation and the average linkage clustering.

In Wielki Staw, group A consist of communities found at depths of 0.5m and 2m. They are characterized by acidophilous, epilithic taxa including *Achnanthes marginulata, Eunotia incisa* and *E. rhomboidea* . Group B comprises communities at depths of 5m and 7m, characterized by acidophilic, epilithic taxa such as *Achnanthes marginulata, E. incisa, E. pseudopectinalis.* Group C is represented by communities at depths of 10, 12, 17m, characterized by acidophilous and acidobionthic, epilithic taxa *Achnanthes marginulata, Eunotia exigua* and acidophilous, planktonic taxa ones such as *Aulacoseira alpigena, A. pfaffiana, A. perglabra* .

In Mały Staw group D comprises communities present at depths of 0.5m and 1m, characterized by acidophilous, epilithic taxa such as *Tabellaria flocculosa* and the acidophilous, planktonic taxon *Aulacoseira distans.* Group E differing somewhat from the others, comprises communities at a depth of 2m. This group is characterized by acidophilous, epilithic species *Eunotia tenella* and planktonic *Aulacoseira alpigena..* Group F comprises communities at depths of 4m, 5m, and 7m characterized by acidophilous, planktonic taxa *Aulacoseira alpigena* and *A. lirata.*

*pH reconstruction comparison*

The measured pH value of the water (Table 1) was close to that calculated by some equations (Fig. 7). The maximum difference between the mean values of measured and calculated pH values was about 0.8 of a pH unit, the minimum difference was 0.1. However, the four different equations used to assess the pH of the

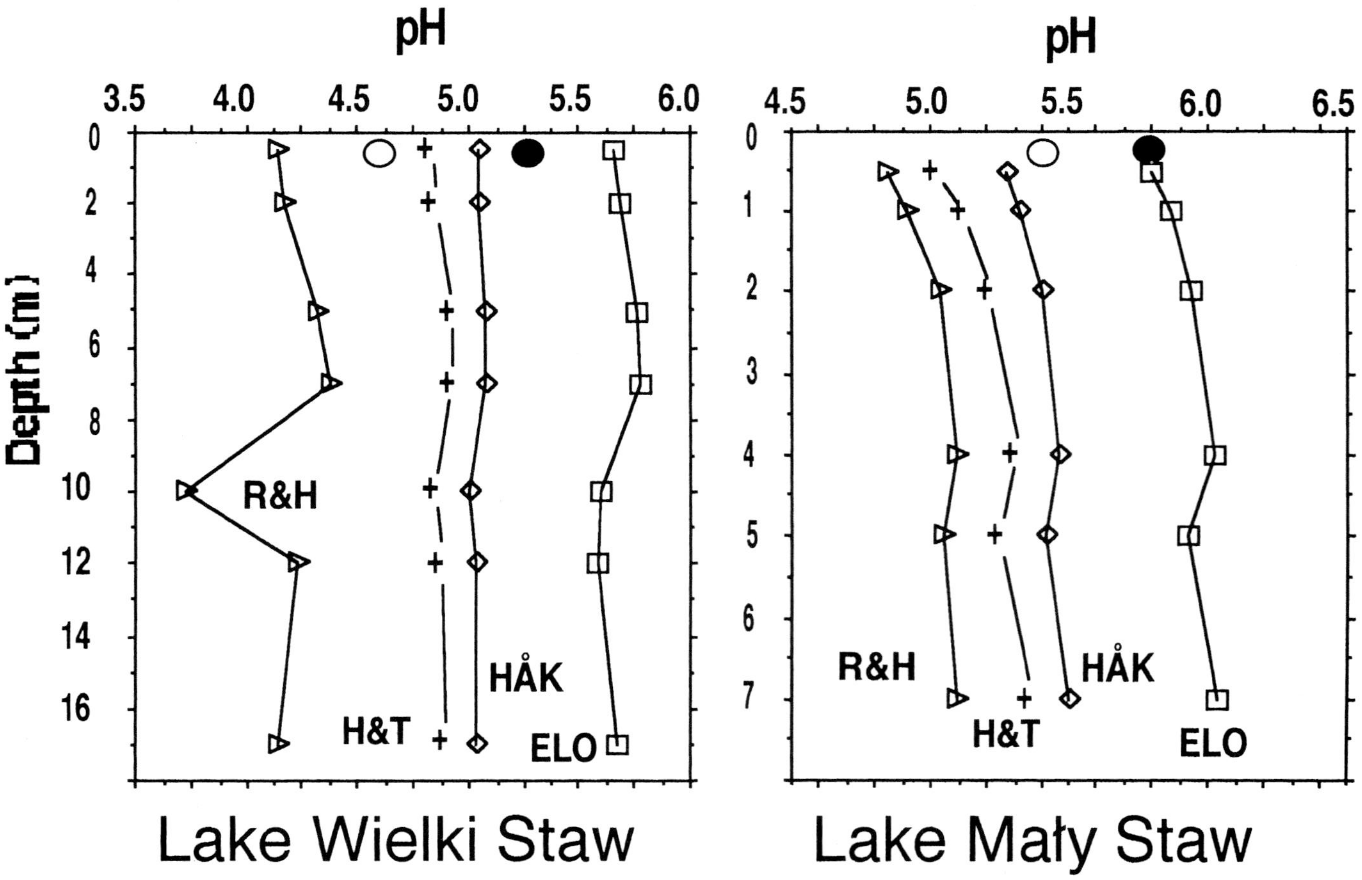


Fig. 7. Reconstructed pH values at different depths of the studied lakes using four different equations (R & H = Renberg & Hellberg 1982. H & T = Huttunen & Turkia 1990, ELO = Eloranta 1990, HAK = Håkansson 1993, o = measured pH Sept.1986, • = measured pH, July 1986.

water gave significantly different pH values. It appears that the lowest calculation of pH was obtained using the Index B (pH ranged from 3.75 to 4.38 in Wielki Staw, and from 4.8 to 5.1 in Mały Staw).

**Discussion**

In lakes Mały and Wielki Staw the most abundant taxa were acidophilous ones such as *Achnanthes marginulata*, *Aulacoseria alpigena*, *A. lirata*, *A. perglabra*, *A. pfaffiana*, *Eunotia incisa*, *E. pseudopectinalis*, *E. rhomboidea*, *E. tenella*, *Tabellaria flocculosa* and acidobiontic *Eunotia exigua*. This diatom flora with the dominance of acidophilous species was similar to that of moderately acidic soft-water (pH 5–6) lakes and pools in Europe (Meriläinen 1967; Van Dam *et al.* 1981; Flower 1986a; Kawecka & Eloranta 1994) and forested, montane acidic lakes of North America (Charles 1985).

Although in the study lakes many species are common to all stations, the diatom communities varied in their composition, number of species and dominance, and share of ecological groups, especially between two lakes but also between shallow and deep sites in each lake. The differences are mainly caused by pH factor.

In Mały Staw, with a higher pH (5.4–5.8) when compared with Wielki Staw (4.6–5.2), there were a greater number of species usually classified as indifferent (mainly *Achnanthes saxonica* and *Fragilaria virescens*) and also some alkaliphilous forms. The proportion of ecological groups with respect to habitats were also different.

Benthic diatom communities in lakes may contain forms originating from other habitats. The relative share of the planktonic and periphytic groups depends on many factors, such as quality of the water, turbulence, type of littoral, morphometry of the lakes and the structure of the shore, type of substrate and seasonal cycle (Eloranta 1990).

In the lakes the most important benthic flora was the epilithic and, with increasing depth, sedimented planktonic species increased. In the shallower lake Mały Staw, penetration of light (Table 1) also created greater possibilities for growth on the surface sediment for epipelic diatoms. In the deeper Wielki Staw the deep-water sediment communities consisted of deposited epilithic and planktonic algae. More apparent also is the presence in the benthic communities of planktonic species in Mały Staw and the epilithic community in Wielki Staw. It seems that the difference in the proportion between the two lakes may also be connected with the pH of the water, whose value in the two lakes differed by about one pH unit. Moreover, in Wielki Staw, surrounded by peat bogs, the pH of the water may be locally still lower than that measured on the water surface where the samples were collected.

In acidic, oligotropic lakes, owing to limited access to silica and phosphorus in the water and high concentration of metals, especially Al, plankton is a minor component of the algal communities or even absent (Kawecka & Eloranta 1994). In lakes at pH 5.6, Charles (1985) recorded a very low population of planktonic diatoms, and DeNicola (1986) observed a considerable decrease in planktonic diatoms in waters at pH 4.9 and 5.6.

In the present study the measured pH of the water was close to that calculated by some equations (Fig. 7). However, estimation of pH values of the water from the benthic diatom communities using 4 different equations gave significantly different pH values. The differences in the predictions may be caused by the type of material collected from various habitats (periphytic, surface of sediment and deep sediment), that is, pure communities were not sampled nor were contaminants removed before estimation of predicted pH values. Communities of diatoms and pH values of water have temporal and spatial changes (Jones & Flower 1986), hence the measurements conducted during one season in this investigation are not entirely adequate. It is obvious that more samples collected from different season should be examined and the contribution of species from contaminanting communities should be taken into account.

However, in this preliminary study, the structure of benthic diatom communities, cluster analysis, and predictive relationship for assessing pH all confirm that diatoms are sensitive indicators of acidity and acidification as documented by several authors.

## Acknowledgements

This studied material was collected by SCUBA diving during an expedition organised by the Section of Free Diving of the Cracov Branch of the Polish Society of Friends of Earth Sciences and Institute of Freshwater Biology of Sciences.

I thank the divers for collecting samples which facilitated this work. I also express my thanks to Prof. P. Eloranta from the University of Helsinki for his assistance in preparing this paper.

## References

Charles, D. F. (1985). Relationships between surface sediment diatom assemblages and lakewater characteristics in Adirondack lakes. *Ecology*, **66**(3), 994–1011.

Cholnoky, B. J. (1968). *Ökologie der Diatomeen in Binnengewässen.* 699 pp. J. Cramer. Lehre. Germany

DeNicola, D. M. (1986). The representation of living diatom communities in deep water sedimentary diatom assemblages in two Maine (U.S.A.) lakes, p.73–85. In: *Diatoms and lake acidity* (J. P. Smol, R. W. Battarbee, R. B. Davis and J. Meriläinen, eds). Dr W. Junk Publishers, Dordrecht.

Eloranta, P. (1988). Periphytic diatoms as indicators of lake acidity. *Verhandlungen der internationalen Vereinigung für theoretische und angewandte Limnologie*, **23**, 470–473.

Eloranta, P. (1990). Periphytic diatoms in the acidification project lakes. In: *Acidification in Finland* (P. Kauppi, P. Anttila & K. Kenttämies, eds), 985–994. Springer, Berlin.

Flower, R. J. (1986a). The relationship between surface sediment diatom assemblages and pH in 33 Galloway lakes, some regression models for reconstructing pH and their application to sediment cores. *Hydrobiologia*, **143**, 93–103.

Håkansson, S. (1993). Numerical methods for the inference of pH variations in mesotrophic and eutrophic lakes in southern Sweden – A Progress report. *Diatom Research*, **8**(2), 349–370.

Hustedt, F. (1937–39). Systematische und ökologische Untersuchungen über die Diatomeen-flora von Java, Bali und Sumatra. *Archiv für Hydrobiologie, Supplement*, **16**, 274–394.

Huttunen, P. & Turkia, J. (1990). Surface sediment diatom assemblages and lake acidity. In: *Acidification in Finland* (P. Kauppi, P. Anttila & K Kenttämies, eds), pp. 995–1008). Springer, Berlin.

Jahn, A. (1985). *Karkonosze polskie*. Polish Academy of Sciences, Karkonskie Towarzystwo Naukowe w Jeleniej Górze. Ossolineum, Wrocław. 566 pp.

Jones, V. J. & Flower, R. J. (1986). Spatial and temporal variability in periphytic diatom communities: Paleolimnological significance in an acidified lake. In: *Diatoms and lake acidity* (J. P. Smol, R. W. Battarbee, R. B. Davis and J. Meriläinen, eds), pp. 87–94. Dr W. Junk Publishers, Dordrecht.

Kawecka, B. (1980). Sessile algae in European mountain streams 1. The ecological characteristics of communities. *Acta Hydrobiologica*, **22**, 361–420.

Kawecka, B. & Eloranta, P. (1994). Zarys ekologii glonów wód słodkich i środowisk ldowych. Wydawnictwo Naukowe PWN, Warszawa. 256 pp.

Komar, T. (1985). Wody powierzchniowe. In: *Karkonosze polskie* (A. Jahn, ed.), pp. 165–190. Polish Academy of Sciences, Karkonskie Towarzystwo Naukowe w Jeleniej Górze. Ossolineum, Wrocław.

Krammer, K. & Lange-Bertalot, H. (1986–1991). *Bacillariophyceae. Süßwasserflora von Mitteleuropa*. **2** (1–4). G. Fischer, Jena.

Lowe, R. L. (1974). *Environmental requirements and pollution tolerance of freshwater diatoms*. National Environmental Research Center Office of Research and Development U. S. Environmental Protection Agency, Cincinnati, Ohio. 334 pp.

Meriläinen, J. (1967). The diatom flora and the hydrogen-ion concentration of the water. *Annales Botanici Fennici*, **4**, 51–58.

Nygaard, G. (1956). Ancient and recent flora of diatoms and Chrysophyceae in Lake Gribso. *Folia Limnologica Scandinavica*, **8**, 32–99, 254–262.

Renberg, I. & Hellberg, T. (1982) The pH history of lakes in southwestern Sweden, as calculated from the subfossil diatom flora of the sediments. *Ambio*, **11**, 30–33.

Van Dam, H. (1993). Diatom investigation In: *Proceedings of the Twelfth International Diatom Symposium* (H. Van Dam, ed.). Kluwer Academic Publishers, Dordrecht/Boston/London. 540 pp.

Van Dam, H., Suurmond, G. & Ter Braak, C. J. .F. (1981). Impact of acidification on diatoms and chemistry of Dutch moorland pools. *Hydrobiologia*, **83**, 425–459.

# Planktonic diatoms in the Vlasinsko Jezero reservoir, Serbia (Yugoslavia)

Radoje Laušević and Mirko Cvijan

*Institute of Botany, Faculty of Biology, University of Belgrade*
*43 Takovska, 11000 Belgrade, Yugoslavia*

## Abstract

Vlasinsko Jezero reservoir is a temperate, dimictic, neutral to slightly alkaline waterbody with soft water and very low mineralization. The most probable status is mesotrophic. Diatoms predominated both qualitatively and quantitatively in the phytoplankton during almost the whole investigated period (seasonally, 1991–1993), with the maximum in early summer.

Distinct spring and smaller, autumn, diatom maxima were recorded with dominance of large diatoms (*Synedra acus*, *Asterionella formosa* and *Fragilaria crotonensis*). Dense diatom populations persisted during spring, but large diatoms were replaced by small centric species (*Cyclotella radiosa*, *Stephanodiscus* sp.) in summer. During late summer, large dinoflagellates (*Ceratium*, *Peridinium* and *Peridiniopsis*) also play an important role in building up the plankton community. Distinct metalimnetic peaks in cell numbers and Chlorophyll-*a* concentration occurred during late summer 1992, produced by *Mallomonas* sp. and large chlorococcal algae which also persist into the autumn.

Regarding the vertical distribution, meta- and hypolimnetic maxima of large diatoms (mainly *Synedra acus*, *S. acus* var. *angustissima* and *Fragilaria crotonensis*) were noted in early summer. In respect to trophic status, our findings indicate a process of gradual eutrophication of the Vlasinsko Jezero reservoir.

## Introduction

Diatoms are valuable indicators of environmental conditions and change. Their indicator value is based on their well-defined ecological tolerances which have been demonstrated in many papers (see Willén 1991). Also, the composition of the aquatic diatom assemblages reflects the water quality and certain environmental conditions in the investigated water bodies which are not deducible, for instance, from macrophyte vegetation surveys or infrequent chemical analysis (Denis & Van Straaten 1992).

The present study deals with the planktonic diatom assemblages in the Vlasinsko Jezero reservoir.

Limnological investigations of that reservoir have been undertaken during the period of its formation (Milovanović & Živković 1953, 1956) as well as during the maturation period (Milovanović & Živković 1958; Milovanović 1973) with a break until the summer of 1989 (Cvijan & Laušević 1991a, 1991b) and again continued from August 1991. However, none of these early studies concentrated particularly on the diatom assemblages.

*Study area*

Vlasinsko Jezero reservoir is situated in the south-east part of Serbia, in the central part of the plateau at an altitude of 1213 m (Fig. 1). It was formed when a peat-bog, the largest in Yugoslavia, floated downstream after the construction of the dam on the river Vlasina in 1949, 2.2 km downstream from the place where it flowed out from the peat-bog. The filling of the depression was completed in 1954 when the lake was formed.

The morphometric data for the lake are given in Table I. The catchment area of the reservoir has been artificially enlarged and, therefore, about 80% of the inflowing water comes from collecting channels. Through only one of them, $74.7 \times 10^6$ $m^3$ of water is drawn from the Lisina pump-accumulation plant annually, from the Lisinsko Jezero reservoir.

From the Vlasinsko Jezero reservoir water goes through an outflow in the southern part of the lake (Fig. 1) into the water system of the four hydro-electric power plants. Their working regime, together with the Lisina pump-accumulation plant, as well as the amount of atmospheric precipitation, causes great variability in the water level of the reservoir, which can vary more than 4 m.

The climate in the region is a typical continental one. The winter period is long, from November to April (Milovanović 1973) and the reservoir is covered by ice usually for two or three months (from December to March). The east shore is pasture land, while the west coast is partly forest, mainly birch and planted pine.

The southern part of the reservoir is shallow with an average depth of 5 m and with many floating peat islands, some of which are very large. The origin of these islands is mainly from the former peat deposits.

The aquatic vegetation, which grows most during summer and autumn, is rich only in the zone lying at 4 to 6 meters (below the water fluctuation level) along the west shore, where *Potamogeton obtusifolium* Mert. et Koch dominates. In the shallow bays, at the mouth of small natural tributaries and in the southern part of the reservoir *Ranunculus aquatilis* L. and *Polygonum amfibium* L. dominate (Blaženčić & Blaženčić 1991).

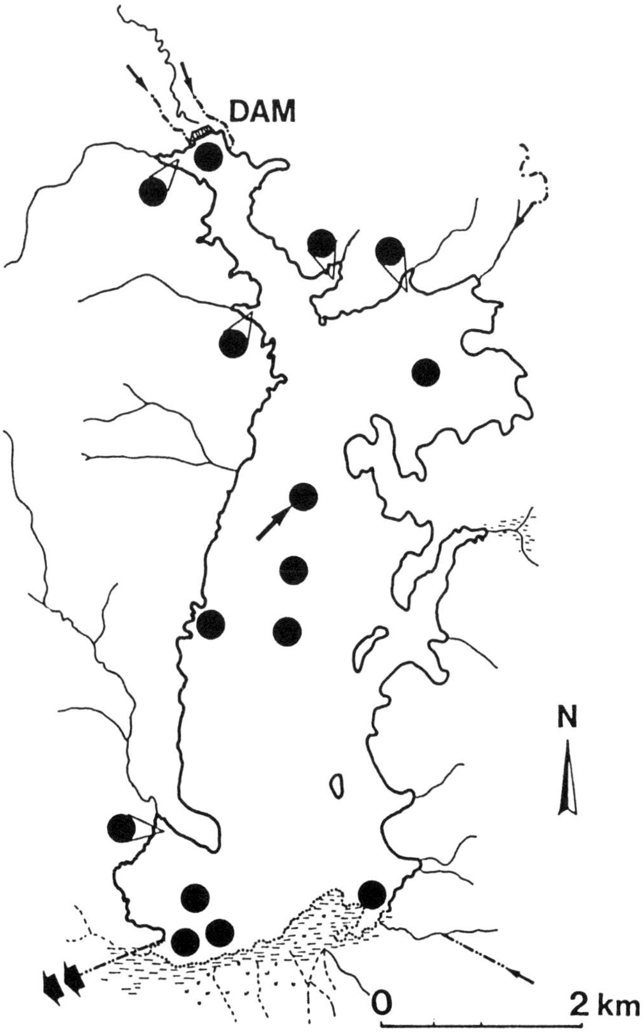


Figure 1. Vlasinsko Jezero reservoir with the sampling stations. Pelagic sampling station is arrowed in black.

Table I. Morphometric data of the Vlasinsko Jezero reservoir.

| | |
|---|---|
| Altitude (m above see level) | 1213 |
| Maximal surface area ($km^2$) | 16.5 |
| Minimal surface area ($km^2$) | 5.6 |
| Volume ($m^3$) | $163 \cdot 10^6$ |
| Maximal depth (m) | 21 |
| Mean depth (m) | 10.5 |
| Catchment area ($km^2$) | 50 + 97.5 |

## Material and Methods

The water temperature was measured directly from the sampling bottle by a thermometer. The transparency was determined by a Secchi disc, 25 cm diameter. The chlorophyll-*a*, corrected for phaeophytin, was determined according to Standard Methods (1985). The limited chemical data, representing only the summer periods in 1991–93, obtained by the Serbian Meteorological and Hydrological Bureau and the Serbian Institute of Public Health, probably only reflect the most striking differences in the chemical characteristics of the water.

The phytoplankton samples for qualitative analyses were collected by hauling a net (36 µm mesh size and 25 cm opening) from the bottom to the surface at three points (the dam, central and southern shallow part), also at points around the reservoir (Fig. 1). 41 samples, preserved immediately in 4% formaldehyde, were collected in August and October 1991, and in April and June 1992. Diatom frustules were cleaned with sulphuric acid (Patrick & Reimer 1966) or oxidized with $H_2O_2$ for lightly silicified species. Slides were made using Naphrax and observed on a Reichart Diastar microscope with a Photostar camera system at ×1200 magnification. All determined taxa were photographed. Both the diatom slides and black-white photographs are kept in the Institute of Botany, Faculty of Biology, Belgrade. Two slides from each sample were counted and the relative abundance was determined according to a six-point scale (Pantle & Buck 1955). At least two hundred frustules were counted on each slide.

Based on the results of the qualitative analyses a sampling station in the central part of the lake was chosen as a representative for the quantitative sampling (the pelagic sampling station, Fig. 1). At two month intervals, from April 1992 to October 1993, samples were collected every 2 m from the surface to the bottom by means of a Hydro–Bios bottle. The phytoplankton samples were preserved in 4% formaldehyde solution and stored in 1 litre plastic bottles. The samples were counted using a Zeiss inverted microscope, according to Utermöhl (1958). Diatoms were determined mainly according to Hustedt (1930), Krammer & Lange-Bertalot (1986, 1988, 1991a, 1991b) and Lange-Bertalot (1993).

## Results

The temperature fluctuated from 5.1°C April 1993 (spring season) to 21.8°C in August 1992 (autumn season). The vertical distribution during a seasonal cycle showed spring and autumn circulation and a summer stratification (Fig. 2). The circulation pattern revealed an isothermal period (5.1°C and 7.5°C) in spring and between 11.4°C and 14.6°C in autumn. The epilimnion increases in thickness from early to late summer, and its thickness fluctuates between 5 m and 7 m (Fig. 2).

The transparency ranged from 2.5 m to 8.3 m and increased with the developing epilimnion thickness (Fig. 2). The circulation in spring and autumn was followed by low transparency and the appearance of visible humic and peat particles in the lake water. The average Secchi depth is 4.4 m.

Some chemical variables are listed in Table II. Chlorophyll-*a* concentrations ranged between 0.25 and 32.6 $\mu g\ l^{-1}$ showing clear seasonal and vertical fluctuations. The highest concentrations were established in the metalimnion during summer stratification and near the surface during circulation. In all seasons, the highest concentrations correspond to the Secchi depth (Fig. 2). The average planktonic algal chlorophyll-*a* is 6.36 $\mu g\ l^{-1}$. During spring and summer it is in the range between 4.16 and 6.34 $\mu g\ l^{-1}$ and between 8.63 and 9.62 $\mu g\ l^{-1}$ for the autumn.

In the water with the above described characteristics, diatoms predominated both qualitatively and quantitatively in the phytoplankton during almost the whole investigated period, with the maximum in early summer. Populations of chlorococcal algae (*Planktococcus* sp.) dominate over diatoms only in the metalimnion in late summer and in the whole water column in autumn 1992. Just two more populations are significant, *Mallomonas* sp. in the lower metalimnion in late summer 1992 and dinoflagellates (*Peridinium* sp., *Peridiniopsis* sp. and *Ceratium hirundinella* (O. F. Müll.) Ehr.) in late summer 1993. Chlorophyll-*a* metalimnetic peaks (Fig. 2) correspond with the maximum development of the non-diatom populations (Fig. 3).

In the 41 qualitative spatial samples, 159 diatom taxa were identified. However, many of them are not planktonic, and were established in, or only in, shallow bays, among the floating peat islands or among the aquatic macrophytes, which indicate that they have been resuspended from benthic or epiphytic communities. In Table III the most frequent and abundant planktonic diatoms are listed.

The results of the quantitative analyses show that the diatoms in the pelagic zone formed dense populations only in spring and early summer (Fig. 3). Fig. 4 shows that only a few diatom taxa formed numerous populations while all others are sparse.

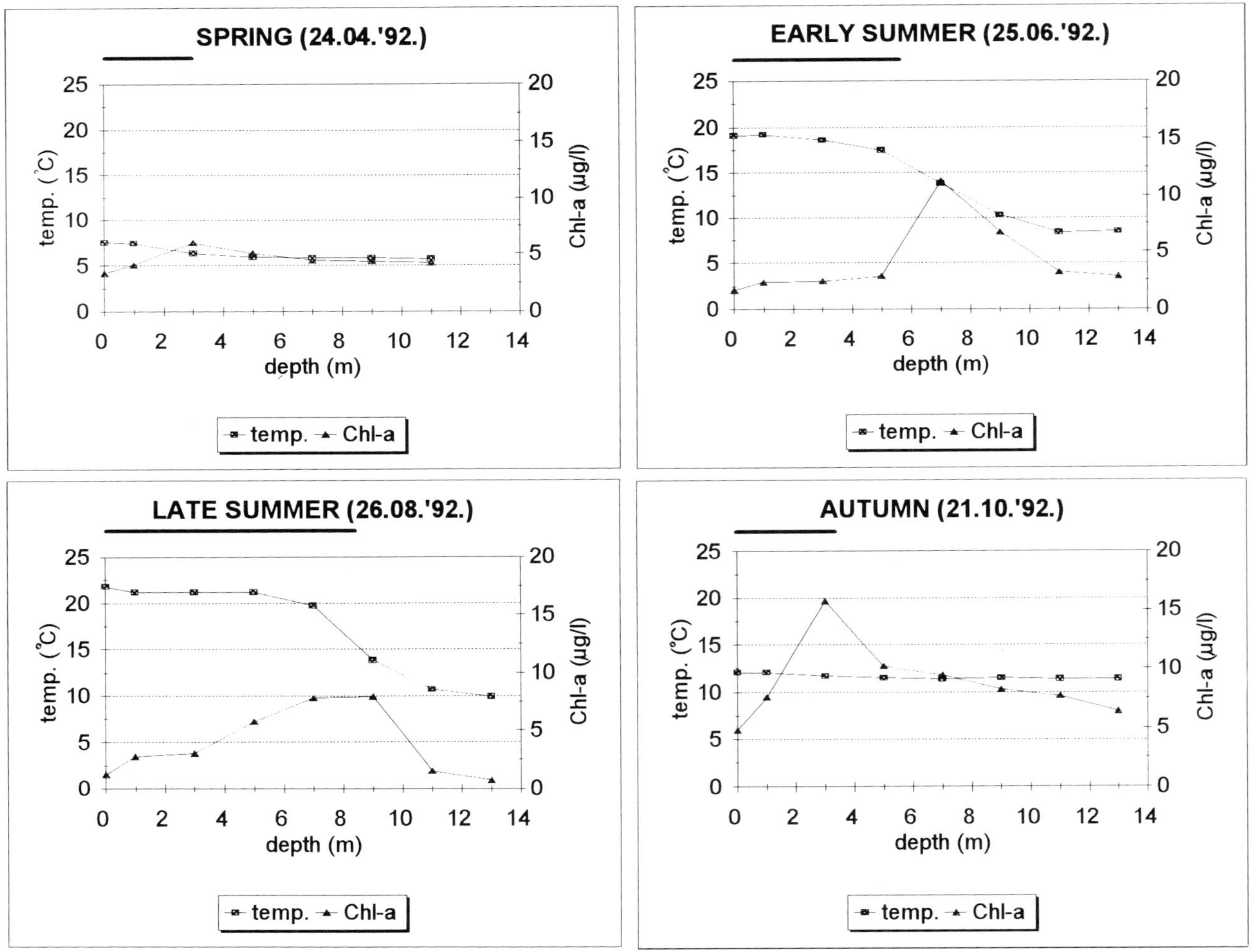


Figure 2. Seasonal and vertical distribution of temperature, planktonic algal Chlorophyll-*a* (µg $l^{-1}$), corrected for phaeophytin, and Secchi depths. Notice different scale of Chlorophyll-*a* concentration for August 1992.

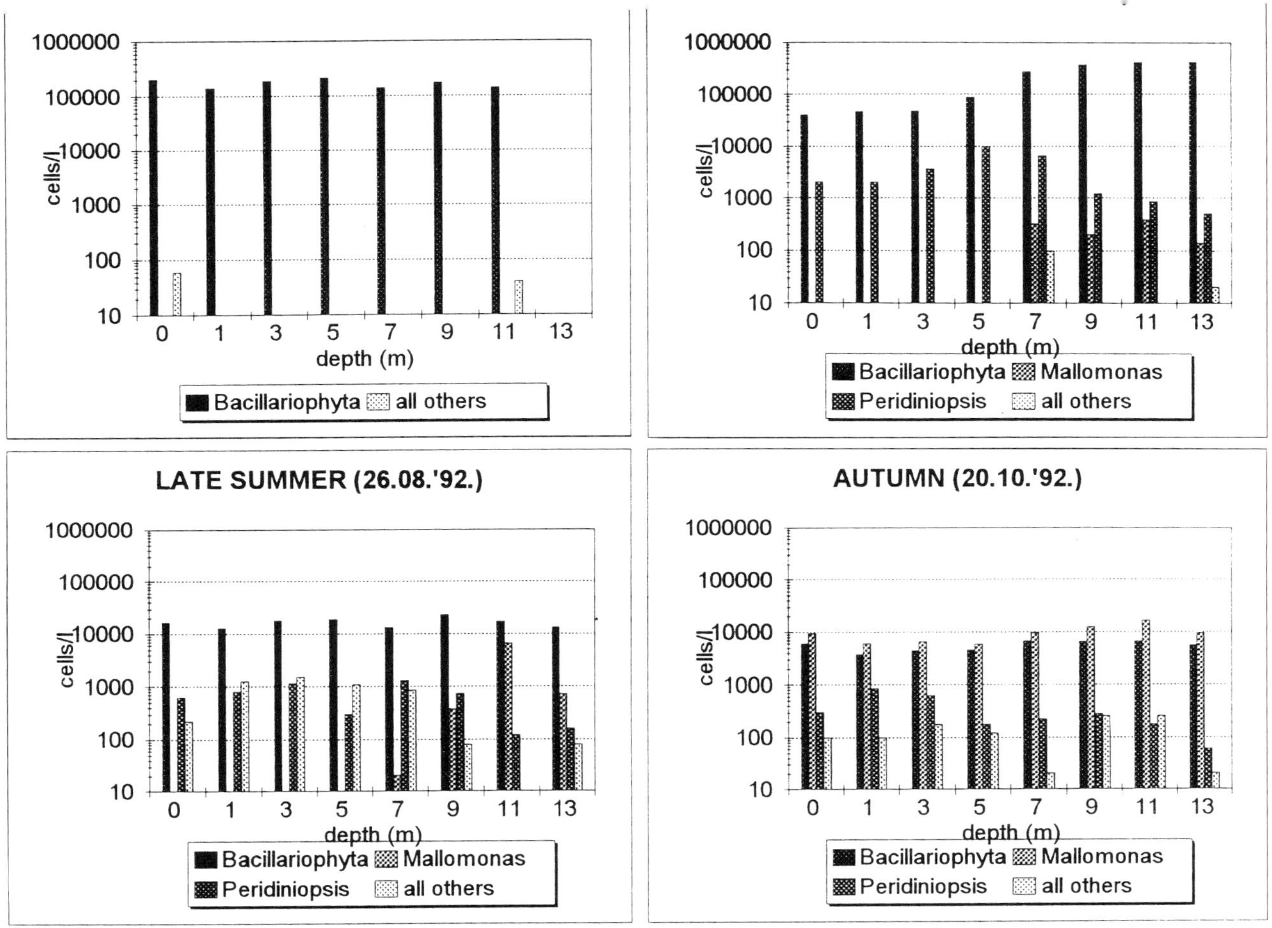


Figure 3. Seasonal and vertical distribution of the population densities, plotted on a $\log_{10}$ scale, of the main phytoplankton groups (BACI – Bacillariophyta, PYRR – Pyrrophyta: *Ceratium* sp., *Peridinium* sp. and *Peridiniopsis* sp., MALL – *Mallomonas* sp., CHLO – chlorococcal green algae, OTHE – all other algae).

Table II. Chemical characteristics of Vlasinsko Jezero reservoir at the pelagic sampling station. Values are calculated for the summer period 1991–93 from unpublished data sets supplied by the Serbian Meteorological and Hydrological Bureau and the Serbian Institute of Public Health.

| | max | min | mean |
|---|---|---|---|
| pH | 8 | 7.1 | 7.6 |
| Alakalinity (mE/l) | 1 | 0.75 | 0.8 |
| Conductivity (μS/cm) | 103 | 72 | 92 |
| Total hardness (°dH) | 3 | 1.7 | 2.5 |
| Ammonia (N mg/l) | 0.026 | 0.01 | 0.016 |
| Nitrite (N mg/l) | <0.001 | <0.001 | <0.001 |
| Nitrate (N mg/l) | 0.67 | 0.17 | 0.41 |
| Ortho phosphates (P mg/l) | 0.006 | <0.001 | 0.002 |
| Total phosphorus (P mg/l) | 0.06 | 0.003 | 0.016 |
| Silicon ($SiO_2$ mg/l) | 4.0 | 3.0 | 3.1 |
| COD (mg/l $KMnO_4$) | 11 | 5.58 | 8.6 |
| BOD (mg/l $O_2$) | 7.35 | 1 | 3.8 |
| $O_2$ (mg/l) | 12 | 4.6 | 9 |
| Saturation ($O_2$ %) | 138 | 38.3 | 97 |

In spring 1992, *Synedra acus* var. *angustissima* dominated the whole water column, at temperature ranging between 5.7 and 7.5°C with populations of 1.46–3.34 · $10^5$ cells $l^{-1}$. *S. acus* var. *angustissima* declined in early summer after *S. acus* reached a population maximum of 9.43 · $10^5$ cells $l^{-1}$ in the metalimnion at a temperature range between 8.5 and 10.3°C. These populations did not persist for the whole summer as they were succeeded by the centric diatom *Cyclotella radiosa* which formed a late summer population of only 1.5–6.4 · $10^4$ cells $l^{-1}$. In autumn 1992 diatoms did not develop dense populations.

Table III. Floristic composition of planktonic diatoms in the Vlasinsko Jezero reservoir at the pelagic sampling station (August 1991 – October 1993).

| |
|---|
| *Asterionella formosa* Hass. |
| *Campylodiscus hibernicus* Ehr. |
| *Ceratoneis arcus* (Ehr.) Kütz. |
| *Cyclotella bodanica* Grun. var. *affinis* Grun. |
| *Cyclotella radiosa* Grun. |
| *Cymbella minuta* Hilse |
| *Fragilaria capucina* Desm. |
| *Fragilaria construens* (Ehr.) Grun. var. *venter* (Ehr.) Grun. |
| *Fragilaria crotonensis* Kitton |
| *Gomphonema gracile* Ehr. |
| *Rhizosolenia longiseta* Zach. |
| *Stephanodiscus hantzschii* Grun. |
| *Stephanodiscus minutulus* (Kütz.) Cl. & Müller |
| *Synedra acus* Kütz. |
| *Tabellaria fenestrata* (Lyngb.) Kütz. |
| *Tabellaria floculosa* (Roth.) Kütz. |

In 1993, a spring bloom of *Asterionella formosa* ($0.35–1.12 \cdot 10^5$ cells $l^{-1}$) was noticed in the whole water column at a temperature range between 5.1–8.6°C; it was succeeded by *Cyclotella radiosa* in early summer. *C. radiosa* produced dense populations ($14.74 \cdot 10^5$ cells $l^{-1}$) in the surface layer at a temperature of 18.7°C and it persisted in late summer and autumn, but with sparse populations. *A. formosa* reappeared in autumn in the whole water column (temperature range 13.0–14.6°C) with a maximum of $4.0 \cdot 10^4$ cells $l^{-1}$ near the surface (Fig. 4).

In terms of relative abundance, *Fragilaria crotonensis* was recorded during August and October 1991 as the dominant pelagic species, but did not reappear in significant numbers during 1992–1993.

## Discussion

Vlasinsko Jezero reservoir is a moderately warm type of lake such as is commonly found in southern and central Europe with a clearly marked summer stratification and differentiated layers of epi-, meta- and hypolimnion, and with isothermal conditions in spring and autumn.

The average Secchi depth (4.4 m) is characteristic for oligotrophic-mesotrophic lakes (Jones & Lee 1982) and much higher compared the period of reservoir formation (2.12 m, calculated from Milovanović 1973). The chemical data allow us to describe this ecosystem as a neutral to slightly alkaline one with soft waters and very low

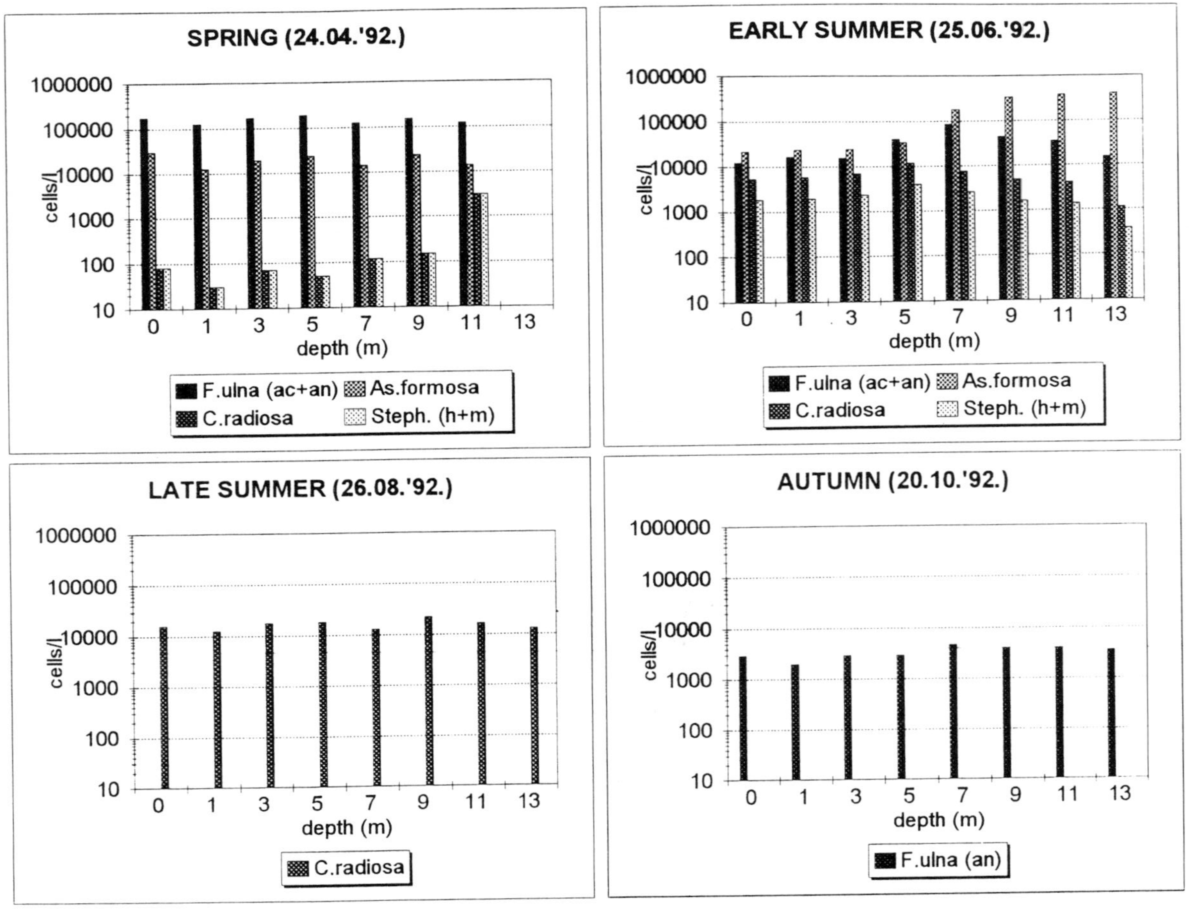


Figure 4. Seasonal and vertical distribution of the most numerous diatom populations, plotted on a $\log_{10}$ scale (AFOR – *Asterionella formosa*, CRAD – *Cyclotella radiosa*, FCRO – *Fragilaria crotonensis*, SACU – *Synedra acus*, SANG – *Synedra acus* var. *angustissima*, OTHE – all other diatoms).

mineralization. The average silicon corresponds with the maximum epilimnion values in 172 Swedish oligotrophic lakes and with many other moderate sized lakes of various nutrient state in Sweden, a number of large central Finnish lakes (Willén 1991), or Loch Leven in Scotland (Bailey-Watts 1976). The average total phosphorus indicates a mesotrophic status of the reservoir, according to the trophic classification of reservoirs (Jones & Lee 1982), but bearing in mind the available data it is hard to explain the maximal levels of total phosphorus. Peat mineralization and the large amount of inflow water from the Lisinsko Jezero reservoir through the pump-accumulation plant probably cause nutrient enrichment, due to the eutrophic status of the Lisinsko Jezero reservoir (Laušević, unpublished data). Further measurements are required. Frequent disturbances in the thermal stratification, distribution of nutrients and phytoplankton, weather conditions and the working regime of a hydroenergetic system, as pointed out by Milovanović (1973), result in a complicated system.

The chlorophyll-*a* concentrations and vertical distribution in relation to the temperature throughout the annual cycle (Fig. 2) are in agreement with the hypothesis of a metalimnion chlorophyll maximum as described by Moll & Stoermer (1982). The distribution pattern of chlorophyll-*a* down the depth gradient corresponds to the theoretical distribution for oligo-mesotrophic lakes (Moll & Stoermer 1982). The average planktonic algal chlorophyll-*a* during spring and summer season is characteristic of mesotrophic lakes (Jones & Lee 1982). Higher concentrations in the autumn (Fig. 2), are characteristic for mesotrophic-eutrophic lakes. These results show how difficult it is to give a definite classification of the trophic status of this reservoir, but the most probable one is the mesotrophic status.

The total number of the diatom taxa in the qualitative samples indicates a great floristic diversity, but many of them have been resuspended from benthic or epiphytic communities, especially in the shallow bays. Cvijan & Laušević (1991b) noticed that small parts of the peat bog have formed again in places alongside the lake, particularly in the southern part. Those places have a specific flora of macrophytes (Milovanović & Zivković 1958; Milovanović 1973; Blaženčić & Blaženčić 1991) and harbour a rich flora of diatoms. However, those with a mostly planktonic life-span are in the minority (Tab. III). The floristical composition of the dominant planktonic diatoms (Fig. 4) is very similar to those in Schlachtensee (Gervais 1991). Our findings for the most abundant taxa basically are in agreement with the published data of their ecological characteristics (Krammer & Lange-Bertalot 1991a, 1991b).

In spite of infrequent sampling, it is possible to follow up the main trends in the diatom community on both seasonal and spatial bases. Distinct spring diatom maxima (Fig. 3) were recorded with dominance of large, pennate diatoms *Synedra acus* and *Asterionella formosa* in 1993 and *Fragilaria crotonensis* in 1994 (Fig. 4). Milovanović (1973) also recorded a spring diatom peak for the period from 1954 to 1964, but she reported the dominance of the epizoic diatom *Synedra cyclopum* Brutschi, and the colonial taxa *Tabellaria* and *Fragilaria*, which differs from our findings. Diatoms, as a group, are known to be especially good competitors during turbulence or the onset of stratification in low light, low temperature and not too nutrient-deficient environment (Willén *et al.* 1992; Sommer *et al.* 1986). However, in contrast to the first statement of the model developed by the Planktonic Ecological Group (PEG – Sommer *et al.* 1986) in Vlasinsko Jezero reservoir large, pennate diatoms dominate in the spring. Sommer *et al.* (1986) restrict their dominance to lakes of moderate depth in which algae start to grow during the spring overturn, which fits for the Vlasinsko Jezero reservoir. Reynolds (1984) stressed their high photosynthetic efficiency and capacity, and high ratio of cell surface area to cell volume which enables them to take up nutrients effectively at low temperature, as the reason for their spring dominance. Furthermore, as mentioned by Gervais (1991), dominance of the genus *Synedra* is predicted by resource competition theory under the conditions of high Si:P ratio and low temperature. *Asterionella formosa* also grows well at low temperature and high Si:P ratio (Lund 1949, 1950a, 1950b; Tilman 1977), as does *Fragilaria crotonensis* (Gervais 1991). *A. formosa* had appeared in the reservoir in 1956 with maximum development in the spring and autumn season (Milovanović 1973) and since then always occurs in the open water, as noted by Lund (1949) for some British lakes. There is no evidence for the appearance of other diatoms, now dominant, in the reservoir from 1949 to 1964 (Milovanović & Živković 1956, 1958; Milovanović 1973).

The absence of dense populations of herbivorous crustacean zooplankton (Ostojić, personal communication) prevents decline of the diatom spring bloom as predicted in statement 2 of the PEG model (Sommer *et al.* 1986). As silicon is available in sufficient quantities and the water still has a certain amount of turbulence, the diatoms have extended their occurrence and proved to be efficient competitors with other algal groups, as already reported by Willén (1988) for lake Mälaren.

Replacement of the large diatoms by small centric species (*Cyclotella radiosa*, *Stephanodiscus* sp.) in summer probably cause silica-depletion (see Willén 1991) but unfortunately we do not have proof from the results of chemical analysis. During summer, large dinoflagellates (genera *Ceratium*, *Peridinium* and *Peridiniopsis*) play an important role in building up plankton community as described in statement 12 of the PEG model (Sommer *et al.* 1986). However, during late summer 1992 distinct metalimnetic peaks in cells numbers and Chlorophyll-*a* concentration (Fig. 2) was produced by *Mallomonas* sp. and large chlorococcal algae which also persist in the autumn. This cannot be explained with available data set.

The autumn plankton community is again dominated by large diatoms (*Asterionella formosa* and *Fragilaria crotonensis*), together with chlorococcal algae in 1992. It is well known that in a strongly mixed water column many species coexist (Carney *et al.* 1988). An autumnal dominance of large diatoms, corresponding to the onset of full overturn, can be understood with the help of the PEG model (statement 18 – Sommer *et al.* 1986).

With regard to the vertical distribution, formation of meta- and hypolimnion maxima of large diatoms (mainly *Synedra acus*, *S. acus* var. *angustissima* and *Fragilaria crotonensis*) were noted in early summer (Fig. 4). This might be explained by the fact that live cells sedimented to layers where there was probably a better Si supply, as suggested by Gervais (1991). Davey & Heany (1989) have shown that diatom sinking velocity depends upon the nutritional status of the algal cells.

In respect to trophic status, our findings indicate a process of gradual eutrophication, because Milovanović (1973) classified Vlasinsko Jezero reservoir as low productive and oligotrophic. Bearing in mind present plans for building up large tourists facilitates on the western shore of the reservoir that trend could accelerate in near future.

## Acknowledgements

We gratefully acknowledge improvements of the manuscript from two anonymous reviewers. The research is supported by Serbian Ministry of Science and Technology Contract No. 0321. R. Laušević is granted by UNESCO MAB Young Sci. Grant No. SC/RP204112.1 for 1992 year. We are grateful to Prof. J. Blaženčić, on her coordination of limnological investigations of Vlasinsko Jezero reservoir, to Mr M. Ristić, director of the hydro electric power plant, on obtaining logistic support for the field research and to Mr M. Vasiljević on assistance in obtaining chemical data.

## References

Bailey-Watts, A. E. (1986). The ecology of planktonic diatoms, especially Fragilaria crotonensis, associated with artificial mixing of a small Scotish loch in summer. *Diatom Research*, **1**(2), 153–168.

Blaženčić, J. & Blaženčić, Ž. (1991). Macrophytes of Vlasina lake. *Bulletin of Natural History Museum, Belgrade*, B **46**, 71–85.

Carney, H. J., Richerson, P. J., Goldman, C. R. & Richards, R. C. (1988): Seasonal phytoplankton demographic processes and experiments on interspecific competition. *Ecology*, **69**(3), 664–678.

Cvijan, M. & Laušević, R. (1991a). Floristical study of algae of Vlasinsko lake (Yugoslavia). *Bulletin of Natural History Museum, Belgrade*, B **46**, 57–69.

Cvijan, M. & Laušević, R. (1991b). Desmids of Vlasinsko lake – from peat bog to lake. *Archiv für Protistenkunde*, **139**, 21–37.

Davey, M. C. & Heaney, S. I. (1989): The control of sub-surface maxima of diatoms in a stratified lake by physical, chemical and biological factors. *Journal of Plankton Research*, **11**(6), 1185–1199.

Denis, L. & Van Straten, D. (1992). A survey of the acid water diatom assemblages of two heathland relics in the Belgian Northern Campine (Groot & Klein Schietveld, Brasschat) with an assessment of their conservational value. *Diatom Research*, **7**(1), 1–13.

Gervais, F. (1991). Which factors controlled seasonal and spatial distribution of phytoplankton species in Schlachtensee (Berlin, F.R.G.) 1987?. *Archiv für Hydrobiologie*, **121**(1), 43–65.

Hustedt, F. (1930), Bacillariophyta. In: *Süßwasserflora von Mitteleuropa* (A. Pascher, ed.), **10**, 466 pp. Fischer, Jena.

Jones, R. A. & Lee, G. F. (1982). Recent advances in assessing impact of phosporus loads on eutrophication-related water quality. *Water Research*, **16**, 503–515.

Krammer, K. & Lange-Bertalot, H. (1986). Bacillariophyceae. 1. Teil: Naviculaceae. In: *Süßwaserflora von Mitteleuropa* (H. Ettl, J. Gerloff, H. Heynig & D. Mollenhauer, eds), **2**, 876 pp. Fischer, Stuttgart.

Krammer, K. & Lange-Bertalot, H. (1988). Bacillariophyceae. 2. Teil: Bacillariaceae, Epithemiaceae, Surirellaceae. In: *Süßwasserflora von Mitteleuropa* (H. Ettl, J. Gerloff, H. Heynig & D. Mollenhauer, eds), **2**, 596 pp. Fischer, Stuttgart.
Krammer, K. & Lange-Bertalot, H. (1991a). Bacillariophyceae. 3. Teil Naviculaceae. In: *Süßwasserflora von Mitteleuropa* (H. Ettl, J. Gerloff, H. Heynig & D. Mollenhauer, eds), **2**, 576 pp. Fischer, Stuttgart.
Krammer, K. & Lange-Bertalot, H. (1991b). Bacillariophyceae. 4. Teil Achnanthaceae, Kritische Erganzungen zu Navicula (Lineolatae) und Gomphonema Gesamtliteraturvetzeichnis. In: *Süßwasserflora von Mitteleuropa* (H. Ettl, J. Gerloff, H. Heynig & D. Mollenhauer, eds), **2**, 437 pp. Fischer, Stuttgart.
Lange-Bertalot, H. (1993). *85 New Taxa and much more than 100 taxonomic clarifications supplementary to Süßwasserflora von Mitteleuropa* Vol. **2**/1–4. 454 pp. J. Cramer, Berlin, Stuttgart.
Lund, J. W. G. (1949). Studies on *Asterionella*. I. The origin and nature of the cell producing seasonal maxima. *Journal of Ecology*, **37**(2), 389–419.
Lund, J. W. G. (1950a). Studies on *Asterionella*. II. Nutrient depletion and the spring maximum. Part I. Observations on Windermere, Esthwaite Water and Blelham Tarn. *Journal of Ecology*, **38**(1), 1–15.
Lund, J. W. G. (1950b). Studies on *Asterionella*. II. Nutrient depletion and the spring maximum. Part II. Discussion. *Journal of Ecology*, **38**(1), 15–35.
Milovanović , D. (1973). Fitoplankton Vlasinskog jezera u periodu 1949–64. (Phytoplankton of the Vlasina lake during the period 1949–1964. In Serbian, with Summary in English). *Archives of Biological Sciences, Belgrade*, **25**(3–4), 177–194.
Milovanović, D. & Živković, A. (1953). Prvo saopštenje o ispitivanju planktonske produkcije u novom baražnom jezeru na Vlasini (First report on plankton production in the new lake formed by the Vlasina dam. In Serbian, with Abstract in English). *Periodicum biologorum*, II/B(7), 266–267.
Milovanović, D. & Živković, A. (1956). Limnološka ispitivanja baražnog jezera na Vlasini (Limnological studies of the artificial lake on Vlasina. In Serbian, with Summary in English). *Zbornik radova Instiuta za ekologiju i biogeografiju Beograd*, **7**(5), 1–47.
Milovanović, D. & Živković, A. (1958). Novi prilog proučavanju planktonske produkcije u baražnom jezeru na Vlasini (An additional study of plankton production in the lake Vlasina. In Serbian, with Summary in English). *Zbornik radova Biološkog Instituta NR Srbije Beograd*, **2**(7), 1–12.
Moll, R. A. & Stoermer, E. F. (1982). A hypothesis relating trophic status and subsurface chlorophyll maxima of lakes. *Archiv für Hydrobiologie*, **94**(4), 425– 440.
Patrick, R. & Reimer, C. W. (1966). *The diatoms of the United States*, **1**(13). 688 pp. Monographs of the Academy of Natural Sciences, Philadelphia.
Pantle, R. & Buck, H. (1955). Die Biologische Terwachung der Gewasser und due Darstellung der Ergebnise. *Gas und Wasserfach*, **96**(18), 604.
Reynolds, C. S. (1984). *The ecology of freshwater phytoplankton*. Cambridge University Press, Cambridge. 384 pp.
Sommer, U., Glivicz, Z. M., Lampert, W. & Duncan, A. (1986). The PEG–model of seasonal succession of planktonic events in fresh waters. *Archiv für Hydrobiologie*, **106**(4), 433–471.

Standard Methods for the Examination of Water and Wastewater. (1985). American Public Health Association, Washington, DC 20005.
Tilman, D. (1977): Resource competition between planktonic algae: an experimental and theoretical approach. *Ecology*, **58**, 338–348.
Utermöhl, H. (1958). Zur vervolkmmung der quantitativen phytoplankton methodik. *Mitt. Int. Verein. Limnol.*, **9**, 1–38.
Willén, E. (1988): Diatoms and reversed eutrophication in lake Mälaren, central Sweden, 1966–1985. In: *Proceedings of Nordic Diatomist Meeting, Stockholm*, 1987. University of Stockholm, Department of Quaternary Research (USDQR) Report 12: 103–109.
Willén, E. (1991). Planktonic diatoms – an ecological review. *Algological Studies*, **62**, 69–106.
Willén, E., Hajdu, S. & Pejler, Y. (1992): Long-term changes in the phytoplankton of large lakes in response to changes in nutrient loading. *Nordic Journal of Botany*, **12**, 575–587.

# Diatom biogeography: some preliminary considerations

D. M. Williams

*Department of Botany, The Natural History Museum, Cromwell Road, London SW7 5BD, UK*

## Abstract

Biogeography is the study of where species live and how they got there. Some diatomists believe that the biogeographic distribution of modern diatoms is almost cosmopolitan. Is this a fair assessment? Given a group of species where information on distribution is known a hypothesis relating distribution to other factors can be proposed. Relevant taxa, from the perspective of diatom distribution, occur within the freshwater araphid diatoms and an example is given from the genus *Tetracyclus*. The general questions of how these groups relate to each other and how these relationships inform about the geographical dimension are addressed.

## Introduction

Diatoms have recently been used to reconstruct pH changes from sediment cores largely due to their application in assessing the problem of lake water acidification by recent acid deposition (e.g. Smol *et al.* 1986). The relationship between diatom occurrences (measured in time by the stratigraphic record and space by their geography) and pH has been shown on a number of occasions to be strongly correlated. That is, there is a correlation between environment and time with species presence. More generally, there is a correlation between ecology, the fossil record and taxonomy. A question that has seemed pertinent, to some at least, is: What are the causes of this correlation? Answers have been sought in the study of diatom community structure. Clearly if a particular group of species frequently occurs together and is correlated with particular environmental conditions, the causes which influence the composition of that community and its structure should be studied. Thus attempts to relate ecology, stratigraphy and taxonomy through physiology become paramount. When viewed in terms of physical and chemical factors, it is assumed that any diatom distribution can be satisfactorily explained – or so this particular argument goes. For instance, Battarbee & Charles (1987) drew attention to two possible means of study relating directly to diatom distribution: (1), there is a lack of research into community structure and population dynamics of diatoms in acid lakes; and (2), the causes of change are little understood in these communities as lakes acidify. They pointed out that each of these areas are being actively studied with the expectation that they will "help to explain why some sites fail to produce expected diatom assemblages for a given pH, and why certain taxa have regionally discontinuous distributions" (Battarbee & Charles 1987, 574).

**Biogeography and the origin of taxa**

The key question in Battarbee and Charles' paper, "why certain taxa have regionally discontinuous distributions", is central to an alternative viewpoint, which represents another approach to the understanding of taxon distributions.

Physiological and ecological investigations will certainly help to clarify some aspects of the problem of species distribution. However, they omit one vital area of consideration: history and the origin of taxa. It is possible (but, perhaps, not necessary) to view the study of physiology and ecology by presupposing that every species can get to any part of the globe and its presence or absence explained by prevailing ecological parameters (see the recent comment of Lange-Bertalot in Schmid & Crawford 1993: 485). According to this view, the origin of taxa is more or less irrelevant to the problems studied. Species arise from a single centre of "creation" and travel, by some means or other, to their appropriate place(s) on the globe; appropriate in that the necessary ecological requirements prevail for it to survive. Paradoxically, the history of biogeography (Nelson 1978, 1983; Browne 1983) suggests that this view was quickly superseded by naturalists who began to investigate plants and animals collected from far away lands and the idea of a single centre of creation was quickly replaced by the notion that there may very well be several centres of creation. The issue, at that time, was whether there was indeed one or several centres of creation. Wallace and Darwin settled on a single centre of creation for each species and migration to explain species wide (and possibly disjunct) distributions. Of course, in the 1850s they were unaware that continents move as well. Recent appreciation of that fact puts biogeography in a new and different light.

I wish to consider an alternative view that sees "biological communities" in terms of the sum total of species contained within rather than a community as an integrated functioning unit. If certain communities on different parts of the globe are absolutely identical in their species composition, interpretation may not be a problem. Community in this sense will equate more or less with the idea of biogeographic regions (Browne 1983). From the perspective of diatoms, there remains an old, unanswered question: Do diatom communities (broadly construed) from different geographic regions but with more or less similar ecological conditions have different species, or more precisely different floras? Bradbury & Krebs (1982: 106 see also Bradbury *et al.* 1985: 45), for instance, argued that "[T]he biogeographic distribution of modern diatoms is very nearly cosmopolitan"; and Bonik (1982) suggested that there were only extremely local endemics "buried", as it were, among cosmopolitan, widespread species. These two sets of authors offered drastically different explanations that tied their observations into an historical viewpoint that "explained" the origin of diatom taxa. Explanations aside, there remains the underlying empirical question: Are diatom distributions "very nearly cosmopolitan"? Numerous regional diatom floras testify to a large number of species peculiar to their geographic region. Thus, the first part of any general solution to biogeographic problems is to establish the limits of distribution of taxa and express them as a pattern (or perhaps series of patterns) extending, ramifying and intertwining around the globe. Perception of this pattern, of course, is dependent upon two factors: taxa and their meaning (such that the units of distribution can be defined precisely) and a common understanding of the extent of distributions (such that the units of endemism can be defined precisely). Both topics are currently under much discussion in the modern systematic literature (e.g. Ladiges *et al.* 1991). Even with these issues to one side, it should, in principle, be possible to chart the distribution of every species, resulting in a series of maps perhaps equal to the number of taxa studied.

The question that has re-occurred consistently since the time of Linnaeus is whether there is a general explanation for all taxon distributions or whether each and every separate distribution requires separate explanation. In a sense, this preoccupation with either a single centre of creation or many centres of creation is rendered somewhat obsolete by the theory of continental plate movements. Biogeography, then, is the study of where species live and how they got there. For instance, two species can occur in the same area but may have arrived there by different routes: one may have evolved elsewhere and dispersed, that is migrated by some means, into that area; whilst the other may have evolved *in situ*. The centre or centres of creation may be irrelevant.

Can we separate out different explanations? It may be that migration evokes ecology: a taxon disperses into a particular area because the conditions were right for its survival. It may be that the alternative evokes history: a taxon occupies a particular area because that is where is originated. Clearly there will be a mix of the two. However, the general question appeals not just to ecology but to history: how did those species get there in the first place. Ecological studies alone will not tell us how species originated in any particular area. What is required is consideration of the historical dimension: that is, research into the systematics (or taxonomy, if you prefer) of relevant taxa. Systematics, in its modern form, presents taxa as summaries of its characters expressed as a branching diagram (Figure 1). It may be possible to relate this branching diagram to a similar one but for areas instead of taxa. If historical interrelationships of areas correspond with the historical interrelationships of taxa there may be a common cause explanation. With this view in mind, ecology can determine whether a particular species can live in an area, while history can determine whether it does or could have.

Most studies in diatom distribution assume that many, if not all, species are cosmopolitan and their occurrence is determined solely by possible migration to the relevant ecological conditions; endemics are usually limited to a few rather infrequent occurrences or special places (such as the Rift Valley Lakes in East Africa, Ross 1983; Kociolek & Stoermer 1991). Systematics can contribute to teasing out historical from

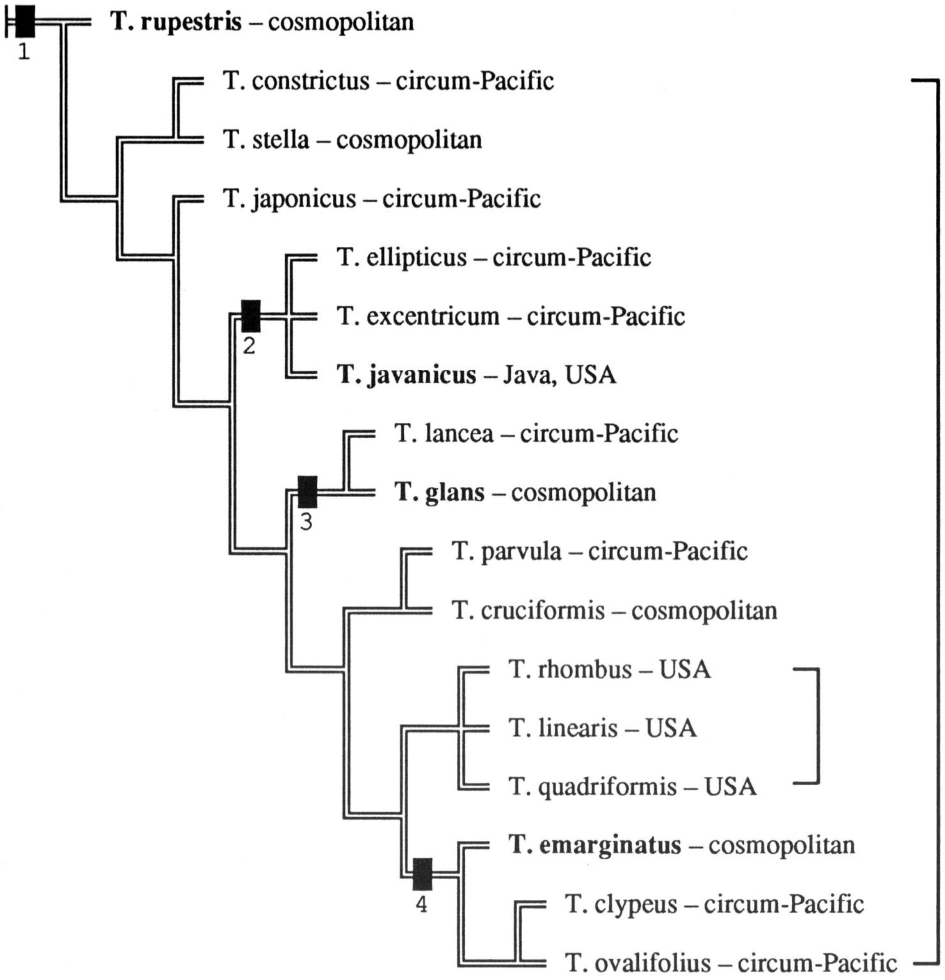


Figure 1: Results from cladistic analysis of morphological characters using three-item statement matrices with fractional weighting (Williams, submitted). Fractional weighting yielded 9 trees (length 2955, CI 80, RI 76) of which the strict consensus tree length 2955 = minimal tree (13 nodes) and thus the preferred tree for this particular matrix. Other analyses differed in their placement of certain taxa (details in Williams, submitted). Extant species in bold. Nodes to which extant taxa "emerge" are labelled 1–4 as small boxes along branches.

ecological explanations and tackle distributional history as a scientific problem relevant to all of life on this earth.

### *Tetracyclus*: an example

Consider the freshwater genus *Tetracyclus*, its taxonomy and distribution. In several recent revisionary studies (Li 1982, 1984; Li & Qi 1984; Williams 1987, 1989, 1990, submitted; Williams & Li 1990) a total of 33 species are now recognised of which only 5 are extant, the remaining 28 known only from fossil deposits (Table 1).

Cladistic results among the species of *Tetracyclus* are ambiguous, yielding between 9 and 113 trees, depending upon approach (Figure 1, details of analyses in Williams, submitted). However, the distribution of species largely centre around the Pacific rim. This poses some wide ranging questions concerning the relationship these taxa may have with Pacific rim geology. For instance, progress has been made in determining the historical relationships among the various terranes that comprise the western United States (Minkley *et al.* 1985; Hendrickson 1986), but is there a correspondence between that history and the one derived from species of *Tetracyclus*? Initial estimates seem to suggest not – however, poor resolution of the interrelationships of *Tetracyclus* may render serious judgment defective. More taxonomic work is required.

Suppose, however, we consider a change of focus, a pre-Jurassic origin for *Tetracyclus* could be postulated on the basis of the proximity of the Pacific terranes at some time in the past. The most primitive species (basal in the cladogram in Figure 1) of *Tetracyclus* is *T. rupestris*, a morphologically simple as well as widespread (cosmopolitan for the Old World, Williams 1987, submitted) species. Yet most species of *Tetracyclus* are extinct and limited to the Pacific rim implying that their extinction (rather than their origin) was bought about by volcanic and tectonic activity in those regions, especially in what is now the west coast of North America, and the east coasts of Japan and China (see Hendrickson 1986 for detailed map). The distribution of species of *Tetracyclus* may have been established at a much earlier time than is usually considered for diatoms as a whole.

It is possible to suggest that the Pacific fragments, when viewed in the past, are a "centre-of-origin" (Nelson & Platnick 1981, 1984) with extant species (*emarginatus*, *rupestris*, *glans*, see Figure 1, nodes 1–4) all now widespread, having migrated over time into the further reaches of the globe. "Centres-of-origin" are tricky, if not impossible, to pin down (Nelson & Platnick 1981, 1984). Explanation of any kind will only have purchase if more taxa with similar distributions show similar geographical displacement and all Pacific rim terranes can be related in one way. In any case, the distributional pattern exemplified by *Tetracyclus* has been recognised for nearly 150 years (Ehrenberg 1849, 1850) yet remains unexplained.

### A way forward

The distribution of *Tetracyclus* and its understanding in a biogeographical context may seem to require consideration of stratigraphy, geology, geography and ecology. It is

Table 1: Species of *Tetracyclus* described in Williams (submitted). There is a total of 33 species of which 5 are extant the remaining 28 known from fossils alone. Available data are appended as either valve and/or girdle. In some instances, especially the fossils, data are only partially known (=p). Species marked as [1] were considered in the character analysis. Details of morphology and analyses are contained in Williams (submitted).

| Species | Valve | Girdle |
|---|---|---|
| *T. glans* [1] | + | + |
| *T. javanicus* [1] | + | +(p) |
| *T. chilensis* | + | +(p) |
| *T. rupestris* [1] | + | + |
| *T. emarginatus* [1] | + | + |
| *T. cruciformis* [1] | + | –(p) |
| *T. inflata* | + | – |
| *T. liensis* | + | – |
| *T. pagesi* | +(p) | – |
| *T. parvula* [1] | +(p) | +(p) |
| *T. quadriformis* [1] | + | –(p) |
| *T. stella* [1] | + | +(p) |
| *T. constrictus* [1] | + | + |
| *T. peragalli* | + | +(p) |
| *T. clypeus* [1] | + | + |
| *T. divisium* | + | – |
| *T. excentricum* [1] | + | + |
| *T. subclypeus* | + | – |
| *T. subdivisium* | + | – |
| *T. boryanus* | + | – |
| *T. ellipticus* [1] | + | +(p) |
| *T. lancea* [1] | + | + |
| *T. lata* | + | – |
| *T. linearis* [1] | + | + |
| *T. maxima* | +(p) | – |
| *T. ovale* | +(p) | – |
| *T. ovalifolius* [1] | + | + |
| *T. rhombus* [1] | + | – |
| *T. shangduensis* | +(p) | +(p) |
| *T. castellum* | + | +(p) |
| *T. japonicus* [1] | + | + |
| *T. pseudocastellum* | + | +(p) |

possible to allow one or more of these factors to assume overall importance: for some it is stratigraphy, for others it is ecology. However, the perspective I have given here suggests that taxonomy is at the core of such investigations: how are the taxa interrelated among themselves? In this light, then, consider some of the following more general questions that emerge from this preliminary account, all, as yet, unanswerable:

(1) Will studies of diatom taxonomy contribute to a general view of biogeography and the history of life on earth?

The prospects look promising, given appropriate data (taxa) from appropriate regions and sufficient resolving power of those data to be meaningful. That is, a dependency on our ability to recover species interrelationships accurately and robustly enough. Little of this work is being done at present (but see the accounts of Kociolek & Stoermer 1991).

(2) Will studies of diatom taxonomy lead to contributions on the early origin of the group *in the absence of appropriate fossil taxa*?

This study seems to suggest that such may also be the case. These considerations have been expressed on several occasions as several authors have already suggested a Precambrian origin for diatoms (e.g. Round & Crawford 1981). This depends on a particular view of the fossil record (not the fossils). Specialised arguments have been used to suggest that the fossil record of diatoms, as known today, may reflect an accurate if not true situation (Sorhannus 1994); and even further, those involved in molecular studies have searched for some calibration factor for any molecular clock-like changes that may be evident (Medlin *et al.* 1993): yet each depends upon the fossil record as recovered. Whatever the outcome of these studies, diatom taxonomy coupled with a (palaeo)biogeographic aspect may allow a different, and perhaps independent, assessment, of diatom origin across all taxa.

In summary, my expectation is that work in progress will identify many more taxa pertinent to the problem of diatom distribution and as that happens we will be able to say something meaningful about diatom distributions and explain, in the words of Battarbee & Charles (1987: 574), " ... why certain taxa have regionally discontinuous distributions." These preliminary results suggest that the historical aspect will provide the necessary dimension sought after but perhaps unfairly ignored by ecologists seeking explanation.

**Acknowledgements**

My grateful thanks to Drs Sandy Knapp, Chris Humphries, Elliott Schubert, Dick Crawford, David Mann, Frank Round and Eileen Cox for critically reading the manuscript; Dr Pat Kociolek and CAS for providing facilities at their institute to study their magnificent freshwater fossil diatom collections; Dr Kociolek also provided much needed inspiration for addressing biogeographical problems as well as drawing my attention to the Ehrenberg (1849, 1850) references.

## References


Battarbee, R. W. & Charles, D. F. (1987). The use of diatom assemblages in lake sediments as a means of assessing the timing, trends and causes of lake acidification. *Progress in Physical Geography*, **11**, 552–580.

Bonik, K. (1982). Gibt es Arten bei Diatomeen? Eine evolutionasbiologische Deutung am Beispiel der Gattung *Nitzschia*. *Senckenbergiana Biologica*, **62**, 413–434.

Bradbury, J. P. & Krebs, W. N. (1982). Neogene and quaternary lacustrine diatoms of the western Snake river basin, Idaho–Oregon, USA. *Acta Geologica. Academiae Scientiarum Hungaricae*, **25**, 97–102.

Bradbury, J. P., Dieterich, K. V. & Williams, J. L. (1985). Diatom flora of the Miocene lake beds near Clarkia in northern Idaho. In: *Late Cenozoic History of the Pacific Northwest* (J. C. Smiley, ed.), 33–59. American Association for the Advancement of Science, San Francisco.

Browne, J. (1983) *The Secular Ark. Studies in the History of Biogeography*. Yale University Press, New Haven & London. 273 pp.

Ehrenberg, G. C. (1849). Über das mächtigste bis jetzt bekannt gewordene (angeblich 500 Fufs mächtige) Lager von mikroscopischen reinen kieselalgen Süsswasser-Formen am Wassfall-Flusse im Oregon. *Bericht über die zur Bekanntmachung geeigneten Verhandlungen der Königlichen Pruessischen Akademie der Wissenschaften zu Berlin*, **1849**, 76–87.

Ehrenberg, G. C. (1850). On Infusorial Deposits on the River Chutes in Oregon. *American Journal of Science*, 2nd ser. **9**, 140.

Hendrickson, D. A. (1986). Congruence of Bolitoglossine biogeography and phylogeny with geologic history: Paleotransport on displaced suspect terranes. *Cladistics*, **2**, 113–129.

Kociolek, J. P. & Stoermer, E. F. (1991). New and interesting *Gomphonema* species from East Africa. *Proceedings of the California Academy of Sciences*, **47**, 275–288.

Ladiges, P. Y., Humphries, C. J. & Martinelli, L. W. (eds) (1991). *Austral Biogeography*. Special issue of *Australian Systematic Botany*, **4**, 227 pp.

Li, Jia Ying (1982). Genus *Tetracyclus* and its stratigraphic significance. *Bulletin of the Institute of Geology. Chinese Academy of Geological Science*, **5**, 149–166.

Li, Jia Ying (1984). Some new species and varieties of the genus *Tetracyclus* Ralfs (Bacillariophyta). *Acta Phytotaxonomia Sinica*, **22**, 231–236.

Li, Jia Ying & Qi Yuzao (1984). Neogene diatom assemblages in China. In: *Proceedings of the 8th International Diatom Symposium* (M. Ricard, ed.), 699–711. O. Koeltz, Koenigstein.

Medlin, L. K., Williams, D. M. & Sims, P. A. (1993). The evolution of the diatoms (Bacillariophyta). I. Origin of the group and assessment of the monophyly of its major divisions. *European Journal of Phycology*, **28**, 261–275.

Minkley, W. L., Hendrickson, D. A. & Bond, C. E. (1985). Geography of Western North American freshwater fishes: Description and relationships to intracontinental tectonism. In: *Zoogeography of freshwater fishes of North America* (C. Hocutt & E. O. Wiley, eds), 519–613. Wiley Interscience, New York.

Nelson, G. J. (1978). From Candolle to Croizat: Comments on the history of biogeography. *Journal of the History of Biology*, **11**, 269–305.

Nelson, G. J. (1983). Vicariance and cladistics: Historical perspectives with implications for the future In: *Evolution, Time and Space: The Emergence of the Biosphere* (R. W. Sims, J. H. Price & P. E. S. Whalley, eds), 469–492. Academic Press, London.

Nelson, G. J. & Platnick, N. I. (1981). *Systematics and Biogeography: Cladistics & Vicariance.* Columbia University Press, New York.

Nelson, G. J. & Platnick, N. I. (1984). *Biogeography.* Carolina Biology Reader, no. **119**.

Ross, R. (1983). Endemism and cosmopolitanism in the diatom flora of the East African Great Lakes. In: *Evolution, Time and Space: The Emergence of the Biosphere* (R. W. Sims, J. H Price & P. E. S. Whalley, eds), 157–177. Academic Press, London.

Round, F. E. & Crawford, R. M. (1981). The lines of evolution of the Bacillariophyta. I. Origin. *Proceedings of the Royal Society of London* **B 211**, 237–260.

Schmid, A.-M. & Crawford, R. M. (1993). Recent events: 7. Deutschsprachiges Diatomologentreffen, Innsbruck, Austria. 26–28 March 1993. *Diatom Research,* **8**, 485–487.

Sorhannus, U. (1994). Relative-rate tests versus paleontological divergence data for diatoms and vertebrates. *Acta Palaeontologica Polonica,* **38**, 199–214.

Smol, J. P., Battarbee, R. W., Davis, R. B. & Merilainen, J. (eds) (1986). *Diatoms and lake acidity. Reconstructing the pH from siliceous algal remains in lake sediments. Developments in Hydrobiology*, **29**.

Williams, D. M. (1987). Observations on the genus *Tetracyclus* Ralfs (Bacillariophyta) I. Valve and girdle structure of the extant species. *British Phycological Journal*, **22**, 383–399.

Williams, D. M. (1989). Observations on the genus *Tetracyclus* Ralfs (Bacillariophyta) II. Morphology and taxonomy of some species from the genus *Stylobiblium. British Phycological Journal*, **24**, 317–327.

Williams, D. M. (1990). Examination of auxospore valves in *Tetracyclus* from fossil specimens and the establishment of their identity. *Diatom Research*, **5**, 189–194.

Williams, D. M. (submitted). Revision of the *Tetracyclus* ellipticus (Bacillariophyceae) complex with notes on their interrelationships and palaeobiogeography. *Bulletin of the British Museum (Natural History), Botany.*

Williams, D. M. & Li, Jia Ying (1990). Observations on the genus *Tetracyclus* Ralfs (Bacillariophyta) III. Description of two new species from Chinese fossil deposits. *British Phycological Journal*, **25**, 335–338.

# Distribution of attached diatoms in inorganic acid lakes in Japan

Sakiko Yoshitake* and Hiroshi Fukushima**

* *Shonan Junior College, Yokosuka, Kanagawa, Japan*
** *Institute of Phycology, Kunitachi, Tokyo, Japan*

## Abstract

Samples of attached algae were collected from 7 inorganic strongly acid waters. In total 75 samples were examined.

The pH at the sampling stations ranged between 2.2 and 6.2. There was considerable variation in the cell numbers recorded in the 7 waters with values ranging from 6 × 10 (Lake Osoresanko) to 17.5 × $10^3$ cells/$mm^2$ (Pond Shunuma). The dominant species also varied. The most dominant taxa in terms of relative frequency were *Eunotia exigua* (19.7%), *Aulacoseira distans* (17.5%), *Pinnularia braunii* (9.3%) and *Anomoeoneis brachysira* (8.3%). In order of increasing pH, they were *Pinnularia braunii*, *Nitzschia capitellata*, *Eunotia septentrionalis*, *Anomoeoneis brachysira*, *Eunotia exigua* and *Aulacoseira distans*. Seasonal changes in the dominant species of attached algae were investigated in Lake Ohnumaike and Lake Mishakaike. However, clear seasonal changes could not be recognised.

Species diversity was generally low with a range from 0.45 (Pond Katanuma) to 2.50 (Lake Akanuma).

Correlation of species diversity with pH was calculated for each sample. The correlation coefficient (r) was 0.299 indicating a very small positive correlation.

## Introduction

There are many volcanic acid lakes distributed throughout Japan. Six of them (inorganic acid lakes) were investigated in order to survey the characteristics of algal composition at various pH levels. In addition to these lakes, a small pond, Shunuma, was studied, in which heavy metals, especially copper from disused mine effluent contribute to a low pH. Epilithic, epipelic and epiphytic algal communities were studied from both taxonomic and ecological viewpoints in each lake and pond.

## Materials and Methods

Samples of attached algae were taken in six acidic lakes and one pond in Japan (Fig. 1). Some environmental factors are shown in Table 1. 21 samples were taken in Lake Mishakaike, 12 in Lake Ohnumaike, 10 in Lake Osoresanko and Lake Tazawako, 9 in Lake Katanuma, 7 in Lake Akanuma and 6 in Pond Shunuma. Fig. 2 shows the distribution of the number of samples in each range of the pH values. Attached algae were taken from the surface of stones, rocks, bottom mud or sand, and sometimes from the surface of fallen leaves or aquatic plants.

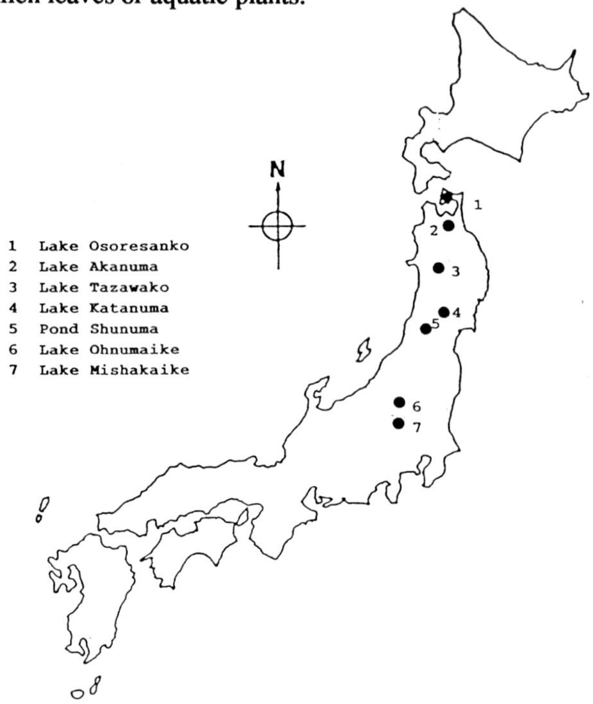


Fig. 1. Sampling location of inorganic acid lakes in Japan.

For the quantitative analysis, 25 $cm^2$ or 100 $cm^2$ of attached algae were taken in addition to the samples for qualitative analysis. Material was fixed with formalin solution. The qualitative samples were cleaned with sulphuric acid and hydrogen peroxide and mounted in pleurax. Identification of diatoms was performed using light micro-photographs at 2000x magnification. For the quantitative analysis, counts were limited to the cells which contained a certain amount of protoplasm, and standing crops were expressed as cell number per $mm^2$ of surface of the substrata. Counts were also examined for the qualitative samples to calculate the diversity index and relative frequencies of each algal taxon. With regard to the determination of dominant species, 90% confidence intervals were calculated for each taxon, and the taxa which represented a larger range than the mean relative frequency (100/total number of taxa at each sampling station), were considered to be the dominant species.

Species diversity was calculated using the Shannon-Weaver Diversity Index (Shannon & Weaver 1949).

Table 1. pH values and water temperature in each lake and pond

| Location | sampling date | no. of sampling station | pH values | water temperature (°C) |
|---|---|---|---|---|
| L. Osoresanko | 1993 Jun. | 5 | 3.0–6.2 | 10.9–22.5 |
| L. Akanuma | 1993 Jun. | 2 | 4.5–4.6 | 16.5–17.3 |
| L. Tazawako | 1994 Jun. | 4 | 5.2–5.3 | 13.0–16.0 |
| L. Katanuma | 1994 May | 6 | 2.2–2.5 | 16.5–21.0 |
| P. Shunuma | 1994 May | 6 | 2.4–3.0 | 12.0–25.0 |
| L. Ohnumaike | 1970 Jun. | 2 | 4.3 | 10.8–12.0 |
| | 1970 Sep. | 2 | 4.1–4.2 | 19.5–20.0 |
| | 1970 Oct. | 2 | 4.3–5.1 | 10.3–11.0 |
| L. Mishakaike | 1993 Apr. | 5 | 3.8 | 6.7–9.3 |
| | 1993 Sep. | 5 | 3.8 | 15.5–16.4 |
| | 1993 Oct. | 4 | 4.3–6.2 | 18.0–21.0 |

## Observations

### *Cell numbers*

For the quantitative analysis, the cell numbers of attached algae growing on stone surfaces were counted. But occasionally (when stones were absent at a site) surface mud, fallen trees or leaves (e.g. Pond Shunuma) were investigated.

Table 2 shows the mean cell number per $mm^2$ in each lake and pond. There was considerable variation in the cell number recorded in the 7 waters with values ranging from $6 \times 10$ to $17.5 \times 10^3$ cells/$mm^2$. Very low cell counts (below 100) were observed in Lake Osoresanko and Lake Akanuma. The largest cell number was observed in Pond Shunuma where the pH was very low (pH 2.6). Though little work has so far been carried out on the use of cell numbers in estimating standing crops in inorganic acid lakes, some studies reported the standing crops (cell numbers) of attached algae in inorganic acid rivers in Japan (Fukushima *et al.* 1951; Fukushima 1957; Fukushima *et al.* 1967). In these reports, the standing crops of attached algae were low between a pH range 3–4 in which the deposition of iron oxide is induced in the rivers of Japan. Moreover, under these circumstances the standing crops in lentic are generally higher than those in lotic situations. In the pH range from approximately 4.0–6.0 in which aluminium deposits, attached algae were poorly developed in a river (Fukushima & Ko-Bayashi 1968). Low standing crops in Lake Mishakaike were probably caused by

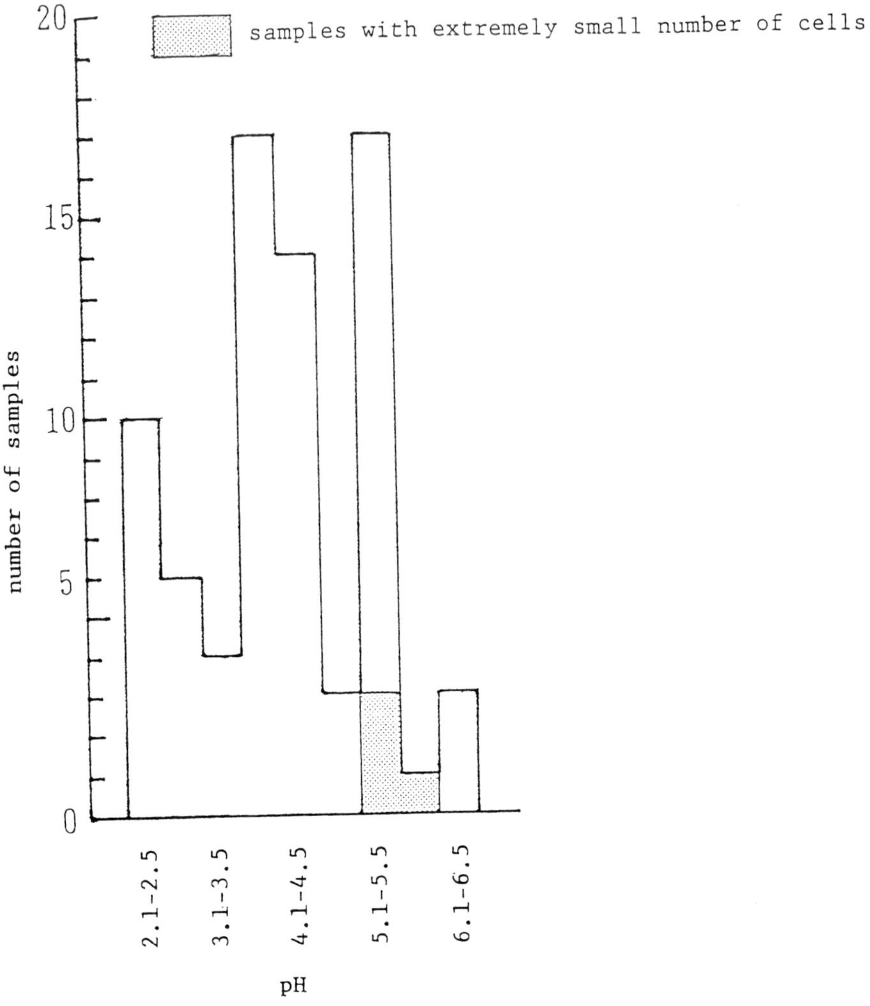


Fig. 2. Number of samples in each range of pH. The following is the detail of samples belonging to each range of pH. pH 2.1-2.5: 9 samples in L. Katanuma and 1 sample in P. Shunuma; pH 2.6-3.0: 5 samples in P. Shunuma and 1 sample in L. Osoresanko. pH 3.1-3.5: 4 samples in L. Osoresanko. pH 3.6-4.0: 17 samples in L. Mishakaike. pH 4.1-4.5: 11 samples in L. Ohnumaike, 3 samples in L. Akanuma and 2 samples in L. Mishakaike. pH 4.6-5.0: 3 samples in L. Akanuma. pH 5.1-5.5: 10 samples in L. Tazawako, 2 samples in L. Osoresanko and 1 sample in L. Akanuma, L. Mishakaike and L. Ohnumaike. pH 6.1-6.5: 3 samples in L. Osoresanko and 1 sample in L. Mishakaike.

▒: samples with extremely small number of cells.

the deposition of iron oxide; a thin layer of iron oxide covered the lake bottom when the lake dried up. On the other hand, the standing crops by cell number were calculated in two pH-neutral lakes, Lake Yunoko (Fukushima & Migita 1969) and Lake Ashinoko (Fukushima *et al.* 1977a). Comparison of standing crops between inorganic acid lakes and pH-neutral lakes were attempted to find some characteristics of standing crops in acidic waters. In a study of Lake Yunoko, the mean value was $18.3 \times 10^3$ (cell/mm$^2$) in March and $8.3 \times 10^3$ in July 1967, while in Lake Ashinoko, it was $1 \times 10^3$ in August 1974 and $2.7 \times 10^3$ in September 1975. Though algal community composition in general becomes simpler (i.e. species diversity declines) with decreasing pH, it is clear that cell numbers of attached algae varied regardless of pH when the data for these pH-neutral lakes were compared with our data (Fig. 3 and Table 2).

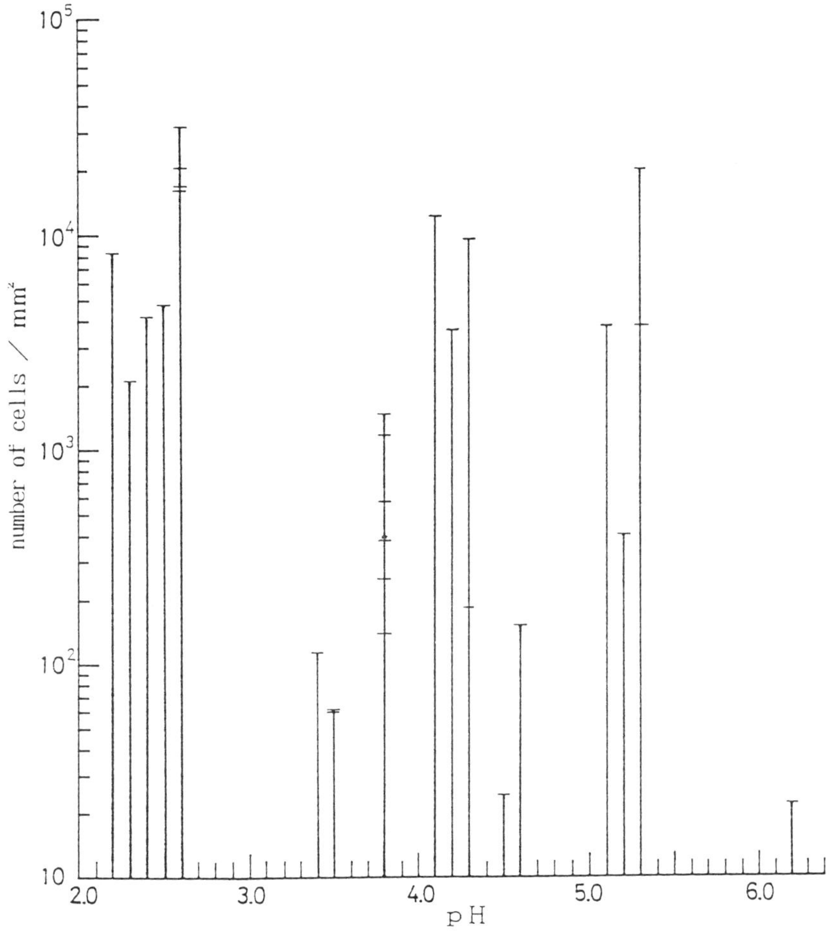


Fig. 3. Interrelation between standing crops (cell number/mm$^2$) and pH.

Table 2. Mean values of standing crop (cells/$m^2$) and species diversity index of attached algae in each lake and pond.

| | Standing crop (cells/$mm^2$) | | | Diversity Index | | |
|---|---|---|---|---|---|---|
| Sampling location | pH range | no. of quantitative samples | mean cell number | pH range | no. of samples | mean D.I. (bit) |
| L. Osoresanko | 3.4–6.2 | 4 | 60 | 3.0–6.2 | 10 | 1.76 |
| L. Akanuma | 4.5–4.6 | 2 | 90 | 4.5–4.6 | 7 | 2.50 |
| L. Tazawako | 5.2–5.3 | 3 | 8020 | 5.2–5.3 | 10 | 1.09 |
| L. Katanuma | 2.2–2.5 | 5 | 4680 | 2.2–2.5 | 9 | 0.45 |
| P. Shunuma | 2.6 | 4 | 17470 | 2.4–3.0 | 6 | 1.35 |
| L. Ohnumaike | 4.1–5.1 | 5 | 5865 | 4.1–5.1 | 12 | 1.83 |
| L.Mishakaike | 3.8 | 6 | 650 | 3.8–6.2 | 21 | 0.70 |

*Dominant species*

Of the 75 samples, four had extremely small cell number and thus it was not possible to determine species. The number of samples in which the dominant species consisted of only one taxon was 43, while in 26 samples they consisted of two taxa, and in 2 samples of three taxa. Of the 30 taxa which appeared as dominant species, 12 were diatoms. The order of relative frequency is *Eunotia exigua* (19.7%), *Aulacoseira distans* (17.5%), *Pinnularia braunii* (9.3%), *Anomoeoneis brachysira* (8.3%), *Chlamydomonas* sp. (7.1%) and *Hormidium subtile* var. *planktonica* (3.0%). Fig. 4 shows the interrelation between pH values and relative frequencies of dominant species with 90% confidence intervals. It is clear that *Pinnularia braunii* frequently appeared in the most strongly acidic waters in the pH range from 2.2 to 3.0. Though *Nitzschia capitellata* was second to *P. braunii* in its tolerance of acidic water, it appeared in only one sampling station with a pH of 3.0. Therefore more data is needed to determine the relation between its distribution and pH. *Eunotia septentrionalis* was distributed in a narrow pH range of 3.0–3.7, while *Anomoeoneis brachysira* appeared as a dominant species in a pH range of 3.3–4.3. In contrast with *E. septentrionalis, E. exigua* was distributed widely in a pH range from 3.4 to 6.2, but it was detected most frequently at a pH of around 3.8. Though the range of pH in *Aulacoseira distans* (pH 3.8–5.1) is similar to that in *E. exigua,* it was most often found in a pH range of 3.8–4.5. In general, many species appearing as dominant ones in this study prevailed in a pH range of 3.8–4.3.

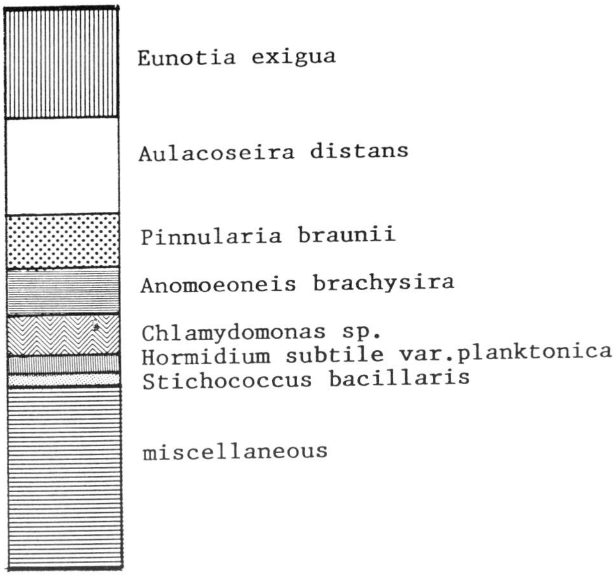


Fig. 4. Interrelation between pH values and relative frequencies of dominant species expressed by 90% confidence intervals.

Table 3. Seasonal change of dominant species of attached algae in Lake Ohnumaike.

| Sampling station | Substrata | Spring 2 VI 1970 | | | Summer 6 IX 1970 | | | Autumn 18 X 1970 | | |
|---|---|---|---|---|---|---|---|---|---|---|
| | | Water temp. (°C) | pH | Dominant species | Water temp. (°C) | pH | Dominant species | Water temp (°C) | pH | Dominant species |
| st. 1 | mud | 12.0 | 4.3 | *Ulothrix* sp. – *Anomoeoneis brachysira* | 19.5 | 4.1 | *Anomoeoneis brachysira* | 11.0 | 4.3 | *Anomoeoneis brachysira* |
| st. 2 | mud | 10.8 | 4.3 | | 20.0 | 4.2 | *Aulacoseira distans* | 10.3 | 5.1 | *A. brachysira* – *A. distans* |
| | stone | | | *Pinnularia gibba* – *Anom. brachysira* | | | *A. distans* – *Surirella amphioxys* | | | |
| | fallen tree | | | *Chlamydomonas* sp. – *Anom. brachysira* | | | *Spirogyra* sp. | | | |

Table 4. Seasonal change of dominant species of attached algae in Lake Mishakaike.

| Sampling station | Substrata | Spring 2 VI 1993 | | | Summer 13 IX 1993 | | |
|---|---|---|---|---|---|---|---|
| | | Water temp. (°C) | pH | Dominant species | Water temp. (°C) | pH | Dominant species |
| st. 1 | stone | 6.8 | 3.8 | *Eunotia exigua* | 15.5 | 3.8 | *Aulacoseira distans* |
| st. 2 | grass | 6.7 | 3.8 | *E. exigua* | 16.4 | 3.8 | *E. exigua* |
| | moss | | | *E. exigua* | | | *E. exigua* |
| st. 3 | stone | 9.3 | 3.8 | *E. exigua* | 15.7 | 3.8 | *A. distans – E. exigua* |
| st. 4 | stone | 7.7 | 3.8 | *E. exigua* | 16.0 | 3.8 | *A. distans – E. exigua* |
| st. 5 | grass | 8.9 | 3.8 | *E. exigua* | 16.2 | 3.8 | *E. exigua – A. distans* |

The standing crop in these acidic waters in Japan is as high as that of other neutral waters in Japan if estimated by cell number. Though the phytoplankton study in the LaCloche (Ontario) lakes with a pH range of 4.2–7.0 by Stokes & Yung (1986) indicated that from high to low pH there was a general trend of decreasing total biomass measured by cell volume, our data is in agreement with the report on L. Katanuma by Watanabe & Asai (1993) who stated that the standing crop of this lake was not so low, despite the very low pH.

There are considerable differences in the dominant species observed in the 7 waters. In L. Osoresanko, *Eunotia exigua*, *E. septentrionalis* and *Nitzschia capitellata*; in L. Tazawako, *Stichococcus bacillaris* and *Ulothrix* sp.; in L. Katanuma, *Pinnularia braunii*; in Pond Shunuma, *Chlamydomonas* sp., *Synechococcus* sp., *Pinnularia braunii* and *Euglena* sp.; in L. Ohnumaike, *Aulacoseira distans*, *Anomoeoneis brachysira* and *Chlamydomonas* sp. and in L. Mishakaike, *Eunotia exigua* and *Aulacoseira distans*. Counts were too low to give a reliable estimate for L. Akanuma.

An estimate of the prevailing dominant species throughout all samples is problematic because of variable sample numbers between lakes. Algal taxa ranked in order of frequency, from high to low, are *Eunotia exigua* (19.7%), *Aulacoseira distans* (17.5%), *Pinnularia braunii* (9.3%) and *Anomoeoneis brachysira* (8.3%). Negoro (1944) recorded 6 taxa of *Pinnularia*, 4 taxa of *Eunotia* and 2 taxa of *Nitzschia* in his comprehensive studies of inorganic strongly acid waters in Japan. Furthermore, he observed (Negoro 1985) that *Pinnularia braunii*, *P. acoricola* and *Eunotia septentrionalis* were not only widely distributed but also abundant in inorganic acid waters. Watanabe & Asai (1993) investigated waters with a pH below 4.0 and pointed out that *Eunotia* and *Pinnularia* were typical genera in such waters. This, to a certain extent, disagrees with our findings. On the other hand, Hustedt (1938, 1939) noted that the diatoms appearing at pH levels below 5 were one taxon of *Anomoeoneis*, *Caloneis*, *Cymbella*, *Desmogonium*, *Melosira* and *Synedra*, 2 taxa of *Frustulia*, 5 taxa of *Navicula*, 7 taxa of *Pinnularia* and 9 taxa of *Eunotia*. In his data, *Anomoeoneis* and *Melosira* (*Aulacoseira*), which have only rarely been observed in acidic waters of Japan, were found to be widely distributed.

No distinct seasonal changes in the dominant species was observed in the two lakes studied. It is considered that in strongly acidic waters, clear seasonal change is hardly recognised, which coincides with the observation of Watanabe & Asai (1993) who suggested that the influence of low pH acted as the strongest limiting factor.

Correlation coefficients between pH and diversity index were calculated ($n = 71$, $r = 0.299$) and the results indicate no significant correlation between them. According to the phytoplankton study by Stokes & Yung (1986), species richness and total biomass decreased as pH fell. This observation is different from our findings. It is unclear whether this discrepancy is due to the differences of ecological behaviour between phytoplankton and attached algae. Further study is needed to clarify this point.

## References


Fukushima, H. (1957). Quantitative analyses of algae at water course of the River Chikuma. Report on the changes of water quality accompanied by a development of sulphur mine in Yatsugatake, 2, 53-63.

Fukushima, H. & Ko-Bayashi, T. (1968). Diatom flora in the River Matsukawa (inorganic acid river), Nagano prefecture. *Bulletin, Yokohama City University Society of Natural History*, **19**(1), 1-8.

Fukushima, H. & Migita, S. (1969). Preliminary report on epilithic algae of Lake Yunoko, Central Japan. *Bulletin, Yokohama City University Society of Natural History*, **20**(2), 1–30.

Fukushima, H., Ko-Bayashi, T. & Fukushima, S. (1977a). Attached algae in Lake Ashinoko. *Algal Flora and Water Pollution*, **6**, 129–133, 214–221.

Fukushima, H., Ko-Bayashi, T. & Fukushima, S. (1977b). Attached algae in Lake Sagamiko and Lake Tsukuiko. *Algal Flora and Water Pollution*, **6**, 65–66, 100–101, 164–165,189–191.

Fukushima, H., Ohno, M. & Yokoyama, N. (1967). Diatom flora and bottom fauna at water course of the River Kitakami. *Journal, Yokohama City University, Series C-54*, **173**, 1-49.

Fukushima, H., Ishii, S., Furuya, N. & Morimoto, Y. (1951). Algal flora at water course of the River Tone in Gunma prefecture. *Seitaigakushi*, **1**(2), 83-87.

Hustedt, F. (1938–1939). Systematische und ökologische Untersuchungen über die Diatomeen-flora von Java, Bali und Sumatra nach dem Material der Deutschen limnologischen Sunda-Expedition. *Archives of Hydrobiology, Supplement*, **15**, 131–506, 639–790, Taf. 9–43, **16**, 1–394.

Negoro, K. (1944). Untersuchungen über die Vegetation der Mineralogen-Azidotrophen Gewässer Japans. *Science Reports of the Tokyo Bunrica Daigaku, section B*. No. **101**, 6, 231–374.

Negoro, K. (1985). Diatom flora of the mineralogenous acidotrophic inland waters of Japan. *Diatom*, **1**, 1–8.

Shannon, C. E. & Weaver, W. (1949). *The Mathematical Theory of Communication*. Urbana, University of Illinois Press, 117 pp.

Stokes, P. M. & Yung, Y.-K. (1986). Phytoplankton in selected LaCloche (Ontario) lakes, pH 4.2–7.0, with special reference to algae as indicators of chemical characteristics. In: *Diatoms and Lake Acidity* (J. P. Smol, R. W. Battarbee, R. B. Davis & J. Meriläinen, eds), 57–72. Dr W. Junk Publishers, Dordrecht.

Vymazal, J. (1984). Short-term uptake of heavy metals by periphyton algae. *Hydrobiology*, **119**, 171–179.

Watanabe, T. & Asai, K. (1993). Common organisms in strongly acidic running water (pH 1.6) and standing water (pH 1.9). In: *Abstract of the 14th Symposium of the Japanese Society of Diatomology*.

# Morphological variation in widely distributed diatom taxa: taxonomic and ecological implications

Eileen J. Cox

*Department of Botany, The Natural History Museum, Cromwell Road, London, SW7 5BD*

## Abstract

Morphological variation has been recorded within a number of diatom species which grow in a range of habitats, under differing environmental conditions. Where close correlation between morphology and ecological range have been demonstrated, particular morphs can be considered indicative of those conditions, whatever the underlying cause of the morphological differences. Other work has shown that morphologically distinct, but sympatric, populations of infraspecific taxa can be reproductively isolated.

Representatives of the widespread diatom, *Navicula gregaria*, were examined from both freshwater and brackish habitats. Consistent differences between freshwater and brackish populations were observed. Experimental studies of clonal material of *N. gregaria* and other *Navicula* spp. revealed that taxa collected from freshwater and marine sites were able to grow over a wide range of salinities but showed some phenotypic plasticity with salinity regime, reminiscent of that observed in field material.

Difficulties in quantifying perceived differences are discussed in relation to current forms of diatom description and diagnosis. The inclusion of additional characteristics which could improve descriptions and facilitate accurate identification, particularly of naviculoid taxa, are advocated. The results are also discussed in relation to ecomorph and ecotype concepts, and to the use of diatoms as environmental indicators.

## Introduction

A number of diatom taxa, e.g. *Caloneis amphisbaena* (Bory) Cleve, *Anomoeoneis sphaerophora* (Ehrenb.) Pfitzer, *Craticula cuspidata* (Kütz.) D.G. Mann, *Cymbella amphicephala* Naegeli in Kützing, exhibit morphological variation under different environmental conditions, such as changing salinity regimes (Schmid 1976), whereas intraspecific morphological variation in other taxa, e.g. *Sellaphora pupula* Mereschk., has been shown to be an expression of discrete breeding populations (gamodemes) (Mann 1988, 1989). In other cases, pairs of taxa may be morphologically similar, e.g. *Surirella ovalis* Brébisson and *Surirella brebissonii* Krammer & Lange-Bertalot, but occur in different habitats; the former is predominantly marine to brackish, the latter freshwater.

The recognition of taxa with restricted ecological ranges can improve the predictive power of diatoms in water quality monitoring, palaeoecological and climate change studies (Anderson *et al.* 1993; Koppen 1975; Hürlimann & Straub 1991). Taxa with wide ecological amplitudes are usually considered of poor indicator value (Descy 1984), but whether widely distributed taxa are unusually tolerant, or comprise series of ecologically and genetically discrete taxa or ecotypes, has rarely been investigated. If clones of such taxa are isolated and grown under defined conditions in the laboratory it should be possible to establish whether the taxon is eurytopic or comprises several ecologically, or physiologically, discrete taxa. The clones of an eurytopic species should be capable of growth over the ecological range. On the other hand, a series of ecotypes should show contrasting growth along the ecological gradient, according to their original locality. Any effects of environment on morphology can be investigated concomitantly.

Consistent differences have been observed between specimens of *Navicula gregaria* Donkin from marine and freshwater sites (Cox 1987), although they would all be included by the classical definition of the species. This study was therefore undertaken to establish whether *N. gregaria* is euryhaline (Simonsen 1962) and exhibits morphological plasticity, or whether it comprises at least two morphologically stable taxa of narrower ecological amplitude. The morphology of *N. gregaria* specimens from both freshwater and marine sites was investigated while the growth of one isolate was monitored under different salinity regimes. The growth and morphology of three other small *Navicula* species were also studied under different salinity regimes.

## Materials and Methods

Samples collected from a range of freshwater and brackish sites were examined with both light and scanning electron microscopy after cleaning in 50% nitric acid, mounted in Naphrax or coated with gold-palladium accordingly. Photographs were taken on a Zeiss Universal photomicroscope (Planapo x100 objective, differential interference contrast optics) and a high resolution Hitachi S-800 scanning electron microscope, using Kodak T Max and Ilford HP4 film respectively. Valves from each field site and experimental condition (see below) were examined by light microscopy for the following features: valve length and breadth; length and breadth of apices; stria density; visibility of stria pores; stria pattern opposite the central raphe endings; thickness of the apical rib. Electron microscopy was used to examine the raphe fissures and details of the pores.

variation along a particular environmental gradient is the prerequisite. Whether the variation is genotypically determined or not is irrelevant to the ecological exercise, although nevertheless of taxonomic and evolutionary interest. It is as erroneous to assume that all morphological variation has a genetic origin, as it is that morphologically similar taxa necessarily have similar ecological requirements.

One of the problems encountered in comparative ecological work is that species identifications may be unreliable, particularly when dealing with taxa which lack obvious distinguishing features. In discussing the shape of naviculoid diatoms, thirty years ago, Hendey (1964) pointed out that, "The difficulty in identifying diatoms from published descriptions arises very largely from the paucity of descriptive terms and the lack of uniformity in their use." However, in spite of technical advances, many descriptions remain stylized and scanty, and descriptive terms are imprecise. This does not mean that subtle morphological differences cannot be discerned, but because the eye is a highly effective discriminator, it is often more effective to compare illustrations than rely on a textual account. Language must be used very precisely and consistently to convey perceived differences as efficiently. Hendey (1964) defined and illustrated his terms in relation to geometrical figures, while illustrations in Barber & Haworth (1981) support the descriptive terminology they use. More formalized developments of such an approach are used very effectively elsewhere, e.g. pollen grain shape and surface ornamentation (Moore *et al.* 1991; Erdtman 1992; Punt *et al.* 1994), mushroom gill insertion and leaf shape (cf. Stearn 1983), but such practice has not been generally adopted for diatoms. Yet reference to particular published figures encourages consistent terminology and minimizes misunderstanding.

In addition to the lack of consistency in terminology, or even agreement on how to measure some features, stylized, minimalist descriptions offered for taxa in floristic works have often been carried over as descriptions of new species, even where space constraints are fewer. Thus, accounts of naviculoid diatoms usually comprise a descriptor of basic outline, modifier to include apical outline, length and breadth ranges, stria arrangement and density (although over which part of the valve is rarely stated). Width and symmetry of axial and central areas may be given, as defined by the absence of striae, but qualitative comparisons are frequently lacking. There may be no information on the distinctness of pores or the refractivity of ribs or valve margins, yet such detail reflects underlying structural variation, e.g. in pore size and spacing. In *N. gregaria* the striae are more distinct in marine specimens, while slight changes in valve outline can also be used to distinguish specimens from brackish/marine and freshwater habitats. Thus, even with a single clone, it is possible to infer growth conditions from variation in valve morphology, once the link between phenotype and environment has been demonstrated. However, full exploitation of the predictive value of diatoms will require closer attention to detail and more strenuous attempts to describe the perceived differences in sufficient detail, using consistent terminology.

## Acknowledgements

Thanks are due to Peter York for assisting with the light photomicroscopy and to Clive Moncrieff for stimulating discussion over shape recognition.

## References

Anderson, D. S., Davis, R. B. & Ford, M. S. (1993). Relationships of sedimented diatom species (Bacillariophyceae) to environmental gradients in dilute northern New England lakes. *Journal of Phycology*, **29**, 264–277.

Barber H. G. & Haworth, E. Y. (1981). *A guide to the morphology of the diatom frustule*. FBA Scientific Publication No.44. 112pp. Titus Wilson & Son Ltd., Kendal.

Cox, E. J. (1987). Studies on the diatom genus *Navicula* Bory. VI. The identity, structure and ecology of some freshwater species. *Diatom Research*, **2**, 159–174.

Cox, E. J. & Ross, R. (1981). The striae of pennate diatoms. In: *Proceedings of the Sixth Symposium on Recent and Fossil Diatoms, Budapest, 1980*, (R.ROSS, ed.) 267–278. Koeltz, Koenigstein.

Descy, J. P. (1984). *Ecologie et distribution de diatomées benthiques dans le bassin belge de la Meuse*. Documents de travail no 18. 25pp. Institut royale des Sciences naturelles de Belgique, Brussels.

Egan, P. F. & Trainor, F. R. (1989a). The role of unicells in the polymorphic *Scenedesmus armatus* (Chlorophyceae). *Journal of Phycology*, **25**, 65–70.

Egan, P. F. & Trainor, F. R. (1989b). The effect of media and inoculum size on the growth and development of *Scenedesmus communis* Hegew. (Chlorophyceae) in culture. *Archiv für Hydrobiologie*, **117**, 77–95.

Erdtman, G. (1992). *Handbook of Palynology*. 2nd Edition. 580pp. Munksgaard, Copenhagen.

Hendey, N. I. (1964). *An introductory account of the smaller algae of British Coastal Waters. Part V: Bacillariophyceae (Diatoms)*. Fisheries Investigations, Series IV. 317pp. HMSO, London.

Hürlimann, J. & Straub, F. (1991). Morphologische und ökologische Charakterisierung von Sippen um den *Fragilaria capucina*-Komplex *sensu* Lange-Bertalot 1980. *Diatom Research*, **6**, 21–47.

Koppen, J. D. (1975). A morphological and taxonomic consideration of *Tabellaria* (Bacillariophyceae) from the northcentral United States. *Journal of Phycology*, **11**, 236–244.

Lange-Bertalot, H. (1980). Zur systematischen Bewertung der bandförmigen Kolonien bei *Navicula* und *Fragilaria*. *Nova Hedwigia*, **33**, 723–787.

Mann, D. G. (1988). The nature of diatom species: analyses of sympatric populations. In: *Proceedings of the 7th International Diatom Symposium, Philadelphia* (F. E. Round, ed.), 113–141. Koeltz, Koenigstein.

Mann, D. G. (1989). The species concept in diatoms: evidence for morphologically distinct, sympatric gamodemes in four epipelic species. *Plant Systematics and Evolution*, **164**, 215–237.

McLachlan, J. (1973). Growth media – marine. In: *Handbook of Phycological Methods: Culture methods and growth measurements*. (J. R. Stein, ed.), 25–51. Cambridge University Press.

Moore, P. D., Webb, J. A. & Collinson, M. E. (1991). *Pollen analysis*. 2nd Edition. 216pp. Blackwell Scientific Publications, Oxford.

Nichols, H. W. (1973). Growth media – freshwater. In: *Handbook of Phycological Methods: Culture methods and growth measurements.* (J. R. Stein, ed.), 7–24. Cambridge University Press.

Punt, W., Blackmore, S., Nilsson, S. & Thomas, A. Le. (1994). *Glossary of pollen and spore terminology.* 71pp. LPP Foundation, Utrecht.

Schmid, A.-M. (1976). Morphologische und physiologische Untersuchungen an Diatomeen des NeusiedlerSees: II. Licht- und rasterelektonenmikroskopische Schalenanalyse der Umweltabhängigen Zyklomorphose von *Anomoeoneis sphaerophora* (KG.) Pfitzer. *Nova Hedwigia*, **28**, 309–351.

Schultz, M. E. (1971). Salinity-related polymorphism in the brackish-water diatom *Cyclotella cryptica. Canadian Journal of Botany*, **49**, 1285–1289.

Simonsen, R. (1962). Untersuchungen zur Systematik und Ökologie der Bodendiatomeen der westlichen Ostsee. *Internationale Revue der gesamten Hydrobiologie, Systematisches Beiheft*, **1**, 1–144.

Stace, C. A. (1980). *Plant taxonomy and biosystematics.* 279pp. Edward Arnold, London.

Stearn, W. (1983). *Botanical Latin.* 3rd Edition. 566pp. David & Charles Ltd., Newton Abbot, Devon.

Trainor, F. R. (1991). The format for a *Scenedesmus* monograph. *Archiv für Hydrobiologie, Supplement*, **88** (*Algological Studies*, **61**), 47–53.

Trainor, F. R. & Egan, P. G. (1990). The implications of polymorphism for the systematics of *Scenedesmus. British Phycological Journal*, **25**, 275–279.

# A morphometric and geographical analysis of two races of *Diploneis smithii / D. fusca* (Bacillariophyceae) in Britain

Stephen J. M. Droop

*Royal Botanic Garden Edinburgh, Edinburgh EH3 5LR, UK*

## Abstract

There is now considerable evidence that the occurrence of morphologically distinct races (morphotypes) within diatom species is widespread; evidence from breeding behaviour, where this is available, shows that reproductive barriers exist between races of a species, and suggests that they should be delimited and named as normally functioning sexual species.

A previous study identified 11 morphotypes within the *Diploneis smithii / D. fusca* complex in one sample from Oban, Western Scotland. Two of these were very similar, differing only in dimensions and shape. The present analysis was initiated to test the constancy of these two morphotypes in other locations around the coast of Britain, and to test the usefulness of rectangularity as a measure of shape.

The results demonstrate that the morphotypes are separate in other British samples and maintain their diagnostic characters with little overlap. One of them is shown to be more variable than the other, and some suggestions are made that might explain the variation. Rectangularity discriminates well between these and other pennate diatoms with simple outlines.

## Introduction

It was shown by Droop (1994) that a single sample of epipelic marine diatoms (from Ganavan, near Oban on the west coast of Scotland – Fig. 1) contained 11 separate entities, which would all be identified as belonging to the complex of taxa containing *Diploneis smithii* (Brébisson) Cleve, *D. fusca* (Gregory) Cleve and *D. vacillans* (A.S.) Cleve using standard reference works such as Hustedt (1927–1966). The empirical groupings identified subjectively based on all the valve characters visible with the light microscope were confirmed by Principal Component Analysis and Hierarchical Cluster Analysis of variation in the six most easily scorable characters:

length, width, striation density, shape of central external raphe endings, proportion of valve with longitudinal ribs between each pair of transapical ribs, and rectangularity (an aspect of the shape of the valves). Each morphotype demonstrated what seems to be a normal size-reduction sequence, and it was suggested that the simplest explanation for the variation pattern is that each morphotype is a normal allogamous species.

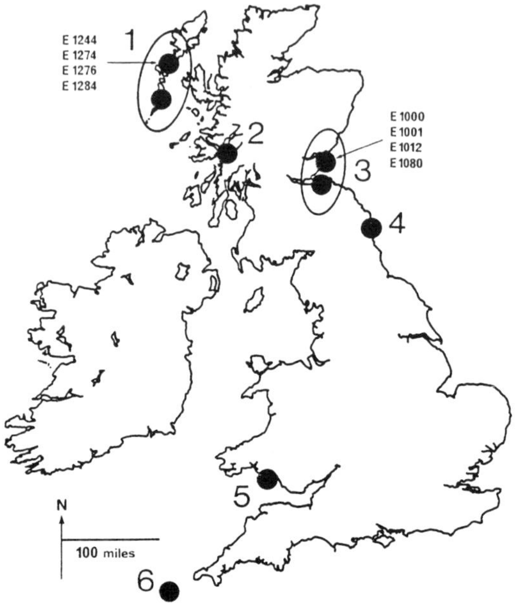


Fig. 1. Map showing the collecting localities of samples and sample groups. See Table 1 for more details about the samples. Where a spot covers the collecting localities of more than one sample, the individual samples are listed. Sample 2 represents Oban, the source of the material studied by Droop (1994).

Two of the morphotypes were more similar to each other than were any other pair, and differed only in their dimensions and valve shape. These were referred to as Morphotypes 3 and 4 (Droop 1994, figs 8–10, 11–14). In all other characters they were more or less indistinguishable from one another, but clearly different from each of the other morphotypes in the sample. Droop (1994, especially figs 8–14, 57–61 and table 2) describes how Morphotypes 3 and 4 differ from the other morphotypes in the same sample, and illustrates the relevant characters in more detail than has been done here.

The following description combines Morphotypes 3 and 4 from Oban and includes the characters that, together, separate them from other morphotypes in the complex; features common to all members of the *D. smithii* / *D. fusca* group are excluded. Morphotypes 3 and 4 are medium- to large-sized elliptical to oblong-elliptical diatoms with striation density ca. 9–10/10 µm; "*vacillans*"-structure (pores arranged in adjacent pairs between each pair of transapical ribs) occupying about three-quarters of the valve surface abaxial to the longitudinal canals (the remainder is occupied by "*smithii*"-structure, i.e. pores arranged alternately between transapical ribs; not illustrated here); abaxial edges of longitudinal canals enclosing a smoothly lanceolate shape ("a", Fig. 16) rather than a rhombic shape; longitudinal canals wider than central nodule ("b", Fig. 16); raphe sternum furrow parallel-sided ("c", Fig. 16), not noticeably constricted close to the central nodule; central external raphe endings hooked, bilaterally asymmetrical ("d", Fig. 16).

The characters of length, width and shape are sufficient to enable complete separation of the two morphotypes from each other when used in combination, although not on their own (Figs 4 and 5): Morphotype 3 has large wide valves that are almost elliptical in outline; the valves of Morphotype 4 are smaller and narrower, and more oblong. However, this distinction, although real, is based on only one sample. It is conceivable that intermediates may exist outside the sample that would mean that the Oban morphotypes had only local significance and were not applicable elsewhere. It was known that diatoms similar to Morphotypes 3 and 4 also occurred elsewhere in Britain, so this study was initiated to test their distinctness in other parts of Britain. The secondary purpose of the analysis was to test the usefulness of rectangularity as a taxonomic character.

## Material and methods

The 14 samples used in the analysis were collected around the shores of Britain between 1858 and 1991 (Table 1), and are housed at the Royal Botanic Garden Edinburgh, UK. Most were from sandy shores between the limits of high and low tide. Two (E1244 and E1284, both from the Outer Hebrides) are from high shore pools. The samples are divided geographically into six groups (Fig. 1). Only Groups 1 and 3 (Outer Hebrides and South-East Scotland respectively) comprise more than one sample.

Collections (except BH307 and WA582) were made using an adaptation of the Eaton & Moss (1966) lens tissue method, then acid-cleaned and mounted in Naphrax. BH307 was collected by Mr Bernard Hartley, probably as washings from sand, and was subsequently filtered to remove small diatoms and debris. Smaller individuals from the populations of the morphotypes under study may have been lost from the sample at this stage, although none were found in a slide made from the material that had passed through the filter (sample BH307<). WA582 was collected by a Rev. R. Taylor and sent to Dr G. Walker Arnott in Glasgow sometime around May 1858, and is probably washed from sand. The mountant Walker Arnott used for the slides was probably Canada Balsam, which has a lower refractive index than modern mountants like Naphrax or Hyrax – diatoms in this mountant show less contrast than those in Naphrax and appear more finely silicified as a result (Figs 29–30).

Photographs were taken on Kodak Technical Pan using a Reichert Polyvar photomicroscope. Measurements were made on a Zeiss Axiophot photomicroscope fitted with a video camera connected to image analysis equipment as detailed in Droop (1994).

Table 1: The sources of material examined. The sample and sample group numbers in the left hand column refer to numbers on the map (Fig. 1).

| | | | |
|---|---|---|---|
| **Sample group 1 Outer Hebrides** | E1244 | 7.6.91 | Newtonferry, North Uist, NF 896 781 Sheltered deep channel at top of sandy shore |
| | E1274 | 8.6.91 | West shore, Berneray, NF 902 824 Sandy shore: mid tidal zone |
| | E1276 | 9.6.91 | Loch Borve, Berneray, NF 907 803 Sandy shore: lower-mid tidal zone |
| | E1284 | 9.6.91 | Loch Borve, Berneray, NF 903 805 Mud over sand in marsh pool at top of shore |
| | BH307 | 8.6.71 | Kilbride Bay, South Uist, ca. NF 7514 Sandy shore |
| **Sample 2 Western Scotland** | E1014 | 1.3.91 | Ganavan Bay, Oban, NM 862 328 Sandy shore: lower tidal zone |
| **Sample group 3 South-East Scotland** | E1000 | 21.1.91 | St Andrews, Fife, NO 504 176 Sandy shore: upper-mid tidal zone |
| | E1001 | 21.1.91 | St Andrews, Fife, NO 504 177 Sandy shore: lower tidal zone |
| | E1012 | 21.1.91 | Cambo Sands, Fife, NO 605 124 Sandy shore: mid tidal zone |
| | E1080 | 19.4.91 | Crail Harbour, Fife, NO 612 073 Sandy shore: lower tidal zone |
| | E1036 | 20.3.91 | Fisherrow Harbour, Lothian, NT 334 733 Sandy shore: lower tidal zone |
| **Sample 4 North-East England** | WA582 | ca.5.1858 | Cresswell, Northumberland Sandy shore |
| **Sample 5 South Wales** | E1301 | 4.7.91 | Three Cliffs Bay, Gower Peninsula, SS 535 877 Sandy shore: lower-mid tidal zone |
| **Sample 6 Isles of Scilly** | B106 | 1979 | St Martin's, Isles of Scilly Sandy shore: intertidal |

For measurement via image-analysis, broken diatoms and those lying obliquely (more than about 1 µm between the vertical focus of opposite sides or ends) were rejected (about 25% of the total) except in two specific cases: in E1080 a single very large valve was lying too obliquely to be measured accurately (Fig. 28), and in E1301 the incomplete valve in Fig. 34 was only the second valve of the narrow morphotype (see below) to be found in the sample. In both cases the figures used in the graphs are likely to be less accurate than the others, but it was felt that they were such significant valves that they should be included. In any case, the inaccuracy is probably only a few percent – the arrows in Figs 6, 7, 10 and 11 indicate the valves in question.

The graphs were created using SigmaPlot (Jandel Scientific, PO Box 7005, San Rafael, CA 94912-8920, USA).

*Characters used in the analysis*

This study has dealt only with those characters that separated Morphotypes 3 and 4 at Oban (Droop 1994): dimensions and rectangularity.

In pennate diatoms the reduction in size during the asexual phase of the life cycle (the MacDonald–Pfitzer rule: MacDonald 1869; Pfitzer 1869) is usually associated with a change in aspect ratio, since length tends to reduce faster than width (see Cox 1993). The aspect ratio therefore decreases and diatoms become more isodiametric. However, this variation in dimensions tends to occur between limits that are more or less fixed for each genotype (Geitler 1932), and so can yield valuable taxonomic information.

Differences in diatom shape can easily be detected by eye, but taxonomic assessment of shape is difficult, especially where differences are slight and where objectivity is required. As in Droop (1994), rectangularity was used as a measure of shape: rectangularity is the area enclosed by the outline of a diatom valve expressed as a proportion of the area of its enclosing rectangle (as defined by the length of the diatom's apical and transapical axes). Using image-analysis, the outline of each diatom was detected semi-automatically by the computer. The diatom's apical axis was taken as the line between the two points furthest apart on the outline; the length of its transapical axis was taken as the sum of the distances from the apical axis to the most distant point on each side. The area enclosed within the outline was taken as the area of the diatom and this area was divided by the product of the diatom's length and width to calculate its rectangularity. The resulting dimensionless number can range from near zero (for a cross-shape) to one (for a rectangle). See below for a fuller discussion of the use of this character.

## Results

513 valves were measured from the 14 samples, including the 115 valves of Morphotypes 3 and 4 studied by Droop (1994). Most of the samples from the diatom herbarium at the Royal Botanic Garden Edinburgh that were found to contain

individuals of these morphotypes were included in the study, and wherever possible all valves were measured (excluding those rejected because they were broken or oblique).

All the valves included in this study are from the taxonomic group that includes Morphotypes 3 and 4 from Oban (Droop 1994). No valves were in any way intermediate between these and any of the other morphotypes occurring at Oban, and there was never any doubt whether a valve did or did not belong to the taxonomic complex in question.

*Descriptions of samples and sample groups*

Each of the geographical sample groups (numbered 1–6 in Fig. 1) contained individuals of two apparently separate populations of valves from the complex that includes Morphotypes 3 and 4 from Oban (Droop 1994). The valves in one population in each sample/sample group were smaller (and especially narrower) than the other, and tended to be more oblong. The only exception was Sample WA582 (4 in Fig. 1), where there seemed to be only one population.

Figs 2–13 are plots of length, width and rectangularity for each individual sample or sample group: even figure numbers are plots of width against length (the upper plot of each pair); odd figure numbers are plots of rectangularity against width. Valves of the smaller, narrower group are shown by solid symbols – the others (including those from WA582) by hollow symbols – throughout Figs 2–15. Each valve is represented by the same symbol throughout the study and equivalent axes have the same range in each plot, so individual points can be traced between the individual plots (Figs 2–13) and the compilations (Figs 14–15), and between width/length plots and the equivalent rectangularity/width plots for the same sample or sample group.

Fig. 14 is a plot of width against length for all valves measured; Fig. 15 is a plot of rectangularity against width for all valves measured. Samples and sample groups are not differentiated, but valves of the narrower morphotype in each sample are represented by filled symbols – the others by hollow symbols.

Figs 16–38 are photographs of large and small examples of each morphotype from the six samples or sample groups. Sample and sample group numbers refer to Fig. 1.

***Sample group 1: Outer Hebrides.*** *Five samples: E1276, BH307, E1244, E1274 and E1284. Figs 2–3, 16–19. 57 valves measured*

This sample group very clearly contains two separate morphotypes (see Figs 16–17, 18–19). BH307 (squares in Figs 2–3) contains both, but is lacking small individuals, probably because it had been filtered. E1276 is the only other sample in this group that contains valves of the narrower morphotype, and these, with the narrow valves of BH307, form a well-defined size-reduction sequence in Fig. 2 (Figs 18–19). One hollow square in Fig. 2 closely approaches the narrower morphotype in dimensions, but its low rectangularity (Fig. 3) clearly shows it to be a small individual of the wider morphotype.

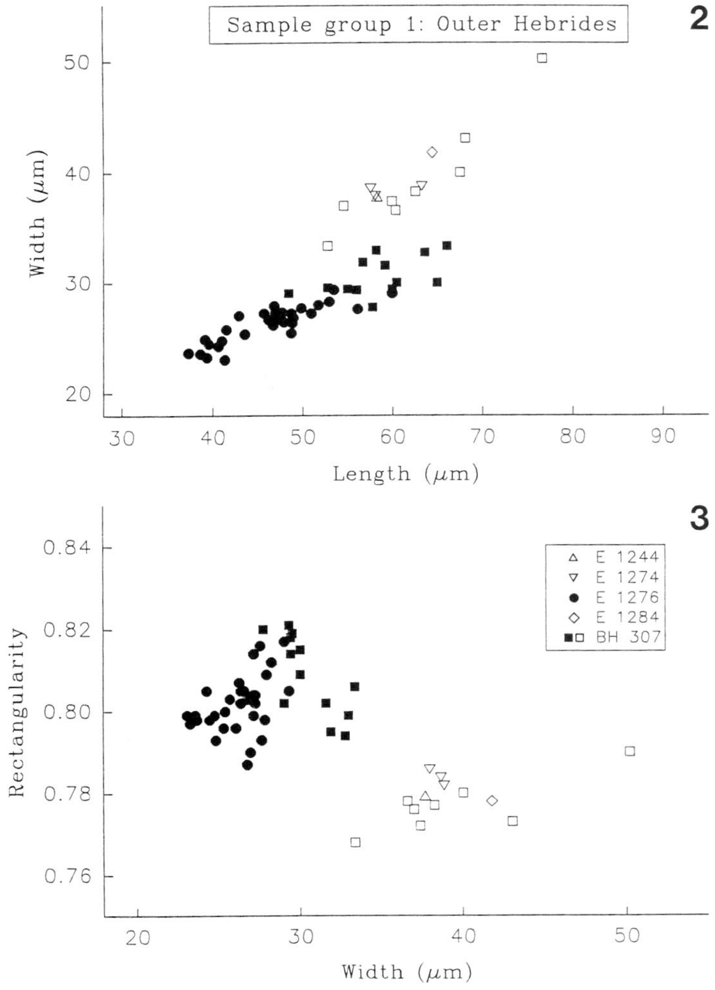

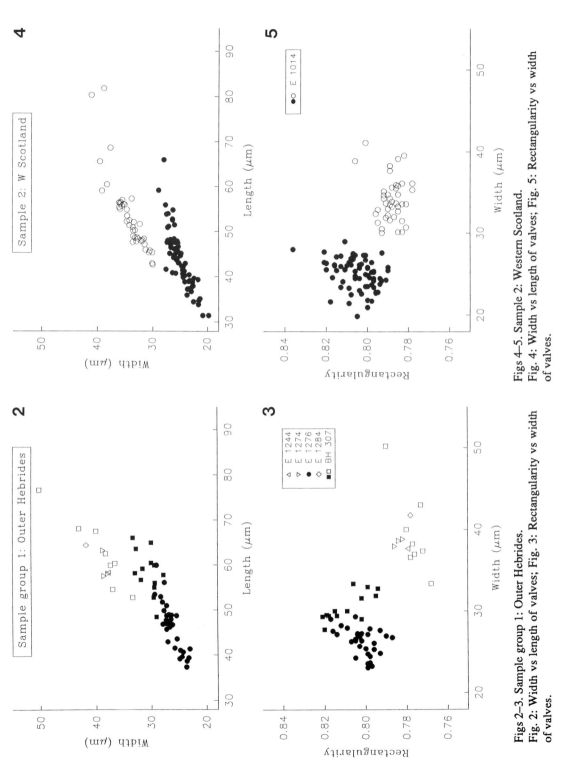

Figs 2–3. Sample group 1: Outer Hebrides.
Fig. 2: Width vs length of valves; Fig. 3: Rectangularity vs width of valves.

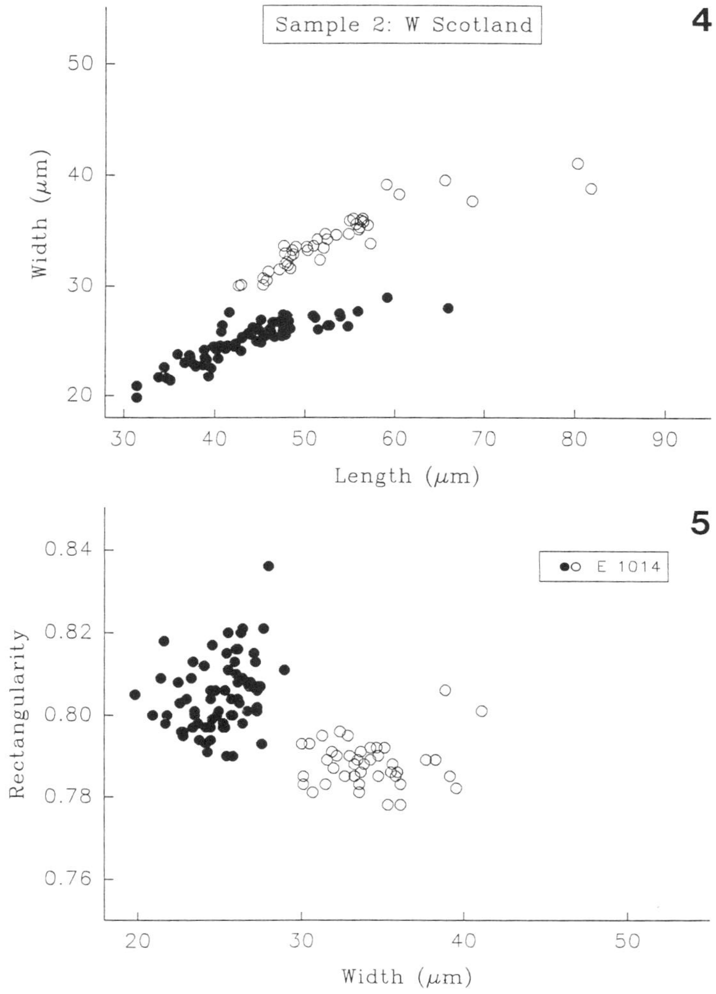


Figs 4–5. Sample 2: Western Scotland.
Fig. 4: Width vs length of valves; Fig. 5: Rectangularity vs width of valves.

The wider morphotype, similarly, forms a well-defined unit (Figs 16–17) composed primarily of valves from BH307, plus three from E1274 and one each from E1244 and E1284. The angle of slope is rather steep, however (Fig. 2) – the widest valve, in particular, seems almost too wide to be part of the same population as the others.

***Sample 2: Oban, Western Scotland.*** *One sample: E1014. Figs 4–5, 20–23. (See also Droop 1994.) 115 valves measured*

This is the sample analysed by Droop (1994) and contains the original Morphotypes 3 and 4. There is a very clear distinction between the two morphotypes (see Figs 20–21, 22–23), with a distinct gap between the widths of valves of each morphotype at a particular length (Fig. 4). Both morphotypes show very well-defined size-reduction sequences (Fig. 4) and rectangularity/width clusters (Fig. 5).

***Sample group 3: South-East Scotland.*** *Five samples: E1036, E1080, E1000, E1001, E1012. Figs 6–7, 24–28. 47 valves measured*

The group of samples from South-East Scotland (and three out of five of the individual samples comprising it) contains two clearly-distinguishable morphotypes (see Figs 24–25, 28, 26–27, but the difference in width for a particular length is somewhat less well marked than in Sample group 1 and Sample 2 (compare Fig. 6 with Figs 2 and 4). The distinction becomes very clear, though, when rectangularity is included as a character (Fig. 7). The very small valve of the wider morphotype (the hollow triangle to the left of Fig. 6) has a very low rectangularity (Fig. 7) and is clearly not part of the cluster of filled symbols that it approaches in its dimensions.

The narrower morphotype forms a close-knit cluster with a clear size-reduction sequence. The wider morphotype forms a similarly coherent cluster except for one valve (arrowed in Figs 6 and 7 and illustrated in Fig. 28).

***Sample 4: Cresswell, North-East England.*** *One sample: WA582. Figs 8–9, 29–30. 50 valves measured*

This sample seems to contain only one morphotype (Figs 29–30), the valves of which form a coherent size-reduction sequence (Fig. 8). The cluster in Fig. 9 is not quite so neat, with one valve exhibiting a particularly low rectangularity. It may or may not be significant that this sample was collected more than 110 years before any of the others in this study; if diatoms evolve very rapidly, this time lapse could conceivably help explain the difference in pattern of variation between the diatoms at Cresswell and those at the other sites studied.

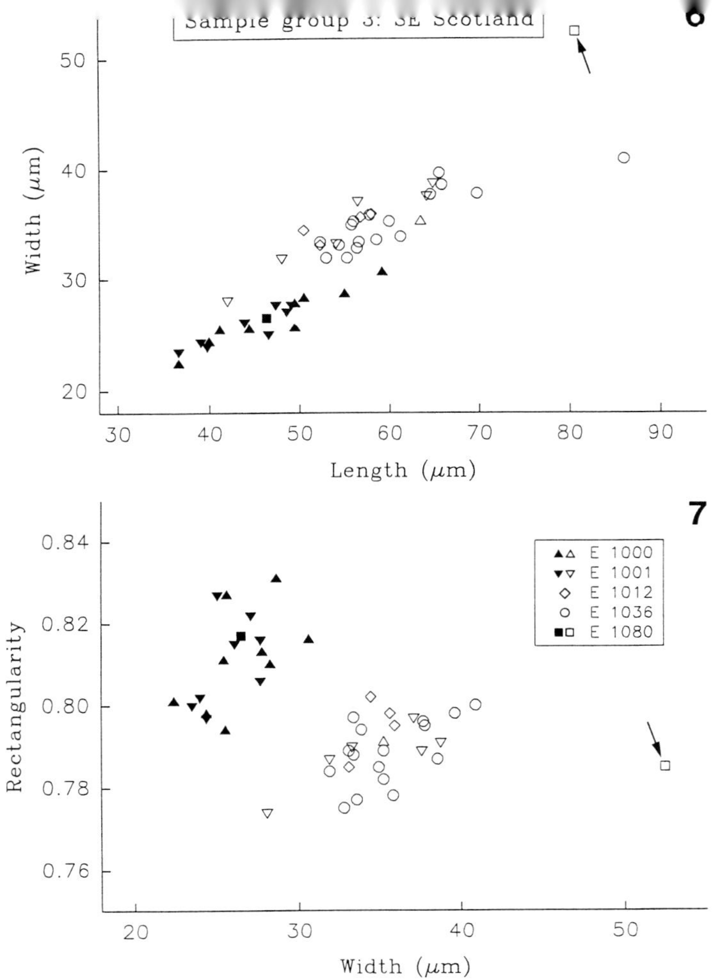

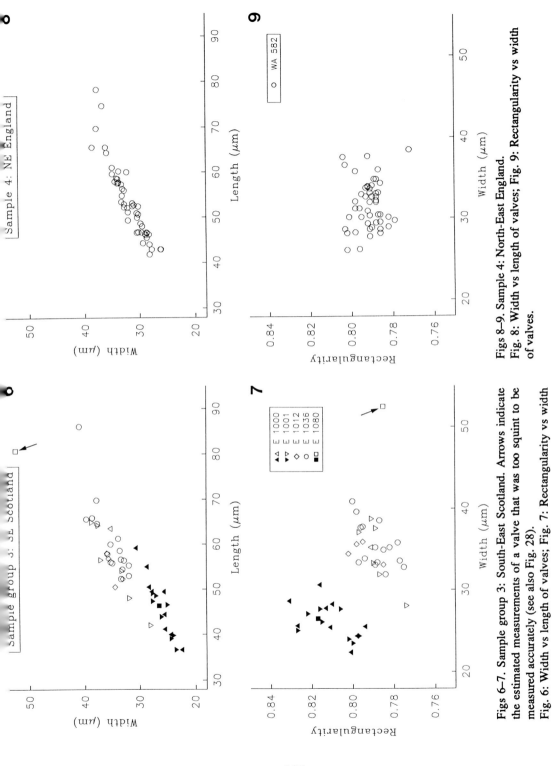

Figs 6–7. Sample group 3: South-East Scotland. Arrows indicate the estimated measurements of a valve that was too squint to be measured accurately (see also Fig. 28).
Fig. 6: Width vs length of valves; Fig. 7: Rectangularity vs width of valves.

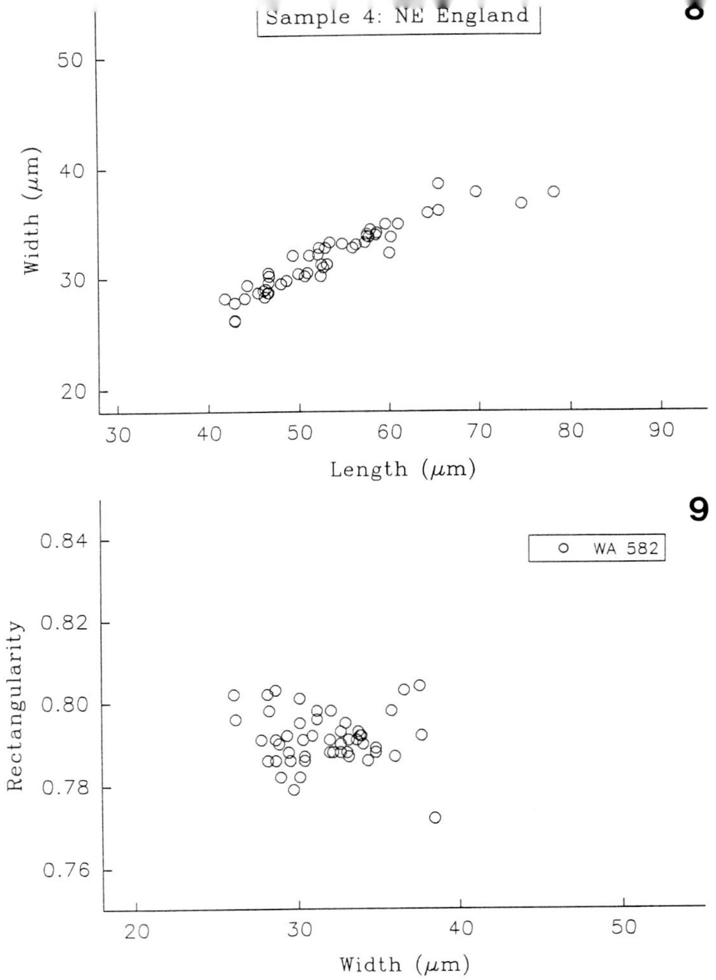


Figs 8–9. Sample 4: North-East England.
Fig. 8: Width vs length of valves; Fig. 9: Rectangularity vs width of valves.

***Sample 5: Gower Peninsula, South Wales.*** *One sample: E1301. Figs 10–11, 31–34. 46 valves measured*

The sample from South Wales was thought at first to contain only one morphotype, except for one valve that was much too narrow (Fig. 33, which might have been a contaminant). However, careful searching uncovered a further, broken valve (Fig. 34) of a similarly narrow diatom. The presence of this fragment lessened the likelihood that the first was a contaminant; it was important to get a series of measurements from it, but this necessitated using the edit facility in the image analysis program to extrapolate the outline to complete half the valve. Measurements of length and rectangularity were taken unaltered; the width was doubled to estimate the width of the whole valve. The resulting points are arrowed on Figs 10 and 11.

The two morphotypes present in the sample are clearly distinct (see Figs 31–32, 33–34) in both dimensions (Fig. 10) and rectangularity/width (Fig. 11). The wider morphotype forms a well-defined size-reduction sequence. There are too few of the narrower valves to allow a size-reduction sequence to be reconstructed, but by analogy with the other samples analysed, there is every reason to believe that the two valves found are indeed derived from the same population.

***Sample 6: St Martin's, Isles of Scilly.*** *One sample: B106. Figs 12–13, 35–38. 198 valves measured*

The Scilly sample contains two separate morphotypes (Figs 35–36 and 37–38), although there is relatively little spatial separation between the two in Fig. 12 and especially in Fig. 13. There is no overlap, though, and never any appreciable doubt which group a particular valve belongs to. A possible exception is the smallest valve of the larger morphotype (Fig. 12, left hand side), which could conceivably have been misidentified, although its rectangularity is low (Fig. 13). The narrower morphotype is rare in the sample, and many valves had to be measured to find a large enough sample.

Both morphotypes viewed separately form well-defined size-reduction sequences.

*All samples taken together*

The analyses of individual samples or sample groups (except WA 582 from Cresswell) all suggest that there are two morphotypes, as originally found at Oban (Droop 1994). However, Figs 14 and 15 show that the distinction between the morphotypes is not absolute, although the overlap between them is slight in both the length/width (Fig. 14) and width/rectangularity (Fig. 15) plots. Note that, if the populations of the wider and narrower morphotypes were not distinguished in Figs 14 and 15, the combined data would appear as a continuum, with no obvious discontinuities. It is only the fact that two morphotypes are separable *within each sample* that indicates that two distinct entities exist, worthy of taxonomic recognition.

In Fig. 14 the combined narrower populations from each sample/sample group (filled circles) form a well-defined size-reduction sequence, with few outliers. In Fig.

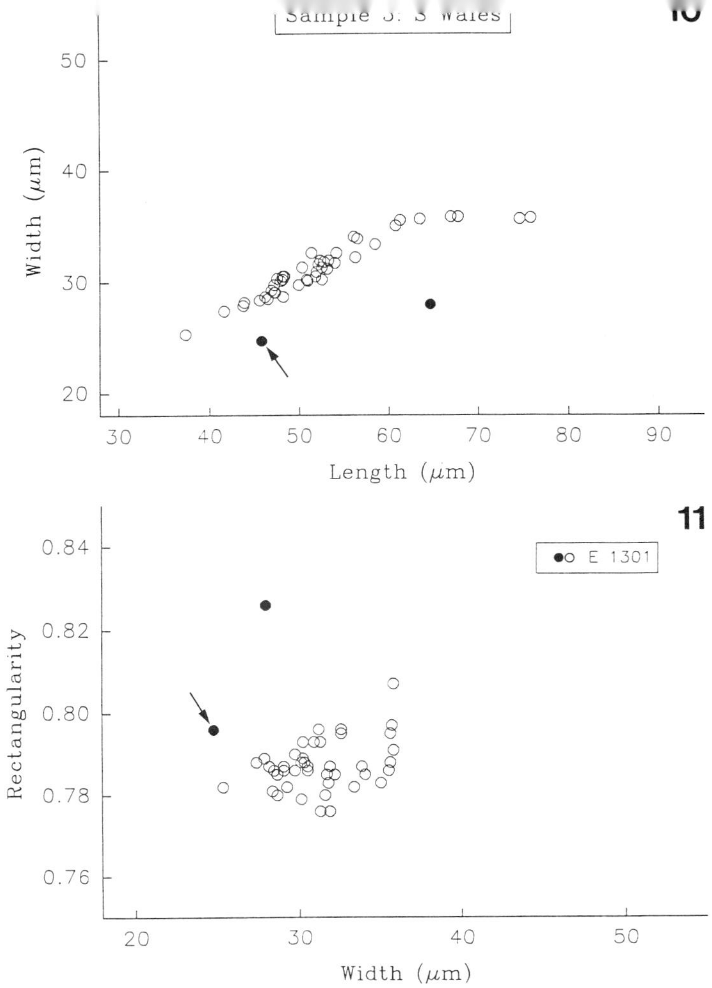


Figs 10–11. Sample 5: South Wales. Arrows indicate the estimated measurements of a broken valve (see also Fig. 34).
Fig. 10: Width vs length of valves; Fig. 11: Rectangularity vs width of valves.

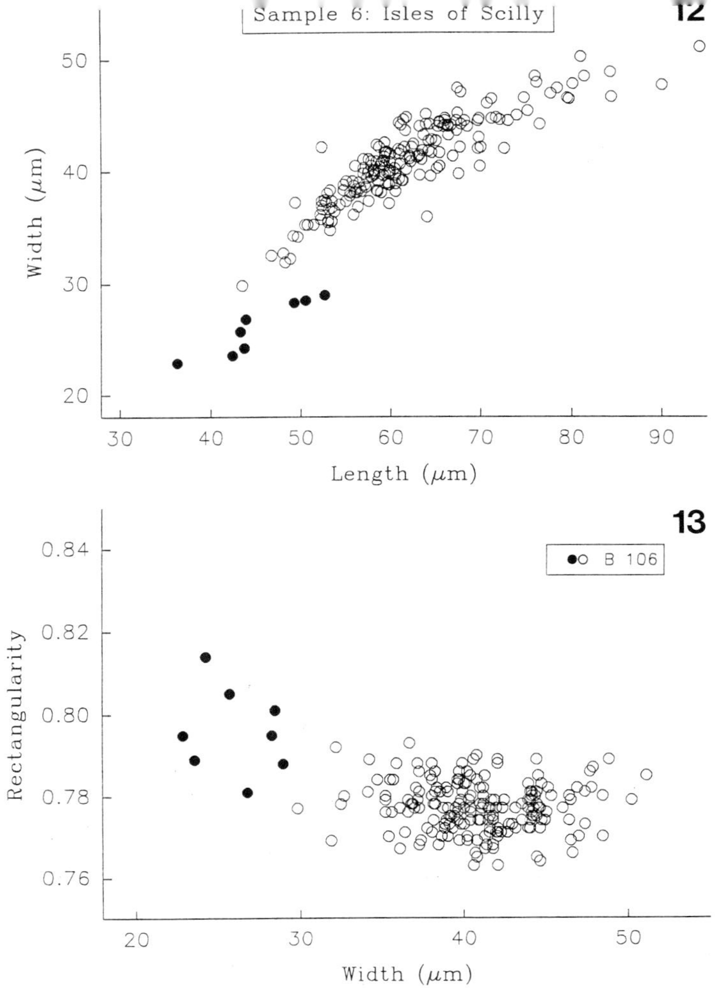

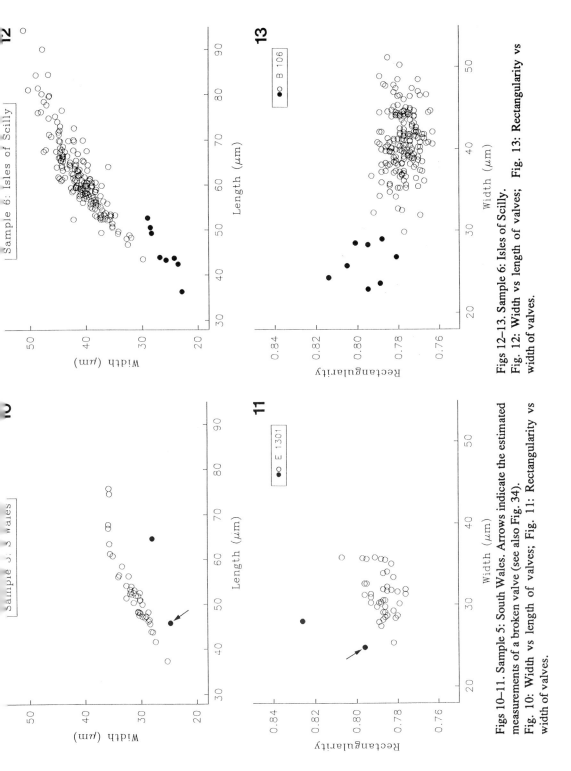

Figs 12–13. Sample 6: Isles of Scilly.
Fig. 12: Width vs length of valves; Fig. 13: Rectangularity vs width of valves.

15, too, these same valves form a reasonably tight cluster with a high average rectangularity. There is no evidence of any heterogeneity that might indicate the presence of more than one narrow morphotype.

It is less easy to say the same about the larger and wider valves in the various samples and sample groups. There is a considerable spread of width for a given length, especially among very large diatoms (Fig. 14), and the spread is rather uneven; the combined data (hollow symbols in Fig. 14) hint that there may be two elongate clusters, representing two separate size-reduction series: one narrower than the other for a given length (either side of the dashed line in Fig. 14). The evidence for this, however, is weak. Thus, for instance, the valves from Sample group 3 and Sample 2 in fact, cross the dashed line between the two "separate" wide morphotypes in Fig. 14. The populations of the wider morphotype are much less variable individually (with respect to width at a particular length) than when they are all combined in Fig. 14: the valves from Samples 4 and 5 (Figs 8 and 10) are narrowest for a given length and some even overlap with the other (narrow) morphotype (Fig. 14). Valves from Sample group 1 and Sample 6 (plus the individual arrowed valve from Sample group 3; Figs 6 and 28) are the widest for a given length. Sample 2 and Sample group 3 are intermediate.

As with the length/width plot (Fig. 14), there is a slight hint of two separate clusters in the rectangularity/width plot (Fig. 15) within the wider morphotype (open symbols), one containing narrower valves with slightly higher rectangularities (either side of the dashed line, Fig. 15). There is little difference in valve rectangularity of the wider morphotype between individual populations, except that there is a slight tendency for populations with narrower valves to have a higher average rectangularity. Sample 4, for example, has relatively narrow valves (overlapping slightly with the narrow morphotype) with relatively high rectangularity (also overlapping slightly with the narrower morphotype). Sample 6 on the other hand has wide valves with low rectangularities.

## Discussion

### *Rectangularity as a measure of shape*

Shape has been important almost from the beginning of diatom taxonomy. One reason for this is undoubtedly that, although the resolution of early microscopes was good enough to see the shape of a diatom's outline, it was not good enough to see very much of the fine detail enclosed within it. The other, more significant, reason for the importance of shape as a character was its perceived constancy within a species. Thus Gregory (1857) began his description of *Amphora spectabilis, n.sp.* with "Form nearly rectangular, broad, with rounded angles; occasionally sub-elliptic". The implication is that if something isn't "nearly rectangular, broad, etc." then it isn't *Amphora spectabilis*!

In an effort to reduce the subjective element in shape assessment, systematists have developed tables or charts as a basis for a standardized descriptive terminology (e.g. the Systematics Association chart of descriptive terminology reproduced in Stearn

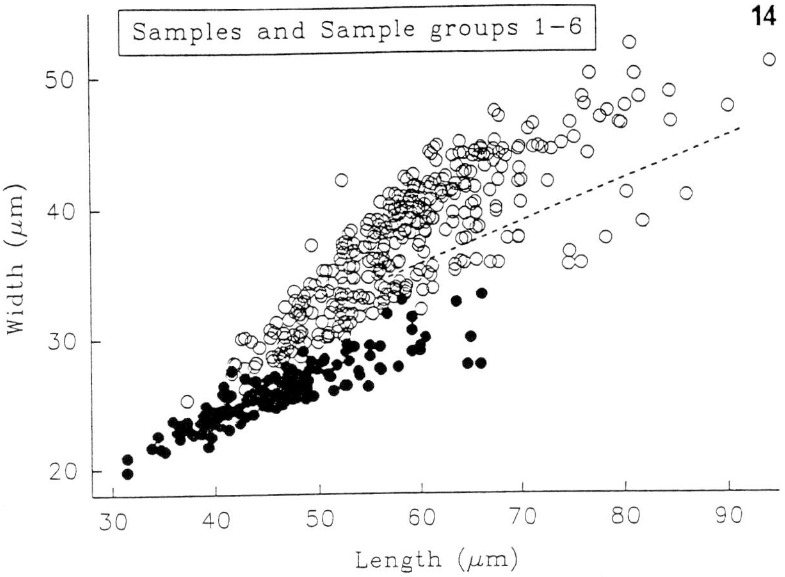


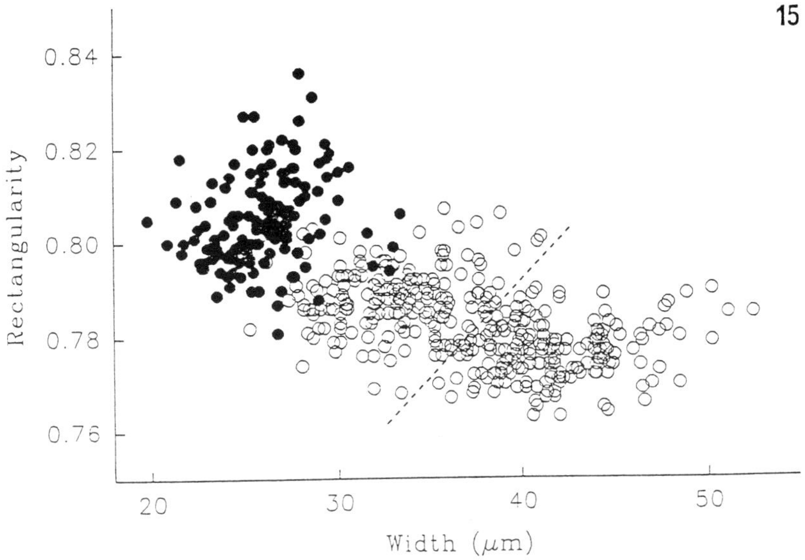


Figs 14–15. Plots of all data from Samples and Sample groups 1–6. Filled circles indicate valves of the smaller, narrower population in each Sample/Sample group (see Figs 2–13); hollow circles indicate valves of the larger, wider population in each Sample/Sample group. Dashed lines indicate a possible subdivision of the hollow circles.
Fig. 14: Width vs length of valves; Fig. 15: Rectangularity vs width of valves.

1992, pp. 310–311). Each different shape is defined graphically with prefixes for variants: "anguste ellipticus", for example, defines an ellipse with an aspect ratio of between 3:1 and 6:1. These charts could theoretically be as complex and as comprehensive as the variation they are designed to help classify. In practice, though, their use is limited; any chart that is easy to use will be so crude (i.e. have so few categories) as to be more or less useless for careful analysis. Conversely, any chart with sufficient detail or resolution will be more or less impossible to use.

In order to analyse shape objectively, efforts have been made to reduce its description to numbers that can be treated statistically. There are many ways of doing this, some involving quite complex mathematics (see Mou & Stoermer 1992 for a brief review). The method most often used for diatoms involves fitting outlines to a polynomial series. Then multivariate statistical analysis can be performed, using the coefficients of successively higher order terms of the polynomial as descriptors of shape.

The most commonly used polynomial series is that of Legendre. This method has been used successfully on several diatom species: *Gomphoneis herculeana* (Ehrenb.) Cleve (Stoermer & Ladewski 1982), *Tabellaria flocculosa* (Roth) Kütz. (Theriot & Ladewski 1986), *Didymosphenia geminata* (Lyngbye) M. Schmidt (Stoermer *et al.* 1986), *Eunotia pectinalis* (O.F. Müller) Rabenhorst (Steinman & Ladewski 1987), and *Surirella fastuosa* Ehrenb. (Goldman *et al.* 1990). The method is computationally complex, although modern image-analysis techniques can do the hard work. It is possible that it will become much more widely used in the future, and may even become a routine tool for the systematist.

However, in some cases it is possible to obtain quantitative measures of shape without the complicated mathematics associated with polynomial expansions, using methods that are nevertheless capable of resolving fine differences between the diatoms under study. Rectangularity is one such method. For many naviculoid diatoms, variation in rectangularity expresses more or less all of the variation in shape. Details such as shape of poles and whether or not the cells are constricted are excepted, but even these can affect rectangularity to the extent that differences can be demonstrated. For example *Mastogloia meisteri* Hustedt (Hustedt 1927–66, fig. 970) is almost rhombic and would have a rectangularity of close to 0.5, but perhaps not quite as low as *M. elegantula* Hustedt (Hustedt 1927–66, fig. 975), which is almost rhombic but with slightly rostrate poles that increase the length of the diatom without greatly increasing its area. Rectangularity could probably also distinguish between populations of *M. elegantula* and *M. lemniscata* Leudiger-Fortmorel (Hustedt 1927–66, fig. 976), the latter having shorter rostrate poles.

Rectangularity is not a true measure of shape since it is not invertible: the shape of a diatom cannot be reconstructed from its rectangularity. Nevertheless, it has many advantages for the diatomist as a crude measure. Firstly, it is understandable: a high rectangularity index for example indicates a diatom that nearly fills its enclosing rectangle. *Cistula lorenziana* (Grun.) Cleve, for example, would have a rectangularity of close to 1. Secondly, it is quickly and easily measured, since it can be calculated from only three parameters: length, width and area. The first two are routinely measured for diatoms in any case. The increasing availability of lower price image-analysis systems and programs could make the measurement of area almost as easy as length and width, but even in the absence of image-analysis, area can be measured accurately by drawing or photographing the diatom, and then counting squares on calibrated paper or cutting round the outline and weighing.

The greatest advantage of rectangularity is that it is logically independent of size and aspect ratio (the ratio of length to width): a rectangle has a rectangularity of 1 whether it is long and thin or short and fat; a rhombus has a rectangularity of 0.5 whatever its shape, and a circle or an ellipse always has a rectangularity of $\pi/4$ (ca. 0.785). Cox (1993) pointed out the dangers of relying too much on shape in diatoms because our perception of shape is so dependent on aspect ratio. However, a population of diatoms that exhibits a large range of sizes and aspect ratios in its size-reduction cycle could exhibit a much narrower range of rectangularity indices. This is reflected in the rather narrow limits of shape generally given in species descriptions. However, many of the complications that can affect rectangularity tend to be lost as a diatom progresses through its size-reduction sequence, perhaps due to the effects of cell turgor (Mann 1994b). In this case, then, small and large individuals of the same species would differ in rectangularity, but this would have no taxonomic significance.

Care must also be taken over the automatic measurement of rectangularity by computer-aided image analysis, since the enclosing rectangle is likely to be defined by the computer by the longest axis and the one perpendicular to it, unless the axes are indicated by the user. In most pennate diatoms the apical axis corresponds to the longest axis, but there are some (e.g. *Cistula lorenziana*) in which the longest axis is diagonal to the axis of symmetry. The rectangularity index of this diatom calculated using the wrong axes would be much too low.

The data presented here demonstrate the usefulness of rectangularity as a taxonomic character in a morphometric analysis. Where dimensions alone did not adequately define the difference between the morphotypes in a sample, a plot of rectangularity against width clearly showed the true relationships of ambiguous points (Figs 2, 3, 6, 7, 12, 13).

*Variation within and between the samples*

The previous study (Droop 1994) identified two very similar but distinct morphotypes in one sample collected in Oban. Each of these occurred in a range of sizes that seemed to indicate a size-reduction sequence typical of a normal diatom life-cycle.

In several other samples, from elsewhere in Britain, two apparently separate populations co-exist, just as in the Oban sample. Each morphotype in each sample exists in a range of sizes that would be typical of an orthodox life-cycle.

The combined data from all the samples (Figs 14 and 15) show that the smaller, narrower diatoms from each sample or sample group (filled symbols) are all very similar. They exhibit a range of sizes (Fig. 14) that suggests very strongly an orthodox life-cycle of size-reduction alternating with size-restitution via auxospores. The tightness of the cluster in Fig. 15 (rectangularity/width) shows that they are all of a similar shape (insofar as rectangularity is a measure of shape) and strengthens the suggestion that they form one entity which, for convenience, we can call Narrow.

The large amount of variation in width for a given length among the combined populations of larger and wider diatoms from each sample (hollow circles, Fig. 14) does not argue quite so convincingly for a single discrete taxon. However, it could be reasonable to ascribe that level of variation to simple genotypic diversity. The other possibility hinted at by the results (Figs 14, 15) is that the combined wider populations actually comprise two separate taxa, which could be referred to as: QuiteWide and VeryWide (delimited in Figs 14–15 by the dashed line). In that case, Sample group 1 would comprise only VeryWide (as well as Narrow described above); Sample 2 would include Narrow and VeryWide, except that the largest valves would fall more clearly into QuiteWide (compare Figs 4 and 14); Sample group 3 would include Narrow, QuiteWide and one valve of VeryWide (Fig. 6, arrow; Fig. 28); Sample 4 would include only QuiteWide; Sample 5 would include Narrow and QuiteWide; Sample 6 would include Narrow and VeryWide.

This suggestion does not attract me for several reasons. Firstly, there is no clear-cut morphological distinction between QuiteWide and VeryWide, either in dimensions or rectangularity; the extremes are easy to tell apart, but many valves are more or less intermediate, especially in Sample 2 and Sample group 3 where the main bulk of the wider populations cross the dashed line in Fig. 14 between VeryWide and QuiteWide.

Secondly, if VeryWide and QuiteWide were different (and assuming both were different from Narrow discussed above and represented in Fig. 14 by filled circles), one might expect to find samples containing valves of all three morphotypes VeryWide, QuiteWide and Narrow. As it is, if samples do contain more than one morphotype, they contain Narrow plus either QuiteWide or VeryWide. No individual sample contains all three (although Sample group 3 contains one very wide valve besides many of QuiteWide and Narrow); and none contains valves of more than one form of Wide in separable populations. Altogether, then, there appear to be just two morphotypes: Narrow and Wide.

How, then, can the variation in width (for a given length) and rectangularity in Wide be explained? The inheritance of shape in diatoms from one asexual generation to the next has a well-established basis (see Mann 1994b): in most diatoms the outline of the valve is physically determined by the shape of the parental cell, since the valves and girdle of the new cell form immediately inside those of the parent. This can be seen especially well in lineages of misshapen cells derived from one abnormal ancestor, in which a teratology has been caused by osmotic or nutrient stress, damage by grazers or parasites, or some failure in the cytokinetic mechanism. Such abnormalities can often be followed for many generations (for example Geissler 1970, Round 1993). Besides causing teratologies, culturing conditions can also alter some aspects of shape more subtly, as has been demonstrated in clones grown under different laboratory conditions. Jahn (1986) found that the widths of valves of *Gomphonema*

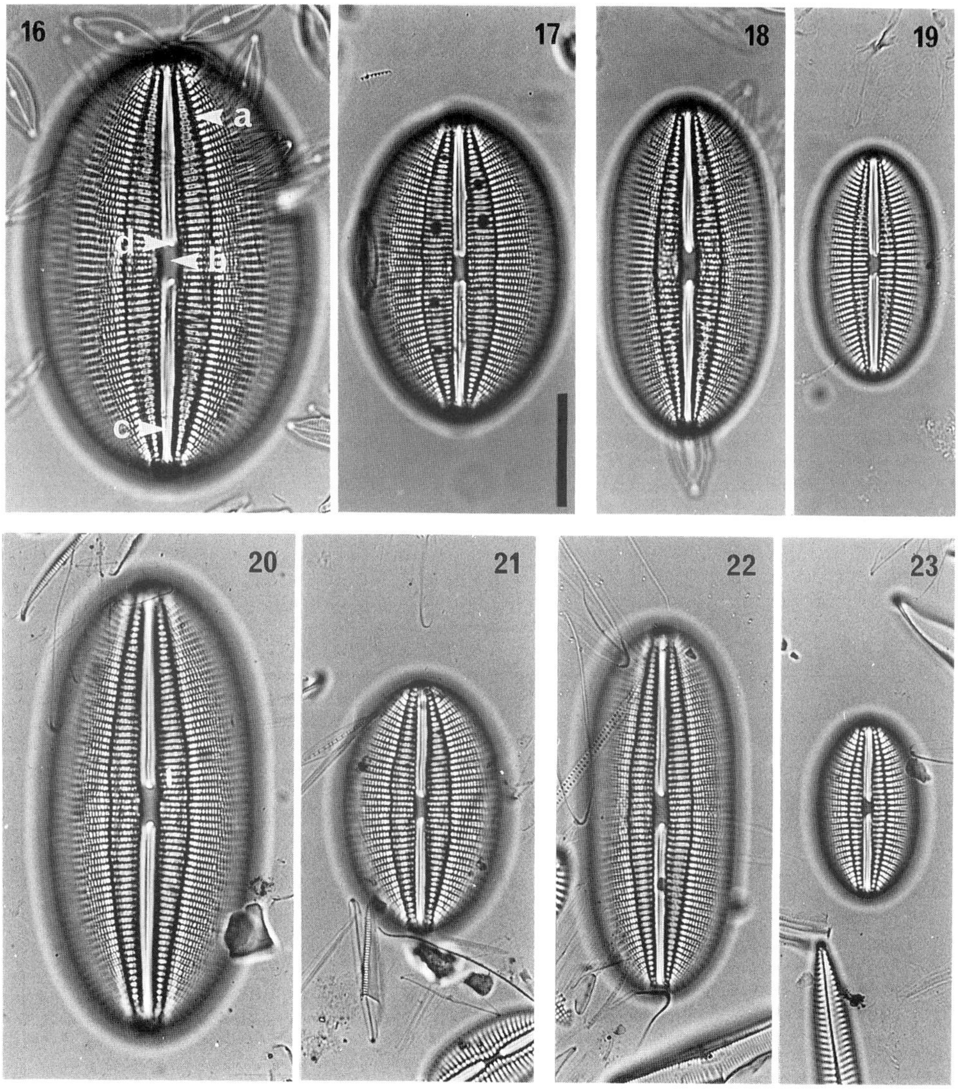


Figs 16–19. Sample group 1: Outer Hebrides. Scale bar = 20 µm.
Fig. 16: large valve of the wider population (BH307); arrows: *a:* abaxial edge of one longitudinal canal; *b:* central nodule; *c:* raphe sternum furrow; *d:* hooked central raphe ending; Fig. 17: small valve of the wider population (BH307); Fig. 18: large valve of the narrower population (E1276); Fig. 19: small valve of the narrower population (E1276).
Figs 20–23. Sample 2: Western Scotland (E1014). Scale bar (in Fig. 17) = 20 µm.
Fig. 20: large valve of the wider population; Fig. 21: small valve of the wider population; Fig. 22: large valve of the narrower population; Fig. 23: small valve of the narrower population.

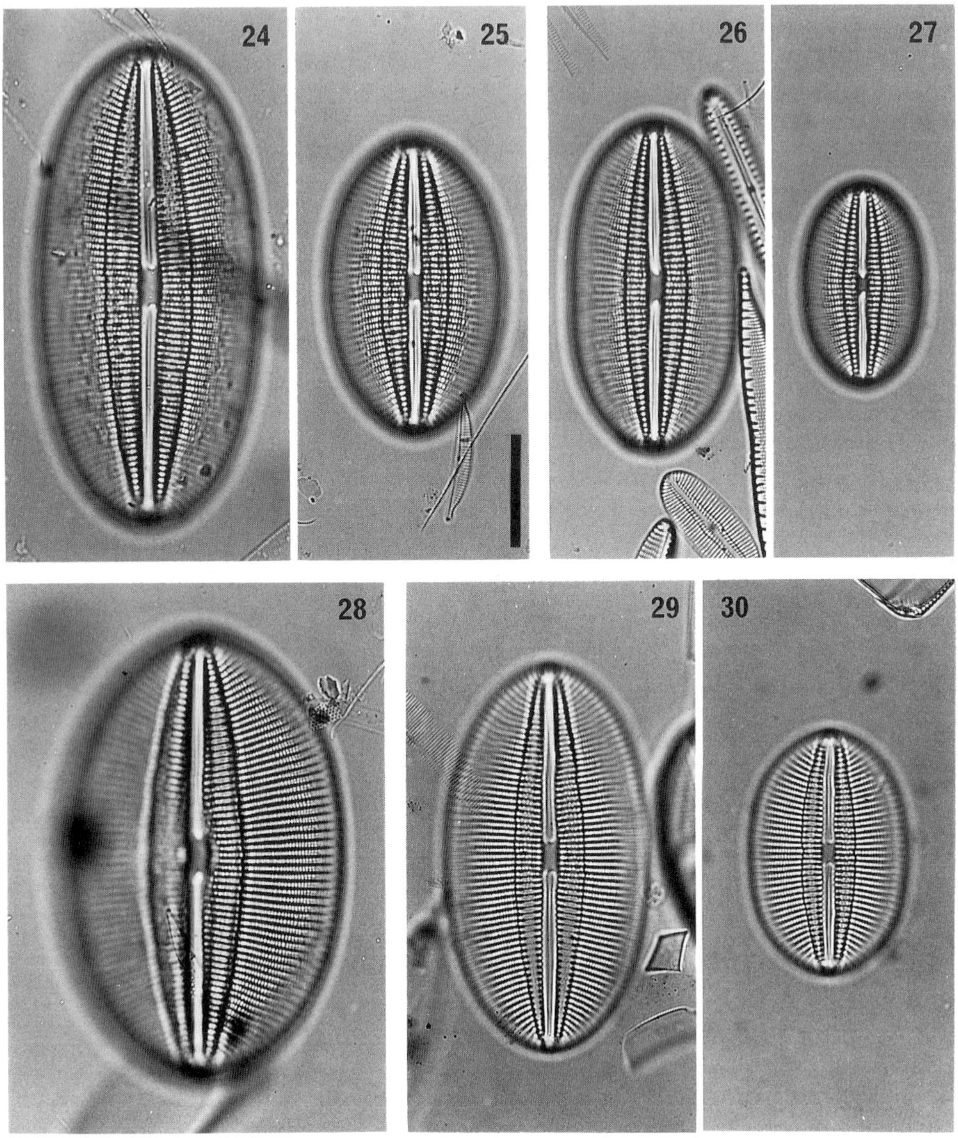


Figs 24–28. Sample group 3: South-East Scotland. Scale bar = 20 µm.
Fig. 24: large valve of the wider population (E1036); Fig. 25: small valve of the wider population (E1036); Fig. 26: large valve of the narrower population (E1000); Fig. 27: small valve of the narrower population (E1000); Fig. 28: very large, squint valve of the wider population (E1080).
Figs 29–30. Sample 4: North-East England (WA582). Scale bar (in Fig. 25) = 20 µm.
Fig. 29: large valve; Fig. 30: small valve.

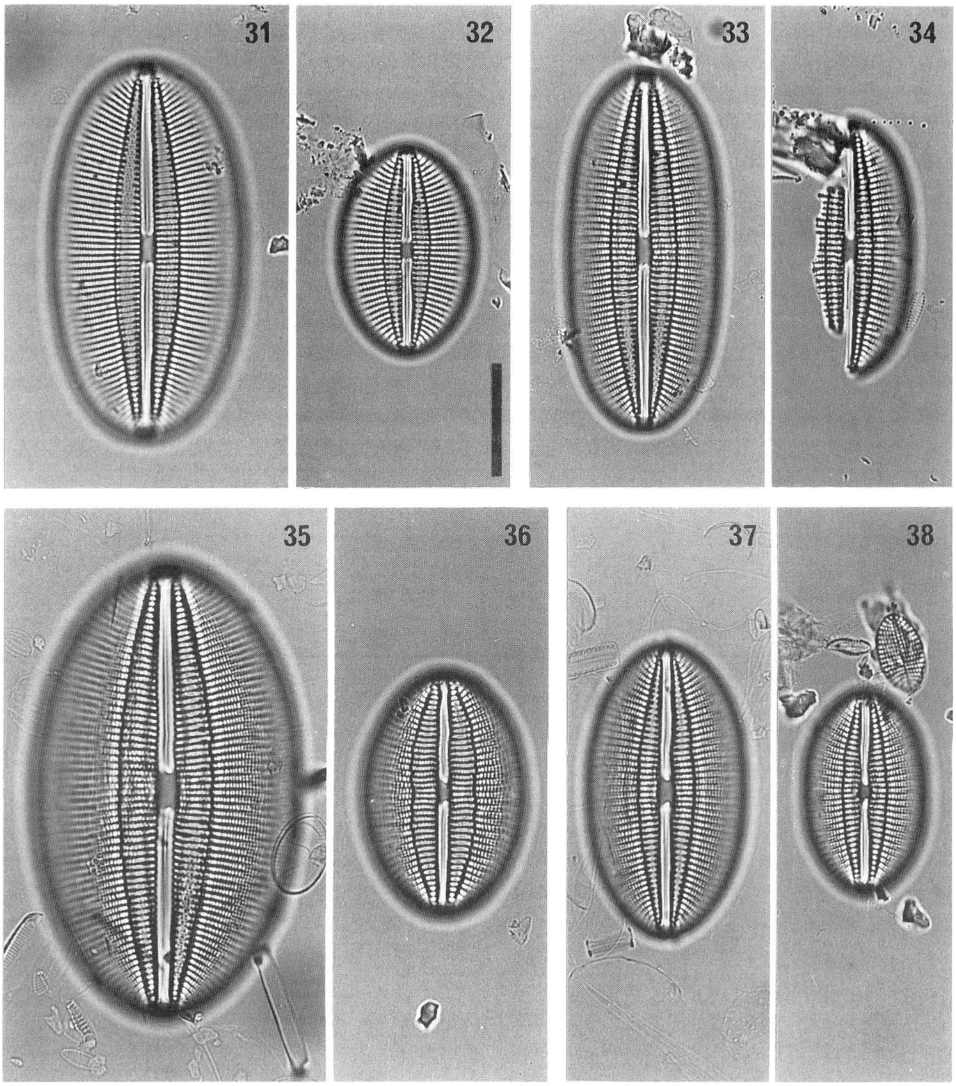


Figs 31–34. Sample 5: South Wales (E1301). Scale bar = 20 µm.
Fig. 31: large valve of the wider population; Fig. 32: small valve of the wider population; Fig. 33: large valve of the narrower population; Fig. 34: small, broken valve of the narrower population.
Figs 35–38. Sample 6: Isles of Scilly (B106). Scale bar (in Fig. 32) = 20 µm.
Fig. 35: large valve of the wider population; Fig. 36: small valve of the wider population; Fig. 37: large valve of the narrower population; Fig. 38: small valve of the narrower population.

*augur* Ehrenb. were much less in high salinity media than in the controls or the natural populations they were taken from; valve length varied much less.

Although most individual diatom cells are the products of mitosis, during which the inheritance of shape is largely physical, there are episodes in the life histories of most diatom species where shape is generated *de novo*, i.e. during the formation of auxospores after the fusion of meiotic products. The inheritance of shape from one sexual generation to the next must be under genetic control, but again there can be environmental effects: osmotic stress, for example, can cause the production of abnormal initial cells (Mann 1994b). In some freshwater diatoms (e.g. *Meridion* and *Synedra*: Geitler 1940a, b), distorted auxospores are quite common, but the distortions are usually lost during size-reduction, perhaps because of the effects of cell turgor (Mann 1994b). Schmid (1979) has observed differences in the shape and structure of initial cells of *Surirella peisonis* Pant. grown in different salt solutions; the different forms could be related to four different varieties of *S. ovalis* and *S. ovata*.

Thus, the variation within the Wide morphotype could be partly phenotypic, reflecting the interaction between genotype and the environment during either the mitotic or the sexual phase of the life cycle. Alternatively, the variation may have a purely genotypic basis. In either case, if two very similar morphotypes exist side by side in the same localities, it is possible that factors in the environment that affect the morphology of one (through phenotypic plasticity or selection) could have a similar effect on the morphology of the other. This does seem to be happening in *Diploneis*; compare Figs 3 and 5: both morphotypes, Wide and Narrow, are more rectangular (higher on the graph) and narrower (further to the left) in Western Scotland (Fig. 5) than in the Hebrides (Fig. 3). The same is true between the morphotypes in South-East and Western Scotland (Figs 7 and 5): they are more rectangular and narrower in Western Scotland (Fig. 5). Similarly, although to a lesser extent, both morphotypes are more rectangular in South-East Scotland (Fig. 7) than the equivalent morphotypes in the Isles of Scilly (Fig. 13). In this case, though, dimensions are not affected to the same extent in both morphotypes: the solid symbols are positioned roughly the same horizontally in Figs 7 and 13; the hollow symbols are further to the right (wider valves) in Fig. 13 than in Fig. 7.

There is at least one other possible cause of the large variation in size and shape within Wide: polyploidy. Mann & Stickle (1991) reported the occasional occurrence of trikaryotic auxospores in *Craticula cuspidata* (Kütz.) D.G. Mann. These were noticeably wider than diploid auxospores. Similarly, in *Dickieia ulvacea* Berkeley *ex* Kütz., polyploid auxospores formed by the fusion of three or four gametes were abnormally large relative to diploid auxospores, especially in their width (Mann 1994a), although the initial cells formed from these large auxospores were sometimes well within the normal diploid size limits, as a result of the contraction of the proptoplast during initial valve formation. However, there is as yet no evidence in any diatom that polyploid initial cells give rise to viable polyploid clones that can survive alongside their diploid progenitors. But even if polyploidy has not played a major part in the evolution of these *Diploneis* populations, it could perhaps be responsible for individual anomalous valves such as the very wide valve in Sample group 3 (Fig. 28).

*Unanswered questions and future research*

The data presented in this paper have shown that the two morphotypes (3 and 4, = Wide and Narrow) first identified from Oban and discussed by Droop (1994), occur in other samples from around the shores of Britain. There is no overlap between the morphotypes when they are sympatric and very little overlap between the two morphotypes when data from all the samples are combined; this evidence suggests that the two are separate species that should be given separate binomials. Each seems to exhibit a normal life-cycle, with auxosporulation alternating with size reduction. However, a number of questions remain to be answered:

1. How much of the variation between and within the morphotypes can be accounted for by phenotypic plasticity? This could be answered in part by autecological and culture studies similar to those of Jahn (1986) and Schmid (1979).

2. Are the morphotypes reproductively isolated? Both Geitler (e.g. 1968 and 1982) and Mann (1984, 1988, 1989) have demonstrated reproductive isolation between morphotypes of some common freshwater species. Similar studies with these *Diploneis* morphotypes would indicate whether or not they were the same biological species.

3. How different are the morphotypes genotypically and what is their evolutionary history? It is only very recently that molecular methods have been applied at the species level and below in diatoms (Medlin *et al.* 1991, Zechman *et al.* 1994). Such approaches are essential if we are to quantify the level of genotypic diversity within and between *Diploneis* morphotypes, since there are too few morphological markers to allow further morphometric or cladistic analysis.

**Acknowledgements**

Part of this work was done with a NERC grant (GR3/7923) and I am grateful for their financial support. My thanks, too, to Anna Bradshaw for assistance with some of the image analysis, to Bernard Hartley for his diatom sample from South Uist, and to David Mann for his support and encouragement, and for his review of the manuscript at several stages.

## References

Cox, E. J. (1993). Diatom systematics – a review of past and present practice and a personal vision for future development. *Nova Hedwigia, Beiheft* **106**, 1–20

Droop, S. J. M. (1994). Morphological variation in *Diploneis smithii* and *D. fusca* (Bacillariophyceae). *Archiv für Protistenkunde*, **144**, 249–270

Eaton, J. W. & Moss, B. (1966). The estimation of numbers and pigment content in epipelic algal populations. *Limnology and Oceanography*, **11**, 584–595

Geissler, U. (1970). Die Variabilität der Schalenmerkmale bei den Diatomeen. *Nova Hedwigia*, **19**, 623–773

Geitler, L. (1932). Der Formwechsel der pennaten Diatomeen (Kieselalgen). *Archiv für Protistenkunde*, **78**, 1–226

Geitler, L., (1940a). Die Auxosporenbildung von *Meridion circulare*. *Archiv für Protistenkunde*, **93**, 288–294

Geitler, L., (1940b). Gameten- und Auxosporenbildung von *Synedra ulna* im Vergleich mit anderen pennaten Diatomeen. *Planta*, **30**, 551–566

Geitler, L. (1968). Kleinsippen bei Diatomeen. *Österreichische Botanische Zeitschrift*, **115**, 354–362

Geitler, L. (1982). Die infraspezifischen Sippen von *Cocconeis placentula* des Lunzer Seebachs. *Archiv für Hydrobiologie, Suppl.* **63**, 1–11

Goldman, N., Paddock, T. B. B. & Shaw, K. M. (1990). Quantitative analysis of shape variation in populations of *Surirella fastuosa*. *Diatom Research*, **5**, 25–42

Gregory, W. (1857). On some new forms of marine Diatomaceae, found in the Firth of Clyde and in Loch Fine. *Transactions of the Royal Society of Edinburgh*, **21**, 473–542

Hustedt, F. (1927–66). Die Kieselalgen Deutschlands, Österreichs und der Schweiz unter Berücksichtigung der übrigen Länder Europas sowie der angrenzenden Meeresgebiete. In: *Dr Rabenhorst's Kryptogamen-Flora von Deutschland, Österreich und der Schweiz* **7**. Leipzig.

Jahn, R. (1986). A study of *Gomphonema augur* Ehrenberg: the structure of the frustule and its variability in clones and populations. In: *Proceedings of the eighth International Diatom Symposium, Paris, August 27–September 1, 1984* (M. Ricard, ed.), 192–204 (labelled 191–203). Koeltz, Koenigstein.

MacDonald, J. D. (1869). On the structure of the diatomaceous frustule and its genetic cycle. *Annals and Magazine of Natural History, ser. 4*, **3**, 1–8

Mann, D. G. (1984). Observations on copulation in *Navicula pupula* and *Amphora ovalis* in relation to the nature of diatom species. *Annals of Botany*, **54**, 429–438

Mann, D. G. (1988). The nature of diatom species: analysis of sympatric populations. In: *Proceedings of the ninth International Diatom Symposium, Bristol, August 24–30, 1986* (F. E. Round, ed.), pp.317–327. Bristol and Koenigstein.

Mann, D. G. (1989). The species concept in diatoms: evidence for morphologically distinct, sympatric gamodemes in four epipelic species. *Plant Systematics and Evolution*, **164**, 215–237

Mann, D. G. (1994a). Auxospore formation, reproductive plasticity and cell structure in *Navicula ulvacea* and the resurrection of the genus *Dickieia* (Bacillariophyta). *European Journal of Phycology*, **29**, 141–157

Mann, D. G. (1994b). The origins of shape and form in diatoms: the interplay between morphogenetic studies and systematics. In: *Shape and form in plants and fungi* (D. S. Ingram & A. J. Hudson, eds), 17–38. Academic Press, London

Mann, D. G. & Stickle, A. J. (1991). The genus *Craticula*. *Diatom Research*, **6**, 79–107

Medlin, L. K., Elwood, H. J., Stickel, S. & Sogin, M. L. (1991). Morphological and genetic variation within the diatom *Skeletonema costatum* (Bacillariophyta): evidence for a new species, *Skeletonema pseudocostatum*. *Journal of Phycology*, **27**, 514–524

Mou, D. & Stoermer, E. F. (1992). Separating *Tabellaria* (Bacillariophyceae) shape groups based on Fourier descriptors. *Journal of Phycology*, **28**, 386–395

Pfitzer, E. (1869). Über Bau und Zellteilung der Diatomeen. *Botanische Zeitung*, **27**, 774–776

Round, F. E. (1993). A *Synedra* (Bacillariophyta) clone after several years in culture. *Nova Hedwigia, Beiheft* **106**, 353–359

Schmid, A.-M. M. (1979). Influence of environmental factors on the development of the valve in diatoms. *Protoplasma*, **99**, 99–115

Steinman, A. D. & Ladewski, T. B. (1987). Quantitative shape analysis of *Eunotia pectinalis* (Bacillariophyceae) and its application to seasonal distribution patterns. *Phycologia*, **26**, 467–477

Stoermer, E. F. & Ladewski, T. B. (1982). Quantitative analysis of shape variation in type and modern populations of *Gomphoneis herculeana*. *Nova Hedwigia, Beiheft* **73**, 347–386

Stoermer, E. F., Qi, Y. Z. & Ladewski, T. B. (1986). A quantitative investigation of shape variation in *Didymosphenia* (Lyngbye) M. Schmidt (Bacillariophyta). *Phycologia*, **25**, 494–502

Theriot, E. & Ladewski, T. B. (1986). Morphometric analysis of shape of specimens from the neotype of *Tabellaria flocculosa* (Bacillariophyceae). *American Journal of Botany*, **73**, 224–229

Zechman, F. W., Zimmer, E. A. & Theriot, E. C. (1994). Use of ribosomal DNA internal transcribed spacers for phylogenetic studies in diatoms. *Journal of Phycology*, **30**, 507–512

# Morphological variability of *Fragilaria vaucheriae* (Kuetzing) Boye-Petersen var. *perminuta* (Grunow) Kobayasi on King George Island (Antarctica)

Hiroshi Fukushima* , Tsuyako Ko-Bayashi ** , Harue Fujita and Sakiko Yoshitake **

* *Institute of Phycology, Kunitachi, Tokyo, Japan*
** *Shonan Junior College, Yokosuka, Kanagawa, Japan*

## Abstract

Overall form of the valve in our specimens is similar to the original illustrations of *Synedra (vaucheriae* var.*) perminuta*. However, three distinct types are recognizable, with respect to the striae at the centre in our specimens. (1) Striae at one side of the margin are clear, while those at the other side are lacking; this is in agreement with the original description of *Synedra perminuta*. (2) Striae at both sides are obscure. (3) Striae are lacking at both sides. So, the nature of the striae at the centre cannot be used as the criterion for classification in this taxon. On the other hand, our specimens closely resemble *Synedra vaucheriae* in their overall form of the valve, but differ in their narrower breadth of valve and denser striae. Therefore, our specimens are placed as *Fragilaria vaucheriae* (Kuetzing) Boye-Petersen var. *perminuta* (Grunow) Kobayasi.

## Introduction

There have been many taxonomic views concerning the classification of *Fragilaria vaucheriae* (Kuetzing) Boye-Petersen var. *perminuta* (Grunow) Kobayashi (Van Heurck 1880–1881, Kobayashi 1955, Watanabe 1971, VanLandingham 1978, Negoro & Gotoh 1983, 1986, Krammer & Lange-Bertalot 1991). While Van-Landingham (1978) has synonymized *Synedra* (*vaucheriae* var. ?) *perminuta* with *Fragilaria vaucheriae* var. *vaucheriae*, Grunow (1881) considered these two taxa as different ones. On the other hand, Krammer & Lange-Bertalot (1991) have considered *Synedra* (*vaucheriae* var. ?) *perminuta* as a variety of *Fragilaria capucina*. The specimens of *F. vaucheriae* var. *perminuta* taken from King George Island were surveyed to make clear their taxonomic position comparing with the data of the specimens of *F. vaucheriae* var. *vaucheriae* from two different sampling places, Japan (Fukushima *et al.* 1992a) and Kenya (Fukushima *et al.* 1992b).

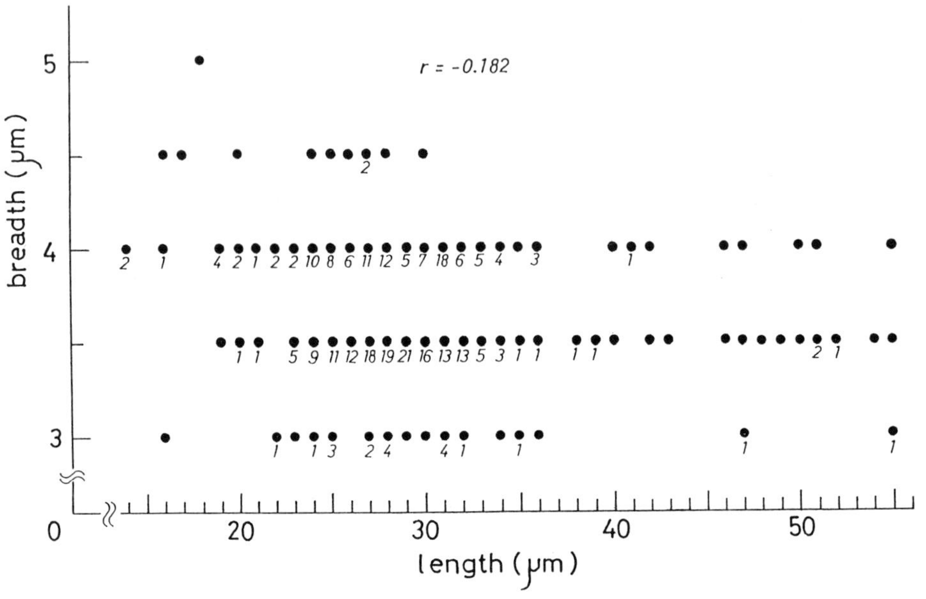


Fig. 1. Diagrams of correlation of breadth with length of valve in *Fragilaria vaucheriae* var. *perminuta*. Numbers under closed symbols are the numbers of specimens that have the same breadth and length (ex. = • one specimen, •1 = two specimens, •2 = three specimens).

## Materials and Method

The investigation was carried out on material from a stream on King George Island (Antarctica, South Shetland Islands). The samples were cleaned by sulphuric acid and hydrogen peroxide treatment and mounted in Pleurax. Almost 370 valves were identified at 2000× magnification on light microphotographs.

## Observations

On the basis of the form of the valve, four types were distinguished. (1) the broad-lanceolate type (Plate 3, figs A–F), (2) the lanceolate type (Plate 3, figs G–P), (3) the prolonged lanceolate type (Plate 3, figs Q–Y) and (4) the linear type (Plate 3, figs Z–f). Their relative frequencies are broad-lanceolate (4.3%), lanceolate (26.4%), prolonged lanceolate (16.0%), linear (51.4%). Apices are acute (Plate 3, figs h & I), rostrate (Plate 3, figs j–t) or capitate (Plate 4, figs A–I). Specimens with rostrate apices are found most frequently with a relative frequency of 84.0%, while those with acute or capitate apices are found at frequencies of 7.0% and 7.2% respectively. Four types are distinguished from the margins of the valve. (1) the nearly parallel type (Plate 4, figs J–P), (2) the slightly convex type (Plate 4, figs Q–W), (3) the slightly concave at the centre of the dorsal side type and (4) the slightly concave at the centre of one side type. Specimens with nearly parallel margins are commonest (83.9%), and those with slightly convex margins are second most common (13.9%). As for the central area, this usually extends to both margins of the valve (Plate 5, figs A–P) or sometimes develops on only one side of the valve (Plate 4, figs X–l). Their relative frequencies are 76.7% and 23.3% respectively. The number of striae at the central area is commonly four in the latter case, although up to 8 can be found. The relative frequencies of 5, 6, 7 and 8 striae at the central area are 4.3%, 9.4%, 7.5% and 1.3% respectively (Plate 6, figs C–o). The regression equation between length and breadth is shown in Figure 1. The coefficient of correlation (r) is negative at -0.182. Studies on *Fragilaria vaucheriae* from samples taken in Lake Shikotsu (northern part of Japan) (Fukushima et al. 1992a) and Lake Nakuru (Kenya) (Fukushima et al. 1992b) revealed similar patterns of morphological variation.

## Discussion

The original illustration of *Synedra* (*vaucheriae* var. ?) *perminuta* Grunow is on Plate 40, Fig. 23 of Van Heurck (1880–1881), which is reproduced on Plate 1, fig. S in this paper. In our samples, some specimens closely resemble the original illustration in their overall form of the valve (Plate 1, figs T–Z), apart from the striae at their centre. Though in the original illustration, striae at the centre are lacking at only one side of the margin, our samples additionally include specimens in which striae at both sides of the margins are obscure or totally lacking. Accordingly, the characteristics of the striae at the centre cannot be used as the criterion for classification.

Krammer & Lange-Bertalot (1991) found specimens which should belong to *Fragilaria capucina* var. *perminuta* (Plate 109, fig. 1; or Plate 1, fig. A in this paper) among the specimens of the syntype of *Synedra famelica* (not lectotype, Herbar Kuetzing 841) from the samples of Lake Zurich. Similar specimens were found in our sample (Plate 1, figs B–G). Some specimens which are similar to the specimens identified as *F. capucina* var. *perminuta* by Krammer & Lange-Bertalot (Plate 1, figs 3–4; Plate 1, figs H and N in this paper), are found also in our sample (Plate 1, figs I–M and O–R). Therefore, our samples contain the taxa previously identified as *Synedra* (*vaucheriae* var. ?) *perminuta,* originally illustrated by Grunow and *Fragilaria capucina* var. *perminuta* identified by Krammer & Lange-Bertalot, and a gradual intergrading of these and various intermediate shapes of valve can be recognized. Therefore these should all be considered as belonging to the same taxon. Krammer & Lange-Bertalot (1991) considered this taxon as *Fragilaria capucina* Desmazieres var. *perminuta.*They pointed out that this taxon was distinguished from *F. capucina* var. *vaucheriae* by its denser striae and its linear-lanceolate valve with more-or-less

Plate 1. A–F: Valves – broad-lanceolate (4.3%); G–P: Valves – lanceolate (26.4%); Q–Y: Valves – prolonged lanceolate (16.0%); Z–f: Valves – linear (51.4%); g: Abnormal form of valve; h, i: Valves with acute apices (7.0%); j–t: Valves with rostrate apices (84.0%).

Plate 2. A–I: Valves with capitate apices (7.2%); J–P: Valves with nearly parallel margins (83.9%); Q–W: Valves with slightly convex margins (13.9%); X–l: Central area – developing on only one side of valve (23.3%).

Plate 3. A–P: Central area – extending to both margins of valve (76.7%); Q–U: Central area – developing on only one side of valve with obscure striae (97.1%); V–Z: Central area – developing on only one side of valve (2.9%); a–o: Central area – extending to both margins of valve in which obscure striae are recognized (97.6%).

Plate 4. A, B: Central area – lacking striae at both sides of margins (2.4%); C, D: Number of short striae at centre of one side of margin – 5 (4.3%); E–I: Number of short striae at centre of one side of margin – 6 (9.4%); J–N: Number of short striae at centre of one side of margin – 7 (7.5%): O: Number of short striae at centre of one side of margin – 8 (1.3%).

Plate 5. A: *Fragilaria capucina* var. *perminuta* = syntype of *Synedra famellica* collected in Lake Zürich (not lectotype, Herbar Kuetzing 841) (Krammer & Lange-Bertalot 1991, Pl. 109, Fig. 1); B–G: Valves which are similar to overall form of A; H–N: *Fragilaria capucina* var. *perminuta* identified by Krammer & Lange-Bertalot (Krammer & Lange-Bertalot 1991, Pl. 109, Figs 3 & 4); I–M: Valves which are similar to overall form of H; O–R: Valves which are similar to overall form of N; S: Original illustration of *Synedra* (*vaucheriae* var. ?) *perminuta* (Van Heurck 1881, Pl. 40, Fig. 23); T–Z: Valves which are similar to overall form of S; a–e: Valves which are similar to overall form of S, with obscure striae at one side of margin at centre; f–l: Valves which are similar to overall form of S, without striae at one side of margin at centre.

Plate 6. A–O: Valves which are similar to overall form of S in Plate 1 and their margins of valve at centre are nearly parellel; P–W: Intermediate form of valve between that of S in Plate 1 and that of *Synedra rumpens*; X–e: Valves which are similar to overall form of *Synedra rumpens*.

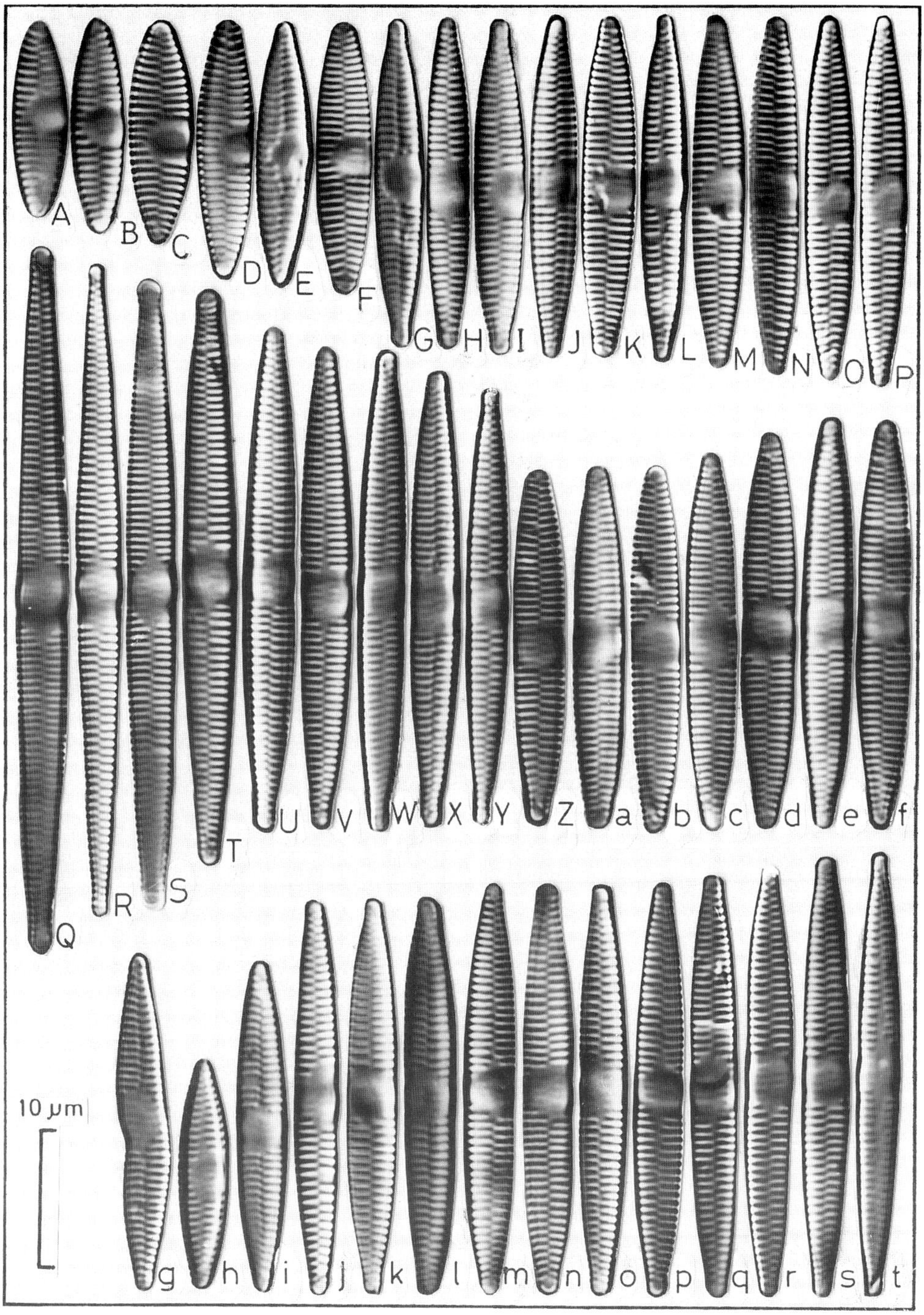
A B C D E F G H I J K L M N O P
Q R S T U V W X Y Z a b c d e f
g h i j k l m n o p q r s t
10 μm

Plate 1.

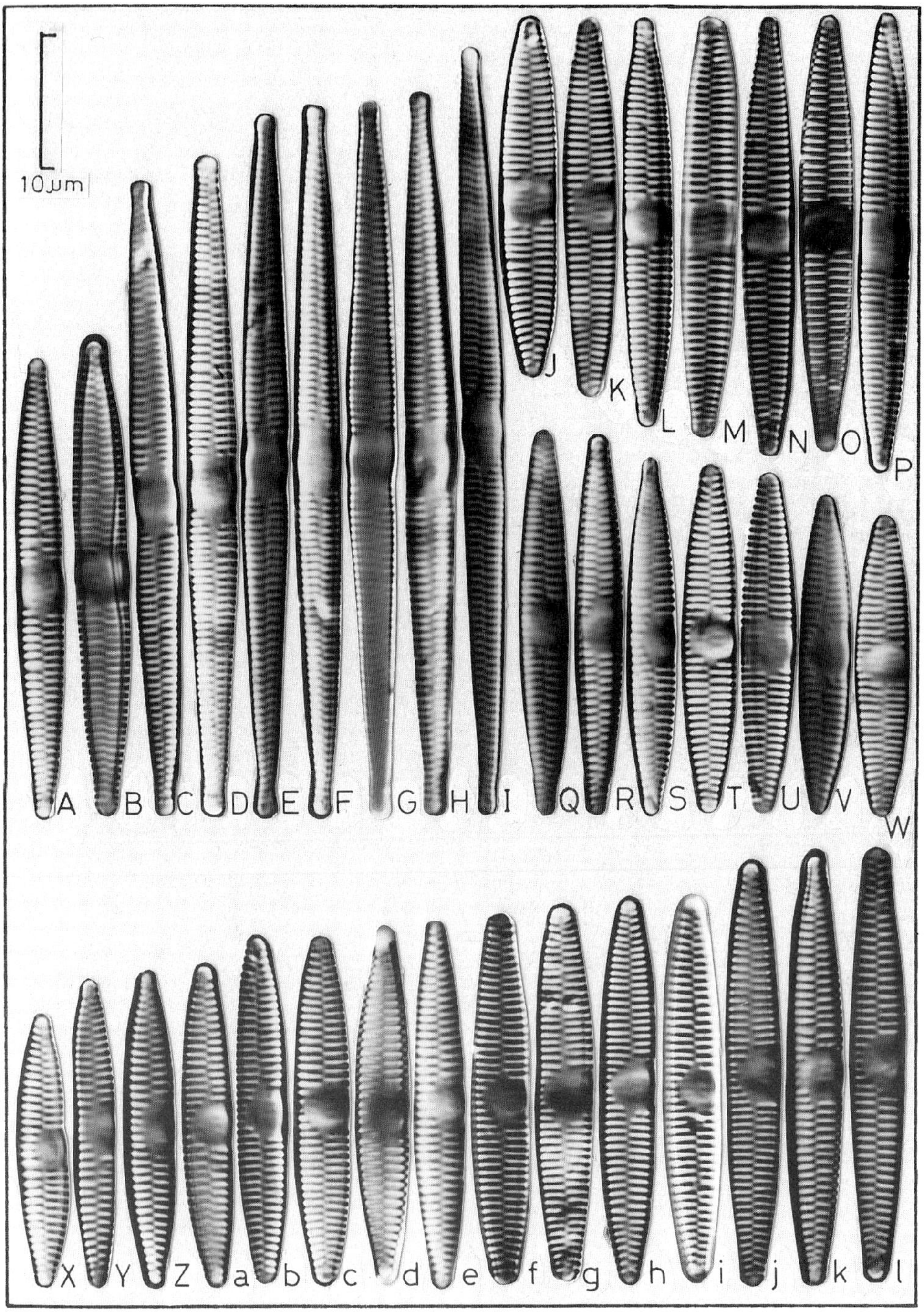


Plate 2.

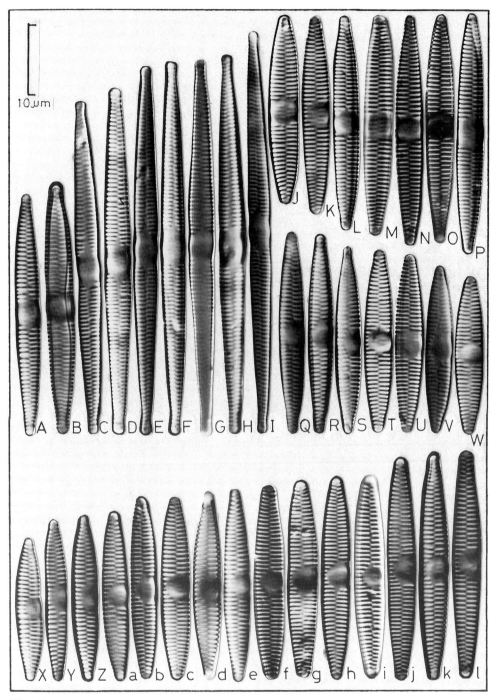

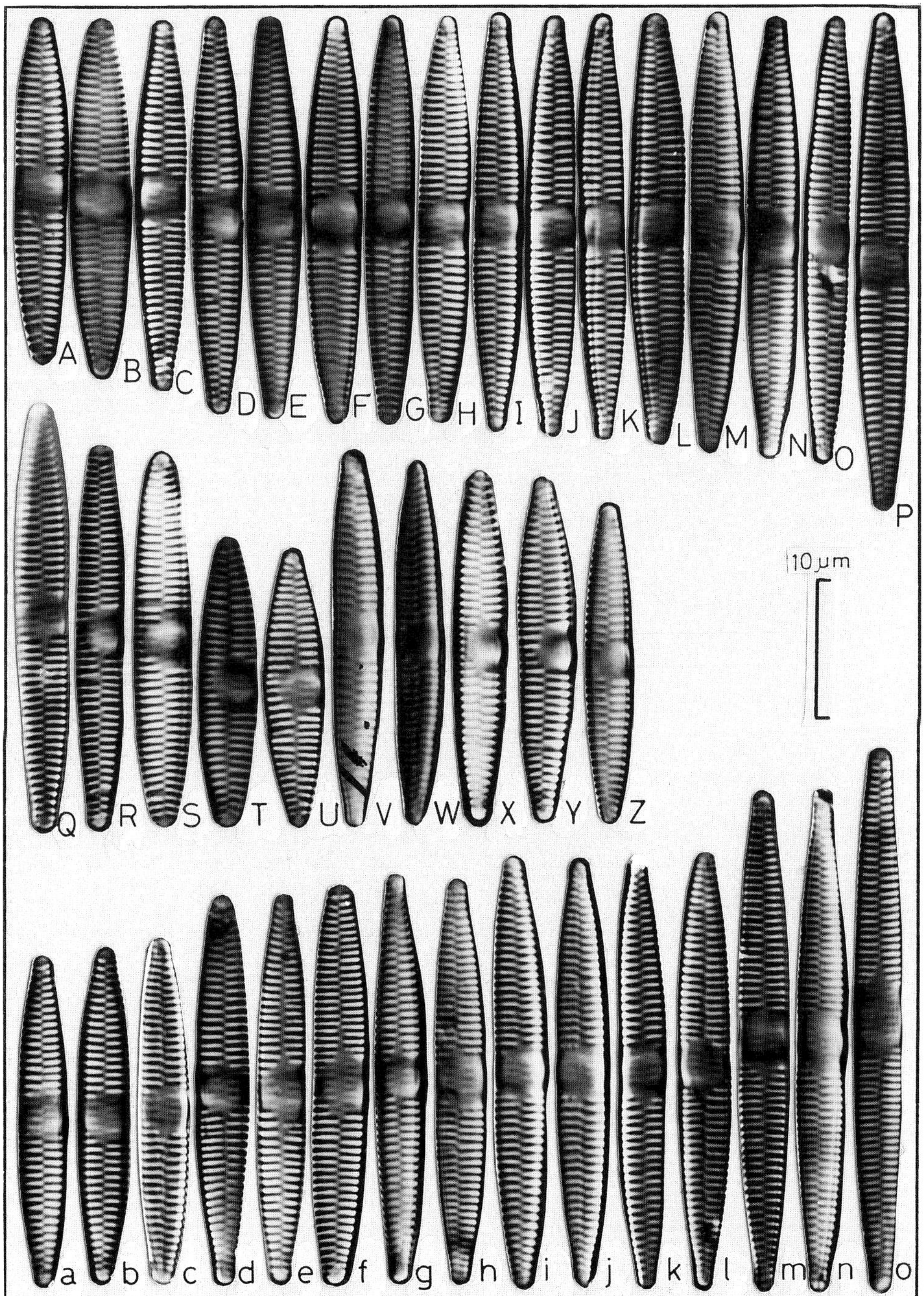


Plate 3.

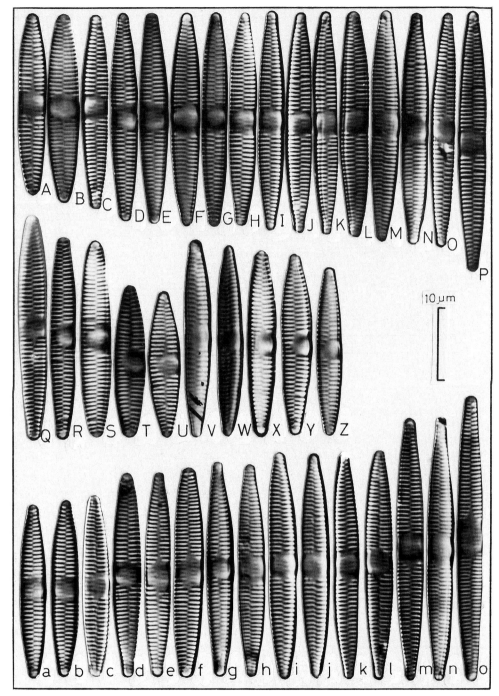

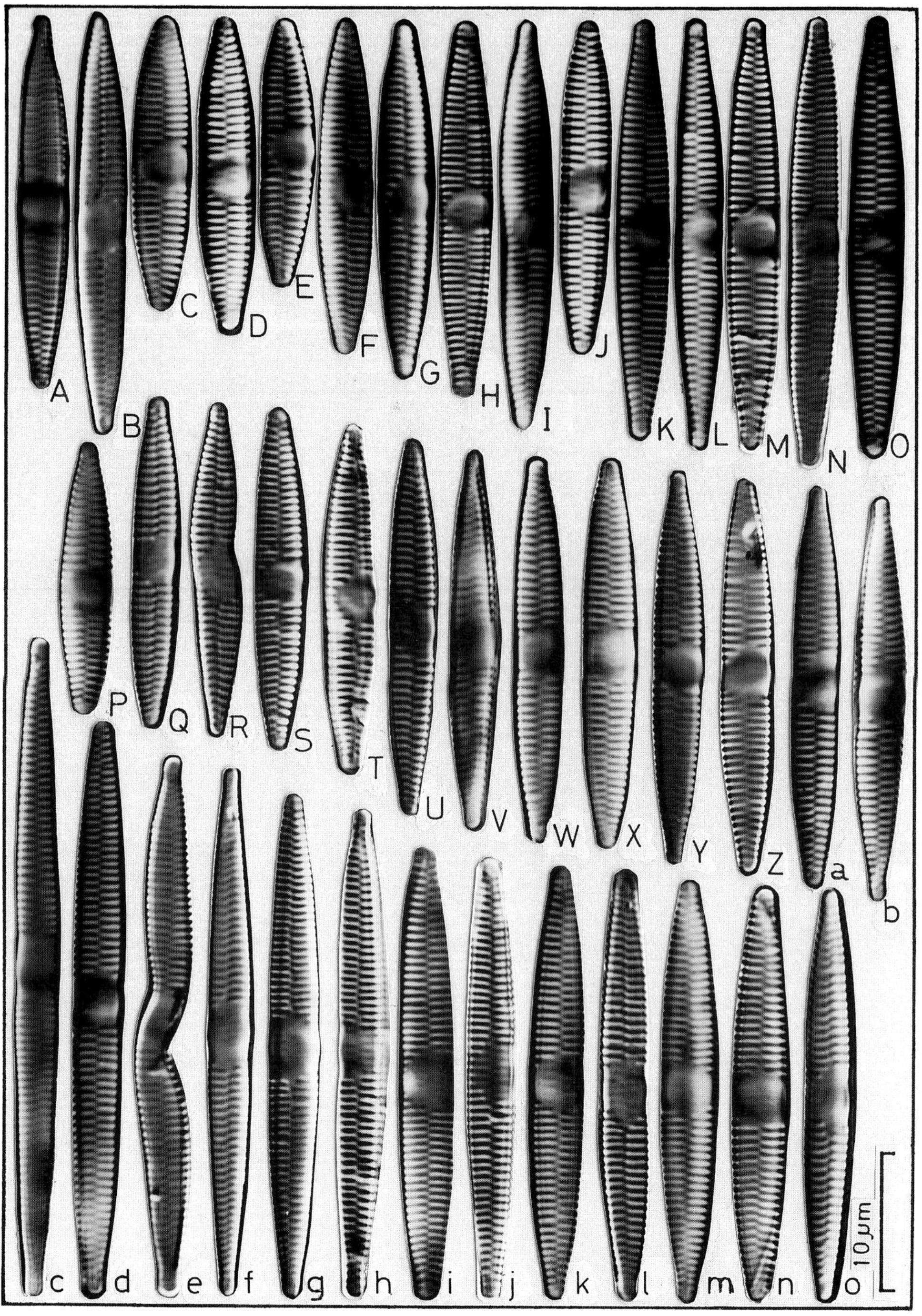


Plate 4.

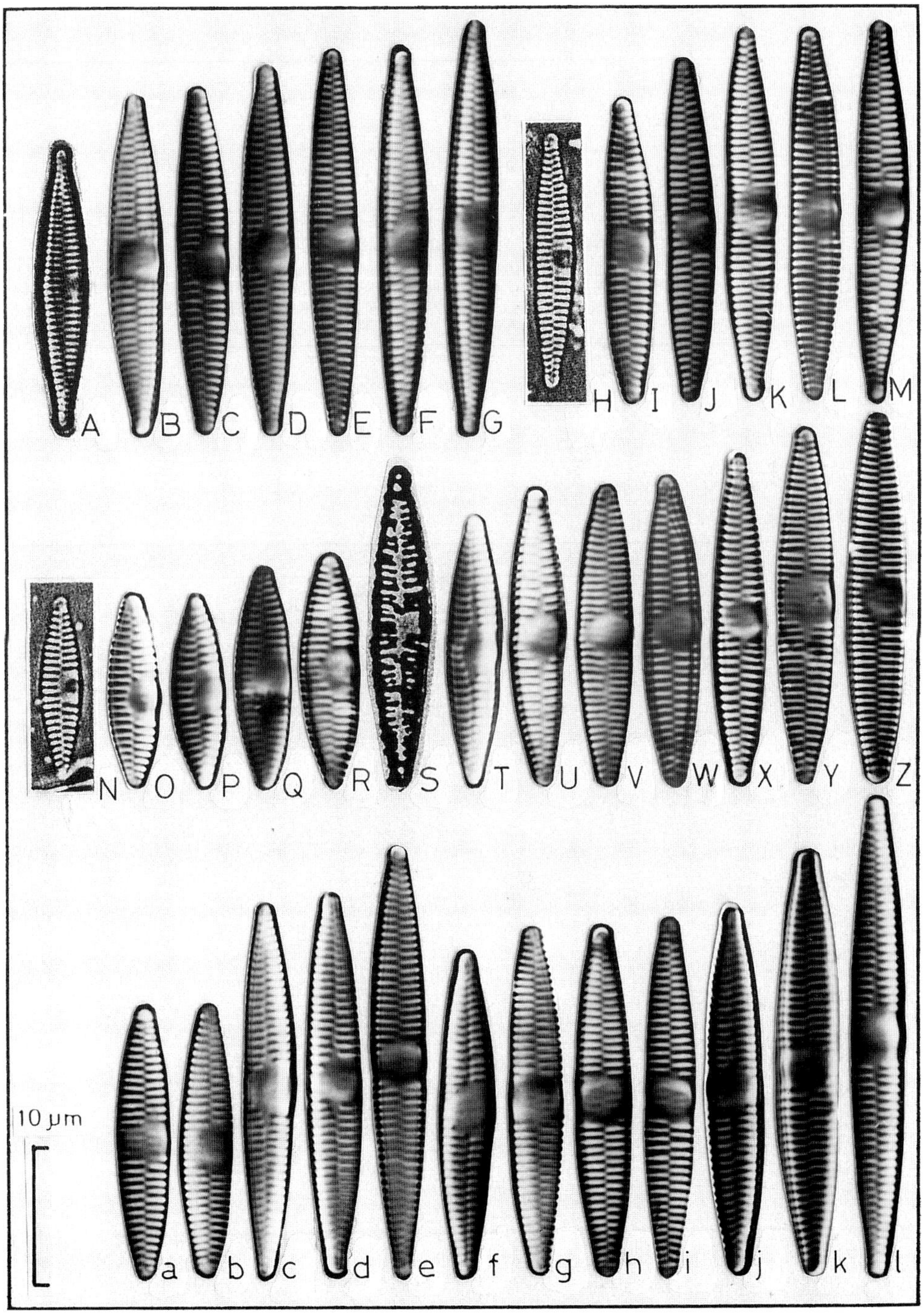


Plate 5.

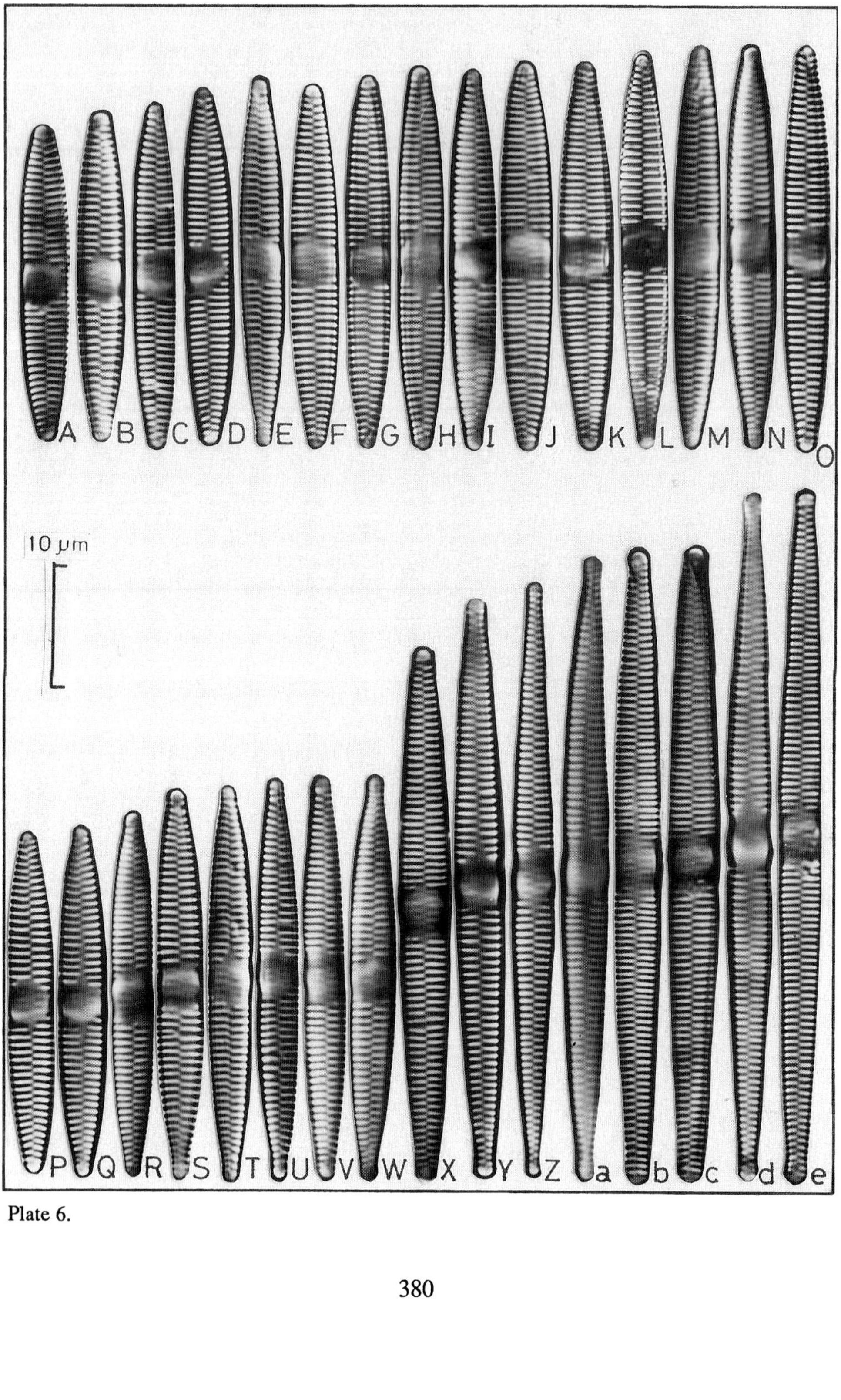


Plate 6.

protruding apices. Though they indicated that the breadth of the valve of *F. capucina* var. *perminuta* was 3–3.5 (rarely 4) μm, in our specimens it is 3–5 μm; the mode is at 3.5 μm (Fig. 2), which is again a larger value than seen in their data. As to the breadth

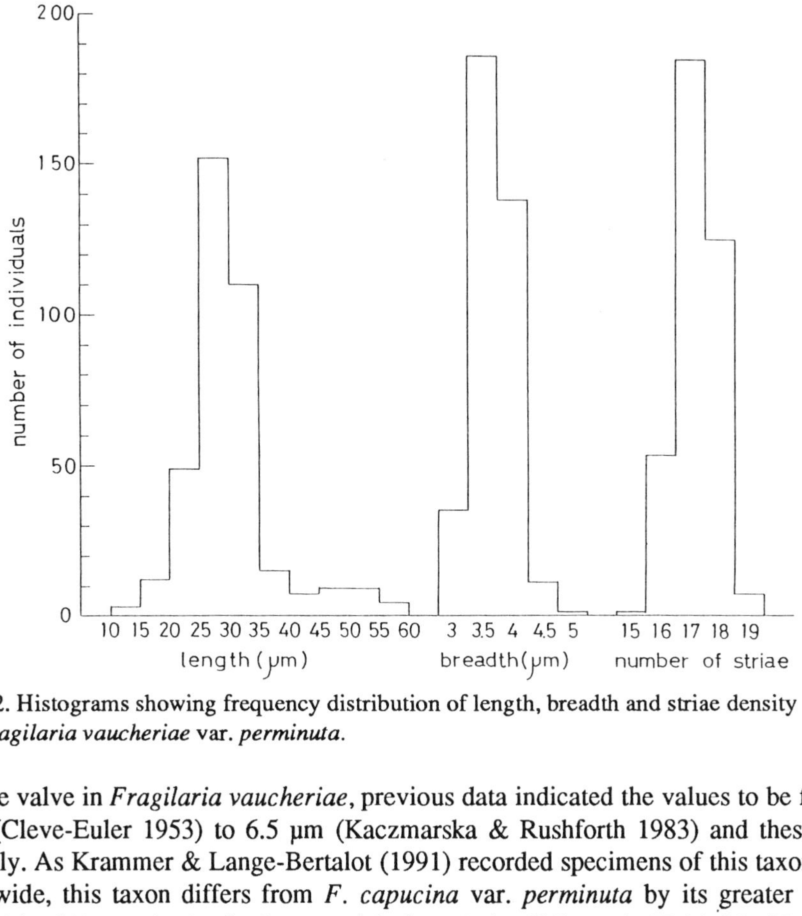

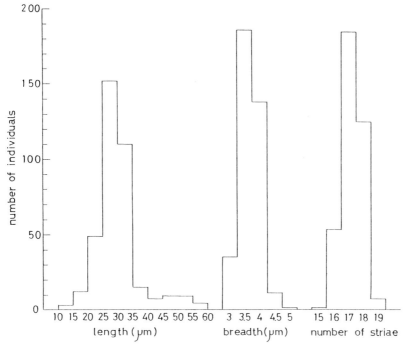

Fig. 2. Histograms showing frequency distribution of length, breadth and striae density in 10 μm of *Fragilaria vaucheriae* var. *perminuta*.

of the valve in *Fragilaria vaucheriae*, previous data indicated the values to be from 1.5 μm (Cleve-Euler 1953) to 6.5 μm (Kaczmarska & Rushforth 1983) and these varied widely. As Krammer & Lange-Bertalot (1991) recorded specimens of this taxon as 4–5 μm wide, this taxon differs from *F. capucina* var. *perminuta* by its greater breadth. Breadth of *F. vaucheriae* in the materials from Lake Shikotsu in Hokkaido (Fukushima *et al.* 1992a) is 3–6 μm while the mode is at 5 μm; breadth is 3–5 μm and mode at 4 μm in the specimens from Lake Nakuru in Kenya (Fukushima *et al.* 1992b). It is clear from these data that the breadth of the valve in *F. construens* var. *perminuta* is less than in *F. vaucheriae* as Krammer & Lange-Bertalot (1991) recorded. Therefore it is clear that these two taxa may be distinguished by the breadth of their valves. Though Negoro & Gotoh (1983) reported the breadth in *F. vaucheriae* to be less than that of *F. capucina* var. *perminuta*, this result might be incorrect. Previous reports have indicated that striae in 10 μm at the centre varied between 9 (Cleve-Euler 1953) and 21 (Watanabe 1971) as for *F. vaucheriae* and this value clearly has a wide range. As to the materials from Lake Shikotsu and Lake Nakuru, striae in 10 μm vary between 9 and 19 the mode is 11 in Lake Shikotsu while in Lake Nakuru, striae distribute between 12 and

15 and the mode is 13, and these ranges are similar to the data (9–14) of Krammer & Lange-Bertalot (1991), and the modes are also within their range. In the specimens from King George Island it varies between 15 and 19; the mode is 17, *i.e.* similar to the record of *F. capucina* var. *perminuta* (17–21) of Krammer & Lange-Bertalot (1991) and our specimens in this study should be regarded as *F. capucina* var. *perminuta* on the basis of the density of the striae. According to Negoro & Gotoh (1983), striae in *F. capucina* var. *perminuta* are denser than *F. vaucheriae* and Gotoh (1986) further reported that the range was 17–20 in 10 μm in the former and 14–15 in the latter and that the two taxa were distinguishable on the basis of their differing distribution. VanLandingham (1978) considered *F. capucina* var. *perminuta* to be synonymous with *F. vaucheriae*. From these observations, we believe that *F. capucina* var. *perminuta* should be named *F. vaucheriae* var. *perminuta.*

The scientific name of this taxon is then as follows:

*Fragilaria vaucheriae* (Kuetzing) Boye-Petersen var. *perminuta* (Grunow) Kobayasi, *Bull. Chichibu Museum of Natural History*, 6, 48, pl. 1, fig. 6 (1955).

Synonym:

*Synedra* (*vaucheriae* var. ?) *perminuta* Grunow, in Van Heurck's *Synop. Diat. Berg.* pl. 40, fig. 23 (1881).

*Fragilaria vaucheriae* (Kuetzing) Boye-Petersen, VanLandingham, *Catalogue of the fossil and recent genera and species of diatoms and their synonyms* 3931 (1978).

On the other hand, our sample contains specimens which should be identified as *Synedra rumpens* (Plate 2, fix X–e) and in addition, intermediate shapes between these specimens and those which are similar to the original illustrations of *Synedra* (*vaucheriae* var. ?) *perminuta* (Plate 1, figs T–l), are recognized (Plate 2, figs P–W). Though we refer here to the relationship between *Synedra rumpens* and *Fragilaria vaucheriae* var. *perminuta*, more study will be needed in order to clarify the distinction.

## Acknowledgement

We are grateful to Dr Shuji Ohtani, Shimane University for providing the samples from Antarctica.

## References

Cleve-Euler, A. (1953). Die Diatomeen von Schweden und Finnland. 2. *Kungl. Svenska Vet. Akad. Handl. Ser.* 4, **4** (1), 1–158, figs 292–483.

Fukushima, H., Ko-Bayasi, H., Kurihara, M. & Fujita, H. (1992a). Morphological variability of *Fragilaria vaucheriae* (Kuetzing) Boye-Petersen. *Japan Journal Water Treat. Biol.*, **28** (1), 33–50.

Fukushima, H., Ko-Bayasi, H., Fujita, A. & Ohno, M. (1992b). Morphological variability of *Fragilaria vaucheriae* (Kuetzing) Boye-Petersen (Further notes). *Ibid.*, **28** (2), 15–34.

Gotoh, T. (1986). Diatom community of Kumano-gawa River estuary. *Diatom*, **2**, 1103–115.
Kaczmarska, I. & Rushforth, S. R. (1983). The diatom flora of Blue Lake Warm Spring, Utah, USA. *Bibl. Diatom*, **2**, 1–49.
Kobayasi, H. (1955). Diatom flora of Chichibu Tama National Park 1. *Bull. Chichibu Museum of Natural History*, **6**, 47–53. Pls 1–3.
Krammer, K. & Lange-Bertalot, H. (1991). Bacillariophyceae 3l Ettle, E. *et al.* Süsswasserflora von Mitteleuropa 2/3, 1–14, 1–576. Gustav Fischer Verlag., Stuttgart, Jena.
Negoro, K. & Gotoh, T. (1983). Diatom vegetation of the less polluted river, the U-kawa River, Kyoto Prefecture. *Japan Journal of Limnology*, **47** (1), 77–96.
Negoro, K. & Gotoh, T. (1986). Diatom vegetation of the River Yura. *Japan. Mem. Facul. Agric. Kinki Univ.*, **16**, 67–118.
Van Heurck, H. (1880–1881). Synopsis des diatomees de Belgique. *Atlas*, 1–132. Texte 235 pp. Anvers.
VanLandingham, S. L. (1978). *Catalogue of the fossil and recent genera and species of diatoms and their synonyms*,7, 3606–4241. J. Cramer, Gaduz.
Watanabe, T. (1971). Attached diatoms from the Takami River, Nara Prefecture. *Annual Rep. Noto Marine Lab.*, **11**, 9–20. Pls 3–8.

# Genetic variation in *Fragilaria capucina* clones along a latitudinal gradient across North America: a baseline for detecting global climate change

Kyle D. Hoagland, Stephen G. Ernst, Susan I. Jensen, Raymond J. Lewis, Virginia I. Miller and Dean M. DeNicola*

*Department of Forestry, Fisheries and Wildlife, University of Nebraska, Lincoln, Nebraska 68583–0814, U.S.A.*

## Abstract

A major objective of this research has been to provide data for detecting global warming effects on freshwater lakes in central North America. The filamentous diatom *Fragilaria capucina* was chosen because it is widely distributed and often dominant in the phytoplankton of lakes throughout this region. A total of 147 clones of *F. capucina* have been isolated and characterized genetically using PCR-RAPD methodology. A majority of these clones exhibited unique combinations of genetic markers, demonstrating that this approach is capable of distinguishing clones at several spatial scales. Within-site banding patterns appear more similar than among-site patterns from different localities. The results should provide a baseline for detecting shifts in the distributions of fingerprinted *F. capucina* populations, which are hypothesized to shift as isotherms migrate northward.

## Introduction

Although inland waters cover less than two percent of the earth's surface, they represent a finite resource critical to the maintenance of terrestrial ecosystems including man (Wetzel 1983). Significant effects of global warming on inland waters in North America are expected (Schindler *et al.* 1990; Meisner 1990; Gorham 1991), particularly in the Great Plains Region (Matthews & Zimmerman 1990; Poiani & Johnson 1991). A report from the Experimental Lakes Area of northwestern Ontario indicated that increases in annual air temperatures over the past 20 years have resulted in an increase in mean and maximum water temperatures of lakes in that region (Schindler *et al.* 1990). This study provided the

* Present address: Dept. of Biology, Slippery Rock University, Slippery Rock, PA. 16057, U.S.A.

first empirical evidence for changes in algal species diversity and productivity due to increases in water temperatures. Byron & Goldman (1990) have also predicted dramatic increases in freshwater algal productivity and shifts in species composition, using a model based on long-term limnological data from Castle Lake, California. Thus, global climatic change may have a direct influence on the base of the food web in freshwater systems.

Temperature is a major determinate of the biogeographical distribution of freshwater planktonic diatoms (Reynolds 1988). Temperature sets the maximum growth rate for algae when conditions are optimal (Eppley 1972) and also influences the degree to which other environmental factors (e.g. nutrients and light) can limit algal growth (Li 1980; Rhee 1982). The optimal and extreme temperatures for growth differ among planktonic diatom species and often correspond to the natural temperature regime in which the taxa occur (Talling 1955; Braarud 1961; Thomas 1966; Seaburg & Parker 1983). Several studies have shown that clonal strains or ecotypes of a single morphologically defined algal species isolated from regions of cold or warm water differ in their extreme and optimum temperatures for growth (Braarud 1961; Guillard & Ryther 1962; Fogg 1969; Hulbert & Guillard 1968; Mosser *et al.* 1977; Johansen *et al.* 1990). Seaburg & Parker (1983) isolated and measured the temperature tolerances of 49 strains for several species of freshwater phytoplankton from warm and cold water environments in Virginia. Based on their results, the authors predicted that a 5°C increase in maximum *in situ* temperatures would decrease the percentage of isolates capable of growth at the maximum temperature by 27%. In addition, they suggested that the distribution of freshwater phytoplankton would be substantially altered not only by excluding growth of cold water forms, but also by permitting warm water forms to grow in colder seasons. Based on these studies, it is clear that long-term ambient environmental temperatures influence the extreme and optimum temperatures for growth in freshwater phytoplankton.

Morphological features continue to be the primary taxonomic determinant in diatom systematic and ecological studies. However, morphological features cannot be used to reliably document the extent of intraspecific variation with regard to thermal ecotypes. While there are several reports in which DNA markers have been used to assess interspecific variability, intraspecific DNA polymorphic markers have been reported for diatoms only by Zechman *et al.* (1994). Soudek & Robinson (1983) provided a report which utilized isozymes to study intraspecific population genetic structure in a freshwater diatom (*Asterionella*). Studies at the intraspecific level are needed to provide DNA fingerprints to assess the genetic basis of the selection response to temperature, because morphological markers do not adequately assess genotype. RAPD (Random Amplified Polymorphic DNA) methodology is well suited to assaying intraspecific variability and assignment of DNA fingerprints for each isolate, and offers a promising approach to providing a baseline for detection of the effects of global warming. In our experiments, we have used DNA fingerprints generated by RAPD analysis to ensure strain integrity, and later for monitoring latitudinal shifts in the distribution of strains as well as shifts in the competitive ability of several isolates in mixed populations. Thus, the objective of this research has been to provide genetic markers and information on genetic variability among diatom clones across a broad latitudinal temperature gradient traversing the Great Plains of North America.

## Materials and Methods

*Fragilaria capucina*, a widely distributed filamentous diatom often dominant in planktonic systems, was isolated from a sample site (i.e. a lake or reservoir) in each of the following localities, along a north-south latitudinal transect: Manitoba, North Dakota, South Dakota, Nebraska, Kansas, Oklahoma and Texas (Fig. 1). Samples were taken by multiple net tows at 1–3 locations on each lake and stored in coolers near ambient lake water temperatures until they were returned to the laboratory. Single filaments of *F. capucina* were isolated into standard soil-water medium (Hoshaw & Rosowski 1973) supplemented with Alga-gro nutrients (Carolina Biological Supply) and maintained in growth chambers on a 12:12 light:dark cycle near the temperature of the lake or reservoir from which they were collected.

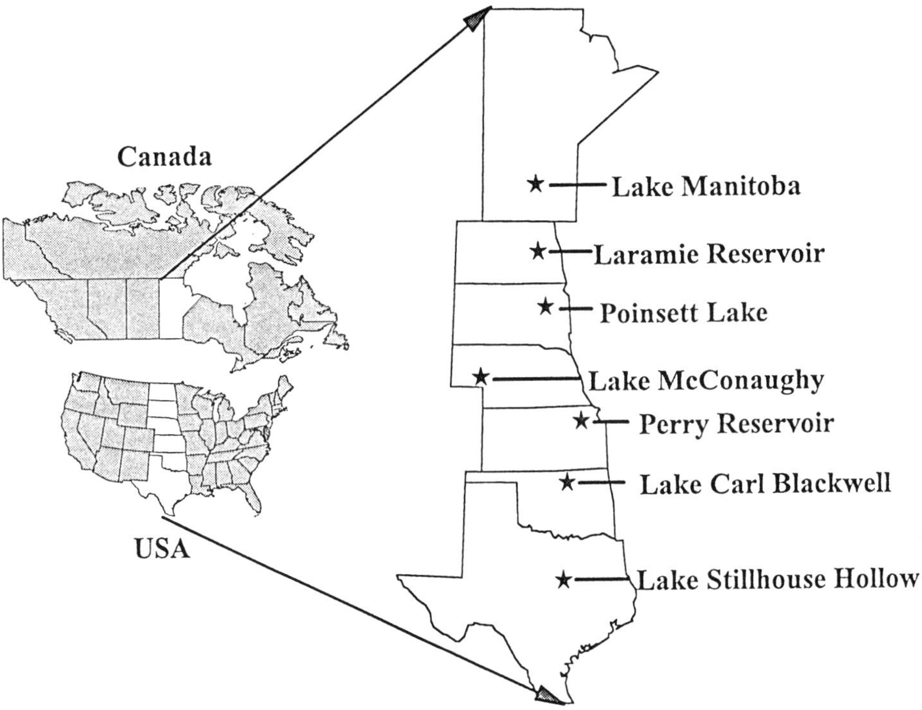


Fig. 1. Location of sites from which *Fragilaria capucina* was collected.

Each clone was grown in 200 mL WC medium (Guillard 1975) plus selenium (Price *et al.* 1987). The diatoms were harvested by filtration and frozen at -80°C until DNA extraction. Bacteria were cultured from the diatom cultures on nutrient agar plates, further grown in liquid nutrient broth, harvested by centrifugation and stored at -80°C for DNA extraction. Total DNA was extracted from cultured clonal isolates of *F. capucina* and bacteria using the method of Varadarajan & Prakash (1991), modified for the small quantities used in this study. Approximately 0.1 g fresh weight of frozen diatom material was ground to a fine powder in liquid nitrogen with a mortar and pestle. The frozen ground powder was suspended in 1.5 mL isolation buffer (100 mM Tris HCl (pH 7.4), 50 mM EDTA (pH 8), 500 mM NaCl). Each sample was transferred to two 1.5 mL microfuge tubes, 10.5 μL β-mercaptoethanol and 50 μL 20% SDS were added to each tube, and the tubes shaken vigorously and incubated at 65°C for 15 min. After incubation, 250 μL of 5 M potassium acetate was added, followed by vigorous shaking and placed at -20°C for 10 min. The samples were microfuged at 14,000 rpm at 4°C for 20 min. The supernatant was transferred to a clean tube, 500 μL of ice-cold isopropanol was added, and the sample was mixed by inverting before being incubated at -20°C for 30 min. The samples were microfuged as before for 20 min., pelleting the DNA. The supernatant was discarded and the pellet dried briefly before dissolving it in 70 μL of TE-1 buffer (50 mM Tris HCl (pH 8), 10 mM EDTA (pH 8)). The samples were microfuged as before for 10 min. The supernatant was carefully transferred to a clean microfuge tube, with the duplicate samples combined, and 15 μL 3 M sodium acetate and 100 μL ice-cold isopropanol added. The samples were microfuged as before for 4 min, pelleting the DNA. The supernatant was poured off, and the pellet washed twice with 100 μL of ice-cold 80% ethanol and dried for 30 min. The dry pellet was resuspended in TE-2 buffer (10 mM Tris HCl (pH 8), 10 mM EDTA (pH 8)). DNA concentration was quantified by spectrophotometer (Beckman DV640) readings at 260 and 280 nm.

RAPD-Polymerase Chain Reaction (PCR) was carried out according to the method of O'Malley (pers. comm.). A reaction volume of 13 μL contained 0.7 U *Taq* DNA polymerase (Perkin Elmer), *Taq* DNA polymerase buffer (Perkin Elmer), 408 μM each of deoxyadenosine triphosphate, deoxycytosine triphosphate, deoxyguanosine triphosphate and deoxyribosyl thymine triphosphate (Gibco BRL), 1.5 mM $MgCl_2$, 0.35 μM primer and 3 ng diatom DNA. PCR amplification was performed on a Perkin-Elmer Cetus DNA Thermal Cycler programmed for 40 cycles at 92°C for 1 min, 37°C for 1 min and 72°C for 2 min. The PCR products were separated by electrophoresis using 1.8% agarose gels and stained with ethidium bromide for visualization with UV light (Maniatis *et al.* 1982). The gel was photographed and the DNA fragments analyzed versus size markers. Bacteria isolated from the cultures were also routinely run, so that confounding bands could be omitted.

A set of 100 PCR primers (401–450) were obtained from Dr. John E. Carlson, Biotechnology Laboratory, University of British Columbia, Canada. A random subset of the *F. capucina* clones across three sample sites was screened initially using primers 401–450 to determine which RAPD primers would provide useful information.

## Results and Discussion

From 12 to 40 clones of *F. capucina* were isolated from each site along the north-south gradient, for a total of 147 clones (Table 1). Of the 50 primers that were screened using selected *F. capucina* clones, seven primers were found to yield amplification products that included polymorphic genetic markers. At present, 36 distinct markers have been identified for use in fingerprinting the various clones and assessing their relatedness. Figure 2 illustrates an example of a gel, showing banding patterns for ten clones using a single primer. A total of seven discrete bands can be seen, with a unique banding pattern for each clone. A total of 115 of 131 clones analyzed to date exhibited unique combinations of genetic markers, demonstrating that this methodology is robust and capable of distinguishing genetic entities at several levels. For example, the four Kansas banding patterns (left) are clearly distinct from those of Manitoba (right), although they do have some bands in common (Fig. 2). Furthermore, within-site banding patterns appear more similar than among-site patterns from different localities.

Table 1. Number of clones of *Fragilaria capucina* isolated, extracted and analyzed by RAPD-PCR from each location along the north-south latitudinal gradient.

| Site/Location | Number of clones |
|---|---|
| Lake Manitoba, Manitoba | 13* |
| Laramie Reservoir, North Dakota | 17 |
| Lake Poinsett, South Dakota | 40 |
| Lake McConaughy, Nebraska | 26 |
| *Perry Reservoir, Kansas | 25 |
| Lake Carl Blackwell, Oklahoma | 14 |
| Lake Stillhouse Hollow, Texas | 12 |

*One clone of *F. capucina* var. *capucina*, 12 clones of *F. capucina* var. *mesolepta*.

The within-site differences are particularly interesting, not only for determining population heterogeneity, but also in light of how little is known regarding sexual reproduction in *Fragilaria*. While isogamy occurs in the closely related genus *Synedra* (Drebes 1977), sexual reproduction has apparently not been described for *Fragilaria*. High within-site diversity may indicate a greater degree of sexual recombination than previously thought. Further analysis of RAPD data may also provide insight into diatom distribution rates and colonization of new habitats.

Among-site banding patterns may be functional in inferring global warming patterns. A major goal of this research has been to provide a baseline for future detection of shifts in clonal distributions over the next half century, as a result of shifts in isotherms northward and coincident northerly migrations of southern clones. Thus, if future *Fragilaria* clones from the more northern reservoir sites exhibit a significant shift toward the banding patterns observed previously in southern sites, it would be indicative that the global warming process is driving diatom redistributions.

The RAPD results are encouraging as a tool for demonstrating genetic differences among individual clones or groups of clones along this latitudinal transect. Clearly, our approach will require concomitant physiological characterizations of selected clones to

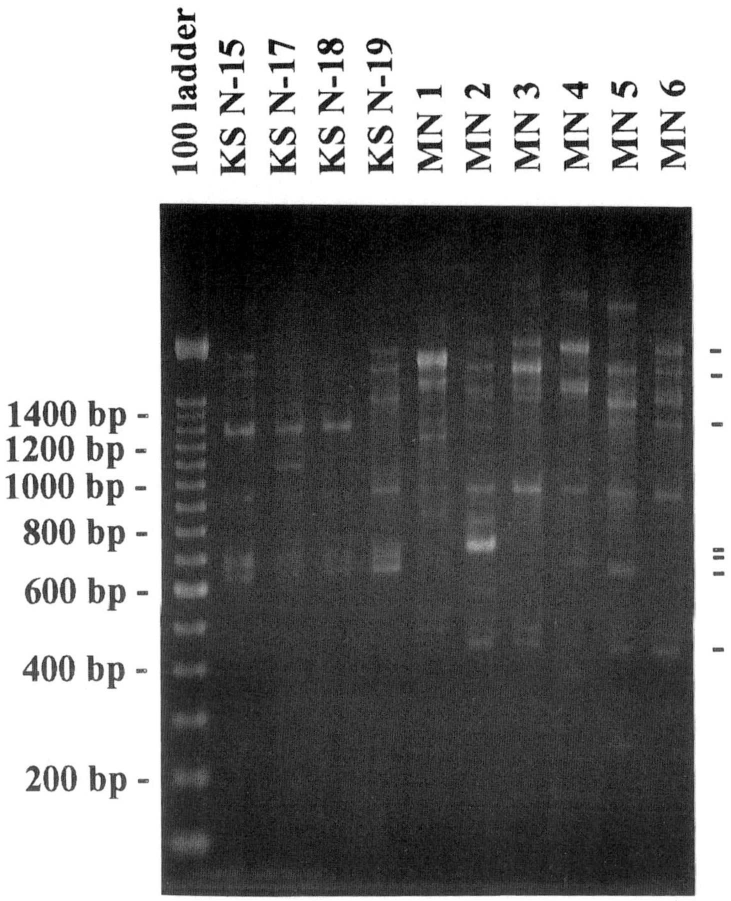


Fig. 2. An example of RAPD with *F. capucina* DNA using primer #411. KS refers to clones collected from Kansas, and MN refers to Manitoba. The numbers on the left side refer to sizes, in base pairs (bp), obtained from the 100 bp ladder marker. Ticks on the right side refer to sizes of RAPD markers that were found in one or more clones in this example. Unmarked bands were excluded because of their similarity to bacterial bands.

determine growth responses to a thermal gradient. In addition, this approach will require that we determine whether or not clonal banding patterns are conservative through time, by extracting DNA for RAPD analysis of cultured clones at several points in time. We have shown that isolated DNA (stored for two months to two years) from selected clones exhibited consistent banding patterns, demonstrating that this methodology is reproducible.

## Acknowledgements

The authors thank J. Holz and R. Spawn for their assistance in collecting and isolating *Fragilaria*, and J. Becker for help with culturing. This research was funded by the National Institute for Global Environmental Change, U.S. Department of Energy. This is Journal Series no. 10921 of the Agricultural Research Division at the University of Nebraska.

## References

Byron, E. R. & Goldman, C. R. (1990). The potential effects of global warming on the primary productivity of a subalpine lake. *Water Resources Bulletin*, **26**, 983–989.

Braarud, T. (1961). Cultivation of marine organisms as a means of understanding environmental influences on populations. In: *Oceanography* (M. Sears, ed.), 271–298, AAAS Pub. **67**, Washington, D.C.

Drebes, G. (1977). Sexuality. In: *The Biology of Diatoms*, pp. 250–281. University of California Press, Berkeley, California

Eppley, R. W. (1972). Temperature and phytoplankton growth in the sea. *Fishery Bulletin*, **70**, 1063–1085.

Fogg, G. E. (1969). Survival of algae under adverse conditions. *Symposium of the Society of Experimental Biology*, **23**, 123–142.

Gorham, E. (1991). Northern peatlands: role in the carbon cycle and probable responses to climatic warming. *Ecological Applications*, **1**, 182–195.

Guillard, R. R. L. (1975). Culture of phytoplankton for feeding marine invertebrates. In: *Culture of Marine Invertebrate Animals*, pp. 29–60. Plenum Publishing Corporation, New York

Guillard, R. R. L. & Ryther, J. H. (1962). Studies of marine plankton diatoms. I. *Cyclotella nana* Hustedt, and *Detonula confervacea* (Cleve) Grun. *Canadian Journal of Microbiology*, **8**, 229–239.

Hoshaw, R. W. & Rosowski, J. R. (1973). Methods for microscopic algae. In: *Handbook of Phycological Methods*, pp. 53–67. Cambridge University Press, Cambridge, England

Hulbert, E. M. & Guillard, R. R. L. (1968). The relationship of the distribution of the diatom *Skeletonema tropicum* to temperature. *Ecology*, **49**, 337–339.

Johansen, J. R., Barclay, W. R. & Nagle, N. (1990). Physiological variability within ten strains of *Chaetoceros muelleri* (Bacillariophyceae). *Journal of Phycology*, **26**, 271–278.

Li, W. K. W. (1980). Temperature adaptation in phytoplankton: cellular and photosynthetic characteristics. In: *Primary Productivity of the Sea*, Vol. **19** (P. G. Falkowski, ed.), pp. 259–279, Plenum Press, NY.

Maniatis, T., Fritch, E. F. & Sambrook, J. (1982). *Molecular Cloning: A Laboratory Manual*. 545 pp. Cold Spring Harbor Laboratory, Cold Spring Harbor, New York

Matthews, W. J. & Zimmerman, E. G. (1990). Potential effects of global warming on native fishes of the southern Great Plains and the Southwest. *Fisheries*, **15**, 26–32.

Meisner, J. D. (1990). Potential loss of thermal habitat for brook trout, due to climatic warming, in two southern Ontario streams. *Transactions of the American Fisheries Society*, **119**, 282–291.

Mosser, J. L., Mosser, A. G. & Brock, T. D. (1977). Photosynthesis in the snow: the alga *Chlamydomonas nivalis* (Chlorophyceae). *Journal of Phycology*, **13**, 22–27.

Poinai, K. A. & Johnson, W. C. (1991). Global warming and prairie wetlands. *BioScience*, **41**, 611–618.

Price, N. M., Thompson, P. A. & Harrison, P. J. (1987). Selenium: an essential element for growth of the coastal marine diatom *Thalassiosira pseudonana*. *Journal of Phycology*, **23**, 1–9.

Reynolds, C. S. (1988). Functional morphology and adaptive strategies of freshwater phytoplankton. In: *Growth and Reproduction of Freshwater Phytoplankton* (C. D. Sandgren, ed.), Cambridge Press, Cambridge

Rhee, G -Y. (1982). Effects of environmental factors and their interactions on phytoplankton growth. *Advances in Microbial Ecology*, **6**, 33–74.

Schindler, D. W., Beaty, K. G., Fee, E .J., Cruikshank, D. R., Debruyn, E. R., Findlay, D. L., Linsey, G. A., Shearer, J. A., Stainton, M. P. & Turner, M. A. (1990). Effects of climatic warming on lakes of the Central Boreal Forest. *Science*, **250**, 967–970.

Seaburg, K. G. & Parker, B. C. (1983). Seasonal differences in the temperature ranges of growth of Virginia algae. *Journal of Phycology*, **19**, 380–386.

Soudek, D., Jr. & Robinson, G. G. C. (1983). Electrophoretic analysis of the species and population structure of the diatom *Asterionella*. *Canadian Journal of Botany*, **61**, 418–433.

Talling, J. F. (1955). The relative growth rate of three plankton diatoms in relation to underwater radiation and temperature. *Annals of Botany* (n.s.), **19**, 329–341.

Thomas, W. H. (1966). Effects of temperature and illuminance on cell division rates of three species of tropical oceanic phytoplankton. *Journal of Phycology*, **2**, 17–22.

Varadarajan, G. S. & Prakash, C. S. (1991). A rapid and efficient method for the extraction of total DNA from the sweet potato and its related species. *Plant Molecular Biology Reporter*, **9**, 6–12.

Wetzel, R. G. (1983). *Limnology*. 2nd ed. 767 pp. Saunders College Pub., Philadelphia

Zechman, F. W., Zimmer, E. A. & Theriot, E. C. (1994). Use of ribosomal DNA internal transcribed spacers for phylogenetic studies in diatoms. *Journal of Phycology*, **30**, 507–512.

# Observations on the valve morphology of *Thalassionema nitzschioides* (Grunow) Hustedt

Jose Luis Moreno-Ruiz* and Sergio Licea**

** Departamento de Hidrobiología, C.B.S., UAM–I, Apdo. Postal 55–535, México, 03940 D.F.*

*** Instituto de Ciencias del Mar y Limnología, Universidad Nacional Autónoma de Mexico, Apdo. Postal 70–305, México, 04510 D. F.*

## Abstract

Fossil material of marine sediments from 11 different localities in the Mexican Pacific and net samples from the southern Gulf of Mexico and the coastal lagoon of Sontecomapan were studied. This paper describes the morphology and the great variability of this species.

Observations by light and electron microscopy of the valve structure, specially areolae and rimoportulae ultrastructure were made in order to detect any variation within the species. One thousand valve measurements were made. In addition, four varieties of this species were identified, there are three new combinations, and two emendations are proposed on the basis of valve outline and morphometry.

## Introduction

While investigating the recent and fossil diatom flora of material from the Mexican Pacific and southern Gulf of Mexico, we found many samples containing abundant valves which have morphological characters similar to those reported for *Thalassionema nitzschioides* (Grunow) Hustedt 1932.

The range of morphological variation and ultrastructure of *T. nitzschioides* have not been fully investigated, though some data are given by Hasle (1960), Hasle & De Mendiola (1967), Akiba (1982), Hallegraeff (1986) and Moreno-Ruiz & Carreño (1993). VanLandingham (1978) listed 11 varieties belonging to this species, even so, delimitation of this taxon and its varieties is sometimes difficult. Therefore, it is neccesary to look for characters to separate infraspecific taxa. The varieties of this species are distinguished by the heteropolar or isopolar apices, the fine structure of the areolae and valve outline.

This paper will discuss and illustrate the morphological and structural characters of nine varieties studied.

## Materials and Methods

Fossil material has been collected from the following areas:

Valle de Diatomita, Baja California, 27°11'29"N, 114°20'55"W, late Miocene (Moreno-Ruiz & Carreño 1994). San Roque, Baja California, located 14 km NW of Valle de Diatomita, late Miocene (age determined by Moreno-Ruiz). Punta Quebrada, Baja California, 27°18.23'N, 115°0.6'W, late Miocene (Helenes-Escamilla 1980). La Ventana, Baja California, 26°07'00"N, 112°11'02"W, early Miocene (age determined by Moreno-Ruiz). El Cien, Baja California, 24°49.8'N, 111°9.2'W, late Miocene (Carreño 1992). DSDP, Leg 63, site 472, Baja California, 23°00.35'N, 113°59.71'W, late Miocene (Barron 1981). San Felipe, Gulf of California, 31°06'29"N, 114°56'25"W, late Miocene (Bohem 1982). Santiago, Gulf of California 23°29'N, 109°41'W, late Pliocene (McCloy 1984). Arroyo Hondo, Maria Madre Island, 21°39'11"N, 106°35'05"W, late Miocene (Carreño 1985). DSDP, Leg 66, sites 490 and 493, 15°59.8'–16°31.3'N, 98°16.1'–99°27.9'W, late Miocene (age determined by Dr. John A. Barron).

The living material was collected by net hauls through the routine sampling programmes, of the following oceanographic cruises in the southern Gulf of Mexico: COSMA 71–04 (28 March–2nd Apr. 1971); COSMA 71–18 (21–26 Sep. 1971); COSMA 72–02 (28 Jan. 2nd Feb. 1972); COSMA 72–12 (7–15 Aug. 1972); FBC 82–03 (20–26 March 1982); ENDEAVOR–010 (July 1977); OGMEX I (24 Febr.–10 March 1987); OGMEX II (27 July–6 Aug. 1987); OGMEX III (29 Nov.– 4 Dec. 1987); OGMEX V (1–10 Aug. 1988) and OGMEX VIII (6–18 Sep. 1989). Additional samples were examined from the lagoon of Sontecomapan located in southwestern coast in the Gulf of Mexico (April 1992 May and Oct. 1993).

All samples were treated by the method of Barron (1981) modified by Moreno-Ruiz & Carreño (1993). The preparations were investigated in the light microscope (LM), scanning (SEM) and transmission electron microscopes (TEM).

The terminology followed is that of Hendey (1964), Anonymous (1975), with the additions of Akiba (1982), Hallegraeff (1986) and Moreno-Ruiz & Carreño (1993).

Studied specimens were deposited at the diatom collection of the Instituto de Ciencias del Mar y Limnología, Universidad Nacional Autónoma de México, México 04510 D.F.

## Observations and Discussion

A summary of valve morphology of nine varieties of *T. Nitzschioides* is given in Table 1.

Table 1. Summary of valve morphology of 10 varieties of *Thalassionema nitzschioides* (Grun.) Hust. The total number of specimens examined was 1,000.

| Taxa | Length µm | Width µm | Areolae in 10 µm: central | Areolae in 10 µm: terminal | Valve outline | Other distinctive characters |
|---|---|---|---|---|---|---|
| *Thalassionema nitzschioides* | 10–84 | 2.5–4 | 7–13 | 7–14 | narrow linear, lanceolate margin | isopolar rounded apices exceptionally pointed or capitate |
| var. *antiqua* | 12–153 | 4–6.5 | 10–14 | 10–14 | linear margin to very slightly convex | isopolar apices, strongly rounded |
| var. *capitulata* | 34–69 | 2.8–6 | 10–12 | 10–13 | margin slightly convex toward the centre of valve | thin valves, isopolar elongated apices, slightly capitate |
| var. *claviformis* | 5–94 | 1.9–7 | 8–13 | 9–13 | linear lanceolate margin, semi-concave or semiconvex | heteropolar rounded apices, one apex wider than the other |
| var. *incurvata* | 8–25 | 2.3–4 | 10–13 | 10–13 | concave valve margin in the middle of valve | isopolar apices, strongly rounded |
| var. *inflata* | 16–43 | 3.3–6.5 | 9–12 | 9–13 | convex margin in the middle | thin to heavy silicified valves, isopolar short apices, semicapitate to capitate |
| var. *lanceolata* | 26–117 | 3.5–8.7 | 9–12 | 9–12 | lanceolate margin | thin to heavy silicified valves, isopolar apices, strongly rounded |
| var. *parva* | 5–<10 | 2.3–4 | 9–12 | 9–12 | linear margin | isopolar rounded apices |
| var. *robusta* | 27–47 | 4.5–6 | 10–12 | 9–12 | semilinear margin, slightly convex | isopolar apices, strongly rounded |
| var. *schraderi* | 15–37 | 6–8.5 | 10–11 | 10–11 | linear to convex margin | very heavily silicified valves, isopolar apices, strongly rounded, capitate to semicapitate |

*Thalassionema nitzschioides* (Grunow) Hustedt 1932. (Figs 1–3, 34–38).

Illustrations: Hustedt 1932, p. 244, Fig. 725; Cupp 1943, p. 182, Fig. 133; Hasle & De Mendiola 1967, p. 111, Figs 5, 13–17, 27–34, 39–44; Schrader 1973, p. 712, pl. 23, Figs 2, 6, 9–10; Barron 1975, p. 157, pl. 13, Figs 8–9; Schrader & Fenner 1976, p. 1001, pl. 5, Fig. 8, pl. 24, Fig. 10; Abbott & Andrews 1979, p. 253, pl. 6, Fig. 10; Akiba 1986, p. 445, pl. 21, Figs 11,19; Hallegraeff 1986, p. 58, Figs 1–4; Moreno-Ruiz & Carreño 1993, Figs 1–4.

Valves narrow, linear-lanceolate outline. Apices rounded, isopolar, exceptionally pointed or capitated, two slit-like rimoportulae, one on each pole and they are slightly oblique or parallel to the mid-line of the valve. Axial area broad, areolae covered with triradiate struts sometimes with lateral ramifications. Length 10–84 µm, width 2.5–4.0 µm, 7–13 central areolae in 10 µm, 7–14 terminal areolae in 10 µm.

Affinity: The type of areolae of *T. nitzschioides* show a similarity to those of *T. hirosakiensis* (Schrader) Akiba and *T. schraderi* Akiba (Akiba & Yanagisawa 1986, Pl. 49, Fig. 2; Pl. 50, Figs 9–10), however it can be distinguished from these species by its lightly silicified valves. The valves of *T. nitzschioides* are wider than 3 µm, and are intermediate with forms *T. antiqua* Schrader (Schrader 1973, Pl. 23, Fig. 28), which has a linear margin, a broad transapical axis and rounded apices; and to *T. robusta* Schrader (Akiba 1982, Pl. 1, Fig. 14) with its semiconvex margins and widely rounded apices. Therefore, from the detail of the marginal areolae, surmounted externally by triradiate struts which may carry irregular side-branches (Hallegraeff 1986, Figs 3a, 4b, 4d), or areolae surmounted by tetraradiate struts with side-branches (Hallegraeff op. cit., Fig. 3a near the end; Moreno-Ruiz & Carreño 1993, Fig. 2; Hasle & Semina 1987, Fig. 61 near the end), together with valve outline variations (Moreno-Ruiz & Carreño 1993, Figs 5–7, 10–11) recorded for *T. nitzschioides*, it would appear that all the above taxa have too few morphologic characters to justify separate species.

General distribution: Baja California, Gulf of California and southern Gulf of México (this study). Cosmopolitan species.

Stratigraphic distribution: Early Miocene to Recent (in agreement with above authors).

*Thalassionema nitzschioides* var. *antiqua* (Schrader) Moreno-Ruiz *comb. nov.* (Figs 4–5, 39–41).

Original description: *Thalassionema antiqua* Schrader 1973, p. 711, pl. 23, Fig. 28.
Synonyms: *Thalassionema nitzschioides* Gombos 1975, p. 322, pl. 2, Fig. 6; *Thalassionema hirosakiensis* Akiba 1982, Figs 3–5; *Thalassionema* aff. *hirosakiensis* Yanagisawa *et al.* 1989, p. 452, pl. 5, Fig. 17.
Illustrations: Schrader & Fenner 1976, p. 1018, pl. 5, Fig. 8.

Valves weakly or heavily silicified, margin linear to very slightly convex. Isopolar apices, strongly rounded, areolae covered with triradiate and tetraradiate struts sometimes with lateral ramifications. Broad axial area. Length 12–153 µm, width 4.0–6.5 µm, 10–14 central and terminal areolae in 10 µm.

Schrader (1973) erected *T. antiqua*, however in his description we found characters that overlap with *T. nitzschioides* (valves linear, broadly rounded ends; 12–10 µm long; apices isopolar, at one apex a very short median spine; axial area linear, narrow, about 2 µm wide; transapical stria about 11 in 10 µm) and *T. nitzschioides* var *incurvata* Heiden (valves sometimes constricted at the middle, about 5 µm wide). On account of the similarities in nearly all features, Schrader's specimens do not justify a separate species. Only one of his specimens (Pl. 23, Fig. 28) has a weakly silicified valve, linear margin and is wider than 4 µm; characters that in part define *T. nitzschioides* var. *antiqua* (Schrader) comb. nov. To the variety *antiqua* also belong three specimens identified by Akiba (1982, Figs 3–5) as *T. hirosakiensis*, and one specimen of *Thalassionema* cf. *hirosakiensis* of Yanagisawa *et al.* (1989, Pl. 5, Fig. 17).

Affinity: This variety is similar to linear valves of *T. schraderi*, but they differ from the very heavily silicified valves of *T. schraderi.*

General distribution: Baja California, Gulf of California, Mexican Pacific. South of Alaska (Schrader 1973), central Pacific (Gombos 1975), Norway basin (Schrader & Fenner 1976); Owasawa, Japan (Akiba 1982); Futaba, Fukushima, Japan (Yanagisawa *et al.* 1989).

Stratigraphic distribution: Early Miocene to Pleistocene.

*Thalassionema nitzschioides* var. *capitulata* (Castracane) Moreno-Ruiz comb. nov. (Figs 6–7, 42–43)

Original description: *Synedra capitulata* Castracane 1886, p. 52, pl. 25, Fig. 13 (*vide* Simonsen 1992).

Synonyms: *S. capitata* Heiden in Heiden & Kolbe 1928, p. 565, pl. 5, Fig. 119 (*vide* Simonsen 1992, pl. 22, Figs 7–10); *T. capitulata* Hustedt 1958, p. 139; *T. nitzschioides* Hasle & De Mendiola 1967, p. 117, Figs 11–12; *T. lineatum* Schrader & Fenner 1976, p. 1001, pl. 5, Fig. 5.

Illustrations: Hasle 1960, p. 16, Fig. 3; Schrader 1973, p. 712, pl. 23, Fig. 8; Akiba 1982, p. 48, pl. 13, Fig. 12; Whiting & Schrader 1985, p. 264, pl. 6, Fig. 28; Akiba & Yanagisawa 1986, p. 549, pl. 48, Fig. 15.

Valves thin, margins slightly convex toward the middle. Apices elongated, isopolar, slightly capitate, widely rounded, two slit-like rimoportulae, one on each pole and they are slightly oblique or parallel to the mid-line of the valve. Axial area broad, areolae covered with triradiate struts sometimes with lateral ramifications. Length 34–69 µm, width 2.8–6.0 µm, 10–12 central areolae in 10 µm, 10–13 terminal areolae in 10 µm.

Hustedt (1958) includes the *Spinigera capitata* Heiden as a synonym of *Thalassionema capitulata* (Castracane) Hustedt. However our observations show that *T. capitulata* has areolae covered with triradiate struts with lateral ramifications and rimoportulae similar to *T. nitzschioides*, differing only by the valves being slightly capitate in var. *capitulata* therefore its characteristics do not permit distinction as a separate species.

Hasle (1960, Fig. 3) shows a specimen as *T. capitulata* Hustedt, with parallel margins in the middle of the valve and capitate apices, because of this we infer intermediate characters with *T. nitzschioides*. Therefore we propose to retain this taxon as *T. nitzschioides* var. *capitulata*.

Ferreira (1982, Figs 7e, 7g) identified two specimens as *T. nitzschioides* var. *obtusa*, whose valves show one apex slightly spatulate, while the other is rounded. These characters differ from the distinctive features of *T. nitzschioides* var. *obtusa*, which has elongated poles rather than capitate poles (Hasle 1960, Figs 5b, e, Pl. 4, Figs 38–39, 43–44). Therefore we believe Ferreira's specimens belong to *T. nitzschioides* var. *capitulata*.

Affinity: This variety is similar to *T. inflata* Heiden, except for the short apices from the latter species.

General distribution: Baja California, Gulf of California, Mexican Pacific, southern Gulf of Mexico. Salpendarm and Simonsbay (Simonsen 1992); Antarctic (Hustedt 1958); tropical Pacific (Hasle 1960); Canarias Islands (Hasle & De Mendiola 1967); north of California (Schrader 1973); Norway basin (Schrader & Fenner 1976); Japan (Akiba 1982, Akiba & Yanagisawa 1986); Oregon State (Whiting & Schrader 1985).

Stratigraphic distribution: Early Miocene to Recent.

*Thalassionema nitzschioides* var. *claviformis* (Schrader) Moreno-Ruiz *in* Moreno-Ruiz & Carreño 1993 (Figs 8–13, 44–46).

Original description: *Thalassionema claviformis* Schrader 1973, p. 711, pl. 23, Figs 11, 15.

Synonyms: *T. nitzschioides* Schrader 1973, p. 712, pl. 23, Figs 1, 12–13, 34; *T. hirosakiensis* Schrader 1976, p. 636, pl. 1, Fig. 16; *Thalassiothrix pseudonitzschioides* Schuette & Schrader 1982, p. 216, pl. 13, Figs 1,5, 9; pl. 2, Figs 10–14; pl. 3, Figs 15–20.

Illustrations: Schrader 1974, p. 556, Fig. 5.4; Schrader & Fenner 1976, p. 1001, pl. 24, Fig. 9; Barron 1985, p. 455, pl. 8, Fig. 15; Hallegraeff 1986, p. 60, Fig. 1a; Yanagisawa *et al.* 1989, p. 452, pl. 5, Fig. 22; Moreno-Ruiz & Carreño 1993, p. 143, Figs 5–9.

Valves linear-lanceolate, semiconcave or semiconvex, symmetric about the apical axis, asymmetric about the transapical axis. Apices rounded, heteropolar, one apex wider than the other. Sometimes one apex is slightly capitate, two slit-like rimoportulae, one on each pole, slightly oblique or parallel to the mid-line of the valve. Axial area broad. Areolae covered with triradiate and tetraradiate struts sometimes with lateral ramifications. Length 5–94 µm, width 1.9–7.0 µm. 8–13 central areolae in 10 µm, 9–13 terminal areolae in 10 µm.

Hendey (1964) pointed out that "The valves are heteropolar, symmetrical about the apical axis, but asymmetric about the transapical axis". Hallegraeff (1986) suggests "Heteropolar valves with an arrow-head shaped spine at one pole were common... These cells appeared in the same colony as isopolar frustules carrying spines at both valve poles and isopolar frustules without spines... and the present observation of isopolar and heteropolar cells in mixed colonies clearly shows that *Thalassiothrix pseudonitzschioides* is a form of *Thalassionema nitzschioides*". Therefore, the presence of one spine at one pole is not a reliable diagnostic character. Rather valve symmetry together with triradiate to tetraradiate struts sometimes with lateral ramifications are characters that define *T. nitzschioides* var. *claviformis*.

Affinity: The semiconcave margins relate this taxon to *T. nitzschioides* var. *incurvata* (Simonsen 1992, Pl. 21, Figs 13–15). The presence of one slightly capitate apex relates it to *T. nitzschioides* var. *capitulata* (Simonsen 1992, Pl. 22, Figs 7–10), however, it differs from both taxa by its heteropolar valves.

General distribution: Mexican Pacific, southern Gulf of Mexico, Sontecomapan. Baja California (Schrader 1974), Gulf of California (Schuette & Schrader 1982), north of California and south of Alaska (Schrader 1973), Norway basin (Schrader & Fenner 1976), east equatorial Pacific (Barron 1985), Australia (Hallegraeff 1986); Futaba, Japan (Yanagisawa *et al.* 1989).

Stratigraphic distribution: Late Eocene to Quaternary

*Thalassionema nitzschioides* var. *incurvata* Heiden *in* Heiden & Kolbe 1928 (*vide* Simonsen 1992). (Figs 16–19, 50–51).

Synonyms: *Thalassionema antiqua* Schrader 1973, p. 711, pl. 23, Figs 26–27, 29–30; *T. nitzschioides* var. *parva* Koizumi & Tanimura 1985, p. 291, pl. 6, Fig. 11.

Illustrations: Simonsen 1992, pl. 21, Figs 13–15.

Valves slightly concave. Apices strongly rounded, isopolar, two slit-like rimoportulae, one on each pole and slightly oblique or parallel to the mid-line of the valve. Axial area broad. Length 8–25.0 µm, width 2.3–4.0 µm, 10–13 central and terminal areolae in 10 µm.

Our observations of the interior valve view show the same kind of rimoportula as for *T. nitzschioides*, which allow recognize the variety *incurvata*. Schrader (1973, Pl. 23, Figs 26–27, 29–30) illustrate four specimens with concave margins and designated these as *T. antiqua*. Koizumi & Tanimura (1985) illustrate one specimen (Pl. 6, Fig. 11) identified as *T. nitzschioides* var. *parva*; however, this specimen has margins slightly concave which is a character of *T. nitzschioides* var. *incurvata* to which Schrader's specimens belong.

Affinity: The variety *incurvata* relates to *T. nitzschioides* var. *claviformis*, because of some valves with one slightly concave margin in the variety *claviformis*. It differs since *T. nitzschioides* var. *claviformis* is heteropolar.

General distribution: Baja California, Gulf of California, Mexican Pacific, southern Gulf of Mexico. South of Alaska (Schrader 1973); west elevation of Shatsky, Japan (Koizumi & Tanimura 1985); Austral Ocean (Simonsen 1992).

Stratigraphic distribution: Late Miocene to Recent.

*Thalassionema nitzschioides* var. *inflata* Heiden in Heiden & Kolbe 1928 (*vide* Simonsen 1992). (Figs 14–15, 20–22, 47–49).

Synonyms: *Fragilaria hirosakiensis* Kanaya 1959, p. 104, pl. 9, Figs 11a–14 (*vide* Akiba 1982); *Thalassionema nitzschioides* Gombos 1975, p. 322, pl. 2, Fig. 4; *T. hirosakiensis* Schrader 1973, p. 711, pl. 23, Figs 31–33; Akiba 1986, p. 472, pl. 21, Figs 2–3.

Illustrations: Akiba 1982, p. 49, Figs 1–2; Barron 1985, p. 445, pl. 8, Fig. 13; Akiba & Yanagisawa 1986, p. 497, 549, pl. 48, Figs 6–9, 13–14; pl. 49, Figs 1–8; Yanagisawa *et al.* 1989, p. 452, pl. 5, Figs 14–16; Simonsen 1992, p. 24, pl. 21, Figs 5–9.

Valves thin to heavily silicified, margin convex in the middle. Apices isopolar slightly extended, rounded, semicapitate to capitate, two slit-like rimoportulae, one on each pole, slightly oblique or parallel to the mid-line of the valve. Axial area broad in the middle, areolae covered with triradiate and tetraradiate struts sometimes with lateral ramifications. Length 16–43 µm, width 3.3–6.5 µm, 9–12 central areolae in 10 µm, 9–13 terminal areolae in 10 µm.

Barron (1985, Pl. 8, Fig. 13) shows one specimen identified as *T. nitzschioides*; Akiba (1986, Pl. 21, Figs 2–3), Akiba & Yanagisawa (1986, Pl. 48, Figs 7–8, 13–14; Pl. 49, Figs 1–2, 6) and Yanagisawa *et al.* (1989, Pl. 5, Figs 14–16) illustrations, show specimens semicapitate to capitate and thin to heavy valves which are similar to the specimens of this study. On the other hand, the structure of areolae covered with triradiate to tetraradiate struts with lateral ramifications, valve outline and the form of rimoportula confirm its position as the variety *inflata*. Therefore all specimens cited above should be included in *T. nitzschioides* var. *inflata*. Akiba (1982) emended the description of *T. hirosakiensis*, pointing out that the close array of marginal areolae (puncta) are the best diagnostic character. However our observations show that this feature (10–12 areolae in 10 µm) is within the range shown by *T. nitzschioides* (7–13 central areolae in 10 µm). Akiba did not mention either valve outline or the heavily silicified valves, only that its morphology is similar to *T. schraderi*. What we observe from Akiba's illustrations (Akiba 1982, Figs 1–2) is that his specimens have heavy valves, convex margins and capitate to semicapitate slightly extended apices, i.e. of *T. nitzschioides* var. *inflata*. Therefore we propose to include *T. hirosakiensis* in *T. nitzschioides* var. *inflata*.

Affinity: This variety shows similarity with *T. schraderi*, *T. nitzschioides* var. *capitulata* and *T. nitzschioides* var. *claviformis*. It differs from *T. schraderi* by its very heavily silicified valves, from the variety *capitulata* by its elongated slightly capitate apices and from the variety *claviformis* by its heteropolarity.

General distribution: Baja California, Nayarit, Gulf of California, Mexican Pacific, southern Gulf of Mexico. Gulf of Alaska; Owasawa, Japan (Akiba 1982); Atsunai, west Hokkaido, Japan (Akiba 1986, Akiba & Yanagisawa 1986); Equatorial central and east Pacific (Gombos 1975, Barron 1985); trench of Japan (Akiba 1986), east Hokkaido, Japan (Akiba & Yanagisawa 1986).

Stratigraphic distribution: Middle Miocene to Recent.

*Thalassionema nitzschioides* var. *lanceolata* (Grunow) Peragallo & Peragallo 1897–1908. (Figs 23–24, 52–56).

Original description: *Thalassiothrix nitzschioides* var. *lanceolata* Grunow *in* Van Heurck 1881, pl. 43, Figs 8–9.

Synonyms: *Spinigera lanceolata* Heiden 1928 (*vide* Simonsen 1992, p. 25, pl. 21, Figs 16–17); *T. nitzschioides* var. *gracilis* Heiden 1928 (*vide* Simonsen 1992, p. 24, pl. 21, Figs 1–4); *T. robusta* Schrader 1973, p. 712, pl. 23, Figs 35–37; *T. obtusa* Abbott & Andrews 1979, p. 253, pl. 6, Fig. 11; *Thalassiothrix robusta* Akiba 1986, p. 441, pl. 21, Fig. 4.

Illustrations: Peragallo & Peragallo 1897–1908, p. 321, pl. 81, Fig. 19; Akiba 1982, p. 48, pl. 1, Fig. 15; Fenner 1978, p. 533, pl. 33, Fig. 21; Yanagisawa *et al.* 1989, pl. 5, Figs 24–26; Yanagisawa *et al.* 1989, p. 452, pl. 5, Fig. 24.

Valves lanceolate, thin to heavy. Apices isopolar, rounded, two slit-like rimoportulae, one on each pole and they are slightly oblique or parallel to the mid-line of the valve. Axial area broad. Areolae covered with triradiate to tetraradiate struts with lateral ramifications. Length 26–117 µm, width 3.5–8.7 µm, 9–12 central and terminal areolae in 10 µm.

Peragallo & Peragallo (1897–1908) point out the presence of lanceolate valves as the main character, however the areolae and rimoportulae structure seen in the present observations verifies its location in the variety *lanceolata.*

*T. nitzschioides* var. *lanceolata* is identical to *Thalassionema robusta,* since Schrader (1973) established *T. robusta* with the lancelate character of the variety *lanceolata.* He points out that *T. robusta* has linear valves, a feature not shown in his illustrations (Pl. 23, Figs 35–37). He includes heavy valves, however this character is variable.

Akiba (1986, Pl. 21, Fig. 4) established the combination *Thalassiothrix robusta,* in which he observes one lanceolate valve with isopolar rounded apices and includes the specimens illustrated by Schrader (1973, Pl. 23, Figs 35–37) and Akiba (1982, Pl. 13, Fig. 15). However his combination is inconsistent because *Thalassiothrix* has at least one subterminal rimoportula and more than one row of areolae at the apices (Hallegraeff 1986, Figs 16–17, 27a–b, 32d–e, 35, 36c–d, 37a, c; 38a, c; 39a–45a; Hasle & Semina 1987, Figs 5–10, 15–22, 34–37, 53–57), features not observed in Akiba's illustration.

Abbott & Andrews (1979, Pl. 6, Fig. 11) show one specimen identified as *Thalassionema obtusa,* with lanceolate valve outline, thin valve and one row of marginal areolae, which are identical features of *T. nitzschioides* var. *lanceolata.* Consequently there are not sufficient characters to separate this specimen as an independent species.

Affinity: This variety relates to heavy valves of *T. nitzschioides* var. *claviformis.* It differs from variety *claviformis* by its isopolar valves.

General distribution: Baja California, Gulf of California, Mexican Pacific. North of California and south of Alasaka (Schrader 1973), plateau of Argentina (Fenner 1978), Southern Carolina (Abbott & Andrews 1979), Tenpoku, Hokkaido, Japan (Akiba 1982, 1986); Futaba, Japan (Yanagisawa *et al.* 1989); Villefranche (Peragallo & Peragallo 1897–1908), austral ocean (Simonsen 1992).

Stratigraphic distribution: Early Miocene to Recent.

*Thalassionema nitzschioides* var. *parva* (Heiden) Moreno-Ruiz emend. (Figs 25–27, 57–58).

Original description: *Thalassionema nitzschioides* var. *parva* Heiden *in* Heiden & Kolbe 1928 (*vide* Simonsen 1992).

Valves small, linear. Apices rounded, isopolar, two slit-like rimoportulae, one at each pole, slightly oblique or parallel to the mid-line of the valve. Axial area broad in the middle, areolae covered with triradiate and tetraradiate struts sometimes with lateral ramifications. Length: 5–9.5 µm, width 2.3–4.0 µm, 9–12 central and terminal areolae in 10 µm.

This variety has been cited with very variable lengths, Hasle (1960) and Barron (1985) show specimens 12 µm long, Fenner (1978) refers to lengths of 21 to 28 µm. All our specimens, fossil and living, were fairly constant in size (< 10 µm). Therefore we propose less than 10 µm as an upper limit size for this taxon.

Affinity: *T. nitzschioides* var. *incurvata* has some valves less than 10 µm length, however its valve outline is concave, differing from *T. nitzschioides* var. *parva.*

General distribution: Baja California, Gulf of California, Mexican Pacific, southern Gulf of México.

Stratigraphic distribution: Late Miocene and Recent.

*Thalassionema nitzschioides* var. *robusta* (Schrader) Moreno-Ruiz emend. (Figs 28–30, 59–62).

Original description: *Thalassionema robusta* Schrader 1973, p. 712, pl. 23, Figs 24, 35–37.

Synonyms: *Thalassionema robusta* Akiba 1982, p. 48, pl. 1, Fig. 14. *Thalassiothrix robusta* (Schrader) Akiba 1986, p. 441, pl. 21, Fig. 4.

Illustrations: Barron 1985, p. 445, pl. 8, Fig. 14.

Valves semilinear, margin slightly convex. Apices isopolar, rounded, two slit-like rimoportulae, one at each pole, slightly oblique or parallel to the mid-line of the valve.

Axial area broad in the middle, areolae covered with triradiate and tetraradiate struts sometimes with lateral ramifications. Length: 27–47 µm, width 4.5–6.0 µm. 10–12 central areolae in 10 µm, 9–12 terminal areolae in 10 µm.

Schrader (1973) erected *T. robusta*, with characters belonging to *T. nitzschioides* var. *lanceolata* (Valves narrow, linear, lanceolate with broadly rounded apices; 55–60 µm long, about 8 µm wide. Apices isopolar, one apex with an obtuse median spine. Axial area wide, lanceolate. One marginal line of punctae, about 9 in 10 µm), consequently its taxonomic position is unprecise. However, Akiba (1982, pl. 1, Fig. 14) illustrates one specimen with a linear valve rather than lanceolate; this we interpret as semilinear with isopolar rounded apices, which differs from the lanceolate shape previously illustrated (Schrader op. cit.). In spite of Akiba's poor description of *T. robusta* this is the basis for its taxonomic position in *T. nitzschioides* var. *robusta*.

Affinity: This variety shows similarity to both *T. nitzschioides* var. *lanceolata* (Grun.) Per. & Per. and *T. nitzschioides* var. *antiqua* (Schr.) comb. nov. It differs from variety *lanceolata* by its lanceolate valves and thin poles, and from variety *antiqua* in linear margins very slightly convex.

General distribution: Baja California. Tenpoku, Hokkaido, Japan (Akiba 1982); east equatorial Pacific (Barron 1985).

Stratigraphic distribution: Middle Miocene to late Miocene.

*Thalassionema nitzschioides* var. *schraderi* (Akiba) Moreno-Ruiz comb. nov. (Figs 31–33, 63–69).

Original description: *Thalassionema schraderi* Akiba 1982, p. 50, Figs 6–11, 16–18.
Synonym: *Thalassionema hirosakiensis* (Kanaya) Schrader 1973, pl. 23, Figs 31–33.
Illustrations: Barron 1980, p. 677, pl. 2, Fig. 6; Barron 1981, p. 536, pl. 5, Fig. 4; Akiba 1986, p. 445, pl. 21, Figs 1–3; p. 472, pl. 21, Figs 13–16; Akiba & Yanagisawa 1986, p. 497, pl. 48, Figs 1–5, 10–12; pl. 50, Figs 1–10.

Valves very heavily silicified, margin linear to convex in the middle. Apices isopolar, strongly rounded, capitate to semicapitate, two slit-like rimoportulae, one at each pole, slightly oblique or parallel to the mid-line of the valve. Axial area broad in the middle, areolae covered with triradiate and tetraradiate struts sometimes with lateral ramifications. Length 15–37 µm, width 6.0–8.5 µm, 10–11 central and terminal areolae in 10 µm.

Akiba (1982) erected to *T. schraderi*, without mentioning its slightly capitate apices observed in his figures 8–9 (upper apices), Akiba (1986, Pl. 21, Figs 13–15) and Akiba & Yanagisawa (1986, Pl. 48, Figs 1–3, 11). In addition the illustrations of Akiba & Yanagisawa (1986, Pl. 50, Figs 9–10) show areolae covered with triradiate and tetraradiate struts with lateral ramifications, similar to *T. nitzschioides*. It is concluded there are not significant differences between the taxa, therefore we propose *T. nitzschioides* var. *schraderi* as a comb. nov.

Affinity: This variety is related to *T. nitzschioides* var. *inflata*. It differs from it in its valve thickness.

General distribution: Baja California. East of Guadalupe Island, (Barron 1981); east Hokkaido Formation, diatomite Shinzan, Monterey, northern of Honshu, Japan; west Gulf of Alaska, northwestern Pacific (Akiba 1982, Akiba 1986, Akiba & Yanagisawa 1986).

Stratigraphic distribution: Late Miocene.

## Conclusions

The study of the morphology and structure of the nominate species shows that the areolae and rimoportulae fine structure are the best consistent characters. This research have been undertaken to define the range of morphological variability in *T. nitzschioides*, and as can be seen from the nine varieties studied, the infraspecific differentiation of *T. nitzschioides* can be based on the valve outline, length, width and silicification of the valve. Table 1 summarizes these conclusions.

## Acknowledgements

We wish to thank two anonymous referees. We are grateful to John A. Barron who provided samples from DSDP, Leg 63, 66 sites 472, 490, 493; Ana Luisa Carreño, Vicente Ferreira, Luis Rafael Segura, Miguel Angel Lara, and Jorge Lodigiani who made material available. Facilities for use of the SEM were obtained through Alfonso Carabez, Tomas Cruz and Margarita Reyes. A. Fernández and B. Jon for providing laboratory facilities. Sonia Ocampo, Edgardo Flores, Sergio López, Raúl Jurado and Wintilo Vega provided technical assistance. Partial support has been provided by Dirección General de Asuntos del Personal Académico, Programa de Apoyo a las Divisiones de Estudios de Posgrado, UNAM, Universidad Autónoma de Baja California Sur and Universidad Autónoma Metropolitana-Iztapalapa.

## References

Abbot, W. H. & Andrews, G.W. (1979). Middle Miocene marine diatoms from the Hawthorn Formation of the Ridgeland Trough, South Carolina and Georgia. *Micropaleontology*, **25**, 225–271.

Akiba, F. (1982). Taxonomy and biostratigraphic significance of a new diatom, *Thalassionema schraderi*. *Bacillaria*, **5**, 43–61.

Akiba, F. (1986). Middle Miocene to Quaternary diatom biostratigraphy in the Nankai trough and Japan trench, and modified lower Miocene trough Quaternary diatom zones for middle-to high latitudes of the North Pacific. *Initial Reports Deep Sea Drilling Project*, **87**, 393–481.

Akiba, F. & Yanagisawa, Y. (1986). Taxonomy, morphology and phylogeny of the Neogene diatom zonal marker species in the middle-to high latitudes of the North Pacific. *Initial Reports Deep Sea Drilling Project*, **87**, 483–557.

Anonymous (1975). Proposals for a standardization of diatom terminology and diagnoses. *Nova Hedwigia Beiheft*, **53**, 323–354.

Barron, J. A. (1975). Late Miocene-early Pliocene marine diatoms from Southern California. *Paleontographica*, **151** (4–6), 97–170.
Barron, J. A. (1980). Lower Miocene to Quaternary diatom biostratigraphy of Leg 57, off Northeastern Japan, Deep Sea Drilling Project. *Initial Reports Deep Sea Drilling Project*, **57** (2), 641–685.
Barron, J. A. (1981). Late cenozoic diatom biostratigraphy and paleoceanography of the middle-latitude eastern North Pacific, Deep Sea Drilling Project Leg 63. *Initial Reports Deep Sea Drilling Project*, **63**, 507–538.
Barron, J. A. (1985). Late Eocene to Holocene diatom biostratigraphy of the equatorial Pacific Ocean, Deep Sea Drilling Project Leg. 85. *Initial Reports Deep Sea Drilling Project*, **85**, 413–456.
Bohem, M. C. F. (1982). Biostratigraphy, lithostratigraphy, and paleoenvironments of the Miocene-Pliocene San Felipe marine sequence, Baja California Norte, Mexico. A thesis submitted to the Department of Geology and the Commitee on graduate studies of Stanford University in partial fulfillment of the requirements for the degree of Master of Science, 1–326.
Carreño, A. L. (1985). Biostratigraphy of the late Miocene to Pliocene on the Pacific island Maria Madre, Mexico. *Micropaleontology*, **31** (2), 139–166.
Carreño, A. L. (1992). Early Neogene Foraminifera of the Cerro Tierra Blanca Member (El Cien Formation), Baja California Sur, and associated microfossils. Universidad Nacional Autónoma de México, Instituto de Geología, Paleontología Mexicana, no. 59, pt. 2
Cupp, E. E. (1943). Marine plankton diatoms of the west of North America. *Bulletin, Scripps Institution of Oceanography*, **5** (1), 1–238.
Fenner, J. (1978). Cenozoic diatom biostratigraphy of the equatorial and southern Atlantic Ocean. *Initial Reports Deep Sea Drilling Project*. supplement, **39** (2), 491–623.
Ferreira, B. V. (1982). Variaciones de tamaño y abundancia en fitoplancton opalino preservado en sedimentos laminados: posible respuesta ecológica al cambio del clima oceánico. Tesis Licenciatura. 1–67. 51 figs.
Gombos, A. M. Jr. (1975). Fossil diatoms from Leg 7, Deep Sea Drilling Project. *Micropaleontology*, **21** (3), 306–333.
Hallegraeff, G. M. (1986). Taxonomy and morphology of the marine plankton diatoms *Thalassionema* and *Thalassiothrix*. *Diatom Research*, **1** (1), 57–80.
Hasle, G. R. (1960). Phytoplankton and ciliate species from the tropical Pacific. *Skrifter Det Norske Videnskaps-Akademi: I. Matematisk-Naturvidenskapelig Klasse*, **2**, 1–50.
Hasle, G. R. & De Mendiola, B. R. E. (1967). The fine structure of some *Thalassionema* and *Thalassiothrix* species. *Phycologia*, **6** (2–3), 107–125.
Hasle, G. R. & Semina, H. J. (1987). The marine planktonic diatoms *Thalassiothrix longissima* and *Thalassiothrix antarctica* with comments on *Thalassionema* spp. and *Synedra reinholdii*. *Diatom Research*, **2** (2), 175–192.
Helenes-Escamilla, E. J. (1980). Stratigraphy depositional environments and foraminiferal of the Miocene Tortugas Formation, Baja California Sur, Mexico. *Boletin de la Sociedad Geológica Mexicana*, **41** (1–2), 47–67.
Hendey, N. I. (1964). *An intoductory account of the smaller algae of british coastal waters*. Part 5, *Bacillariophyceae (Diatoms)*. H.M.S.O., London. 1–317.
Hustedt, F. (1932). Die Kieselalgen Deutschlands Osterreichs und der Schweiz unter Berücksichtigung der übrigen Lander Europas sowie der angrenzenden Meeresgebiete. In: *Die*

*Kryptogamen-Flora von Deutschland, Osterreich und der Schweiz* (L. Rabenhorsts, ed.), **7** (2), 1–846. Acad. Verlag., Leipzig.

Hustedt, F. (1958). Diatomeen aus der Antarktis und dem Südatlantik. Dtsch. Antark. Exped. 1938–1939, 2.

Koizumi, I. & Tanimura, Y. (1985). Neogene diatom biostratigraphy of the middle latitude western North Pacific, Deep Sea Drilling Project Leg 86. *Initial Reports Deep Sea Drilling Project*, **86**, 269–300.

McCloy, C. (1984). Stratigraphy and depositional history of the San Jose del Cabo trough, Baja California Sur, Mexico. In: *Geology of the Baja California Peninsula.* (V. A. Frizzell, ed.), **39**, 267–273. Pacific Section Society of Economic Paleontologists and Mineralogists, Los Angeles, California.

Moreno-Ruiz, J. L. & Carreño, A. L. (1993). Morfología de *Thalassionema nitzschioidez* var. *claviformis* (Schrader) Moreno-Ruiz nov. comb. *Facultad de Ciencias del Mar, Universidad Católica del Norte, Coquimbo, Chile.* Serie Ocasional, **2**, 141–148.

Moreno-Ruiz, J. L. & Carreño, A. L. (1994). Diatom biostratigraphy of Bahia Asuncion, Baja California Sur, Mexico. *Revista Mexicana de Ciencias Geológicas*, **11** (2), 243-252.

Peragallo, H. & Peragallo, M. H. (1897–1908). *Diatomées marines de France, et des districts maritimes voisins.* 1–493. Grez sur Loing.

Schrader, H. (1973). Cenozoic diatoms from the northeast Pacific, Leg 18. *Initial Reports Deep Sea Drilling Project*, **18**, 673–797.

Schrader, H. (1974). Revised diatom stratigraphy of the experimental Mohole drilling, Guadalupe site. *Proceedings of the California Academy of Sciences*, 4th series, **34** (23), 517–562.

Schrader, H. (1976). Cenozoic planktonic diatom biostratigraphy of the southern Pacific Ocean. *Initial Reports Deep Sea Drilling Project*, **35**, 605–671.

Schrader, H. & Fenner, J. (1976). Norwegian Sea Cenozoic diatom biostratigraphy and taxonomy, part I: Norwegian Sea Cenozoic diatom biostratigraphy, part II: diatoms at Leg 38, taxonomic references. *Initial Reports Deep Sea Drilling Project*, **38**, 921–1099.

Schuette, G. & Schrader, H. (1982). *Thalassiothrix pseudonitzschioides* sp. nov.: a common pennate diatom from the Gulf of California. *Bacillaria*, **5**, 213–223.

Simonsen, R. (1992). The diatom types of Heinrich Heiden in Heiden & Kolbe 1928. *Bibliotheca Diatomologica*, **24**, 1–100.

Van Heurck, H. (1880–1885). *Synopsis des diatomées de Belgique*. Anvers, Texte 235 p., Atlas 132 pls.

VanLandingham, S. L. (1978). Catalogue of the fossil and recent genera and species of diatoms and their synonyms. Part 7. *Rhoicosphenia* through *Zygoceros*. *Lehre*: J. Cramer, **7**, 3606–4235.

Whiting, M. C. & Schrader, H. (1985). Late Miocene to early Pliocene marine diatom and silicoflagellate floras from the Oregon coast and continental shelf. *Micropaleontology*, **31** (3), 249–270.

Yanagisawa, Y., Nakamura, K., Susuki, Y., Sawamura, K., Yoshida, F., Tanaka, Y., Honda, Y. & Tanahashi, M. (1989). Tertiary biostratigraphy and subsurface geology of the Futaba District, Joban Coalfield, northeast Japan. *Bulletin of the Geological Survey of Japan*, **40** (8), 405–467.

Figures 1–33, transmitted light photomicrographs (BF = bright field, PC = phase contrast). Scale bar = 10 µm. ATBA, VD = Valle de Diatomita; SR, SRM = San Roque; END = ENDEAVOR, southern Gulf of Mexico; SONT = Lagoon of Sontecomapan; SANT = Santiago; AH = Arrogo Houdo.

1. *T. nitzschioides* (Grunow) Hustedt 1/VD17–3A/R37VDTH–27, BF.
2. *T. nitzschioides* 1/ATBA14–3/R23–16, BF.
3. *T. nitzschioides* 2/VD16–1/R35VDTH–5, BF.
4. *T. nitzschioides* var. *antiqua* (Schrader) Moreno-Ruiz comb. nov. 3/VD15–1/R38VDTH–29, BF.
5. *T. nitzschioides* var. *antiqua* 2/VD16–1/R35VDTH–5, BF.
6. *T. nitzschioides* var. *capitulata* (Hustedt) Moreno-Ruiz comb. nov. 15/VD3–1/R16VDTH–20, BF.
7. *T. nitzschioides* var. *capitulata* 13/VD5–1/R33VDTH–20, BF.
8. *T. nitzschioides* var. *claviformis* (Schrader) Moreno-Ruiz 16/VD2–7/R17VDTH–8, BF.
9. *T. nitzschioides* var. *claviformis* 8A/VD10'–2/R14VDTH–2, PC.
10. *T. nitzschioides* var. *claviformis* 10/VD8–2/R17VDTH–17, BF.
11. *T. nitzschioides* var. *claviformis* 16/VD2–7/R17VDTH–18, PC.
12. *T. nitzschioides* var. *claviformis* 17/VD1–1A/R17VDTH–32, PC.
13. *T. nitzschioides* var. *claviformis* 2/VD16–1/R34VDTH–22, BF.
14. *Thalassionema nitzschioides* var. *inflata* Heiden 17/VD1–1 B/R27VDTH–32, PC.
15. *T. nitzchioides* var. *inflata* 1/VD17–3A/R37VDTH–26, BF.
16. *T. nitzschioides* var. *incurvata* Heiden 5/VD13–1/R12VDTH–30, BF.
17. *T. nitzschioides* var. *incurvata* 3/VD15–2/R35VDTH–11, BF.
18. *T. nitzschioides* var. *incurvata* 17/VD1–1A/R18VDTH–3, PC.
19. *T. nitzschioides* var. *incurvata* 17/VD1–1B/R27VDTH–29, PC.
20. *T. nitzschioides* var. *inflata* 16/VD2–7/R1VDTH–9, BF.
21. *T. nitzschioides* var. *inflata* 1/VD17–3A/R36VDTHALA–1, BF.
22. *T. nitzschioides* var. *inflata* 2/VD16–1/R34VDTHALA–23, BF.
23. *T. nitzschioides* var. *lanceolata* (Grunow) Peragallo & Peragallo 3/VD15–2/R35VDTH–22, BF.
24. *T. nitzschioides* var. *lanceolata* 13/VD5–1/R33VDTH–18,BF.
25. *T. nitzschioides* var. *parva* (Heiden) Moreno-Ruiz emend. 2/VD16–1/R34VDTH–30, BF.
26. *T. nitzschioides* var. *parva* 2/VD16–1/R34VDTH–29, BF.
27. *T. nitzschioides* var. *parva* 6/VD12–2/R34VDTH–17, PC.
28. *T. nitzschioides* var. *robusta* (Schrader) Moreno-Ruiz emend. 13/VD 5–1/R25VDTH–4, PC.
29. *T. nitzschioides* var. *robusta* 13/VD5–1/R25VDTH–5, BF.
30. *T. nitzschioides* var. *robusta* 13/VD5–1/R16VDTHALA–10, BF.
31. *T. nitzschioides* var. *schraderi* (Akiba) Moreno-Ruiz comb. nov. SRM2–1/R2ROSS–27, BF.
32. *T. nitzschioides* var. *schraderi* SRM1–3/R2ROSS–21, BF.
33. *T. nitzschioides* var. *schraderi* SRM2–1/R2ROSS–25, BF.

Figures 34–69, Electron microscopy (TEM = transmission, SEM = scanning). Figs 34, 44, 47, 59 scale bars = 10 μm; Figs 35–43, 45–46, 48–58, 60–69 scale bars = 1 μm.

34. *T. nitzschioides* (Grunow) Hustedt END 010–EST26–753, TEM.
35. *T. nitzschioides* END 010–EST26–754, TEM.
36. *T. nitzschioides* SONT1/3052, SEM.
37. *T. nitzschioides* SR1/3885, SEM.
38. *T. nitzschioides* SR1/3886, SEM.
39. *T. nitzschioides* var. *antiqua* (Schrader) Moreno-Ruiz comb. nov. SR2/3955, SEM.
40. *T. nitzschioides* var. *antiqua* SR2/3956, SEM.
41. *T. nitzschioides* var. *antiqua* SR2/3957, SEM.
42. *T. nitzschioides* var. *capitulata* (Castracane) Moreno-Ruiz comb. nov. 17/VD17–20/3373, SEM.
43. *T. nitzschioides* var. *capitulata* SANT1–1/3591, SEM.
44. *T. nitzschioides* var. *claviformis* (Schrader) Moreno-Ruiz 17/VD17–20/3066, SEM.
45. *T. nitzschioides* var. *claviformis* 17/VD17–20/3072, SEM.
46. *T. nitzschioides* var. *claviformis* 17/VD17–20/3071, SEM.
47. *T. nitzschioides* var. *inflata* Heiden SR2/3949, SEM.
48. *T. nitzschioides* var. *inflata* SR2/3950, SEM.
49. *T. nitzschioides* var. *inflata* 17/VD17–20/3174, SEM.
50. *T. nitzschioides* var. *incurvata* Heiden SR2/3927, SEM.
51. *T. nitzschioides* var. *incurvata* SR2/3928, SEM.
52. *T. nitzschioides* var. *lanceolata* (Grunow) Peragallo & Peragallo 17/VD17–20/3382, SEM.
53. *T. nitzschioides* var. *lanceolata* 17/VD17–20/3383, SEM.
54. *T. nitzschioides* var. *lanceolata* SR2/3922, SEM.
55. *T. nitzschioides* var. *lanceolata* SR2/3923, SEM.
56. *T. nitzschioides* var. *lanceolata* SR1/3903, SEM.
57. *T. nitzschioides* var. *parva* (Heiden) Moreno-Ruiz emend. AH–16E/3444, SEM.
58. *T. nitzschioides* var. *parva* 17/VD17–20/3350, SEM.
59. *T. nitzschioides* var. *robusta* (Schrader) Moreno-Ruiz emend. 17/VD17–20/3374, SEM.
60. *T. nitzschioides* var. *robusta* 17/VD17–20/3375, SEM.
61. *T. nitzschioides* var. *robusta* 17/VD17–20/3377, SEM.
62. *T. nitzschioides* var. *robusta* SR2/3916, SEM.
63. *T. nitzschioides* var. *schraderi* (Akiba) Moreno-Ruiz comb. nov. SR1/3904, SEM.
64. *T. nitzschioides* var. *schraderi* SR2/3929, SEM.
65. *T. nitzschioides* var. *schraderi* SR2/3932, SEM.
66. *T. nitzschioides* var. *schraderi* SR2/3110, SEM.
67. *T. nitzschioides* var. *schraderi* SR2/3111, SEM.
68. *T. nitzschioides* var. *schraderi* SR2/3952, SEM.
69. *T. nitzschioides* var. *schraderi* SR2/3954, SEM.

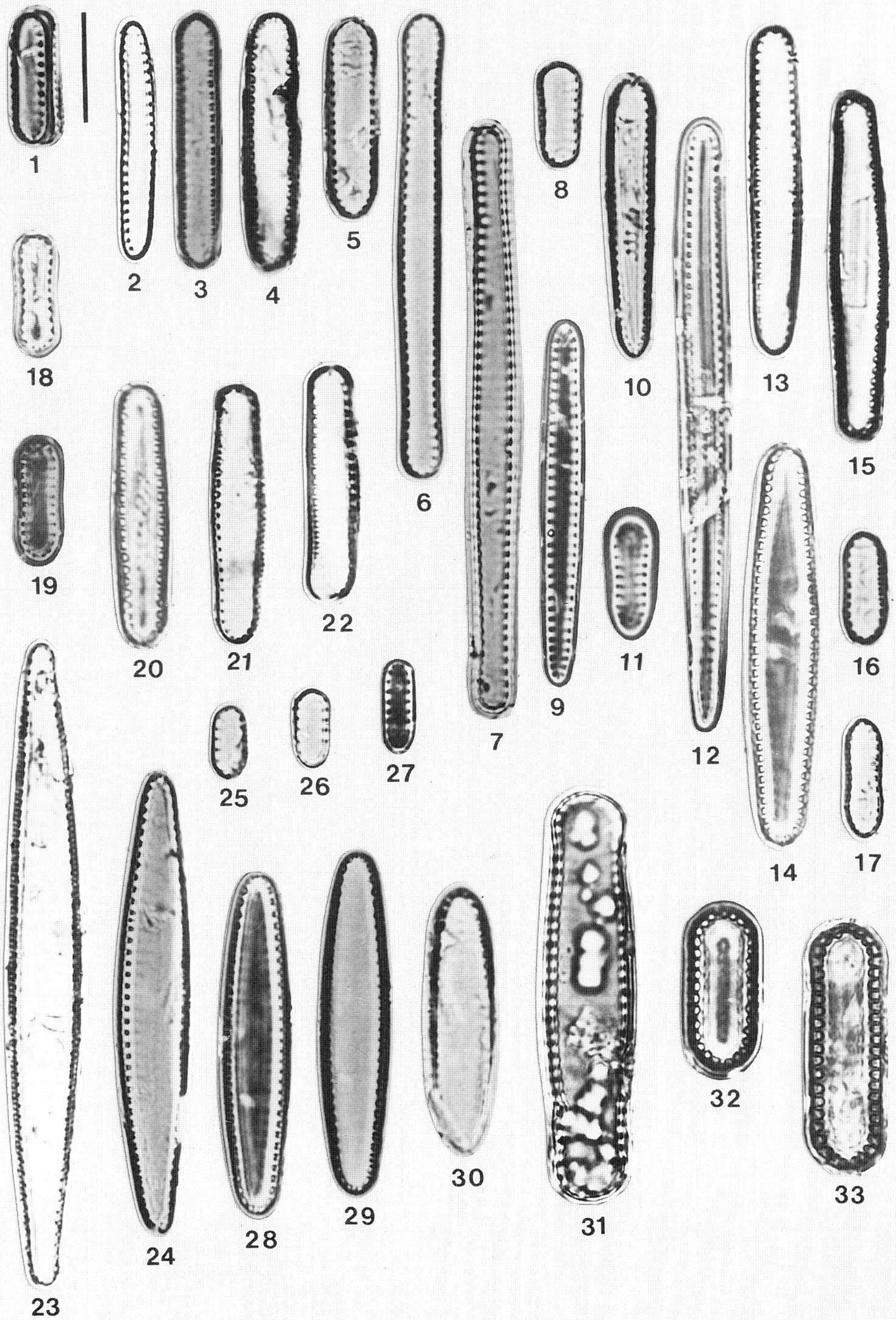
1
2
3
4
5
6
7
8
9
10
11
12
13
14
15
16
17
18
19
20
21
22
23
24
25
26
27
28
29
30
31
32
33

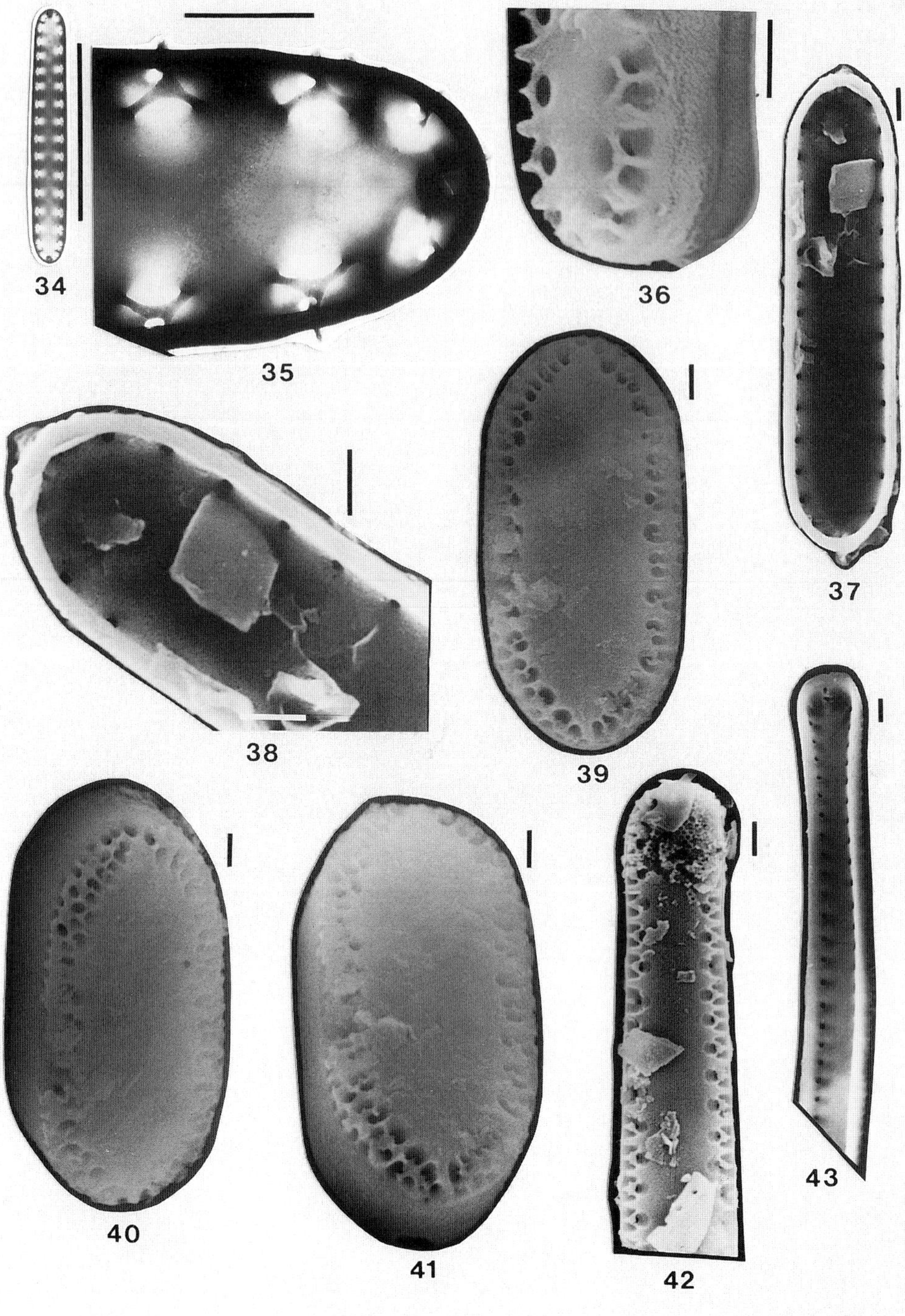
34
35
36
37
38
39
40
41
42
43

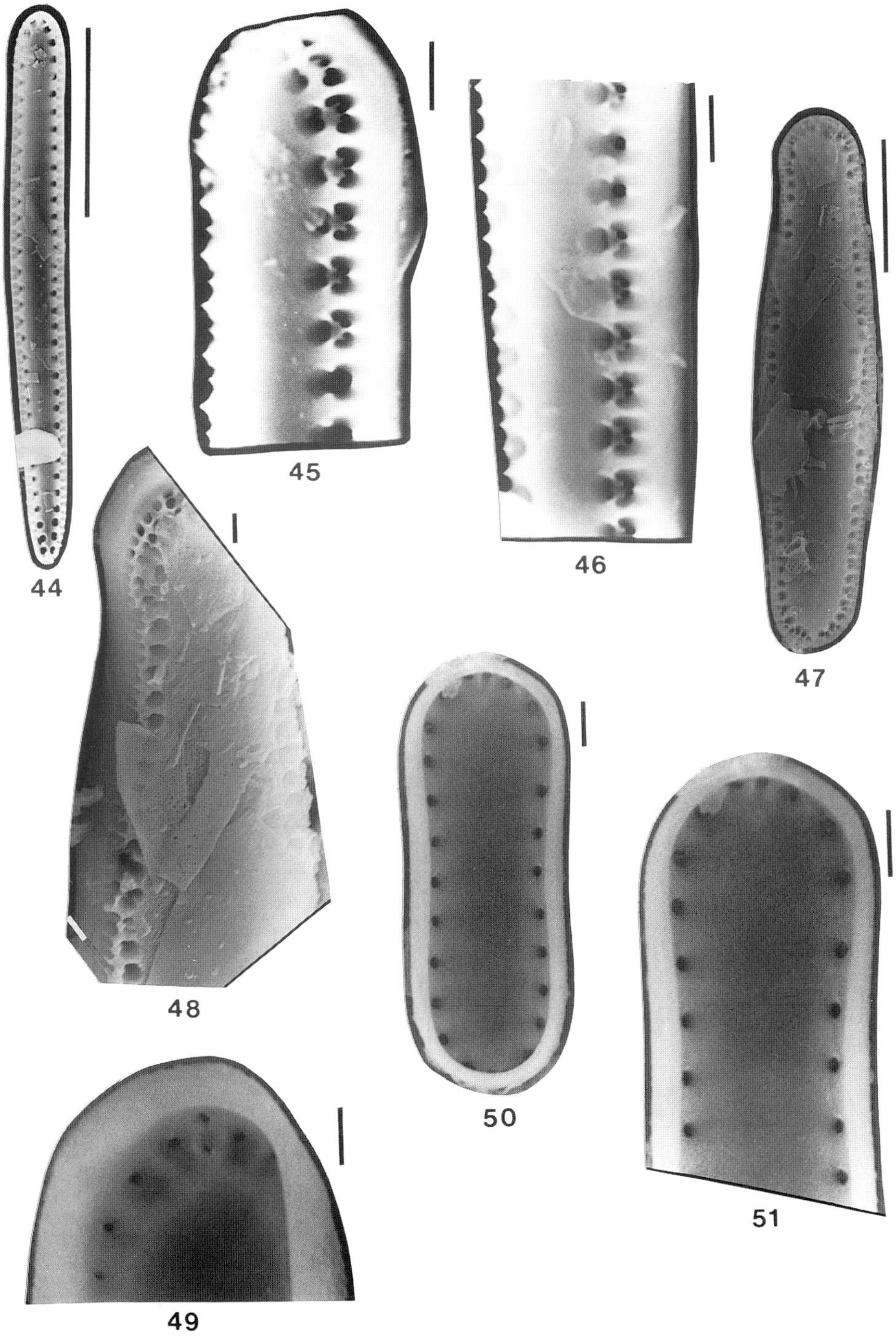
44
45
46
47
48
49
50
51

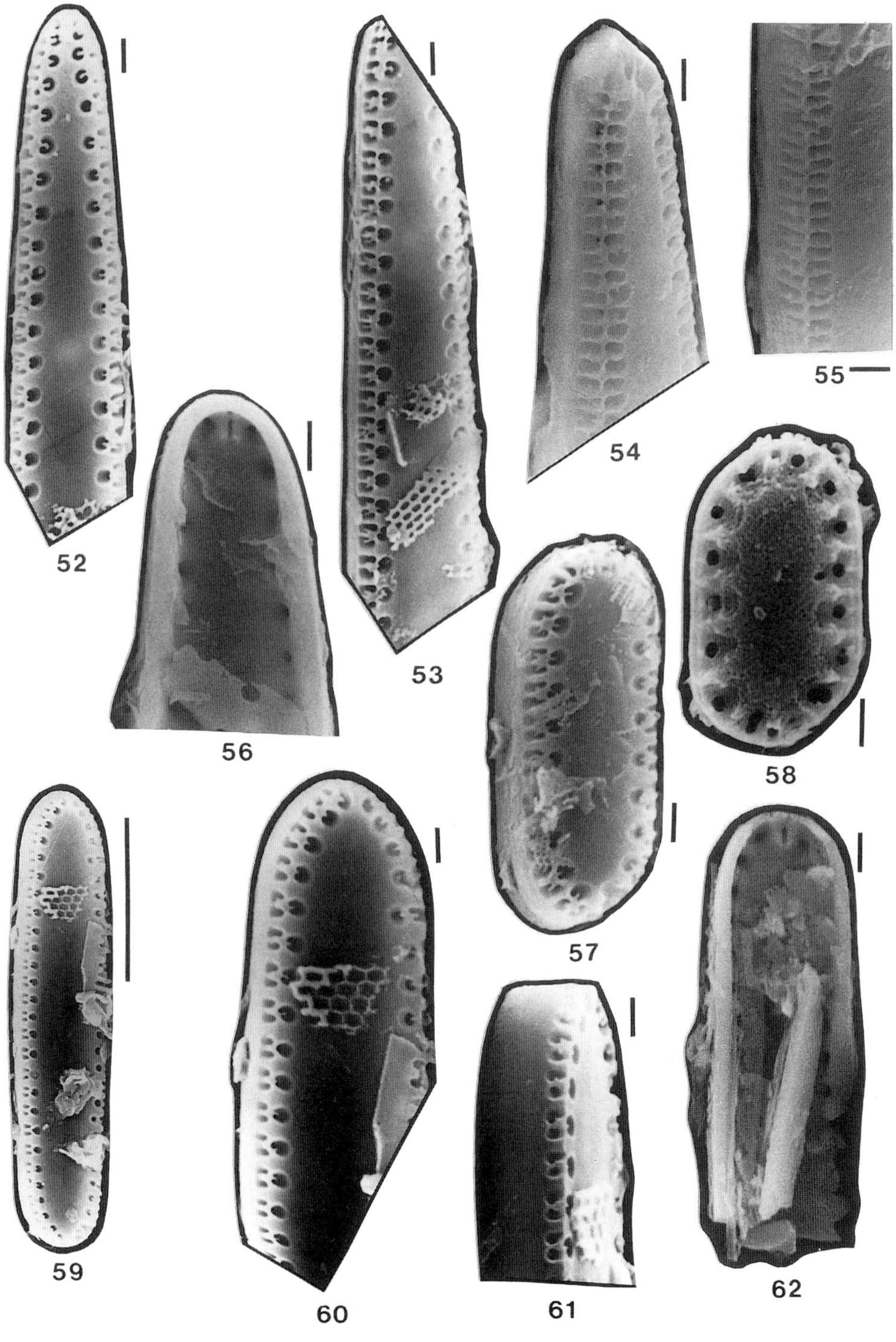
52
53
54
55
56
57
58
59
60
61
62

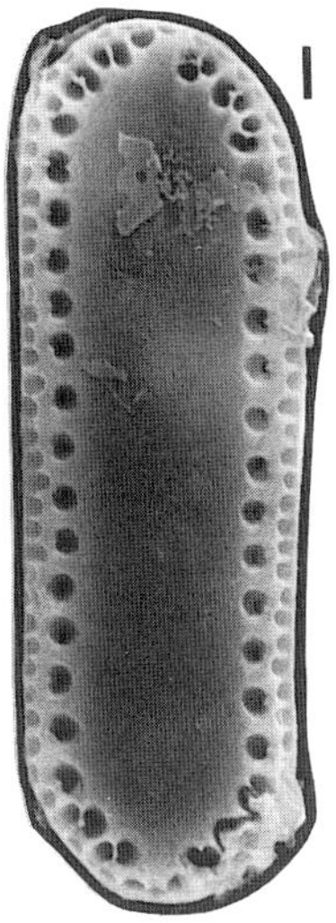
63

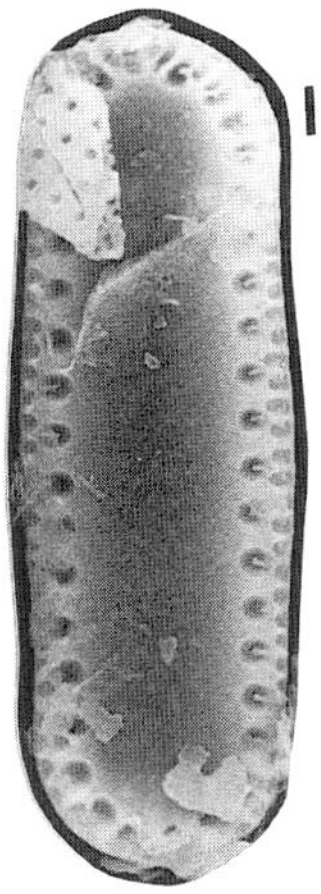
64

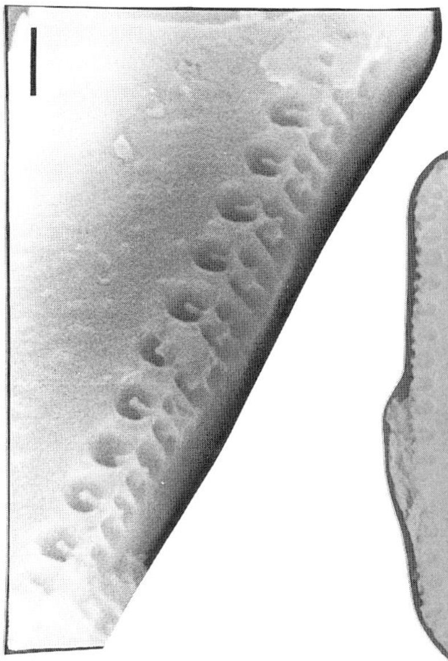
65

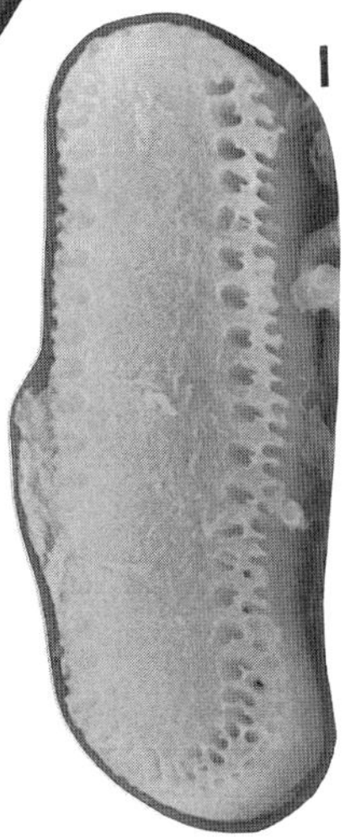
66

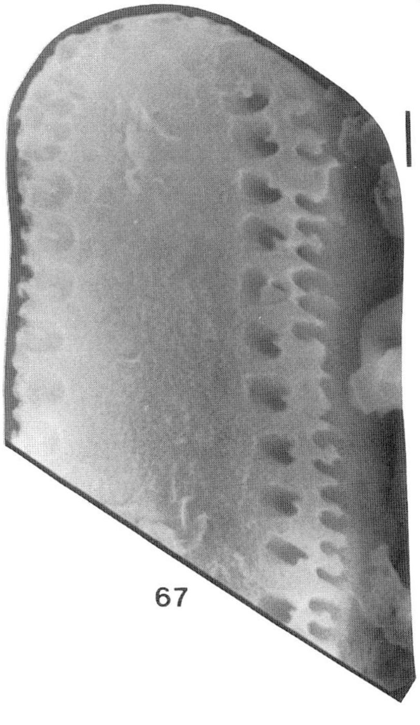
67

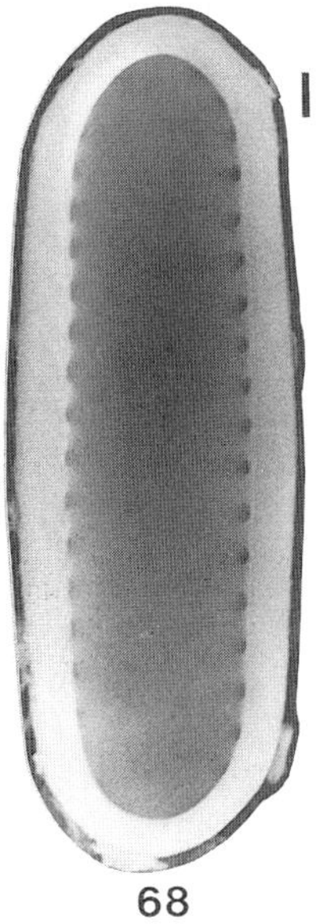
68

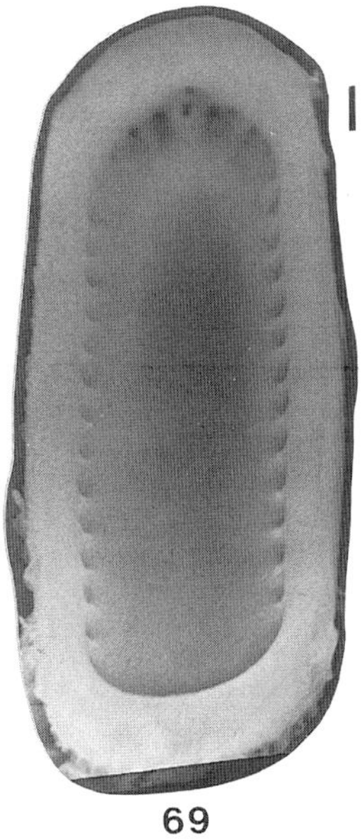
69

# A new species *Azpeitia elmorenii* sp. nov. from Late Miocene sediments of Baja California, Mexico

*Jose Luis Moreno-Ruiz, **Hermilo Santoyo and *Wintilo Vega-Murillo

**Depto. Hidrobiología, C.B.S., UAM–I, Apdo. Postal 55–535, México, D.F. 03940, México.*

***Depto. Biología Marina, UABCS. Carretera al Sur km. 5.5 A. P. 19–B C. P. 23080 La Paz, B. C. S.*

## Abstract

From material of latest Miocene of San Felipe diatomite, Baja California, Norte, the morphological variation and ultrastructure of the diatom *Azpeitia elmorenii* Moreno-Ruiz sp. nov. is described as revealed by light and scanning electron microscopy (SEM).

The species is characterised by a central siliceous ring, with a variable number of areolae in between, one eccentric rimoportula, slightly projected, presence of one marginal ring of elliptic rimoportulae, cribra with a variable number of poroids, two rows of areolae on the valve mantle and a valve face with irregular areolae.

SEM observations indicate this species is related to *A. praenodulifera* (Barron) Sims and Fryxell. A variety of the nominate species *A. elmorenii* var. *ferreira* Moreno-Ruiz var. nov. was found, its main difference being the extension of the external aperture of the rimoportulae.

## Introduction

The genus *Azpeitia* is a group that has been problematic in its taxonomic delimitation. Initially it was described in the genus *Coscinodiscus* which is also the case for the fossil taxa: *A. aeginensis* (Schmidt) Sims, *A. biradiata* (Greville) Sims, *A. elegantula* (Greville) Sims, *A. endoi* (Kanaya) Sims and Fryxell, *A. grovei* (Grove ex Schmidt) Sims, *A. oligocenica* (Jousé) Sims, *A. praenodulifera* (Barron) Sims and Fryxell, *A. salisburyana* (Lohman) Sims, *A. tuberculata* (Greville) Sims, *A. tuberculata var. atlantica* (Gleser and Jousé) Sims, *A. vetustissima* (Pantocsek) Sims and *A. voluta* (Baldauf) Sims. The recent species with fossil representatives have not escaped from this error: *A. africana* (Janisch ex Schmidt) Fryxell and Watkins, *A. barronii* Fryxell and Watkins, *A. neocrenulata* (VanLandingham) Fryxell and Watkins, *A. nodulifera* (Schmidt) Fryxell and Sims, *A. tabularis* (Grunow) Fryxell and Sims (Fryxell *et al.* 1986). In the San Felipe Diatomite, some studies have been made in which no *Azpeitia* representatives are mentioned (Hertlein 1968, Mandra and Mandra 1972). However, Bohem (1982) indicates the presence of *Coscinodiscus nodulifer* (=*A. nodulifera*) while Moreno-Ruiz (1990) on studying some species of the genus *Coscinodiscus* from this location also observed specimens of *Azpeitia* which differ from the species described at present.

## Materials and Method

The material comes from the San Felipe Diatomite, Baja California, México, located between 31°06'29" N and 114°56'25" W. 23 samples were collected, encompassing 5.5 m of diatomaceous sediments.

The samples were treated according to a mixed technique suggested by Moreno-Ruiz and Carreño (1993). Preparations for light microscope and SEM were obtained from this material. Samples were studied using a Zeiss IM–35 microscope and a Jeol–35C scanning electron microscope. 124 valves were studied.

The terminology used to designate the morphological characters is the one proposed by Anonymous (1979), Fryxell *et al.* (1986) and Sims *et al.* (1989).

## Results

*Azpeitia elmorenii* Moreno-Ruiz sp. nov.

Type locality: San Felipe Diatomite Member of the Llano el Moreno Formation, Baja California Norte.
Collector: Ferreira, V. & Moreno-Ruiz, J. L.
Holotype: LF–ICMyL slide SFII/3–1, Fig. 1, deposited at the Diatom Herbarium, Laboratorio de Fitoplancton, Instituto de Ciencias del Mar y Limnología, UNAM, México.
Iconotypes: Figs 2–7.
The species takes its name from Llano el Moreno, where the diatomitic sequence is located.

Valve circular, slightly convex (Figs 1–2). The center of the valve has a distinctive annulus surrounding 2–22 areolae. There is a circular external opening to the rimoportula in a slightly eccentric position (Figs 1–3). Diameter of the valve is 27.1–95.65 µm. 4–9 central to 3–9 marginal areolae in 10 µm. Rows of areolae, radiate in a slightly decusate pattern. The cribra are reticulate with a variable number of poroids (Fig. 3). The mantle is shallow, slightly tilted and formed by two alternate rings of areolae (10–11 in 10 µm). The proximal ring are in general vertical with smaller areolae, those of the distal ring larger and extended vertically (Figs 4–5). The striae (10–12 in 10 µm) structured by the chambers of the areolae in the distal ring (Figs 4–5). There are 10–26 rimoportulae (2 in 10 µm) between the margin and the mantle each with a rounded opening (Fig. 4). The valvocopula is wide and a thin copula and a third thin band can be seen (Figs 4–5). In internal valve view the eccentric rimoportula can be observed with a short, smooth neck, the rimoportulae between the margin and the mantle have a short neck and an elliptical aperture surrounded by a thin band (Fig. 7). The internal foramena are circular (Figs 6–7).

*Azpeitia elmorenii* var. *ferreira* Moreno-Ruiz var. nov.

Type locality: San Felipe Diatomite Member of the Llano el Moreno Formation, Baja California Norte.
Collector: Ferreira, V. & Moreno-Ruiz, J. L.
Holotype: LF–ICMyL slide SFII/15 ß–1, Fig. 10, deposited at the Diatom Herbarium, Laboratorio de Fitoplancton, Instituto de Ciencias del Mar y Limnología, UNAM, México.
Iconotypes: Figs 8–9, 11–16.

This variety is named in honor of Vicente Ferreira from the Centro de Investigación Científica y Estudios Superiores de Ensenada, because he provided diatomaceous sediments and helped in their collection.

Valve similar to the nominal species (Figs 8, 10–11), annulus with 2–18 areolae. The external aperture of the rimportula is eccentric and slightly elongated to elongated and from slightly wavy to wavy (Figs 8–12). Valve diameter 29.7–92.8 µm, 4–7 central areolae, 4–8 marginal aerolae in 10 µm. Radial rows of areolae with large and small areolae in a slightly decusate pattern (Figs 8, 10–11). The mantle is narrow slightly tilted, formed by two alternate rings of areolae (11–12 in 10 µm). The proximal ring near the margin has smaller areolae and is vertical; the areolae of the distal ring are larger and tilted (Fig. 13). Striae (10–12 in 10 µm), structured by the chambers of aerolae of the distal ring (Fig. 13). There are 10–24 rimoportulae lying between the margin and the mantle (2 in 10 µm). In the internal valve view a wide, circular hyaline area can be seen with a variable number of areolae between it and the elliptical aperture of the rimportulae (Figs 14–15). The rimoportulae between the margin and the mantle have an elliptical aperture surrounded by thin band (Figs 14, 16). The foramena are circular (Figs 14–16).

## Discussion

*A. elmorenii* is similar to *A. praenodulifera* (Barron) Sims & Fryxell, it differs in that the latter has the proximal ring of areolae tilted, the distal ring vertical, and the base of the external aperture of the rimportulae with a bridge. On the other hand, the internal valve view presents an eccentric, semicircular hyaline area (Sims *et al.* 1989,

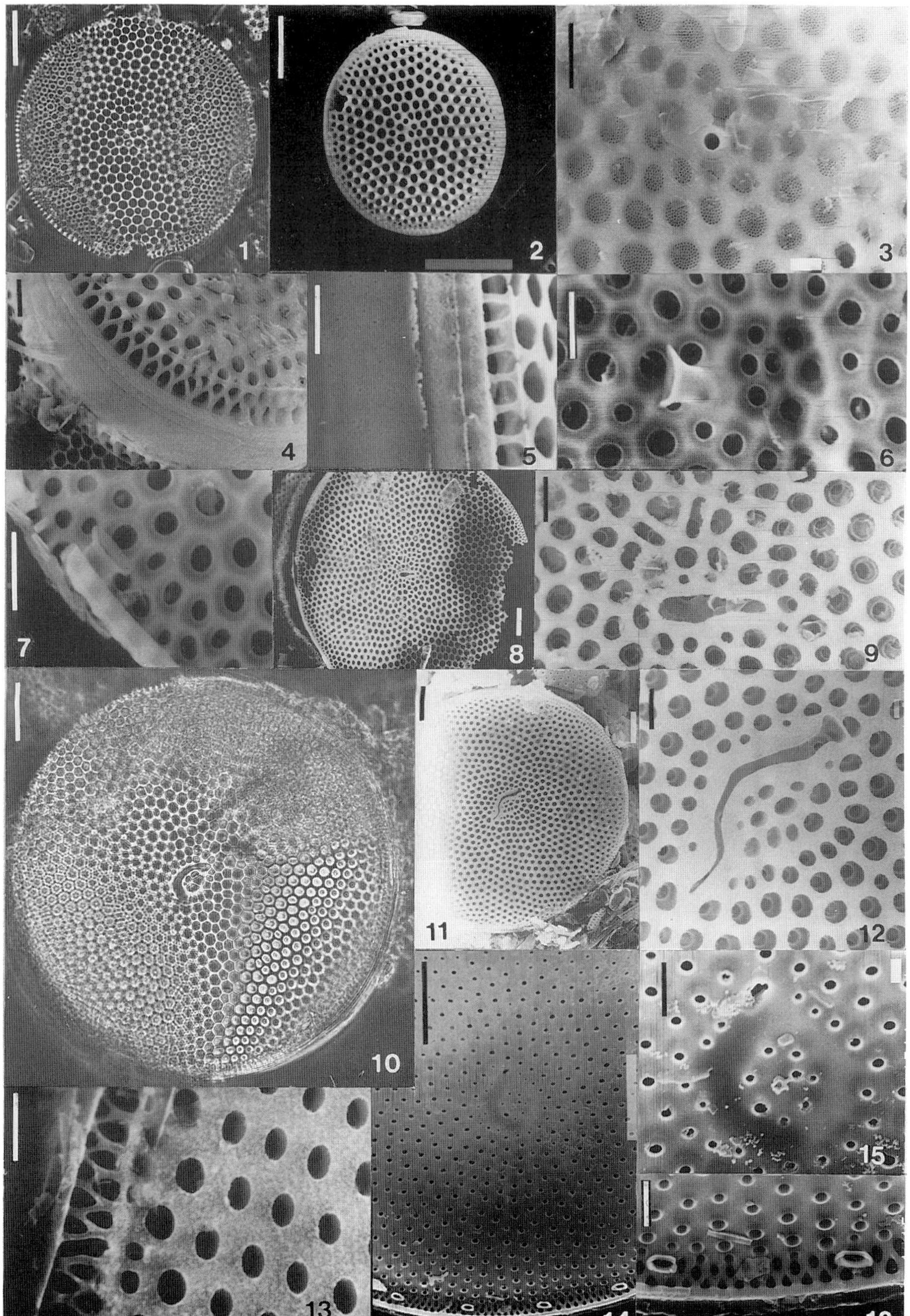
1
2
3
4
5
6
7
8
9
10
11
12
13
14
15
16

Pl. 1, Figs 10, 12). It is also close to *A. nodulifera* (Schmidt) Fryxell & Sims, but the latter presents the remains of a small rimoportulae over the eccentric rimoportula (Fryxell *et al.* 1986, Fig. 17. 1C). With *A. nodulifera* fo. *cyclopa* (Jousé) Sims, the main difference lies in the small rimoportula over the eccentric rimoportula (Sims *et al.* 1989, Pl. 2, Fig. 10).

*A. elmorenii* var. *ferreira* differs from the nominal species by the external aperture being semi–elongate to elongate and in the presence of a wide hyaline area with a variable number of areolae internally.

It is similar to *A. aeginensis* (Schmidt) Sims, but it differs in that the latter has a wide hyaline area around the external aperture of the eccentric rimoportula. In the mantle, the proximal ring of areolae is tilted and the distal ring is vertical (Sims *et al.* 1989, Pl. 1, Fig. 5).

*Geographical distribution*

Both taxa are present in the San Felipe diatomite, in the north of Baja California. As far as we know, they have not been previously recorded, so distribution is unknown.

*Stratigraphic distribution*

In the samples analysed, *Nitzschia reinholdii* Kanaya ex Barron and Baldauf and *Thalassiosira praeoestrupii* Dumont, Baldauf and Barron are present at 6.1 Ma (Barron 1992) in latest Miocene and due to the absence of other index diatoms, these sediments can be dated about this age.

---

Figs 1–7. *Azpeitia elmorenii* Moreno-Ruiz sp. nov. Figs 8–16. *A. elmorenii* var. *ferreira* Moreno-Ruiz var. nov. Figs 1, 10. Light photomicrographs (PC = phase contrast), scale bars = 10 µm (Figs 2–9, 11–16 scanning electron microscopy [SEM]), 10 µm (Figs 2, 8, 11, 14), 2 µm (Figs 3–7, 9, 12–13, 15–16). Fig. 1. Valve outline, annulus and marginal rimoportulae SFII/3–1/R25TAzp–29, PC. Holotype. Fig. 2. Annulus and radial row of areolae SFX2/2281, SEM. Fig. 3. External aperture of eccentric rimoportula and cribra SF15d/1864, SEM. Fig. 4. External aperture marginal rimoportulae, proximal and distal rings of areolae and three cingulum bands SF2+50d/2078, SEM. Fig. 5. Valvocopula, copula and third band SF3/3527, SEM. Fig. 6. Rimoportula SFX2/2280, SEM. Fig. 7. Marginal rimoportula SF2+50c/2023, SEM. Fig. 8. Valve outline SF16–2/1629, SEM. Fig. 9. Elongated external aperture of eccentric rimoportula SF16–2/1630, SEM. Fig. 10. Valve outline, annulus and eccentric rimoportula SFII/15ß1/R26TAzp–4, PC. Holotype. Fig. 11. Valve outline, annulus and eccentric rimoportula SF16–2/1620, SEM. Fig. 12. Weavy external aperture of eccentric rimoportula SF16–2/1621, SEM. Fig. 13. Proximal and distal ring of areolae SF16–2/1626, SEM. Fig. 14. Central area and marginal rimoportulae SFII–1/16–1/1543, SEM. Fig. 15. Annulus and broken eccentric rimoportula SFII–1/16–1/1544, SEM. Fig. 16. Marginal rimoportulae SFII–1/16–1/1545, SEM.

*A. elmorenii* has a distinctive morphology which can probably serve as an index species. The total biostratigraphic range of this species was not determined because the base of the section is covered and does not have a previous register to 6 Ma (Boehm 1982), and there is a possibility that it has a restricted extension which could reinforce its usefulness as a species with stratigraphic value.

In spite of the studies in these sequences (Hertlein 1968; Mandra and Mandra 1972; Boehm 1982; Moreno-Ruiz 1990), most of the species of diatoms are still unknown. It is possible that both taxa are from warm waters as indicated for other species of this genus (Sims *et al.* 1989).

*Affinity*

*A. praenodulifera* because *A. elmorenii* does not possess a small rimoportula over the eccentric rimoportulae and ten areolae in the proximal ring separate one rimoportula from the other. This species has a stratigraphic range from early Miocene to early middle Miocene (Sims *et al.* 1989), which suggest the possibility that *A. elmorenii* arose from the first taxon. The sister species of the latter would correspond to *A. nodulifera* which has its first occurrence in the middle Miocene (Burckle 1978 and Barron 1981b, *vide* Sims *et al.* 1989).

**Acknowledgements**

G. A. Fryxell and S. Licea for the revision and suggestion of the present study. V. Ferreira provided and helped to collect samples of the San Felipe diatomite. J. A. Barron for his giving valuable help on age. M. Reyes, G. Sánchez, A. Carabez and T. Cruz for access to the SEM. A. Fernández and B. Jon for the use of laboratory facilities. G. Gaxiola for his interest in the realitation of this study. Edgardo Flores, Raúl Jurado, Sergio López and Cesar García for their technical assistence. Sonia Ocampo for the translation into English. This research was partially supported by the Dirección General de Asuntos del Personal Académico, UNAM, by PADEP, UNAM., CICESE, Baja California Norte and UABCS.

**References**

Anonymous. (1975). Proposals for a standardization of diatom terminology and diagnoses. *Nova Hedwigia Beih.*, **53**, 323–354.

Barron, J. A. (1992). Neogene diatom datum levels in the equatorial and north Pacific. In: *The Centenary of Japanese Micropaleontology* (T. Saito & K. Ishizaki, eds). Terra Sci. Pub., Tokyo. 413–425.

Bohem, M. C. F. (1982). Biostratigraphy, lithostratigraphy, and paleoenvironments of the Miocene–Pliocene San Felipe marine sequence, Baja California Norte, Mexico. A thesis submitted to the department of Geology and the Commitee on graduate studies of Stanford University in partial fulfillment of the requirements for the degree of Master of Science. 1–326.

Fryxell, G. A., Sims, P. A. & Watkins, T. P. (1986). *Azpeitia* (Bacillariophyceae): Related genera and promorphology. *Sistematic Botany Monographs*, **13**, 1–74.

Hertlein, L. G. (1968). Three late Cenozoic molluscan faunules from Baja California, with a note on the diatomite from west of San Felipe. *Proceedings of the California Academy of Sciences*, 4th Ser., **30** (19), 401–405.

Mandra, Y. T. & Mandra, H. (1972). Paleoecology and taxonomy of silicoflagellates from an upper Miocene diatomite near San Felipe, Baja California, Mexico. *California Academy of Sciences, Occasional Papers*, **99**, 1–35.

Moreno-Ruiz, J. L. (1990). Morfología y sistemática del género *Coscinodiscus* Ehrenberg (Bacillariophyceae), en el sur del Golfo de México. UNAM, Tesis Maestría. 1–113. 12 tabs., 151 figs (unpub.).

Moreno-Ruiz, J. L. & Carreño, A. L. (1993). Morfología de *Thalassionema nitzschioides* var. *claviformis* (Schrader) Moreno-Ruiz nov. comb. *Facultad de Ciencias del Mar, Universidad Católica del Norte, Coquimbo, Chile*, Serie Ocasional, **2**, 141–148.

Ross, R., Cox, E. J., Karayeva, N. I., Mann, D. G., Paddock, T. B. B., Simonsen, R. & Sims, P. A. (1979). An emended terminology for the siliceous components of the diatom cell. In: *Proceedings of the 5th Symposium on Recent and Fossil Marine Diatoms, Antwerp, 1978. Nova Hedwigia Beih.*, **64**, 511–530.

Sims, P. A., Fryxell, G. A. & Baldauf, J. G. (1989). Critical examination of the diatom genus *Azpeitia*: species useful as stratigraphic markers for the Oligocene and Miocene epochs. *Micropaleontology*, **35** (4), 293–307.

# Preliminary efforts towards a diatom type catalogue with an example from *Fragilariforma virescens*

Geraldine Reid, Robert Huxley and David M. Williams

*Department of Botany, The Natural History Museum, Cromwell Road, London SW7 5BD, U.K.*

## Abstract

Each taxon should have unique name; each name, however used in the past or indeed in the future, should ideally be connected to a specimen to act as its name-bearer. We will demonstrate how this requirement can alleviate problems of conflicting taxonomies with reference to the species *Fragilariforma virescens*. Currently, this species is recognised either as the type of the genus *Fragilariforma* or a species of *Fragilaria*. According to VanLandingham's catalogue, there are 54 infraspecific entries and at least 7 synonyms under the name *virescens*. In addition, the species catalogue at the Natural History Museum indicates at least a further 15 names. We present an evaluation of some of these names from the perspective of available type specimens and how this information can link to *all* possible alternative classifications. The importance of specimen collections and their study is emphasised.

## Introduction

It has become almost a cliche to refer to particular taxonomies as being in a state of flux when different classifications compete for some sort of supremacy. It may be true that *all* classifications are constantly in need of revision, constantly in need of accommodating new data and never able to express a final solution. Such "difficulties" are the province of the science of systematics and taxonomy and encompass two areas of study: the theory of classification and the acquisition and understanding of new data. All this may be of little interest to those whose job it is not to ponder these variables but to use the classification presented for application to other areas of science. These persons are presented only with the question: Which of the available classifications do I use? This is, of course, quite a different question from which is the correct classification? Although these issues are central to the progress of taxonomy, we will not dwell on them any further. Instead, we offer a solution of sorts to those who "need a name" for some other purpose. We consider only the specific and infraspecific levels although other taxonomic levels will require future consideration.

This paper is not intended as a discourse on the finer points of taxonomic nomenclature or on the precise requirements for executing such studies. We present a few ideas that will hopefully encourage fellow diatomists to pursue similar avenues of investigation in their work.

Each species must have a unique name. This simple principle avoids confusion between two (or more) possibly different taxa having the same name. It is clear, however, that this simple principle can become complicated as names are used in different senses at different times by different authors. The *International Code of Botanical Nomenclature* has been constructed such to help avoid and prevent such confusions persistently occurring (Greuter 1994). Irrespective of the sometimes complex rules of that Code (Williams 1989; Ross 1994) is one underlying factor: each name, however used in the past or indeed in the future, requires a specimen to act as the name-bearer[1]. The specimen and the name become forever linked, irrespective of how "typical" that specimen turns out to be (Williams 1993). This is the basis of the type system of nomenclature which applies to all living organisms. The central point is that each name now requires typification; that is the name *must* be attached to a specimen and that specimen must be the one intended by the authors to represent that particular taxon.[2]

In diatom taxonomy, typification has not been adhered to with any great consistency in the past, although that trend is changing and practically all new taxa described in the eight volumes of *Diatom Research* (1986–1993) have type slides and their herbaria identified. However, more pertinent is the absence of identification of types for many of the older names and, in most instances, this action will be required retrospectively. This in itself introduces hazards as it will not be the original author

---

[1] The word specimen can be variously interpreted. In the best of all possible situations the specimens should be that which the author based his descriptions upon, which includes the possibility of using illustrations from the original protologue (description). The reader should consult Williams (1989) or Ross (1994) for further details of the ICBN. Since those papers were written a new ICBN has been produced (Greuter 1994). This version is noteworthy in that it introduces the concept of the epitype: "An epitype is a specimen or illustration selected to serve as an interpretative type when the holotype, lectotype or previously designated neotype, or all original material associated with a validly published name, is demonstrably ambiguous and cannot be critically identified for purposes of the precise application of the name of the taxon. When an epitype is designated, the holotype, lectotype or neotype that the epitype supports must be explicity cited" (Art. 9. 7).

[2] "All names published from 1 January 1958 onwards have designated holotypes" and there must be "an indication of the holotype for the name of a genus or taxon of lower rank; this will normally be the identification number of the microscope slide on which it is preserved, and also the herbarium or institution in which it is preserved." (Ross 1994: 432 and 434).

who "connects" taxon with name. Neverthless, if, in an ideal world, all valid names that have appeared in the literature have a reference to a relevant slide then the process of naming will be made considerably easier. If such a catalogue was available then diatom researchers would simply check the list and examine the type slide.

Is such a catalogue realistic? VanLandingham's monumental index (1969–79) indicates that there may be something like 250,000 names to deal with. VanLandingham's stop date for including taxonomic entries was 1964. The nomenclatural index in the Natural History Museum (BM) (incomplete though it is) indicates at least another 50,000 names to contend with since 1964. In addition, we suspect that of the total number of valid names something in the region of only 25% have been typified with appropriate specimens available in extant herbaria. However, although clearly a monumental task, once typification has been completed it no longer needs doing again. As a sobering example of what can be achieved, the Linnean typification project, undertaken largely at the BM, will have typified all Linnaeus' plant names with an appropriate specimen giving plant taxonomy a definitive record of Linnaeus' efforts tied to a fundamental modern database of specimens (Jarvis 1992; Jarvis *et al.* 1993).

How can we proceed with such an ambitious project in diatom taxonomy? Several things are required: (1) An inventory of all names for taxa so far described; (2) A search for slides that are representative of type material for each valid name and, if necessary, correct typification made (see Ross 1994: 434–5 for details); and (3) An indication of which names may never be typified along with proposals for abandoning them until such times as they can be verified. To a certain extent all taxonomic research will require the results of such efforts as well as contribute to its growth.

### *Fragilariforma virescens* and its "varieties"

*F. virescens* (Ralfs) Williams & Round is generally considered a cosmopolitan species (Williams 1995*a*). This may be misleading. For example, a number of distinct specimens described from South and Central America, although clearly related to *F. virescens*, have not been given appropriate attention (Williams, submitted *b*). They have either been identified as *F. virescens* or else as one of its varieties. Given that there is a general interest in diatom distribution (Mann & Droop, in press), it would be pertinent to know how many of these named varieties are actually distinct from the nominate type and have unique distributions. For instance, the taxon known as *F. virescens* var. *exigua* is now considered not only to be a distinct species but also belongs to a different genus (Flower, Jones & Round,. in prep.); its geographic distribution is unclear.

From the perspective of biogeography, there is insufficient data to make firm pronouncements on the distribution of species of *Fragilariforma*. For the most part this is a consequence of current taxonomy. However, our preliminary efforts suggest that there is a sister group relationship (they are each other's closest relatives) between the tropical species *F. strangulata* and a new species yet to be described (Williams, submitted *b*, Williams, in prep.). Both species are confined to the southern continents. This contrasts with the interrelationships among *F. constricta, F. acidobiontica, F. virescens* and *F. floridana* which together form a branch distinct from the *F. strangulata* clade. *F. acidobiontica* and *F. floridana* have been recorded only from the American continents: *F. floridana* occurs in the Southern states of North America and south to Brazil (including Central America, Honduras; Williams, pers. obs.). Yet, *F. constricta* and *F. virescens* are both considered widespread species probably cosmopolitan to the old world. From the perspective of current taxonomy how true is this pronouncement, given the inclination to include varieties within distributional limits and the suspicion that those same varieties may indeed be good species? The solution lies in an examination of the known varieties of *F. constricta* and *F. virescens* and the most pertinent starting point is the type material.

**Some conclusions**

The taxon we have decided to concentrate on is *F. virescens* which is recognised as either the type of the genus *Fragilariforma* (Williams & Round 1987) or a species of *Fragilaria sensu* Lange-Bertalot (Lange-Bertalot 1993). According to VanLandingham (1971 in 1967–1979) there are 54 infraspecific entries and at least 7 synonyms for the name *virescens*. In addition, the species catalogue at the BM indicates at least another 15 taxa that have been named as part of the *virescens* complex. Beyond such questions as what constitutes infraspecific rank, how many of these names have been or can be typified?

We have undertaken an examination of the slides present in the BM. Of the roughly 60 names known, six have material definitely identified from the collections; a further 50 or so are expected to have type material in the BM or at least will be identifiable from other extant collections; but at least eight are considered to be lost from the possibility of typification by a specimen altogether. Given that these figures suggest that over 90% of available names have extant type material, the prospects look good.

An additional complicating factor will be one of interpretation. When faced with a simple line drawing (Figures 1–65) and material which may contain many species from the same genus, associating the drawing and description with an actual specimen may be problematic. A sample of original illustrations are presented as Figures 1–65 to give some idea of how difficult it is to interpret some of these simple line drawings in the absence of specimens.

The significance of any findings are pertinent to both the typification of taxa belonging to *Fragilariforma* as well as ongoing research into the biogeographical distribution of freshwater araphid diatoms (Williams, submitted *b*).

Placed in a broader perspective, the capture of type specimen data from *Fragilariforma* is part of the BM strategy to transfer specimen data to electronic form and provide a searchable integrated core database for earth and life science specimens. It has been policy to record all new diatom accessions in this way since 1988 and

60,000 diatom slides are now recorded on the database. However, given the large number (c. 68 million) of specimens held in BM (that is, all Departments, not just Botany), a system of prioritisation has been used for retrospective capture, and important specimens with immediate value to current and future research, such as that described above, will be given priority. Indeed, it is crucial to understand that while disovery and identification of type specimens are valuable in themselves, the exercise is not hollow as distinct areas of research benefit. In short, the collections illuminate as well as document.

A good proportion of the collections are loaned to other workers during the course of each year. Study on these taxa is rightly published in appropriate scientific journals. Currently, as a matter of policy, we request that additional information gained from study of specimens is added either to the slide label or the herbarium sheet; electronic is now the most desirable form. These data are now to be entered into existing BM databases. Such procedures should become as routine as sending type slides for safe-keeping in any major specimen collection.

## Acknowledgements

Our thanks to Charlie Jarvis for his considerable help with the interpretation of the Tokyo Code and for his advice on type collections, Bob Press for producing material related to various presentations of this work, David Mann and Stephen Droop for sending us a preprint of their paper and the referees of this volume for their detailed comments on various drafts of the manuscript.

## References

Greuter, W. (1994). *International Code of Botanical Nomenclature. (Tokyo Code).* IAPT, International Association for Plant Taxonomy, Koeltz Scientific Books, Königstein, Germany. 389 pp.

Jarvis, C. E. (1992). The Linnean Plant Name Typification Project. *Botanical Journal of the Linnean Society*, **109**, 503–513.

Jarvis, C. E., Barrie, F. R., Allan, D. M. & Reveal, J. L. (1993). *A list of Linnean Generic Names and their Types*. International Association for Plant Taxonomy, Koeltz Scientific Books, Königstein, Germany. 100 pp.

Lange-Bertalot, H. (1993). 85 neue Taxa und über 100 weitere neu deinierte Taxa ergänzend zur Süsswasserflora von Mitteleuropa, vol. 2/1–4. *Bibliotheca Diatomologica*, **27**. 454 pp.

Mann, D. G. & Droop, S. J. M. (in press). Biodiversity, biogeography and conservation of diatoms. *Hydrobiologia*.

Ross, R. (1994). Nomenclature for diatomists. *Diatom Research*, **8**, 429–438.

VanLandingham, S. L. (1967–79). *Catalogue of the fossil and recent genera and species of diatoms and their synonyms*. Vols 1–8. J. Cramer, Vaduz.

Williams, D. M. (1989). Publication of new and revised taxa: A guide to the International Code of Botanical Nomenclature. *Journal of Paleolimnology*, **2**, 55–59.

Williams, D. M. (1993). Diatom nomenclature and the future of taxonomic database studies. *Nova Hedwigia, Beiheft* **106**, 21–31.

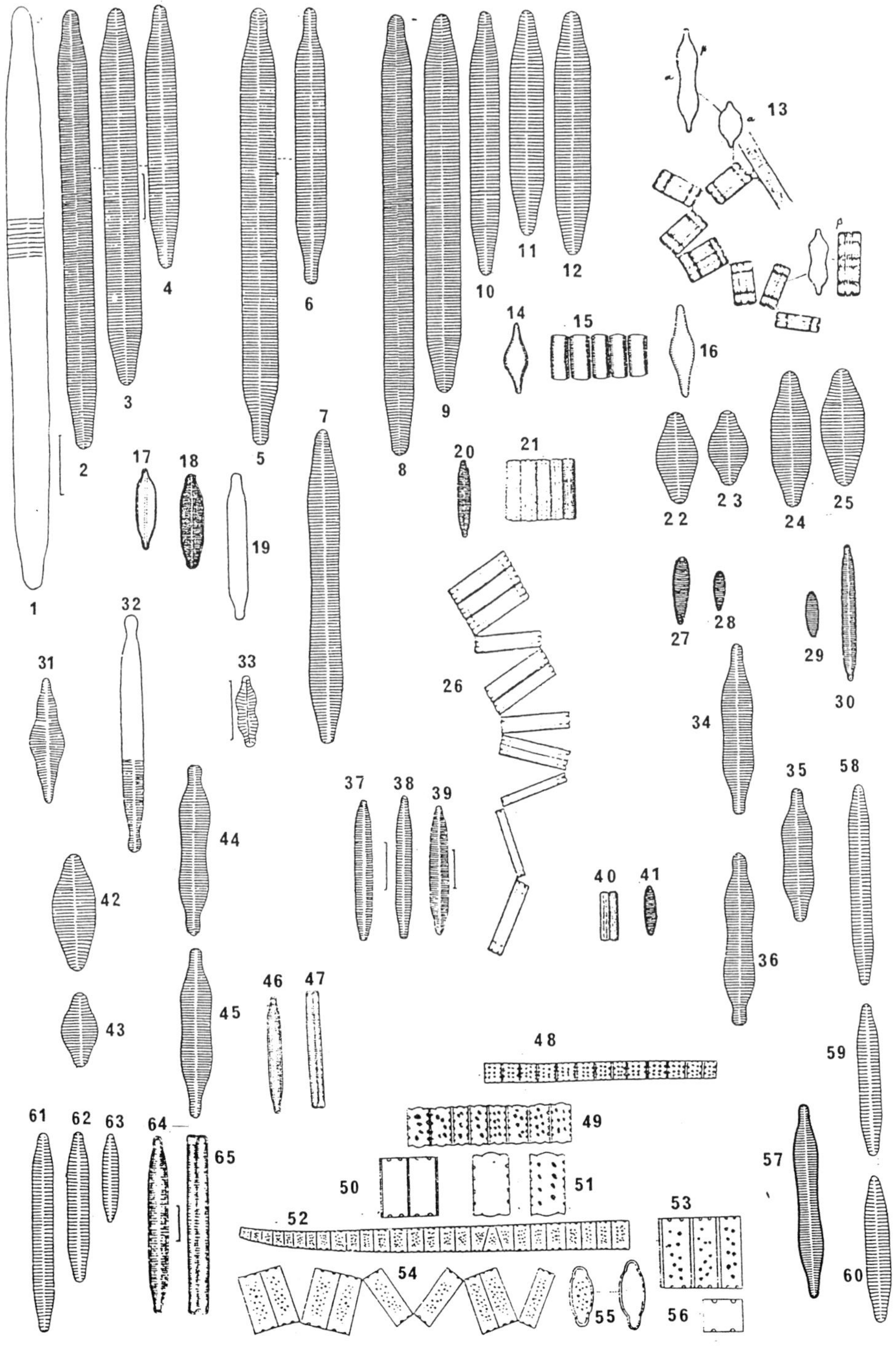
1
2
3
4
5
6
7
8
9
10
11
12
13
14
15
16
17
18
19
20
21
22
23
24
25
26
27
28
29
30
31
32
33
34
35
36
37
38
39
40
41
42
43
44
45
46
47
48
49
50
51
52
53
54
55
56
57
58
59
60
61
62
63
64
65

Williams, D. M. (1995 *a*). Diatom Biogeography: Some Preliminary Considerations. This volume.

Williams, D. M. (submitted, *b*). Notes on the genus *Fragilariforma* (Fragilariophyceae: Bacillariophyta) with a description of a new Miocene fossil species, *Fragilariaforma platensis*. *Nova Hedwigia, Beiheft*.

Williams, D. M. & Round, F. E. (1987). Revision of the *genus Fragilaria*. *Diatom Research*, **2**, 267–288.

---

Figures 1–65. Illustrations of some varieties of *Fragilariforma virescens* taken from original publications.

Figs 1–4. *Fragilaria virescens* f. *curta* Cleve-Euler; Figs 5, 6. *Fragilaria virescens* var. *birostrata* f. *typica* Cleve-Euler; Fig. 7. *Fragilaria virescens* var. *genuina* f. *mesolepta* Cleve-Euler; Figs 8–12, 24, 25. *Fragilaria virescens* var. *genuina* f. *typica* Mayer; Fig. 13. *Fragilaria undata* Wm. Smith; Figs 14–16. *Fragilaria inflata* Pant.; Figs 17–19. *Fragilaria virescens* var. *genuina* Mayer; Figs 20, 21. *Fragilaria producta* var. *bohemica* Grun.; Figs 22, 23. *Fragilaria virescens* var. *genuina* f. *curta* Mayer; Fig. 26. *Fragilaria virescens* var. Wm. Smith; Figs 27, 28. *Fragilaria virescens* f. *clavata* Grun.; Figs 29, 30. *Fragilaria virescens* var. *oblongella* Grun.; Fig. 31. *Fragilaria istvanffii* Pant.; Fig. 32. *Fragilaria virescens* var. *capitata* Krasske; Fig. 33. *Fragilaria virescens* var. *birostrata* f. *minor* Cleve-Euler; Figs 34–36. *Fragilaria virescens* var. *undatiformis* Mayer; Figs 37–39. *Fragilaria producta* var. *bohemica* Grun.; Figs 40, 41. *Fragilaria virescens* var. *subsalina* Grun.; Figs 42, 43. *Fragilaria virescens* f. *curta* Mayer; Figs 44, 45. *Fragilaria virescens* var. *birostrata* f. *undatiformis* Cleve-Euler; Figs 46, 47. *Fragilaria aequalis* var. *producta* Lagerstedt; Figs 48–56. *Fragilaria virescens* Ralfs; Fig. 57. *Fragilaria mesolepta* Rabenhorst; Figs 58–60. *Fragilaria producta* f. *genuina* Mayer; Figs 61–65. *Fragilaria producta* var. *genuina* Cleve-Euler.

# *Actinoptychus* Ehrenb. and kindred genera in Cretaceous and Paleogene deposits

Nina I. Strelnikova

*Department of Botany, Biological Faculty, St. Petersburg State University, Universitetskaya Emb. 7/9, St. Petersburg, 199034, Russia*

## Abstract

Eight species of *Actinoptychus* are illustrated (*A. heterostrophus, A. simbirskianus, A. tenuis, A. pericavatus, A. packii, A. seductilis, A. intermedius* and *A. senarius*). Scanning electron micrographs of these illustrate two different morphological types within this genus. The related genera *Glyphodiscus*, *Lepidodiscus* and *Corona* are also illustrated and certain issues of phylogeny are discussed.

## Introduction

*Actinoptychus* Ehrenb., *Glorioptychus* Hanna, *Lepidodiscus* Witt and *Corona* Witt are characteristically listed as occurring in Cretaceous and Paleogene floras. However, this group has not received particular attention from either morphological taxonomists or biostratigraphers, although some of the species have been studied separately by SEM.

Different taxonomic systems disagree over the classification of the aforementioned genera (Schütt 1896; Hustedt 1930; Simonsen 1979). In "The Diatoms of the USSR" (Makarova 1988), *Actinoptychus* and *Lepidodiscus* are placed in the family Heliopeltaceae, in the order Coscinodiscales, while *Corona* is listed within the family Auliscaceae, in the order Auliscales. Round *et al.* (1990) also place *Actinoptychus* and *Lepidodiscus,* along with *Glorioptychus*, in the family Heliopeltaceae in the order Coscinodiscales, but do not include *Corona* in their classification.

*Glorioptychus*, *Lepidodiscus* and *Corona* are extinct genera with few species. *Actinoptychus*, on the other hand, is still evolving; it includes about 150 fossil and contemporary species and infraspecific taxa. Nearly 50 of these species have been found in Cretaceous and Paleogene deposits. Only a few of them, including *A. senarius,* have survived up to the present.

During the last 25 years several morphological studies (SEM) of species of these four genera have been published (Wornardt 1970, 1971; Fryxell & Hasle 1973; Ross & Sims 1973; Andrews 1979; Round *et al.* 1990).

This study is a preliminary to a revision of the genera *Actinoptychus*, *Glorioptychus*, *Lepidodiscus* and *Corona* on the basis of the morphological characteristics of their valves as shown by the SEM. Its aim is also to disclose phylogenetic relationships within and between them.

## Material and Methods

Specimens were picked individually from the samples listed below. These specimens were positioned in such a way that both the interior and exterior views, as well as broken valves, could be observed. Specimens were processed for SEM using the technique described by Nikolaev (1982), and were examined using a JEOL–JSM 35C microscope.

The samples examined were from:

Russia

- exposure XI (14) from Tiltim in the basin of River Synya, cores from sites 22, 82 near Ust'–Manya in the basin of River Severnaya Sos'va, both in the Prepolar Urals, Western Siberia, Upper Cretaceous (Campanian).
- Povolzh'e. Exposures "Sengilei", Inza, Ulyanovsk district, Paleocene.
- Exposures Kantemirovka, Voronezhsky district. Middle Eocene (for details, see Strelnikova 1974, 1992).

Norwegian Sea Basin.

Leg 38 Glomar Challenger, Site 340 – 7– 6. Eocene (for details, see Dzinoridze *et al.* 1976).

USA.

Moreno Gulch, Fresno County, California (Collection received from G.D. Hanna). Upper Cretaceous (Maastrichtian).

All the samples listed were cleaned and prepared by the author and stored in the collection of diatoms of the Department of Botany, St. Petersburg State University.

## Observations

The genera *Actinoptychus*, *Glorioptychus* and *Corona* have many morphological features in common: the complicated relief of the valves is similar, as is the structure of the poroid areolae or passage pores, and the rimoportulae tend to be situated near the margin. The most obvious similarity is between *Actinoptychus* and *Glorioptychus* and between *Lepidodiscus* and *Corona*. The relationships within these pairs will be discussed further and the possible lines of evolution suggested.

### *Actinoptychus* Ehrenberg (1843: 400)

Radial undulation is characteristic of the genus. The valves consist of alternating raised and depressed sectors. There are three raised (convex) and three depressed (concave) sectors in the species described in this paper. However, other representatives of the genus may have up to 20 sectors, or even more.

When studying *Actinoptychus* species one needs to take into account the structures of the external and internal surfaces of both the convex and the concave sectors of the valves. Some species of *Actinoptychus*, for example *A. packii* Hanna and *A. pericavatus* Brun, show a tendency towards a more elaborate relief. In the closely related genus *Glorioptychus*, with only one species, *G. callidus* Hanna (figs 47–53), the relief is even more elaborate. In the marginal part of the valve there is an alternation of raised and depressed sectors in a more intricate pattern than in *Actinoptychus* (figs 47, 48).

The valves of the first group of species [*A. heterostrophus* A. Schmidt (figs 1, 3), *A. simbirskianus* A. Schmidt (figs 7, 8)] have smooth external and internal surfaces (monolaminate) with the structure consisting solely of poroid areolae (fig. 2) or passage pores.

The external and internal surfaces of the valves in the second group are different (bilaminate): the internal surface is smooth, while the external surface has *tubercles* [(*A. tenuis* Streln. (figs 11, 12), *A. pericavatus* Brun (fig. 16), *A. seductilis* A. Schmidt (figs 21, 23)]; these tubercles fuse to produce a cellular network giving the external surface a corrugated look. There are also some species in the same group [*A. intermedius* A. Schmidt (figs 27, 28), *A. senarius* Ehrenb. (fig. 32)] with tubercles united into ribs fused together to produce a honeycomb pattern (figs 30, 33).

Within the genus there appears to be a tendency towards a more elaborate relief of the sectors with accessory bulges on the raised sectors and concavities on the depressed sectors, e.g. in *A. packii* Hanna (figs 17, 18), and in *A. pericavatus* Brun (figs 13, 14). Rimoportulae also increase in number. In most of the species illustrated here there is one rimoportula located near the valve margin in the middle of each raised sector (*A. simbirskianus*, *A. packii*, *A. pericavatus*, etc.), but in *A. seductilis* there are three rimoportulae per raised sector (figs 22, 23). *A. heliopelta* Grun. has multiple rimoportulae (Andrews 1979).

The structure of the pore apparatus is a difficult problem as an intact velum was encountered in only a few cases [*A. heterostrophus* A. Schmidt (fig. 20), *A. senarius* (Ehrenb.) Ehrenb. (figs 33, 34)]; in these it was on the external surface. Passage pores without a velum on either the external or the internal surface appear to be more characteristic of this genus. It is of course possible that cribrate areolae have been lost during diagenesis. Initiation of pseudolocular areolae with internal vela seems to be probable in the course of evolution from cellular network structures formed of ribs with a silica basal layer pierced by passage pores, e.g. in *A. intermedius* A. Schmidt (figs 30, 31).

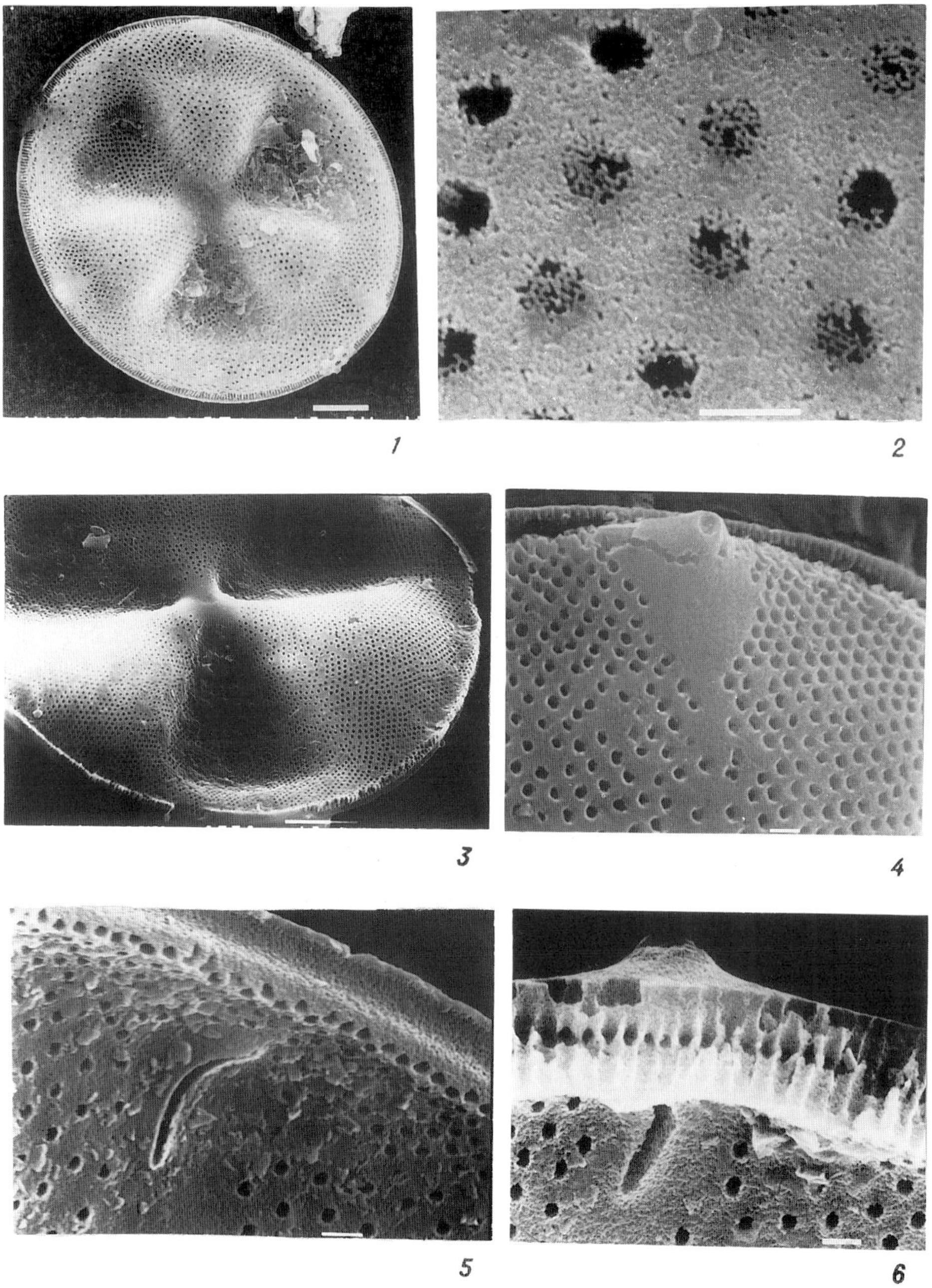

Figs 1–6. *Actinoptychus heterostrophus* A. S. Figs 1–3, 6. Inza, fig. 4. Tiltim, fig. 5. Sengilei. Figs 1, 2, 4. Valve exterior; Figs 3, 5, 6. Valve interior; Fig. 2. External cribrum; Fig. 4. External opening of rimoportulae; Figs 5, 6. Internal fissure of rimoportulae.

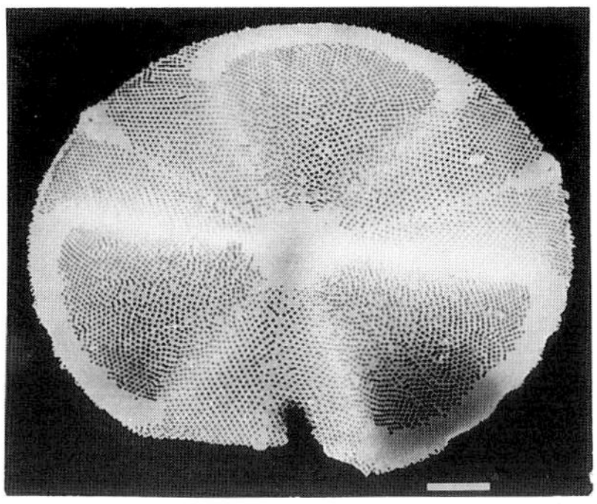

7

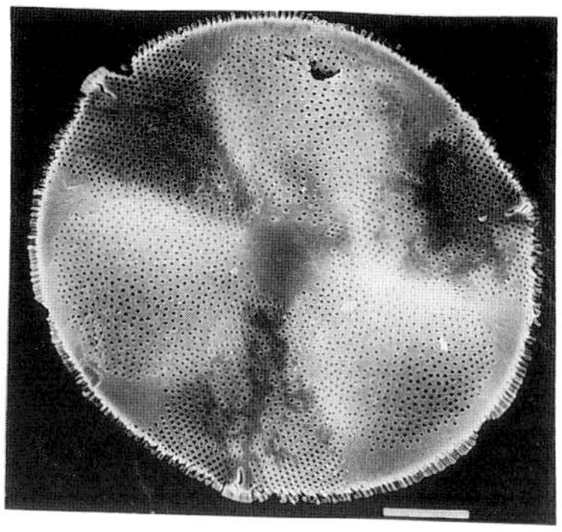

8

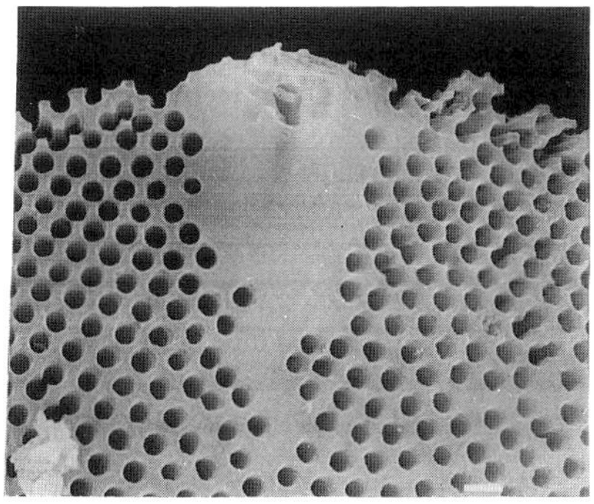

9

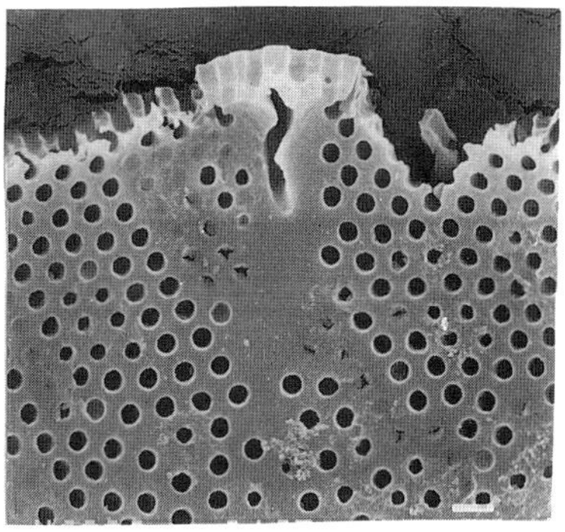

10

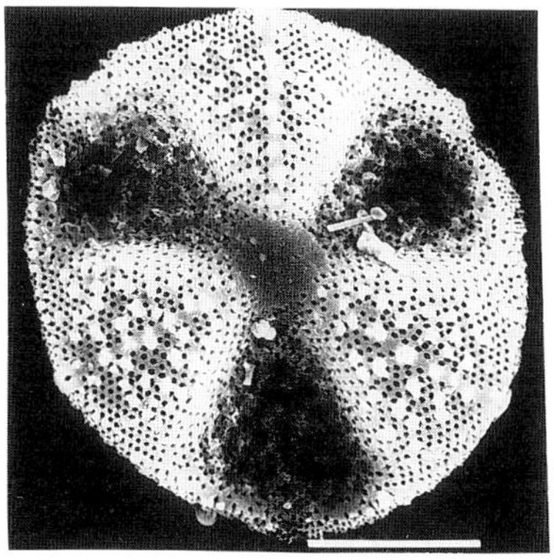

11

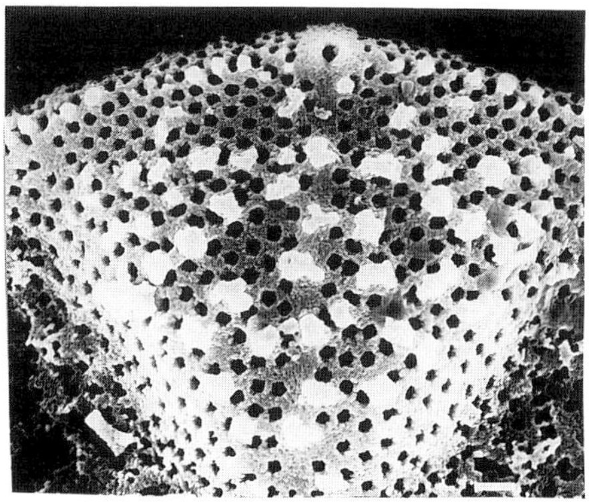

12

Figs 7–10. *Actinoptychus simbirskianus* A. S. Tiltim. Figs 7, 9. Valve exterior; Figs 8, 10. Valve interior; Fig. 9. External opening of rimoportula; Fig. 10. Internal fissure-shaped opening of rimoportula. Figs 11, 12. *A. tenuis* Streln. Ust' – Manya. Exterior of the valve with minute tubercles.

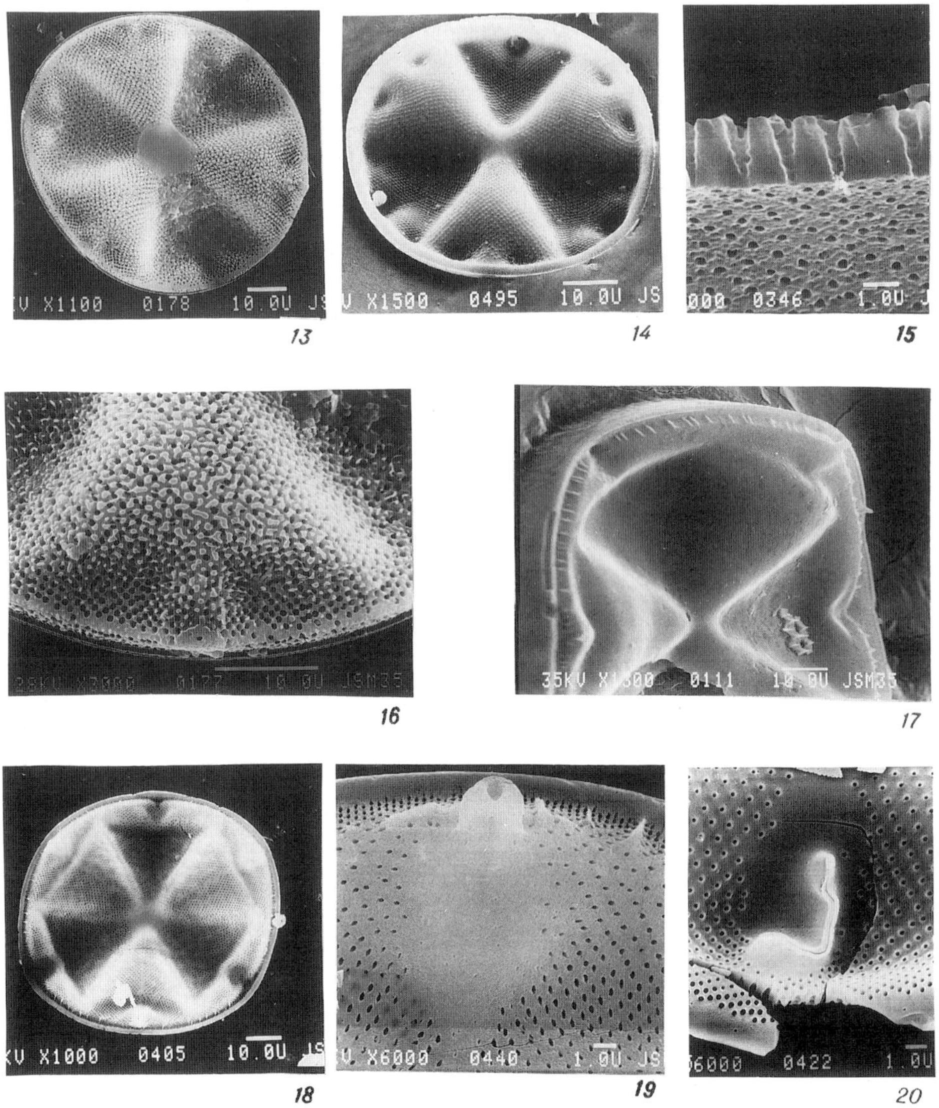


Figs 13–16. *Actinoptychus pericavatus* Brun. Kantemirovka. Figs 13, 16. Valve exterior. Fig. 14. Valve interior. Fig. 15. Fracture face: cone-shaped passage pores. Fig. 16. External openings of rimoportulae. Figs 17–20. *A. packii* Hanna. Moreno Shale. Figs 17, 18. Valve exterior. Fig. 19. External opening of rimoportula. Fig. 20. Internal fissure of rimoportula.

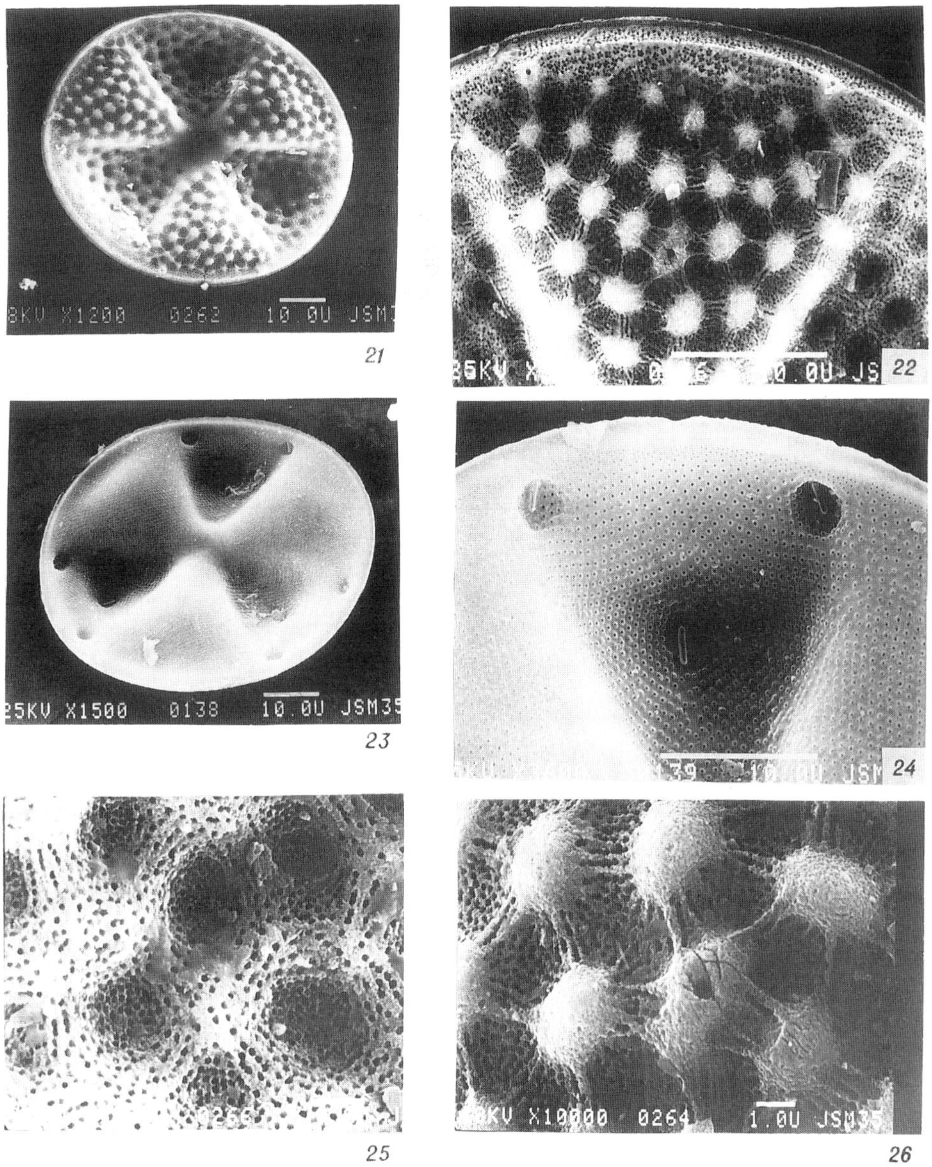


Figs 21–26. *Actinoptychus seductilis* A. S. Inza. Figs 21, 22, 25, 26. Valve exterior with fused tubercles. Fig. 22. External openings of rimoportulae. Fig. 24. Internal fissures of 3 rimoportulae.

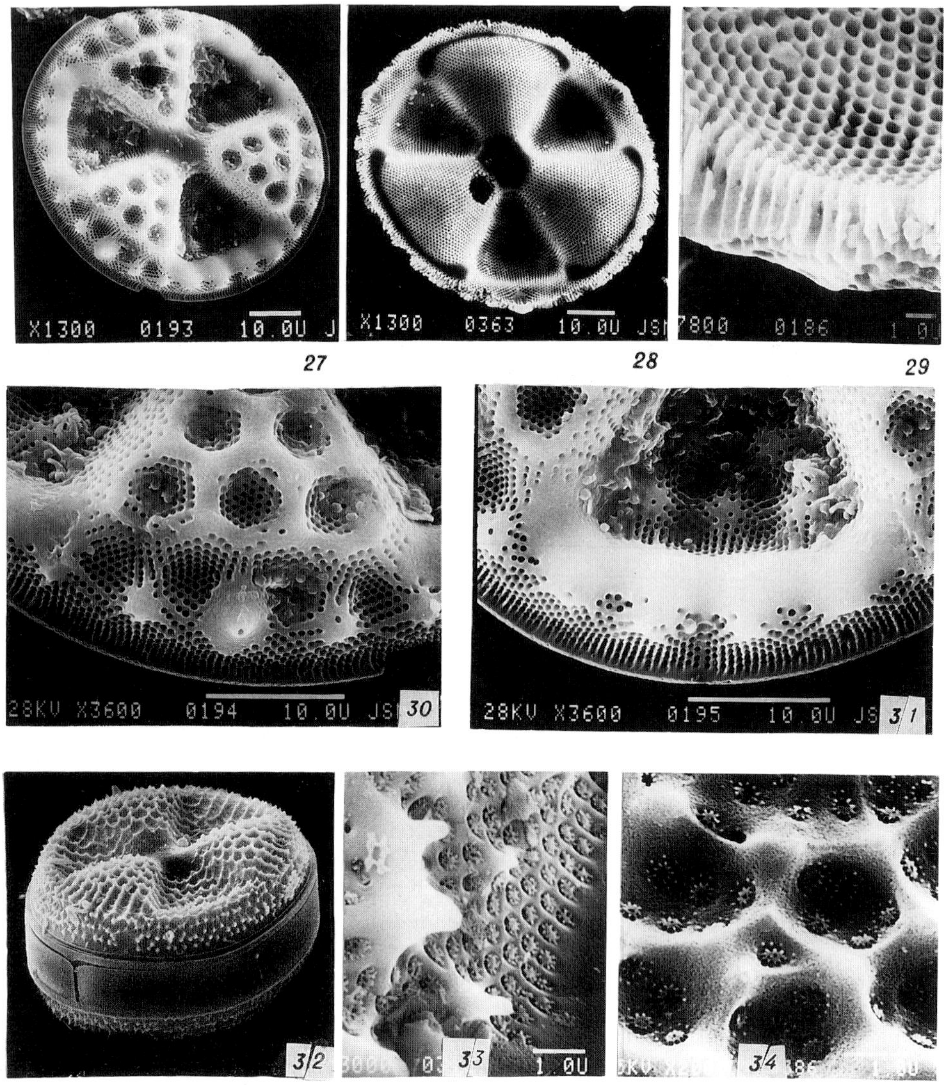


Figs 27–31. *Actinoptychus intermedius* A. S. Kantemirovka. Figs 27, 30, 31. Valve exterior. Fig. 28. Valve interior. Fig. 29. Fracture face: passage pores. Fig. 30. A network of ribs on the raised sector and an external opening of rimoportula. Fig. 31. Fused tubercles on the depressed (concave) sector, a hyaline bolster at the base of the sector. Figs 32–34. *A senarius* Ehr. Norwegian Sea. Valve exterior. Figs 33, 34. Star–shaped velum of the rota type on the external surface of poroid areolae.

### *Glorioptychus* Hanna (1927): 19, tab. 2, figs 6, 7)

In the closely related genus *Glorioptychus*, with only one species, *G. callidus* Hanna (figs 35, 36), the relief is even more elaborate. In the marginal part of the valve there is an alternation of raised and depressed sectors in a more intricate pattern than in *Actinoptychus*. However, the division of the valve into six sector still persists. There is a network of ribs on the external surface, and the regular rows of passage pores (fig. 37). (See also Round *et al.* 1990, p. 202).

### *Lepidodiscus* Witt (1866: 163, tab. 7, fig. 6)

An intricate concentric and radial undulating relief is characteristic of this genus. Three species are described in the literature; however, further examination of the material indicates that *Lepidodiscus* is a monotypic genus with only one species, *L. elegans* Witt. It is characteristic of late Cretaceous and Paleocene deposits (Strelnikova 1974). The raised and depressed sectors can be seen clearly in oblique view. The external surface is divided by a series of tubercular irregular hyaline ribs oriented radially. On the internal side of the valve the marginal raised sectors are separated by narrow hyaline "buttresses" (Round *et al.* 1990).

There are radial hyaline spaces on the internal surface of the valve corresponding to the radial ribs on the external surface; between these there are radial rows of passage pores. There are also oblique rows of passage pores on the lateral sides of the raised sectors and vertical parallel rows on the valve mantle (figs 38, 39). There is a rimoportula situated at the valve in the middle of each sector; each opens to the externa by a short, thick, stubby tube (fig. 40) surrounded by a silica ridge, and to the internal by a narrow fissure, straight or slightly curved and radially oriented. (See Round *et al.* 1990, p. 204).

### *Corona* (Lefébure & Chenevière 1938: 10, pl. 1, fig. 1)

Concentric and radially undulate relief is characteristic of this monotypic genus. Its only species, *C. magnifica* Lef. & Chen. (fig. 42), [*Craspedoporus magnifica* (Lef. & Chen.) Hendey & Sims according to Hendey & Sims (1987)], is characteristic of mid-Eocene deposits in the European part of Russia. Raised sectors along the valve margin look like truncated cones with a foramen at the top surrounded by a thick hyaline rim (fig. 43). The foramina closely resemble ocelli, but even the most careful SEM studies of numerous valves have not revealed thin plates with porelli. Rimoportulae are located near the margin of the valve between the raised sectors (figs 43, 44), each opening to the outside by a short conical tube and to the inside by a short narrow fissure enclosed by a thick rim. There is a network of coarse ribs with irregularly shaped spaces between them on the external surface; the internal surface is smooth. Passage pores are arranged in radial or fan-like rows on the internal surface.

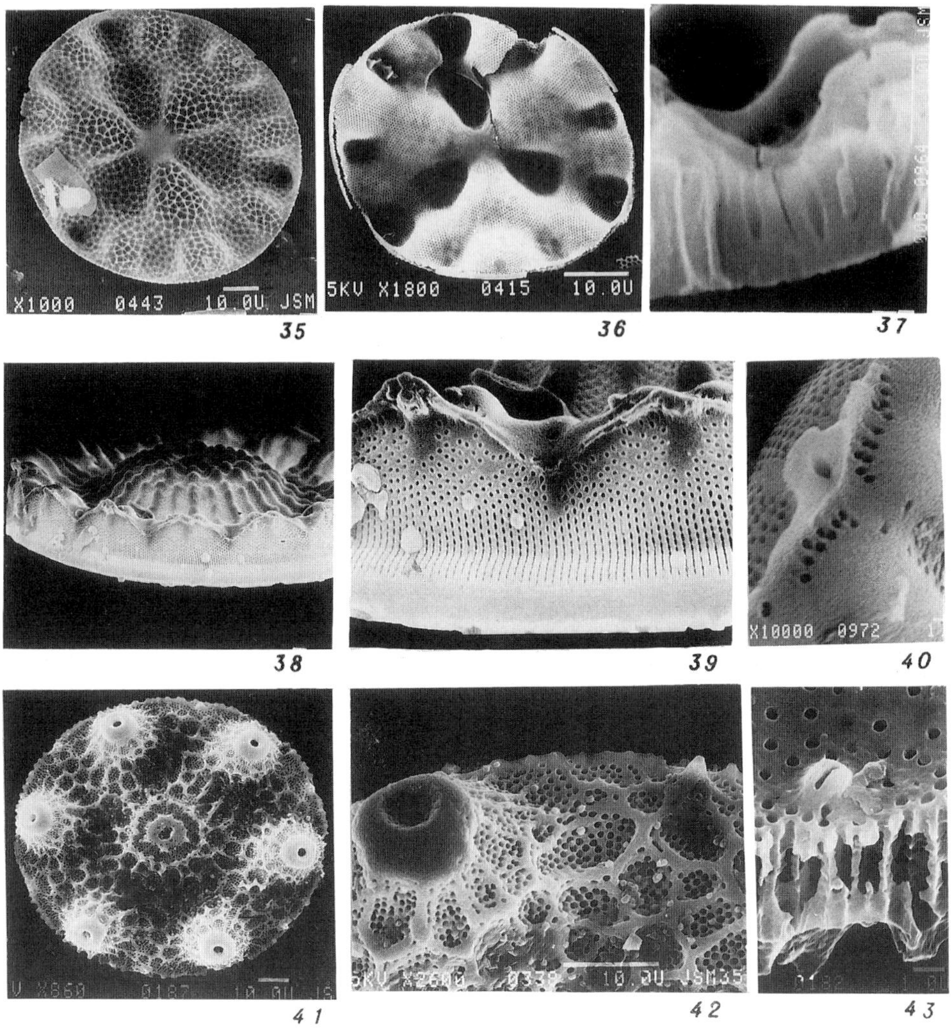


Figs 35–37. *Glorioptychus callidus* Hanna. Moreno Shale. Fig. 35. Valve exterior with a network of ribs. Fig. 36. Valve interior. Fig. 37. Fracture face: passage pores. Figs 38–40. *Lepidodiscus elegans* Witt. Ust'- Manya. Valve exterior. Figs 38, 39. The mantle of the valve with rimoportulae. Fig. 40. External opening of rimoportula. Figs 41–43. *Corona magnifica* Lef. et Chen. Kantemirovka. Figs 41, 42. Valve exterior with a network of ribs, cone-shaped eminence and rimoportulae. Fig. 43. Valve interior with a fissure of rimoportula.

It was due to the dissimilarities of the structure of the external and internal surfaces of the valve that *C. retinervis* Sheskukova & Gleser (1964: 87, tab. IV, figs 1, 2) was erroneously described as a separate species.

## Discussion

All the four genera described above form a coherent group with numerous features in common (the relief of the valves, the configuration of the pore apparatus, the shape and fine structure of the rimoportulae). The forms with a simple valve structure with poroid areolae but without tubercles or ribs (like *A. heterostrophus*) appear to be the primitive type. The second group of *Actinoptychus*, along with the more complicated forms, *Glorioptychus*, *Lepidodiscus* and *Corona*, have tubercles or ribs on the external surface and poroid areola or passage pores.

*Glorioptychus*, *Lepidodiscus* and *Corona* appear to be dead-end evolutionary branches; the first became extinct during the Cretaceous, the second in the Paleocene, whilst the third persisted for a short time during the middle Eocene. However, the group of species with a more complicated (bilaminar) structure within the genus *Actinoptychus* continues its evolution with representatives, including *A. senarius*, which still survive today.

The genera *Actinoptychus, Glorioptychus* and *Lepidodiscus* ought to be classified in a single family, the Heliopeltaceae. On the position of the genus *Corona* there are several opinions. Hendey & Sims (1987) treat it as a synonym of *Craspedoporus* and the new combination *Craspedoporus magnifica* (Lef. et Chen.) Hendey et Sims is made. They include it in subfamily within the Eupodiscaceae following for Simonsen (1979). In "The Diatoms of the USSR" (1988) *Corona* and *Craspedoporus* belongs to the family Auliscaceae, Order Auliscales. The question is what is the real position of this genus.

*Polymyxus* L. W. Bailey, a monotypic extant genus, is intermediate between *Actinoptychus* and *Aulacodiscus* Ehrenb. according to Ross & Sims (1973, p. 106). I have no other information about *Polymyxus* but, if this is so, it indicates that *Aulacodiscus* should be included in the Heliopeltaceae, as in "The Diatoms of the USSR" (Makarova 1988) rather than placed in a separate family, as in Round *et al.* (1990).

The ratio of the number of species of Coscinodiscaceae to Heliopeltaceae within the order Coscinodiscales increases with time because of the diminishing number of the Heliopeltaceae; this is an indication that the latter is an ancient family with a tendency towards extinction.


## Acknowledgements

The study was supported by the Russian Foundation for Basic Research (Grant 036).

Thanks are expressed to L. A. Kartseva and N. V. Tchintsova for their assistance with SEM, and to the photographer N. A. Ablyamitova for her help in processing the plates.


## References


Andrews, G. W. (1979). Morphological variation in the Miocene diatom *Actinoptychus heliopelta* Grunow. *Nova Hedwigia, Beiheft* **64**, 79–98.

Diatomovyanaliz (1949). v. 2. Gosgeolisdat, Leningrad, 239 pp.

Dzinoridze, R. N., Jouse, A. P., Koroleva-Golikova, G. S., Kozlova, G. E., Nagaeva, G. S. Petrushevskaya, M. G. & Strelnikova, N. I. (1976). Diatom and radiolarian Cenozoic stratigraphy, Norwegian Basin, DSDP Leg 38. In: *Initial Reports DSDP* (M. Talwani, G. Udintsev *et al.* eds), **38**, 289–427, Washington.

Ehrenberg, C. G. (1843). Verbreitung und Einflusses mikroskopischen Lebens in Sud- und Nord America. *Abh. Kongl. Akad. Wiss. Berlin*, **1841**, 291–445.

Fryxell, G. & Hasle, G. (1973). Coscinodiscineae: some consistent patterns in diatom morphology. *Nova Hedwigia, Beiheft* **45**, 69–97.

Hanna, G. D. (1927). Cretaceous diatoms from California. *Occasional Papers, California Academy of Science*, **13**, 5–49.

Hendey, N. I. & Sims, P. A. (1987). Examination of some fossil Eupodiscoid diatoms with descriptions of two new species of *Craspedoporus* Greville. *Diatom Research*, **2**, 23–34.

Hustedt, F. (1930). Die Kieselalgen. In: *Dr L. Rabenhorsts Kryptogamen Flora Deutschlands, Österreich und der Schweiz*, **I**, 925 S., Leipzig.

Lefebure, P. & Cheviere, E. (1938). Description et iconographie des diatomées rares ou nouvelles. *Bulletin Societe Française Microscopie*, **7**, 2.

Makarova, J. V. (ed.). (1988). *The Diatoms of the USSR, fossil and recent*. II, 1, 115 pp. Nauka, Leningrad.

Nikolaev, V. A. (1982). K metodike prigotovlenya preparatov diatomovyh vodorosley dlya svetovogo I skaniruyuschego elektronnogo microskopov [Manual on prepration of diatom specimens for light microscopy and SEM*]*. *Botanicheskii Zhurnal*, **67**, 12, 1677–1679.

Ross, R. & Sims, P. A. (1973). Observation on family and generic limits in the Centrales. *Nova Hedwigia, Beiheft* **45**, 97–128.

Round, F. E., Crawford, R. M. & Mann, D. G. (1990). *The Diatoms. Biology and morphology of the genera*. Cambridge University Press, Cambridge. 747 pp.

Schütt, F. (1896). Bacillariales (Diatomeae). In*: Die Naturlichen Pflanzenfamilien nebst ihren Gattungen und wichtigeren Arten* (A. Engler & K. Prantl, eds), Teil 1, Abt. 1, 31–153. Leipzig.

Sheshukova-Poretzkaja, V. S. & Gleser, S. I. (1964). Novii vidi morskih paleogenovih diatimovih vodorosley USSR. [Recent species of marine Paleogene diatoms of the Ukraine]. *Novosti Systematiki Nizshikh Rastenii*, 78–92, Nauka, Moscow, Leningrad.

Simonsen, R. (1979). The diatom system: ideas on phylogeny. *Bacillaria*, **2**, 9–71.

Strelnikova, N. I. (1974). *Diatomei pozdnego mela (Zapadnaya Sibir)*. [Diatoms of late Cretaceous in Western Siberia], 203 pp. Nauka, Moscow.

Strelnikova, N. I. (1992). *Paleogenovie diatomovie vodorosli* [Paleogene diatoms]. Izdatelstvo SPbU, St. Petersburg. 310 pp.

Witt, O. (1886). Uber den Polierschifer von Archangelsk-Kurojedovo in Goub. Simbirsk. *Verh. Russ. Mineral. Ges*., ser. 2, **22**, 137–177. St. Petersburg.

Wornardt, W. (1970). Diatom research and the scanning electron microscope. *Nova Hedwigia, Beiheft* **31**, 355–376.

Wornardt, W. (1971). Eocene, Miocene and Pliocene marine diatoms and silicoflagellates studied with the scanning electron microscope. In*: Proceedings of the II Planktonic Conference, Roma, 1970* (A. Farinacci, ed.), 1277–1300. Edizioni Tecnol. Seienza, Roma.

# Recording environmental changes in the Southern Baltic Sea – current results from a diatom study within Project ODER

Elinor Andrén

*Department of Geology and Geochemistry, Stockholm University, S–106 91 Stockholm, Sweden*

## Abstract

Two short sediment cores from the southern Baltic Sea were analysed with respect to their siliceous microfossil content. The aim was to detect and date changes in the composition of the diatom flora and if possible interpret the causes. The study shows that a significant change occurred in the diatom assemblages at the turn of the century. The shift from periphytic taxa to a predominance of small planktonic taxa indicates that the depth of the photic layer has decreased, probably due to eutrophication of the Baltic Sea. The result indicates that the eutrophication began to affect the investigated area approximately 100 years ago.

## Introduction

The River Oder drains highly polluted areas of western Poland and eastern Germany through a series of shallow water lagoons, the Oder estuary, and flows out into the Arkona and Bornholm Basins (Fig. 1). These lagoons, termed "boddens" and "haffs", with only restricted connections to the coastal waters, function as effective traps for contaminants (Leipe *et al.* 1989). The supply of agricultural nutrients and high light levels in the summer months results in hypertrophic conditions and the deposition of organic-rich sediments which often become anoxic (Brügmann & Bachor 1990).

The aim of the present study was to detect and quantify changes in the composition of the diatom flora in the southern Baltic Sea, caused by natural and anthropogenic input into the Oder System.

Since the 1920's, diatom stratigraphic analysis has been used successfully to distinguish the different salinity stages of the Baltic Sea (Miller 1986). The relationship between diatom assemblages and anthropogenic discharge into the sea is not as clear as the link to variations in salinity. In recent years, several papers on the use of diatoms as

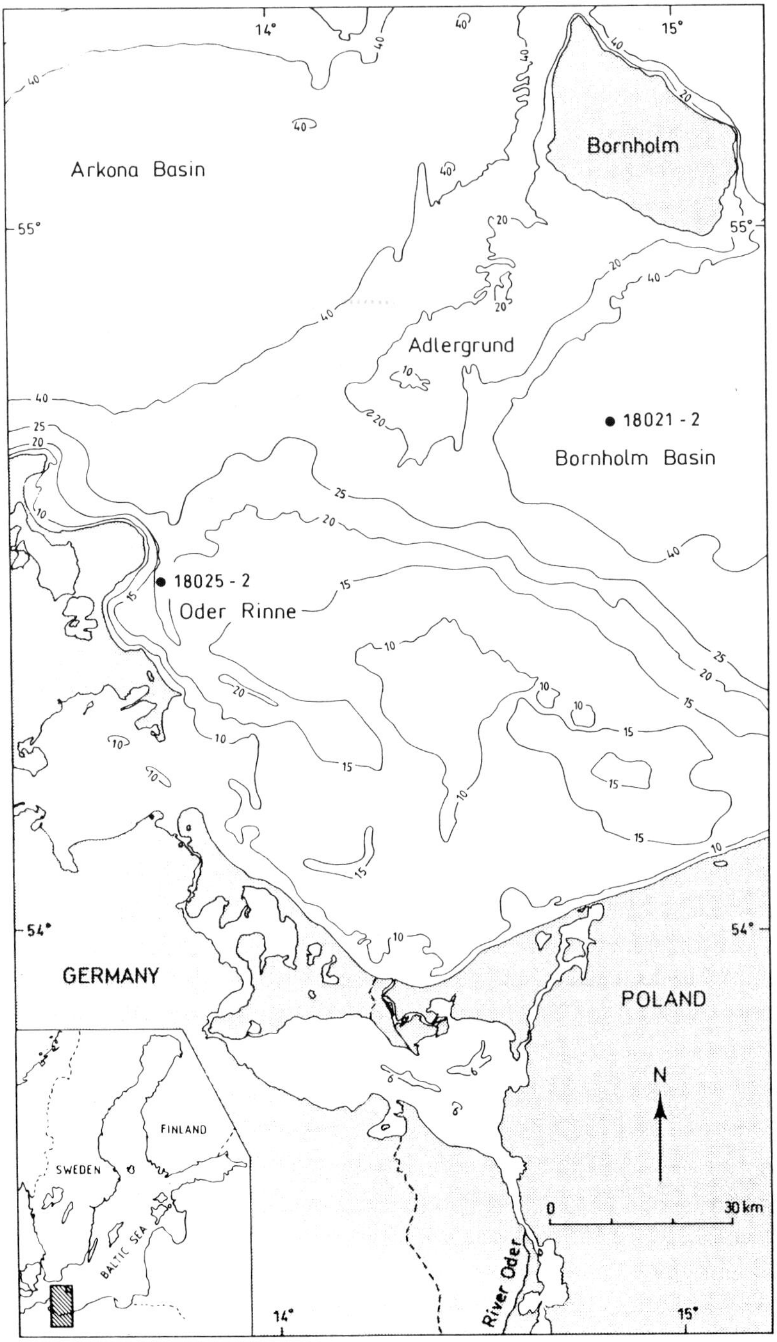


Fig. 1. Location of the sediment stations investigated in the southern Baltic Sea.

indicators of ecological change in the Baltic Sea have been published (Leskinen & Hällfors 1990; Miller & Risberg 1990; Risberg 1990; Wendker 1990; Grönlund 1993; Sakson & Miller 1993; Witkowski 1994). However, there is a shortage of diatom data from the area investigated in this study. Hällfors & Niemi (1975) published a study of samples from the uppermost sediment layer in the Bornholm Basin and several long cores were investigated with respect to diatoms by Abelmann (1985).

In this paper, two superficial sediment cores; 18021–2 from the Bornholm Basin and 18025–2 from Oder Rinne, will be compared with respect to the composition of the diatom flora and a possibly reason for changes in the stratigraphy discussed.

This investigation was carried out within the project "ODER – Oder Discharge Environmental Response" which was initiated during 1993 as a component of the EC Environment Programme. The objective of project ODER is to evaluate how the River Oder estuary and the connected deep basins have been influenced by the environmental changes occurring in the drainage area during the last centuries. Methods applied were *e.g.* radio nuclides (University of Edinburgh), organic and heavy metal geochemistry (University of Kiel and Baltic Sea Research Institute, Rostock-Warnemünde, respectively), diatom stratigraphy and benthic foraminifera analysis (Stockholm University), meiobenthos (Academy of Agriculture, Szczecin).

## Material and Methods

Two approximately 65 cm long sediment cores, 18021–2 from the Bornholm Basin (54°43.54' N, 14°50.49' E) at a water depth of 54 m and 18025–2 from Oder Rinne (54°29.53' N, 13°43.02' E) at a water depth of 20 m, were sampled from the research vessel Alkor (Fig. 1). The sampling equipment used was a 1 m long gravity corer. The collected sediment consisted of silty clay and clay gyttja with an organic carbon content ranging from *c.* 1 to 5 %. The sub sampling interval varied between 1 and 5 cm. The sediment samples were enriched with respect to siliceous microfossils according to Battarbee (1986). To estimate the concentration of diatom valves per unit weight of dry sediment (absolute abundance), a random settling technique adopted for microfossils was used (Laws 1983, Granlund 1984 and Bodén 1991). The diatom absolute abundance, which gives information about both changes in the production and preservation of diatom frustules, has been calculated at each level analysed.

The cover glasses were mounted in Naphrax (refraction index $n_D$=1.73). Quantitative analyses were carried out under light microscope with a magnification of X1000 using oil immersion ($n_D$=1.515). The counting convention by Schrader and Gersonde (1978) was used and approximately 300 diatom valves, excluding *Chaetoceros* spp. resting spores, were counted at each level.

The following floras and papers were used for identification and as a source of ecological information: Hustedt (1927–1966, 1930); Cleve-Euler (1951–1955); Mölder & Tynni (1967–1973); Tynni (1975–1980); Krammer & Lange-Bertalot (1986, 1988, 1991a, 1991b); Fryxell & Hasle (1972); Hasle (1977, 1978); Sundbäck (1987); Hasle & Lange (1989, 1992); Kuylenstierna (1989–1990) and Snoeijs (1993).

Dating of the sediment were carried out on parallel cores, using $^{137}Cs$ and $^{210}Pb$, at the University of Edinburgh, Great Britain. The ages of the sediment in core 18021–2 and 18025–2 have been estimated from age models of $^{137}Cs$ and $^{210}Pb$-data, known as the Constant Initial Concentration method (Robbins & Edgington 1975), which is a more accurate way to date cores than using the linear accumulation rates.

## Results

### *Bornholm Basin, 18021–2*

A diagram showing the total composition of siliceous microfossils, with *Chaetoceros* spp. resting spores as a separate group, has been constructed (Fig. 2, left). All identified diatom taxa were grouped according to their different life forms combined with their salinity requirements. A total of 129 taxa divided into 34 genera were recorded in the topmost 39 cm of core 18021–2. A selected diatom diagram (excluding *Chaetoceros* resting spores) showing the relative abundance of taxa were constructed (Fig. 3). The diagram has been divided stratigraphically into three diatom assemblage zones (DAZ) based upon the changes in diatom content and absolute abundance. These zones, from the oldest to the youngest, are described as follows:

DAZ 1. The sediments of this zone attain an age of *c*. 400–250 years according to the calculated age model for this station. Very low numbers of siliceous microfossil absolute abundance is distinctive of the zone.

DAZ 2. The sediments reach an age of *c*. 250–95 years. Significant for this zone is the maximum in diatom abundance, which is comprised mainly of brackish-marine planktonic species *Actinocyclus octonarius* Ehrenberg (including its varieties) and *Pseudosolenia calcar avis* (Schultz) Sundström. Maxima of some brackish-marine periphytic species, e.g. *Grammatophora oceanica* Ehrenberg and *Diploneis stroemi* Hustedt occur, as well as some indifferent taxa, e.g. *Amphora copulata* (Kützing) Schoeman & Archibald and *Epithemia turgida* (Ehr.) Kützing.

DAZ 3. The sediments of zone 3 are characterised by a decrease in diatom abundance, except for a mass occurrence of *Chaetoceros* spp. resting spores. So, there is a change in the diatom flora from a dominance of periphytic taxa to a predominance of planktonic taxa. The brackish-marine planktonic species *Actinocyclus octonarius* and *Thalassiosira hyperborea* v. *lacunosa* (Berg) Hasle show peaks of 60 and 40 percent, respectively. Five planktonic species have their first appearance in this zone: *Cyclotella choctawhatcheeana* Prasad, *Thalassiosira* cf. *levanderi* Van Goor, *T.* cf. *angulata* (Greg.) Hasle, *T. baltica* (Grun.) Ostenfeld and *Coscinodiscus granii* Gough. The age of the sediment in this zone is *c*. 95–0 years.

### *Oder Rinne 18025–2*

The siliceous microfossils analysed were grouped and diagrams constructed in the same way as for core 18021–2 (Figs 2 (right) & 4). 156 taxa divided upon 40 genera were recorded. The diagram was divided into three diatom assemblage zones characterised by the features described below:

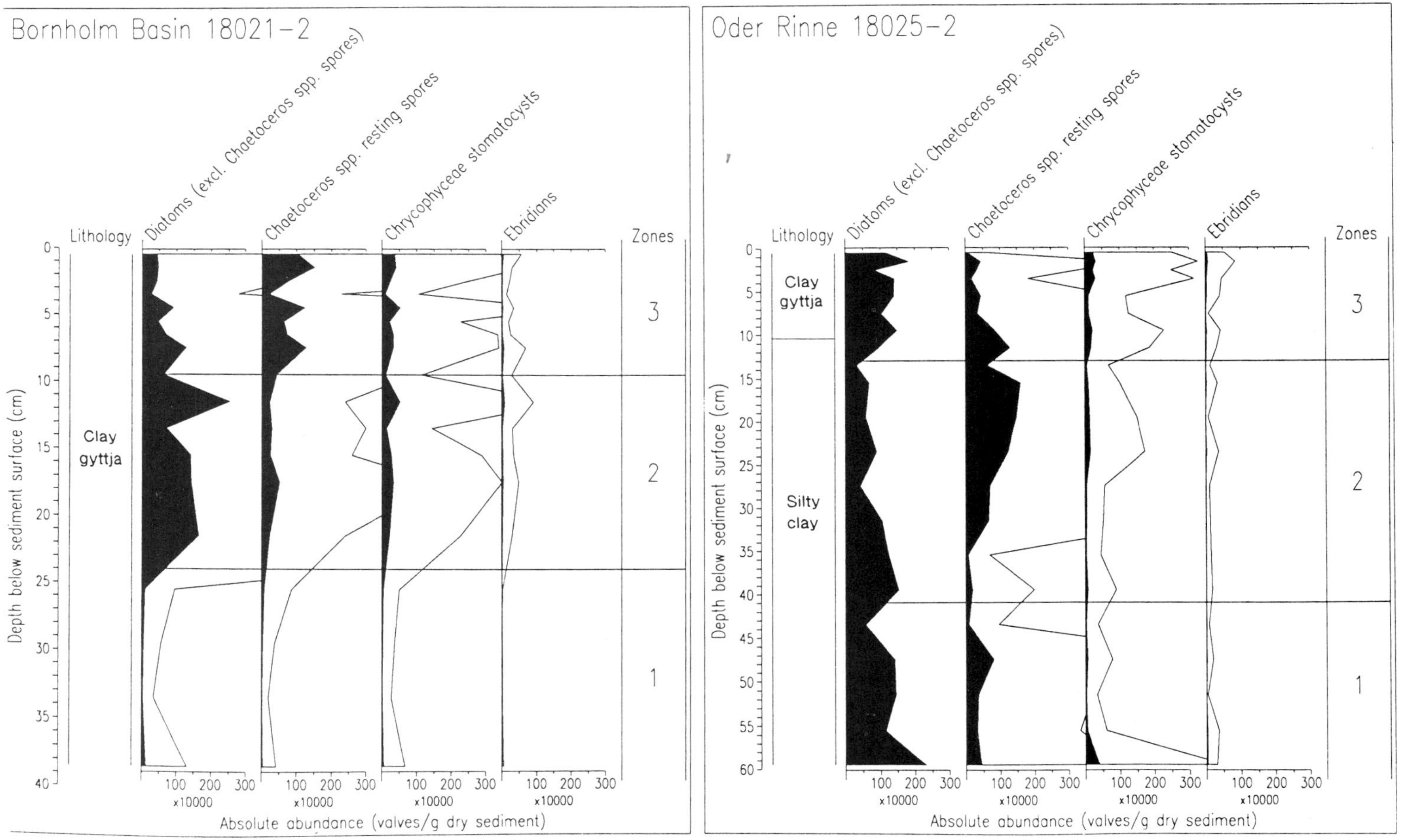


Fig. 2. Diagrams showing the total composition of siliceous microfossils in the sediment cores 18021–2 from Bornholm Basin (left) and 18025–2 from Oder Rinne (right). The black areas show the absolute abundance (valves/g dry sediment) and the solid lines are these numbers exaggerated 10 times.

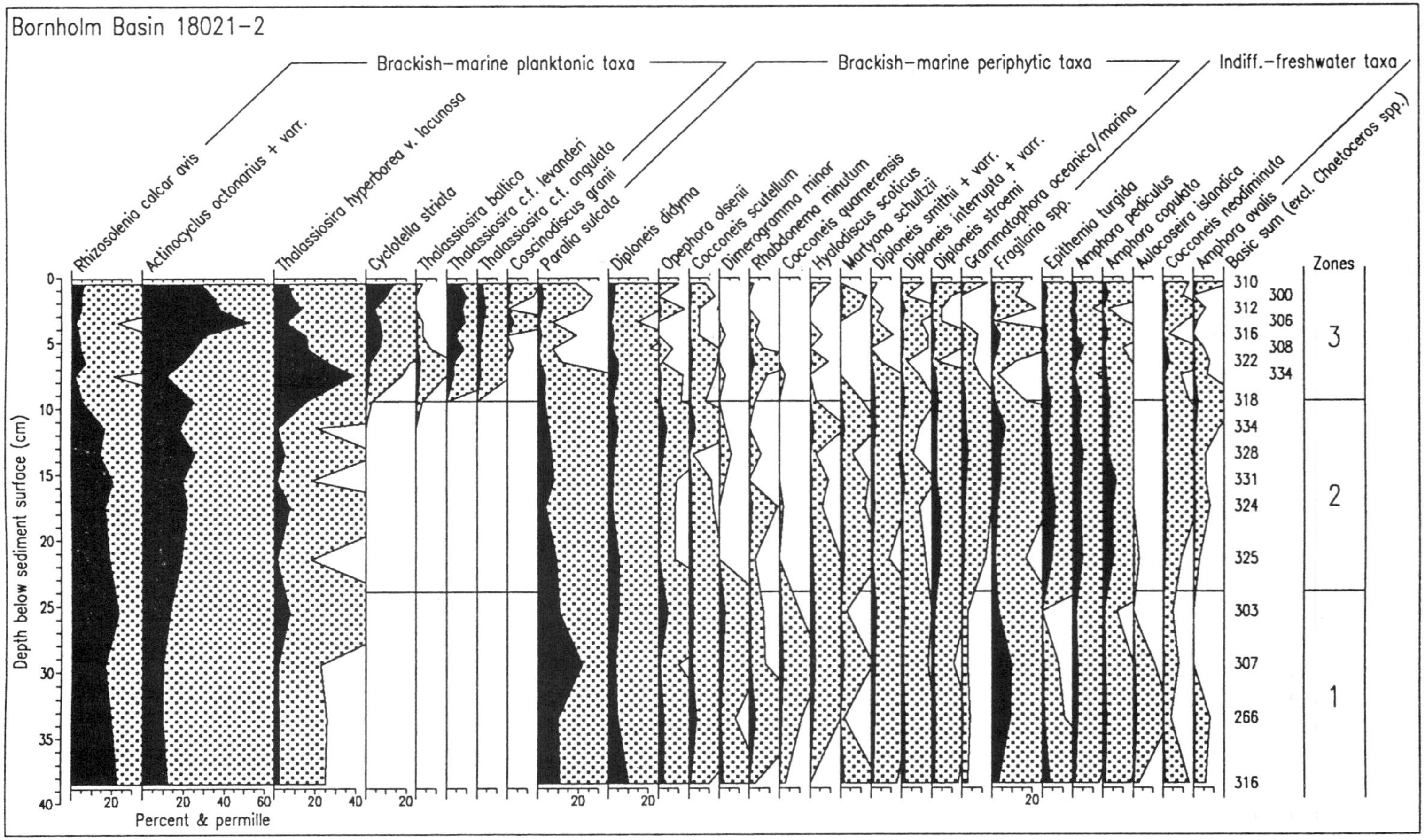


Fig. 3. Diatom diagram from core 18021–2, Bornholm Basin. The black areas show the selected species relative abundance in percent and the dotted areas in permille. The diagram has been divided stratigraphically into 3 diatom assemblage zones.

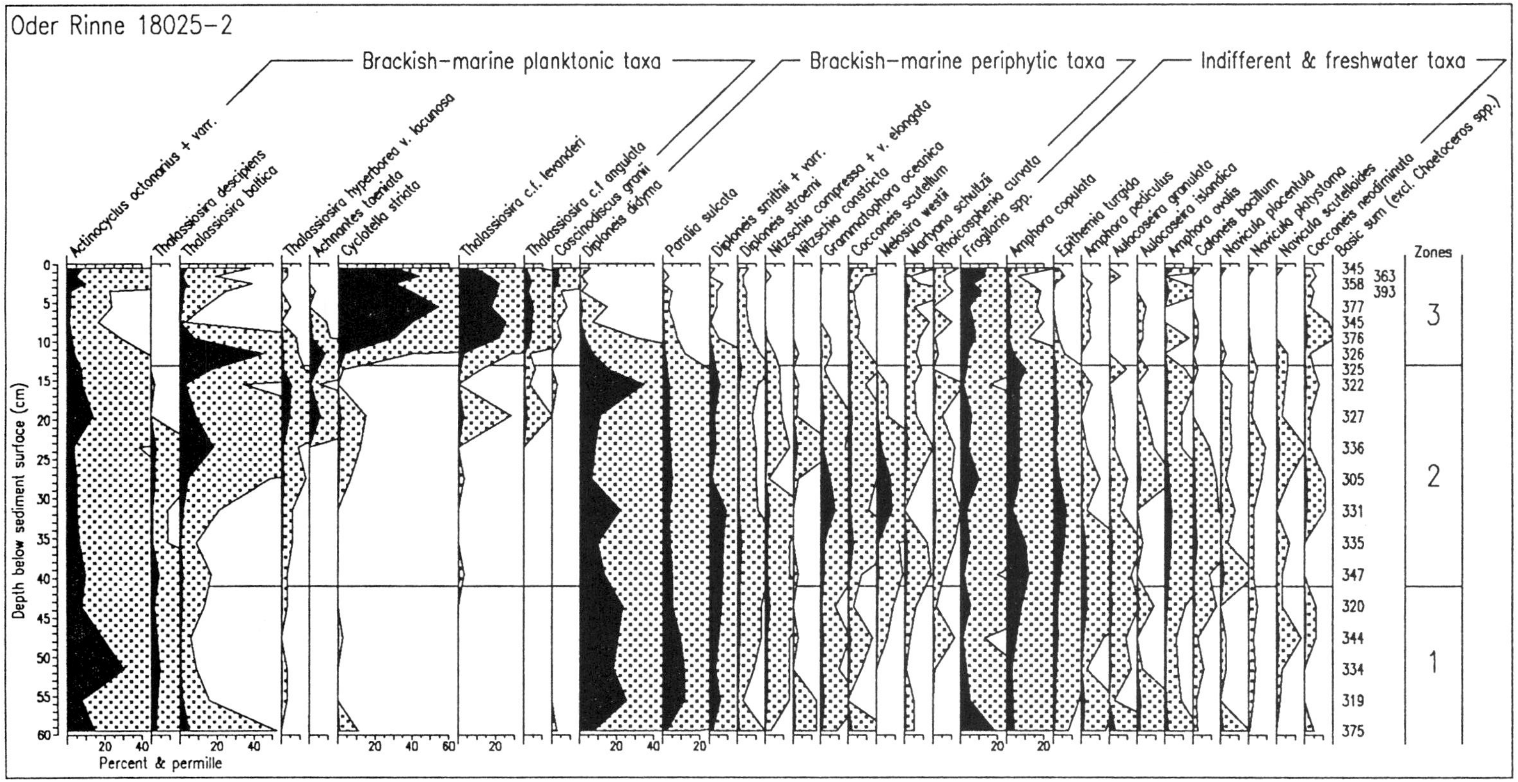


Fig. 4. Diatom diagram from core 18025–2, Oder Rinne. The black areas show the selected species relative abundance in percent and the dotted areas in permille. The diagram has been divided stratigraphically into 3 diatom assemblage zones.

DAZ 1. The siliceous microfossils mainly consist of diatoms, composed by brackish-marine periphytic taxa occurring in high percentages e.g. *Diploneis didyma* (Ehr.) Ehrenberg, *Paralia sulcata* (Ehr.) Cleve and *Diploneis smithii* (Brébisson) Cleve (including varieties). The indifferent and freshwater taxa amount to *c*. 30% and are mainly composed of *Fragilaria* spp. and *Amphora copulata*. There are low percentages brackish-marine planktonic taxa, except for a peak in the occurrence of *Actinocyclus octonarius*. The bottom of this zone is too old to be covered by the age model of $^{210}$Pb-data, but the minimum age is *c*. 600 years.

DAZ 2. The sediments of zone 2 reach an age of *c*. 470–90 years. The absolute abundance of diatoms decreased except for a mass occurrence of *Chaetoceros* spp. resting spores. The diatom flora is dominated by a similar brackish-marine periphytic taxa as in zone 1, complemented by peaks of *Grammatophora oceanica* and *Melosira westii* Smith. Some indifferent taxa reach their maximum percentages, e.g. *Amphora copulata, A. ovalis* (Kütz.) Kützing and *Epithemia turgida*. There is still a low share of planktonic taxa, with small peaks in the occurrence of *Thalassiosira hyperborea* v. *lacunosa, Achnantes taeniata* Grunow and an increase of *T. baltica*.

DAZ 3. The zone is dominated by diatoms. The absolute abundance of *Chaetoceros* spp. resting spores decreases. The essential characteristics that separates this zone from zone 2 is the transition into a predominance of brackish-marine planktonic taxa. These taxa consist of the mass occurrence of *Cyclotella choctawhatcheeana, Thalassiosira* cf. *levanderi* and a peak in the appearance of *T. baltica*. The age of the sediment in this zone is *c*. 90–0 years.

**Discussion**

The changes which separate the diatom flora into the different zones have about the same age in both analysed cores, even if the responding species are not always the same. The most striking change is the transition in DAZ 2 to 3 into a predominance of brackish-marine planktonic taxa, mainly composed of the small planktonic *Cyclotella choctawhatcheeana* and *Thalassiosira* cf. *levanderi*. At the same level the number of species was decreasing, especially the brackish-marine periphyton, which almost halved from DAZ 1 and 2 into DAZ 3.

Since the turn of the century, the Baltic phosphorus load has increased about eightfold, and the nitrogen load about fourfold (Larsson *et al.* 1985). Most of the increased input of nutrients is likely to have taken place after 1950 (Elmgren 1989). Primary production in summer has approximately doubled in the southern Baltic proper since the 1970's (Cederwall & Elmgren 1990). Increased primary production reduces the depth of the photic layer and, consequently, also the water depth to which the periphytic diatoms can survive.

Station 18025 is situated close to the German coast at a water depth of 20 m. The coastal living periphytic taxa which are dependent upon light penetration in the water column are clearly decreasing in number from DAZ 1 and 2 to DAZ 3. Effected species are for instance *Diploneis didyma, D. smithii, Tryblionella punctata* W. Smith, *Grammatophora oceanica* and several *Navicula* species.

Station 18021 is situated in the southern Bornholm Basin at a water depth of 54 m. This is too deep for colonisation by periphytic taxa but a shallow sand bank in the vicinity, Adler grund with a water depth of less than 10 m, can be the source for most of the brackish-marine periphytic taxa found in core 18021–2. Species that show a similar tendency to decrease in the core from Oder Rinne are, e.g. *Diploneis didyma, Dimerogramma minor* (Greg.) Ralfs, *Hyalodiscus scoticus* (Kütz.) Grunow and *Martyana schultzii* (Brockmann) Snoeijs. The decrease in periphytic taxa seems to expand the possibilities for the planktonic taxa because they not only increase in share but also in number of taxa.

According to Grönlund (1993), based upon a diatom study of a superficial sediment core from the Gotland Basin, eutrophication began some 200 years ago. Witkowski (1994) studied the diatom flora of the Gulf of Gdansk and found a distinct increase of centric taxa *c.* 90 years ago. In this study the alterations of the diatom flora occurred approximately at the turn of the century, which correspond to the results reported by Witkowski (1994). There are difficulties in comparing investigations from different sites within the Baltic Sea, because the flora mainly reflects the local sedimentation of diatoms (Hällfors & Niemi 1975) and the coring stations are situated at varying water depth.

The diatom absolute abundance curves show different patterns in the two analysed cores. There is an uncertainty to interpret the absolute abundance to be a direct reflection of the production, as changed sedimentation rates and dissolution of frustules are not considered.

The influx of freshwater diatoms amounted to *c.* 20 relative percent at some levels in the core from Bornholm Basin and as much as 40 % in the core from Oder Rinne. The main constituents were periphytic taxa such as *Fragilaria* spp. (sensu Hustedt), *Epithemia turgida* and *Amphora copulata* which probably have been transported out into the Baltic by river outflow. The Oder Rinne station is situated in the direct plume of outflowing river water from the River Oder into the Arkona Basin. Consequently the large influx of freshwater taxa most likely originates from the River Oder. The input into the south-western Bornholm Basin derives from various sources of which the main one is probably the River Oder.

There is confusion concerning the identification of the small planktonic *Thalassiosira* species. The species bear a close resemblance to *T. levanderi*, which has not yet been properly described.

*Achnantes taeniata* is a planktonic diatom taxon that thrives in sea ice and is accordingly likely to reflect the ice cover conditions in the Baltic (Hasle & Syvertsen 1990). Risberg (1990) has proposed that the taxon is favoured by the increased input of nutrients into the Baltic. The frustules of this species are very weakly silicified and are less likely to be preserved in the sediment. Nevertheless, the species is recorded in the core from Oder Rinne (18025–2) as a peak in the diatom diagram around the border between DAZ 2 and 3. This corresponds quite well with the interpretation by Risberg

(1990), but as the ice cover situation in the area has not been investigated, it could well match the results reported by Hasle & Syvertsen (1990).

It is difficult to interpret the abundance of *Chaetoceros* spp. resting spores in the cores analysed. No vegetative cells of *Chaetoceros* were found in the sediment and this makes it impossible to classify the resting spores. The lack of vegetative cells could be due to dissolution of their very weakly silicified frustules in the water column. The occurrence of resting spores in the Baltic has been discussed, but the cause of formation is still unknown. Miller & Risberg (1990) and Risberg (1990) propose that the formation of resting spores is favoured by an increased access of nutrients, or by unfavourable living conditions for the vegetative cells. Hajdu *et al.* (in manus.) interpret the development of vegetative cells to be dependent upon the weather situation and not only the eutrophication, especially as they seem to be favoured by sea ice formation. Core 18021–2 from Bornholm Basin shows mass occurrence of *Chaetoceros* resting spores in the topmost 10 cm which is the same pattern as reported by Miller & Risberg (1990) and Risberg (1990). The core 18025–2 from Oder Rinne shows corresponding mass occurrence 10–25 cm under the sediment surface.

Chrysophyceae stomatocysts are considered to be widely distributed in freshwater, but some species have also been found living in marine environments (Abelmann 1985). In this study no attempt has been carried out to identify the cysts and they are mainly interpreted as being redeposited river outwash. There is a slight increase in the absolute abundance of cysts upwards in both sediment cores, which contradicts the fact that the outwashed freshwater diatoms are less abundant in the top sediment. Possibly the interpretation of the cysts as all freshwater is a source of error in this case.

The ebridian *Ebria tripartita* (Schumann) Lemmermann, a unicellular marine phytoplankton abundant in nutrient rich cool waters (Lipps 1979), was recorded in both cores analysed. The abundance was very low with a slightly increase towards the sediment surface.

## Conclusions

This study shows that a significant change occurred in the diatom flora of the southern Baltic Sea at the turn of the century. The change, recorded as a shift from a periphytic taxa to a predominance of small planktonic taxa, indicates that the photic layer must have decreased in depth probably due to eutrophication of the Baltic. The conclusion is that the eutrophication began to affect the investigated area approximately 100 years ago.

## Acknowledgements

I am grateful to my supervisors: Assoc. Prof. Otto Hermelin for involving me into this project and for reading the manuscript and Assoc. Prof. Urve Miller for valuable discussions concerning the interpretation as well as reading the manuscript. I sincerely thank all participants of Project Oder and the crew of r/v Alkor. The $^{137}Cs$ and $^{210}Pb$

datings, and the age models were carried out by Dr Graham Shimmield and Mr Timothy Brand at the University of Edinburgh. Figure 1 was drawn by Inger Arnström. Grants were received from the European Community Environment Programme.

## References

Abelmann, A. (1985). Palökologische und ökostratigraphische Untersuschungen von Diatomeenassoziationen an holozänen Sedimenten der zentralen Ostsee. *Berichte Reports, Geologisch-Paläontologisches Institut der Universität Kiel,* **9**.

Battarbee, R. W. (1986). Diatom analysis. In: *Handbook of Holocene Palaeoecology and palaeohydrology* (B. E. Berglund, ed.), 527–570. John Wiley & Sons. Chichester.

Bodén, P. (1991). Reproducibility in the random settling method for quantitative diatom analysis. *Micropaleontology,* **37**(3), 313–319.

Brügmann, L. & Bachor, A. (1990). Present state of the baltic coastal waters off Mecklenburg-Vorpommern, Germany. *GeoJournal* **22**(2), 185–194.

Cederwall, H. & Elmgren, R. (1990). Biological effects of eutrophication in the Baltic Sea, particularly the coastal zone. *Ambio,* **19,** 109–112.

Cleve-Euler, A. (1951–1955). Die Diatomeén von Schweden und Finnland. *Kungliga Vetenskapsakademiens Handlingar I–V.* 4:e serien **2**:1 (1951) 163 pp., **3**:3 (1952) 153 pp., **4**:1 (1953) 158 pp., **4**:5 (1953) 255 pp., **5**:4 (1955) 232pp.

Elmgren, R. (1989). Man's impact on the ecosystem of the Baltic Sea: energy flows today and at the turn of the century. *Ambio,* **18**, 326–332.

Fryxell, G. A. & Hasle, G. R. (1972). *Thalassiosira eccentrica* (Ehrenberg) Cleve, *T. symmetrica* sp. nov., and some related centric diatoms. *Journal of Phycology,* **8**, 297–317.

Granlund, A. (1984). An approach to a statistically random settling technique for microfossils. *Stockholm Contributions in Geology,* **39**(3), 119–125.

Grönlund, T. (1993). Diatoms in surface sediments of the Gotland Basin in the Baltic Sea. *Hydrobiologia,* **269/270**, 235–242.

Hajdu, S., Larsson, U. & Skärlund, K. (in manus). Växtplankton. *In*: Himmerfjärdsrapporten (Ragnar Elmgren, ed.). *Swedish Environmental Protection Agency.* (In Swedish).

Hällfors, G. & Niemi,Å. (1975). Diatoms in surface sediment from the deep basins in the Baltic Proper and the Gulf of Finland. *Merentutkimuslait. Julk.* **240**, 71–77.

Hasle, G. R. (1977). Morphology and taxonomy of *Actinocyclus normanii* f. *subsalsa* (Bacillariophyceae). *Phycologia,* **16**(3), 321–328.

Hasle, G. R. (1978). Some *Thalassiosira* species with one central process (Bacillariophyceae). *Norwegian Journal of Botany,* **25**, 77–110.

Hasle, G. R. & Lange, C. B. (1989). Freshwater and brackish water *Thalassiosira* (Bacillariophyceae): taxa with tangentially undulated valves. *Phycologia,* **28**(1), 120–135.

Hasle, G. R. & Lange, C. B. (1992). Morphology and distribution of *Coscinodiscus* species from the Oslofjord, Norway, and the Skagerak, North Atlantic. *Diatom Research,* **7**(1), 37–68.

Hasle, G. R. & Syvertsen, E. E. (1990). Arctic Diatoms in the Oslofjord and the Baltic Sea – a bio- and palaeogeographic problem. In: *Proceedings of the 10th International Diatom Symposium* (H. Simmola, ed.), 285–300. Koeltz Scientific Books, Koenigstein.

Hustedt, F. (1927–1966). Die Kieselalgen Deutschlands, Österreichs und der Schweiz unter Berücksichtigung der übrigen Länder Europas sowie der angrenzenden Meeresgebiete. In: *Dr Rabenhorst's Kryptogamen-Flora von Deutschland, Österreich und Schweiz,* **7**. Akademische Verlagsgesellschaft. Leipzig.

Hustedt, F. (1930). *Die Süsswasser-flora Mitteleuropas. Heft 10: Bacillariophyta (Diatomeae).* 468 pp. Verlag von Gustav Fischer. Jena.

Krammer, K. & Lange-Bertalot, H. (1986). Bacillariophyceae. 1. Teil Naviculaceae. In: *Süsswasserflora von Mitteleuropa* (H. Ettl, J. Gerloff, H. Heynig & D. Mollenhauer, eds), 857 pp. Gustav Fischer Verlag. Stuttgart, New York.

Krammer, K. & Lange-Bertalot, H. (1988). Bacillariophyceae. 2. Teil Bacillariaceae, Epithemiaceae, Surirellaceae. In: *Süsswasserflora von Mitteleuropa* (H. Ettl, J. Gerloff, H. Heynig & D. Mollenhauer, eds), 596 pp. Gustav Fischer Verlag. Stuttgart, New York.

Krammer, K. & Lange-Bertalot, H. (1991a). Bacillariophyceae. 3. Teil Centrales, Fragilariaceae, Eunotiaceae. In: *Süsswasserflora von Mitteleuropa* (H. Ettl, J. Gerloff, H. Heynig & D. Mollenhauer, eds), 576 pp. Gustav Fischer Verlag. Stuttgart, Jena.

Krammer, K. & Lange-Bertalot, H. (1991b). Bacillariophyceae. 4. Teil Achnantaceae. In: *Süsswasserflora von Mitteleuropa* (H. Ettl, J. G. Gärtner, Gerloff, H. Heynig & D. Mollenhauer eds), 437 pp. Gustav Fischer Verlag. Stuttgart, Jena.

Kuylenstierna, M. (1989–1990). Benthic algal vegetation in the Nordre Älv Estuary (Swedish West Coast), **1** (text, 1990), 244 pp., **2** (plates, 1989), 76 pls. *Department of Marine Botany. University of Göteborg.*

Larsson, U., Elmgren, R. & Wulff, F. (1985). Eutrophication and the Baltic Sea: causes and consequences. *Ambio*, **14**, 9–14.

Laws, R. A. (1983). Preparing strewn slides for quantitative microscopical analysis: A test using calibrated microspheres. *Micropaleontology*, **29**(1), 60–65.

Leipe, T., Brügmann, L. & Bittner, U. (1989). On the distribution of heavy metals in recent brackish-water-sediments of estuaries of the G.D.R. *Chemie der Erde* **49**, 21–38. (In German).

Leskinen, E. & Hällfors, G. (1990). community structure of epiphytic diatoms in relation to eutrophication on the Hanko Peninsula, South Coast of Finland. In: *Proceedings of the 10th International Diatom Symposium* (H. Simola, ed.), 323–333. Koeltz Scientific Books, Koenigstein.

Lipps, J. H. (1979). Ebridians. In: *The Encyclopedia of Palaeontology*, **276** (R. W. Fairbridge & D. Jablonski, eds). Dowden, Hutschinson & Ross, Stroudsburg, Pennsylvania.

Miller, U. (1986). Ecology and palaeoecology of brackish water diatoms with special reference to the Baltic Sea. In: *Proceedings of the 8th International Diatom Symposium* (M. Rickard, ed.), 601–611. Koeltz Scientific Books, Koenigstein.

Miller, U. & Risberg, J. (1990). Environmental changes, mainly eutrophication, as recorded by fossil siliceous micro-algae in two cores from the uppermost sediments of the north-western Baltic. *Beiheft zur Nova Hedwigia*, **100**, 237–253.

Mölder, K. & Tynni, R. (1967–1973). Über Finnlands rezente und subfossile Diatomeen I–IV. *Comtes Rendus de la Société géologique de Finlande N:o XXXIX*, 199–217 (1967), *Bulletin of the Geological Society of Finland* **40**, 151–170 (1968), *Bulletin* **41**, 235–251 (1969), *Bulletin* **42**, 129–144 (1970), *Bulletin 43*, 203–220 (1971), *Bulletin* **44**, 141–149 (1972), *Bulletin 45*, 159–179 (1973).

Risberg, J. (1990). Siliceous microfossil stratigraphy in a superficial sediment core from the northwestern part of the Baltic Proper. *Ambio*, **19**(3), 167–172.

Robbins, J. A. & Edgington, D. N. (1975). Determination of recent sedimentation rates in Lake Michigan using $^{210}$Pb and $^{137}$Cs. *Geochimica et Cosmochimica Acta*, **39**, 285–304.

Sakson, M. & Miller, U. (1993). Diatom assemblages in superficial sediments from the Gulf of Riga, eastern Baltic Sea. *Hydrobiologia*, **269/270**, 243–249.

Schrader, H.-J. & Gersonde, R. (1978). Diatoms and silicoflagellates. *Utrecht Micropaleontological Bulletin* **17,** 129–176.

Snoeijs, P., ed. (1993). Intercalibration and distribution of diatom species in the Baltic Sea, **1**. *The Baltic Marine Biologists Publication* No. **16a**. 129 pp. Opulus Press, Uppsala.

Sundbäck, K. (1987). The episammic marine diatom *Opephora olsenii* Möller. *Diatom Research*, **2,** 241–249.

Tynni, R. (1975–1980). Über Finnlands rezente und subfossile Diatomeen VIII–XI. *Geological Survey of Finland Bulletin* **274,** 55 pp. (1975), *Bulletin* **284**, 37 pp., 17 plates (1976), *Bulletin* **296,** 55 pp., 17 plates (1978), *Bulletin* **312,** 93 pp. (1980).

Wendker, S. (1990). Untersuchungen zur subfossilen und rezenten Diatomeenflora des Schlei-Ästuars (Ostsee). *Bibliotheca Diatomologica,* **20**, 230 pp. J. Cramer, Berlin, Stuttgart.

Witkowski, A. (1994). Recent and fossil diatom flora of the Gulf of Gdansk, Southern Baltic Sea. *Bibliotheca Diatomologica,* **28**, 313 pp. J. Cramer, Berlin, Stuttgart.

# Stratigraphical studies of the Holocene deposits in Älgpussen, a small bog in eastern Sweden

Jonas Björck*, Jan Risberg*, Sven Karlsson* and Per Sandgren**

*Stockholm University, Department of Quaternary Research, Odengatan 63, S–113 22 Stockholm, Sweden

**Lund University, Department of Quaternary Geology, Tornavägen 13, S–223 63 Lund, Sweden

## Abstract

The bog Älgpussen is situated approximately 30 km southwest of Stockholm. The bog occupies a depression in the landscape, probably a kettle hole. The surroundings consist of hilly glaciofluvial deposits. The isolation threshold of the depression, when it was separated from the Baltic Sea, cuts through silt and has been levelled to +24.5 m a.s.l. A 400 cm long core from the central part of the bog has been analysed for organic carbon and mineral magnetic parameters and the siliceous microfossils and pollen stratigraphy has been determined. In addition, six tandem accelerator datings were carried out.

The results show that the stratigraphical sequence covers the last c. 8000 years. The Baltic stages of the Ancylus Lake, the Mastogloia Sea and the Litorina Sea are represented. The Litorina 4 transgression is reflected as a peak in *Melosira westii* f. *parva* Brander dated to around 3500 BP. The basin was isolated c. 2900 BP. After that, as indicated by the presence of *Gomphonema parvulum* (Kützing) Grunow, a polluted lake prevailed for almost 1000 years. The pollution was most likely caused by human activities, such as waste deposition and/or cattle breeding. The lake was filled in and a *Carex* fen, and later on a *Sphagnum* bog, developed. Three periods of intense anthropogenic activities have been identified: 2800–1800 BP, 1400–1300 BP and 800–500 BP.

The taxonomy of a *Melosira* species, found in the uppermost *Carex* peat and tentatively determined as *M. nygaardii* Camburn, is discussed.

## Introduction

Älgpussen is a small bog situated c. 30 km southwest of Stockholm in eastern Sweden (59°05' N, 17°34' E, Fig. 1). The site is located in a Nature Conservation area

"The River Moraån valley". Archaeological excavations in the vicinity of Älgpussen have revealed finds from Late Neolithic onwards (Fig. 2). The surface area of the bog covers c. 1300 $m^2$ and the drainage area is c. 18,000 $m^2$. The bog lies in a depression, probably a kettle hole. The surroundings consist of hilly glaciofluvial deposits. The aim of this investigation was to study Holocene palaeoenvironmental changes, such as shore displacement, vegetation successions and human impact, by stratigraphical analysis of the deposits in the bog. Methods used were siliceous microfossil (mainly diatoms) and pollen analyses, mineral magnetic measurements, organic carbon measurements and radiocarbon dating (AMS). This paper presents results and interpretations of the siliceous microfossil analyses. Results of the other analyses will be published elsewhere.

The isolation threshold cuts through silt and has been levelled to +24.5 m. According to Miller in Brunnberg *et al.* (1985) and Risberg *et al.* (1991) this altitude corresponds to isolation ages of 3000 BP and 3700 BP, respectively. These ages are compared with the isolation age of the Älgpussen basin, 2900 BP, in the discussion. According to the shore displacement model constructed by Miller (in Brunnberg *et al.* 1985) it is also possible that the Litorina transgression 4 (L 4) can be documented in the sedimentary strata.

*Field work and lithostratgraphy*

The fieldwork consisted of corings, with a Russian peat sampler (Jowsey 1987), to ascertain the lateral extension of various accumulated strata. The stratigraphy revealed an unusual feature; a repeated sequence of gyttja clay and clay (Fig. 3). A 400 cm long core from the central part of the bog was sampled and analysed. The layers investigated consist of c. 200 cm minerogenic sediment overlain by c. 200 cm organic material (clay gyttja and peat, Table 1). The isolation threshold was levelled and related to the altitude of a known benchmark in the Swedish height system RH00.

## Methods

The organic carbon content was measured in an ELTRA Metalyt 80 W and calculated as % of dry sample. Definitions of mineral magnetic measuring techniques, parameters and ratios, and examples of applications are described in e.g. Thompson & Oldfield (1986), Sandgren & Risberg (1990), Higgit *et al.* (1991) and Sandgren & Fredskild (1991). Pollen preparation techniques and analyses follow Berglund & Ralska-Jasiewiczowa (1986). The radiocarbon datings have been carried out at The Svedberg Laboratory, Uppsala University, using the AMS technique (Possnert 1990). Ages presented refer to uncorrected and uncalibrated years before present (1950) based on the half-life $T_{1/2} = 5568 \pm 30$ years.

The preparation of the slides for siliceous microfossil analysis was carried out according to Battarbee (1986). After the separation of siliceous microfossils from the sediment and peat, the residual was mounted in Naphrax ($n_D$ = 1.73). The analyses were carried out under a light microscope with a magnification of X1000 and oil immersion.

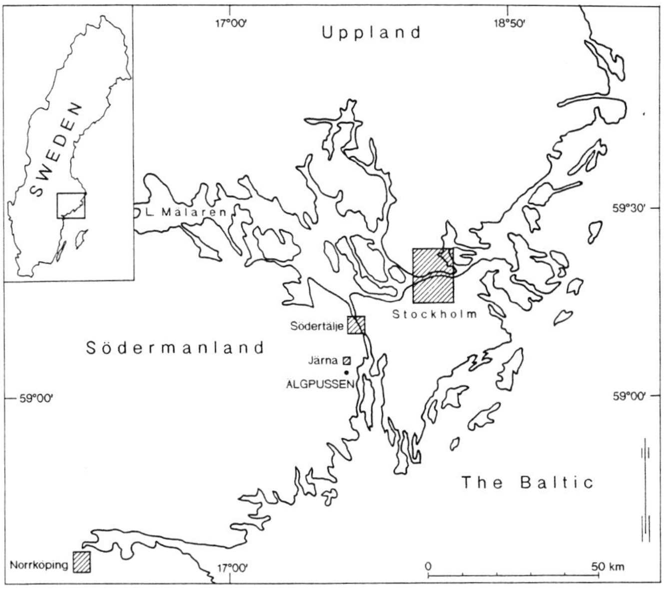


Fig. 1. Location map showing eastern Svealand, comprising the counties of Södermanland and Uppland, separated by Lake Mälaren. Älgpussen is situated c. 30 km southwest of Stockholm.

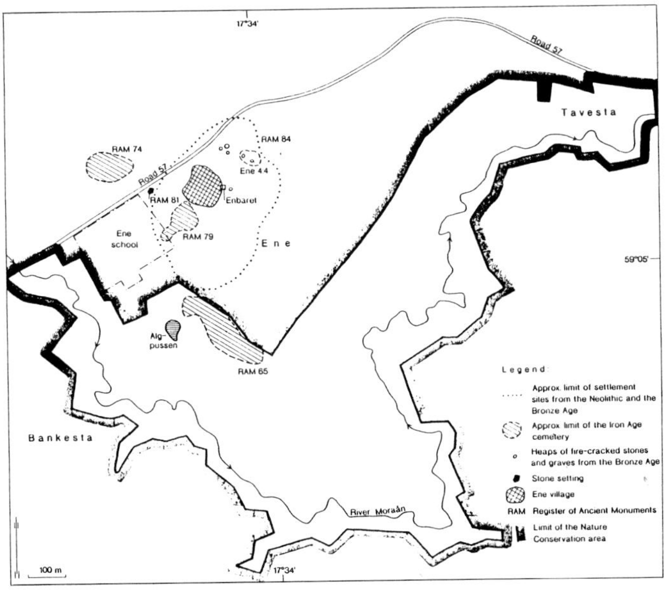


Fig. 2. Map showing Älgpussen within the Nature Reserve "The River Moraån Valley" and the surrounding archaeological excavation areas.

The diatom floras of Cleve-Euler (1951–1955) and Krammer & Lange-Bertalot (1986, 1988, 1991a,b) were used for identifications and ecological data.

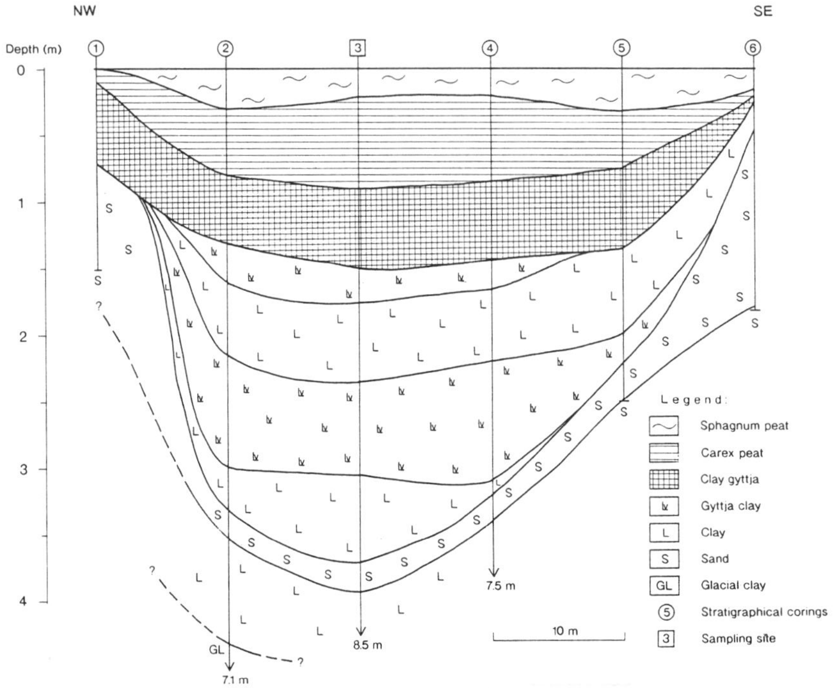


Fig. 3. The lateral extention of accumulated strata in Älgpussen. The shape of the bog is semicircular and the diameter c. 60 m.

Table 1. Stratigraphical description of the analysed core (sampling site 3, cf. Fig. 3).

| Depth (cm) | Lithological description |
|---|---|
| 0–20 | *Sphagnum* peat |
| 20–90 | *Carex* peat with *Equisetum* and *Eriophium* |
| 90–150 | Clay gyttja |
| 150–175 | Gyttja clay |
| 175–235 | Clay |
| 235–305 | Gyttja clay |
| 305–373 | Clay |
| 373–395 | Silty sand |
| 395–›850 | Clay |

## Results and interpretation

The relative abundances of siliceous microfossils, i.e. diatoms, phytoliths, Chrysophyceae stomatocysts, sponge spicules and ebridians, are shown in Fig. 4. In Fig. 5 the diatoms have been grouped according to the halobion systems of Kolbe (1927), Hustedt (1957) and Simonsen (1962). In addition, aerophilous taxa and species common in the Ancylus Lake have been included. Selected species of major importance for the intepretation of the stratigraphical layers are shown in Fig. 6. The results from the pollen and charcoal analysis, together with the organic carbon content, are shown in Fig. 7. The results of the mineral magnetic analysis are shown in Fig. 8. According to the radiocarbon dates (Table 2) a chronology has been established using the primary axis in Figs 4–8. Definitions of six stratigraphic zones (ÄLG–1 to ÄLG–6) were based on the results from all analyses.

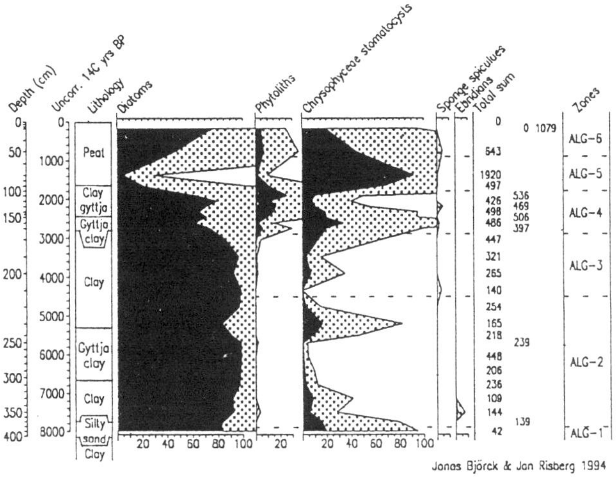


Fig. 4. Total siliceous microfossil diagram from Älgpussen. Solid areas represent percentages and stippled areas per milles.

Table 2. Radiocarbon datings (uncorrected $^{14}C$ years BP) from Älgpussen.

| Lab. no | Material | Depth (cm) | $^{14}C$ yrs BP | Fraction |
|---|---|---|---|---|
| Ua–3762 | *Carex* peat | 31 | 190 ± 55 | SOL |
| Ua–3761 | *Carex* peat | 71 | 1410 ± 50 | SOL |
| Ua–3760 | clay gyttja | 125 | 2105 ± 55 | SOL |
| Ua–3899 | gyttja clay | 168 | 2845 ± 70 | SOL |
| Ua–3759 | gyttja clay | 175 | 2820 ± 55 | SOL |
| Ua–3758 | gyttja clay | 235 | 5415 ± 60 | SOL |

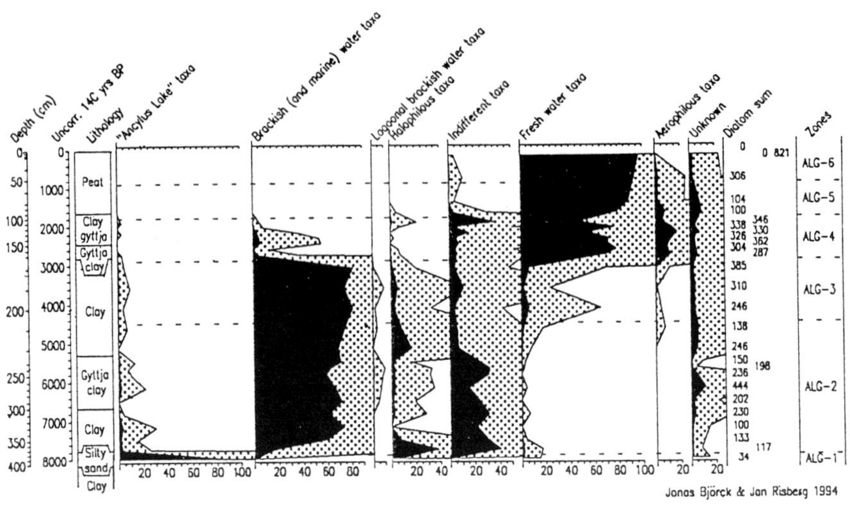

Fig. 5. Diatoms grouped according to the halobion system with the addition of species typical for the Ancylus Lake and aerophilous taxa.

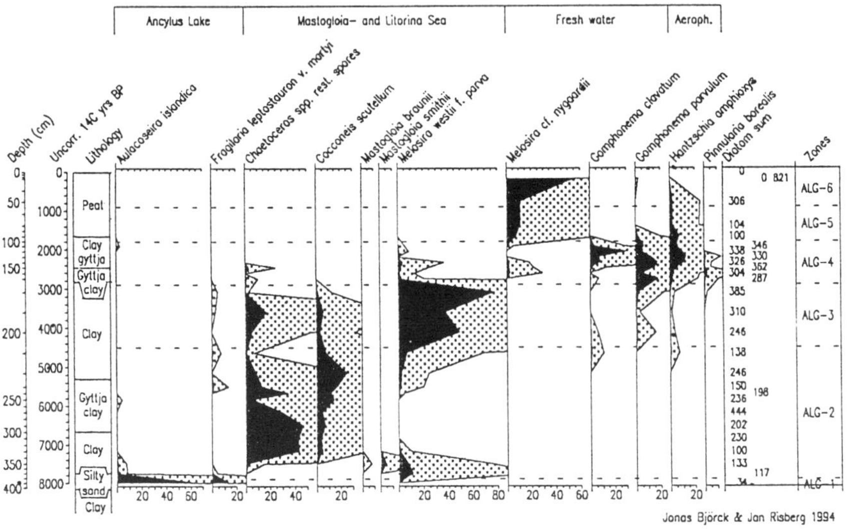

Fig. 6. Selected species important for the interpretation of palaeoenviromental changes within and around Älgpussen.

**ÄLG–1.** 400–380 cm. 8200–7900 BP. The magnetic susceptibility ($\chi$), HIRM and SIRM show high but decreasing values. Pollen of *Pinus* and *Betula* dominated but also *Alnus* and *Quercus* are present. Among the few diatoms found, e.g. *Aulacoseira islandica* + ssp. *helvetica* (O. Müller) Simonsen, *Ellerbeckia arenaria* (Moore)

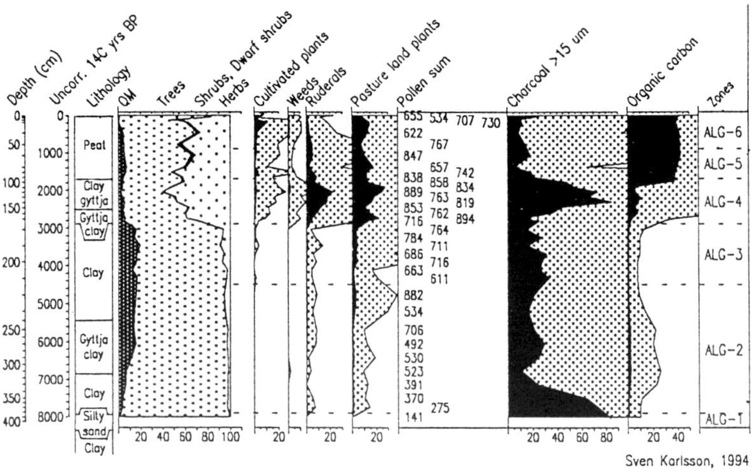


Fig. 7. Simplified pollen diagram with different plant groups, charcoal particles and organic carbon content, from Älgpussen.

Crawford and *Fragilaria leptostauron* v. *martyi* (Héribaud) Lange-Bertalot, taxa are typical for the Ancylus Lake. The frustules found are considered as not being reworked. Therefore, the sediments are interpreted to have been deposited during the Baltic Ancylus Lake stage.

**ÄLG–2.** 380–215 cm. 7900–4500 BP. Magnetic concentration parameters show stable values. *Quercetum mixtum* display increasing values from the middle of the zone (6500 BP), which represents the Holocene climatic amelioration. The brackish (and marine) diatom taxa *Nitzschia compressa* v. *elongata* (Grun.) Lange-Bertalot, *Chaetoceros* spp. resting spores and *Cocconeis scutellum* Ehr. dominate. In the beginning of the zone *Mastogloia* spp., together with the planktonic *Thalassiosira* spp. and *Melosira westii* f. *parva* Brander (Plate 1a), show relatively high percentages. This part of the zone includes the transition phase from the Ancylus Lake to the Litorina Sea, i.e. the brackish Mastogloia Sea.

**ÄLG–3.** 215–175 cm. 4500–2900 BP. At the transition ÄLG–2/3 the $\chi$ display slightly increased values while there is a marked drop in the end of the zone. The first signs of cultivated plants can be seen in the pollen assemblages (7300 BP). Brackish (and marine) species dominate in the diatom diagram, especially the planktonic *Melosira westii* f. *parva* that occurs in very high abundances (up to 80%), indicating an increase in water depth. This increase is correlated with the L 4 transgression, which reached its maximum level around 3500 BP (Miller in Brunnberg *et al.* 1985). Phytoliths and Chrysophyceae stomatocysts were increasing. Aerophilous diatoms, e.g. *Hantzschia*

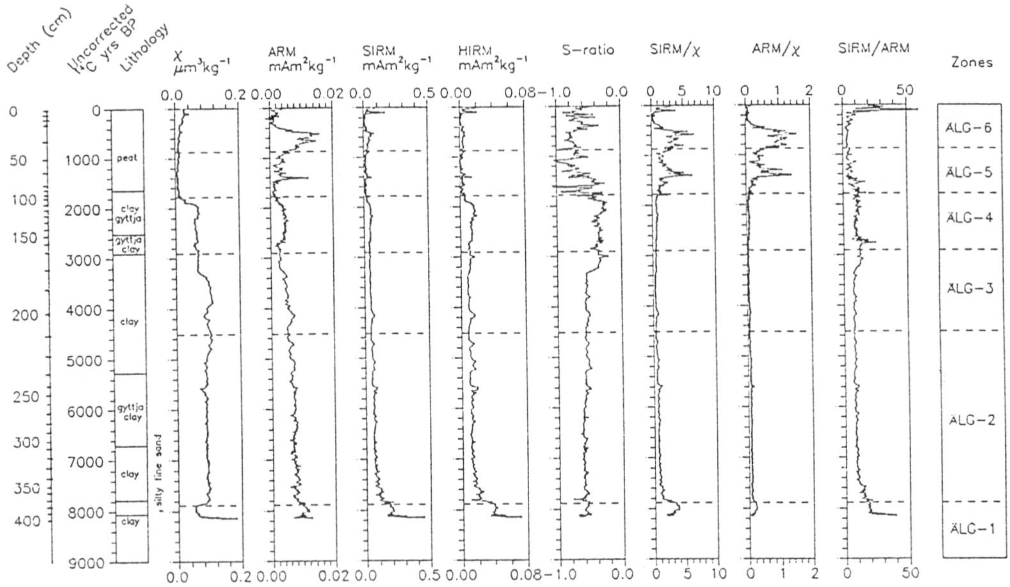

Fig. 8. Mineral magnetic parameters and ratios from Älgpussen.

*amphioxys* (Ehr.) Grun. (Plate 1b) and *Pinnularia borealis* (Ehr.) and fresh water species also increased. This, together with the pollen composition, suggests the first signs of anthropogenic activities in the vicinity.

**ÄLG–4.** 175–100 cm. 2900–1800 BP. ARM shows relatively high values throughout the zone, compared to the end of ÄLG–3, indicating an inwash of very fine grained, soil derived magnetic particles (cf. Mullins 1977). There are an increasing number of charcoal particles and pollen from cultivated plants, weeds, ruderals and pasture land plants. Dominating cultivated plants were *Hordeum* and *Triticum. Linum usitatissimum* were found in the beginning and the end of the zone. The basin was isolated c. 2900 BP as reflected by an increase in fresh water diatom taxa and a subsequent decrease in brackish (and marine) water taxa. *Gomphonema parvulum* (Kütz.) Grun. (Plate 1c), phytoliths and Chrysophyceae stomatocysts increased. Aerophilous taxa show their greatest abundance, with *Hantzschia amphioxys* as the dominating taxon. This is interpreted as a period (2800–1800 BP) of intense human activities in the vicinity of Älgpussen. The growth of *Gomphonema parvulum* can be related to a "Massenentwicklung in alphamesosaproben und mässig polysaproben Gewässern" (Krammer & Lange-Bertalot 1986, p.359). At the end of the zone there is a peak in *Gomphonema clavatum* Ehr. (Plate 1d), a taxon that is "sensibel gegen organische Belastung" (Krammer & Lange-Bertalot 1986, p.367). This is interpreted as an initial

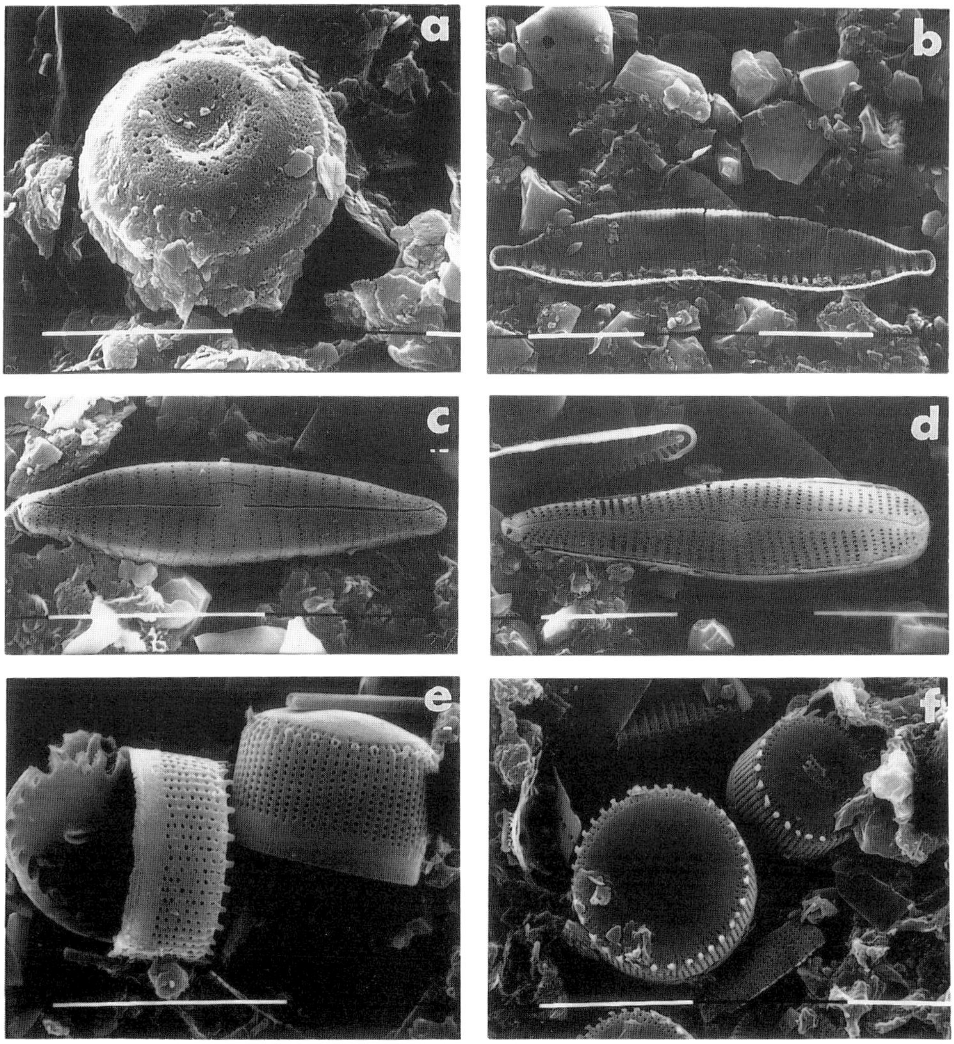


Plate 1 a–f. Examples of fossil diatoms found in Älgpussen and mentioned in the text, SEM. 1 scale bar = 10 μm. **a.** *Melosira westii* f. *parva* Brander, **b.** *Hantzschia amphioxys* (Ehr.) Grun., **c.** *Gomphonema parvulum* (Kütz.) Grun., **d.** *Gomphonema clavatum* Ehr., **e–f.** *Melosira cf. nygaardii* Camburn. On plate 1 **e**, note the valve mantle which contains 20–25 striae/10 μm that are parallel to the pervalvar axis. Striae are finely punctate, puncta c. 32/10 μm. On plate 1 **f**, note the discus that have only marginally arranged areolae.

high load of organic material within the lake but at the end of the zone the water quality improved.

**ÄLG–5.** 100–50cm. 1800–900 BP. At the transition ÄLG–4/5 the $\chi$ decreases markedly. Around 1300 BP ARM, SIRM/$\chi$, and ARM/$\chi$ display peaks. S-ratio fluctuate markedly. The peaks coincided with a peak in pollen from pasture land plants.This indicates a second period of intense anthropogenic activity in the vicinity (c. 1400–1300 BP). There is a dominance of Chrysophyceae stomatocysts in the peat. The phytoliths and aerophilous taxa decrease. The dominating diatom genera is *Eunotia* spp. which suggests a natural acidification of the basin.

**ÄLG–6.** 50–0 cm. 900–0 BP. High values of ARM, SIRM/$\chi$ and ARM/$\chi$ together with high frequencies of herb pollen, with a peak of cultivated plants (mainly *Secale*), indicate the third and last intense period of anthropogenic activity (800–500 BP). *Melosira* cf. *nygaardii* Camburn (Plate 1e–f) dominated, and a few aerophilous species were present. The presence of *M.* cf. *nygaardii* indicates that the site experienced a wet stage. The increasing acidification is reflected by high numbers of *Eunotia* spp. and also by *Tabellaria* spp. Chrysophyceae stomatocysts decreased. The palaeoenvironmental changes within and around the Älgpussen sedimentary basin are summarised in Fig. 9.

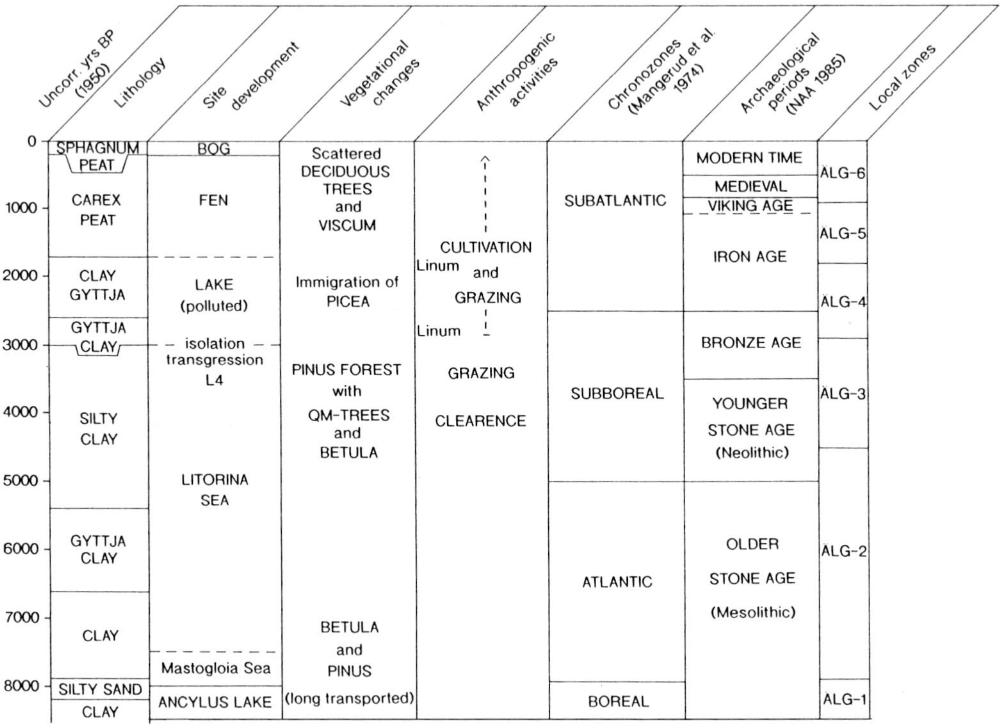


Fig. 9. Summary of the most important palaeoenvironmental changes, based on stratigraphical investigations of mineral magnetic parameters, pollen, and siliceous microfossils, within and around the Älgpussen sedimentary basin.

**Discussion**

In general, the stratigraphy of the Älgpussen sedimentary basin seems to represent the last 8000 years. Within the interval 5400–2800 BP, however, there are indications of a slow sedimentation rate. Probably, there is a short hiatus at the transition clay/gyttja clay (ÄLG–3/4). In time, this corresponds to the regression after the L 4 maximum (c. 3200–3000 BP) when water depth became shallow enough to establish an erosional and/or transportation bottom.

The isolation age of Älgpussen (2900 BP) corresponds well with the shore displacement model of Miller (in Brunnberg *et al.*1985). The Älgpussen bog represents the same type of overgrown basins as Miller mainly used to construct her shore displacement model. Therefore, the sediments dated from Älgpussen will exhibit a similar degree of contamination by rootlets. The shore displacement model presented by Risberg *et al.*(1991) is based only on radiocarbon dates of lake sediments showing a varying degree of reservoir ages. In comparison with the latter model the isolation age of Älgpussen is c.700 radiocarbon years too young.

The repeated sequence of clay/gyttja clay is indicative of a transgressive phase. Compared to the shore displacement model of Miller (in Brunnberg *et al.* 1985) both L 3, with a maximum at 4500 BP, and L 4, with a maximum at 3400 BP, should be documented within the clay layer (5400–2800 BP, cf. Table 2). There is a slight increase in magnetic susceptibility around 4500 BP that might correspond to L 3. At a slightly younger age (4200 BP) there is also a peak in the planktonic *Melosira westii* f. *parva* (cf. Miller 1973). Other indications of L 3 are, however, weak probably depending on a too deep water (c. 10 m) at the sampling site. We consider the pronounced peak in *M. westii* f. *parva* at 3200 BP to be the result of the L 4 transgression. At that time water depth was shallower and had more effect on the ratio planktonic/periphytic taxa.

The pollution indicative taxon *Gomphonema parvulum* show high percentages coinciding with high values of ARM, phytoliths and pollen of apophytes. The high organic load might be the result of anthropogenic activities such as waste deposition and/or cattle breeding. At the end of ÄLG–4 *Gomphonema parvulum* decreases and *G. clavatum* increases. According to Krammer & Lange-Bertalot (1986, p. 367) the latter taxon prefer "Quellewässeren auf Silikatböden", to be sensitive to organic load and to indicate an oligosaprob environment. This indicates an improved water quality in the end of the zone ÄLG–4. There are at least two possible explanations for this change. The first is related to a marked change in anthropogenic activities, *e.g.* a re-organisation of the cattle breeding. The second explanation might be a re-routing of the ground water flow within the glaciofluvial deposit that surrounds Älgpussen, bringing "clean" water into the basin.

*Gomphonema parvulum* is difficult to separate from *G. angustatum* Agardh because of the very broad variations in the morphology of the frustules (Krammer & Lange-Bertalot 1986; Wasell & Håkansson 1992). The frustules that have been identified as *G. parvulum* in this investigation were 17–22 µm long, 4–5 µm broad and had 12–18 striae in 10 µm (Plate 1e). These data are considered to correspond to the description of *G. parvulum* in Krammer & Lange-Bertalot (1986 p. 358).

The frustules of *Melosira cf. nygaardii* found in this investigation are a small form, 6–9 µm in diameter and a mantle height of 4–5µm. The valve mantle contains striae that are parallel to the pervalvar axis and striae 20–25/10 µm (Plate 1g–h). Striae are finely punctate and puncta c. 32/10 µm. The discus have only marginally arranged areolae and a shallow pseudoseptum is present. According to Camburn & Kingston (1986 p.26) the valve diameter of *Melosira nygaardii* is within the interval (5.0) 6.9–11.0 µm and the mantle height vary between (4.0) 7.0–17.0 (19.0) µm. Striae are 20–22/10 µm and finely punctated with puncta 25–30/10 µm. The valve mantle striae on the frustules found in this investigation show a finer areolation and the mantle height is within the lowermost interval of *M nygaardii*. We have, however, tentatively identified the frustules found in the *Carex* peat in Älgpussen as *Melosira nygaardii* Camburn.

## Acknowledgements

Financial support was received from Stockholm County Council, Secretariat for Structural Development (Swedish: Stockholms Läns Landstings Miljövårdsfond), Södertälje Municipality, National Board of Antiquities and the Swedish Natural Science Research Council. We thank B.A. Bengt Elfstrand, National Board of Antiquities, Linköping, for the archaeological information in Fig. 2. M.Sc. Elinor Andrén, M.Sc. Per Westman and Mr Johan Pettersson assisted in the field work.

## References

Battarbee, R. W. (1986). Diatom analysis. In: *Handbook of Holocene Palaeoecology and Palaeohydrology* (B. E. Berglund, ed.). John Wiley & Sons. Chichester, 527–570.

Berglund, B. E. & Ralska-Jasiewiczowa, M. (1986). Pollen analysis and pollen diagrams. In: *Handbook of Holocene Palaeoecology and Palaeohydrology*. (B. E. Berglund, ed.). John Wiley & Sons. Chichester, 455–484.

Brunnberg, L., Miller, U. & Risberg, J. (1985). Project Eastern Svealand: Development of the Holocene landscape. *ISKOS*, **5**, 85–91.

Camburn, K. E. & Kingston, J. C. (1986). The genus *Melosira* from soft-water lakes with special reference to northern Michigan, Wisconsin and Minnesota. In: *Diatoms and Lake Acidity* (J. P. Smol, R. W. Battarbee, R. B. Davis & J. Meriläinen, eds). Dr. W. Junk Publishers, Dordrecht, 17–34.

Cleve-Euler, A. (1951–1955). Die Diatomeen von Schweden und Finnland. *Kungliga Svenska Vetenskapsakademiens Handlingar* **I–V**. 4:e serien 2:1 (1951) 163 pp, 3:3 (1952) 152 pp. 4:1 (1953) 158 pp. 4:5 (1953) 255 pp. 5:4 (1955) 232 pp.

Higgitt, S. R., Oldfield, F. & Appelby, P. G. (1991). The record of land use change and soil erosion in the late Holocene sediments of the Petit Lac dÁnnecy, eastern France. *The Holocene*, **1**, 14–28.

Hustedt, F. (1957). Die Diatomeenflora des Fluss-Systems der Weser im Gebiet der Hansenstadt Bremen. *Abhandlungen des Naturwissenschaftlichen Vereins*. Bremen, **34**, 181–440.

Jowsey, P. C. (1987). An improved peat sampler. *New Phytologist,* **65**, 245–248.

Kolbe, R. W. (1927). Zur Ökologie, Morphologie und Systematik der Brackwasser-Diatomeen. Die Kieselalgen des Sperenberger Salz-Gebiets. *Pflanzenforschung,* **7**, 146 pp. Jena.

Krammer, K. & Lange-Bertalot, H. (1986). Bacillariophyceae. 1. Teil. Naviculaceae. In: *Süsswasserflora von Mitteleuropa.* (H. Ettl, J. Gerloff, H. Heynig & D. Mollenhauer, eds). Band, **2/1.** 857 pp. Gustav Fischer Verlag. Stuttgart.

Krammer, K. & Lange-Bertalot, H. (1988). Bacillariophyceae. 2. Teil. Bacillariaceae, Epithemiaceae, Surirellaceae. In: *Süsswasserflora von Mitteleuropa.* (H. Ettl, J. Gerloff, H. Heynig & D. Mollenhauer, eds). Band **2/2.** 596 pp. Gustav Fischer Verlag. Stuttgart.

Krammer, K. & Lange-Bertalot, H. (1991a). Bacillariophyceae. 3. Teil. Centrales, Fragilariaceae, Eunotiaceae. In: *Süsswasserflora von Mitteleuropa.*(H. Ettl, J. Gerloff, H. Heynig & D. Mollenhauer, eds). Band **2/3.** 576 pp. Gustav Fischer Verlag. Stuttgart.

Krammer, K. & Lange-Bertalot, H. (1991b). Bacillariophyceae. 4. Teil. Achnanthaceae, Kritische Ergänzungen zu *Navicula* (Lineolate) und *Gomphonema.* In: *Süsswasserflora von Mitteleuropa.* (H. Ettl, J. Gerloff, H. Heynig & D. Mollenhauer, eds). Band **2/4.** 437 pp. Gustav Fischer Verlag. Stuttgart.

Mangerud, J., Andersen, S. T., Berglund, B. E. & Donner, J. (1974). Quaternary stratigraphy of Norden, a proposal for terminology and classification. *Boreas,* **3**, 109–128.

Miller, U. (1973). Belägg för en subboreal transgression i Stockholms-trakten. *Lund University. Department of Quaternary Geology.* Report, **3**, 96–104.

Mullins, C. E. (1977). Magnetic susceptibility of the soils and its significance in Soil Science: A review. *Journal of Soil Science,* **28**, 223–246.

NAA (1985). *Nordic Archaeological Abstracts,* 389 pp. Højbjerg-Viborg.

Possnert, G. (1990). Radiocarbon dating by the accelerator technique. *Norwegian Archaeological Review,* **23**, 30–37.

Risberg, J., Miller, U. & Brunnberg, L. (1991). Deglaciation, Holocene shore displacement and coastal settlements in eastern Svealand, Sweden. *Quaternary International,* **9**, 33–37.

Sandgren, P. & Fredskild, B. (1991). Magnetic measurements recording Late Holocene man-induced erosion in S. Greenland. *Boreas,* **20**, 315–331.

Sandgren, P. & Risberg, J. (1990). Magnetic minerology of the sediments in Lake Ådran, eastern Sweden, and an interpretation of early Holocene water level changes. *Boreas,* **19**, 57–68.

Simonsen, R. (1962). Untersuchung zur Systematik und Ökologie der Bodendiatomeen der Westlishen Ostsee. *Internationale Revue gesamt. Hydrobiologie,* Berlin, Systemat. Beihefte **1**. 1–144.

Thompson, R. & Oldfield, F. (1986). *Environmental magnetism.* 227 pp. Allen & Unwin, London.

Wasell, A. & Håkansson, H. (1992). Diatom stratigraphy in a lake on horseshoe island, Antartica: a marine-brackish-fresh water transition with comments on the systematics and ecology of the most common diatoms. *Diatom Research,* **7** (1), 157–194.

# The diatom record of a core from the seaward part of the coastal plain of Belgium

Luc Denys

*Departement Biologie, Universitair Centrum Antwerpen (RUCA), Groenenborgerlaan 171, B–2020 Antwerpen, Belgium*

## Abstract

Holocene fossil diatom assemblages in a core taken in the more seaward part of the western Belgian coastal plain, and close to two major former tidal channels, are discussed. The bulk of the taphocoenoses consists of easily transported tychoplanktonic and sessile species (mainly *Cymatosira belgica* and *Rhaphoneis minutissima*). The contribution of epipelon of strictly local origin becomes considerable only when high-intertidal deposits are reached. Shifts in assemblage composition and local palaeoenvironment are closely linked to the lateral development of the channel systems, and it is likely that events recorded at this site are linked to phenomena of more regional importance. At least this appears to be the case for events occurring in the periods 6500–6000 and 4230–3830 cal BP.

## Introduction

Diatom analysis has become a widely used method in palaeoenvironmental studies of coastal and nearshore deposits, and is being used to address questions regarding relative sea-level change, crustal movements, coastal genesis, hazard assessment, sediment transport, archaeology, pollution, etc. The more sheltered parts of coastal areas, where accumulative conditions predominate, and where siliciclastic deposits alternate with organic beds, offer better possibilities for most palaeo-environmental studies than areas where the sequence consists entirely of either minerogenic sediments or peat (Streif 1980). It has also been recognized that environmental resolution of fossil diatom assemblages, or taphocoenoses, is considerably better in the higher intertidal zone than in the lower intertidal and adjacent subtidal (e.g. Brockmann 1940; Denys 1994; Nelson & Kashima 1993). In the latter, hydrodynamic and sedimentary conditions are more dynamic, resulting in more homogeneous assemblages, whereas the contribution of diatoms of strictly local origin may be small (see e.g. works of Brockmann 1950; Vos & de Wolf 1993a). Consequently, accurate palaeoenvironmental reconstructions by means of diatoms may be difficult for sediments deposited here (e.g. Vos & de Wolf 1993a, 1994).

In a palaeoecological diatom study of the Holocene deposits in the western part of the coastal plain of Belgium (Denys 1993) concerning coastal development, sea-level history, and the relations between palaeoenvironments and diatom taphocoenoses, the focus was therefore primarily on sequences with mainly low-energy lithofacies. Nevertheless, some of the records originate from deposition under energetic conditions, or – as in the core discussed here – relate to sites more strongly influenced by nearby tidal channels. The results presented in this paper illustrate some of the difficulties, and possibilities, in interpreting the diatom record from such environments. Also their relation to the variations in marine influence in the study area is considered.

*Study area*

Holocene coastal deposition in the western part of the Belgian coastal plain (Fig. 1A) goes back to *c.* 9400 cal BP. The deposits were formed in the course of the postglacial sea-level rise and rest upon Pleistocene material of different age and facies, or where these are absent, on Eocene clay. During the initially very rapid sea-level rise, a marine influence penetrated into the area through pre-existing NW–SE directed palaeovalley systems (Fig. 1B). By 6500–6000 cal BP, tidal environments stretched out almost as far as the present limits of the western coastal plain. Within this time, a protective barrier system – most likely a barrier island chain – had been built up, which except in the extreme west was situated a few km offshore from the present coastline (Baeteman & Denys, in press). Its development enabled extensive fen and bog peat growth in nearly the entire plain in the period from about 5000–4750 to 3000–2000 cal BP. Hereafter tidal sedimentation predominated, as the protective barrier degraded. The present shoreline configuration was reached at *c.* 1200 cal. BP. The plain silted-up

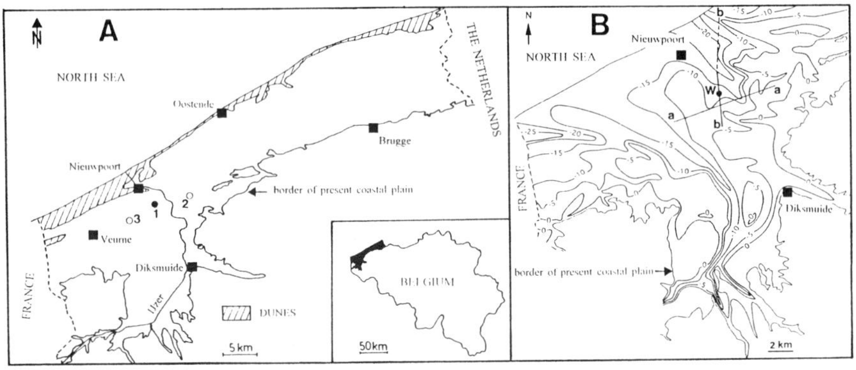


Fig. 1. A. Map of the coastal plain of Belgium, showing location of the coring site Wolvenest (1), and the sites Spermalie 2 (2) and Booitshoeke (3) mentioned in the text. B. Morphology of the Pleistocene surface in the western coastal plain (after Baeteman 1993), showing coring site Wolvenest (W), and the stratigraphic sections a–a and b–b shown in Fig. 2. Contours are in m TAW.

forming salt marshes and salt meadows. Within a few centuries, dykes were constructed, transforming the area into a polder.

The 17 m long core considered here, named Wolvenest, was drilled mechanically (10 cm ϕ). It is located in the more seaward part of the study area (51°06'40" N, 2°46'45" E; Fig. 1), close to a large channel system south-west of Nieuwpoort, which remained active throughout most of the build-up of the sequence. A branch of this system passes a few km S of the site (Figs 1B–2). A second channel system is present to the north-east.

*Core description and chronology*

The surface at the coring site lies at 3.00 m TAW (the Belgian ordnance datum).

Holocene sediments occur above a depth of 14,78 m (Fig. 2 A–B). Below 5,76 m, the infill is entirely sililiciclastic, with mainly fine-sandy and silty sediments in the lower part, followed by clayey deposits. This part of the sequence is barren of diatoms. Judging from the lithological characteristics and age/depth position, it represents a channel infill consisting of presumably subtidal deposits in its lower part, covered by mud-flat sediments.

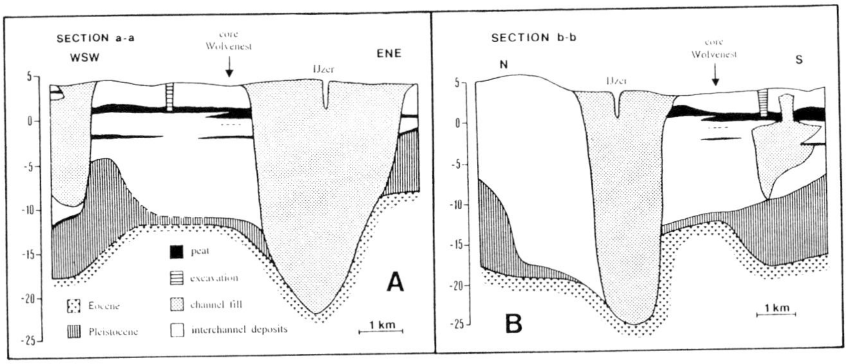


Fig. 2. A. Stratigraphic section a–a (see Fig. 1B.). B. Stratigraphic section b–b (*idem*). Altitudes in m TAW. Drawn after Baeteman & Van Strijdonck (1989).

An equally diatom-barren *Phragmites* peat occurs from 5.76 to 5.61 cm (Fig. 3), which was dated at *c*. 7300–7020 cal BP (Table 1). Immediately above the lowermost peat, many small detrital peat fragments, irregular silt accretions, and (mainly fragmented) *Hydrobia* shells occur. The top of the peat appears to be scoured. The possibility of a depositional hiatus therefore cannot be excluded, and the dates obtained from the top of the peat may not represent the termination of its growth. The lateral extent of this peat bed is limited, and towards the north it has been eroded.

Above the lowermost peat, clayey and silty sediments are found until 3.52 m. From 4.42 to 4.26 m a decalcified muddy clay with *Phragmites* rhizomes underlies a 2 cm thick peaty clay layer with vegetation remains. This is followed again by a muddy clay containing rhizomes, becoming more organic near the top, and grading into the second peat layer at 3.52 m. This part of the sequence contains diatom remains.

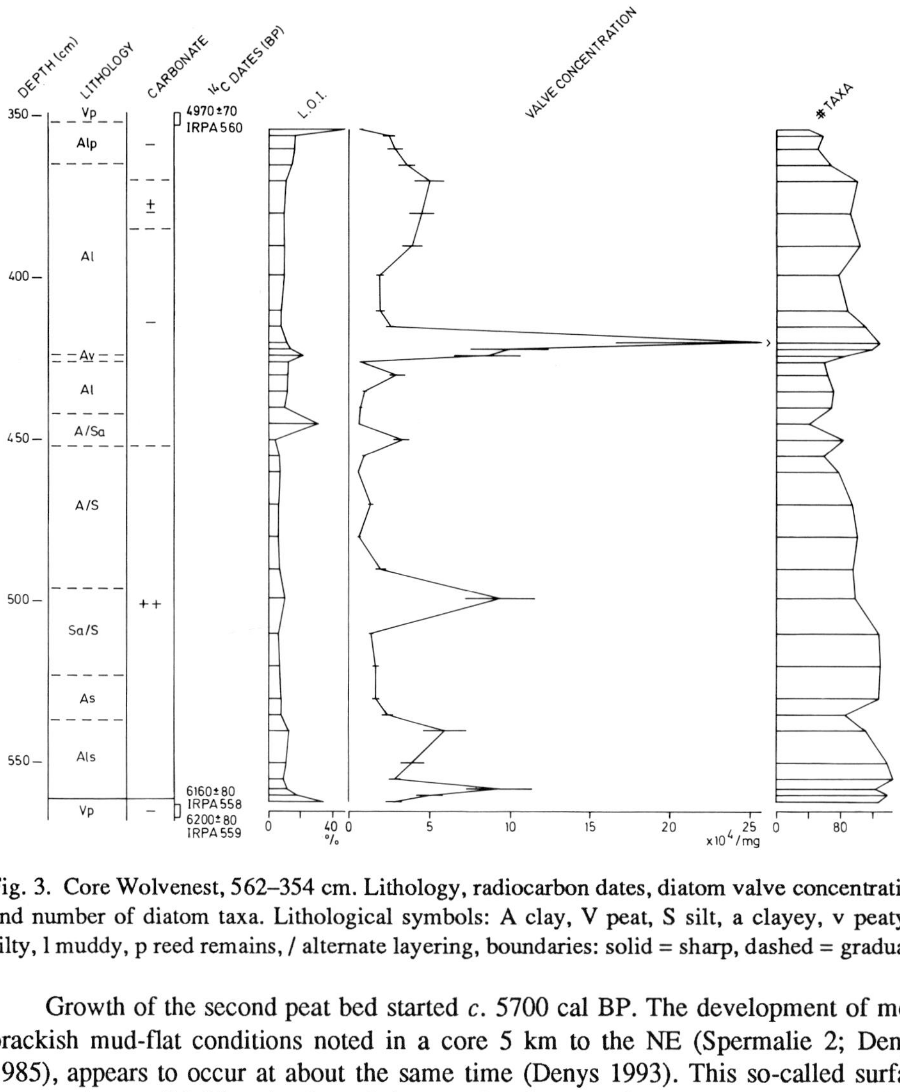


Fig. 3. Core Wolvenest, 562–354 cm. Lithology, radiocarbon dates, diatom valve concentration, and number of diatom taxa. Lithological symbols: A clay, V peat, S silt, a clayey, v peaty, s silty, l muddy, p reed remains, / alternate layering, boundaries: solid = sharp, dashed = gradual.

Growth of the second peat bed started *c.* 5700 cal BP. The development of more brackish mud-flat conditions noted in a core 5 km to the NE (Spermalie 2; Denys, 1985), appears to occur at about the same time (Denys 1993). This so-called surface peat reaches until 2.28 m depth, and shows a development from reed peat to amorphous peat (from 3.47 m) and finally *Sphagnum*/*Eriophorum* peat (3 m). Some more clayey lenses occur between 2.88 and 2.85 m, and diatom remains are present only from 2.93 to 2.84 m (Fig. 4). The most likely time interval for this part of the core was determined as *c.* 4230–3830 cal BP.

The peat is covered by a slightly silty and peaty muddy clay, containing lumps of peat and little to no lime; this extends up to 1,79 m (Fig. 5). The transition from peat to clay is very sharp and oblique. The date for the top of the peat (*c.* 2780 cal BP) is likely to predate considerably the onset of minerogenic sedimentation (*cf. infra*). The muddy

Table 1. Radiocarbon dates from the core Wolvenest. Calibrated ages were obtained with CALIB 3.0.3 (Stuiver & Reimer 1993), method A. References: 1 Dauchot-Dehon & Van Strydonck (1989), 2 Dauchot-Dehon *et al.* (1986).

| depth (cm) | age BP | cal BP | cal BP 1σ | cal BP 2σ | lab. no. | material | ref. |
|---|---|---|---|---|---|---|---|
| 228–232 | 2710±60 | 2780 | 2860–2760 | 2940–2740 | IRPA 859 | peat | 1 |
| 282–285 | 3550±60 | 3830 | 3900–3720 | 3980–3650 | IRPA 860 | peat | 1 |
| 292–295 | 3830±70 | 4230 | 4400–4090 | 4420–3990 | IRPA 825 | peat | 1 |
| 349–353 | 4970±70 | 5710 | 5850–5620 | 5900–5590 | IRPA 560 | reed peat | 2 |
| 561–563 | 6160±80 | 7020 | 7170–6900 | 7210–6810 | IRPA 558 | wood | 2 |
| 561–565 | 6200±80 | 7090 | 7200–7000 | 7230–6880 | IRPA 559 | reed peat | 2 |
| 573–577 | 6420±80 | 7280 | 7390–7120 | 7420–7170 | IRPA 561 | reed peat | 2 |

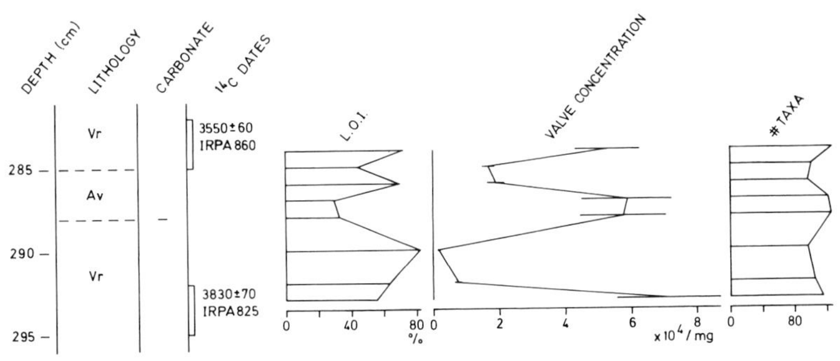


Fig. 4. As Fig. 3, 293–284 cm. Lithological symbols: see Fig. 3, r ombrotrophic.

clay is followed abruptly by a strongly bioturbated sticky clay, containing carbonates, which grades at *c.* 1.42 m into a compact clay with numerous silt lenses and small plant remains. These deposits were suitable for diatom analysis as well, except for the upper 0.4 m which were anthropogenically disturbed.

Mean accumulation rates for the core (uncorrected for compaction), obtained by linear interpolation, are shown in Fig. 6. These indicate a considerably slower minerogenic accretion after *c.* 6000 cal BP than before.

Formerly, the deposits below and above the surface peat would have been classified as Calais and Dunkerque deposits respectively.

## Methods

Loss on ignition (L.O.I.) of the sediments was determined according to Dean (1974). Presence of carbonates was tested with hydrochloric acid.

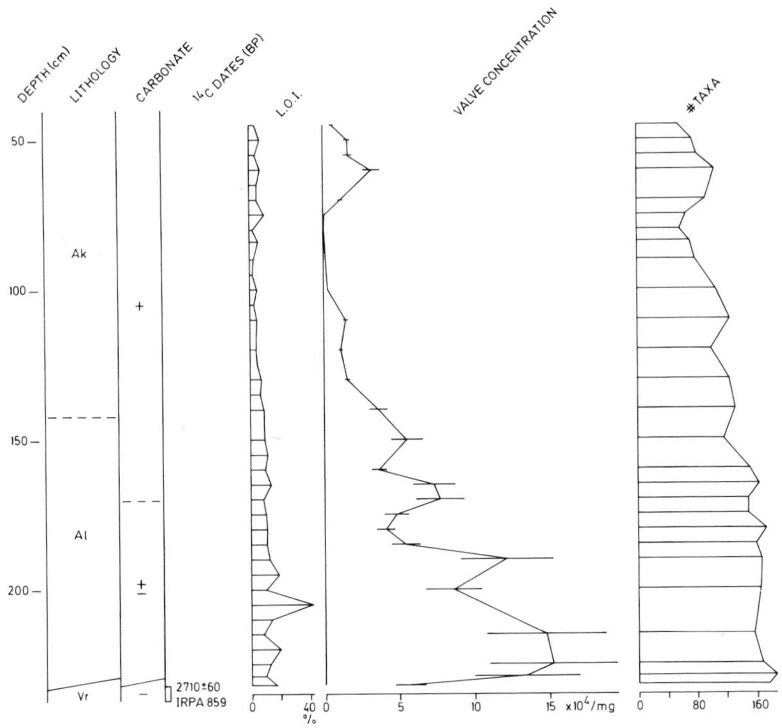


Fig. 5. As Fig. 3, 227–45 cm. Lithological symbols: see Figs 3–4, k compact.

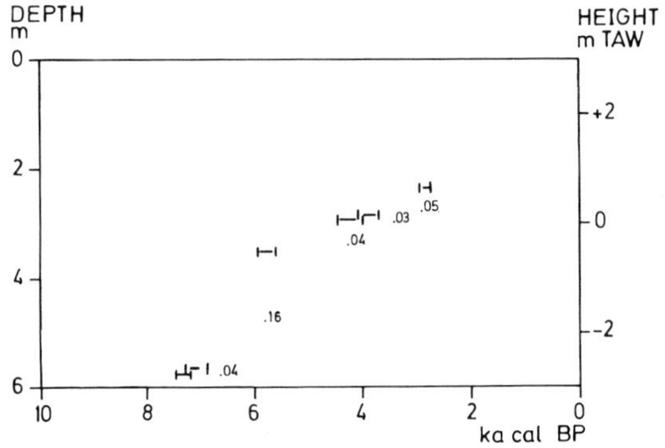


Fig. 6. Age/depth diagram of radiocarbon dated levels in the core Wolvenest (–1σ, I I 2σ), with average accumulation rates (in cm/year).

For diatom analysis, 0.1 g of dried sediment was treated with hydrogen peroxyde and hydrochloric acid. Tablets with *Lycopodium* spores (Stockmarr, 1971) were added, to allow estimates of microfossil concentrations. Coarse sand was removed by careful decanting. Mounts were made using pleurax. Generally, 500 to 700 valves were counted at 1250× magnification (Denys 1984). At three levels with very low valve concentrations a smaller number was counted.

Taxa were grouped into categories on the basis of an extensive literature survey, allowing ecological and taphonomic spectra to be calculated according to the codes and classifications listed by Denys (1991). Only relative abundance diagrams concerning the major taxa are reproduced here. Estimated standard errors on concentrations follow Stockmarr (1971), and 95% confidence limits on relative abundances are calculated according to Mosimann (1965, p. 643). Taxonomic names follow Denys (1991).

Zonation of the species diagrams was established by visual inspection of relative abundance data, stratigraphically constrained clustering (Grimm 1987), DCA ordination (ter Braak 1987), and calculation of the distance measures SIMI (McIntire & Moore 1977) and D (Patrick 1977) between adjacent samples.

## Results

*Cymatosira belgica*, a polyhalobous tychoplanktonic-benthic species, is the principal diatom in zone 1 (562–510 cm; Fig. 7). It is accompanied here mainly by sessile taxa: small forms of *Achnanthes delicatula*, *Opephora marina* and *Rhaphoneis* (*Delphineis*) *minutissima*. A number of less frequent species (*Achnanthes* sp. 1, *Catenula adhaerens*, *Cocconeis molesta*, *C. placentula*, *Navicula perminuta*, *N. salinicola*, *Nitzschia frustulum*) also occur. This community lives attached to siliciclastic grains and various other sorts of particles. The abundance of potentially epipsammic species remains well below that observed in sand-flat sediments (*cf.* Vos & de Wolf 1993b), nevertheless indicates still mildly energetic conditions, and suggests the existence of sandy flats nearby (along the channels). The common occurrence of taxa with an adnate (*Achnanthes*, *Cocconeis*, *Rhaphoneis*, small *Navicula* spp.) and a prostrate growth form (*Fragilaria*, *Nitzschia*, *Opephora*) may reflect the varying intensity of sediment displacement; this is also suggested by the alternation of silty with more clayey sediments. Concerning *Rhaphoneis minutissima*, it must be remarked that – in agreement with Vos & de Wolf (1993a) – it should not be considered a species typical of the highly energetic sand flats, although several reports do indicate that it will readily grow on sand grains (e.g. Amspoker 1977; Round & Mann 1980; Round *et al.* 1990, with a striking photograph on p. 10, Fig. 10b). Indeed it is generally more frequent in rather fine-grained deposits than in very sandy ones (Denys 1993; Vos & de Wolf 1988, 1993a, 1994), which may be explained by its common adherence to detrital flocs, pellets, clay aggregates and silt-sized particles (see also Sabbe 1993). Moreover, it was noted that the adhesive capacity of *Rhaphoneis minutissima* is rather limited (de Jonge 1985), making it quite susceptible to winnowing from the sediments in more energetic areas, and subsequent transport. As long as these properties are kept in mind in interpreting high abundances of *R. minutissima*, it appears of no great importance whether this species (like taxa such as *Delphineis surirella* and *Rhaphoneis amphiceros*) is classified as tychoplanktonic (Vos & de Wolf 1993a) or as sessile (Denys 1991). The delimitation of life-form categories remains a purely pragmatic tool,

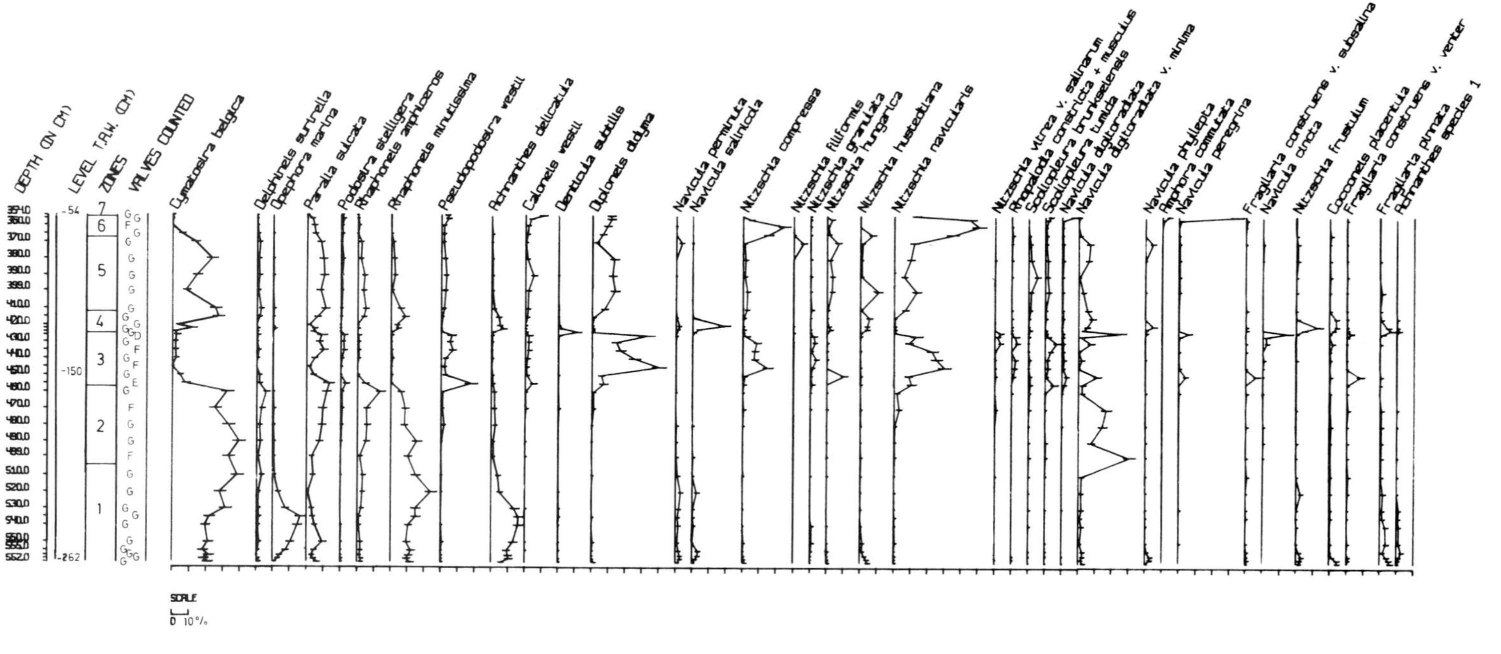


Fig. 7. Core Wolvenest, 562–354 cm. Relative abundance of major taxa. Valves counted: D 300–399.5, E 400–499.5, F 500–599.5, G 600–799.5.

and cannot reflect the full range of behaviour of individual taxa (*cf.* van den Hoek *et al.* 1979). Characteristic high-intertidal taphocoenoses are lacking at the transition from the peat to the overlying clay, but several species with more brackish affinities are slightly more abundant in the lower part of the zone (*Cocconeis placentula, Fragilaria pinnata, Navicula phyllepta, Nitzschia frustulum, N. hustedtiana*). Also, a larger number of freshwater taxa were noted in this zone. This may point to a somewhat lower or more variable salinity initially, but may also be attributed in part to some reworking of brackish and freshwater sediments. In general, polyhaline conditions appear to have prevailed throughout the zone. The epipelon is poorly represented (*c.* 20% above the peat, and rapidly decreasing to < 10%), and euplankton is very scarce. Rapid sedimentation under still intertidal conditions may be assumed most likely. The local transition from reed marsh to tidal flat is linked to the extension of the channel lithofacies 1–2 km to the S (Fig. 2B).

In zone 2 (499–460 cm), the situation remains more or less unchanged as far as salinity is concerned. The decrease of *Achnanthes delicatula* and *Opephora marina*, in favour of *Paralia sulcata* and especially the epipelic *Navicula digitoradiata* var. *minima,* leaves little doubt that the environment became calmer. In view of the abrupt nature of this evolution it is unlikely that this resulted merely from silting-up. Rather, channel migration or inactivation may be assumed (*cf.* the movement of the channel to the south; Fig. 2B). The increased relative abundance and concentration of the sessile *Rhaphoneis amphiceros* near the top of the zone is most likely due to increased sediment-associated import, and not to local growth.

A brackish, high tidal flat, with less favourable conditions for preservation of diatom valves, develops in zone 3 (455–426 cm). Only a few cm of sediment were deposited. Several epipelic and euryhaline mesohalobous species persist forming a typical assemblage for this type of deposit (*cf.* Denys 1985, 1989, 1993). *Caloneis westii, Diploneis didyma, Nitzschia compressa, N. navicularis* and *Scoliopleura tumida* are most common though numbers of valves remain low. The peaks of *Fragilaria construens* var. *subsalina* and var. *venter* at 450 cm indicate the proximity of freshening salt marshes (*cf.* Denys 1990). At the same level the high frequency of *Pseudopodosira westii* (*cf.* Denys 1993, 1994), and the restricted development of *Caloneis amphisbaena* f. *subsalina, Navicula digitoradiata, N. peregrina, N. pygmaea* and *Nitzschia hungarica*, show that the sedimentation level approached the local mean high water mark closely at the site. A small peak in L.O.I. is notable here (Fig. 3). In the middle part of the zone, the salt marsh aspect disappears again, to return modestly near 430 cm, where a slightly increased proportion of presumably epiphytic *Achnanthes brevipes, Cocconeis placentula, C. scutellum* var. *parva* and *Rhopalodia* spp., and of the epipelic *Gyrosigma balticum, Navicula cincta, Nitzschia vitrea* var. *salinarum* and *Scoliopleura tumida* is observed.

The salt marsh character becomes more pronounced in the lower part of zone 4 (424–420 cm), as *Fragilaria construens* var. *venter, F. pinnata, Nitzschia scalpelliformis* (2.4%) and the "aerophilous" *Denticula subtilis, Navicula cincta, N. soehrensis* var. *muscicola* (1.4%; often associated with rather dry mosses!), and *Nitzschia vitrea* var. *salinarum* show peaks at 424 cm. The eurytopic *Navicula digitoradiata* var. *minima* also becomes dominant. The strong increase in valve concentration (Fig. 3) indicates prolific diatom growth, a low influx of minerogenic material, or both. L.O.I. again shows a slight increase. Diatom assemblages related to the taphocoenoses encountered in zones 3 and 4 were described by Round (1960) from open stands of *Halimione* and *Puccinellia*, and by Körber-Grohne (1967) from *Juncus gerardi* and *Scirpus maritimus* vegetation.

The shifts occurring above 424 cm include a marked rise of *Cymatosira belgica*, and an increase of potentially epipsammic species, such as *Achnanthes delicatula*, *Cocconeis molesta*, *C. placentula*, *Navicula clamans*, *N. perminuta*, *N. salinicola*, *Nitzschia frustulum*, *Opephora marina* and *Rhaphoneis minutissima*. They indicate a resumption of marine influence, continuing in zone 5 (415–370 cm), with the tides again supporting an important contribution of polyhalobous tychoplankton (*Cymatosira*, *Paralia*) and the *Rhaphoneis* spp. The more local epipelic assemblage, including *Diploneis didyma*, *Caloneis westii*, *Nitzschia hungarica*, *N. hustedtiana*, *N. navicularis* and *Scoliopleura brunkseiensis*, reveals the persistence of essentially brackish conditions on this tidal flat, which received rather large amounts of sediment.

With silting-up proceeding, the influx of tychoplankton decreases dramatically in zone 6 (365–356 cm), and an epipelic assemblage with mainly *Diploneis didyma*, *Nitzschia compressa* and *N. navicularis* remains. In the uppermost sample of the zone, *Amphora commutata*, *A. copulata*, *Caloneis amphisbaena* f. *subsalina* and *Navicula peregrina* reveal the transition to salt marsh, occurring definitely in zone 7 (354 cm). Here *Navicula peregrina* predominates. This holoeuryhaline species, reaches optimal development at ß-mesohaline salinities (e.g. Brockmann 1950; Budde 1931; Koppen & Crow 1978), as does *Amphora commutata* (*cf.* also Legler & Krasske 1940; Simonsen 1962). The presence of some freshwater species (*Amphora copulata*, *Pinnularia viridis*) and of numerous stomatocysts points to rapid freshening of the waters. *Navicula digitoradiata* would be sensitive to drying according to Hopkins (1964), and typical subaerial taxa are scarce. This agrees well with the presumed ecology of *Navicula peregrina* and the *Amphora* species, indicating a permanently wet situation. Nevertheless valve concentration decreases and preservation becomes poor as the formation of reed peat starts.

The assemblage composition of zone 8 (293–284 cm; Fig. 8), corresponding to the more clayey part within the peat, is quite homogeneous. *Cymatosira belgica* and *Paralia sulcata* are the most prevalent taxa. Some sessile species are present in low quantities (*Achnanthes delicatula*, *Delphineis surirella*, *Rhaphoneis* spp.), as are a limited number of brackish epipelic taxa. The latter show a small increase in the uppermost part of the zone (8B, 286–284 cm). Concentration curves for individual taxa are very similar to the one for total valve concentration, and more or less opposite to that of L.O.I. (Fig. 4), suggesting a strong relation to sediment influx. Most of the valves from this zone are to be considered allochthonous. Yet, the radiocarbon dates, lithology, and the slight relative increase of *Pseudopodosira westii*, *Diploneis didyma*, *Nitzschia navicularis* and *Scoliopleura tumida* in the upper part indicate that the intercalation of clay sediments does not result from a very shortlived event. In view of the tychoplankton predominance, it is likely that a more prolonged period of stronger tidal influence occurred, during which the bog was flooded from a nearby channel.

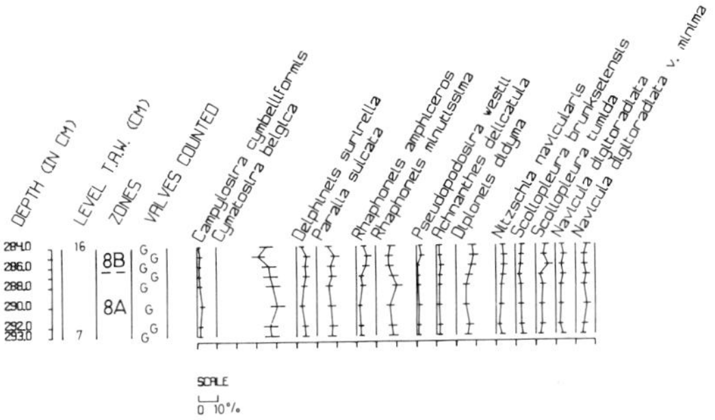


Fig. 8. As Fig. 7, 293–284 cm.

Above the peat there is an assemblage with codominance of marine tychoplankton (*Cymatosira*, *Paralia*) and the sessile *Rhaphoneis minutissima* (zone 9, 232–110 cm; Fig. 9). As in the case of zone 1, a limited, more brackish component is present in the lowermost peaty-clayey part (*Amphora coffeaeformis*, *Navicula agnita*, *N. cryptocephala*, *N. phyllepta*, *Nitzschia hungarica*, *Synedra fasciculata*), whereas a pure high-intertidal assemblage is lacking. It is unlikely that significant erosion of the peat could occur under the low energy conditions prevailing initially on this mud flat. Nevertheless, an erosive contact surface is observed, and reworked peat occurs in the overlying sediments. Moreover, the radiocarbon date from the top of the peat is too old to be consistent with the estimated termination of peat growth elsewhere in the area (Denys 1993). It is likely that either a period of erosion and non-deposition occurred, or that part of the peat was removed by peat diggers before the flooding. In this area the upper part of the peat is of good burning quality, and traces of peat digging have been found nearby (Stockmans & Vanhoorne 1954). If the behaviour of *Navicula parva*, *N. perminuta*, *N. salinicola*, *Nitzschia frustulum* and *Opephora marina* is considered, a slightly calmer environment might be conceivable above 170 cm, where less reworked peat is present as well (*cf.* L.O.I., Fig. 5). Although their abundance remains modest, the tychoplanktonic *Campylosira cymbelliformis* and *Plagiogrammopsis vanheurckii* and the neritic plankton species *Thalassiosira decipiens* and *Thalassionema nitzschioides* are characteristic. They suggest a relatively strong influence from the open sea, and sedimentation rather near to a major channel. Valve concentrations show a significant decline throughout the zone, probably due to increased dilution by clastics.

Signs of enhanced silting become apparent in zone 10 (100–84 cm), where the mesohalobous mud-flat epipelon starts to increase. As is often the case (Denys 1993, 1994), this is accompanied by an improved representation of transport-resistent tychoplanktonic species of more polyhalobous character (*Actinoptychus splendens*, *Paralia sulcata*, *Podosira stelligera*, *Pseudopodosira westii*).

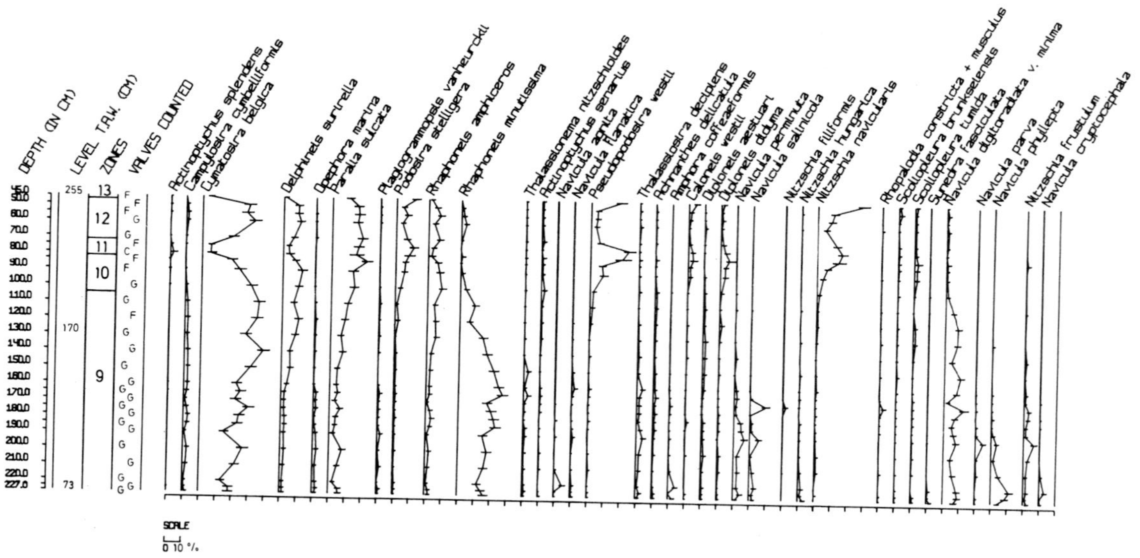


Fig. 9. As Fig. 7, 227–45 cm. Valves counted: see Fig. 7, C 200–299.5.

The increase in abundance of these taxa culminates in the next zone (11, 80–75 cm), where they outnumber the autochthonous epipelic community, which is dominated by *Nitzschia navicularis*.

Hereafter, a marked reactivation of the tidal influence takes place, resulting in the resurgence of *Cymatosira belgica* to former abundance levels (zone 12, 70–50 cm). The uppermost sample (zone 13, 45 cm) shows the restoration of a brackish situation fairly close to the local mean high water, again with dominance of *Nitzschia navicularis* and *Pseudopodosira westii*.

**Discussion**

Preservation of diatom remains in siliciclastic coastal deposits is often difficult to understand, and an entirely satisfactory explanation for their complete absence in the lower part of the core cannot be given. The good preservation immediately above the lower peat makes a purely post-sedimentary cause unlikely. Presumably diatom concentrations were already very low in these sediments from the beginning, enhancing dissolution of any valves present. As elsewhere, the lack of diatom valves in coastal peats, other than those influenced by tides or freshwater run-off, is a general feature in the Belgian area. Silica depletion of the pore water as a result of leaching (fluctuating watertables, ombrogenic conditions) and uptake by living organisms, and improved dissolution of biosilica due to defficiency of solution-retarding metal ions and biological activity, may be held responsible for this (Denys 1993; Vos & de Wolf 1994).

The intertidal diatom taphocoenoses present in the core reflect the generally high sediment supply to the site, favoured by its relatively seaward position and the proximity of major channel systems. There is a low representation of a locally developing epipelic component, in spite of the predominantly clayey-silty character of the tidal flats. High mud-flat assemblages are poorly represented. The bulk of the fossil assemblages consists of easily transported tychoplankonic and sessile taxa (*Cymatosira*, *Delphineis*, *Rhaphoneis*, *Paralia*). Only when the influence of the tidal channels decreases, or when local silting-up results in a high-intertidal environment, and sediment supply becomes reduced, does epipelon development improves.

*Cymatosira belgica* and *Rhaphoneis minutissima* often predominate in marine inter- and subtidal deposits in the southern North Sea region (e.g. Denys 1989, 1993; Vos & de Wolf 1993a). Vos & de Wolf (1988; 1993a; 1994) suggest that these species originate from the shallow open sea and are allochthonous in the tidal area. However, it may be more likely that they come from the (lower) intertidal community as well, and that their "source area" cannot be delimited as strictly as these authors propose. Both species appear similar in the way they are deposited in sediments. In Belgian coastal deposits high joint abundances appear to be related particularly to processes of rapid infilling. Vos & de Wolf (1988, 1994) note that in North Holland, *C. belgica* generally is more abundant in Atlantic and Subboreal deposits than in the younger "Dunkerque" sediments, and hence that this species may be used as an (eco)stratigraphic marker. In the Belgian coastal deposits, the relative abundance of this taxon is lower than the highest values mentioned by these authors, but the situation appears to be reversed, with a more important representation in Subatlantic sediments, except for the aforementioned cases of very rapid infilling. This suggests that the proportion of *C. belgica* reflects the more open coastal configuration in Subatlantic times, enabling more substantial sediment transport from the fore- and nearshore to the intracoastal tidal areas. Stronger tidal currents and more intense resuspension and transport of sediments may have allowed a better proliferation of the species in the intertidal zone as well, by creating a rather large area of silty to fine-sandy tidal flats and a higher density of gullies and small channels. The possible increase of tidal amplitudes on the North Sea (e.g. Franken 1987) should be considered as well.

Due to the high sensitivity of diatom assemblages to environmental changes related to local marine influence in coastal deposits, their record may contribute to the refinement of chronologies of marine activity ("sea-level tendencies"; Shennan *et al.* 1983) (Denys 1993). Although the sediments in this study provide a fairly low temporal resolution this is illustrated at the local level by the detection of variations in the periods of marine influence which are only partially observed from the lithofacies changes. The dating of these events, occurring in largely siliciclastic beds, remains less reliable than that of the commonly used trans- and regressive overlaps, which may be dated directly by $^{14}C$ methods. Age estimates based on interpolations between radiocarbon dated levels will suffer from (unknown) variations in accumulation rate and compaction of the sediment. In this specific case, the lower part of the core – where moreover the lowermost date available most likely refers to an erosive contact – would appear especially susceptible to error. Using calibrated 2$\sigma$ limits, linear interpolations would yield an age between *c.* 6550 and 6000 cal BP for the two periods of more pronounced silting-up recorded between 450 and 424 cm in the core. Based on evidence from the entire coastal plain, periods of reduced marine influence were suggested from *c.* 6450 to 6300 and from about 6220 to 6140 cal BP (Denys 1993). Less tentative is the correlation of the 4230–3830 cal BP inundation of the peat bog, to a series of similar phenomena observed in the area (Baeteman & Van Strijdonck 1989; Denys 1993, 1994), such as near Booitshoeke to the SW of the site (Fig. 1), pointing to a distinct period of increased tidal activity and limited reactivation of at least the main channel crossing the coastal plain SW of Nieuwpoort. Ecostratigraphic indications of multiple sedimentation cycles or channel reactivations are uncommon in the tidal deposits formed after the surface peat (Denys 1993). Unfortunately, an age estimate cannot be proposed for the lateral shifts of the tidal flat subenvironments observed in the Wolvenest core. This precludes a comparison with further data or the presumed chronology of the much debated "Dunkerque transgressive phases". On the whole, however, the observed changes in assemblage composition and local palaeoenvironment are closely related to the lateral development of extensive channel systems. Hence, it is likely that events recorded at this site are linked to phenomena of more regional importance.

## Acknowledgements

Part of the study was supported by an IWONL grant. Coring and dating were commissioned by the Belgian Geological Survey. C. Baeteman is acknowledged for appreciated comments.

## References

Amspoker, M. C. (1977). The distribution of intertidal epipsammic diatoms on Scripps Beach, La Jolla, California, USA. *Botanica Marina*, **20**, 227–232.

Baeteman, C. (1993). The western coastal plain of Belgium. In: *Quaternary shorelines in Belgium and The Netherlands. Guidebook fieldmeeting 1993 of the Subcommission on Shorelines of northwestern Europe* (C. Baeteman & W. de Gans, eds), 1–24, 43–55. Ministry of Economic Affairs, Brussel.

Baeteman, C. & Denys, L. (in press). Holocene shoreline and sea-level data from the Belgian coast. *Paläoklimaforschung*.

Baeteman, C. & Van Strijdonck, M. (1989). Radiocarbon dates on peat from the Holocene coastal deposits in West Belgium. In: *Quaternary sea-level investigations from Belgium* (C. Baeteman, ed.). *Professional Paper Belgische Geologische Dienst*, **241**, 59–91.

Brockmann, C. (1940). Diatomeen als Leitfossilien in Küstenablagerungen. *Westküste*, **2**, 150–181.

Brockmann, C. (1950). Die Watt-Diatomeen des schleswig-holsteinischen Westküste. *Abhandlungen Senckenbergischen Naturforschenden Gesellschaft*, **478**, 1–26.

Budde, H. (1931). Die Algenflora westfälischer Salinen und Salzgewässer. 1. Teil. *Archiv für Hydrobiologie*, **23**, 462–490.

Dauchot-Dehon, M. & Van Strydonck, M. (1989). Institut Royal du Patrimoine Artistique radiocarbon dates XIII. *Radiocarbon*, **31**, 187–200.

Dauchot-Dehon, M., Van Strydonck, M. & Heylen, J. (1986). Institut Royal du Patrimoine Artistique radiocarbon dates XI. *Radiocarbon*, **28**, 69–77.

Dean, W. E. (1974). Determination of carbonate and organic matter in calcareous sediments and sedimentary rocks by loss on ignition: comparison with other methods. *Journal of Sedimentary Petrology*, **44**, 242–248.

De Jonge, V. N. (1985). The occurrence of 'epipsammic' diatom populations: a result of interaction between physical sorting of sediment and certain properties of diatom species. *Estuarine, Coastal and Shelf Science*, **21**, 607–622.

Denys, L. (1984). Diatom analysis of coastal deposits: methodological aspects. *Bulletin van de Belgische Vereniging voor Geologie*, **93**, 291–295.

Denys, L. (1985). Diatom analysis of an Atlantic-Subboreal core from Slijpe (western Belgian coastal plain). *Review of Palaeobotany and Palynology*, **46**, 33–53.

Denys, L. (1989). Observations on the transition from Calais deposits to surface peat in the western Belgian coastal plain – results of a paleoenvironmental diatom study. In: Quaternary sea-level investigations from Belgium (C. Baeteman, ed.). *Professional Paper Belgische Geologische Dienst*, **241**, 20–43.

Denys, L. (1990). *Fragilaria* blooms in the Holocene of the western coastal plain of Belgium. In: *Proceedings of the 10th Diatom Symposium* (H. Simola, ed.), 397–406. Koeltz Scientific Books, Koenigstein.

Denys, L. (1991). A check-list of the diatoms in the Holocene deposits of the western Belgian coastal plain with a survey of their apparent ecological requirements. I. Introduction, ecological code and complete list. ***Professional Paper Belgische Geologische Dienst*, 246,** 1–41.

Denys, L. (1993). ***Paleoecologisch diatomeeënonderzoek van de holocene afzettingen in de westelijke Belgische kustvlakte.*** 830 pp. Unpublished doctoral thesis, Universitaire Instelling Antwerpen.

Denys, L. (1994). Diatom assemblages along a former intertidal gradient: a palaeoecological study of a Subboreal clay layer (western coastal plain, Belgium). *Netherlands Journal of Aquatic Ecology*, **28**, 85–96.

Franken, A. F. (1987). Rekonstruktie van het paleo-getijklimaat in de Noordzee. *Waterloopkundig Laboratorium Delft Rapport*, X 0029–00, 1–71.

Grimm, E. C. (1987). CONISS: a FORTRAN 77 program for stratigraphically constrained cluster analysis by the method of incremental sum of squares. *Computers & Geoscience*, **13**, 13–35.

Hopkins, J. T. (1964). A study of the diatoms of the Ouse Estuary, Sussex II. The ecology of the mud-flat flora. *Journal of the Marine Biological Association U. K.*, **44**, 333–341.

Koppen, J. D. & Crow, J. H. (1978). Some midsummer diatom assemblages along the saline gradient of a small coastal stream in Kachemak Bay, Alaska. *Botanica Marina*, **21**, 199–206.

Körber-Grohne, U. (1967). *Geobotanische Untersuchungen auf der Feddersen Wierde*. 358 pp. Franz Steiner Verlag, Wiesbaden.

Legler, F. & Krasske, G. (1940). Diatomeen aus dem Vansee (Armenien). Beiträge zur Ökologie der Brackwasserdiatomeen. I. *Beihefte Botanisches Centralblatt*, **60B**, 334–345.

McIntire, C. D. & Moore, W. W. (1977). Marine littoral diatoms: ecological considerations. In: *The biology of diatoms* (D. Werner, ed.), 333–371. University of California Press, Berkeley.

Mosimann, J. E. (1965). Statistical methods for the pollen analyst. In: *Handbook of paleontological techniques* (B. Kummel & D. Raup, eds), 636–673. W.H. Freeman and Co., San Francisco – London.

Nelson, A. R. & Kashima, K. (1993). Diatom zonation in southern Oregon tidal marshes relative to vascular plants, foraminifera, and sea level. *Journal of Coastal Research*, **9**, 673–697.

Patrick, R. (1977). Ecology of freshwater diatoms – diatom communities. In: *The biology of diatoms* (D. Werner, ed.), 284–332. University of California Press, Berkeley.

Round, F. E. (1960). The diatom flora of a salt marsh on the River Dee. *New Phytologist*, **59**, 332–348.

Round, F. E., Crawford, R. M. & Mann, D. G. (1990). *The diatoms. Biology and morphology of the genera*. 747 pp. Cambridge University Press, Cambridge.

Round, F. E. & Mann, D. G. (1981). *Psammodiscus* nov. gen. based on *Coscinodiscus nitidus*. *Annals of Botany*, **46**, 367–373.

Sabbe, K. (1993). Short-term fluctuations in benthic diatom numbers on an intertidal sandflat in the Westerschelde estuary (Zeeland, The Netherlands). In: *Proceedings of the Twelfth International Diatom Symposium* (H. van Dam, ed.). *Hydrobiologia*, **269/270**, 275–284.

Shennan, I., Tooley, M. J., Davis, M. J. & Haggart, B. A. (1983). Analysis and interpretation of Holocene sea-level data. *Nature*, **302**, 404–406.

Simonsen, R. (1962). Untersuchungen zur Systematik und Ökologie der Bodendiatomeen der westlichen Ostsee. *Internationale Revue der gesamten Hydrobiologie, Systematische Beihefte*, **1**, 8–144.

Stockmans, F. & Vanhoorne, R. (1954). Etude botanique du gisement de tourbe de la région de Pervijze (Plaine Maritime belge). *Mémoires de l'Institut royal des Sciences Naturelles de Belgique*, **130,** 1–144.

Stockmarr, J. (1971). Tablets with spores used in absolute pollen analysis. *Pollen et Spores*, **13,** 615–621.

Streif, H. (1980). Cyclic formation of coastal deposits and their indications of vertical sea-level changes. *Oceanis*, **5** *Fasc. Hors-Série*, 303–306.

Stuiver, M. & Reimer, P. J. (1993). Extended $^{14}C$ data base and revised CALIB 3.0 $^{14}C$ age caibration. In: *Calibration 1993* (M. Stuiver, A. Long & R. S. Kra, eds). *Radiocarbon*, **35,** 215–230.

Ter Braak, C. J. F. (1987). *CANOCO a FORTRAN program for canonical community ordination by (partial) (detrended) (canonical) correspondence analysis, principal components analysis and redundancy analysis (version 2.1)*. 95 pp. ITI–TNO, Wageningen.

Van Den Hoek, C., Admiraal, W., Colijn, F. & De Jonge, V. N. (1979). The role of algae and seagrasses in the ecosystem of the Wadden Sea: a review. In: *Flora and vegetation of the Wadden Sea* (W. J. Wolff, ed.), 9–118, 172–1198. Report 3 of the Wadden Sea Working Group, Texel.

Vos, P. C. & De Wolf, H. (1988). Paleo-ecologisch diatomeeën onderzoek in de Noordzee en Provincie Noord-Holland in het kader van het kustgenese project, taakgroep 5000. *Rijks Geologische Dienst Rapport*, **500,** 1–144.

Vos, P. C. & De Wolf, H. (1993a). Reconstruction of sedimentary environments in Holocene coastal deposits of the southwest Netherlands; the Poortvliet boring, a case study of palaeoenvironmental diatom research. In: *Proceedings of the Twelfth International Diatom Symposium* (H. van Dam, ed.). *Hydrobiologia*, **269/270**, 297–306.

Vos, P. C. & De Wolf, H. (1993b). Diatoms as a tool for reconstructing sedimentary environments in coastal wetlands; methodological aspects. In: *Proceedings of the Twelfth International Diatom Symposium* (H. van Dam, ed.). *Hydrobiologia*, **269/270**, 285–296.

Vos, P. C. & De Wolf, H. (1994). Palaeoenvironmental research on diatoms in early and middle Holocene deposits in central North Holland (The Netherlands). *Netherlands Journal of Aquatic Ecology*, **28**, 97–115.

# Eemian diatom floras in the Amsterdam glacial basin

H. de Wolf and P. Cleveringa

*Geological Survey of the Netherlands, P.O. Box 157, 2000 AD Haarlem, The Netherlands.*

## Abstract

Recent multidisciplinary studies on sea-level changes in the Quaternary and the corresponding sedimentation patterns, processes and environments offer starting-points for re-interpretation of Harting's (1852) and Brockmann's (1928) diatom data. Eemian sediments from cores from the Amsterdam glacial basin, dated by pollen analysis, were re-examined for their diatom associations.

Based on the diatom associations, the infilling of the Amsterdam basin can be divided into a freshwater (clay and gyttja) and a marine (clay) phase. In addition, the depth of the water changed from shallow to deep during infilling of the basin. The infilling of the basin started under cold conditions, when the tree-line had not receded. *Navicula jaernefeltii* Hust. is the dominant species in the clay at the base of the infilling. There is a hiatus between the clay *(Navicula jaernefeltii* zone) and the overlying gyttja (*Cyclotella-Stephanodiscus* zone). The first marine influence coincided with a *Fragilaria* bloom, when the lake became shallower.

The marine phase itself is characterized by three diatom floras. Firstly, in the lower portion of the infilling, the flora indicates shallow and clear marine waters. Harting identified several species (such as *Hyalodiscus scoticus* (Kütz.) Grun., *Grammatophora oceanica* Grun.) of this flora, which most closely resembles the contemporary Baltic Sea flora. Secondly, the "classic" Eemian flora, consisting of species regarded as indicative of warmer and southerly waters (*Stephanopyxis turris* (Grev. & Arnott) Ralfs, *Cocconeis sp.*), and species indicating a supply of oceanic waters (*Chaetoceros* and *Rhizosolenia spp.*). Thirdly, in the uppermost Eemian section there is a flora showing a marked similarity to the holocene North Sea flora (*Cymatosira belgica* Grun., *Rhaphoneis amphiceros* Ehr.).

These floras have also been found at locations elsewhere in Europe, so that their distribution is not merely regional, but appears to be wider.

## Introduction

Over a century ago, Harting (1874) studied marine sediments in the area of the Eem River, in the vicinity of Amersfoort, The Netherlands. Previously Harting (1852), had already described similar sediments from the subsoil of Amsterdam. In both cases, the sediments were deposited in an ice-tongue basin ("glacial valley") inside ridges pushed up by the ice. The sediments of these infillings owe their name to their occurrence in bore-holes from the glacial valley of Amersfoort near the Eem River (Harting 1874; Lorié 1906, 1916). Subsequently the stratotype near the city of Amersfoort was documented by Zagwijn (1963).

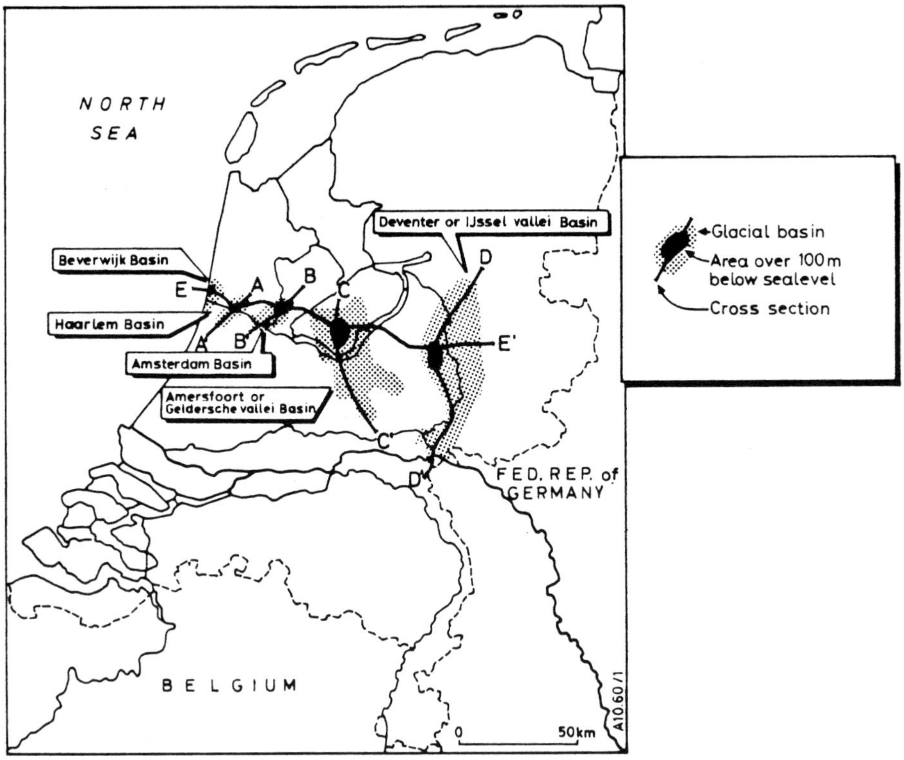


Fig. 1. Location of the major glacial basins in The Netherlands and position of the cross-section (after De Gans *et al.* 1986).

In his earlier study, Harting (1852) examined some cores (made for water-wells) from what we know now as another glacial valley system, the so-called Amsterdam Glacial Basin. In the subsoil he recognized a clay layer that contained a rich freshwater diatom flora as well as brackish and marine species in its uppermost portion, situated underneath a sandy deposit. Sometimes a layer consisting almost exclusively of diatoms was found. Subsequently this layer was assigned separate status in the literature as "Laag van Harting" (Harting's layer) (Zagwijn 1983). The "Laag van Harting" has become a collective term especially for the gyttjas at the base of glacial basin infillings.

In his study on the marine diatoms of the Dutch Quaternary, Brockmann (1928) provided a survey of the diatoms including those from the "Laag van Harting".

Van der Werff analysed the samples from the Valkenweg core (unpublished data of the Geological Survey) and gave an analysis that included diatoms from the gyttja bed. His approach was qualitative rather than quantitative and resulted in a list of species and an interpretation of the environmental conditions. Van der Werff's slides (see under Methods and materials) also formed the basis of the present study. Mainly on the basis of Van der Werff's data, Zagwijn (1983) typified the fresh, brackish and marine sediments from the Valkenweg borings. Up till now, the Amsterdam Basin is the only known basin with a freshwater diatom flora at the base of the marine Eemian deposits. Zagwijn (1983) published the relative heights of the sea-level during the Eemian, with dating based on pollen-analysis.

The object of this paper is to present local diatom assemblage zones for the Amsterdam Glacial Basin and to construct a regional paleoecological framework for the Eemian Interglacial in The Netherlands. For the (relative) dating framework, data from pollen analysis were used.

### *Geological setting*

Figure 1 shows that the Amsterdam Glacial Basin is one of a series of basins in The Netherlands, situated at the southern fringe of the maximum extent of the Saale Glaciation (De Gans *et al.* 1986; Van den Berg & Beets 1986). The depth of the basins increases from West to East (Fig. 2). This is the reason why the Eemian sediments vary in thickness – with a maximum of over 60 metres. In general, this infilling consists of a diatom-bearing gyttja layer with overlying marine clays. The sedimentation is terminated by a layer of sand, sometimes rich in shells (Zagwijn 1983). In many places the Eemian deposits are superimposed on the ground-moraine of the Saale glacial and are in turn covered by deposits from the Weichsel glacial. The depth at which the top of the marine deposits is encountered ranges from 8 metres (at the fringe of the tongue basins) to about 30 metres (northern North Holland). The depth of the basins varies even more, depending on the scouring effects of ice-lobes during the Saalian and on the topography over which the Eemian sea transgressed.

## Materials and methods

Slides of three cores from the Amsterdam basin in the collections of the Geological Survey were examined: namely Amsterdam Valkenweg 25E255 (AVW), Broek in Waterland 25E172 (BIW) and Amsterdam E7-119 25G594 (E7). The core samples had been treated with HCl and $H_2O_2$ and a qualitative analysis had been made. A quantitative analysis of the slides was made during 1993. The Zeeburgereiland core 25G922 (ZBE) was sampled in 1991 and subsequently analysed.

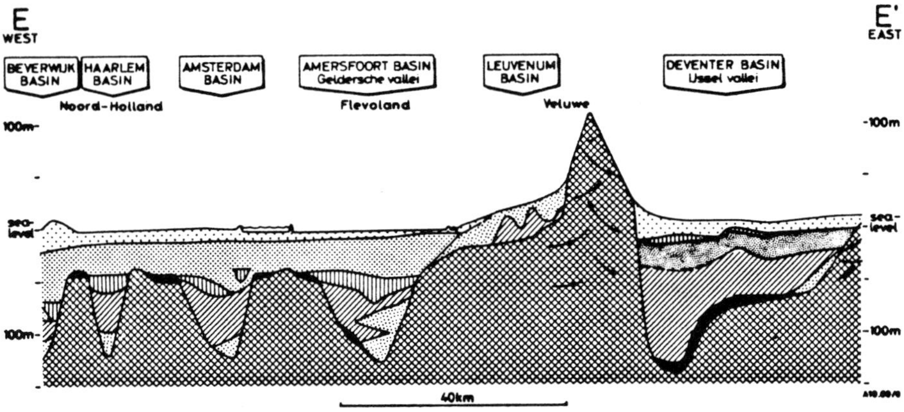


Fig. 2. Cross-section over the major glacial basins (after De Gans *et al.* 1986).

Of the samples examined, the Valkenweg core yields the most complete record. The results are, therefore, discussed in detail while the results of diatom analysis of the other three borings are dealt with in more general terms.

Determinations of the diatoms were made using literature from Cleve-Euler (1951, 1952, 1953a, 1953b, 1955), Brockmann (1928), Grönlund (1988, 1991a, 1991b) and Harting (1852).

In the absence of precise dating of the Eemian, the "classic" pollen analysis dating is used (Jessen & Milthers 1928; Zagwijn 1961). Therefore, a brief outline of the development of vegetation in the Eemian is necessary because pollen zonation is used as a global time-scale in the discussion.

## Observations

*Valkenweg (AVW)*

In zone A, consisting of clay and decalcified material towards the top (De Wolf 1994b) (Fig. 3), *Navicula jaernefeldtii* Hustedt was the principal species. This diatom is found in shallow pools and was restricted to relatively little disturbed areas (Stoermer 1980). *N. jaernefeldtii* appears to be more common in higher latitudes like the Shetland Islands, Scotland and Scandinavia, areas with only scattered shrubs and dwarf trees.

Other species found in this zone were *Achnanthes calcar* Cl., *Gomphocymbella ancyli* (Cl.) Hust. and *Fragilaria pinnata* Ehr. These species have also a northerly distribution.

The calcareous gyttja from zone B, dominated by species like *Melosira italica* (Ehr.) Kütz. and *Stephanodiscus minutulus* (Kütz.) Cl. & Möller, with *Cyclotella kuetzingiana* Thwaites appeared at a later stage and was deposited in a fresh, shallow mesotrophic lake.

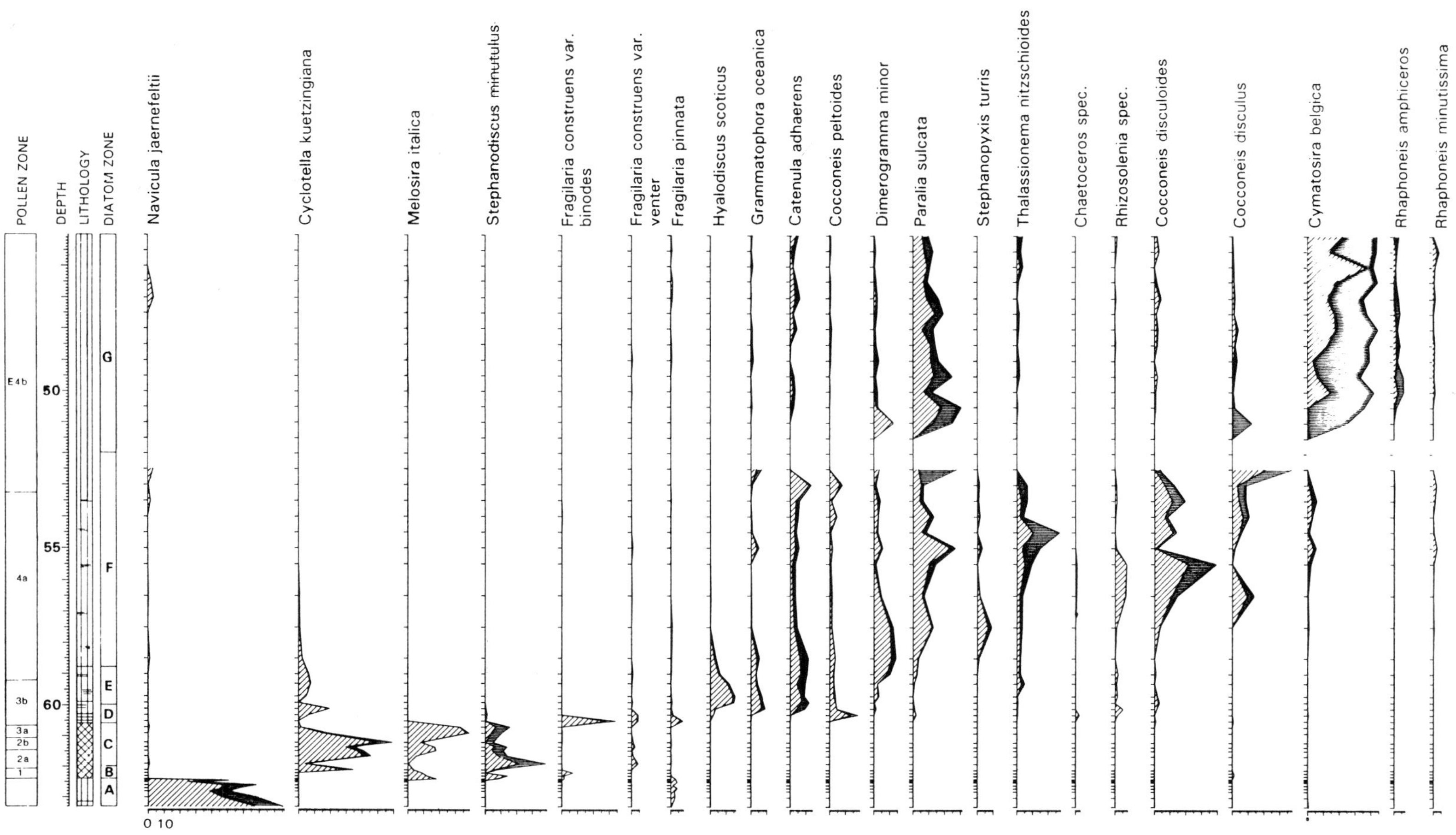


Fig. 3. Percentage diagram of the most important diatom species from the Valkenweg core; slanting lines: % unbroken diatoms, horizontal lines: % broken diatoms. Legend lithological column, see Fig. 6.

In zone C (gyttja), with species like *Cyclotella kuetzingiana*, *Melosira italica* and *Stephanodiscus minutulus*, the lake became more eutrophic and deeper, but it was still a freshwater environment.

Finely laminated clay-gyttja from zone D provided the first evidence of marine influence. The presence of *Fragilaria* spp. indicates that the lake passed through a shallower phase.

In the finely laminated clay of zone E, *Grammatophora oceanica* and *Hyalodiscus scoticus* were observed; *Catenula adherens* Mereschkovsky, *Dimeregramma minor* (Greg.) Ralfs and *Cocconeis peltoides* Hust. made their appearance. The last mentioned species, together with *Thalassionema nitzschioidesm* Grun., *Chaetoceros* resting-spores and peaks of *Rhizosolenia* fragments, were also encountered in zone F.

Apart from *Hyalodiscus scoticus* and *Grammatophora oceanica*, *Catenula adherens* and *Dimeregramma minor* provide a good indication of the sedimentation environment. These species are indicative of a shallow sea with growths of seaweeds and a reduced salinity, little or no tidal action and clear water. *Thalassionema* and the *Chaetoceros* spores are allochthonic and indicative of a nutrient-rich oceanic influence.

The samples of zone F consist of finely laminated clay. Like the previous zone, they contain a fair number of epipsammic species such as *Catenula adherens* and *Dimeregramma minor*. In addition this zone contains various *Cocconeis* species such as *C. peltoides*, *C. disculus* (Schum.) Cl. and *C. disculoides* Hust. Other diatoms found in this zone included *Thalassionema nitzschioides*, *Stephanopyxis turris*, resting-spores of *Chaetoceros*, peaks of *Rhizosolenia* fragments and the species mentioned by Brockmann (1928, 1940) as "forms characteristic for the Eemian" (Fig. 5).

The laminated clay indicates stratification of the water. *Chaetoceros* spores and *Thalassionema nitzschioides* are indicative of high nutrient levels (Van Iperen *et al.* 1993). The presence of resting spores of *Chaetoceros* indicates the occurrence of sudden shocks such as nutrient deficiency (Pitcher 1986).

The autochthonic environment is represented by the *Cocconeis* species, *Catenula adherens* and *Dimeregramma minor*. *Hyalodiscus scoticus* and *Grammatophora oceanica* were, however, absent. This indicates a shallow sea with low salinity, little or no tidal action and clear water. Apparently, the substratum on which *Hyalodiscus* and *Grammatophora* live had disappeared. *Thalassionema*, *Stephanopyxis* and the *Chaetoceros* spores are allochthonic and indicate a nutrient-rich oceanic influence. At the transition between zone F and zone G some samples were found to contain few diatoms, if any.

In zone G, *Cymatosira belgica* is the dominant species. Together with the *Rhaphoneis* species, this constitutes a flora resembling that of the recent North Sea coast. In contrast to the ampotixene *Cocconeis* species in zone E, *Cymatosira belgica* is indifferent to tides and *Rhaphoneis amphiceros* is pseudo-ampotiphilic (Simonsen 1962). The sediment of this zone was deposited in a tidally influenced coastal area.

Results from pollen analysis are presented in the next section and summarized in Fig. 6.

*Broek in Waterland (BIW)*

The deepest portion of the core contains the *Cyclotella-Stephanodiscus* flora (De Wolf 1994) (zone C, Fig. 4) which probably indicates that in this core infilling began in a eutrophic freshwater lake. Sediments with species from zones A and B are absent. The freshwater lake deposit is followed by an association of two floras also observed in the Valkenweg core: species of zone C and of zone E. The presence of *Grammatophora oceanica* and *Hyalodiscus scoticus* indicates a marine influence, which has resulted in reworking of zone C. The *Fragilaria* phase (zone D) is absent in the deeper portion of the basin. Zone F is represented by species such as *Thalassionema nitzschioides*, *Catenula adherens* and *Cocconeis disculoides*. The *Rhizosolenia sp.* peak appears to be characteristic of the basis of this zone and is compatible with the aspect of zone F in the Valkenweg core.

The dominance of *Cymatosira belgica* in the uppermost sample is compatible with zone G.

Pollen analysis was not carried out on this core.

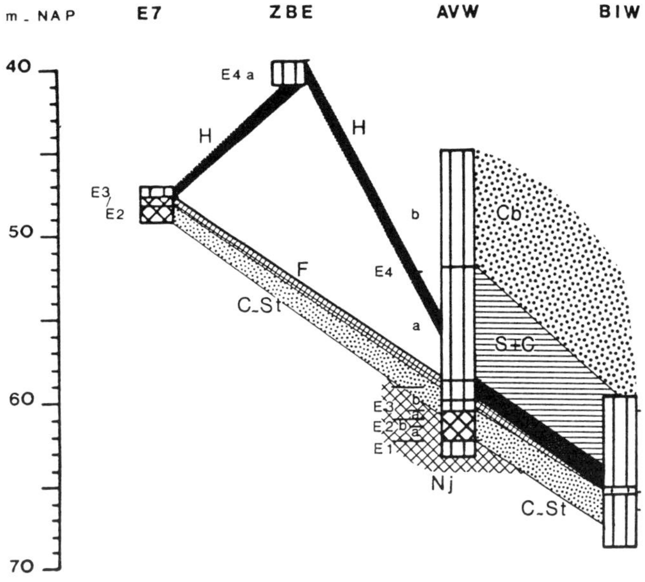


Fig. 4. Profile with diatom zones and pollen zones of the cores and their depth in the basin. Legend lithological column, see Fig. 6. E7 – Amsterdan E7 core; ZBE – Zeeburgereiland core; AVW – Amsterdam Valkenweg core; BIW – Broek in Waterland core. E1 – E4 -polenzones. Cb = *Cymatosira belgica* zone; S+G = *Stephanopyxis–Cocconeis* zone; H = *Hyalodiscus* zone; F = *Fragilaria* zone; C-St = *Cyclotella – Stephanodiscus* zone; Nj = *Navicula jaernefeltii* zone.

| | | | Eemformation | alluv. Nordsee | Litorinasee |
|---|---|---|---|---|---|
| *Hyalodiscus scoticus* . . . . . . . . . . . | M. | 2 | + | | + |
| *Stephanopyxis Turris* . . . . . . . . . . | M. | 1× | + | | + |
| *Trinacria Regina* . . . . . . . . . . . | M. | 1× | + | | |
| *Cocconeis distans* . . . . . . . . . . . | M. | 1× | + | | + |
| „ *quarnerensis* . . . . . . . . | M. | 1× | + | | + |
| *Terpsinoe americana* . . . . . . . . . . | M. | 1× | + | r | + |
| *Biddulphia pulchella* . . . . . . . . | Br. | ? | + | r | + |
| *Synedra crystallina* . . . . . . . . . . | M. | 1× | + | r | + |
| *Endictya oceanica* . . . . . . . . . . . | M. | 2 | + | | |
| *Melosira ornata* . . . . . . . . . . . | M. | 1× | + | | |
| *Biddulphia spinosa* . . . . . . . . . . | M. | 1× | + | | |
| „ *nobilis* var. *pentagona* . . . . . | M. | 1× | + | | |
| „ *Rhombus* v. *trigona* . . . . . . | M. | 1× | | + | |
| *Eupodiscus Argus* . . . . . . . . . . . | M. | 1× | | + | |
| *Melosira Westii* . . . . . . . . . . . | M. | 1× | + | + | |
| *Hyalodiscus stelliger* . . . . . . . . . | M. | 1× | + | + | |
| *Actinoptychus splendens* . . . . . . . . | M. | 1× | + | + | |
| *Biddulphia Rhombus* . . . . . . . . . . | M. | 1 | + | + | |
| „ *Favus* . . . . . . . . . . . | M. | 1 | + | + | |

Fig. 5. Diatoms characteristic for the Eemian sediment, the holocene North Sea sediments and the Litorina Sea sediments (after Brockmann 1928).

*Amsterdam E7 (E7)*

The lowermost sample of this section, consisting of 5 sub-samples (De Wolf 1994d), can be assigned to zone C because of the presence of *Cyclotella kuetzingiana* and *Stephanodiscus minutula*. Zones A and B are absent in this core (Fig. 4). Two superimposed samples in which *Fragilaria spp.* are dominant (e.g. *F. construens* (Ehr.) Grun. including the var. *venter* (Ehr.) Grun. and var. *subsalina* Hust., and *F. brevistriata* (Grun.) are assigned to zone D. Other species of interest in this zone are *Amphora pediculus* (Kütz.) Grun. and *Opephora martyi* Héribaud. In the uppermost two samples, the presence of *Hyalodiscus scoticus* and many *Chaetoceros* and *Rhizosolenia spp.* indicates reworking when the findings are compared with those for the Valkenweg (zones E and F).

Pollen analysis indicate that this sediment belongs to pollen zone E2/E3.

*Zeeburgereiland (ZBE)*

This core yielded a thin and laminated marine Eemian section (De Wolf 1994a) The samples that contained *Hyalodiscus scoticus* and *Grammatophora oceanica* indicate sedimentation of zone E. The other zones are absent (Fig. 4). Pollen analysis has also been carried out for this core (Cleveringa 1994). The findings clearly indicate that this sediment belongs to pollen zone E4b (see Fig. 6)

## Vegetational history

The Eemian interglacial vegetational history of The Netherlands is generally known in its outline (Zagwijn 1975). All Eemian pollen sequences of the marine deposits represent long-distance pollen from the mainland, whereas the pollen in the gyttjas of the terrestrial deposits of channel fills as well as the first infillings of the tongue basins are of a more local origin.

The latter are mainly restricted to the earlier part of the Eemian. The vegetation succession started with a *Betula*-dominated phase (zone E1), when the forests were already of the closed type. In the next phase *Pinus* spread to the area (zone E2), with a simultaneous increase in the proportion of *Quercus*, *Fraxinus* and *Ulmus* in the forest vegetation (zone E3). *Corylus* also arrived in the area. *Fraxinus* and *Ulmus* probably grew on river banks as solitary trees. *Tilia* populated the more loamy parts of the river landscape. *Salix* was also very common in this environment, some areas of which were occasionally wet.

After the *Quercus* maximum, *Corylus* reached its optimum (zone E4a). Just before, the pollen diagrams indicate an increase of the proportion of *Alnus*. After the *Corylus* maximum, *Taxus* and *Alnus* became common trees in the vegetation (zone E4b).

During the *Alnus-Corylus-Taxus* phase, migration of *Carpinus* and *Picea* (zone E5) started. *Corylus*, *Fraxinus*, *Tilia* and *Quercus* – trees of the temperate deciduous

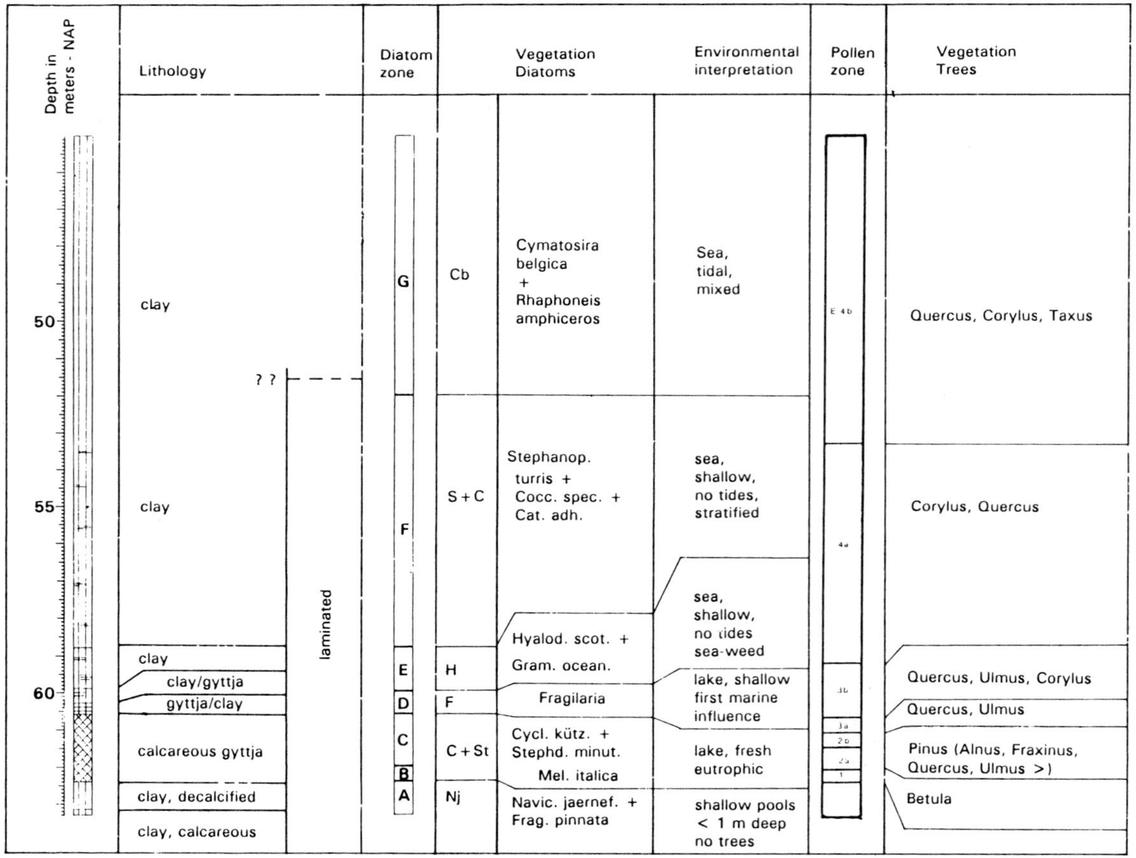


Fig. 6. Summary of the results of the Valkenweg core. Pollen-zones after Zagwijn (1961, 1975, 1993).

forest – declined. Therefore, *Carpinus* and *Picea* became common species in the forests of the second part of the Eemian. The latter possibly also grew on *Sphagnum* bogs. In the final phase of the Eemian, *Picea* was still prevalent, together with *Alnus*, *Betula* and *Pinus* (zone E6a, b). At present, forests of the *Pinus-Betula* type are characteristic of the boreal vegetation zone in which *Picea* is an admixed species and *Alnus* grows in favourable damp places.

Although a comparison of pollen diagrams of the Eemian revealed marked differences between pollen values of trees and herbs, and also in the thickness of the sediments of the individual zones, the conclusion is nevertheless warranted that basically they all indicate the same type of forest development. The differences are probably caused by local factors. The very clear succession of forest phases during the Eemian interglacial is uniform over large areas in northern and continental Europe (Menke & Tynni 1984; Behre 1989; Mangerud 1991; Zagwijn 1993).

The developmental history of the forests and the occurrence and distribution of some plants are good indicators of the paleoclimate. The presence of *Hedera* and *Viscum* pollen in the first part of the Eemian pollen stratigraphy indicate high summer temperatures (at least 15.5°C according to Iversen 1944) together with mild winters (Zagwijn 1993). Under the influence of a more oceanic climate during the second half of the Eemian, mull soils gradually changed into acid litter and competitiveness of the temperate deciduous trees was diminished (Andersen 1966). During this phase of soil development, *Ilex* became an important factor in the flora. A striking phenomenon is that the occurrence of *Ilex* pollen in the pollen association is closely related to diminishing percentages for *Taxus*. Clearings in the *Taxus-Corylus* vegetation favoured the growth and pollination of *Ilex*.

The *Betula-Pinus-Picea* pollen assemblage may reflect the cooling of the climate at the end of the Eemian.

In this paper, the stratigraphic position of the cores examined and the diatom floras observed are linked with the pollen zonation described (above). This zonation is valid for a large portion of Europe. Pollen stratigraphy thus makes correlation over long distances possible. The choice of major changes in vegetation as time boundaries does not impede a relative placement in time of the changes in the diatom flora. Fig. 6 shows this schematic zonation according to major changes.

## Discussion

A comparison with the results of diatom and pollen analysis shows that the pollen findings differ for the trajectories of zone A and zone B. Unpublished pollen data (Zagwijn 1993) for the clay underneath the gyttja show the presence of trees such as *Betula*, *Pinus*, *Salix* and *Juniperus*, while for the herbs *Artemisia* reaches values of up to 10%. In contrast, the gyttja itself has high percentages of *Betula* and *Pinus*, whereas *Artemisia* is practically absent. The pollen analysis suggests that the clay was probably deposited in a tundra-like situation. There were occasional patches of willow, birch and juniper dwarf shrub and in frost-disturbed places there was possibly a local abundance of *Artemisia* plants. The vegetation changed completely during the deposition of the gyttja (zone B). Birch and pine were the dominant trees, it became a more or less closed forest.

The available data on *Navicula jaernefeldtii*, and also on *Fragilaria pinnata*, tends to indicate an ideal habitat for both species in areas north of the tree-line. The data from pollen analysis of the lower portion of the infilling of the basin do not contradict this conclusion. The abrupt transition and north of tree-line – closed forest, means that there is a stratigraphic hiatus between the clay and the gyttja. Zagwijn (1983) already suspected the existence of such a hiatus, because he emphasized that the top layer of the lacustrine glacial clay is decalcified. It should be noted that in this basin a presumed gradual passing of the tree-line has not been demonstrated in the cores examined. Further study will be necessary to determine whether this is a local phenomenon or whether it occurs in several or even all basins.

The *Fragilaria* association (zone B) is, like the diatom association of zone A, not present throughout the lake. Because of the striking similarity to the Holocene, it seems possible that the fossilization of the *Fragilaria* flora is related to stress in the system. For the Holocene, fossilization of *Fragilaria* took place in shallow marginal zones of the lagoon (Vos & De Wolf 1994). In the situation where lagoon-type deposits are formed, seepage water from the higher Pleistocene subsoil and stagnant fresh water supplied by the rivers are both present. With increasing marine influence, *Fragilaria* spp. are seen to flourish and to be covered by sediment.

A similar situation occurred in the Eemian, although the circumstances were different. In the Holocene the sea transgressed over a relatively flat coast, whereas during the Eemian the sea penetrated into a "fjord-like landscape", the result of the Saale glaciation. The basins, some of which probably had a threshold, were initially filled with fresh water. On first contact with the sea, a stressed environment developed locally. In all probability, the flourishing *Fragilaria* populations were also fossilized here.

From the changes in the species composition of zones A, B and C, a change in the water depth can also be deduced. Exact changes are difficult to demonstrate in the case of the Amsterdam Basin because infilling of the basin was not continuous (hiatus). The problem of discontinuity should be solved by a study of additional basins whose morphology is accurately known.

Diatom layer E was present in all borings. This means that the highest level in the infilling of the basin (Zeeburgereiland) indicates that a threshold, if at all present, must have lain at ca 40 metres below Dutch Ordnance Level (NAP) or even lower. The concurrence of "oceanic species" and *Hyalodiscus scoticus* indicates that the basin was within reach of the sea. Little sediment reached the system, however, because *Hyalodiscus scoticus* grows in clear water on higher algae (such as *Antithamnion*, Edsbagge 1966).

The presence of laminated sediments indicates slight tidal action. Temperature stratification is the possible reason why lamination develops and is fossilized. When the sea-water reaches the basin, the way in which it happens leads to mixture or separation of fresh and marine waters and possibly further contributes to the development and preservation of rhythmic lamination. The nature of sampling does not permit additional conclusions. It should be noted, however, that the changing counts for *Chaetoceros* and *Rhizosolenia* species in the higher layers of the infilling indicate changes in the supply of fresh water. Possibly there was large-scale development of algal mats (Sancetta 1994).

The decrease of *Hyalodiscus scoticus* in zones E and F, in favour of *Cocconeis* spp. and *Catenula adherens* is provisionally ascribed to an increasing water depth leading to disappearance of host species (sea-weed and higher algae) on which *Hyalodiscus scoticus* lives. The spacing of samples does not allow us to ascertain whether continuing rise in the sea-level (Zagwijn 1983) was of decisive significance in this respect.

The section of samples poor in, or devoid of, diatoms at the transition of zones F and G probably reflects the transition from a practically non-tidal inland sea to a fully marine situation. A tidal sea was the result, leading to increased mixing and exchange. The "intermediate zone" practically devoid of diatoms, then constitutes a phase of transition. Afterwards, *Cymatosira belgica* and *Rhaphoneis amphiceros* characterise the diatom flora.

The massive occurrence of *Cymatosira belgica* in marine sediments is also known from the Holocene (Vos & De Wolf 1994). A morphological situation of an open coast prior to development of barriers and dunes led to a phase during the Holocene in which *Cymatosira belgica* fossilized in massive numbers. From sedimentological data it becomes apparent that in this phase sediment was transported from the sea inland via large East-West channels (Beets *et al.* 1992). Sediment transport also occurred in the case of the Eemian, with the notable difference that the channels were already present. namely the old courses of the Vecht and Rhine (Zagwijn 1975a, b).

An important result of the investigation was that the sediments that contain *Cymatosira belgica* and *Rhaphoneis amphiceros,* only occur in the deepest portions of the basin. All the evidence indicates that the basin reached its present shape during their deposition. Possible mixing of floras characterizing the diatom zones can be attributed to the way in which infilling of a tidally affected basin occurred. Undercurrents, dominant wind direction, height of the water column, as well as other factors, largely determine the distribution and thus the displacement of sediment in lakes and basins.

## Conclusions

Infilling of the basin began when the climate was cold. The area investigated occurred to the North of the tree-line. The water was oligo- to mesotrophic, and the basin was shallow. Later the water became richer in nutrients and deeper and can be called a lake (zones B, C).

Between the phase of a shallow basin and a lake a hiatus is present in the record (decalcification). During this period the tree-line passed and closed forests with pine and birch developed.

The first contacts with the sea resulted in a *Fragilaria* flora. The samples that contain *Fragilaria* also contain the first marine species, although only sporadically. From the profile it becomes apparent that the *Fragilaria* level is especially present in the littoral zone of the lake – in the deepest portion this species is rare or absent.

In the truly marine portion, three clearly distinct phases can be distinguished.

Firstly, a phase with a flora indicative of shallow and clear marine water (zone E). *Hyalodiscus* and *Grammatophora* species are epiphytic on higher algae and occurred when the depth of the water was between about 3 and 15 metres. A depth of 25 metres, as given by Edsbagge (1966) for *Hyalodiscus* and *Antithamnion plumula*, is probably an exceptional case. These species occur at a total salt-content of 8-30 ‰. The environment was characterized by little or no current and an undisturbed sedimentation. This flora, only encountered in zone E, compares most closely with the contemporary Baltic flora and it is not limited to the Eemian. This flora has also been found in the Eemian of North Germany (Menke & Tynni 1984; Heck & Brockmann 1950; Brockmann 1932; Von der Brelie 1959), Poland (Cheremisinova 1965), Estonia (Liivrand 1987) and Finland (Grönlund 1991).

Secondly, a phase containing the "Eemian flora" which was already known from the Netherlands (Brockmann 1928). No recent equivalent of this flora is known. This flora consists of species regarded as "warm" and/or "southerly" and species indicative of nutrient-rich oceanic waters. The latter elements of the flora were already scarcely present from the first marine contact onwards.

Thirdly, a phase with a flora closely resembling the marine-littoral flora known from the Dutch coastal areas. In the Holocene deposits it is even better represented than in the recent situation.

In the case of gradual sedimentation the pollen and diatom zones do not coincide and it even becomes apparent that environmental changes are reflected in the diatom association at an earlier moment than in the pollen spectrum.

In the presence of hiatuses, the boundaries of the pollen and diatom zones coincide.

The presence or absence of a threshold in a glacial basin results in a specific environment. Fossilization is closely linked to this point. Obviously this defines why the fresh and marine floras are so well preserved. We may presume, that in the other basins the transitions we missed in the cores of this basin are well developed.

In the Amsterdam basin the stratigraphic record is incomplete. The portion lacking may be present elsewhere. For an accurate and complete survey, additional basins must be examined (see Fig. 1).

The diatom layer observed by Harting (1852), subsequently called "Laag van Harting", contains freshwater and marine species. Results of the present study showed that Harting's layer consists of a combination of zones D and E. In view of the local occurrence of the diatom-rich sediment, special conditions prevailing at certain moments within the basin led to a massive development, death and fossilization of diatoms.

## Acknowledgements

We are indebted to the Director of the Geological Survey of the Netherlands for permission to publish the results from this study. We appreciate the constructive suggestions by Adri Burger, Tuulikki Grönlund, Tom Meijer and Dirk Beets.

## References

Andersen, S. T. (1966). Interglacial vegetational succession and lake development in Denmark. *The Paleobotanist*, **15**, 117-127.

Beets, D. J., Van der Valk, L., Stive, M. J. F. (1992). Holocene evolution of the coast of Holland. *Marine Geology*, **103**, 423-443.

Behre, K. E. (1989). Biostratigraphy of the last glacial period in Europe. *Quaternary Science Reviews*, **8**, 25-44.

Brockmann, C. (1928). Die Diatomeen im marinen Quartär Holands. *Abhandlungen der Senckenbergischen Naturforschenden Gesellschaft*, **41**, 3-73.

Brockmann, C. (1932). Die Diatomeen aus dem Interglazial von Oldenbüttel. In: Die Eem- und ihre begleitenden Junginterglazial-Ablagerungen bei Oldenbüttel in Holstein (H. L. Heck, ed.). *Abhandlungen der Preußischen Geologischen Landesanstalt Neue Folge, Heft*, **140**, 45-50.

Brockmann, C. (1940). Diatomeen als Leitfossilien in Küstenablagerungen. *Westküste*, **2**, H. 2/3, 150-181.

Cheremisinova, Ye. A. (1965). Age of the marine interglacial (Joldian) clays of Gda_sk bay (Poland). *Doklady Akad. Nauk SSSR*, Vol. **155**, 64-66.

Cleve-Euler, A. (1951). Die Diatomeen von Schweden und Finnland. *Kungliga Svenska Vetenskapsakademiens Handlingar. Fjärde Serien*, **2**,1, 1-161.

Cleve-Euler, A. (1952). Die Diatomeen von Schweden und Finnland V. *Kungliga Svenska Vetenskapsakademiens Handlingar. Fjärde Serien*, **3**,3, 1-153.

Cleve-Euler, A. (1953a). Die Diatomeen von Schweden und Finnland II. *Kungliga Svenska Vetenskapsakademiens Handlingar. Fjärde Serien*, **4**,1, 1-158.

Cleve-Euler, A. (1953a). Die Diatomeen von Schweden und Finnland III. *Kungliga Svenska Vetenskapsakademiens Handlingar. Fjärde Serien*, **4**,5, 1-254.

Cleve-Euler, A. (1955). Die Diatomeen von Schweden und Finnland IV. *Kungliga Svenska Vetenskapsakademiens Handlingar. Fjärde Serien*, **5**,4, 1-230.

Cleveringa, P. (1994). Pollenanalytisch onderzoek van de boring Zeeburgereiland 25G922. *Internal Report No. 1169, Department Paleobotany Kaenozoicum, Geological Survey of the Netherlands.*

de Gans, W., De Groot, Th., Zwaan, H. (1986). The Amsterdam basin, a case study of a glacial basin in The Netherlands. In: *INQUA Symp. on the Genesis and Lithology of glacial deposits* (J. J. M. van der Meer, ed.), 205-216. A. A. Balkema, Rotterdam.

De Wolf, H. (1993). History of diatom research in The Netherlands and Flanders. In: *Proceedings of the Twelfth International Diatom Symposium* (H. van Dam, ed.), 1-9 Kluwer Academic Publishers, Dordrecht.

De Wolf, H. (1994a). Diatomeeenonderzoek van de boring Zeeburgereiland 25G922. *Internal Report No. 572, Diatom Department, Geological Survey of the Netherlands.*

De Wolf, H. (1994b). Diatomeeenonderzoek van de boring Amsterdam Valkenweg 25E255. *Intern Rapport No. 574, Diatom Department, Geological Survey of the Netherlands.*

De Wolf, H. (1994c). Diatomeeenonderzoek van de boring Broek in Waterland 25E172. *Intern Rapport No. 580, Diatom Department, Geological Survey of the Netherlands.*

De Wolf, H. (1994d). Diatomeeenonderzoek van de boring E7-119 te Amsterdam, 25G495. *Intern Rapport No. 581, Diatom Department, Geological Survey of the Netherlands.*

Edsbagge, H. (1966). Zur Ökologie der marinen angehefteten Diatomeen. *Botanica Gothoburgensia,* **VI**, 136 pp.

Grönlund, T. (1988). The diatom flora of the Eemian deposit at Haapavesi, western Finland. *Geological Survey of Finland, Report of Investigation,* **79**, 14 pp.

Grönlund, T. (1991a). New corings fron Eemian interglacial marine deposits in Ostrobothnia, Finland. *Geological Survey of Finland, Bulletin,* **352**, 23 pp.

Grönlund, T. (1991b). The diatom stratigraphy of the Eemian Baltic Sea on the basis of sediment discoveries in Ostrobothnia, Finland. Dissertation. *Geological Survey of Finland, Report of Investigations,* **102**, 26 pp.

Harting, P. (1852). De bodem onder Amsterdam onderzocht en beschreven. *Verhandelingen der eerste klasse van het Koninklijk. Nederlandsche Instituut van Wetenschappen, Letterkund en Schoone Kunsten.* 3e reeks, **5**, 73-232.

Harting, P. (1874). De bodem van het Eemdal. *Verslagen en Mededelingen van de Koninklijke Nederlandse Akademie van Wetenschappen, afdeling Natuurkunde II,* **8**, 292-290.

Heck, H. L. & Brockmann, C. (1950). Eem Ablagerungen bei Lübeck. *Schriften des Naturwissenschaftlichen Vereins für Schleswig-Holstein,* **24**, 80-86.

Iversen, J. (1944). Viscum, Hedera and Ilex as climate indicators. *Geologiska Föreningens i Stockholm Förhandlingar, Band* **66**, 463-483.

Jessen, K. & Milthers, V. (1928). Stratigraphical and paleontological studies of interglacial freshwater deposits in Jutland and Northwest Germany. *Danmarks Geologiske Undersøgelse,* II-**48**, 1-139.

Liivrand, E. (1987). Regional type section of the Eemian marine deposits on Suur-Prangli. *Proceedings of the Academy of sciences of the Estonian SSR. Geology*, 36, 20-26.

Lorié, J. (1906). De geologische bouw der Gelderse Vallei. *Verhandelingen der Koninklijke Akademie van Wetenschappen te Amsterdam. 2e sectie,* **XIII**, 1, 1-100

Lorié, J. (1916). De geologische bouw van de Geldersche Vallei II. *Verhandelingen der Koninklijke Akademie van Wetenschappen te Amsterdam. 2e sectie,* **XIX**, 1, 1-23

Mangerud, J. (1991). The last Interglacial/Glacial cycle in Northern Europe. In*: Quaternary Landscapes* (Linda C. K. Shorne & Edward I Cushing, eds), 38-75. Belhaven Press, London

Menke, B. & Tynni, R. (1984). Das Eeminterglazial un das Weichselfrühglazial von Rederstall/Dithmarschen und ihre Bedeutung für die mitteleuropäische Jungpleistozän-Gliederung. *Geologisches Jahrbuch,* **A76**, 1-109.

Pitcher, G. C. (1986). Sedimentary flux and the formation of resting spores of selected *Chaetoceros* species at two sites in the southern Benguela system. *South African Journal of Marine Science,* **4**, 231-244.

Sancetta, C. (1994). Mediterranian sapropels: Seasonal stratification yields high production and carbon flux. *Paleoceanography,* **9**, 195-196.

Simonsen, R. (1962). Untersuchungen zur Systematik und Ökologie der Bodendiatomeen der westliche Ostsee. Internationale Revue der gesamten Hydrobiologie. *Systematische Beiheft* **1**. Akademie-Verlag, Berlin, 144 pp.
Stoermer, E. F.(1980). Characteristics of benthic algal communities in the Upper Great Lakes. *Environmental Research Laboratory, Duluth Office of Research and Development U.S. Environmental Protection Agency Duluth, Minnesota.*
Van den Berg, M. W. & Beets, D. J. (1986), Saalian glacial deposits and morphology in The Netherlands, In: *INQUA Symposium on the Genesis and Lithology of glacial deposits* (J. J. M. van der Meer, ed.), 235-251. A. A. Balkema, Rotterdam.
Van Iperen, J. M., & Van Bennekom, A. J., Van Weering, T. C. E. (1993). Diatoms in surface sediments of the Indonesian Archipelago and their relation to hydrography. In: *Proceedings of the Twelfth International Diatom Symposium* (H. van Dam, ed.), 113-129 Kluwer Academic Publishers, Dordrecht
Von der Brelie, G. (1959). Watt-Ablagerungen des Eem-Meeres im Raum von Norderney. *Zeitschrift der Deutschen Geologische Gesellschaft*, **111/1**, 1-7
Vos, P. C. & De Wolf, H. (1994). Palaeoenvironmental research on distoms in early and middle Holocene deposits in central North Holland (The Netherlands). *Netherlands Journal of Aquatic Ecology*, **28**, 97-117.
Zagwijn, W. H. (1963). Vegetation, climate and radiocarbon datings in the Late Pleistocene of the Netherlands. *Mededelingen Geologische Stichting Nieuwe, Serie* No. **14**, 15-45.
Zagwijn, W. H. (1975a). De paleogeografische ontwikkeling van Nederland in de laatste drie miljoen jaar. *Koninklijk Nederlands Aardrijkskundig Genootschap Geografisch Tijdschrift*, **IX**, 3, 181-201.
Zagwijn, W. H. (1975b). Chronostratigrafie en biostratigrafie. In: *Toelichting bij geologische overzichtskaarten van Nederland* (W. H. Zagwijn & C. J. van Staalduinen, eds), 109-114, Geological Survey of the Netherlands Haarlem.
Zagwijn, W. H. (1983). Sea-level changes in the Netherlands during the Eemian. *Geologie en Mijnbouw*, **62**, 437-450.
Zagwijn, W. H. (1993). Publikatie over het klimaat van het Eemien. In: *An analysis of Eemian climate in western and central Europe* (C. Turner, ed.), Cambridge University Press, in press

# Evolution of freshwater centric diatoms within the Euroasian continent

Galina Khursevich

*Institute of Geological Sciences, The Academy of Sciences of Belarus, Zhodinskaya street 7, Minsk 220141, Republic of Belarus*

## Abstract

Critical analysis of literary evidence and the results of personal investigations suggest four successive stages in the development of freshwater centric diatoms during the Cenozoic both in Asia and Europe. Evolution of non-marine centric diatoms during the Cenozoic is associated with the processes of extinction and neogeneration due to global geological and palaeoclimatic changes.

## Introduction

During the last twenty years considerable progress has been made in the study of freshwater centric diatoms. The ultrastructural features of many species of *Stephanodiscus*, *Cyclotella*, *Ellerbeckia*, *Aulacoseira* and *Actinocyclus* have been revealed by use of scanning electron microscopy (SEM) and their taxonomic status has been defined more precisely. New freshwater centric genera such as *Cyclostephanos* Round, *Mesodictyon* Theriot & Bradbury, *Concentrodiscus* Khurs., Moiss. & Suchova, *Ectodictyon* Khurs. & Tschern. and *Pliocaenicus* Round & Håkansson within the Order Thalassiosirales, *Alveolophora* Moiss. & Nevretd. and *Pseudoaulacosira* Lupik. et Khurs. within the Order *Aulacosirales*, *Actinostephanos* Khurs., *Undatodiscus* Lupik. and *Lobodiscus* Lupik. & Khurs. within the Order Coscinodiscales have been described from the Upper Cenozoic sediments. The extent of certain families, and even orders, have been reconsidered and new systems of centric diatom taxonomy reflecting the modern level of knowledge have been proposed (Round *et al.* 1990; Makarova [ed.] *et al.* 1992).

The problem dealing with the delineation of *Stephanodiscus*, *Cyclostephanos* and *Cyclotella* – the three most taxonomically diverse freshwater genera of the family Stephanodiscaceae has been discussed in the articles of Round (1970), Theriot *et al.* (1987), Khursevich (1989b), Loginova (1990), Serieyssol & Gasse (1990) and Round & Håkansson (1992). Some questions concerning the evolution and phylogeny of non-marine centric diatoms have been considered by Round & Sims (1981), Van Landingham (1985), Krebs *et al.* (1987), Fourtanier & Gasse (1988), Krebs (1990), Temniskova-Topalova *et al.* (1990), Khursevich (1991a, b, c) and Serieyssol & Gasse (1991). The first attempted zonal subdivision of the Eastern Russia continental Neogene deposits based on diatoms has been undertaken by Moissejeva (1993).

The purpose of this report is to present the evolutionary patterns of freshwater centric diatoms during the Cenozoic both in Asia and Europe.

## Materials and Methods

Numerous facts available in the literature about the freshwater fossil diatom flora from different regions within the Euroasian continent have been generalized. Moreover, results of personal electron-microscopic investigations of various centric diatoms from the Upper Cenozoic lacustrine deposits of Armenia, Belarus, Latvia, Lithuania, the Ukraine, the central areas of Russia, Siberia, Kamchatka, the Primorye region of the Far East, as well as of the Lake Baikal and the Black Sea, have been used. In most cases the type material of species has been examined. For SEM, specimens were sputter coated with gold and observed in a JEOL–35C.

The chronology of diatom-bearing sediments in the regions studied is based on palaeomagnetic evidence or absolute dating (K–Ar, radiometric, track methods) but more often on the stratigraphy and palaeontological record (pollen, spore, leaf and seed flora, molluscs or mammals) known from the literature. My reasons concerning freshwater centric diatom evolution are sometimes of a speculative character.

## Results

Freshwater diatoms of the classs *Centrophyceae* are known from the end of the Late Eocene – Early Oligocene within Asia (Lupikina & Dolmatova 1985; Moissejeva & Nevretdinova 1990). Primarily, they were represented by ancient species of *Aulacoseira, Alveolophora, Gleseria* and *Ellerbeckia.* This diatom community was found in palaeobasins on the northern shore of the Penzhinskaya Bay and the western shore of Kamchatka. The first representatives of *Actinocyclus* and *Melosira* appeared in freshwater deposits of West Siberia at the end of the Early Oligocene – Late Oligocene, and developed together with the *Aulacoseira* and *Alveolophora* species (Rubina & Khursevich 1990).

The systematic composition of lacustrine centric diatoms expanded in the Early Miocene when *Pseudoaulacosira* was characteristic of the Transbaikal area and the genus *Undatodiscus* of West Siberia (Lupikina & Khursevich 1991a; Lupikina *et al.* 1991). Moreover, the area occupied by already existing genera (especially *Alveolophora* and *Actinocyclus*) was considerably expanded at the expense of new species development in the Near East, Transbaikal area and Primorye (Servant-Vildary *et al.* 1988; Khursevich 1990; Moissejeva & Nevretdinova 1990). Only the genus *Gleseria* Lupik. & Dolm. became extinct by the Early Miocene.

The Class Centrophyceae expanded during the Middle Miocene. First of all, representatives of five genera of the Order Thalassiosirales emerged in palaeobasins of Asia. Ancient freshwater species of *Thalassiosira*, *Stephanodiscus* and *Cyclostephanos* were recognized in Middle Miocene sediments of the Vitim plateau, *Concentrodiscus* in the Upper Sulbanian depression, *Cyclotella* in the Tunkin depression (Khursevich 1989b, 1990; Khursevich *et al.* 1989; Nikiteeva 1993). The family Hemidiscaceae of the Order Coscinodiscales was supplemented by a new monotypic genus *Lobodiscus* which becomes highly developed in lakes of the Tunkin depression (Lupikina & Khursevich 1991b).

Maximum generic diversity is fixed in the first half of the Late Miocene (14 genera). At this time one more genus (*Orthoseira* Crawford) appeared in the ancient basins of Asia. A larger extinction of genera took place at the end of the Late Miocene, when the genera *Alveolophora*, *Pseudoaulacosira*, *Undatodiscus*, *Lobodiscus* and *Concentrodiscus* became extinct. The freshwater species of *Actinocyclus* ended their development during the Early Pliocene. A new genus *Ectodictyon*, evolved in the Pliocene lakes of the Chara basin (Khursevich & Tschernjaeva 1989). The Pliocene and Pleistocene were characterized by spatial adaptive radiation of the genera *Stephanodiscus* Ehr. and *Cyclotella* (Kütz.) Bréb. in palaeobasins of the Transcaucasus, the Baikal rift zone, the Central Kamchatka depression, Japan, etc. (Mori 1974; Lupikina 1980; Aleshinskaya & Pirumova 1981; Ozornina 1987; Loginova & Khursevich 1989; Popova *et al.* 1989). Some species of these genera became extinct in the Early Pleistocene or later. However, many representatives of *Stephanodiscus* and *Cyclotella* continue to be dominant, together with *Aulacoseira* and *Cyclostephanos* species, in the phytoplankton of Holocene and modern lakes within Asia.

In general, four major successive stages are revealed in the evolution of freshwater centric diatoms during the Cenozoic in Asia: 1) the Late Eocene – Oligocene stage, with dominant ancient species of *Aulacoseira*, *Alveolophora*, *Gleseria* and *Ellerbeckia* and appearance of the first representatives of *Actinocyclus* and *Melosira* in the second half of this time interval; 2) the Early Miocene stage, with intensive development of not only *Aulacoseira*, *Alveolophora* and *Ellerbeckia*, but also *Pseudoaulacosira*, *Melosira* and *Actinocyclus* as well as a local extension of *Undatodiscus*; 3) the Middle – Late Miocene stage, which is characterized by the highest generic diversity (up to 14) of non-marine centric diatoms and an abundance of both previously existing widespread genera and the local occurrence of *Concentrodiscus*, *Lobodiscus* and *Undatodiscus*, which were extinct by the Miocene – Pliocene boundary; by the first appearance of ancient species of *Thalassiosira*, *Stephanodiscus*, *Cyclostephanos* and *Cyclotella*; 4) the Pliocene – Holocene stage with adaptive radiation and a high frequency of *Stephanodiscus* and *Cyclotella* and considerable importance of *Cyclostephanos* and *Aulacoseira* and the local development of the genus *Ectodictyon* in the Pliocene.

Freshwater centric diatoms in the Cenozoic basins of Europe are known from the Early Miocene in the territory of Czech Republic (Reháková 1965) and the Ukraine (Olshtynskaya 1993). *Aulacoseira*, *Melosira*, *Ellerbeckia* and *Actinocyclus* species developed in that time. The most essential change in the systematic composition of centric diatoms took place during the Middle Miocene, evidenced in the appearance of six new genera in the European lakes. Representatives of *Actinostephanos* inhabited the palaeobasins of Austria, Hungary and Russia (the north-western part of Donbas, Khursevich 1989a). The diatom assemblage, including species of *Alveolophora, Thalassiosira* and *Stephanodiscus*, was characteristic of Middle Miocene lakes of Belarus (Yakubovskaya *et al.* 1989). Ancient species of *Cyclotella, Mesodictyon* and *Stephanodiscus* emerged at the Middle – Late Miocene boundary in palaeobasins of France (Serieyssol & Gasse 1991).

Intense speciation and active development of non-marine species of *Actinocyclus* proceeded in Europe during the Middle and Late Miocene in the territory of Bulgaria, Czech Republic, Germany, Belarus and Russia. In the Late Miocene two more genera (the genus *Orthoseira* in the Czech Republic, and the genus of *Pseudoaulocosira* in Belarus) appeared in the European lakes. The ancient species of *Stephanodiscus, Cyclostephanos* and *Cyclotella* evolved in the waters of the West and Eastern Paratethys (the western part). Moreover, the area occupied by *Cyclotella* species expanded to Greece and Spain, *Mesodictyon* species to Belarus and *Actinostephanos* taxa to the Czech Republic. However, the latter genus became extinct by the end of the Late Miocene.

The extinction of *Pseudoaulacosira, Mesodictyon* and *Actinocyclus* freshwater representatives took place during the Early Pliocene. A new genus, *Pliocaenicus,* developed in the Pliocene palaeobasins of Germany (Round & Håkansson 1992). Also the systematic diversity of *Stephanodiscus, Cyclostephanos* and *Cyclotella* increased considerably in the lacustrine phytoplankton of Europe during the Pliocene. Some endemics of these genera, known from the Pliocene non-marine bottom deposits of the Black Sea (Khursevich 1989b), the Upper Pliocene diatomaceous sediments of the Kama river drainage-basin (Loseva 1982), etc. disappeared at the boundary with the Pleistocene. The abundance and diversification of *Cyclotella* and *Stephanodiscus* species in various regions of Europe falls in the Pleistocene–Holocene time interval. *Aulacoseira* taxa also retained a high frequency.

So, four major stages can be distinguished in the evolution of freshwater centric diatoms during the Cenozoic in Europe: 1) the Early Miocene stage documented by ancient representatives of *Aulacoseira, Melosira, Ellerbeckia* and *Actinocyclus*; 2) the Middle Miocene stage with active development of *Actinocyclus, Actinostephanos, Aulacoseira, Alveolophora* and accompanying species of *Thalassiosira* and *Stephanodiscus*; 3) the Late Miocene stage with abundance of *Aulacoseira, Pseudo-aulacosira, Actinocyclus* and *Actinostephanos* as well as the systematic diversity of *Cyclotella* and *Mesodictyon*; 4) the Pliocene–Holocene stage with dominance of *Stephanodiscus, Cyclotella, Cyclostephanos* and *Aulacosira* and the local expansions of *Pliocaenicus* and *Actinocyclus* in the Pliocene.

Dominant and characteristic species of freshwater centric diatoms from deposits of Cenozoic age of Asia and Europe are illustrated in Figs 1–44.

Figs 1–12. SEM. Figs 1, 2. *Alveolophora areolata* (Moiss.) Moiss., specimens from the Lower Miocene deposits in the Tajgonos peninsula. Fig. 1. Internal valve view showing alveolae on the mantle (left) and external view of complete valve (right). Fig. 2. Internal valve view showing the different structure of the valve face surface and mantle. Figs 3–7. *A. bifaria* Nevretd. & Moiss., specimens from the Upper Eocene – Lower Oligocene sediments on the northern shore of the Penzhinskaya Bay. Fig. 3. External view showing areolae cribra on the valve face surface. Fig. 4. Internal valve view showing pseudoalveolae on the mantle. Fig. 5. Internal view showing different structures of the valve face surface and mantle. Fig. 6. Sibling valves with interlocking spines. Fig. 7. Detail of the internal valve mantle surface with rimoportulae in the base of pseudoalveolae. Figs 8–12. *A. jouseana* (Moiss.) Moiss., specimens from the Middle Miocene deposits in Belarus. Fig. 8. Internal valve mantle surface with two rimoportulae. Figs 9, 10. External views of complete valves. Figs 11, 12. Internal views showing pseudoseptae and pseudoalveolae on the valve mantle. Scale bars = 10 μm in Figs 1, 3–6; 1 μm in Figs 2, 7–12.

Figs 13–22. SEM. Figs 13, 15. *Aulacoseira canadensis* (Hust.) Simon., specimens from the Lower Miocene sediments in the Tajgonos peninsula. Fig. 13. External views of the valve mantle. Fig. 15. Internal valve mantle surface with a rimoportula (arrow). Figs 14, 16. *A. praedistans* f. *curvata* Moiss., specimens from the Upper Miocene deposits of West Kamchatka, sibling valves with interlocking spines Fig. 17. *Ellerbeckia kochii* var. antiqua Khurs., specimen from the Lower – Middle Miocene sediments of the Dzhilinda depression. Structure of tube processes on the valve mantle, internal. Figs 18–21. *Pseudoaulacosira moisseevae* (Lupik.) Lupik. & Khurs., specimens from Upper Miocene deposits in Belarus (Figs 18, 20) and the Upper Miocene sediments of West Kamchatka (Figs 19, 21). Fig. 18. External valve face surface with small "hillocks" in the central part. Fig. 19. Sibling valves with interlocking spines Fig. 20. External view of frustule. Fig. 21. Internal valve surface with areolae cribra and a ring of rimoportulae on the mantle. Fig. 22. *Ellerbeckia arenaria* var. *teres* (Brun) Crawford, specimen from the Lower Miocene deposits of West Siberia. External valve mantle surface with openings of tube processes. Scale bars = 10 μm in Figs 13–16, 21, 22; 1 μm in Figs 17–20.

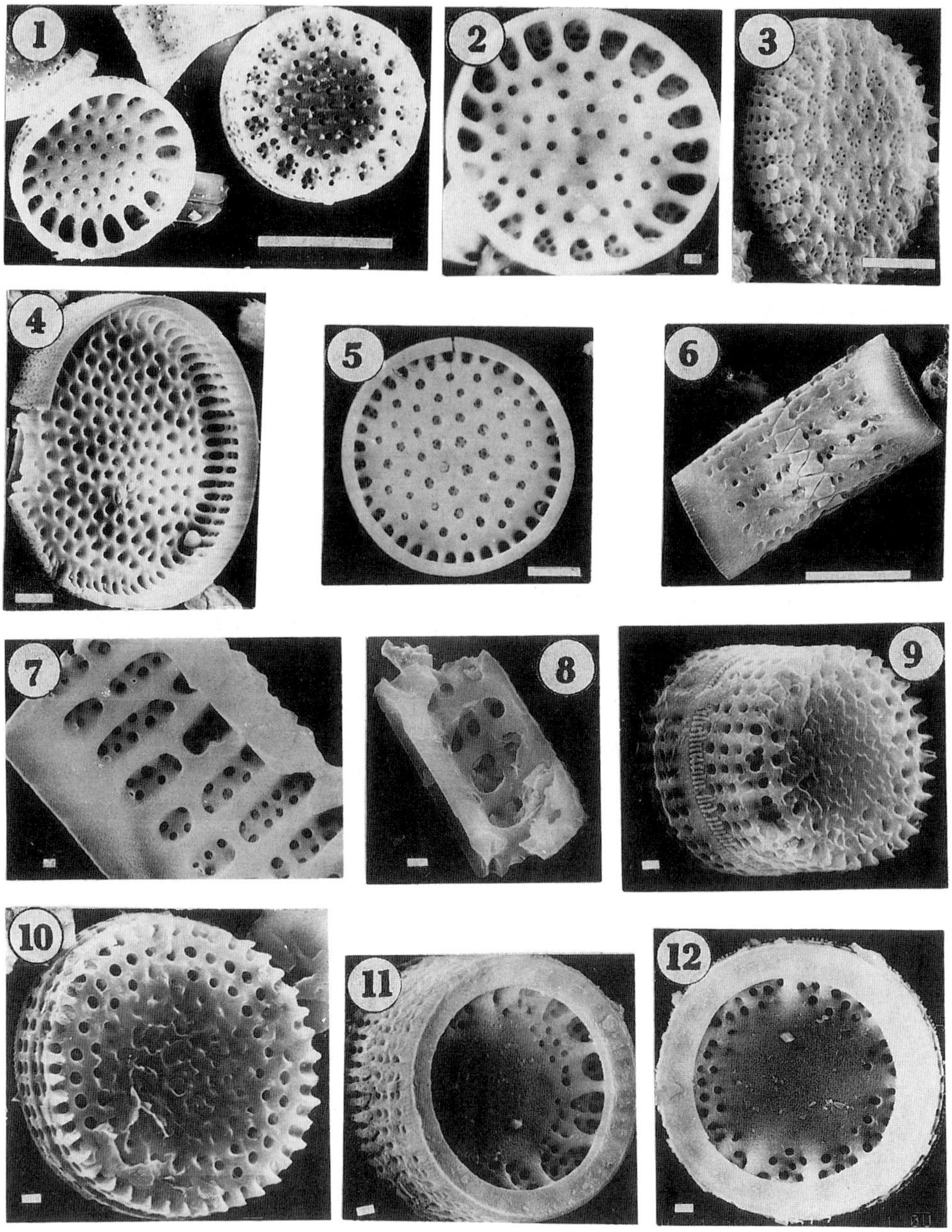
1
2
3
4
5
6
7
8
9
10
11
12

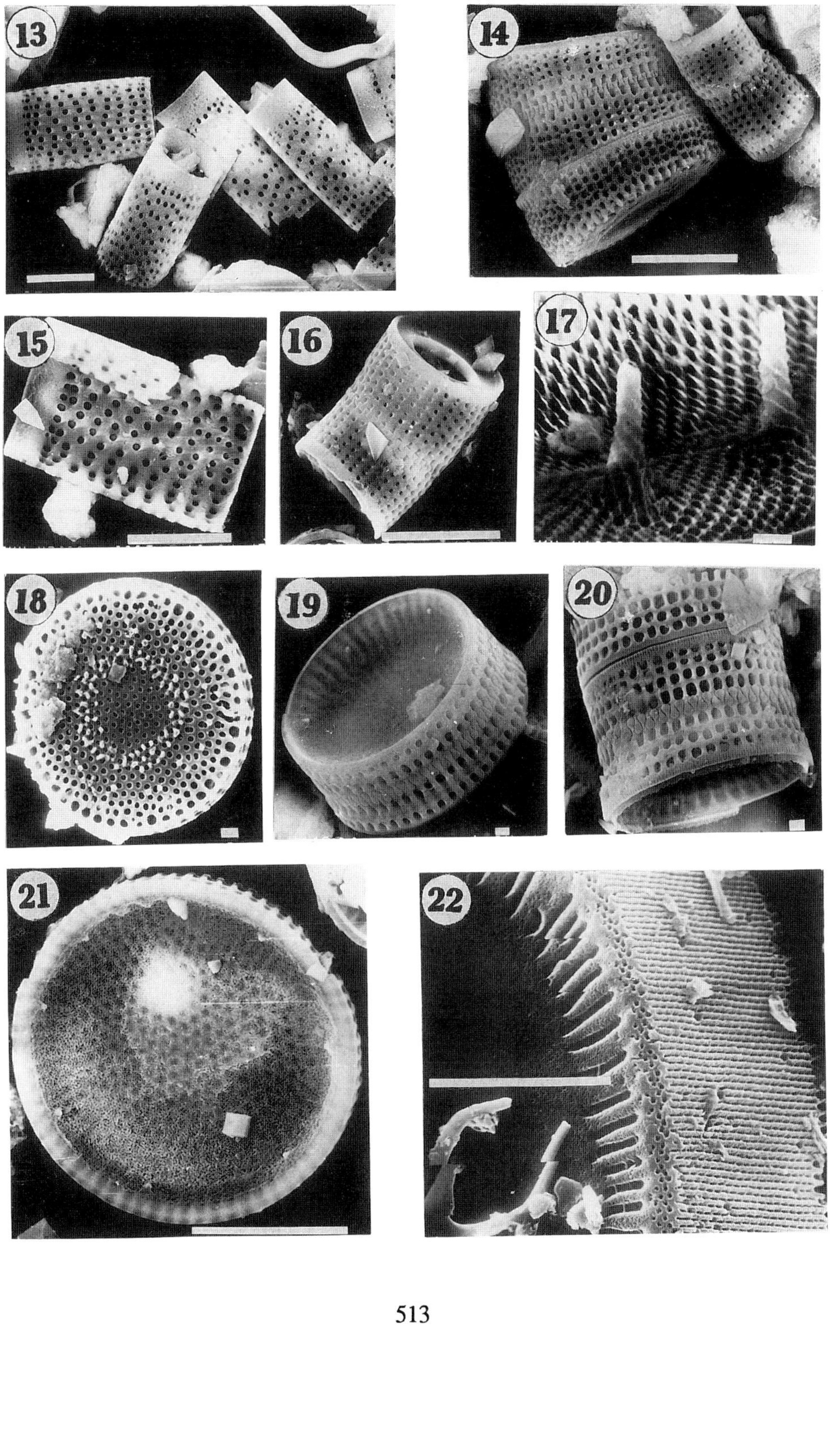
13
14
15
16
17
18
19
20
21
22

Figs 23–32. SEM. Fig. 23. *Actinocyclus trapeziformis* (Rub.) Rub. & Khurs., specimen from the Upper Oligocene sediments of West Siberia. Detail of the external valve surface with a distinct hyaline ring on the boundary of the valve face and mantle with irregularly located solid concise spines. Fig. 24. *A. lobatus* (Rub.) Rub. & Khurs., specimen from the Lower Miocene deposits of West Siberia. External valve surface with a single pseudonodulus (arrow) and lobed spines. Figs 25–27. *Undatodiscus variabilis* (Lupik.) Lupik., specimens from the Upper Miocene sediments of West Kamchatka. Fig. 25. Detail of the external valve surface with three dome-like convexities in the peripherical zone. Fig. 26. Detail of the internal valve surface with rimoportulae located in the concavities. Fig. 27. Internal valve surface with rimoportulae placed on the periphery. Figs 28–30. *Lobodiscus sibericus* (Tscher.) Lupik. & Khurs., specimens from the Middle – Upper Miocene deposits of the Tunkin depression. Fig. 28. Detail of the external valve surface with areolae cribra. Fig. 29. Detail of the valve mantle with the internal part of rimoportula and its external opening on the distal rim of a lobed spine. Fig. 30. Detail of the valve mantle with lobed spines. Figs 31, 32. *Actinocyclus gorbunovii* (Sheshuk.) Sheshuk. & Moiss., specimens from the Upper Miocene sediments of West Siberia. Fig. 31. Detail of the external valve surface with areolae cribra. Fig. 32. Internal valve surface with a single pseudonodulus and the marginal ring of rimoportulae. Scale bars = 10 µm in Figs 23–28, 30–32; 1 µm in Fig. 29.

Figs 33–44. SEM. Figs 33–35. *Concentrodiscus abnormis* Khurs., Moiss. & Suchova, specimens from the Middle–Upper Miocene deposits of the Upper Sulbanian depression. Figs 33, 35. External valve surface with several concentric zones. Fig. 34. Internal valve surface showing areolae cribra, a single rimoportula and the marginal ring of fultoportulae. Figs 36, 37. *Ectodictyon varians* Khurs. & Tschern., specimens from the Pliocene sediments of the Chara depression. Fig. 36. Internal valve surface with a single rimoportula and a marginal ring of fultoportulae. Fig. 37. External valve surface with areolae cribra and openings of fultoportulae on the mantle. Figs 38, 39. *Actinocyclus tubulosus* Khurs., specimens from the Upper Miocene deposits of West Siberia. Fig. 38. External valve surface with areolae cribra and long tubular extensions of rimoportulae. Fig. 39. Internal valve surface with a single pseudonodulus and the marginal ring of rimoportulae. Figs 40, 41. *Mesodictyon nemanensis* (Khurs.) Khurs., specimens from Upper Miocene sediments in Belarus. Fig. 40. Internal valve surface with a marginal ring of fultoportulae. Fig. 41. External view of a complete valve. Figs 42, 43. *Cyclostephanos omarensis* (Kuptz.) Khurs. & Log., specimens from the Upper Pliocene deposits of the Kama river basin. Fig. 42. External view showing areolate striae on the valve face surface which become alveolate on the mantle. Fig. 43. Internal valve surface with alveolae openings on the mantle and fultoportulae with three satellite pores on the valve surface. Fig. 44. *Stephanodiscus* aff. *carconensis* var. *minor* Grun., specimen from Middle Miocene sediments in Belarus. External view of complete valve. Scale bars = 1 µm in Figs 33–37, 40–44; 10 µm in Figs 38, 39.

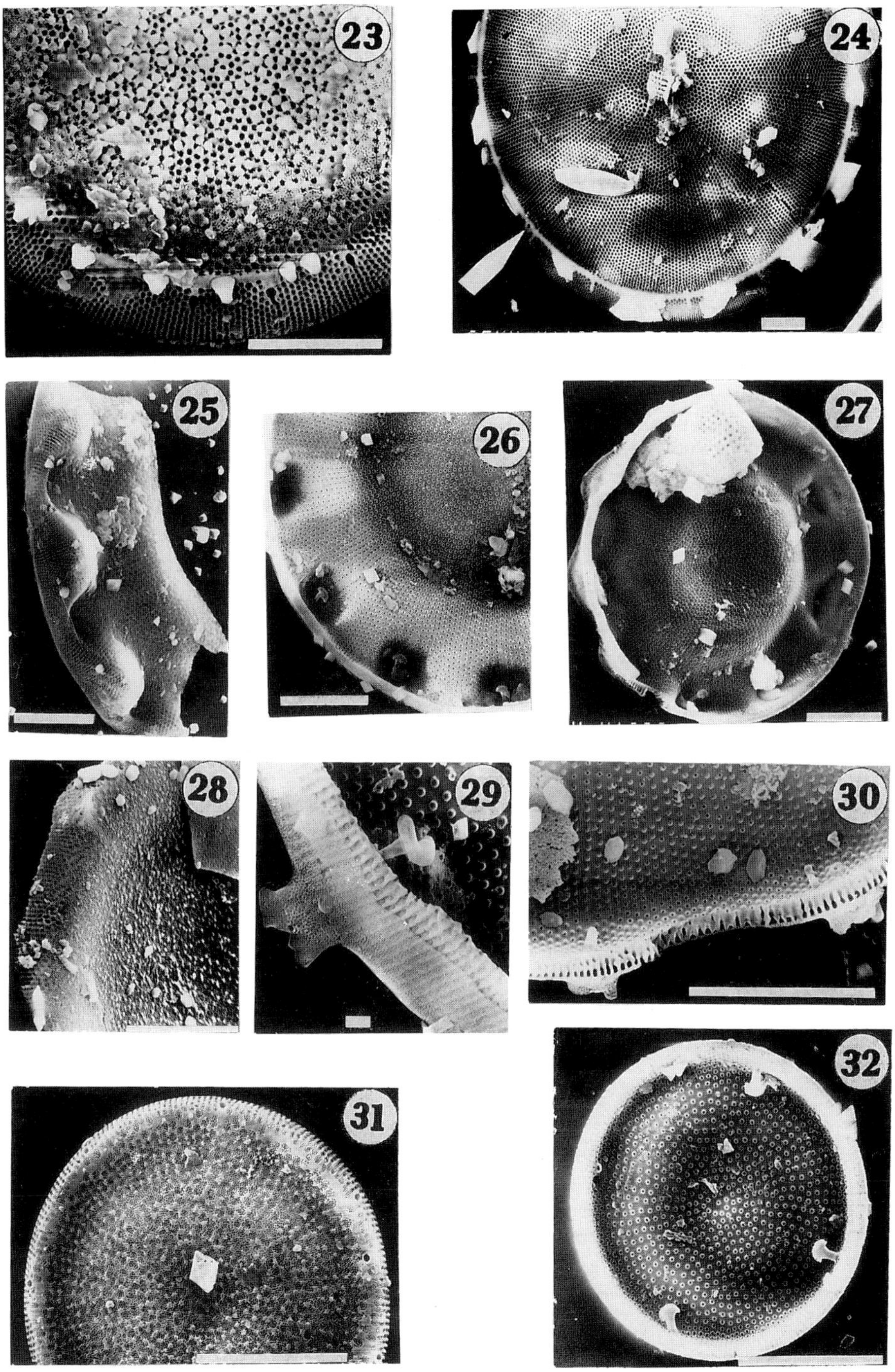
23
24
25
26
27
28
29
30
31
32

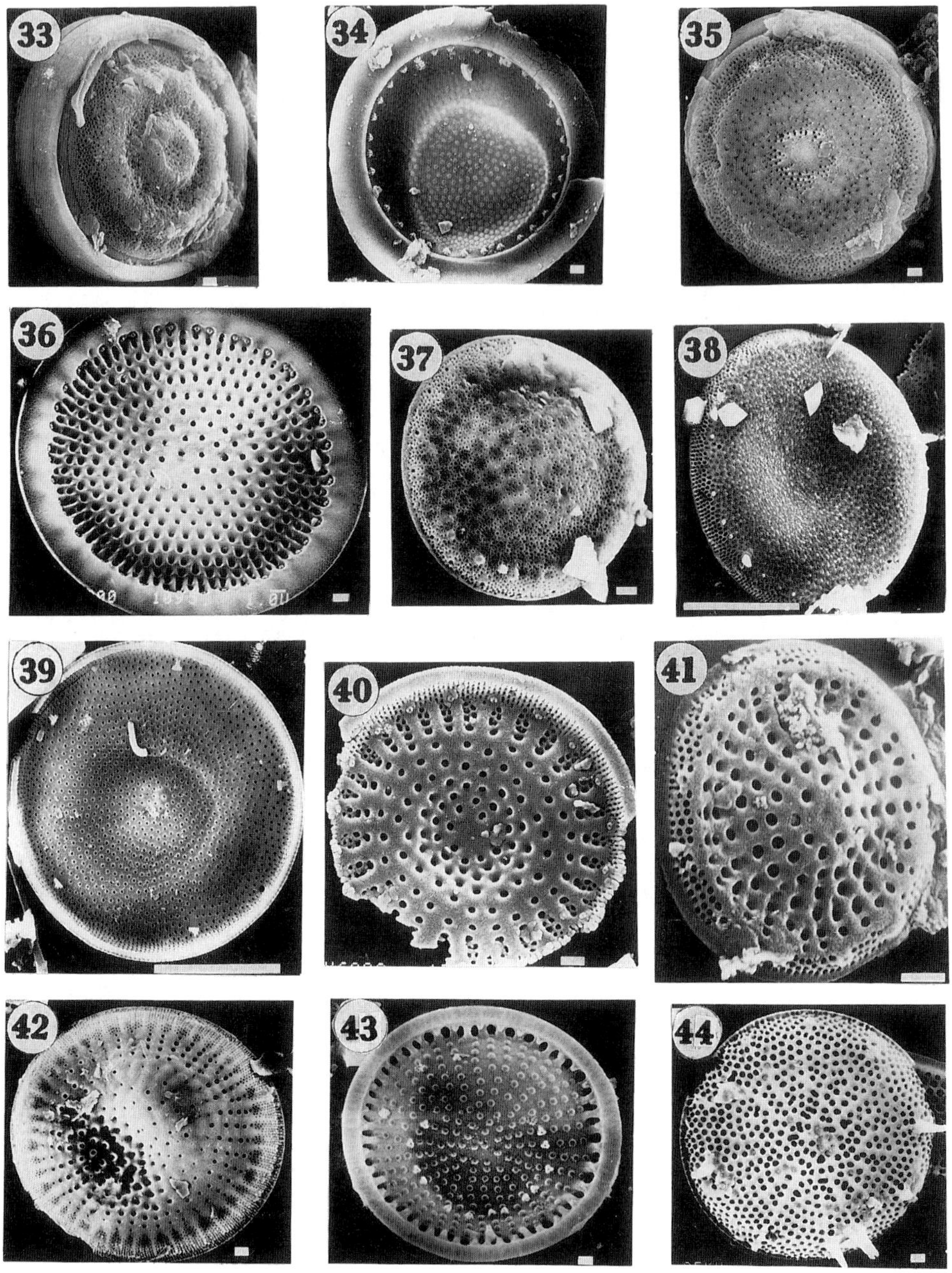
33
34
35
36
37
38
39
40
41
42
43
44

Four stages revealed in the evolutionary development of freshwater centric diatoms both in Asia and Europe reflect global geological and climatic changes. An extensive regression of the ocean in the Oligocene (about 30 my. years ago) gave new ecological niches for development of the continental fauna and flora. Global transgression of the ocean (17–13 my. B.P.), humid climate and intensive development of Alpine orogenesis in the Miocene, accompanied by volcanic activity, stimulated abundant development of diatoms and accumulation of thick, diatomaceous deposits in different regions within the Eurasian continent. Further evolution of centric diatoms proceeded under the influence of increased cooling of the climate during the Pliocene and of repeated cover glaciations interrupted by interglacials in the Pleistocene.

## Discussion

On the whole, representatives of 19 centric genera developed in freshwater palaeobasins within the Euroasian continent. The extinct genera *Gleseria* (end of the Late Eocene–Early Miocene), *Undatodiscus* (Early–Late Miocene), *Lobodiscus* (Middle–Late Miocene), *Concentrodiscus* (Middle–Late Miocene) and *Ectodictyon* (Pliocene) are known, at present, only from Asia. Freshwater species of *Actinocyclus* evolved in the Cenozoic Asian palaeobasins during a rather long time interval (from the end of Early Oligocene through Early Pliocene), in Europe – during Early Miocene –Early Pliocene. The geological range of lacustrine *Actinocyclus* species from the western USA is restricted mainly to Early–Middle Miocene (Krebs 1990). According to VanLandingham (1985) their age range is estimated to be Late Oligocene–Late Miocene (maybe even including the Pliocene). The extinct genus *Alveolophora* was abundant and diversified (6 species) within Asia. Only one species has been found, also in the Miocene, of Europe and North America. However, the species of *Mesodictyon*, especially characteristic for the Late Miocene of North America (USA), Europe (France, Belarus) and Africa, have not been yet discovered in Asia. The genus *Pseudoaulacosira*, on the contrary, developed during a shorter time interval within Europe (Late Miocene – beginning of Early Pliocene) as compared with Asia (Early – Late Miocene). Extinct species of the genus *Actinostephanos* were typical of the Middle and Late Miocene palaeobasins of Europe (Russia, Hungary, Austria and Czech Republics) and have not been yet found in the ancient freshwater deposits of other northern hemisphere regions. Species of *Stephanodiscus*, *Cyclostephanos* and *Cyclotella* were abundant and diversified in European basins as early as from the Late Miocene but in Asian basins mainly occurred in the Late Pliocene.

## Acknowledgements

I would like to thank Dr G. A. Anziferova, Dr V. A. Belova, Ms N. I. Golovenkina, Dr O. P. Kondratene, Dr E. I. Loseva, Ms E. G. Lupikina, Dr A. I. Moissejeva, Dr V. V. Mukhina, Ms T. L. Nevretdinova, Dr S. P. Ozornina, Dr A. P. Olshtynskaya, Dr L. G. Pirumova, Dr Z. Rehakova, Dr N. V. Rubina, Dr E. A.

Tscheremissinova, Dr G. P. Tschernjaeva and Dr D. Temniskova-Topalova for providing material from their collections. I am grateful to Ms A. A. Svirid for all kinds of technical help.

## References

Aleshinskaya, Z. V. & Pirumova, L. G. (1981). Morphological pecularities of diatoms from the Pliocene lacustrine deposits of Armenia. In: *Historical evolution of fossil organisms*. Moskow, 97–109. (In Russian).

Fourtanier, E. & Gasse, F. (1988). Premiers jalons d'une biostratigraphie et évolution des diatomées lacustres d'Afrique depuis 11 M.a. *C. R. Acad. Sci. Paris*, **306** (II), 1401–1408.

Khursevich. G. K. (1989a). New data on morphology and systematic position of *Stephanodiscus matrensis* (Bacillariophyta). *Botanichesky Zhurnal*, **74**(4), 487–489.

Khursevich, G. K. (1989b). Atlas of the *Stephanodiscus* and *Cyclostephanos* species (Bacillariophyta) from the Upper Cenozoic sediments of the USSR. *Nauka I tekhnika*, Minsk. (In Russian). 167 pp.

Khursevich, G. K. (1990). Morphology and Taxonomy of some centric diatom species from the Miocene sediments of the Dzhilinda and Tunkin hollows. In: *Abstracts of the 11th International Diatom Symposium* (E. Fourtanier, ed.), p. 55. San Francisco.

Khursevich, G. K. (1991a). Major stages of the freshwater centric diatoms development in the Cenozoic. *Doklady Akademii Nauk BSSR*, **35**(5), 463–466.

Khursevich, G. K. (1991b). Evolution of the diatoms freshwater genera of the Order *Thalassiosirales* (Bacillariophyta). *Algologia*, **1**(2), 79–84.

Khursevich, G. K. (1991c). Evolution and phylogeny of freshwater representatives of the Order *Coscinodiscales* (Bacillariophyta). *Algologia*, **1**(4), 73–78.

Khursevich, G. K. & Tschernjaeva, G. P. (1989). *Ectodictyonaceae* – a new family in the Class *Centrophyceae* (Bacillariophyta). *Botanichesky Zhurnal*, **74**(7), 1034–1035.

Khursevich, G. K., Moissejeva, A. I. & Suchova, G. A. (1989). *Concentrodiscus* – a new genus of the family *Stephanodiscaceae* (Bacillariophyta). *Botanichesky Zhurnal*, **74**(11), 1660–1661.

Krebs, W. N. (1990). The biochronology of freshwater planktonic diatom communities in Western North America. In: *Abstracts of the 11th International Diatom Symposium* (E. Fourtanier, ed.), p. 67. San Francisco.

Krebs, W. N., Bradbury, J. P. & Theriot, E. (1987). Neogene and Quaternary lacustrine diatom biochronology, Western USA. *Palaios*, **2**, 505–513.

Loginova, L. P. (1990). Evolution and phylogeny of the diatom genus *Cyclotella*. In: *Proceedings of the 10th International Diatom Symposium* (H. Simola, ed.), 47–53. Koeltz, Koenigstein.

Loginova, L. P. & Khursevich, G. K. (1989). Fossil diatom flora of the Lake Baikal. In: *New representatives of fossil fauna and flora of Belarus and other regions of the USSR. Nauka I tekhnika, Minsk*, 167–189. (In Russian).

Loseva, E. I. (1982). *Atlas of the Late Pliocene diatoms of Prikamja. Nauka, Leningrad* (In Russian). 204 pp.

Lupikina, E. G. (1980). Sequence and accumulation conditions of lacustrine volcanic–sedimentary deposits from the data of diatom analysis. In: *Volcanic Centre: structure, dynamics, material composition. Nauka, Moskow* (In Russian). pp. 23–52.

Lupikina, E. G. & Dolmatova, L. M. (1985). About the Paleogene lagoonal diatom flora of Kamchatka. *Paleontologichesky Zhurnal*, **1**, 120–128.

Lupikina, E. G. & Khursevich, G. K. (1991a). *Pseudoaulacosira* – a new genus of the freshwater diatoms in the class Centrophyceae. *Botanichesky Zhurnal*, **76**(2), 290–291.

Lupikina, E. G. & Khursevich, G. K. (1991b). *Lobodiscus* (Tscher.) Lupik. & Khurs. – a new genus in the class Centrophyceae (Bacillariophyta). *Algologia*, **1**(3), 67–70.

Lupikina, E. G., Khursevich, G. K. & Rubina, N. V. (1991). The taxonomic status of the genus *Undatodiscus* (Bacillariophyta). *Botanichesky Zhurnal*, **76**(5), 743–745.

Makarova, I. V. (ed.) (1992). *Diatoms of the USSR, fossil and recent. Nauka, St Petersburg*. (In Russian). 125 pp.

Moissejeva, A. I. (1993). Diatoms in continental Neogene stratigraphy. In: *Abstracts of the 5th workshop on diatom algae* (M. A. Grachev, ed.), 99–101. Irkutsk.

Moissejeva, A. I. & Nevretdinova, T. L. (1990). The new family and genus of freshwater diatom algae (Bacillariophyta). *Botanichesky Zhurnal*, **75**(4), 539–544.

Mori, S. H. (1974). Diatom succession in a core from Lake Biwa. In: *Paleolimnology of Lake Biwa and the Japanese Pleistocene*, **2**, 247–254.

Nikiteeva, T. (1993). New species of *Cyclotella* found in sediments of Lake Baikal and Tunka depression. In: *Abstracts of the 5th workshop on diatom algae* (M. A. Grachev, ed.), 101–103. Irkutsk.

Olshtynskaja, A. P. (1993). New species of the *Actinocyclus* Ehr. genus (Bacillariophyta). *Algologia*, **3**(3), 57–59.

Ozornina, S. P. (1987). Diatoms from the Upper Pliocene deposits of the Central Kamchatka. In: *Materials on the stratigraphy and paleogeography of East Asia and the Pacific Ocean. Vladivostok* (In Russian). pp. 151–155.

Popova, S. M., Mats, V. D., Tschernjaeva, G. P. *et al.* (1989). *Paleolimnological reconstructions (the Baikal rift zone). Nauka, Novosibirsk.* (In Russian). 111 pp.

Reháková, Z. (1965). Fossile Diatomeen der südböhemischen Becker ablagerungen. *Rozpravy Ustredniho ustavu geologickego*, **32**, 1–96.

Round, F. E. (1970). The delineation of the genera *Cyclotella* and *Stephanodiscus* by light microscopy, transmission and reflecting electron microscopy. *Nova Hedwigia*, **31**, 591–604.

Round, F. E. & Håkansson, H. (1992). Cyclotelloid species from a diatomite in the Harz Mountains, Germany, including *Pliocaenicus* gen. nov. *Diatom Research*, **7**, 109–125.

Round, F. E. & Sims, P. A. (1981). The distribution of diatom genera in marine and freshwater environments and some evolutionary considerations. In: *The 6th Symposium on Recent and Fossil Diatoms* (R. Ross, ed.), 301–320. O. Koeltz, Koenigstein,.

Round, F. E., Crawford, R. M. & Mann, D. G. (1990). *The Diatoms. Biology and morphology of the genera.* Cambridge University Press, Cambridge. 747 pp.

Rubina, N. V. & Khursevich, G. K. (1990). New representatives of the genus *Actinocyclus* (Bacillariophyta) from the Late Oligocene of the Western Siberia. *Botanichesky Zhurnal*, **75**(11), 1565–1567.

Serieyssol, K. K. & Gasse, F. (1990). A cladistic analysis of Miocene and Pliocene *Cyclotella* and *Cyclostephanos*. In: *Abstracts of the 11th International Diatom Symposium* (E. Fourtanier, ed.), p. 103. San Francisco.

Serieyssol, K. & Gasse, F. (1991). Diatomées néogènes du Massif Central Français: quelque faits biostratigraphiques. *C. R. Acad. Sci. Paris*, **312**(II), 957–964.

Servant-Vildary, S., Paicheler, J. C. & Semelin, B. (1988). Miocene lacustrine diatoms from Turkey. In: *Proceedings of the 9th International Diatom Symposium* (F. E. Round, ed.), 165–180. Biopress, Bristol & Koeltz, Koenigstein.

Temniskova-Topalova, D., Ognjanova-Rumenova, N. & Valeva, M. (1990). Non-marine diatoms from Neogene sediments of Bulgaria. In: *Proceedings of the 10th International Diatom Symposium* (H. Simola, ed.), 357–363. Koeltz, Koenigstein.

Theriot, E., Stoermer, E. & Håkansson, H. (1987). Taxonomic interpretation of the rimoportula of freshwater genera in the centric diatom family *Thalassiosiraceae*. *Diatom Research*, **2**(2), 251–265.

Yakubovskaya, T. V., Khursevich, G. K. & Loginova, L. P. (1989). The first information on the fossil flora from the Tonezh brown coal field. *Doklady Akademii Nauk BSSR*, **33**(7), 653–656.

VanLandingham, S. L. (1985). Potential Neogene diagnostic diatoms from the western Snake River Basin, Idaho and Oregon. *Micropaleontology*, **31**(2), 167–174.

# Holocene *Nitzschia scalaris* (Ehr.) W. Smith blooms in the coastal glo-lakes of Finland

Atte Korhola

*Laboratory of Physical Geography, P.O. Box 9 (Siltavuorenpenger 20 A), FIN–00014 University of Helsinki, Finland*

## Abstract

Sediment cores were collected from two ancient glo-lake systems in the vicinity of Helsinki, and were studied for diatoms. The sediment sequences were made up of gyttja-clays, clay-gyttjas, gyttjas and peats, deposited in the course of the gradual emergence of the sites from the Baltic Sea between 8000 and 6000 yrs BP. High abundances of *Nitzschia scalaris* were recorded, accounting for as much as 47% of the total flora. These occur mainly in freshwater sediments, deposited after the actual isolation of the sites. Environmental instability and the proximity of the transgressive Litorina Sea probably stimulate mass development of *N. scalaris*.

## Introduction

Along the coast of the northern Baltic, new lake basins are continuously formed as a function of the crustal land uplift. Different successional phases may be distinguished in this process. Glo-lakes represent transitional stages between the Baltic Sea inlets (flads) and independent freshwater lakes (the words "flad" and "glo" are of Swedish dialectal origin). Gloes are shallow water bodies, which only occasionally (during high water periods) have any contact with the saline environment, and their surface often lies above the sea level (Munsterhjelm 1987). Many of the present glo-lakes are meromictic, the main factor leeping them so, being the presence of salt water of higher density (Lindholm 1975).

Alterations in salinity also seem to be the most significant factor determining changes in diatom assemblages in the coastal waters of Finland (Räsänen & Tolonen 1983; Alhonen 1986). The changes in diatom communities in the course of the isolation can successfully be followed using sediment cores (Alhonen 1971). These analyses normally show a progression from marine diatoms, through brackish-water diatoms to freshwater forms, indicating that the sites had once been open bays in the

Baltic and had gradually passed through a series of coastal configurations before final isolation as freshwater lakes. The determination of a lake isolation horizon on the basis of the fossil diatom assemblages is often difficult, however, since some sensitive diatom species react immediately to the variations in salinity while others show optimal development at a somewhat later stage (Florin 1946; Hyvärinen 1982; Räsänen & Tolonen 1983).

*Nitzschia scalaris* is a diatom species which has often created problems in terms of isolation studies, as although it is primarily a brackish taxon and an almost constant member of the saline flora of the Litorina Sea flads ("lagoons"), it can also occur in isolated basin systems regardless of whether these are derived from saline or non-saline environments (Eronen 1974). In this paper I try to relate the succession of *N. scalaris* within the diatom assemblages to the environmental changes that took place.

## Material and Methods

Two mire basins, which are located in the vicinity of Helsinki and are above the highest morphological Litorina shore, were chosen for this study (Fig. 1). Kotasuo (60°15'N, 24°35'E) is a small (14 ha) forested raised bog, surrounded by rock faces and till ridges, and lying at 40-43 m a.s.l. The more Mottisuo (60°19'N, 24°56'E; 114 ha) is an eccentric raised bog with hummocks and hollows, situating on a fairly broad fluvioglacial terrain at an altitude of 44 m a.s.l. Both of the mires have previously been lakes, which is evident from the limnic sediments (gyttja) at their bottoms. The hydroseral development of the Kotasuo bog has been described in detail on a previous occasion (Korhola 1990).

The sediment samples for diatom analysis were taken with a Russian corer from the deepest points of the basins (Fig. 1). The sediments were $^{14}$C-dated at the Dating Laboratory of the University of Helsinki (Hel). Organic content of the sediment was determined by loss on ignition at 550°C. Diatoms were prepared by oxidation using $H_2O_2$ (Renberg 1990). The residues were mounted in Hyrax® and a total of 200–400 valves per sample were counted on random transects using a NIKON OPTIPHOT-2 microscope with phase-contrast optics at 1000×. Diatoms were identified using a range of floras, in particular Möder & Tynni (1967–73), Tynni (1975–80) and Germain (1981). Only the most numerous taxa are considered here. The nomenclature of the diatoms follows Hartley (1986).

## Results

### *Kotasuo*

The sediments at the bottom of the Kotasuo core consist of gyttja-clays and clay-gyttjas. At a depth of 850 cm the sediment changes into a green-greyish fine detritus gyttja, and at c. 750 cm into a brown-greyish coarse detritus gyttja (Fig. 2). The changes are gradual, and for most of the sequence the sediments are relatively uniform. The loss-on-ignition curve (Fig. 2) correlates well with the visual changes in the

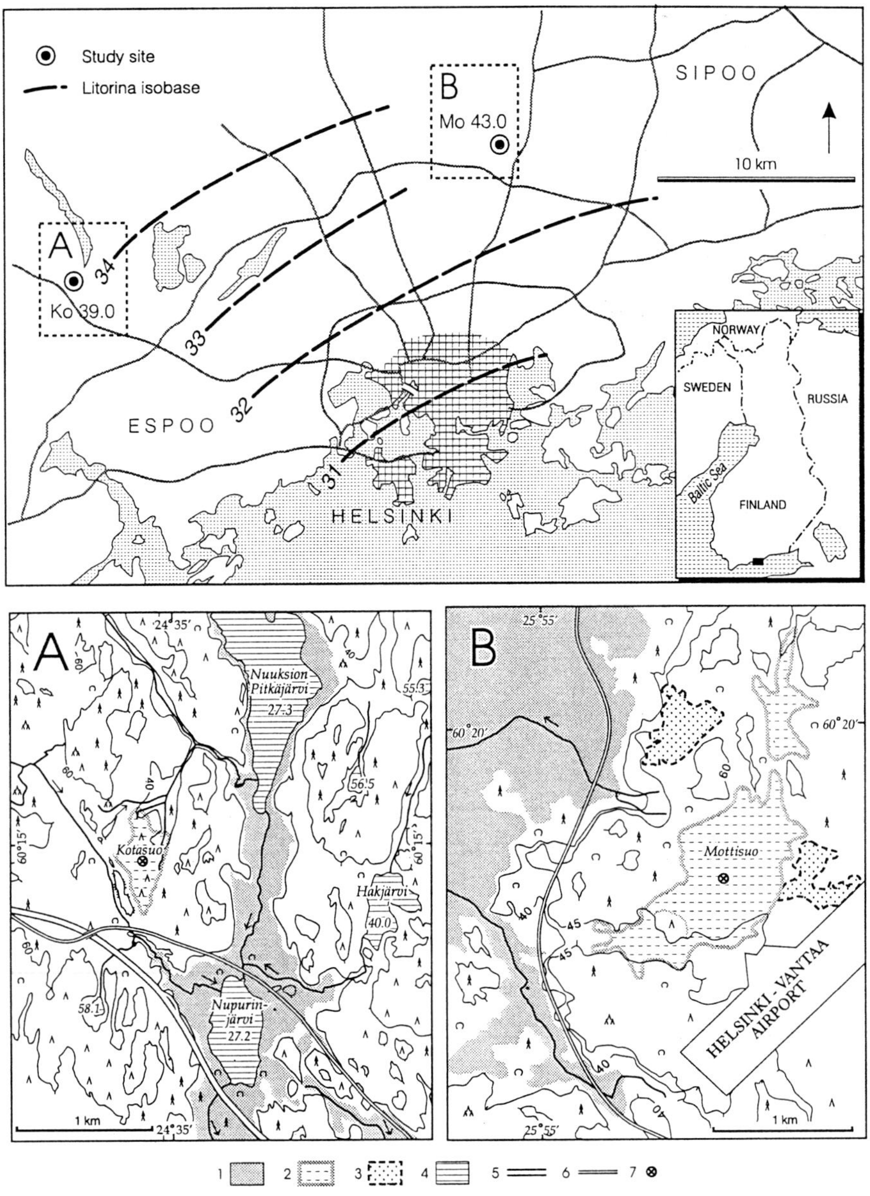


Fig. 1. Map of the Helsinki area, showing locations of the studied sites (above). Isobases for the highest Litorina shore are indicated after Hyvärinen (1982). The numbers denote altitudes above sea level. (A) Kotasuo and its surroundings, (B) Mottisuo and its surroundings. (1) maximum extent of the Litorina Sea, (2) mire, (3) gravel pit, (4) water, (5) motorway, (6) local road, (7) sampling site.

Kotasuo, ( 39.0 m a.s.l. ), diatoms

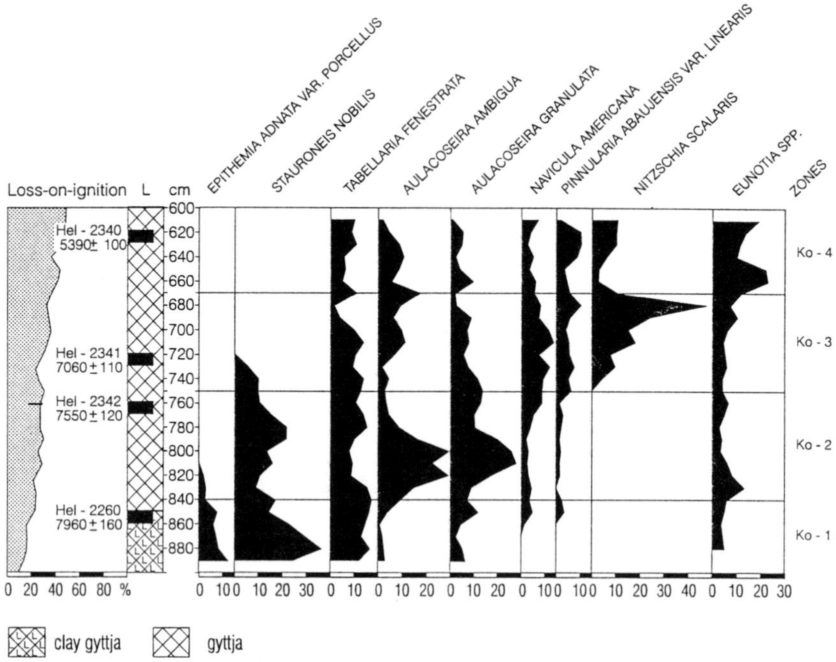


Fig. 2. Diatom diagram of the most numerous species and loss-on-ignition curve for Kotasuo. L = lithostratigraphy.

stratigraphy, showing a slow but gradual increase in organic content from the bottom of the core to values of 30–40% in the upper sections.

According to previous investigations (Korhola 1990) the basin was cut off from the Baltic Sea approx. 8300 yr BP. The isolation is marked in the diatom stratigraphy by a decline in typical Ancylus Lake species and their replacement with taxa indicative of a small freshwater body, the most abundant of which are *Stauroneis nobilis*, *Tabellaria fenestrata*, *Aulacoseira granulata* and *A. ambigua* (zones Ko-1 and Ko-2 in Fig. 2). At a depth of 750–760 cm an entirely new species, *Nitzschia scalaris*, appears in the diatom flora, its percentages increasing rapidly so that by a level of 680 cm it accounts for about 48% of the total diatom spectrum (zone Ko-3). According to the radiocarbon dates the most prominent appearance of this species occurred between 7500–6100 yr BP. *Navicula americana* is also well represented in this section. The uppermost part of the profile is characterised by an increase in values of acidophilous *Eunotia* and *Pinnularia* species (zone Ko-4).

*Mottisuo*

A sharp contact is observed in the Mottisuo core between the basal minerogenic sediment and the above lying peat, as reflected in a jump in the loss-on-ignition values

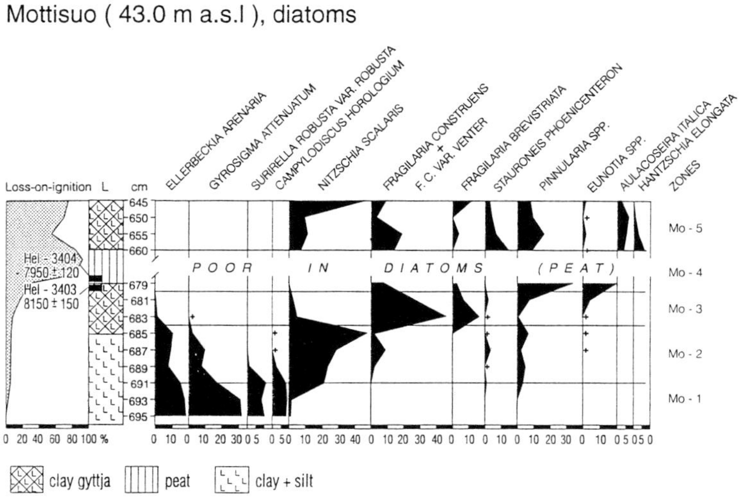


Fig. 3. Diatom diagram of the most numerous species and loss-on-ignition curve for Mottisuo.

within a distance of a few centimetres from below 20% in the gyttja clay and clay gyttja to around 90% in the peat (Fig. 3). The organic content decreases again markedly in the interval 660–640 cm, reflecting the higher propotion of mineral material present. Beyond this point organic matter content returns to figures of 90–98% typical of peat deposits.The gyttja clay at the bottom of the core must originate from the Ancylus stage in the Baltic history, as its diatom flora is dominated by typical "large-lake" oligohalobic forms such as *Ellerbeckia arenaria* and *Gyrosigma attenuatum* (Mo-1). Epithemiae in general and Surirellae are also well represented. The high proportions of epiphytic and benthic diatoms and the low representation of pelagic-planktonic species would suggest shallow littoral conditions. *Nitzschia scalaris* becomes the dominant species in the interval 689-683 cm, exceeding 45% of the total diatom count at its maximum (Mo-2). According to the radiocarbon results the first appearance of *N. scalaris* occurred in pre-Litorina sediments, i.e. >7500 yr BP. The first freshwater diatoms also make their appeararance during this interval. At 683 cm the proportions of *N. scalaris* decline suddenly, these being replaced by the brackish or freshwater tychoplanktonic *Fragilaria* species (Mo-3). The actual isolation of the basin from the Baltic is marked by a change in the diatom assemblage at arround 681 cm, where the freshwater *Eunotia*, *Pinnularia* and *Tabellaria* species become dominant (Mo-4). This isolation event was radiocarbon dated to approx. 8000 yr BP.

The change from lake sediments to peat in the interval 677–660 cm indicates that the basin became overgrown soon after its isolation. The radiocarbon date from the first peat gave an age of 7950 ± 120 (Hel–3404). Diatom frustules disappear from the stratigraphy entirely following this transformation, but reappear at a depth of 660 cm indicating the inundation of the mire basin. The uppermost horizon studied here, features species which had disappeared from the site prior to the onset of mire formation, e.g. *Surirella* spp. and *N. scalaris*, and ones that are complete newcomers such as *Aulacoseira italica*, *Hantzschia elongata* and *Pinnularia* spp. *N. scalaris* becomes very abundant at the last level examined (Fig. 4). In the pollen stratigraphy (Korhola 1995) the *Tilia* pollen curve begins at a depth of 660 cm (T°). This reference point has been dated to approx. 7400 BP in the Helsinki area (see Korhola 1990; Hyvärinen 1984); consequently, the second bloom of *N. scalaris* took place at that time. Diatom frustules disappear completely with the establishment of the terrestrial mire environment at 645 cm.

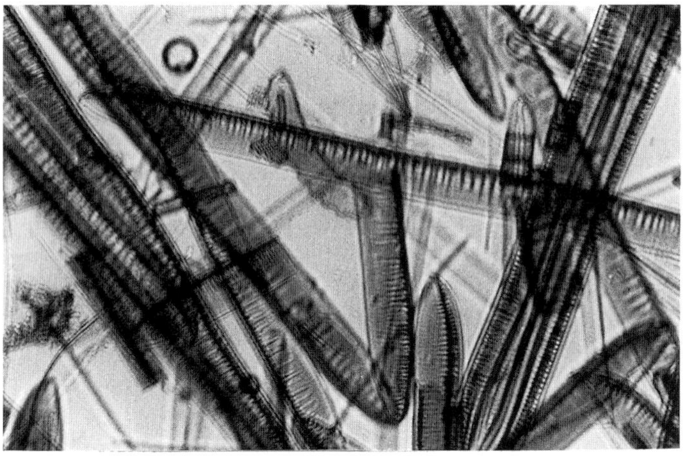

Fig. 4. *Nitzschia scalaris* bloom at the uppermost section of the studied profile in Mottisuo (200×).

## Discussion

There are several diatoms which inhabit both brackish and fresh water (Simonsen 1962). One such species is *Nitzschia scalaris*, which is considered primarily as a halophilous species (Tynni 1980) but can survive also in land waters, particularly in systems that have recently been cut off from the sea. Eronen (1974) discusses numerous cases of this kind in detail in connection with the Holocene shore displacement. Also worthy of mention in this context are the observations from modern coastal lakes of Finland made by Fontell (1926), who found *N. scalaris* in 18 out of 21 small coastal lakes situated at a distance of 100 m to about 2000 m from the sea. However, both in the fossil as well as in the modern material, the frequences of *N. scalaris* in post-isolation sediments were usually low.

The mass occurrence of *N. scalaris* in the sediment profiles studied here is exceptional in many respects. Especially interesting is that the species made its appearance in Kotasuo *de novo* at about 7500 BP, a long time after the isolation of the basin. Moreover, in both cases the described changes in the fossil biocommunity are of such a magnitude, and are so rapid and sufficiently simultaneous that they can scarcely be attributed to autogenous trends. An allochtonous imput of some kind would seem a more probable explanation. and in this case the most natural cause would be that the basins were influenced by the rising Litorina Sea.

The present profiles nevertheless carry no direct evidence of the Litorina transgression, as is to be expected since the basins are located clearly above the highest Litorina shore, but it should be noted that the shore of the Baltic remained very close to the sites, as long arms of the sea extended some distance inland along fault-line valleys (Fig. 1). Thus it is quite possible that some saline water could have reached the basins by an indirect route, e.g. as a result of exceptionally high winds and fierce wave action, or through groundwater seepage. One requirement for such a chain of events would be that there was some degree of relative rise in sea level as well – a proposal that is consistent with the recent findings by Ristaniemi & Glückert (1988) but contradicts Hyvärinen's (1982) view of a constant sea level in the Helsinki region during the Litorina stage (Korhola 1995). The postulated transgression is also supported by the re-inundation of the mire basin at Mottisuo, which must reflect a rise in water levels in the surrounding area. It is possible that a small inwash of saline sea weater could have raised the content of electrolytes in these basins, thus increasing their nutrient level.

One should also consider the possibility that the high frequences of *N. scalaris* observed here could simply be artifacts of differential preservation of diatom frustules, as meromictic lakes, in particular, are special cases in which diatom preservation is frequently poor (Meriläinen 1971). Furthermore, instances are documented in Korhola (1990, 1992) of frustules being dissolved due to increased silica uptake with the advancing hydroseral development. In the above cases the effects were universal, however, rather than selective. It should also be mentioned that usually large, highly silicified frustules are more resistant to dissolving and mechanical erosion than small, poorly silicified ones (Flower 1993). In this sense the clear increase in the numbers of the small *Fragilaria* and *Eunotia* species together with the large *N. scalaris* forms, strongly suggests that it is not only a question of the dissolution-effect but of a true ecological succession.

It is significant that the two cores contain parallel occurrences of an alkaline (± brackish water) flora and a nutrient-poor flora typical of dystrophic waters, which also tend to suggest an allochtonous influx of nutrients rather than an inherently nutrient-rich depositional environment. In Mottisuo, in particular, the occurrence of the *N. scalaris* communities together with tychoplanktonic *Fragilaria* species and benthic-epiphytic, and epontic taxa such as *Cocconeis placentula*, *Rhopalodia gibba*, *Amphora ovalis*, *Epithemia* spp. and *Pinnularia* spp. points to a high-energy, mineral-poor wetland system with brackish water influence (Stabell 1985; Denys 1989). This diatom assemblage greatly resembles, in a structural sense, those occurring in the humus-rich marsh sediments deposited under brackish water conditions on the coast of Belgium and Holland (the "darg" deposits; Vos & De Wolf 1993). We may thus conclude that environmental instability and the proximity of the transgressive Litorina Sea stimulated mass development of *N. scalaris* in these two glo-lakes.

## Acknowledgements

I thank H. Seppä and J. Weckström for their help in the field and K. Lehto for preparing the diagrams. The research was funded by the Academy of Finland (Project no. 101 7383).

## References

Alhonen, P. (1971). The stages of the Baltic Sea as indicated by the diatom stratigraphy. *Acta Botanica Fennica*, **92**, 18 pp.

Alhonen, P. (1986). Holocene lacustrine microfossils and environmental changes. *Bulletin of the Geological Society of Finland*, **58**, 57–69.

Denys, L. (1989). Observations on the transition from Calais deposits to surface peat. Results of a paleoenvironmental diatom study. *Professional Papers of the Belgian Geological Survey*, **241**, 20–43.

Eronen, M. (1974). The history of the Litorina Sea and associated Holocene events. *Commentationes Physico-Mathematicae*, **44**, 79–195.

Florin, M.-B. (1946). Clypeusfloran i postglaciala fornsjölagerföljder i östra Mellansverige. *Geologiska Föreningen i Stockholm Förhandlingar*, **68**, 429–457.

Flower, R. J. (1993). Diatom preservation: experiments and observations on dissolution and breakage in modern and fossil material. *Hydrobiologia*, **269/270**, 473–484.

Fontell, C. W. (1926). Om brak- och saltvattendiatomacéers förekomst i sött vatten i närheten av kusten. *Acta Societatis pro Fauna & Flora Fennica*, **55**, 3–21.

Germain, H. (1981). *Flore des diatomées. Diatomophycées.* Société Nouvelle des Éditions Boubée. Paris. 444 pp.

Hartley, B. (1986). A check-list of the freshwater, brackish and marine diatoms of the British Isles and adjoining coastal waters. *Journal of the Marine Biological Association of the United Kingdom*, **66**, 531–610.

Hyvärinen, H. (1982). Interpretation of stratigraphical evidence of sea-level history: a Litorina site near Helsinki, southern Finland. *Annales Academiae Scientiarum Fennicae A. III.*, **14**, 139–149.

Hyvärinen, H. (1984). The *Mastogloia* stage in the Baltic Sea history: diatom evidence from southern Finland. *Bulletin of the Geological Society of Finland*, **56**, 99–115.

Korhola, A. (1990). Paleolimnology and hydroseral development of the Kotasuo bog, southern Finland, with special reference to the Cladocera. *Annales Academiae Scientiarum Fennicae A. III.*, **155**, 1–40.

Korhola, A. (1992). The Early Holocene hydrosere in a small acid hill-top basin studied using crustacean sedimentary remains. *Journal of Paleolimnology*, **7**, 1–22.

Korhola, A. (1995). The Litorina transgression in the Helsinki region, southern Finland: new evidence from coastal mire deposits. *Boreas* (in print).

Lindholm, T. (1975). Coastal meromictic lakes on Åland (SW Finland). *Aqua Fennica*, **1975**, 24–40.

Meriläinen, J. (1971). The recent sedimentation of diatom frustules in four meromictic lakes. *Annales Botanici Fennici*, **8**, 160–176.

Mölder, K. & Tynni, R. (1967–1973). Über Finnlands rezente und subfossile Diatomeen I–VII. I, *Bulletin de la Commision géologique Finlande*, **29**, 199–207 (1967); II, *Bulletin of the*

*Geological Society of Finland*, **40**, 151–170 (1968); III, *ibid.*, **41**, 235–251 (1969); IV, *ibid.*, **42**, 129–144 (1970); V, *ibid.*, **43**, 203–220 (1971); VI, *ibid.*, **44**, 141–149 (1972); VII, *ibid.*, **45**, 159–179 (1973).

Munsterhjelm, R. (1987). Flads and gloes in the archipelago. *Geological Survey of Finland, Special Paper* **2**, 55–61.

Renberg, I. (1990). A procedure for preparing large sets of diatom slides from sediment cores. *Journal of Paleolimnology*, **4**, 87–90.

Räsänen, M. & Tolonen, K. (1983). Changes in diatom flora deposited annually during the freshening of the impounded, sea-bay of Gennarbyviken in southern Finland. *Hydrobiologia*, **103**, 147–152.

Ristaniemi, O. & Glückert, G. (1988). Ancylus- ja Litorinatransgressiot Lounais-Suomessa. In: *Tutkimuksia geologian alalta* (V. Lappalainen, ed.). *Annales Universitatis Turkuensis C 67*, 129–145.

Simonsen, R. (1962). Untersuchungen zur Systematik und Ökologie der Bodendiatomeen der Westlichen Ostsee. *International Revue des gesamten Hydrologie. Systematische Beiheft* **1**, 144 pp.

Stabell, B. (1985). The development and succession of taxa within the diatom genus *Fragilaria* Lyngbye as a response to basin isolation from the sea. *Boreas*, **14**, 273–286.

Tynni, R. (1975–1980). Über Finnlands rezente und subfossile Diatomeen VIII–XI. VIII, *Geological Survey of Finland, Bulletin*, **274**, 55 pp. (1975); IX, *ibid.*, **284**, 37 pp. (1976); X, *ibid.*, **296**, 55 pp. (1978); XI, *ibid.*, **312**, 93 pp. (1980).

Vos, P. C. & De Wolf, H. (1993). Reconstruction of sedimentary environments in Holocene coastal deposits of the southwest Netherlands; the Poortvliet boring, a case study of palaeoenvironmental diatom research. *Hydrobiologia*, **296/270**, 297–306.

# Diatom stratigraphy of the Mazovian (Holstein, Likhvin) interglacial lacustrine sediments at Biała Podlaska (Eastern Poland)

Barbara Marciniak* and Leszek Lindner**

*Institute of Geological Sciences, Polish Academy of Sciences, Al. Żwirki i Wigury 93, 02-089 Warszawa, Poland

**Institute of Geology, Warsaw University, Al. Żwirki i Wigury 93, 02-089 Warszawa, Poland

## Abstract

At the Biała Podlaska site (eastern Poland) diatoms have been found preserved within lake sediments between two till horizons. Palynological studies of the sediments, dated by the thermoluminescence method as about 209 ka, show that they represent the Mazovian Interglacial s.s. (Holstein s.s., Likhvin s.s.). On the basis of diatom analysis, five local diatom phases have been distinguished (BP1–BP5) which are correlated to pollen analysis. The general trend of the diatom succession, and the presence of *Cyclotella comta* (*radiosa*) var. *lichvinensis* Jouse and in particular *C. vorticosa* Berg, i.e. forms typical for the Mazovian Interglacial in Poland and the Likhvin Interglacial in the Russian Plain, confirm the probable correlation of these interglacials on the basis of diatoms.

## Introduction

Biała Podlaska is situated in southern Podlasie (eastern Poland). The area is a glacial upland, cut by a depression opening towards the South into the Krzna river valley (Fig. 1). The sequence is composed of silts and bituminous shales 5 m thick, overlying sand and gravel deposits (Fig. 2). They are rich in diatoms. Together with the underlying sands with gravel, they represent a lacustrine environment and infill a depression within the older till, dated using the thermoluminescence method (TL) as 230 ± 10 ka (Krupiński 1988b). Their age (the bottom part) has been dated as 209.5 ± 10 ka. The organogenic sediments are covered by a lense of sand, gravel and clay. The whole series, including the overlying till, dated as 170–160 ka (Lindner *et al.* 1991), is glaciotectonically disturbed. In places, a horizon of sand or gravel represents residuum of the till, dated as 18.7 ± 2 ka (Krupiński 1988b).

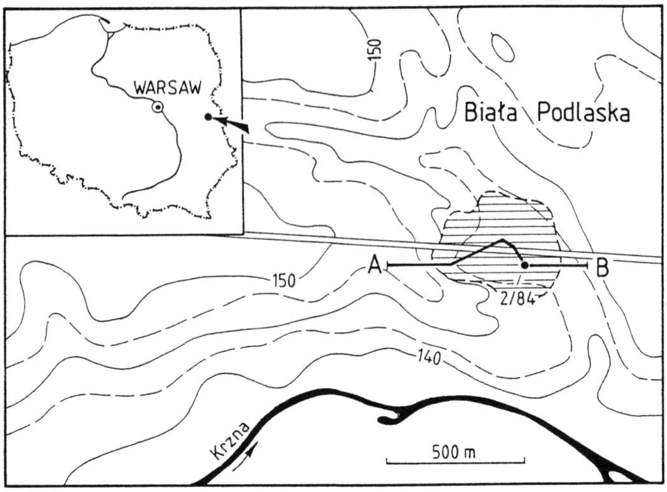


Fig. 1. Location of Biała Podlaska site, borehole 2/84 and geological section (A-B).

Palynological investigations (Krupiński *et al.* 1986, 1988; Krupiński 1988a) have shown that the lacustrine sediments at Biała Podlaska represent the Mazovian Interglacial s.s. (Holstein s.s., Likhvin s.s.) containing the characteristic four periods (I–IV) of vegetation development and 10 local pollen zones (A–K). These sediments (Fig. 2) can be placed above the Sanian 2 = Elsterian 2 till, overlain in part by sands with gravels (Liviecian = Fuhne Glaciation?), in part by clays (Zbójnian = Dömnitz Interglacial?) and in part by the younger till (Odranian = Saalian Glaciation) after Lindner (1988a, b, 1991).

The sediments of the Mazovian s.s. (= Holstein s.s. = Likhvin s.s.) Interglacial are correlated in the European Lowland with the 11th oxygen stage in the deep-sea cores (see Cepek 1986; Lindner 1988a; Zubakov & Borzenkova 1990) dated as 440–367 ka (compare Shackleton & Opdyke 1973), or with the 9th oxygen stage in the deep-sea cores (Zagwijn 1989) dated as 347–297 ka (compare Shackleton & Opdyke 1973). The results of the thermoluminescence measurements of the Biała Podlaska sediments suggest an even younger date of that interglacial and correlation with the 7th oxygen stage dated as 251–195 ka (see Shackleton & Opdyke op. cit.).

*Diatom analysis*

Forty-one samples were analysed. The samples were obtained from the Biała Podlaska profile 2/84 (Figs 1 and 2) from a depth of 1.20–5.05 m. Preparation was done using the method recommended for sediments containing a high content of organic matter (Siemińska 1964). Diatoms were found in 33 samples which were qualitatively and quantitatively analysed. The results, after counting about 500 specimens per sample, are presented in graphic form (Figs 3, 4).

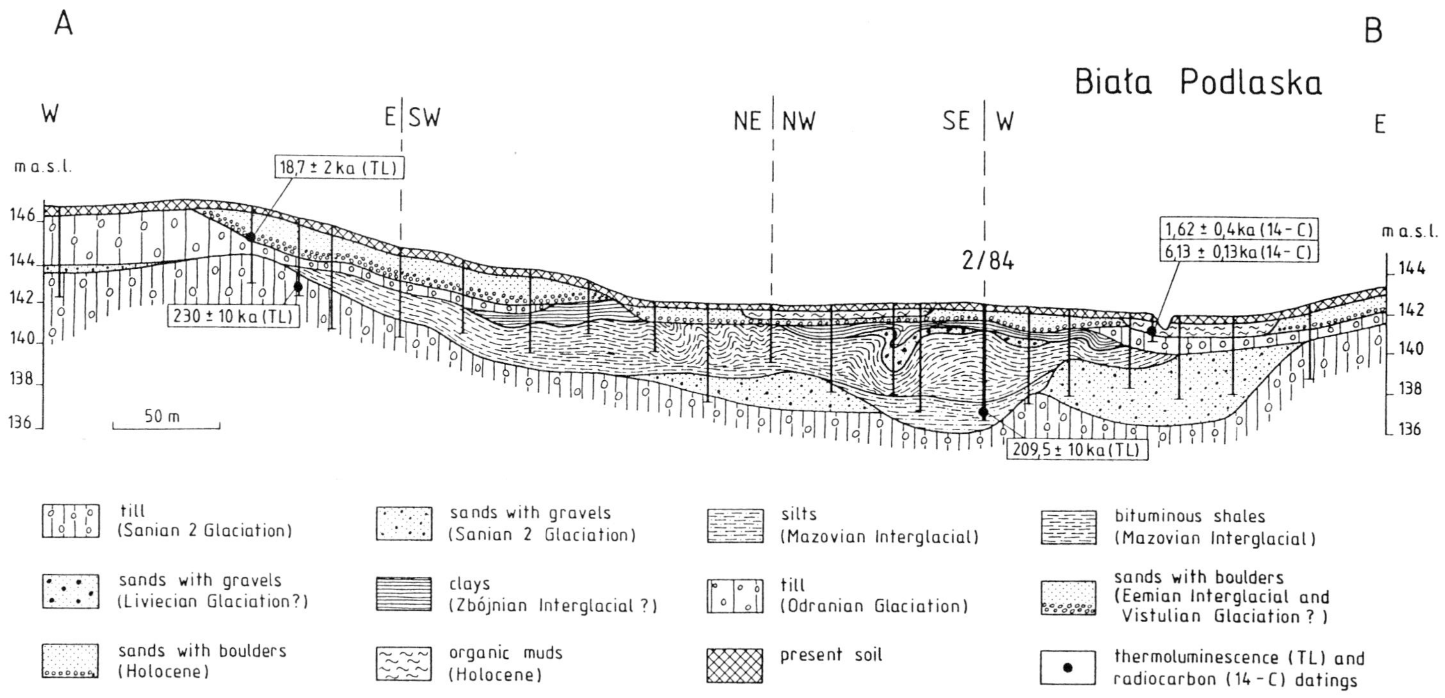


Fig. 2. Geological section A-B of organogenic sediments of the Mazovian Interglacial in the Biała Podlaska area based on Krupiński *et al.* (1988); Krupiński (1988b).

Changes in the composition and relative frequency of diatom taxa (in total 77 taxa) in the Biała Podlaska sediments, profile 2/84 allowed subdivision into five local phases of diatom development BP1–BP5). Subphases have been distinguished within the diatom phases, marked with small letters. They indicate minor changes in diatom occurrence. This subdivision makes it possible to compare the diatom phases with the equivalent periods and local pollen zones from the same profile (compare Figs 3, 4) done by Krupiński (Krupiński *et al.* 1986, 1988; Krupiński 1988a).

*Diatom phase BP1*

Changes in diatom assemblages can be divided into three subphases based on dominant forms (Fig. 3). At the beginning (subphase BP1a) *Fragilaria lapponica* (>32%) and *Cymbella* (*Navicula*) *diluviana* (>15.6%) dominated. Next (in subphase BP1b) there was an increase in *Amphora pediculus* (>27%) and *Navicula scutelloides* (>18%). At the end of this phase (subphase BP1c) a maximum relative frequency of *Navicula scutelloides* (>36%) was noted as well as a considerable proportion of *Fragilaria construens* var. *binodis* (>23%). These changes represent the initial stage of diatom development. These diatoms now live in alkaline lakes, mainly on the lake bottom and within the vegetation of the lake margin. There are very few planktonic diatoms which may point to low temperature and/or low water level during the first stage of lake development.

At that time the first pre-interglacial forest appeared around the lake. This corresponds to period I divided into pollen zones A and B (A–*Betula*, *NAP*, *Salix*, *Juniperus*, B–*Betula*, *Pinus*, *Larix*), after Krupiński *et al.* (1986, 1988).

*Diatom phase BP2*

This phase is characterised (particularly in the subphase BP2a) by the gradual development of *Cyclotella* spp., mainly *C. vorticosa* (>17%) and *C. comta* var. *lichvinensis*, and also *Aulacoseira ambigua* (>15%). Subphase BP2b is characterised by a marked decrease in the number of *Navicula scutelloides*, a slight increase in *Fragilaria pinnata* (>10%) and only small oscillations in the frequency of *F. brevistriata*, and *F. construens* and var. *venter*. A relatively high level of *Fragilaria construens* var. *binodis* (>19%) is noted in this subphase (Fig. 3). The increasing importance of planktonic diatoms (particularly *Cyclotella*) was typical in lakes during the older part of the Mazovian Interglacial (compare Marciniak 1983, 1986, 1990; Bińka *et al.* 1987, Lindner *et al.* 1991). This suggests a slight deepening of the lake and is probably the result of a gradual warming of climate as well as slight improvement of trophic conditions during this phase.

In the pollen spectrum it is period II, the pollen zone C–*Picea*, *Alnus* (*Pinus*, *Betula*), the first stage of interglacial flora development, i.e. forests of boreal type (Krupiński *et al.* 1986, 1988; Krupiński 1988a).

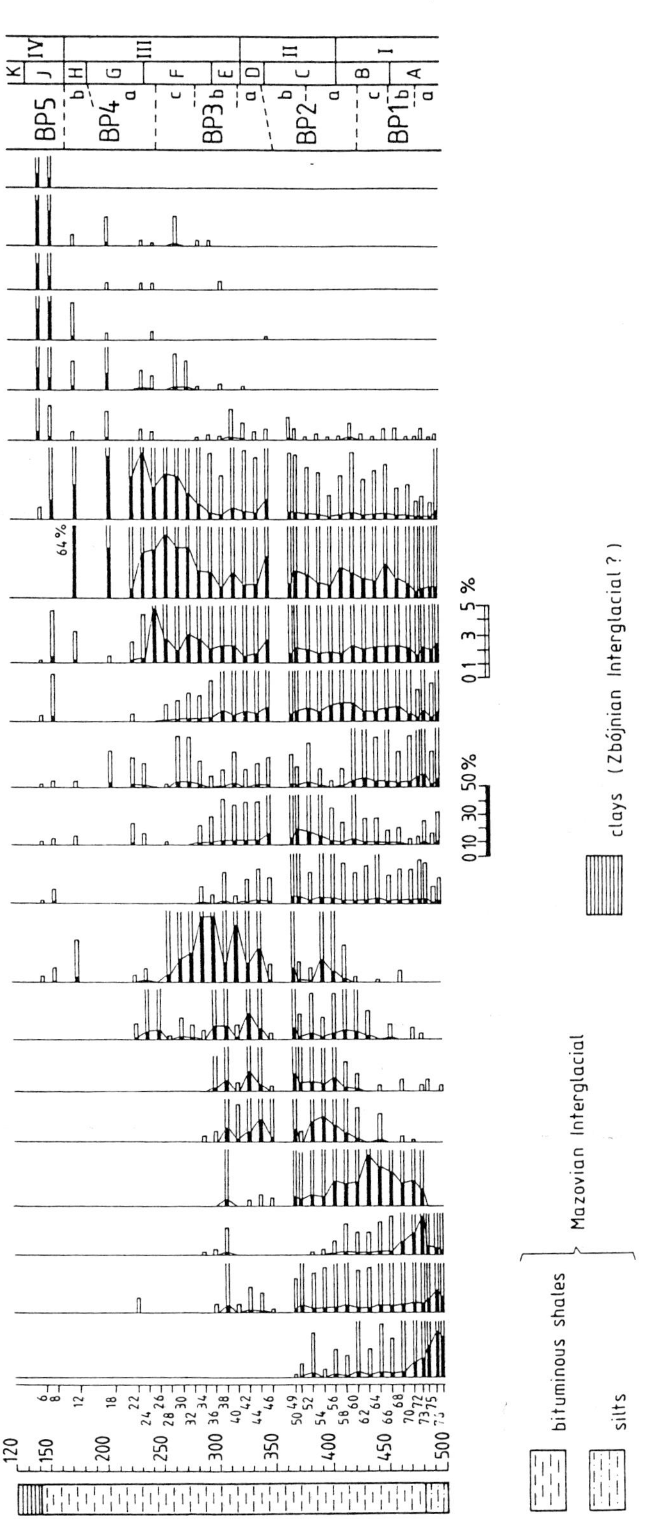


Fig. 3. Diatom percentage diagram of dominant taxa from borehole Biała Podlaska 2/84.

*Diatom phase BP3*

*Aulacoseira ambigua* is the most typical component of the diatom microflora (together with *Cyclotella* spp., *Aulacoseira granulata* and *Stephanodiscus* sp.) in the BP3 phase (Fig. 4). *A. ambigua* increases during subphases BP3a and BP3b and gradually decreases in the BP3c subphase, when *Fragilaria construens* var. *binodis* (44%) and *F. construens* var. *venter* (31%) dominate. The dominance of planktonic diatoms, particularly *Aulacoseira ambigua* (>46%), points to an increase in trophy and a rather high water level in the lake during the earlier part of this phase.

The diatom phase BP3 corresponds to the decline of period II, i.e. the pollen zone D–*Picea, Alnus* and to the lower part of period III embracing the pollen zones E and F (E–*Taxus, Picea, Alnus*, F–*Pinus, Picea, Alnus*) which are associated with forest development during the climatic optimum of the Mazovian Interglacial *sensu lato* (according to Krupiński 1988a).

*Diatom phase BP4*

Two subphases were distinguished (BP4a and BP4b) on the basis of the proportions of *Fragilaria* taxa. In the subphase BP4a *Fragilaria construens* var. *venter* (>46%) is very common, whereas in the subphase BP4b there is a maximum frequency of *Fragilaria construens* var. *binodis* (>64%). Changes in diatom composition (prevalence of periphytic diatoms) , as well as a lack of diatoms in some samples may prove shallowing or significant water level changes in the lake.

In the pollen profile this is the upper part of period III, i.e. pollen zones G and H (G–*Carpinus, Abies, Quercus, Corylus*, H–*Carpinus, Quercus, Corylus, Abies*) with forest flora recording the climatic optimum of the Mazovian Interglacial *sensu stricto* (according to Krupiński 1988a).

*Diatom phase BP5*

This phase is characterised by much damaged, corroded fragments of diatom frustules belonging to the genera *Stauroneis, Pinnularia, Navicula, Neidium* and *Fragilaria*, as well as corroded spore fragments of *Aulacoseira* sp. The above composition and very bad state of preservation of their frustules proves a shallowing of the lake, which had most probably changed into flooded peat. Changes of the diatom population during the fifth and last phase are probably a result of cooling and aridity of climate. These changes reflect the decline of the lake.

In the pollen diagram this phase corresponds to the decline of the Mazovian Interglacial in period IV, the pollen zone J–*Pinus, Betula, Sphagnum* (*Picea, Carpinus, Abies*). This was the transitional period from mixed forest communities to partly dispersed *Pinus* forests with *Betula* and *Abies*. At this time an increasing proportion of *Artemisia* and *Larix* was noted which indicates a more continental climate (Krupiński *et al.* 1986, 1988; Krupiński 1988a).

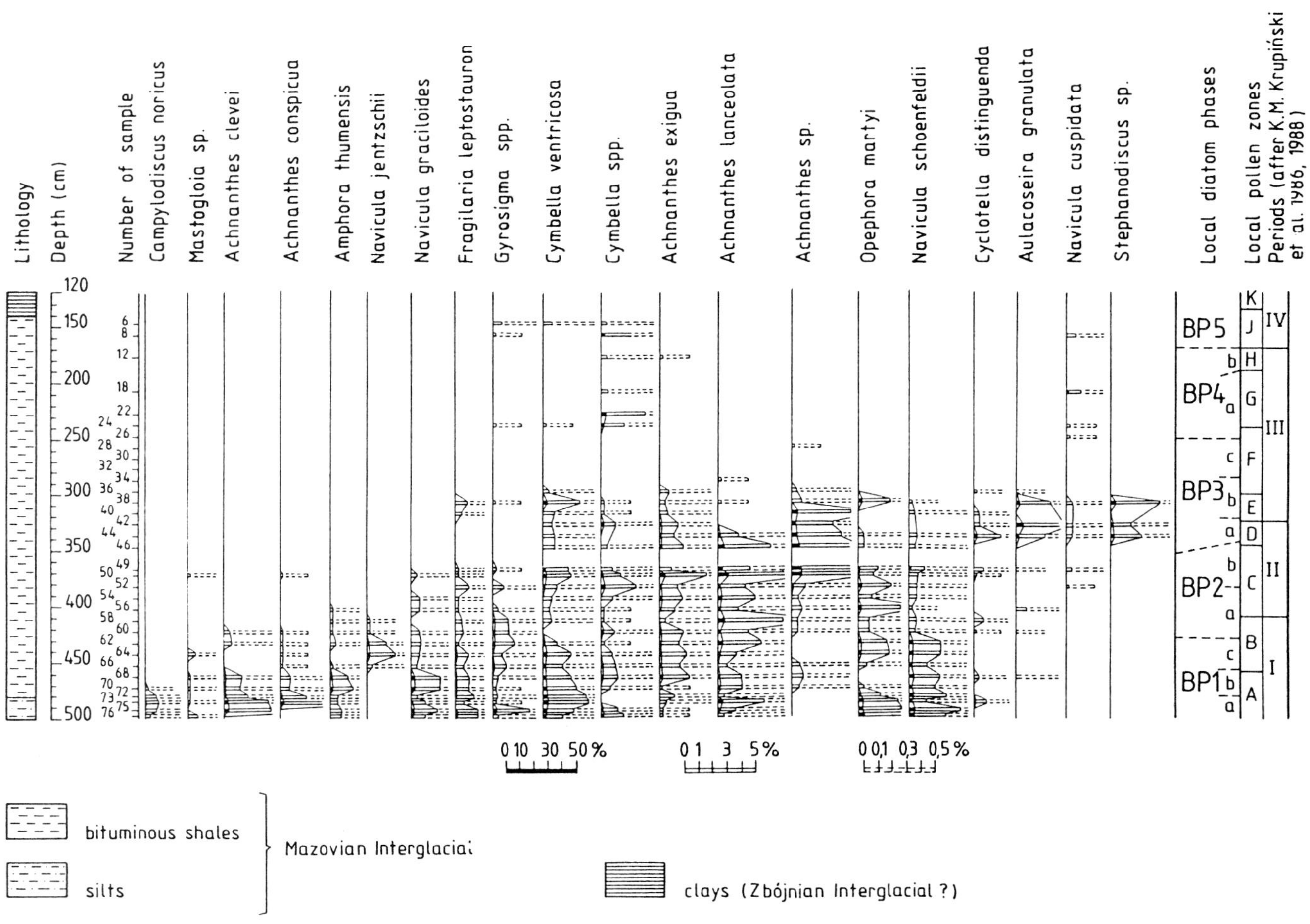


Fig. 4. Diatom percentage diagram of less common taxa from borehole Biała Podlaska 2/84.

## Final remarks

The Mazovian Interglacial of eastern Poland and Likhvin (Aleksandria) of the Russian Plain, have many sites containing similar diatom assemblages despite being far apart. The sites of the Mazovian Interglacial at Adamówka and Krepiec are particularly rich in *Cyclotella* taxa, as are those at Likhvin, Chekalin and Matveev Rov. More diversified floras have been noted at Krepiec and Gvoznitsa, where there is a large proportion of *Stephanodiscus* spp., *Aulacoseira* spp., *Fragilaria crotonensis* and *Asterionella formosa* (compare Loginova 1979, 1981, 1982; Marciniak 1986, 1990; Bińka *et al.* 1987; Lindner *et al.* 1991).

As compared with previously described diatom floras in Poland from interglacial lake profiles at Krepiec and Adamówka, the sediments at Biała Podlaska contain a different type of diatom flora and the composition is more diverse, as is the down core variation, particularly of the dominant forms. In the lower part of these sediments there are numerous *Navicula scutelloides* (>36%). This species is also abundant (>53%) in the Ishkold profile in Byelorussia, where the composition of the flora is still more diverse. Eight stages and twelve phases have been distinguished in the Ishkold lake development, where *Aulacoseira* is very common. This profile represents a climatic period with two optima, Maloaleksandria and Prinyemen (Yelovicheva & Khursevich 1981).

During the climatic optimum described at Biała Podlaska, *Aulacoseira* is also common, mainly *Aulacoseira ambigua.*

The pattern of development obtained here, and the presence of *Cyclotella comta* var. *lichvinensis* and *C. vorticosa*, i.e. forms typical for the Mazovian Interglacial s.s. in the area of Poland and for the Likhvin (=Aleksandria Interglacial) in the Russian Plain, strongly suggests the possibility of correlation of these interglacials on the basis of diatoms.

The need to raise the chronostratigraphic position of the Mazovian Interglacial s.s. on the basis of thermoluminescence dating of the Biała Podlaska sediments seems possible, both in the light of the discussion concerning the chronostratigraphic position of the Holstein s.s. Interglacial in Western Europe (compare Linke *et al.* 1985; Sarnthein *et al.* 1986; Schwarcz & Grün 1988; Balescu *et al.* 1991), and the recent data concerning the position of interglacial warmings of climate within the Pleistocene sediments of the North Sea where the Holstein Interglacial is correlated with the 7th oxygen state (compare Sejrup & Knudsen 1993).

## Acknowledgement

The authors are thankful to an anonymous reviewer for critical reading of the manuscript, valuable comments and correction of the language.

## References


Balescu, S., Packman, S. C. & Wintle, A. G. (1991). Chronological separation of interglacial raised beaches from Northwestern Europe using thermoluminescence. *Quaternary Research*, **35**, 91–102.

Bińka, K., Marciniak, B. & Ziembińska-Tworzydło, M. (1987). Palynologic and diatomologic analysis of the Mazovian Interglacial deposits in Adamówka (Sandomierz Lowland). *Kwartalnik Geologiczny*, **31**, 453–474.

Cepek, A. G. (1986). Quaternary stratigraphy of the German Democratic Republic. In: *Quaternary glaciations in the Northern Hemisphere* (V. Sibrava, D. Q. Bowen & G. M. Richmond, eds). *Quaternary Sciences Reviews*, **5**, 359–364.

Krupiński, K. M. (1988a). Plant succession of the Mazovian Interglacial age at Biała Podlaska (Eastern Poland). *Przeglad Geologiczny*, **11 (427)**, 647–655.

Krupiński, K. M. (1988b). About interstadial flora above the interglacial succession in the site Biała Podlaska. *Przeglad Geologiczny*, **11 (427)**, 665–669.

Krupiński, K. M., Lindner, L. & Turowski, W. (1986). Sediments of the Mazovian Interglacial at Biała Podlaska (Eastern Poland). *Bulletin of the Polish Academy of Sciences, Earth Sciences*, **34** (4), 365–373.

Krupiński, K. M., Lindner, L. & Turowski, W. (1988). Geologic-floristic setting of ther Mazovian Interglacial sediments at Biała Podlaska (E. Poland). *Acta Palaeobotanica*, **28** **(1–2)**, 29–47.

Lindner, L. (1988a). Stratigraphy and extents of Pleistocene continental glaciation in Europe. *Acta Geologica Polonica*, **38** **(1–4)**, 63–83.

Lindner, L. (1988b). Outline of Pleistocene stratigraphy of Biała Podlaska region (eastern Poland) and attempt of correlation with neighbouring area of the Soviet Union. *Przeglad Geologiczny*, **11** (427), 637–647.

Lindner, L. (1991). Problems of correlation of main stratigraphic units of the Quaternary of mid-western Europe. *Przeglad Geologiczny*, **5–6** **(457–458)**, 249–253.

Lindner, L., Krupiński, K. M., Marciniak, B., Nitychoruk, J. & Skompski, S. (1991). Pleistocene lake sediments of the site Hrud I near Biała Podlaska. *Kwartalnik Geologiczny*, **35** (4), 337–362.

Linke, G., Katzenberger, O. & Grün, R. (1985). Description and ESR dating of the Holsteinian interglaciation. *Quaternary Sciences Reviews*, **4**, 319–331.

Loginova, L. P. (1979). *Paleogeographiya likhvinskogo mezhlednikovya sredney polosy vostochno-evropeyskoy ravniny*. 138 pp. Nauka y tekjhnika, Minsk.

Loginova, L. P. (1981). Sravnitielnaya kharakteristika likhvinskoy diatomovoy flory Byelorussi y sopredelnykh rayonov. In: *Geologischeskiye issledovaniya kainozoya Byelorussi*. pp. 121–129. Nauka y tekhnika, Minsk.

Loginova, L. P. (1982). The Likhvin diatom flora from the central part of the East-European Plain, its paleogeographical and stratigraphic significance. *Acta Geologica Academiae Scientiarum Hungaricae*, **25**, 149–160.

Marciniak, B. (1983). Diatoms in the Mazovian Interglacial of the Lublin Upland. *Bulletin de l'Academie Polonaise des Sciences, Serie des Sciences de la terre*, **30** **(1–2)**, 77–85.

Marciniak, B. (1986). Diatoms in the Mazovian (Holstein, Likhvin) Interglacial sediments of south-eastern Poland. In: *Proceedings of the 8th International Diatom Symposium* (M. Ricard, ed.), 483–494. O. Koeltz, Koenigstein.

Marciniak, B. (1990). Dominant diatoms in the interglacial lake sediments of the Middle Pleistocene in Central and Eastern Poland. *Hydrobiologia*, **214**, 253–258.

Sarnthein, M., Stremme, H. E. & Mangini, A. (1986). The Holstein interglaciation.: Time-stratigraphic Position and Correlation to Stable-Isotope Stratigraphy of Deep-Sea Sediments. *Quaternary Research*, **26**, 283–290.

Schwarcz, H. P. & Grün, R. (1988). Comment on M. Sarnthein, H. E. Stremme & A. Mangini "The Holstein interglaciation: Time-Stratigraphic Position and Correlation to Stable-Isotope Stratigraphy of Deep-Sea Sediments". *Quaternary Research*, **29**, 75-79.

Sejrup, H. P. & Knudsen, K. L. (1993). Paleoenvironments and correlation of interglacial sediments in the North Sea. *Boreas*, **22**, 223–235.

Shackleton, N. J. & Opdyke, N. D. (1973). Oxygen isotope and paleomagnetic stratigraphy of equatorial Pacific core V28–238: Oxygen isotope temperatures and ice volumes on a 10/5 and 10/6 years scale. *Quaternary Research*, **3** (**1**), 39–55.

Siemińska, J. (1964). Bacillariophyceae – Okrzemki. In: *Flora słodkowodna Polski*, 6, PWN, Warszawa, 610 pp.

Yelovicheva, Y. K. & Khursevich, G. K. (1981). Ob uslozhneni stratigraphi srednego pleistotsena. In: *Geologicheskiye issledovaniya Kainozoya Byelorussi*, pp. 109–121. Nauka y Tekhnika, Minsk.

Zagwijn, W. H. (1989). The Netherlands during the Tertiary and Quaternary: A case history of Coastal Lowland evolution. *Geologie en Mijnbouw*, **68**, 107–120.

Zubakov, V. A. & Borzenkowa, I. I. (1990). Global palaeoclimate of the Late Cenozoic. *Developments in Palaeontology and Stratigraphy*, **12**, 456 pp. Elsevier, Amsterdam.

# Environment of Zeit Formation and post-Zeit section (Miocene – Pliocene ) in the Gulf of Suez, Egypt

Essam Tawfik[1] and William N. Krebs[2]

[1]*Gulf of Suez Petrolum Co, Maadi, Cairo, Egypt*
[2]*Amoco Production Co., Houston, Texas, USA.*

## Abstract

The Zeit Formation is late Miocene (Tortonian) in age and accumulated in a variety of shallow marine, fresh-water and brackish water settings. In the central and southern portions of the Gulf of Suez, the Zeit Formation is unconformably overlain by Pliocene shallow marine calcareous deposits known as the Ashrafi Formation, while in the northern part, it is unconformably overlain by the Darag Group (Wardan and Zafarana Formations). The Zafarana Formation was deposited in the same Zeit Formation environments, while the Wardan Formation was probably deposited in shallow marine environments.

Two assemblage zones were demonstrated in the present study, one characteristic of the Zeit Formation, the other typical of the post-Zeit section. The results are significant in providing age and environmental information on the shallow sub-surface rocks in the Gulf of Suez.

## Introduction

The Gulf of Suez is located on the northeastern margin of the Arab Republic of Egypt (Fig. 1). It is a complex rift basin that separates the African plate from the Sinai microplate. It originated during the latest Oligocene to earliest Miocene and is filled with as much as 5 km of Neogene and Quaternary sediment (Evans 1988). The latest Neogene-Quaternary rocks of the Gulf of Suez have been divided into several lithostratigraphic units (Figs 2a & b; El-Shafy 1990). Although siliceous microfossils have been recovered from much older rocks in the sub-surface and outcrops of the Gulf of Suez (Colletta *et al.* 1986; Noel & Rouchy 1986; Rouchy 1986; Krebs 1988a, b; Stonecipher *et al.* 1991), the present study is confined to those found in these younger units.

Age and environmental data about the younger evaporites and clastics of the Gulf of Suez are poor because of the lack of fossil recovery. Abdel Salam & El-Tablawy (1970) and Robertson Research (Robinson *et al.* 1982) reported the occurrence of diatoms in the Zeit Formation, while Krebs (1988a, b) documented the presence of marine and non-marine diatoms in the thin, fine-grained beds within these evaporites and clastics.

The purpose of this investigation is to document the stratigraphic distribution of all siliceous microfossils (particularly diatoms) within the Zeit Formation and "post-Zeit" section, to identify forms that have worldwide biochronologic importance, to shed light on the geologic ages of the subject formations, and to establish the environment of deposition of each formation based upon siliceous microfossils.

## Materials and Methods

The data for this study were derived from the well cuttings of six wells that span the length of the Gulf of Suez (see Fig. 1)

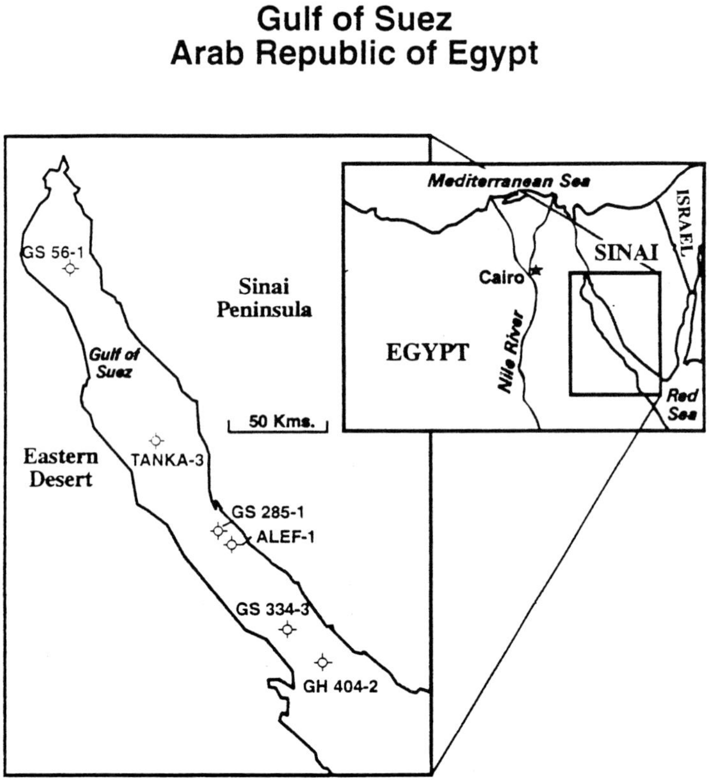


Fig. 1. Location of subject wells.

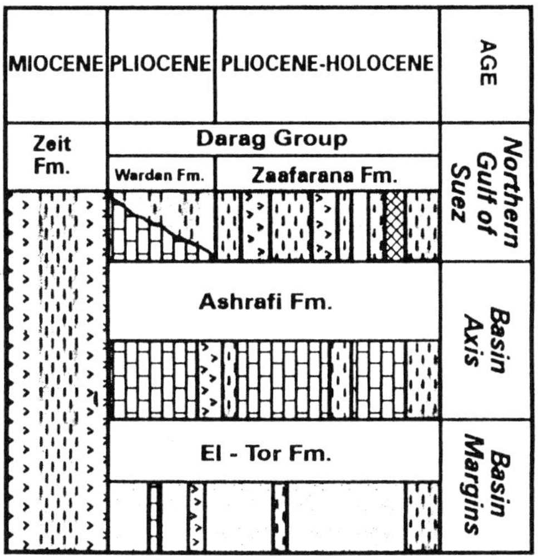


Fig. 2a. Stratigraphy of the post-Miocene section in the Gulf of Suez region (modified from El-Shafy 1990).

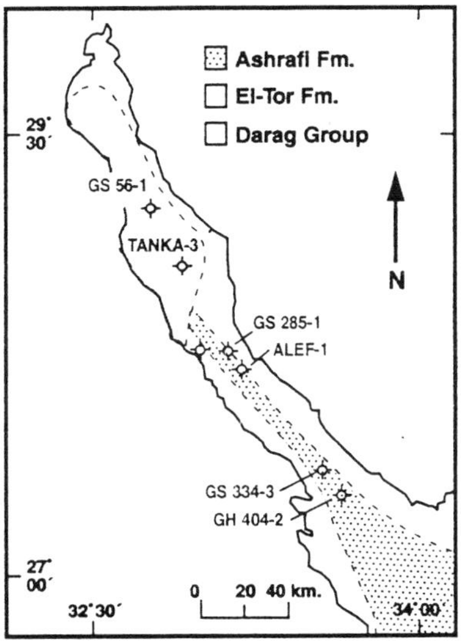


Fig. 2b. Lithofacies map of the post-Miocene section in the Gulf of Suez region (modified from El-Shafy 1990).

Because most of the subject formations consist of evaporites and/or coarse clastics, lithology and other logs were used to select sample intervals of fine-grained clastics where microfossil recovery would be most favourable. Sample processing consisted of treatment in hydrogen peroxide and acids, multiple centrifugation to neutral pH and the preparation of slides of the coarse and fine fractions. Data and preliminary range charts were processed through the RAGWARE system, charts were generated for the final displays (Figs 3 & 4).

## Results and Discussion

### *Recovery*

Diatoms (marine and nonmarine; DI) were the most common siliceous microfossils. Other siliceous microfossils included sponge spicules (PO), and silicoflagellates (SI). The Zeit Formation typically has the most diverse and best preserved diatom assemblages (Fig. 4). The Ashrafi Formation is dominated by marine forms, but there is a minor nonmarine component (Fig. 5). The Zaafarana and Zeit Formations contain a more mixed assemblage of marine and non-marine diatoms. The Wardan Formation, encountered in the TANKA–3 and GS 56–1 wells, is less diatomaceous and contains a predominantly marine flora (Fig. 3). Diatom recovery from the formation which underlies the Zeit Formation, the South Gharib Formation, is very sparse and is characterized by mixed marine and non-marine assemblages. Diatom "ghosts", the product of silica diagenesis, occur in the deeper portions of the Zeit Formation (GS 285–1) and in the South Gharib, Ras el Behar, and Belayim Formations. The presence of these ghosts is significant because they reveal that these units were once diatomaceous and that well-preserved diatoms from these formations may be recovered elsewhere, and that the phase transformation of diatomaceous opaline silica (opal–A) to cristobalite (opal–CT) has begun.

### *Diatom biostratigraphy and geologic ages of the Zeit, Ashrafi Formations and the Darag Group*

The stratigraphic distribution of diatoms within the late Neogene–Quaternary sediments of the Gulf of Suez provides new information on their geologic ages and environments of deposition (Fig. 5).

The occurrence of *Thalassiosira brunii* near the top of the Zeit Formation (Fig. 4) implies that this unit is no younger than Tortonian (late Miocene) in the subject wells. The presence of *Rhizosolenia miocenica, Triceratium condecorum, Thalassiosira temperei s.l.* and the abundance of *Rossiella mediopunctata* within the Zeit Formation corroborate this late Miocene determination.

*Actinoptychus annulatus*, an extant species, is restricted to the Zeit Formation (Fig. 4), and thus probably did not survive the Tortonian in the Gulf of Suez.

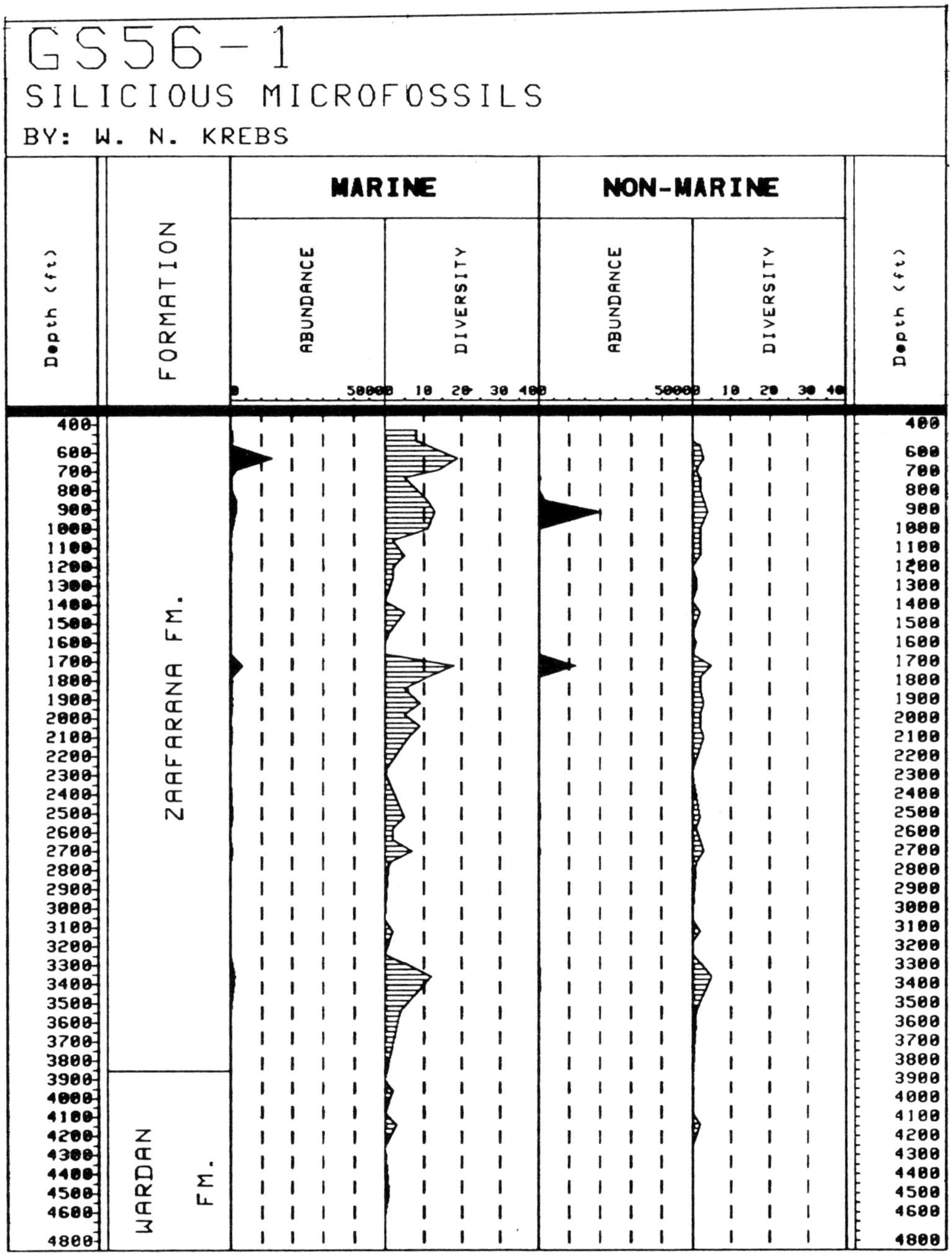


Fig. 3. GS56–1. Siliceous microfossils (W. N. Krebs).

Plate 1.

1. *Actinocyclus ehrenbergii* Ralfs 1861
   Zaafarana Fm GS. 56–1  3360 feet  D = 80 µm
2. *Actinocyclus ehrenbergii* v. *tenella* (Brebisson) Hustedt 1929
   Zeit Fm. GS. 334–3  2800–10 feet  D = 25 µm
3. *Actinocyclus* sp. 4
   Zaafarana Fm. GS. 56–1  630–40 feet  D = 64 µm
4. *Actinocyclus ehrenbergii* Ralfs 1861
   Zeit Fm. GS. 334–3  1900–10 feet  D = 50 µm
5. *Actinocyclus* sp.
   Ashrafi Fm. GS. 404–2  1700 feet  D = 49 µm
6. *Actinocyclus ellipticus* Grunow in Van Heurck 1883
   Zeit. Fm. Alef–1 2840–50 feet  D = 35 µm
7. *Actinocyclus* sp. 2
   Zeit Fm. GS. 334–3  1900–10 feet  D = 33 µm
8. *Actinocyclus curvatulus* Janisch in Schmidt 1876
   Zeit Fm. GS. 334–3  2440–50 feet  D = 39 µm
9. *Thalassiosira oestrupii* (Ostenfeld) Proshkina-Lavrenko 1956
   Zaafrana Fm. GS. 56–1  1780 feet  D = 18 µm
10. *Paralia* sp.
    Zaafrana Fm. GS. 56–1  630–40 feet  D = 38 µm
11. *Grammatophora* sp.
    Zeit Fm. GS. 285–1  2200–50 feet  D = 32 µm
12. *Caloneis* sp.
    Ashrafi Fm Alef–1  800–30 feet  L = 97 µm
13. *Opephora* sp.
    Zeit Fm. Alef–1  2840–50 feet  L = 51 µm
14. *Diatom elongatum* (Lyngbye) Agardh 1824
    Zaafarana Fm. GS. 56–1  630–40 feet  L = 52 µm
15. *Achnanthes* sp. –1
    Ashrafi Fm. GS. 285–1  1550–70 feet  L = 70 µm
16. *Achnanthes* sp. –2
    Zeit Fm. GS. 285–1  2370–400 feet  L = 48 µm
17. *Actinocyclus* sp. –3
    Ashrafi Fm. GS. 285–1  750–70 feet  D = 125 µm
18. *Coscinodiscus subtilis* Ehrenberg 1841
    Ashrafi Fm. Alef–1  800–30 feet  D = 52 µm
19. *Coscinodiscus kurzii* Grunow 1888
    Zeit Fm. GS. 334–3  1900–10 feet  D = 145 µm
20. *Actinocyclus* cf. *motilis* Bradbury & Krebs 1995
    Zeit Fm. GS. 334–3  1960–70 feet  D = 44 µm
21. *Diploneis smithii* (Brebisson in Wm. Smith) Cleve 1894
    Ashrafi Fm. GS. 285–1  1050–100 feet  L = 65 µm
22. *Diploneis* aff. *ovalis* Cleve 1891
    Zeit Fm. GS. 334–3  1900–10 feet  L = 48 µm
23. *Coscinodiscus marginatus* Ehrenberg 1841
    Zeit Fm. GS. 334–3  2380–90 feet  D = 78 µm
24. *Coscinodiscus argus* Ehrenberg 1838
    Ashrafi Fm. Alef–1  2840–50 feet  D = 63 µm
25. *Coscinodiscus* sp.
    Zeit Fm. GS. 285–1  2470–520 feet  D = 40 µm
26. *Diploneis* sp.
    Alef–1  2840–50 feet  D = 40 µm
27. *Nitzschia granulata* Grunow 1862
    Zeit Fm. GS. 334–3  1900–10 feet  L = 35 µm
28. *Navicula halionata* Pantocsek 1886
    Zeit Fm. GS. 334–3  1900–10 feet  L = 35 µm
29. *Navicula humerosa* de Brébisson
    Zeit Fm. GS. 334–3  1900–10 feet  L = 92 µm
30. *Dictyocha* sp.
    Ashrafi Fm. GS. 334–3  1600–10 feet  LD = 49 µm
31. *Mesocena quadrangula* (Ehrenberg) ex Haeckel Bukry 1973
    Zaafarana Fm. GS. 56–1  3360 feet  L = 58 µm
32. *Mesocena diodon borderlandensis* Bukry 1976
    Ashrafi Fm. GS. 285–1  750–70 feet  L = 83 µm
33. *Mesocena hexalitha* Bukry 1976
    Zeit Fm. GS. 334–3  1770 feet  D = 18 µm
34. *Octactis pulchra* Schiller 1925
    Ashrafi Fm. GS. 285–1  1000–20 feet  D = 20 µm.

PLATE 1

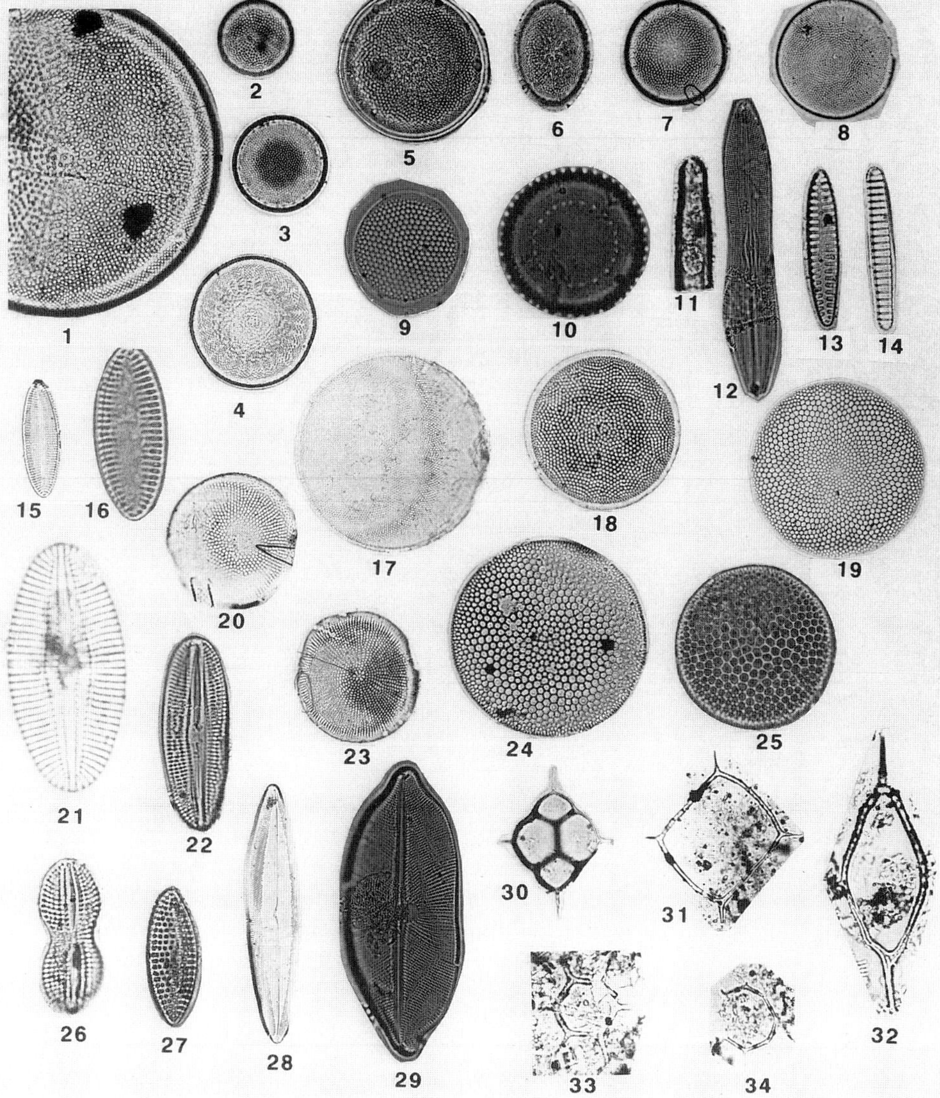

1
2
3
4
5
6
7
8
9
10
11
12
13
14
15
16
17
18
19
20
21
22
23
24
25
26
27
28
29
30
31
32
33
34

## Plate 2

1. *Thalassiosira antiqua* (Grunow) Cleve-Euler 1941
   Ashrafi Fm. GS. 334–3 1600–10 feet D = 43 µm
2. *Thalassiosira brunii* Akiba et Yanagisawa 1985
   Zeit Fm. GS. 334–3 2320–30 feet L = 31 µm
3. *Thalassiosira temperei* var. 2 *sensu* Fourtanier 1987
   Zeit Fm. GS. 285–1 2370–400 feet L = 22 µm
4. *Thalassiosira excentricus* var. *fasciculata* (Hustedt) 1962
   Ashrafi Fm. GS. 334–3 1600–10 feet D = 51 µm
5. *Thalassiosira temperei* s.l. (Brun) Akiba et Yanagisawa 1986
   Zeit Fm. GS. 334–3 2320–30 feet D = 21 µm
6. *Stephanodiscus rotula* (Kützing) Hendey 1964
   Zeit Fm. GS. 56–1 850–80 feet D = 43 µm
7. *Triceratium favus* Ehrenberg 1839
   Zaatarana Fm. GS. 56–1 480 feet L = 113 µm
8. *Triceratium condecorum* Ehrenberg 1944
   Zeit Fm. Alef–1 2840–50 L = 62 µm
9. *Triceratium spinosum* Bailey 1843
   Zeit Fm. GS. 285–1 2850–920 feet L = 102 µm
10. *Triceratium antediluvianum* (Ehrenberg) Grunow 1867
    Zeit. Fm. GS. 285–1 2850–920 feet L = 60 µm
11. *Triceratium antediluvianum pentagona* Hustedt 1929
    Zeit Fm. GS. 285–1 2850–920 feet D = 50 µm
12. *Chaetoceros* spores
    Zeit Fm. GS. 285–1 2430–50 feet D = 16 µm
13. *Bacteriastrum* sp.
    Zeit Fm. GS. 285–1 750–70 feet D = 12 µm
14. *Hyalodiscus* sp.
    Ashrafi Fm. Alef–1 800–30 feet D = 32 µm
15. *Hemidiscus ovalis* Lohman 1938
    Ashrafi Fm. Alef–1 2840–50 feet D = 24 µm
16. *Rossiella mediopunctata* (Hajos) Gersonde et Schrader 1968
    Zeit Fm. Alef–1 2840–50 feet L = 33 µm
17. *Cyclotella* cf. *quillensis* Bailey 1921
    Zaafarana Fm. GS. 56–1 560–80 feet D = 17 µm
18. *Cyclotella* sp.
    Zeit Fm. GS. 334–3 1960–70 feet D = 17 µm
19. *Cyclotella* aff. *kutzingiana* Thwaites (Schmidt *et al.* 1874)
    Zeit Fm. GS. 334–3 1960–70 feet D = 12 µm
20. *Cyclotella iris* Brun et Heribaud 1902
    Zeit Fm. Alef–1 2320–80 feet D = 26 µm
21. *Azpeitia* aff. *salisburyana* (Lohman) Sims 1948
    Zeit Fm. GS. 285–1 2370–400 feet D = 36 µm
22. *Aulacoseira* spp.
    Zeit Fm. Alef–1 1180–90 feet L = 25 µm
23. *Rhaphoneis* aff. *fatula* Lohman 1938
    Ashrafi Fm. GS. 334–3 1600–10 feet L = 41 µm
24. *Rhaphoneis* aff. *sachalinensis* Sheshukova-Poretzkaya 1967
    Zeit Fm. GS. 334–3 1900–10 feet L = 56 µm
25. *Rhaphoneis amphiceros* var. *tetragona* Grunow 1883 in Van Heurck
    Zeit Fm. GS. 285–1 2770–820 feet D = 30 µm
26. *Biddulphia* spp.
    Zeit Fm. GS. 285–1 2770–820 feet L = 124 µm
27. *Rhizosolenia styliformis* Brightwell 1858
    Zeit Fm. GS. 2895–1 3500–570 feet L = 38 µm
28. *Rhizosolenia hebatata* (Bial.) Gran 1904
    Zeit Fm. GS. 285–1 2370–400 feet L = 42 µm
29. *Rhizosolenia miocenica* Schrader 1973
    Zeit Fm. Alef–1 2840–50 feet L = 31 µm
30. sponge spicules
    Zeit Fm. GS. 285–1 3020–70 feet L = 62 µm
31. *Plagiogramma tesselatum* Greville 1859
    Ashrafi Fm. GS. 334–3 1600–10 feet L = 72 µm
32. *Auliscus* sp.
    Zaafarana Fm. GS. 56–1 690–700 feet D = 41 µm
33. *Actinoptychus* sp.
    Ashrafi Fm. GS. 285–1 1750–800 feet D = 80 µm
34. *Actinoptychus annulatus* (Wallish) Grunow 1883
    Zeit Fm. Alef–1 940–50 feet LD = 49 µm
35. *Surirella* sp.
    Zeit Fm. GS. 285–1 2470–520 feet L = 62 µm
36. *Ellerbeckia arenaria* (Ralfs ex Moore) Crawford 1986
    Zeit Fm. GS. 334–3 2500–10 feet D = 67 µm
37. *Asteromphalus* sp.
    Zaafarana Fm. GS. 56–1 630–40 feet D = 67 µm

PLALTE 2

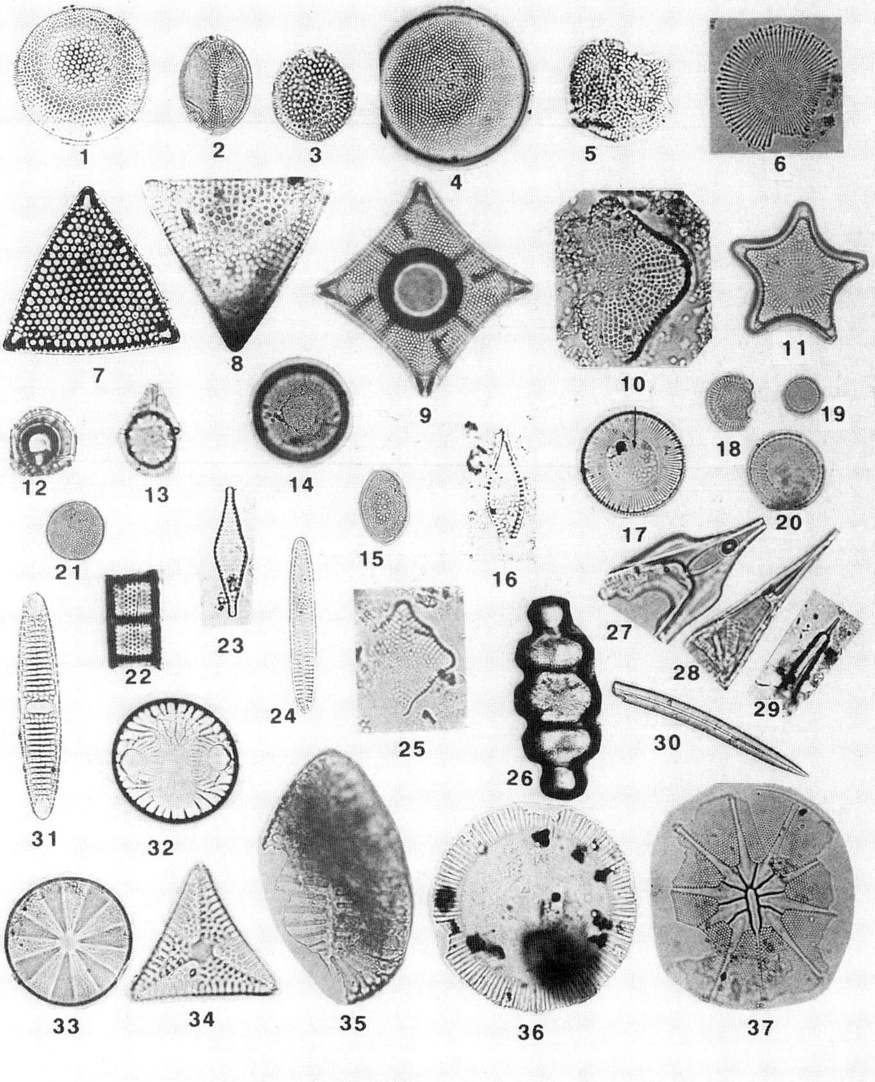
1
2
3
4
5
6
7
8
9
10
11
12
13
14
15
16
17
18
19
20
21
22
23
24
25
26
27
28
29
30
31
32
33
34
35
36
37

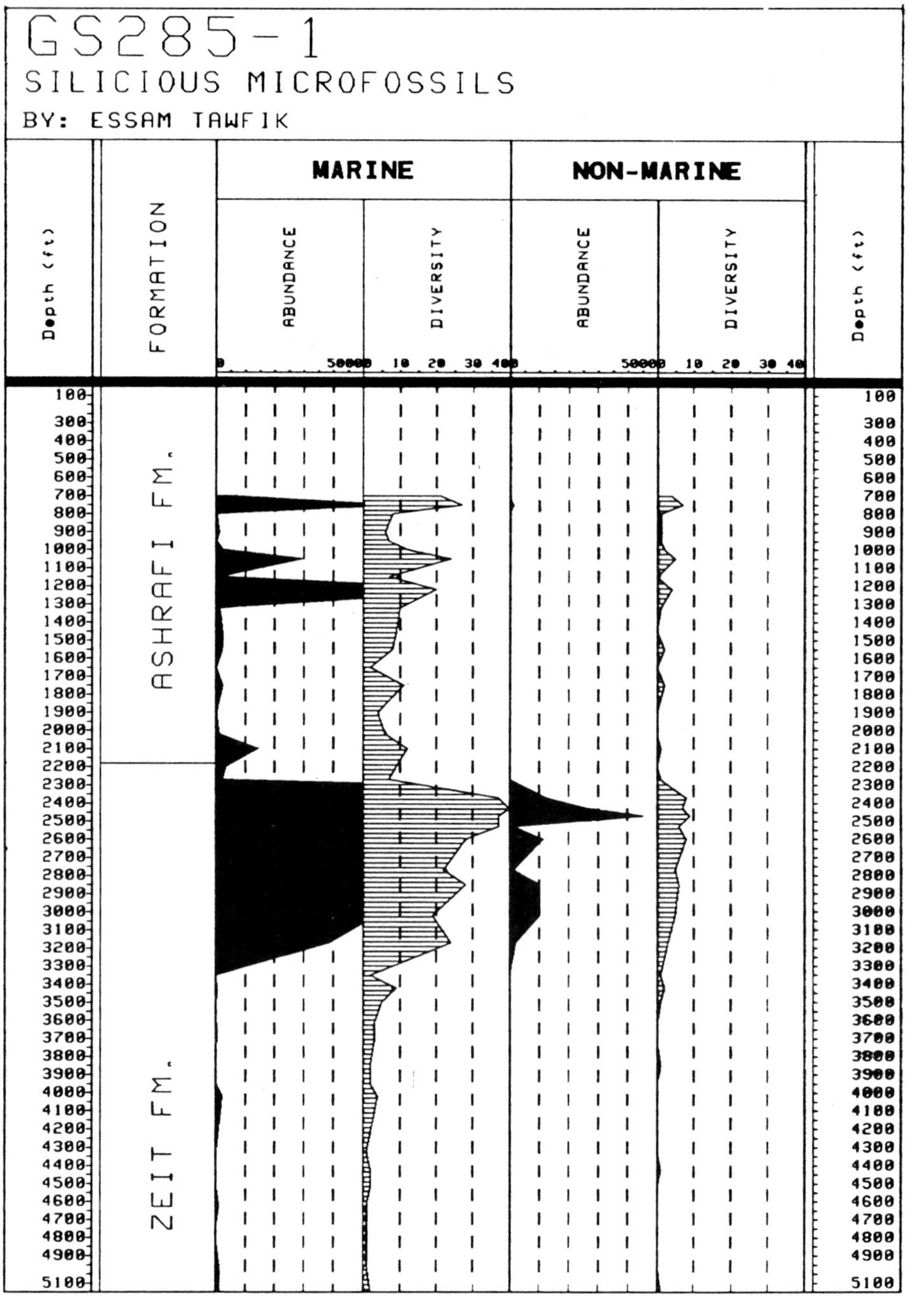


Fig. 4. GS 285–1. Siliceous microfossils (E. Tawfik).

The first downhole occurrence of this distinctive diatom is another important marker for the top of the Zeit Formation.

The overlying Ashrafi, or Wardan-Zaafarana Formations (Darag Group) contain Pliocene to Pleistocene marine and non-marine diatoms (Fig. 3). The last upsection occurrence of *Hemidiscus ovalis* (GS 56–1, GS 285–1), and the presence of *Thalassiosira antiqua* (GS 285–1) suggest that at least a portion of this section is Pliocene. The nonmarine diatoms *Cyclotella* cf. *C. quillensis* and *Stephanodiscus rotula* and the silicoflagellate *Mesocena quadrangula* in the Zaafarana Formation corroborate with the Pliocene–Pleistocene age. The occurrence of *Thalassiosira oestrupii* (GS 56–1) and the silicoflagellate *Octactis pulchra* (GS285–1) within the Zaafarana and Ashrafi Formations, respectively, indicates that both units are not older than Pliocene. There is no evidence from siliceous microfossils in the six subject wells for a late Miocene (Messinian) section. The contact between the Zeit Formation and overlying Wardan or Ashrafi Formation appears to be unconformable (Fig. 5), and comprises the Messinian and a possible portion of the Tortonian and Zanclean (early Pliocene). This unconformity has been corroborated by calcareous nannofossils and ostracods (El-Shafy 1990) and geologic field evidence (Evans 1988).

*Environments of deposition*

The Zeit, Ashrafi, Wardan, and Zaafarana Formations accumulated in a variety of shallow marine, marginal marine, and non-marine environments (Fig. 5). Diatom assemblages in the Ashrafi and Zeit Formations in the GS 285–1 well are typical of both units in the other wells. The Ashrafi Formation is characterized by a shallow marine assemblage. The abundance of *Actinocyclus* spp. corresponds with a near-shore environment, while the abundant and diverse assemblage of benthic diatoms (*Actinoptychus* spp., *Hyalodiscus* sp., *Paralia* sp., *Navicula* spp.) indicates shallow water conditions. *Thalassionema nitzschioides and Thalassiothrix longissima* are also abundant and suggestive of nutrient-rich marine water. There is a very minor non-marine diatom component in the Ashrafi Formation in the GS 285–1 well. Unlike the Zeit and Zaafarana Formations, evaporites are not a significant lithologic component in the Ashrafi Formation, nor are there significant numbers of fresh- and/or brackish-water diatoms. Diatom assemblages indicate that the Ashrafi Formation accumulated beneath nutrient-rich shallow marine water.

The underlying Zeit Formation contains a mixed assemblage of marine, freshwater, and brackish-water diatoms. Many marine species found in the Ashrafi Formation occur in the Zeit Formation as well, although there are some significant first occurrences. The Zeit marine diatoms, however, also lived in shallow water, while the abundance of non-marine diatoms corresponds with fresh- and brackish-water influences. In fact, some intervals within the Zeit Formation have a dominantly freshwater diatom assemblage while others are largely marine. The Zeit Formation appears to have formed in freshwater lacustrine, shallow marine, and lagoonal-estuarine settings. This non-marine influence in the Zeit Formation is corroborated by

| Lithostratigraphy | Age | Diatom Biostratigraphy | Paleoenvironment |
|---|---|---|---|
| Darag Group/ Ashrafi Fm. | Pliocene - Holocene | *Cyclotella* cf. *C. quillensis* - *Triceratium favus* - *Stephanodiscus rotula*<br>*Rhaphoneis* aff. *R. fatula*<br>*Hemidiscus ovalis* | shallow marine (Ashrafi and Wardan Fms.) to brackish water (Zaafarana Fm.) |
| | Messinian | | |
| Zeit Fm. | Late Miocene (Tortonian) | *Thalassiosira brunii* - *Actinoptychus annulatus* - *Thalassiosira temperei* s.l. | freshwater lacustrine, brackish water, shallow marine |

Fig. 5. Diatom biostratigraphy and paleoenvironments in the Zeit and Ashrafi Formations and Darag Group.

the abundance of the freshwater alga *Pediastrum spp.* (D. T. Pocknall, pers. comm.), and is in significant contrast with the overlying Ashrafi and Wardan Formations.

Diatom recovery in the Wardan Formation in the TANKA–3 and GS 56–1 wells is poor. In the former, it is dominantly shallow marine in aspect, and in the GS56–1 well, it may be caved. In any case, unlike the Zeit Formation, the wardan Formation is only slightly diatomaceous.

The Zaafarana Formation, although much younger than the Zeit Formation, formed in similar environments (Fig. 5). Non-marine diatoms, especially *Aulacoseira* spp., *Stephanodiscus rotula,* and *Cyclotella* cf. *C. quillensis,* are a significant component of the Zaafarana diatom assemblages. These non-marine diatoms are often mixed with shallow water marine forms, although some well intervals are dominated by either marine or non-marine diatoms. As in the Zeit Formation, evaporites are a significant constituent of the Zaafarana Formation. The latter appears also to have formed in shallow marine, freshwater, and brackish-water environments.

## Acknowledgements

The authors are grateful for the assistance of E. A. Balcells-Baldwin and Bert Barrett in the creation of the RAGWARE plot range charts that constitute Figs 3 & 4. We also thank David T. Pocknall for providing palynological information, Merrell A. Miller for a critical review of the manuscript and his helpful suggestions, and Jane Leighty, Gamal Mostafa for document wordprocessing services. G. W. McCraw and S. G. Friedrichs processed samples and prepared microscope slides.

## References

Abdel Salam, H. & El-Tablawy, M. (1970). Pliocene diatom assemblages from East Bakr and East Gharib exploratory wells in Gulf of Suez. Seventh Arab Petroleum Congress, Kuwait, paper no. **57** (B–3).

Colletta, B., Moretti, I., Chenet, P. Y., Muller, C. & Gerard, P. (1986). The structure of the Gebel Zeit area: a field example of tilted block crest in the Suez rift. *Total Compagnie Française des Pétroles, Institut Français du Pétrole,* Ref. I.F.P. 34 547.

El-Shafy, A. A. (1990). Miocene-Pliocene boundary in the Gulf of Suez region, Egypt. In: *EGPC Tenth Exploration and Production Conference, Cairo,* November 1990, 21 pp.

Evans, A. L. (1988). Neogene tectonic and stratigraphic events in the Gulf of Suez rift area, Egypt. *Tectonophysics,* **153,** 235–247.

Krebs, W. N. (1988a). Preliminary results – diatom analysis of Gulf of Suez well and outcrop samples. Amoco Production Co., Houston Regional Exploration report, 11 pp.

Krebs, W. N. (1988b). Addendum, Preliminary results – diatom analysis of Gulf of Suez well and outcrop samples. Amoco Production Co., Houston Regional Exploration, memo to R. W. Pierce, June 17 1988, 2 p.

Noël, D. & Rouchy, M. (1986). Transformations minérales *in situ* de frustules de diatomées du Miocène d'Egypte. Double voie de la diagènese: silicification et argilogènese. *C. R. Acad. Sci. Paris,* 303 (II) no. **19,** 1743–1748.

Robinson, E., Winsborough B. & Jiang, N. J. (1982). Biostratigraphic analysis of samples from five wells, Egypt. *Robertson Research (US)* rept. no. 823/157, 11 pp.

Rouchy, J.–M. (1986). Les évaporites miocènes de la Méditerranée et de la mer Rouge et leurs enseignements pour l'intérpretation des grandes accumulations évaporitiques d'origine marine. Bulletin Soc. Géol. France, **8** (II) no. 3, 11–520.

Stonecipher, S., May, J. A., Steinmetz, J. C. & Dyess, J. N. (1991). Evidence for deep-water evaporite deposition in the Miocene Kareem Formation, southwestern Gulf of Suez, Egypt. *Abstract, International Explorationist Dinner Meeting, Houston, Houston Geological Society,* April 17, 1991, 1 p.

# Diatom paleoecology of East Lake, Ontario: a 5400 yr record of limnological change

Jing-Rong Yang* and Hamish C. Duthie

*Department of Biology, University of Waterloo, Waterloo, Ontario, Canada N2L 3G1*

## Abstract

Diatom assemblages in a radiometrically dated sediment core from East Lake, on the northeastern shore of Lake Ontario, Canada, were analysed for the purpose of describing limnological events surrounding isolation from Lake Ontario. The earliest sample, 5350 BP, was dominated by euplanktonic diatoms and others favoring cold oligotrophic water, suggesting connection to Lake Ontario. The isolation process was evidently initiated around 5170 BP by formation of a baymouth bar, reflecting shoreline erosion caused by the rapidly rising water level of Lake Ontario. The process was marked by replacement of euplanktonic diatoms with shallow wetland benthic and epiphytic species in the newly formed lagoon, and a decline in diatom-inferred water depth. The Nipissing Flood increased diatom-inferred water levels in East Lake between 4500 BP and 3700 BP, and was signalled by a major change in diatom community composition. A post-Nipissing decline in water level was evidenced by a second shallow wetland phase. Continuing isostatic adjustment of the Lake Ontario outlet within the past 1000 y has resulted in the modern East Lake. We conclude that while hydrological events in Lake Ontario controlled major changes in water levels in East Lake over much of the mid- and late-Holocene, some lacustrine environmental changes may be attributed to climatic factors.

## Introduction

Diatom paleolimnological studies in Lake Ontario have mainly involved the analysis of short cores covering the recent historical period (Stoermer *et al.* 1985a; Yang *et al.* 1993; Duthie & Sreenivasa 1971). However, a paleo-environmental reconstruction of the Hamilton Harbour basin from analysis of a 8.5 m long core (Duthie *et al.* 1995) demonstrated that

*Present address: Department of Biology, University of Ottawa, 30 Marie Curie, P.O. Box 450 STN A, Ottawa, Ontario, Canada K1N 6N5

diatoms from nearshore sites are especially sensitive to long term water level fluctuations and other environmental changes.

This study addresses the long term development of East Lake, situated on the northeastern shore of Lake Ontario. Our major objective was to investigate Holocene environmental changes, in particular limnological responses to transgressions of Lake Ontario, by using proxy evidence from fossil diatoms in a radiometrically dated sediment core. We use the data to test the hypothesis that hydrological events in Lake Ontario have controlled water level changes in East Lake over the mid- and late-Holocene. The Holocene climatic history of southern Ontario has been described by Edwards and Fritz (1988) and Edwards and Buhay (1994). We anticipated that climatic factors would be signalled in the responses of the diatom record to changes in inferred trophic status, relative temperature and pH. The site near the eastern end of Lake Ontario was chosen for comparison with the Hamilton Harbour site at the western end of the lake (Duthie *et al.* 1995).

## Materials and Methods

East Lake (77°14'W, 43°54'N, Fig. 1) has a surface area of 1150 ha, a maximum depth of 8 m, and is separated from Athol Bay, Lake Ontario, by a permanent baymouth bar. The outlet to Athol Bay is a small river approximately two kilometres in length which passes through the baymouth bar and exits into a sand beach. Total phosphorus in the lake water ranges from 5 to 50 µg/L, and ranges of pH and Secchi disk transparency are 8.1–8.4 and 3.5–6.0 m repectively. Typical summer chlorophyll *a* values are around 6.3 µg/L (Paine 1990).

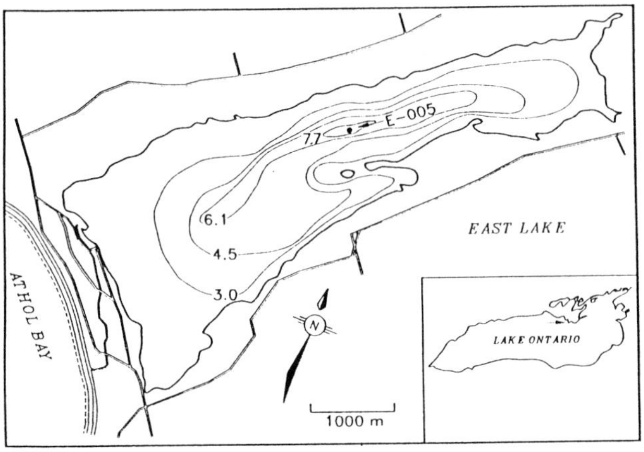


Fig 1. East Lake, showing location, coring site and depth contours (m).

In June of 1990, Core E–005 was collected by L. D. Delorme and A. Cywinska from East Lake at 8 m water depth using a lightweight core (Cywinska 1994). The natural length of the core was 332 cm which translated to a total uncompacted core depth of 388 cm (Turner 1992). The core was immediately subsectioned into 1 cm intervals, and subsamples were withdrawn for specific analyses after freeze-drying. A total of 68 samples were chosen for siliceous microfossil analysis at 5 cm intervals natural depth in the core. Selected samples were dated using $^{210}Pb$ and $^{14}C$ analysis (Turner 1992). The dates given for the diatom zones are estimates based on a constant rate of sedimentation between dated horizons.

Siliceous microfossils were cleaned with $H_2O_2$ and mounted in Hyrax ®. The slides were examined at 1000× magnification with a Nikon Labophot microscope (N.A. 1.25) and all diatoms were identified to the species or varietal level. The main taxonomic references used for identification included Hustedt (1930, 1937–1939, 1957), Cleve-Euler (1951–1955), Patrick & Reimer (1966, 1975) and Germain (1981). Approximately 600 valves were counted from each slide. Chrysophyte cysts, freshwater sponge fossils, and diatoms were counted from the same slides.

Enumeration data were processed through a database management system (StatView 512$^{+TM}$) and Excel. Calculated quantities include estimates of absolute abundance (microfossils per g dry weight of sediment), estimates of relative abundance (percentage of total assemblage) and estimates of assemblage diversity (Shannon–Wiener Index). Multivariate analyses were performed on relative abundance with the SYSTAT statistical package. All species that occurred in at least five samples and at >3% relative abundance in at least one sample were included in the statistical analysis. Cluster analysis was used to examine the pattern of community change with time. This analysis is based on the Euclidean distance measure and used the average distance clustering method (Stoermer *et al.* 1985a, Yang *et al.* 1993).

To characterize the fossil diatom assemblages, all identified diatom taxa were assigned to four habitat groups (euplankton, tychoplankton, epiphyton, benthos) using mainly the criteria of Patrick & Reimer (1966, 1975), and Lowe (1974). Changes in relative abundance of these groups in core samples were used to qualitatively indicate water-level fluctuations in the past. Isolation of the East Lake basin was identified by examining the disappearence of euplanktonic taxa of large lake species. Water depths of the bay before isolation from Lake Ontario were estimated by weighted average calibration (Stevenson *et al.* 1989):

$$ID_i = \sum_{k}^{m} W_k P_{ik} \Big/ \sum_{k}^{m} P_{ik}$$

Where $ID_i$ = inferred water depth at core sample $i_{Wk}$ = weighted average (optimum) value of water depth for each diatom taxon (Yang & Duthie 1995). Water depths of post-isolation East Lake were inferred by using multiple regression:

$D = -941.214 + 9.57Eup\% + 9.496Tyc\% + 9.361Epy\% + 9.406Ben\%$
($r = 0.871$, $r^2 = 0.758$, SE = 1.4 m, P = 0.0001); where D = diatom inferred water depth, Eup = euplankton, Tyc = tychoplankton, Epy = epiphyton, Ben = benthos.

Changes in trophic status were inferred with the empirical equations of Yang & Dickman (1993):

LTSI = 2.643–7.575 log (Index D) (SE = 0.94)

where LTSI = lake trophic status index, Index D = (O% + OM% + M%)/(E% + ME% + M%),O = oligotrophic, OM = oligomesotrophic, M = mesotrophic, ME = mesoeutrophic and E = eutrophic. The assignment of diatom taxa to trophic status categories was based mainly on Great Lakes literature. Listings may be found in Yang *et al.* (1993) and Yang (1994).

pH changes were estimated qualitatively by examining the relative abundance of diatom pH-preference groups of Hustedt (1937–1939), Meriläinen (1967), Cholnoky (1968), Lowe (1974) and Duthie (1989).

The distribution optima of many diatoms with respect to temperature and other climatic parameters have been reported (Lowe 1974, Stoermer & Ladewski 1976, 1978). Based on autecological characteristics from literature reports, we classified the diatom taxa found in this study in five temperature categories using a modified Hustedt (1957) system:- **Oligothermal** (oth) group; cold water forms, occurring at a water temperature no greater than 10°C and with an optimum below 5°C: **Oligomesothermal** (omth) group; cool water forms, occurring between 5–15°C and with optimum between 5–10°C: **Mesothermal** (mth) group; warm water forms, occurring between 10–20°C and with a temperature optimum between 10–15°C: **Mesoeuthermal** (meth) group; warm water forms, occurring between 15–30°C, with an optimum between 15–25°C: **Euthermal** (eth) group; warm water forms, found at temperatures no less than 25°C and with an optimum above 25°C.

## Results

### *Stratigraphy of siliceous microfossils*

In the deepest samples, the absolute abundance of siliceous microfossils declines between 388 and 360 cm and diatom species richness decreases from 39 species to zero (Fig. 2). Chrysophyte cysts are completely absent between 370 cm and 310 cm. Both fossil diatoms and freshwater sponges are absent between 360 and 310 cm except for very low densities at 336 and 346 cm.

All three types of siliceous microfossils increase between 310 and 260 cm. Diatom species richness in this interval fluctuates around 20 species and diversity sharply rises from 310 cm. The stratigraphic patterns of both chrysophytes and freshwater sponges above 300 cm are similar. Stratigraphic patterns of both diatom species richness and diversity above 260 cm are similar; both have high values in the top sample and in the interval between 80 cm and 120 cm.

Relative abundances of four diatom habitat groups change throughout the core (Fig. 3). Diatom assemblages are dominated by euplanktonic and tychoplanktonic groups between 360 cm and 388 cm. The benthic group is abundant in both the top and bottom sample of this interval, but is rare in between; the epiphytic group is absent. In the interval between 310 cm and 270 cm benthic diatoms dominate. Both euplanktonic and epiphytic groups above 270 cm disappear or become rare until 150 cm, and are replaced by the tychoplanktonic and benthic groups. Tychoplanktonic diatoms decrease in relative abundance from 150 cm, and epiphytic diatoms become important. From 85 cm to 50 cm the tychoplanktonic group

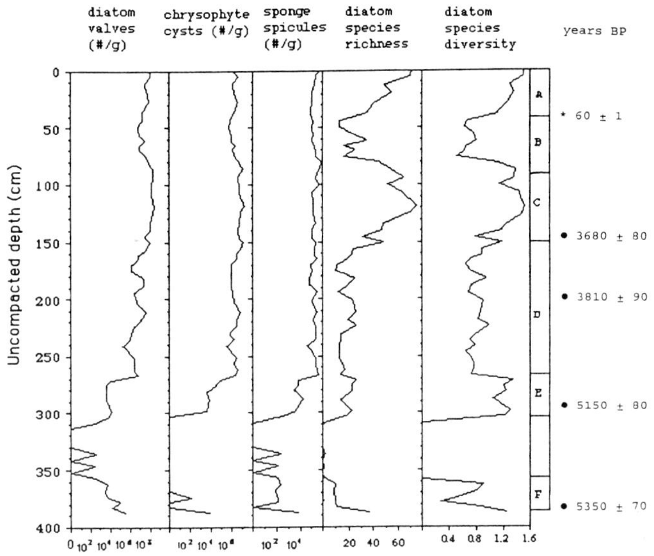


Fig. 2. Absolute abundance of diatom valves, chrysophyte cysts and sponge spicules (No./g dry weight of sediment), and diatom species diversity and species richness in core E–005 from East Lake. Diatom zones A–F are derived from cluster analysis. Radiometric dates (Turner 1992) shown as years before present (BP): • $^{14}C$ date, * $^{210}Pb$ date.

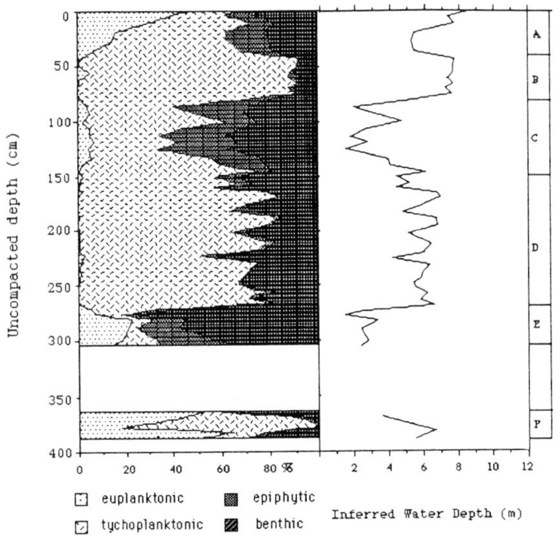


Fig. 3. Relative (%) abundance of diatom habitat types and diatom-inferred water depths in core E–005 in relation to diatom zones A–F. Information between 363 cm and 308 cm is lacking.

completely dominates the assemblages again. However, its relative abundance sharply declines in the top 50 cm, and is replaced by the euplanktonic group.

A total of 148 diatom taxa were found (Yang 1994). Relative abundances of 45 dominant taxa are shown in Fig. 4.

*Aulacoseira italica* is the most abundant species in the bottom sample of the core. It disappears immediately thereafter, not reappearing until 40 cm. *Fragilaria capucina* var. *lanceolata, Synedra rumpens, Stephanodiscus alpinus* and *Frustulia rhomboides* are major components of assemblages in the lower portion of the core below 360 cm.

*Aulacoseira ambigua, Tabellaria fenestrata, Eunotia pectinalis, Cymbella minuta, Synedra ulna* and *Cymbella affinis* are abundant below 360 cm, and then between 310 cm and 270 cm. More species become abundant between 310 and 270 cm, including *Gyrosigma attenuatum, Epithemia turgida, Cymbella laevis, Fragilaria vaucheriae, Eunotia valida, Denticula thermalis, Navicula vulpina, Pinnularia viridis, Navicula lanceolata, Hantzshia amphioxys, Navicula cuspidata* and *Surirella linearis.*

Beginning at 270 cm core depth, a major change occurs. The above-mentioned species become rare or absent, and are replaced by others. *Fragilaria brevistriata* is the most abundant species between 270 cm and 200 cm. It is also important in the top 90 cm of the core. Other major species in the middle portion of the core are *Navicula pupula* var. *elliptica* and *Surirella tenuis. Navicula radiosa* and *Gomphonema subclavatum* have similar stratigraphic patterns, being abundant between 310 cm and 270 cm, and again between 140 cm and 80 cm.

Profiles of both *Navicula pupula* and *Mastogloia smithii* exhibit cyclical patterns in abundance, but *N. pupula* is rare or absent between 270 cm and 160 cm. *N. viridula* is abundant (>10%) only at the middle portion of the core between 280 cm and 160 cm. *N. laterostrata* occurs only as a minor component of assemblages between 270 cm and 70 cm.

*Fragilaria construens* is the most important species between 80 cm and 40 cm. It reaches its maximum abundance at 75 cm (70%). It is also abundant in the core interval between 270 and 130 cm. The distribution of *F. construens* var. *venter* in the core is similar to that of *F. construens.* The stratigraphic pattern of *F. pinnata* is inverse to that of *F. construens* and *F. construens* var. *venter.*

*Cocconeis placentula, Achnanthes linearis* and *Cyclotella bodanica* have their highest abundant values between 140 cm and 70 cm. Beginning at 40 cm, *Amphora ovalis, Achnanthes conspicua, Martyana martyi, Aulacoseira italica* and *Fragilaria crotonensis* become important. In the top 10 cm of the core, the flora is dominated by *Cyclotella comensis, C. krammeri* and *Epithemia turgida.*

*Diatom-inferred (DI) water conditions*

*DI water depth*

DI depth exhibits marked fluctuations throughout the core (Fig. 3). DI depth below 370 cm is around 6 m, dropping sharply above this point. There were not enough diatoms to calculate DI depth for the interval between 310 cm and 358 cm.

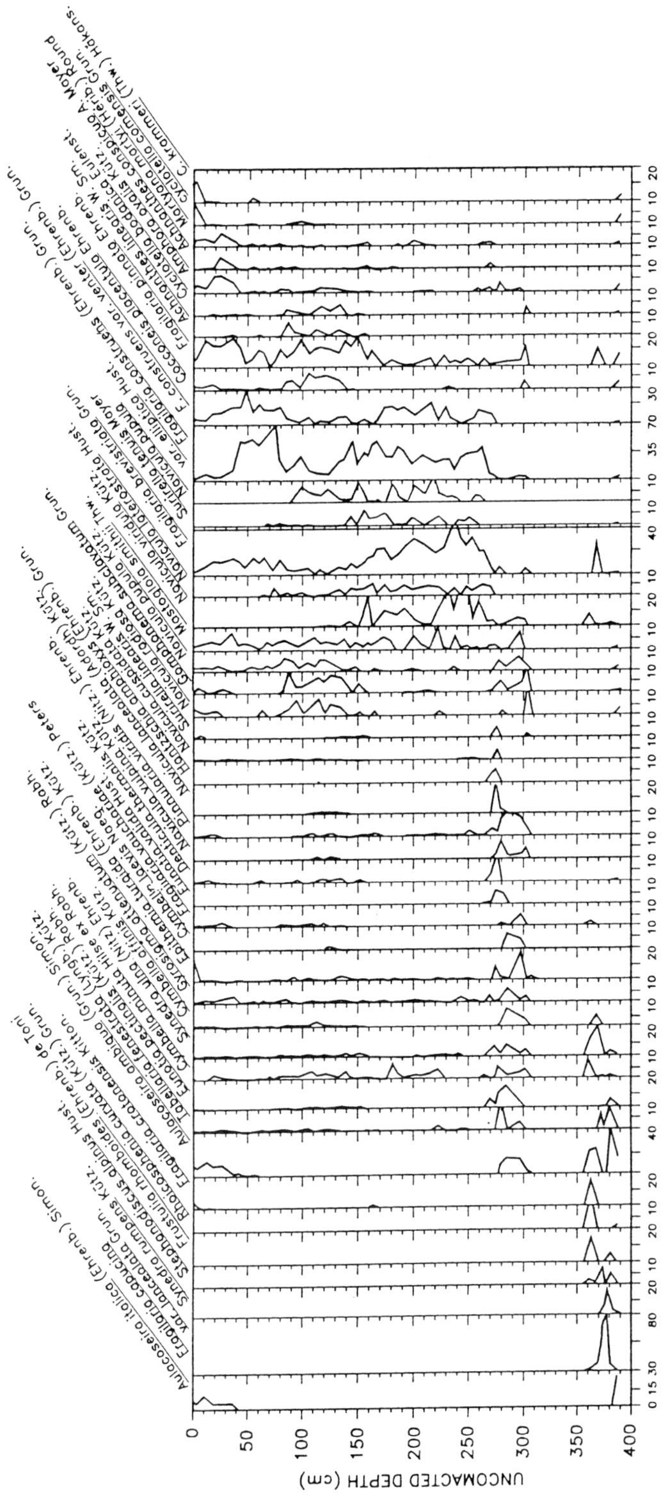


Fig. 4. Relative (%) abundance of 45 abundant diatoms in core E–005.

DI depth between 310 cm and 275 cm is < 3 m. From 275 cm it increases sharply, reaching over 6.5 m at 271 cm and then remains relatively stable around 6 m between 271 cm and 230 cm. Above 230 cm, it fluctuates cyclically between 4 m and 7 m until a core depth of 146 cm.

From 146 cm, DI depth drops sharply to below 1.5 m in the 125 cm sample. It varies around 2 m deep in the interval between 12 cm and 86 cm, except for an isolated maximum of 4 m at 101 cm. It increases sharply again after 86 cm, and remains at 7.5 m until the 44 cm sample. Finally, DI depth reaches 8.5 m in the top sample which compares with a measured depth of 8 m.

*Inferred Trophic status*

In general, mesotrophic diatom species dominate diatom assemblages throughout the core except in the lowest core interval (388 cm–360 cm) where both eutrophic and oligo-mesotrophic categories are abundant (Fig. 5a). The highest abundance of mesoeutrophic diatoms occurs in the interval between 270 cm and 180 cm, reaching a maximum at 235 cm (> 45%). Both oligotrophic and oligomesotrophic groups are rare or absent above 300 cm.

In contrast to the profile of DI depth, the profile of DI lake trophic status index (LTSI) does not strongly fluctuate. Below 360 cm DI trophic status increases upcore from 1.2 (oligotrophic status) to 3.6 (mesotrophic status). LTSI values increase from 300 cm and the maximum value of 5.1 occurs at 240 cm. From this point, LTSI gradually decreases and drops to 2.5 (oligomesotrophic) at 140 cm, persisting around this value until 50 cm. From 50 cm to the top, the LTSI decreases.

*Inferred temperature*

In the lowest sample the diatom assemblage is dominated by the oligothermal group (Fig. 5b). Above this level, but below 360 cm, the proportions of the warm habitat groups (mesothermal, mesoeuthermal and euthermal groups) and cold habitat groups (oligothermal and oligomesothermal groups) are similar. Taxa preferring warmer water dominate assemblages throughout the core above 300 cm, except at 11 cm. A high abundance of warm habitat taxa (>50%) appears at intervals between 270 cm and 130 cm, and between 80 cm and 40 cm.

*Inferred pH*

Profiles of pH preference groups are shown in Fig. 5c. The alkaliphilous group dominates the diatom flora, composing over 50% of the total diatoms in most samples of the core. The acidophilous diatom group is a minor component, significant only at the interval between 300 cm and 270 cm.

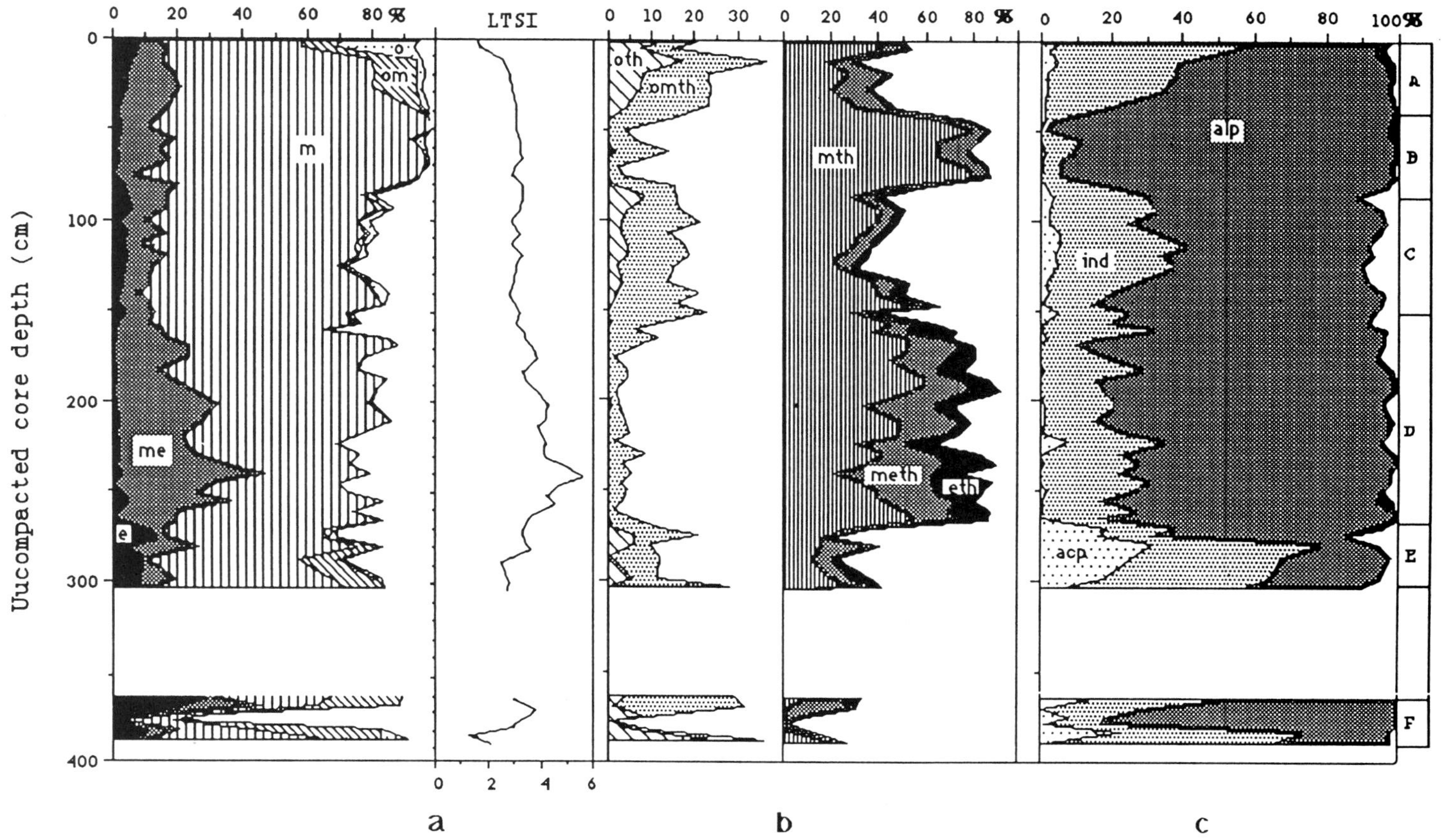


Fig. 5. a) Relative (%) abundance of five diatom trophic preference groups (o = oligotrophic, om = oligomesotrophic, m = mesotrophic, me = mesoeutrophic, e = eutrophic), and the value of diatom-inferred trophic status index (TSI) in relation to diatom zones A – F in core E–005. b) Relative (%) abundance of five diatom temperature groups (oth = oligothermal, omth = oligomesothermal, mth = mesothermal, meth = mesoeuthermal, eth = euthermal). c) Relative (%) abundance of diatom pH preference groups (acp = acidiphilous, ind = indifferent, alp = alkiliphilous).

## Synthesis and Discussion

The siliceous microfossil stratigraphy reveals that the environment of the East Lake basin has experienced major long term changes from an embayment of Lake Ontario to a shallow isolated lake. Based on cluster analysis of the 67 most abundant diatom taxa (Fig. 6), six major lacustrine phases (Zones A–F) may be recognized. A summary is given in Table 1.

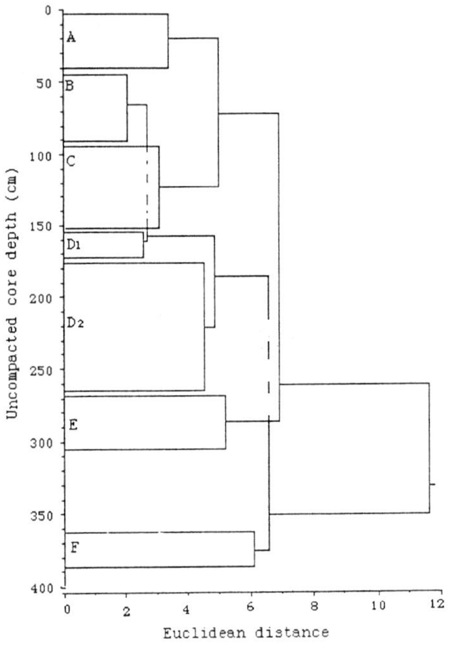


Fig. 6. Cluster analysis based on the 67 most abundant diatom taxa in core E–005.

*Zone F (388 cm – 361 cm, ca. 5350 BP – 5290 BP)*

Diatom assemblages in the bottom samples are dominated by a single euplanktonic species, *Aulacoseira italica.* This species is often abundant in the relatively low nutrient environment of Lake Ontario prior to European settlement (Stoermer *et al.* 1985a). Its dominance in the earliest sample of the core clearly shows that East lake was an embayment of Lake Ontario at ca. 5350 B.P. The low LTSI (< 2) suggests that the bay was in the lower level of oligomesotrophic status.

Above 380 cm, the abrupt decrease in diatom species richness, species diversity and siliceous microfossil concentration reveals a major environmental change, and may signal the beginning of isolation of the East Lake basin from Lake Ontario. *Aulacoseira italica* disappears, diatom assemblages become dominated by other euplanktonic species such as *Aulacoseira ambigua, Fragilaria capucina* var. *lanceolata, Synedra rumpens,*

Table 1. Summary of East Lake diatom zones with approximate dates and inferred environmental conditionsl

| DIATOM ZONE (YEARS BP) | SUMMARY OF DIATOM-INFERRED ENVIRONMENTAL CONDITIONS IN EAST LAKE |
|---|---|
| A (0 – 60) | Zone of human impact. Increase in euplanktonic species of *Cyclotella*. Decline in inferred trophic status. Greatly increased shoreline erosion and sedimentation rate. |
| B (60 – c.1000) | Inferred water depth increases to 7–8 m with continued rise in L. Ontario water level. A cooler, wetter climate also implicated. Return of planktonic diatoms. Increases in mesothermal and in mesotrophic diatoms. The *Ambrosia* pollen horizon, marking the historical period of European settlement (c.150 BP), is near the top of this zone. |
| C (c.1000 – 3690) | Post-Nipissing stabilization. Inferred water depth drops to 1.5 m. Second wetland phase. Benthic and epiphytic taxa dominate. Inferred temperature declines. A cooler, drier climate (Edwards & Buhay 1994). |
| D (3690 – 4480) | Inferred depth increases to about 7 m in early part of zone; coincides with Nipissing maximum in L. Ontario. Assemblages dominated by small species of *Fragilaria* and *Navicula*. High percentage of warm water diatom groups agrees with hypsithermal temperature maximum (Edwards & Buhay 1994). Mesotrophic and meso-eutrophic forms predominate; alkaliphilous diatoms abundant. |
| E (4480 – 5170) | Lagoon; shallow wetland < 2.5 m deep. High species diversity in a relatively warm and moist environment. Inferred trophic status mesotrophic – eutrophic. Alkaliphilous forms increase, acidophilous forms decrease. |
| (5170 – 5290) | Paucity of siliceous fossils limits interpretation. We postulate isolation by baymouth bar formation from an eroding shoreline caused by rapidly rising L. Ontario water levels. |
| F (5290 – 5350) | A zone of rapid change; early dominance of *Aulacoseira italica* implies connection with L. Ontario. Cold, oligotrophic conditions, increasing to mesotrophic. Evidence of acidic marginal wetland. |

*Stephanodiscus alpinus*, and *Tabellaria fenestrata*. These taxa are often abundant in relatively high nutrient environments of the modern Great Lakes (Stoermer *et al.* 1985a, 1985b; Yang 1994). As the East Lake basin became increasingly isolated from Lake Ontario, evidently more nutrients became available. A similar diatom response to parallel isolation processes have been reported by Wolin (1992) and Duthie *et al.* (1995). DI depth at the core location fluctuates from 5 m to 6 m.

At 369 cm, all previously dominant taxa except *Fragilaria crotonensis* were replaced by *Synedra ulna*, *Fragilaria brevistrata* and *Fragilaria pinnata*. In contrast to the previous dominants, these species are usually abundant in small lakes with high nutrient concentrations (Lowe 1974; Beaver 1981; Yang & Dickman 1993). Establishment of these species suggests that a sand bar between East Lake and Lake Ontario was developing, resulting in increased sedimentation. Occurrence of the epiphytic species *Cymbella affinis* (>5%) implies that aquatic macrophyte vegetation was developing.

The fossil diatom assemblage at 363 cm shows that a striking environmental change had occurred. At this level, all previously dominant species except *Aulacoseira ambigua* and *Synedra ulna* disappear. Increasing shallow-habitat benthic and epiphytic diatoms are consistent with a lower water level.

The isolation process is consistent with glacioisostatic uplift of the eastern Lake Ontario basin, which includes the East Lake region (Andrews 1968, Sly & Prior 1984). Our data clearly shows evidence of connection between the East Lake basin and Lake Ontario around 5350 BP. This is at variance with Anderson & Lewis (1985) who estimated the lake level of eastern Lake Ontario around 5000 BP to be about 15 m below its modern level. The major shifts in the diatom record described here imply rapidly changing water levels.

Siliceous microfossils between 363 cm and 308 cm (5290 BP–5170 BP) are almost completely absent; only a few shallow habitat diatom valves and sponge spicules were observed. We believe that the sand bar was formed in this interval. The rapidly rising water level of lake Ontario was likely accompanied by extensive shoreline erosion and baymouth bar formation. The newly isolated lagoon was probably a high sedimentation environment until it stabilized.

*Zone E (305 cm – 276 cm, 5170 BP – 4480 BP)*

Diatom assemblages below 280 cm are characterized by a high percentage of benthic diatoms and extremely low abundances of tychoplanktonic diatoms. DI depth in the early part of this zone is less than 2.5 m. It is notable that several acidophilous diatom species such as *Eunotia pectinalis*, *Eunotia valida*, and *Pinnularia viridis* became important components of the flora. These species are often abundant in shallow marsh environments (Scherer 1988).

By the end of Zone E, the acidophilous species *Eunotia pectinalis* and *Pinnularia viridis* sharply decrease or disappear. Planktonic species also become rare. Alkaliphilous taxa, *Denticula thermalis*, *Navicula lanceolata*, *Hantzshia amphioxys*, *Navicula cuspidata*, *N. radiosa* and *Surirella linearis* become dominant. The percentage of benthic diatoms sharply increases, and DI depth decreases to below 1.5 m. The wetland most likely expanded during this period. At the same time, an increased pH is indicated by an increase in the alkaliphilous group. DI trophic status increases in this zone, and similar patterns of temperature preference groups between Zone F and E indicate similar thermal regimes.

Clearly, diatom assemblages in Zone E imply a low water level period in East Lake. According to Edwards & Buhay (1994) the interval between 5000 BP. and 4500 BP. was a moist hypsithermal period. Although the wetland was maintained by the moist climate, East Lake water levels were mainly controlled hydrologically by the water plane of Lake Ontario acting via the groundwater table across the sandbar.

*Zone D (267 cm – 153 cm, 4480 BP – 3690 BP)*

In comparison with Zone E, Zone D is characterized by a higher concentration of siliceous microfossils and a lower diatom diversity. Diatom assemblages are characterized by small species of *Fragilaria* and *Navicula*. Except for *Cymbella minuta* and *Mastogloia smithii*, all major dominant taxa in Zone E become rare or absent. Small tychoplanktonic *Fragilaria* taxa became dominant. The DI depth during this period fluctuates around 7 m. Relatively warmer water in this period is also suggested by a high percentage of warmer water diatom groups.

The rise of water level in the East Lake basin at the beginning of this period is clearly linked to higher Lake Ontario levels and increased precipitation during the Nipissing Flood (Anderson & Lewis 1982). Maximum values of both temperature and moisture in the Lake Ontario region is proposed at 4000 BP (Edwards & Buhay 1994). According to Anderson and Lewis (1985), the highest water level in eastern Lake Ontario was also found around 4000 BP. In addition, higher water levels are consistent with the pollen data from the the same core where decreases in herbaceous vegetation suggest less extensive emergent vegetation (B. Warner, University of Waterloo, unpublished data).

*Zone C (153 cm – 93 cm, ca. 3690 BP – c. 1000 BP)*

In this zone, diatom assemblages are characterized by abundant epiphytes. The abundance of the previously dominant small forms of *Fragilaria* and *Navicula* decreases. Instead, the shallow benthic taxa *Navicula radiosa*, *N. pupula* and the epiphytic taxa *Gomphonema subclavatum*, *Cocconeis placentula* and *Achnanthes linearis* became important. Among these taxa, *Gomphonema subclavatum*, *Navicula radiosa* and *N. pupula* were also dominant in Zone E, a previous low lake level stage.

The lowering of the water level in the early Zone C reflects stabilization of the water plane in Lake Ontario after the Nipissing Phase. Anderson and Lewis (1985) suggested that the water level in Lake Ontario subsided after the Nipissing Phase as outlet channels adjusted to increased outflow from upper Great Lakes discharge. In addition, an increasingly dry climate in southern Ontario after 4000 BP probably contributed to the decrease in lake level (Swain 1978; Payette 1984; McCarthy and McAndrews 1988; Edwards and Buhay 1994).

The DI water depth in this zone fluctuates strongly. At 146 cm the inferred depth is still 6 m, but drops sharply thereafter, reaching about 1.5 m at the 125 cm sample. A slight increase in euplanktonic diatoms at 134 cm provides some evidence for the Lake Ontario highstand noted by McCarthy and McAndrews (1988) at 3000 BP in Grenadier Pond. However, the brief highstand at 1300 BP (100 cm) in our data was not detected by McAndrews.

Major changes in diatom assemblage structure at the end of Zone C and early Zone B are reflected in sharp declines in diatom species richness and diversity. Furthermore, whereas DI water depth at the end of Zone C is 1.8 m, it is 7.7 m in early Zone B.

*Zone B (93 cm – 42 cm, c. 1000 – 60 BP)*

Diatom assemblages in Zone B were again dominated by small *Fragilaria* taxa, mostly by the alkaliphilous *F. construens*, *F. construens* var. *venter* and *F. brevistrata* (Fig. 4). The occurrence of large lake euplanktonic taxa such as *Aulacoseira italica*, *Cycoltella comensis* and *C. Krammeri* in late Zone B may be a consequence of the backflooding of the Outlet River connection between East Lake and Lake Ontario (Paine 1990), driven by continuing lake level rise with isostatic rebound of the Lake Ontario outlet. Hamilton Harbour, at the western end of Lake Ontario, also showed an increase in DI water depth within the last 1000 years (Duthie *et al.* 1995).

*Zone A (42 cm – surface, 60 BP – present)*

Euplanktonic diatoms sharply increase. *Cyclotella comensis* and *C. krammeri* become major components in this period. DI depth increases to 8.5 m in the top sample, consistent with a measured water depth of 8 m at the core site in 1990.

A sudden increase in the abundance of *Ambrosia* pollen in the topmost sample in Zone B marks the historical period of European settlement (about 150 BP, B. Warner, unpublished). The responses from diatom assemblages were increased diatom species richness and diversity (Fig. 2). Surprisingly, LTSI values in this interval decrease. Although a food cannery operated on the shores of East Lake for a number of years in mid-century, it has left no evidence in the sedimentary diatom record at the coring site.

In conclusion, the long-term development of East Lake has been largely controlled by hydrological changes in adjacent Lake Ontario, either directly by transgression or via the groundwater plane. The core records the Nipissing transgression and bar formation (reflecting rapid shoreline erosion at highstand), followed by the establishment of East Lake, which persisted after the Nipissing phase subsided because of the bar. The subsequent gradual rise in Lake Ontario due to continuing isostatic adjustment of the outlet, has recently resulted in limited influence from Lake Ontario due to back-flooding. East Lake affords an interesting contrast with Hamilton Harbour (Duthie *et al.* 1995) because Burlington Bar already existed prior to connection with Lake Ontario.

## Acknowledgements

We thank L. D. Delorme for providing the core material, for supplying the radiometric dates, and for his advice and interest throughout the project. We thank T. W. Edwards for critically reading the manuscript and for his many useful suggestions. The research was funded by a GLURF grant to HCD.

## References

Anderson, T. W. & Lewis, C. F. M. (1982). The Mid–Holocene Nipissing Flood into Lake Ontario. *America Quaternary Association Program and Abstracts, Seventh Biennial Conference*, Seattle, Washington, page 60.

Anderson, T. W. & Lewis, C. F. M. (1985). Postglacial water-level history of the Lake Ontario Basin. In: *Quaternary Evolution of the Great Lakes*, (P. F. Karrow & P. E. Calkin, eds), 231–251. *Geological Association of Canada Special Paper* **30**.

Andrews, J. T. (1968). Postoglacial Rebound: Similarity and prediction of Uplift Curves. *Canadian Journal of Earth Sciences,* **5**, 39–47.

Beaver, J. (1981). *Apparent ecological characteristics of some common freshwater diatoms.* Water Quality Branch, Ministry of the Environment, Toronto, Ontario. 294pp.

Cholnoky, F. J. (1968). *The Ecology of Diatoms in Inland Waters.* J. Cramer, Lehre (in German), 699 pp.

Cleve-Euler, A. (1951–1955). Die Diatomeen von Schweden und Finland. *Kunglinga Svenska Vetenskaps-Academiens Handlingar,* Serien IV, 2(1), 1–163, 3(3), 1–153, 4(1), 1–158, 4(5), 1–255, 5(4), 1–232.

Cywinska, A. (1994). Stratigraphic analyses of ostracode fossils in East Lake, Ontario; the reconstruction of paleoclimatic and trophic conditions. Ph.D. thesis, University of Waterloo, Ontario. 220 pp.

Duthie, H. C. (1989). Diatom-inferred pH history of Kejimkujik Lake, Nova Scotia, a reinterpretation. *Water Air and Soil Pollution,* **46**, 317–322.

Duthie, H. C. & Sreenivasa, M. R. (1971). Evidence for the eutrophication of Lake Ontario from the sedimentary diatom succession. *Proceedings of the 14th Conference for Great Lakes Research.* International Association for Great Lakes Research, 1–13.

Duthie, H. C., Yang, J.-R., Edwards, T. W. D., Wolfe, B. B. & Warner, B. G. (1995). Hamilton Harbour, Ontario: 8300 years of limnological and environmental change inferred from microfossil and isotropic analyses. *Journal of Paleolimnology.* In press.

Edwards, T. W. D. & Buhay, W. M. (1994). Isotope paleolimnogy in southern Ontario. In: *Great Lakes Archaeology and Paleoecology: exploring interdisciplinary initiatives for the nineties* (R. I. MacDonald, ed.), 23–30. Quaternary Sciences Institute, University of Waterloo, Canada.

Edwards, T. W. D. & Fritz, P. (1988). Stable-isotope paleoclimate records for southern Ontario, Canada: comparison of results from marl and wood. *Canadian Journal of Earth Sciences,* **25**, 1397–1406.

Germain, H. (1981). *Flore des Diatomees Eaux et Saumatres.* Boubee, Paris, 444p.

Hustedt, F. (1930). *Die Süsswasser-Flora Mitteleuropas (Jena).* **10**, Bacillariophyta (Diatomeae), Verrlag von Gustav Fischer, pp. 1–466.

Hustedt, F. (1937–1939). Systematic and ecological investigation of the diatom flora of Java, Bali and Sumatra. *Archive für Hydrobiogie (Stuttgart) Supplement* Band. **15**, 131–177, **16**, 187–295 & **17**, 393–506 (in German).

Hustedt, F. (1957). The diatom flora of the Weser River systems. *Abhandl. Naturwiss. Ver. Bremen*, **34**, 181–440 (in German).

Lowe, R. L. (1974). *Environmental Requirements and Pollution Tolerance of Freshwater Diatoms*. EPA–670/4–74–005, Cincinnati, Ohio. 334 pp.

McCarthy, F. M. G. & McAndrews, J. H. (1988). Water levels in Lake Ontario 4230–2000 years B.P.: evidence from Grenadier Pond, Toronto, Canada. *Journal of Paleolimnology* , **1**, 99–113.

Meriläinen, J. (1967). The diatom flora and the hydrogen ion concentration of water. *Ann. Bot. Fenn.* **4**, 51–58.

Paine, J. D. (1990). East Lake water level management study. Technical Report, Prince Edward Conservation Authority, Ontario, Canada, 1–30.

Patrick, R. & Reimer, C. W. (1966). *The Diatoms of the United States Exclusive of Alaska and Hawaii*, **1**. The Academy of Natural Sciences of Philadelphia, Monograph No. **13**, 688 pp.

Patrick, R. & Reimer, C. W. (1975). *The Diatoms of the United States Exclusive of Alaska and Hawaii*, **2**. The Academy of Natural Sciences of Philadelphia, Monograph No. **13**, 213 pp.

Payette, S. (1984). Peat inception and climate changes in northern Quabec. In: *Climate changes on a yearly to millenial basis* (Morner and Karlen eds). D. Reidel publishing Co. Boston. pp. 173–180.

Scherer, R. P. (1988). Freshwater diatom assemblages and ecology/paleoecology of the Okefenokee Swamp/Marsh Complex, Southern Georgia, U.S.A. *Diatom Research*, **3**, 129–157.

Sly, P. G. & Prior, J. W. (1984). Late-glacial Geology in the Lake Ontario basin, *Canadian Journal of Earth Sciences*, **21**, 802–821.

Stevenson, A. C., Birks, H. J., Flower, R. J. & Battarbee, R. W. (1989). Diatom–based pH reconstruction of lake acidification using canonical correspondence analysis. *Ambio*, **18**, 44–52.

Stoermer, E. F. & Ladewski, T. B. (1976). Apparent optimal temperatures for the occurrence of some common phytoplankton species in southern Lake Michigan.*University of Michigan, Great Lakes Research Division Special Report*, No. **18**. 48 pp.

Stoermer, E. F. & Ladewski, T. B. (1978). Phytoplankton association in Lake Ontario During IFYGL. *University of Michigan, Great Lakes Research Division Special Report*, No. 62. 106 pp.

Stoermer, E. F., Wolin, J. A., Schelske, C. L. & Conley, D. J., (1985a). An assessment of ecological changes during the recent history of Lake Ontario based on siliceous algal microfossils preserved in the sediments. *Journal of Phycology*, **2**, 257–276.

Stoermer, E. F., Wolin, J. A., Schelske, C. L. & Conley, D. J. (1985b). Post settlement diatom succession in the Bay of Quinte, Lake Ontario. *Canadian Journal of Fisheries and Aquatic Sciences*, **42**, 754–767.

Swain, A. M. (1978). Environmental changes during the past 2000 years in north central Wisconsin: Analysis of pollen, charcoal and seeds from varved lake sediments. *Quaternary Research*, **10**, 55–68.

Turner, L. J. (1992). $^{210}$Pb dating of lacustrine sediments from East Lake. Technical Note, RAB–90–22, National Water Research Institute, Lakes Research Branch, Canada Center for Inland Waters, Burlington, Ontario, Canada. 54 pp.

Wolin, J. A. (1992). Paleoclimatic implications of Late Holocene lake-level fluctuations in Lower Herring Lake, Michigan. Ph.D. thesis, University of Michigan, Ann Arbor, Michigan. 155 pp.

Yang, J. (1994). Reconstruction of paleoenvironmental conditions in Hamilton Harbour and East Lake, Ontario from analyses of siliceous microfossils. Ph.D. Thesis, University of Waterloo, Waterloo, Ontario, Canada. 244pp.

Yang, J. & Dickman, M. (1993). Diatoms as indicators of lake trophic status from Central Ontario. *Diatom Research,* **8**, 179–193.

Yang, J. & Duthie, H. C. (1995). Regression and weighted averaging models for inferring water depth from sedimentary diatom assemblages in Lake Ontario. *Journal of Great Lakes Research*, **21**, 84–94.

Yang, J., Duthie, H. C. & Delorme, L. D. (1993). Reconstruction of the recent environmental history of Hamilton Harbour from quantitative analysis of siliceous microfossils. *Journal of Great Lakes Research*, **19**, 55–71.

# CASPIA update on saline lake diatoms: report on a workshop

Laurence R. Carvalho[1] and Sherilyn C. Fritz[2]

[1]*Department of Botany, The Natural History Museum, Cromwell Road, London SW7 5BD, U.K.*

[2]*Department of Earth & Environmental Sciences, Lehigh University, 31 Williams Drive, Bethlehem, PA 18015, U.S.A.*

## CASPIA workshop

The workshop began with an outline of the history of the CASPIA project and its present state. What follows is a summary of this, accompanied by a brief outline of some of the points raised.

Diatom remains in the sediments of saline lakes can provide a direct record of past salinity and an indirect measure of water level and climate change (Fritz *et al.* 1991; Gasse 1987). Quantitative reconstructions of salinity require the development of a transfer function calibrated from a modern data set of diatoms and water chemistry. The CASPIA project (Climate and Salinity Project) was set up in 1991 by a group involved in the collection of these modern data sets from various saline lake regions around the world (Juggins *et al.* 1994). The aim of the project was to merge the different regional data sets into a single data base of diatom and environmental data.

Merging of regional data sets is desirable as it can provide a fuller understanding of species optima and tolerances. It also improves salinity transfer functions where modern analogues of fossil taxa do not occur within the same geographical region. Consistent sampling methodology and taxonomy is a prerequisite before merging the data sets, particularly if weighted-averaging based transfer functions are to be applied, as these rely on precisely quantified, species-rich assemblages (ter Braak & Looman 1986).

Taxonomic consistency is being achieved using methods of taxonomic quality control (TQC) developed during the SWAP (Munro *et al.* 1990) and PIRLA (Kingston *et al.* 1992) projects on lake acidification. Initially this involved the comparison of

[1] Address for correspondence

floras and identification of problem taxa, followed by several workshops where the problem taxa, largely within the genera *Navicula, Nitzschia*, and *Amphora*, were compared, and agreements were made on protocols for identification and nomenclature. A programme of slide exchange was also initiated to test workshop protocols relating to these taxa.

Presently the system of TQC is being taken a step further for the African data sets (East Africa: Gasse *et al.* 1983; Sahara and Sahel: Gasse 1987; Gasse *et al.* 1987; North Africa: Ben Khelifa 1989) and the Northern Great Plains (North America) data set (Fritz *et al.* 1993) by the development of a saline lake flora for these regions. All taxa will be checked against and tied to type material, or voucher material held at the Natural History Museum when type material is not available. The flora will include a diatom iconograph consisting of light and scanning electron microscope images of type specimens and project-based material. Since the flora and iconograph will be a reference tool for micropalaeontologists it will emphasise distinguishing characters visible under the light microscope and contain images of specimens in various dissolution states. The flora will develop concurrently with the publications of accounts of the more important taxa (Table 1). These will initially focus on problematical taxa, taxa that have different names in different regions but are believed to be synonymous, and taxa which are present in core material from one region but only appear to have modern analogues in another region. The first completed study investigated material recorded as *Cyclotella caspia* Grunow, and has shown that its correct name is *Cyclotella choctawhatcheeana* Prasad (Carvalho *et al.* submitted).

A linked taxonomic and ecological data base is also being developed, which will include detailed ecological and distribution data for each taxon, derived from the modern calibration data sets, as well as full taxonomic and nomenclatural history, images, and morphological descriptions. It is hoped that the data base, including images, will eventually be available through the internet to the wider diatom and palaeolimnological community.

In addition to the regional diatom and water chemistry data sets listed in Juggins *et al.* (in press) we were made aware of three others that have been developed, or are being developed, from Turkey (Kashima 1994), Yunnan Plateau, S. China, and Yucatan Peninsula, Mexico (Tom Whitmore).

The importance of publicising the project is to encourage taxonomic and methodological standardization in the development of modern calibration sets, and to include samples from other types of saline environment, such as saline springs or industrially-derived saline waters. We would be pleased to hear from anyone with interests in any aspect of the project.

Table 1: Initial taxa to be investigated from African and North American data sets

| | |
|---|---|
| *Amphora:* | *acutiuscula* Kütz.<br>*coffeaformis* (Ag.) Kütz.<br>*ovalis* var. *pediculus* (Kütz.) Van Heurck<br>*perpusilla* (Grun. *in* Van Heurck) Grun. *in* Van Heurck |
| *Campylodiscus:* | *bicostatus* Wm Smith<br>*clypeus* (Ehrenb.) Ehrenb. |
| *Cyclotella:* | *choctawhatcheeana* Prasad (misidentified as *C. caspia* Grun.)<br>*meneghiniana* Kütz.<br>*quillensis* Bailey |
| *Navicula:* | *bulnheimii* Grun. *in* Van Heurck<br>*capitata* Ehrenb.<br>*cincta* (Ehrenb.) Ralfs *in* Pritchard<br>*cryptocephala* Kütz.<br>*cuspidata* (Kütz.) Kütz.<br>*digitoradiata* (Gregory) Ralfs *in* Pritchard<br>*elkab* Otto Müller<br>*halophila* (Grun. *in* Van Heurck) Cleve<br>*oblonga* (Kütz.) Kütz<br>*pseudohalophila* Cholnoky<br>*radiosa* Kütz.<br>*veneta* Kütz. |
| *Nitzschia:* | *constricta* (Kütz.) Ralfs *in* Pritchard<br>*hungarica* Grun.<br>*punctata* Wm Smith (Grun.)/*compressa* f. *minor* A. Cleve-Euler<br>*elegantula* Grun. *in* Van Heurck<br>*fonticola* Grun. *in* Cleve *et* Möller<br>*inconspicua* Grun. (*frustulum* var. *subsalina* Hustedt)<br>*lacuum* Lange-Bertalot<br>*microcephala* Grun. *in* Cleve *et* Möller<br>*palea* (Kütz.) Wm Smith<br>*subacicularis* Hustedt |
| *Surirella:* | *brightwelli* Wm Smith<br>*crumena* Bréb *in* Kütz.<br>*peisonis* Pantocsek |

## References


Ben Khelifa, L. (1989). *Diatomées continentales et paléomilieux du Sud-Tunisien (PALHYDAF site 1) au Quaternaire supérieur. Approche statistique basée sur les diatomées et les milieux actuels.* Ph.D. Thesis, Université de Paris-Sud, 366 pp.

Carvalho, L. R., Cox, E. J. C., Fritz, S.C., Juggins, S., Sims, P. A., Gasse, F. & Battarbee, R. W. (submitted). Harmonising the taxonomy of salt lake diatoms: "*Cyclotella caspia*" as a case study. *Diatom Research.*

Fritz, S. C., Juggins, S. & Battarbee, R. W. (1993). Diatom assemblages and ionic characterization of lakes of the Northern great Plains, North America: a tool for reconstructing past salinity and climate fluctuations. *Canadian Journal of Fisheries and Aquatic Sciences*, **50**, 1844–1856.

Fritz, S. C., Juggins, S., Battarbee, R. W. & Engstrom, D. R. (1991). Reconstruction of past changes in salinity and climate using a diatom-based transfer function. *Nature (London)*, **352**, 706–708.

Gasse, F. (1987). Diatoms for reconstructing palaeoenvironments and palaeohydrology in tropical semi-arid zones: examples of some lakes in Niger since 12000 BP. *Hydrobiologia*, **154**, 127–163.

Gasse, F., Talling, J. F. & Kilham, P. (1983). Diatom assemblages in East Africa: classification, distribution, and ecology. *Revue d'hydrobiologie tropicale*, **16**, 3–34.

Gasse, F., Fontes, J. C., Plaziat, J. C., Carbonel, P., Kaczmarska, I., Dedeckker, P., Soulié-Marsche, I., Callot, Y. & Dupeuble, P. A. (1987). Biological remains, geochemistry and stable isotopes for the reconstruction of environmental and hydrological changes in the Holocene lakes from North Sahara. *Palaeogeography, Palaeoclimatology, Palaeoecology*, **60**, 1–46.

Juggins, S., Battarbee, R. W., Fritz, S. C. & Gasse, F. (1994). The CASPIA project: diatoms, salt lakes, and environmental change. *Journal of Paleolimnology*, **12**, 191–196

Kashima, K. (1994). Sedimentary diatom assemblages in freshwater and saline lakes of the Anatolia Plateau, central part of Turkey: an implication for reconstruction of paleosalinity change during Late Quaternary. (*This volume*).

Kingston, J. C., Cumming, B. F., Uutala, A. J., Smol, J. P., Camburn, K. E., Charles, D. F., Dixit, S. S. & Kreis Jr., R. G. (1992). Biological quality control and quality assurance: a case study in paleolimnological biomonitoring. *Proceedings International Symposium on Ecological Indicators*. Elsevier, New York.

Munro, M. A. R., Kreiser, A. M., Battarbee, R. W., Juggins, S., Stevenson, A. C., Anderson, D. S., Anderson, N. J., Berge, F., Birks, H. J. B., Davis, R. B., Flower, R. J., Fritz, S. C., Haworth, E. Y., Jones, V. J., Kingston, J. C. & Renberg, I. (1990). Diatom quality control and data handling. *Philosophical Transactions of the Royal Society of London, B*, **327**, 257–261.

ter Braak, C. J. F. & Looman, C. W. N. (1986). Weighted averaging, logistic regression and the Gaussian response model. *Vegetatio*, **65**, 3–11.